Iminosugars

Iminosugars

From synthesis to therapeutic applications

Editors

Philippe Compain and Olivier R. Martin

CNRS, University of Orleans, France

BICENTENNIAL 1807 WILEY 2007 BICENTENNIAL

John Wiley & Sons, Ltd

Telephone (+44) 1243 779777

Email (for orders and customer service enquiries): cs-books@wiley.co.uk
Visit our Home Page on www.wiley.com

Other Wiley Editorial Offices

John Wiley & Sons Inc., 111 River Street, Hoboken, NJ 07030, USA

Jossey-Bass, 989 Market Street, San Francisco, CA 94103-1741, USA

Wiley-VCH Verlag GmbH, Boschstr. 12, D-69469 Weinheim, Germany

John Wiley & Sons Australia Ltd, 33 Park Road, Milton, Queensland 4064, Australia

John Wiley & Sons (Asia) Pte Ltd, 2 Clementi Loop #02-01, Jin Xing Distripark, Singapore 129809

John Wiley & Sons Canada Ltd, 6045 Freemont Blvd, Mississauga, Ontario, L5R 4J3

Wiley also publishes its books in a variety of electronic formats. Some content that appears in print may not be available in electronic books.

Anniversary Logo Design: Richard J. Pacifico

Library of Congress Cataloging in Publication Data

Iminosugars: from synthesis to therapeutic applications / edited by
Philippe Compain and Olivier R. Martin.
p. ; cm.
Includes bibliographical references and index.
ISBN 978-0-470-03391-3 (cloth : alk. paper)
1. Imino sugars—Synthesis. 2. Imino sugars—Therapeutic use. I. Compain, Philippe. II. Martin, Olivier R. [DNLM: 1. Imino Sugars—chemical synthesis. 2. Imino Sugars—therapeutic use. 3. Glycoside Hydrolases—antagonists & inhibitors. QU 54 I32 2007]
QP702.I45I45 2007
612′.01578—dc22

2007023155

British Library Cataloguing in Publication Data

A catalogue record for this book is available from the British Library

ISBN 978-0-470-03391-3

Typeset in 10.5/12.5pt Minion by Integra Software Services Pvt. Ltd, Pondicherry, India
Printed and bound in Great Britain by Antony Rowe Ltd., Chippenham, Wiltshire
This book is printed on acid-free paper responsibly manufactured from sustainable forestry in which at least two trees are planted for each one used for paper production.

Contents

Foreword

The field of carbohydrate chemistry has a long and distinguished history in the annals of organic molecules that dates back to its founder Emil Fischer. Carbohydrates (sugars) are among the most abundant natural products. Their occurrence and importance transcends traditional boundaries separating them from other small molecule gifts provided by Nature such as amino acids, hydroxy acids, terpenes and related compounds. Carbohydrates are constituents of biologically important functional macromolecules such as polysaccharides, glycoproteins and lipopolysaccharides, not to mention RNA and DNA. The role of carbohydrates in molecular recognition at the cellular and subcellular level is the primary basis of immunity in living organisms including man.

Carbohydrates have also been associated with many therapeutically important medicines including antibiotics, antiviral agents and antitumour compounds. As monosaccharides, they have provided a playground for synthetic chemists to explore their exquisitely disposed functionality on conformationally distinct and stereochemically defined cyclic hemiacetals. No other class of natural products offers as much stereochemical diversity embodied in enantiopure chiral compounds as is found in the family of aldopyranose and aldofuranose sugars, as well as their aldulose (ketose) equivalents.

Stereoelectronic effects, exemplified by the anomeric effect, have been a fertile area of study for theoretical as well as experimental chemists with important consequences in the emerging field of glycobiology.

Perhaps the most revealing aspect of a natural carbohydrate analogue was the discovery in 1966 of nojirimycin, a D-glucose molecule containing nitrogen in the ring rather than the traditional oxygen. Its antibiotic properties instigated the synthesis of other congeners such as 1-deoxy-nojirimycin, formally a polyhydroxylated 2-hydroxymethyl piperidine of known 'D-*gluco*' configuration. The isolation of 1-deoxy-nojirimycin from natural sources, and the realization that it had inhibitory properties towards α-glucosidases, started a new class of carbohydrate analogues which have been named iminosugars. Remarkably, iminosugars have demonstrated a range of biological activities that span a wide cross section of diseases. The approval of Glyset™ and Zavesca™, within the last decade, for the treatment of complications associated with type II diabetes and for Gaucher's disease respectively, is a testament to their importance as medicines for unmet medical needs. Clearly, the field of iminosugars has emerged as a fertile area for research on both chemical and biological fronts.

This book covers a wide cross section of iminosugar chemistry, biology and medicinal chemistry in 14 chapters. The Editors have done an excellent job of gathering some of the most active researchers in the field. The coverage of specific topics is extensive and clearly exposed.

It is remarkable that such a small, yet electronically distinct, difference in the replacement of oxygen by the more basic nitrogen, can have dramatically beneficial therapeutic effects. Certainly, the future augurs well to uncover many more fascinating aspects of iminosugars, especially with regard to their biological activities, mode of action and therapeutic potential in treating disease.

Stephen Hanessian
Montréal, Canada
March, 2007

Preface

"The productive scientist must be a traditionalist who enjoys playing intricate games by preestablished rules in order to be a successful innovator who discovers new rules and new pieces with which to play them."

Thomas S. Kuhn in *The Essential Tension* (University of Chicago Press, 1977).

The field of iminosugars is a very exciting area of research. These carbohydrate mimetics were first imagined and synthesized by chemists in the 1960s before being isolated from Nature a few years later. Since the discovery of their biological activity as potent glycosidase inhibitors in the 1970s, iminosugars have been the subject of intense studies at the interface between organic synthesis, glycobiology and medicinal science. In 1999, Arnold Stütz edited the first book devoted to iminosugars (*Iminosugars as Glycosidase Inhibitors: Nojirimycin and Beyond*, Wiley-VCH, New York; 1999). The title of this outstanding monograph reflected the main features of the field at that time and the strong interplay between isolation of natural products such as nojirimycin, synthesis and biochemical research. However, the most striking word of the title is undoubtedly the preposition 'beyond' which was premonitory. The pace of discovery in the area of iminosugars has indeed increased spectacularly over the past decade. The scope of their biological activity has been extended to the inhibition of numerous proteins including enzymes acting on sugars (glycosyltransferases, glycogen phosphorylases, nucleoside-processing enzymes, UDP-Gal*p* mutase) but also metalloproteinases. Glyset™ and Zavesca™, the two first examples of iminosugar-based drugs, have been commercialized for complications associated with diabetes and for the treatment of Gaucher's disease, respectively. New biological and medicinal applications are being uncovered almost every month!

In view of these recent breakthroughs, it appeared to us very timely to publish a new book on the latest developments in the synthesis and the biological evaluation of iminosugars of therapeutic interest with some of the major experts in the area. During the genesis of this project, we have had the opportunity of physically gathering most of the authors at a symposium we organized in August 2005 during the 230th American Chemical Society meeting in Washington DC (*Iminosugars: synthesis and therapeutic potential*, Oral symposium # 7057). The success of this stimulating meeting further reinforced our wish to edit an interdisciplinary book devoted to iminosugars. We are very grateful to the authors, from no less than four continents, for having contributed to the high level of quality of this book. We would like to thank external reviewers for helpful comments, their time and their expertise. Special thanks are also due to

Fiona Woods, Joan Marsh, Andrea Baier and other members of the publishers' team for their faith in our project and for their constant help. Finally, we are most grateful to Stephen Hanessian, a pioneer in the field, for having accepted with enthusiasm our invitation to write the Foreword for this book.

Philippe Compain
Olivier R. Martin
Orleans, France
January, 2007

List of contributors

J. M. F. G. Aerts, Department of Medicine, University of Cambridge, Addenbrooke's Hospital, Cambridge, UK.

N. Asano, Faculty of Pharmaceutical Sciences, Hokuriku University, Kanazawa, Japan.

T. M. Block, Drexel Institute for Biotechnology and Virology Research and Hepatitis B Foundation, Drexel College of Medicine, Doylestown, Pennsylvania, USA.

M. Bols, Department of Chemistry, University of Copenhagen, Universitetsparken, Denmark.

T. D. Butters, Oxford Glycobiology Institute, University of Oxford, Oxford, UK.

L. Cipolla, Department of Biotechnology and Bioscience, University of Milano-Bicocca, Italy.

P. Compain, Institut de Chimie Organique et Analytique, CNRS – Université d'Orléans, Orléans, France.

T. M. Cox, Department of Medicine, University of Cambridge, Addenbrooke's Hospital, Cambridge, UK.

V. Desvergnes, Institut de Chimie Organique et Analytique, CNRS – Université d'Orléans, Orléans, France.

J.-Q. Fan, Department of Human Genetics, Mount Sinai School of Medicine, New York, USA.

B. La Ferla, Department of Biotechnology and Bioscience, University of Milano-Bicocca, Italy.

S. Gerber-Lemaire, Laboratory of Glycochemistry and Asymmetric Synthesis, Swiss Federal Institute of Technology (EPFL), Lausanne, Switzerland.

W. A. Greenberg, Department of Chemistry and the Skaggs Institute of Chemical Biology, The Scripps Research Institute, La Jolla, California, USA.

B. Gu, Drexel Institute for Biotechnology and Virology Research, Drexel College of Medicine, Doylestown, Pennsylvania, USA.

L. Juillerat-Jeanneret, University Institute of Pathology, Centre Hospitalier Universitaire Vaudois (CHUV) and University of Lausanne (UNIL), Lausanne, Switzerland.

V. Liautard, Institut de Chimie Organique et Analytique, CNRS – Université d'Orléans, Orléans, France.

O. Lopez, Department of Chemistry, University of Aarhus, Denmark.

O. R. Martin, Institut de Chimie Organique et Analytique, CNRS – Université d'Orléans, Orléans, France.

M. L. Mitchell, Department of Chemistry and the Skaggs Institute of Chemical Biology, The Scripps Research Institute, La Jolla, California, USA.

F. Nicotra, Department of Biotechnology and Bioscience, University of Milano-Bicocca, Italy.

Y. Nishimura, Bioactive Molecules Research Group, Microbial Chemistry Research Center, Tokyo, Japan

P. A. Norton, Drexel Institute for Biotechnology and Virology Research, Drexel College of Medicine, Doylestown, Pennsylvania, USA.

C. Pillard, Institut de Chimie Organique et Analytique, Université d'Orléans, Orléans, France.

F. M. Platt, Department of Pharmacology, University of Oxford, Mansfield Road, Oxford, UK.

V. L. Schramm, Department of Biochemistry, Albert Einstein College of Medicine, Bronx, New York, USA.

S. Toumieux, Institut de Chimie Organique et Analytique, CNRS – Université d'Orléans, Orléans, France.

P. C. Tyler, Carbohydrate Chemistry, Industrial Research Limited, Lower Hutt, New Zealand.

P. Vogel, Laboratory of Glycochemistry and Asymmetric Synthesis, Swiss Federal Institute of Technology (EPFL), Lausanne, Switzerland.

L. Whalen, Department of Chemistry and Chemical Biology, University of New Mexico, Albuquerque, New Mexico, USA.

C.-H. Wong, Department of Chemistry and the Skaggs Institute of Chemical Biology, The Scripps Research Institute, La Jolla, California, USA.

1

Iminosugars: past, present and future

Philippe Compain and **Olivier R. Martin**

Iminosugars are sugars in which the endocyclic oxygen is replaced by a basic nitrogen atom. This apparently simple substitution raises many synthetic challenges and opens the way to remarkable biological properties. As such, iminosugars undoubtedly form the most attractive class of carbohydrate mimics reported so far. The origin of their therapeutic use goes back to ancient times and traditional Chinese phytomedicines. In Occident, Haarlem oil, the first medication produced on an industrial scale in the 17th century, was recommended for the treatment of diabetes and for whitening the skin. One of the major constituents of Haarlem oil was an extract from leaves of *Morus alba*, the white mulberry, an extremely rich source of iminosugars [1]. The scientific history of iminosugars began in the early 1960s with the almost simultaneous reports of the synthesis of sugar derivatives containing a nitrogen atom in the ring by the groups of Paulsen [2], Jones [3] and Hanessian [4]. At that time, the replacement of the endocyclic oxygen atom in sugars by heteroatoms (N, S, P) [5] to form 'heteroses' [6] was purely an academic exercise. In 1966, Paulsen published the first synthesis of 1-deoxynojirimycin (DNJ) [7] (Figure 1.1). The same year, Inouye *et al.* isolated nojirimycin from bacteria (*Streptomyces*) and identified its antibiotic properties [8]. The first renaissance of iminosugars came from the isolation of DNJ from natural sources and the finding of its biological activity as an α-glucosidase inhibitor by Bayer chemists in 1976. This discovery triggered an enormous amount of interest in imino analogues of carbohydrates [9].

We are currently witnessing a second renaissance in this field. Since the late 1990s, the rate of discoveries has increased dramatically. Original structures have been designed and synthesized such as seven- or eight-membered iminoalditols, conformationally constrained analogues of iminosugars and complex glycoconjugate mimetics (Figure 1.2). Innovative synthetic strategies have been developed, including combinatorial approaches to iminosugar libraries [10]. The asymmetric synthesis of a pyrrolidinol in only four steps in water has been reported [11] as well as concise *de novo* approaches

Iminosugars: From synthesis to therapeutic applications Edited by Philippe Compain and Olivier R. Martin
 ISBN 978-0-470-03391-3

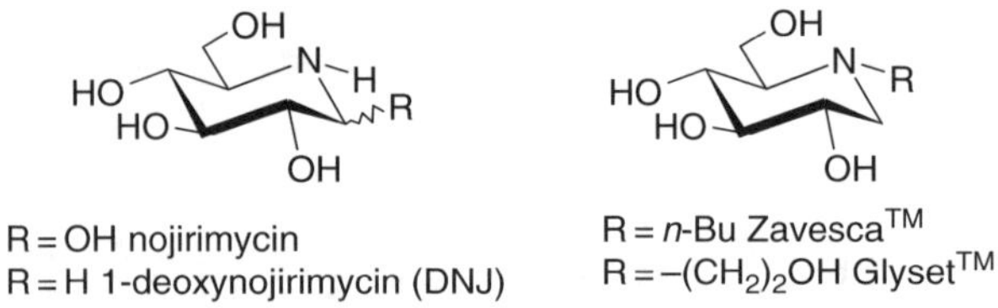

Figure 1.1 Some emblematic iminosugars

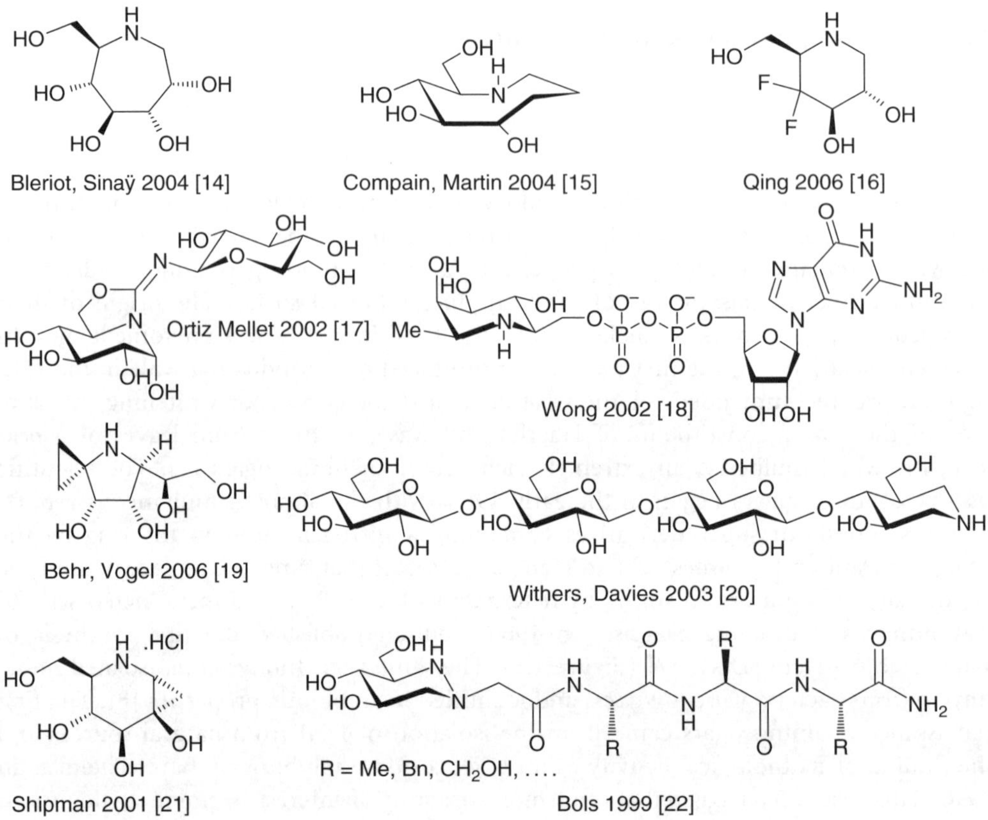

Figure 1.2 Recent examples of original iminosugar structures

by way of enantioselective proline-catalysed reactions [12]. The new concept of dynamic combinatorial chemistry has been applied to accelerate the discovery of iminoalditol-based glycosidase inhibitors [13].

The scope of the biological activity of iminosugars has been extended to the inhibition of a number of enzymes of medicinal interest such as glycosyltransferases [23], glycogen phosphorylases [24], nucleoside-processing enzymes [25], a sugar nucleotide mutase [26] and metalloproteinases [27]. Thanks to iminosugars, significant progress has thus been made in glycobiology in the past 10 years. One of the most spectacular breakthroughs is certainly the discovery that reversible competitive inhibitors could positively influence the folding state of abnormal glycosidases, thus preventing their destruction by quality control in the endoplasmic reticulum (ER) and ER-associated degradation (ERAD) [28]. This new concept, known as chemical chaperone therapy, is being evaluated clinically with a DNJ analogue (Phase II) for the treatment of Fabry disease, highlighting its strong potential as a new therapeutic option for lysosomal diseases. One may also consider the incredibly powerful inhibitors of nucleoside phosphorylases based on iminosugar *C*-nucleosides. These compounds designed as transition-state analogues display inhibition values in the femtomolar range and are among the most powerful inhibitors described for any enzyme to date [29]!

The amazing diversity of enzymes inhibited by iminosugars promises a new generation of medicines in a wide range of diseases such as diabetes, viral infections, lysosomal storage disorders or tumour metastasis (Table 1.1) [30]. Various structures are currently involved in clinical trials and the first successes are being recorded. Recently, two iminosugar-based drugs have been approved: Glyset™ in 1996 for the treatment of complications associated with type II diabetes, and Zavesca™ in 2003 as the first oral treatment for Gaucher disease, a severe lysosomal storage disorder (Figure 1.1).

The aim of this book is to present the multifaceted aspects of iminosugars from their structure to their biological activities, and from synthesis to therapeutic applications. Chapter 2 focuses on naturally occurring iminosugars. Chapter 3 is devoted to recent synthetic strategies and combinatorial approaches towards iminosugar libraries. Important classes of iminosugars and their biological activities are presented in Chapter 4 (imino-*C*-glycosides), Chapter 5 (imino-*C*-disaccharides), Chapter 6 (isoiminosugars, i.e. 1-aza carbasugars) and Chapter 8 (iminosugar *C*-nucleosides). Chapter 7 deals with recent developments in the field of glycosyltransferase inhibitors. Moving closer to therapeutic applications, Chapters 9 and 12 are devoted to iminosugars as antiviral and antitumour agents respectively. The two main strategies for the chemotherapeutic treatment of lysosomal diseases are reviewed in Chapter 10 (substrate reduction therapy) and Chapter 11 (chaperone therapy). Chapter 13 is an overview of the medicinal use of iminosugars, including key reflections on their therapeutic potential. Finally, Chapter 14 provides tables which correlate the structure of more than 600 iminosugars of therapeutic interest with their biological activities, where such data are available. Compounds have been selected for their potential as therapeutic agents. The goal of these at-a-glance tables is to facilitate and to stimulate further research in the area of iminosugar by relating structure to properties.

Table 1.1 Enzymatic and therapeutic targets of iminosugars

Enzymatic targets	Therapeutic targets
Glycosidases mid-1970s [9]	Diabetes mid-1970s
Glycosyltransferases 1992 [23]	Viral diseases 1980s
Nucleoside-processing enzymes 1993 [25]	Cancers 1980s
UDP Gal mutase 1997 [26]	Lysosomal diseases 1990s
Glycogen phosphorylases 1997 [24]	Psoriasis 2000s [31]
Metalloproteinases 2004 [27]	Cystic fibrosis 2006 [32]

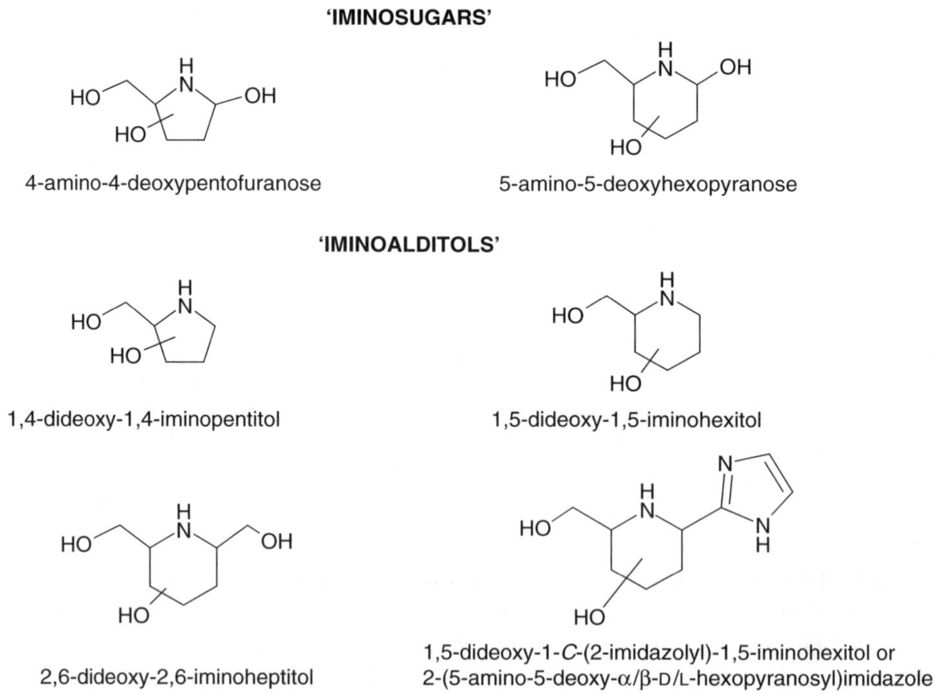

Figure 1.3

Before concluding, we would like to make some remarks concerning the nomenclature of iminosugars and derivatives. For several years, chemists have used the term 'azasugars' to refer to analogues of pyranoses and furanoses in which the ring oxygen atom is replaced by nitrogen. This nomenclature was adopted because it permitted a clear distinction between common amino sugars, such as glucosamine derivatives, and the new class of carbohydrate analogues. However, this nomenclature is evidently incorrect since, strictly, *aza* is to be used when carbon is replaced by nitrogen. Using the standard rules of carbohydrate nomenclature [33], it is possible to name the nitrogen analogues of sugars without resorting to special descriptors (Figure 1.3).

The use of azasugar should therefore be abandoned in preference to the above nomenclature. In addition, we have recently seen an increasing use of the term 'iminocyclitol'. This term should not be used as it relates to a totally different class of compounds based on a carbocyclic ring.

The new class of compounds, referred to as 1-*N*-iminosugars, is even more confusing. Strictly, 1-aza-analogues of sugars represent a little-studied class of compounds in which the C-1 carbon atom is replaced by nitrogen, thus giving a saturated 1,2-oxazine-type structure. The further replacement of the ring oxygen by a carbon atom leads to aza-carba analogues of sugars. Evidently, it is simpler to name such compounds as branched-chain alditols or as substituted pyrrolidines or piperidines. For example, the so-called 'isofagomine' is 1,2,5-trideoxy-2-hydroxymethyl-1,5-imino-D-xylitol (Figure 1.4) – a short name for this family of compounds could be isoiminosugars.

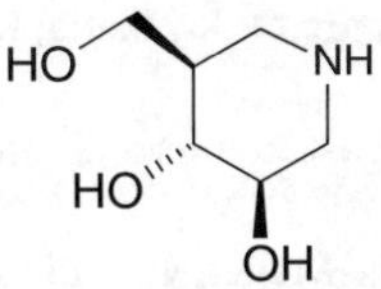

'Isofagomine'

1,2,5-trideoxy-2-hydroxymethyl-
1,5-imino-D-xylitol

Figure 1.4

Regarding the impressive series of discoveries in the field over the past 10 years, we can confidently conclude that it is a 'boom time' for iminosugar chemistry and biology! These multitask molecules have now moved from the laboratory to the clinic and it is likely that so far we have only seen the tip of the iceberg.

References

1. Aerts, J. M. F. G., 2006 *Proc. European Working Group for the Study of Gaucher Disease*; http://www.ewggd.co.uk/EWGGD%20Abstracts.pdf.
2. (a) Paulsen, H., 1962 *Angew. Chem. Int. Ed. Engl.*, **1**, 597. (b) Paulsen, H., 1962 *Angew. Chem. Int. Ed. Engl.*, **1**, 454.
3. (a) Jones, J. K. N. and Turner, J. C., 1962 *J. Chem. Soc.*, 4699–4703. (b) Jones, J. K. N. and Szarek, W. A., 1963 *Can. J. Chem.*, **41**, 636–640.
4. (a) Hanessian, S. and Haskell, T. H., 1963 *J. Org. Chem.*, **28**, 2604–2610. (b) Hanessian, S., 1966 *Chem. Commun.*, 796–798.
5. Paulsen, H. and Todt, K., 1968 *Adv. Carbohydr. Chem.*, **23**, 115–232.
6. Vasella, A. and Veuffray, R., 1982 *Helv. Chim. Acta*, **65**, 1134–1144.
7. (a) Paulsen, H., 1966 *Angew. Chem. Int. Ed. Engl.*, **5**, 495–511. (b) Paulsen, H. and Todt, K., 1967 *Chem. Ber.*, **100**, 3385–3396; 3397–3407.
8. (a) Inouye, S., Tsuruoka, T. and Niida, T., 1966 *J. Antibiot., Ser. A*, **19**, 288–292. (b) Inouye, S., Tsuruoka, T., Koaze, Y. and Niida, T., 1968 *Tetrahedron*, **24**, 2125–2144.
9. Stütz, A. E. (ed.), 1999 *Iminosugars as Glycosidase Inhibitors: Nojirimycin and Beyond* Wiley-VCH, New York; for an historical background, see Chapter 1 by Paulsen, H., The early days of monosaccharides containing nitrogen in the ring.
10. Cipolla, L., La Ferla, B. and Gregori, M., 2006 *Comb. Chem. High Throughput Screen*, **9**, 571–582.
11. Lindström, U. M., Ding, R. and Hidestal, O., 2005 *Chem. Commun.*, 1773–1774.
12. (a) Liao, W. W., Ibrahem, I. and Cordova, A., 2006 *Chem. Commun.*, 674–676. (b) Calderon, F., Doyagüez, E. G. and Fernandez-Mayoralas, A., 2006 *J. Org. Chem.*, **71**, 6258–6261.
13. Gerber-Lemaire, S., Popowycz, F., Rodriguez-Garcia, E., Asenjo, A. T. C., Robina, I. and Vogel, P., 2002 *ChemBioChem*, **3**, 466–470.
14. (a) Li, H., Bleriot, Y., Chantereau, C., Mallet, J.-M., Sollogoub, M., Zhang, Y., Rodriguez-Garcia, E., Vogel, P., Jimenez-Barbero, J. and Sinaÿ, P., 2004 *Org. Biomol. Chem.*, **2**, 1492–1499. (b) Martin, O. R., 1997 in *Carbohydrate Mimics: Concepts and Methods* (ed. Y. Chapleur) Wiley-VCH, New York, p. 259.

15. Godin, G., Garnier, E., Compain, P., Martin, O. R., Ikeda, K. and Asano, N., 2004 *Tetrahedron, Lett.*, **45**, 579–581.
16. Wang, R.-W., Qiu, X.-L., Bols, M., Ortega-Caballero, F. and Qing, F.-L., 2006 *J. Med. Chem.*, **49**, 2989–2997.
17. Garcia-Moreno, M., Diaz-Pérez, P., Ortiz-Mellet, C. and Garcia Fernandez, J. M., 2002 *Chem Commun.*, 848–849.
18. Mitchell, M. L., Tian, F., Lee, L. V. and Wong, C.-H., 2002 *Angew. Chem. Int. Ed. Engl.*, **41**, 3041–3044.
19. Laroche, C., Behr, J.-B., Szymoniak, J., Bertus, P., Schütz, C., Vogel, P., and Plantier-Royon, R., 2006 *Bioorg. Med. Chem.*, **14**, 4047–4054.
20. Varrot, A., Tarling, C. A., Macdonald, J. M., Stick, R. V., Zechel, D. L., Withers, S. and Davies, G. J., 2003 *J. Am. Chem. Soc.*, **125**, 7496–7497.
21. Désiré, J. and Shipman, M., 2001 *Synlett*, 1332–1334.
22. Lohse, A., Jensen, K. B., Lundgren, K. and Bols, M., 1999 *Bioorg. Med. Chem.*, **7**, 1965–1971.
23. (a) Wong, C.-H., Dumas, D. P., Ichikawa, Y., Koseki, K., Danishefsky, S. J., Weston, B. W. and Lowe, J. B. 1992 *J. Am. Chem. Soc.*,**114**, 7321–7322. (b) Compain, P. and Martin, O. R., 2003 *Curr. Top. Med. Chem.*, **3**, 541–560.
24. (a) Bols, M., Hazelle, R. and Thomsen, I. B. 1997 *Chem. Eur. J.*, **3**, 940–947. (b) Heightman, T. D., Vasella, A., Tsitsanou, K. E., Zographos, S. E., Skamnaki, V. T. and Oikonomakos, N. G., 1998 *Helv. Chim. Acta*, **81**, 853–864. (c) Somsak, L., Nagy, V. Hadady, Z., Docsa, T. and Gergely, P., 2003 *Curr. Pharm. Des.* **9**, 1177–1189.
25. (a) Horenstein, B. A., Zabinski, R. F. and Schramm, V. L., 1993 *Tetrahedron Lett.*, **34**, 7213–7216. (b) Schramm, V. L. and Tyler, P. C., 2003 *Curr. Top. Med. Chem.*, **3**, 525–540.
26. (a) Lee, R. E., Smith, M. D., Nash, R. J., Griffiths, R. C., McNeil, M., Grewal, R. K., Yau, W., Besra, G. S., Brennan, P. J. and Fleet, G. W. J., 1997 *Tetrahedron Lett.*, **38**, 6733–6736. (b) Lee, R. E., Smith, M. D., Pickering, L. and Fleet, G. W. J. 1999 *Tetrahedron Lett.*, **40**, 8689–8692.
27. Moriyama, H., Tsukida, T., Inoue, Y., Yokota, K., Yoshino, K., Kondo, H., Miura, N. and Nishimura, S.-I., 2004 *J. Med. Chem.*, **47**, 1930–1938.
28. (a) Fan, J.-Q., Ishii, S., Asano, N. and Suzuki, Y., 1999 *Nature Medicine*, **5**, 112–115. (b) Fan, J.-Q., 2003 *Trends Pharm. Sci.*, **24**, 355–360.
29. Singh, V., Evans, G. B., Lenz, D. H., Mason, J. M., Clinch, K., Mee, S., Painter, G. F., Tyler, P. C., Furneaux, R. H., Lee, J. E., Howell, P. L. and Schramm, V. L., 2005, *J. Biol. Chem.*, **280**, 18265–18273.
30. Martin, O. R. and Compain, P. (eds) 2003 Iminosugars: recent insights into their bioactivity and potential as therapeutic agents, in *Curr. Top. Med. Chem.*, Bentham, Hilversum, Netherlands, Volume 3, issue 5, pp. 471–591
31. Clinch, K., Evans, G. B., Fleet, G. W. J., Furneaux, R. H., Johnson, S. W., Lenz, D. H., Mee, S. P. H., Rands, P. R., Schramm, V. L., Taylor Ringia, E. A. and Tyler, P. C., 2006 *Org. Biomol. Chem.*, **4**, 1131–1139.
32. Norez, C., Noel, S., Wilke, M., Bijvelds, M., Jorna, H., Melin, P., DeJonge, H. and Becq, F., 2006 *FEBS Letters*, **580**, 2081–2086.
33. McNaught, A. D., 1997 *Carbohydr. Res.*, **297**, 1–92.

2

Naturally occurring iminosugars and related alkaloids: structure, activity and applications

Naoki Asano

2.1 Introduction

Great interest in iminosugars (also known as azasugars) has been aroused in recent years because it has been shown that such compounds and their derivatives have enormous therapeutic potential in many diseases such as diabetes, viral infection and lysosomal storage disorders [1–3]. Iminosugars are sugar mimics with a nitrogen atom in place of the ring oxygen and they inhibit glycosidases. The biological properties can be explained by their structural resemblance to the terminal sugar moiety in the natural substrates. Glycosidases are involved in a wide range of important biological processes, such as intestinal digestion, post-translational processing of the sugar chain of glycoproteins, quality-control systems in the endoplasmic reticulum (ER) and ER-associated degradation mechanisms and the lysosomal catabolism of glycoconjugates. Hence, inhibition of these glycosidases can have profound effects on carbohydrate catabolism in the intestine, maturation, transport and secretion of glycoproteins, and can alter cell–cell or cell–virus recognition processes.

Naturally occurring alkaloids inhibiting glycosidases are classified into five structural classes: polyhydroxylated pyrrolidines, piperidines, indolizidines, pyrrolizidines and nortropanes [1, 4]. Furthermore, they also occur in glycosylated forms. However, the structures of natural products are much more diverse. There is no doubt that there are more novel structures waiting to be discovered. This review describes the recent studies on natural inhibitors of α- and β-glucosidases, α- and β-galactosidases and α-mannosidase, as well as their biological applications.

Iminosugars: From synthesis to therapeutic applications Edited by Philippe Compain and Olivier R. Martin
 ISBN 978-0-470-03391-3

2.2 α-Glucosidase inhibitors

2.2.1 Structures and *in vitro* inhibitory activity

Over 40 years have passed since nojirimycin (NJ) (**1**) was discovered as the first natural glucose mimic with a nitrogen atom in place of the ring oxygen [5]. NJ is stored as a bisulphite adduct because it bears a hydroxyl group at the anomeric position (C-1) and is fairly unstable. NJ is a good inhibitor of both α- and β-glucosidases, with IC_{50} values of 9 μM toward human lysosomal α-glucosidase and 19 μM toward human lysosomal β-glucosidase (unpublished data). 1-Deoxynojirimycin (DNJ) (**2**) was originally prepared by catalytic hydrogenation of NJ with a platinum catalyst or by chemical reduction with $NaBH_4$ [6]. DNJ was later isolated from the roots of mulberry trees and called molanoline [7]. Mulberry trees (*Morus* spp.) are cultivated in China, Korea and Japan, and their leaves are used to feed silkworms (*Bombyx mori*). Mulberry leaves have been used traditionally to cure and prevent 'Xiao-ke' (diabetes) in Chinese herbal medicine. The root bark of mulberry trees has been used as a Chinese herbal medicine called 'Sang-bai-pi' (Japanese name 'Sohakuhi') for anti-inflammatory, diuretic, antitussive and antipyretic purposes, while the fruits are used as a tonic and sedative. In 1994, the improvement of the purification procedures using a variety of ion-exchange resins led to the isolation of a number of water-soluble alkaloids from the genus *Morus* (Moraceae) [8, 9]. Fagomine (1,2-dideoxynojirimycin) (**3**) and *N*-methyl-DNJ (**4**) as polyhydroxylated piperidines are also isolated from all parts of mulberry trees [8–10]. Although DNJ is a potent inhibitor of all kinds of α-glucosidases, the deoxygenation of DNJ at C-2 to give fagomine markedly lowered its inhibition toward α-glucosidases [11]. The biosynthesis of oligosaccharide chains in *N*-linked glycoprotein involves the cotranslational transfer of a $Glc_3Man_9(GlcNAc)_2$ precursor from a dolichol carrier onto the asparagine residues in the proper sequon of the protein [12]. Initial processing of the oligosaccharides starts with the removal of the outermost α–1,2-linked glucose residue by α-glucosidase I, and α-glucosidase II successively removes the two remaining α–1,3-linked glucose residues in the endoplasmic reticulum (ER). *N*-Methyl-DNJ is a more potent inhibitor of α-glucosidase I than of α-glucosidase II [13, 14]. Mulberry trees additionally contain 1,4-dideoxy-1,4-imino-D-arabinitol (DAB) (**5**), which was originally isolated from fruits of *Angylocalyx boutiqueanus* (Leguminosae) [15] and found to be a potent inhibitor of yeast α-glucosidase [16]. We later reported that DAB is a good inhibitor with a broad inhibitory spectrum toward mammalian glycosidases, such as ER α-glucosidase II, Golgi α-mannosidases I and II and digestive α-glucosidases [11]. In particular, DAB is a more potent inhibitor of α-1,6-glucosidase (isomaltase) than α-1,4-glucosidase (maltase) [11].

In 1988, α-homonojirimycin (α-HNJ) (**6**) was isolated from the neotropical liana, *Omphalea diandra* (Euphorbiaceae), as the first example of a naturally occurring DNJ derivative with a carbon substituent at C-1 [17]. However, before the isolation of the natural product, the 7-*O*-β-D-glucopyranosyl-α-HNJ (Glc-HNJ) (**7**) had been designed and synthesized as a potential drug for the treatment of diabetes [18, 19]. α-HNJ has been detected in adults, pupae and eggs of the neotropical moth, *Urania fulgens*, whose larvae feed on *O. diandra*, and the level of α-HNJ in pupae was about 0.5 per cent

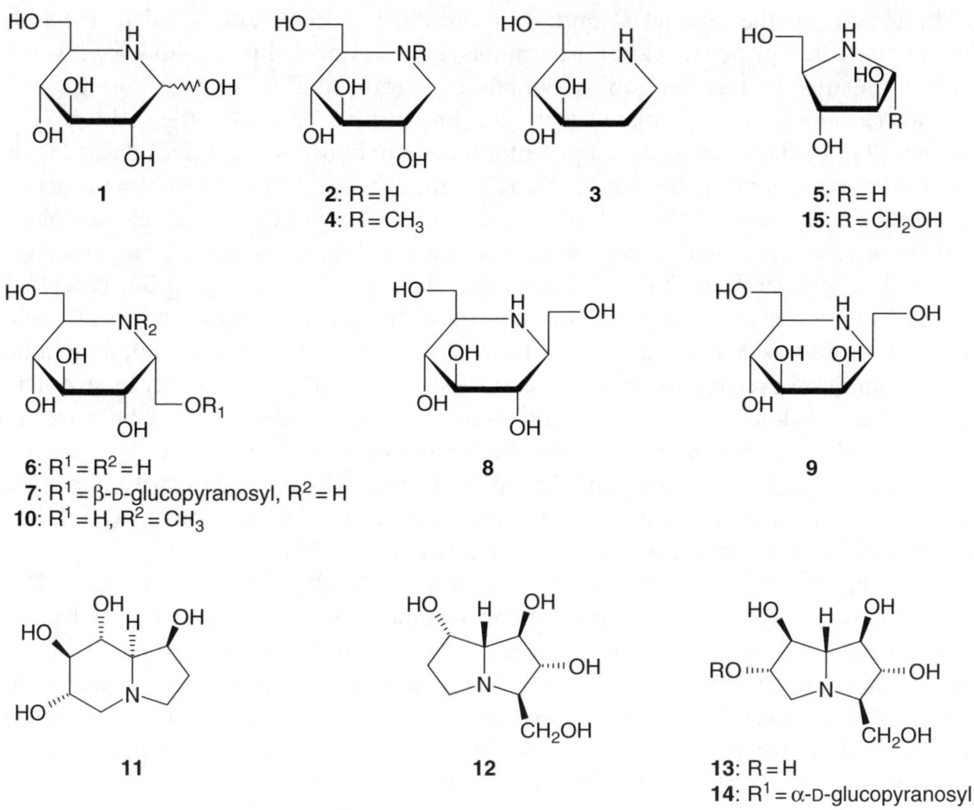

Figure 2.1 Structures of α-glucosidase inhibitors

dry weight [20]. Until 1990, the natural occurrence of α-HNJ had been strictly limited to the Euphorbiaceae family plants. *Aglaonema treubii* (Araceae) is a very common indoor foliage plant and a native to the tropical rainforests of South-East Asia. In 1997, a 50 per cent aqueous EtOH extract of *A. treubii* was found to inhibit α-glucosidase strongly and was subjected to various ion-exchange column chromatographic steps to give a number of polyhydroxylated pyrrolidine and piperidine alkaloids including **6**, **7**, β-homonojirimycin (β-HNJ) (**8**), and β-homomannojirimycin (β-HMJ) (**9**) [21]. α-HNJ is a more specific α-glucosidase inhibitor than DNJ, showing no significant inhibition of other glycosidases [22]. Interestingly, β-HNJ and β-HMJ are not inhibitors of β-glucosidase and β-mannosidase, respectively, but inhibitors of α-glucosidases [22]. α-HNJ and its synthetic *N*-methyl derivative (*N*-methyl-α-HNJ) (**10**) were evaluated for the inhibition of *N*-linked oligosaccharide processing of the viral envelope glycoproteins using an influenza virus-infected Madin–Darby canine kidney (MDCK) cell [23]. By treatment with 100 μg/ml of *N*-methyl-α-HNJ in the medium, essentially all of the *N*-linked oligosaccharide chains of the virus were of the high mannose type with the major structure being characterized as $Glc_3Man_9(GlcNAc)_2$. Similar results were obtained with α-HNJ although this compound was less effective both *in vitro* and *in vivo*.

The toxicity of the legume *Castanospermum australe* for livestock led to the isolation of the toxic principal castanospermine (**11**) [24] and this alkaloid gave rise to a great impetus in research on *N*-containing sugars and its application. In 1981, castanospermine was first isolated from the immature seeds, with the yield of 0.057 per cent [24]. Castanospermine is a potent inhibitor of lysosomal α-glucosidase [25] and disturbs the lysosomal catabolism of glycogen [26]. The syndrome resembles the genetic disorder Pompe disease. When administered to various types of animal cells in culture, castanospermine prevented glycoprotein processing and therefore caused the production of *N*-linked glycoproteins having oligosaccharides mostly of the $Glc_3Man_9(GlcNAc)_2$ type [27–29]. *C. australe* coproduces a pyrrolizidine alkaloid, australine (**12**), which can be regarded as a ring-contracted form of castanospermine [30] but is a much weaker inhibitor of α-glucosidases than castanospermine [31, 32]. A highly hydroxylated pyrrolizidine alkaloid casuarine (**13**) and its 6-*O*-α-D-glucoside (**14**) have been isolated from the bark of *Casuarina equisetifolia* (Casuarinaceae) [33, 34]. Casuarine is a potent inhibitor of rat digestive maltase and *Aspergillus niger* amyloglucosidase, with IC_{50} values in a submicromolar range, and its 6-*O*-α-D-glucoside is a potent competitive inhibitor of porcine kidney trehalase, with an IC_{50} value of 18 nM [32].

In 1976, 2,5-dideoxy-2,5-imino-D-mannitol (DMDP) (**15**), mimicking β-D-fructofuranose, was found in leaves of the legume *Derris elliptica* [35]. DMDP is a more potent inhibitor of yeast α-glucosidase than mammalian α-glucosidases [36, 37]. Recently, the L-enantiomer of DMDP was synthesized from D-gulonolactone and found to be a more powerful and more specific α-glucosidase inhibitor than the natural product DMDP [38]. More interestingly, the natural D-enantiomer is a competitive inhibitor of α-D-glucohydrolases but its synthetic L-enantiomer is a noncompetitive inhibitor [39]. Similarly, the L-enantiomer of DAB was found to be a more potent inhibitor of mammalian digestive α-glucosidases than the enantiomeric natural product DAB and also to be a noncompetitive inhibitor [39]. Recent studies suggest that D-iminosugars are competitive inhibitors of D-glycohydrolases but their L-enantiomers are noncompetitive inhibitors of the enzymes [39, 40].

2.2.2 Applications

The intestinal oligo- and disaccharidases are fixed components of the cell membrane of the brush border region of the wall of the small intestine. These enzymes digest dietary carbohydrate to monosaccharides which are absorbed through the intestinal wall. They include sucrase, maltase, isomaltase, lactase, trehalase and hetero-β-glucosidase. In the late 1970s, it was realized that inhibition of some, or all, of these activities by inhibitors could regulate the absorption of carbohydrate and these inhibitors could be used therapeutically in the oral treatment of the non-insulin-dependent diabetes mellitus (NIDDM or type 2 diabetes) [41].

The strong inhibition of digestive α-glucosidases by DNJ, which are produced by *Bacillus* and *Streptomyces* [41–43], and mulberry trees [8–10], attracted the interest of various research groups and a large number of *N*-substituted DNJ derivatives were prepared in the hope of increasing the *in vivo* activity. Miglitol (**16**) was identified as one of the most favourable candidates showing a desired glucosidase inhibitory profile [44]. Miglitol differs from acarbose in that it is almost completely absorbed from the intestinal tract, and may possess systemic effects in addition to the effects in the intestinal border [45, 46]. In 1996, Glyset (miglitol) tablets were granted market clearance by the US Food and Drug Administration (FDA) and introduced onto the market in 1999 as a more effective second-generation α-glucosidase inhibitor with fewer gastrointestinal side effects. In 2006, it was introduced onto the market in Japan under the brand name Seibule. α-Glucosidase inhibitors are especially suited for patients whose blood glucose levels are slightly above normal and can also benefit those who have high blood glucose immediately after a meal, a condition known as postprandial hyperglycemia. These drugs slow the rate at which carbohydrates are broken down into monosaccharides in the digestive tract and therefore lengthen the digestive process. Other antidiabetic agents such as sulphonylureas and biguanides sometimes are prescribed in combination with α-glucosidase inhibitors to help increase the effectiveness of this therapy. Protective effects of the α-glucosidase inhibitors have been reported for various diabetic complications. Interestingly, α-glucosidase inhibitors are also being studied as a possible treatment for heart disease, a common complication in diabetic patients. Although repetitive postprandial hyperglycemia increases ischemia/reperfusion injury, this effect can be prevented by treatment with α-glucosidase inhibitors [47].

In type 2 diabetes, hepatic glucose production is increased [48]. A possible way to suppress hepatic glucose production and lower blood glucose in type 2 diabetes patients may be through inhibition of hepatic glycogen phosphorylase [49]. In enzyme assays, Fosgerau *et al.* reported that DAB (**5**), which has a broad inhibitory spectrum toward mammalian glycosidases [11], is a potent inhibitor of hepatic glycogen phosphorylase [50]. Furthermore, in primary rat hepatocytes, DAB was shown to be the most potent inhibitor (IC_{50} 1 μM) of basal and glucagon-stimulated glycogenolysis ever reported [51]. Recently, Jakobsen *et al.* have reported that isofagomine (IFG) (**17**) synthesized chemically is a potent inhibitor of hepatic glycogen phosphorylase, with an IC_{50} value of 0.7 μM and, furthermore, is able to prevent basal and glucagon stimulated glycogen degradation in cultured hepatocytes with IC_{50} values of 2–3 μM [52]. However, its *N*-substitution always resulted in a loss of activity compared with the parent compound, and fagomine (**3**) was a weak inhibitor of this enzyme, with an IC_{50} value of 200 μM [52]. Glycogen phosphorylase inhibitors would be a beneficial target to attack in the development of new antihyperglycemic agents.

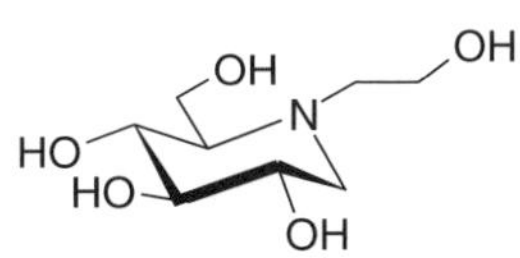

16

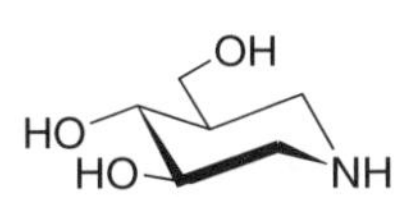

17

Figure 2.2 Structures of miglitol as a commercially available antidiabetic agent and isofagomine as a glycogen phosphorylase inhibitor

The viral envelope glycoproteins are often essential for virion assembly and secretion and/or infectivity. Compounds that interfere with the glycosylation processes of viral glycoproteins can be expected to be antiviral agents. In fact, α-glucosidase inhibitors such as DNJ, *N*-butyl-DNJ (**18**), castanospermine (**11**) and 6-*O*-butanoylcastanospermine (MDL 28574) (**19**) inhibit human immunodeficiency virus (HIV) replication and HIV-mediated syncytium formation *in vitro* [53–56]. The anti-HIV activity (EC_{50}), determined by syncytial counts, of DNJ, *N*-butyl-DNJ, castanospermine, 6-*O*-butanoylcastanospermine are 560, 56, 29 and 1.1 μM, respectively, but these compounds are not as potent as zidovudine ($EC_{50} = 0.1$ μM) [56]. These sugar analogues showing anti-HIV activity have the common property that they are potent processing α-glucosidase inhibitors but not processing α-mannosidase inhibitors. The activities of these inhibitors toward processing α-glucosidase I correlate well with the antiviral effects [56, 57]. The *in vivo* data obtained to date do not promise practical use of processing α-glucosidase I inhibitors as anti-HIV agents. Problems exist in achieving therapeutic serum concentrations of inhibitors needed to inhibit α-glucosidase I sufficiently and side effects such as diarrhoea occur. With respect to diarrhoea, the prodrug of *N*-butyl-DNJ, glycovir (SC 49483) (**20**), was developed as a candidate anti-HIV agent. This prodrug is *N*-butyl-DNJ tetrabutanoate, which will be converted into active *N*-butyl-DNJ after it passes through the intestine, avoiding the diarrhoea [58]. However, it is believed that morphologic changes in various tissue cells were the result of nonspecific inhibition of host α-glucosidases by the prodrug, causing clinically silent perturbation in host cell glycoprotein processing and/or glycoprotein transport [59].

In contrast to the heavily glycosylated HIV envelope glycoproteins, the envelope glycoproteins of the hepatitis B virus (HBV) contain only two glycosylation sites [60]. However, the HBV glycoproteins are sensitive to inhibitors of the *N*-linked glycosylation pathway. In this virus, correct glycosylation appears to be necessary for processes involved in transport of the virus out of the host cell. *In vitro* treatment of HBV with *N*-butyl-DNJ results in a high proportion of virus particles being retained inside the cells [61]. Block *et al.* [62] reported that *N*-nonyl-DNJ (**21**) reduces the viremia in chronically infected woodchucks in a dose dependent manner. *N*-Nonyl-DNJ is 100–200

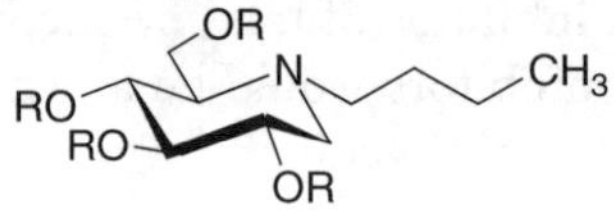

18: R = H
20: R = $COCH_2CH_2CH_3$

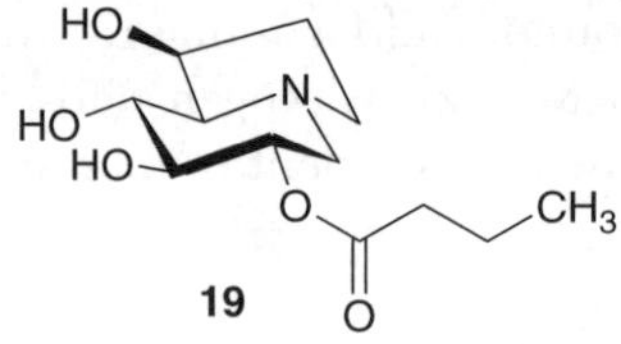

19

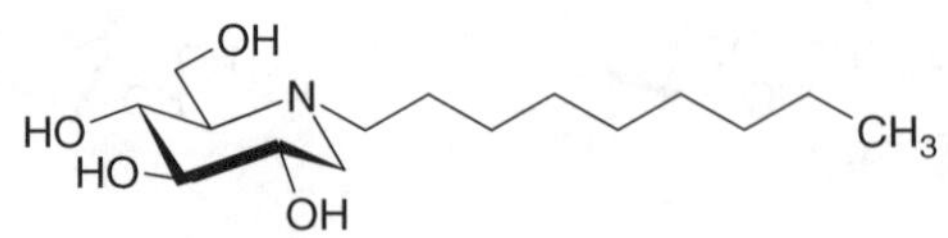

21

Figure 2.3 Structures of α-glucosidase inhibitors as antiviral agents

times more potent than *N*-butyl-DNJ in inhibiting HBV in cell based assays [60]. Furthermore, *N*-nonyl-DNJ, compared with *N*-butyl-DNJ, exhibits a prolonged hepatic retention of bovine viral diarrhoea virus (BVDV), a tissue culture surrogate of the human hepatitis C virus (HCV) [63] (see Chapter 9 in this volume). A single drug against HBV and HCV may be of great therapeutic value. However, when processing α-glucosidase inhibitors are used as antiviral agents, it remains to be determined what effects occur on host cell glycoprotein processing and/or glycoprotein transport.

2.3 β-Glucosidase inhibitors

2.3.1 Structures and *in vitro* inhibitory activity

In mammals, three prominent β-glucosidases have been well characterized: lysosomal glucocerebrosidase (GCase), intestinal lactase-phlorizin hydrolase (LPH) and the cytosolic β-glucosidase present in the liver of mammalian species. LPH is enterocyte-specific disaccharidase crucial for the digestion of dietary lactose and β-glucosides, while GCase hydrolyses glucosylceramide derived from endogenous membrane glycolipids. However, a metabolic role for the cytosolic β-glucosidase has not yet been established.

Nojirimycin (NJ) is a moderate inhibitor of human GCase, with an IC_{50} value of 19 μM. The removal of the anomeric OH group to give DNJ markedly lowered its inhibitory potential toward GCase [64]. Interestingly, the DNJ derivatives with a longer chain than the butyl group at the imino group enhance the GCase activity with increasing chain length [64, 65]. In particular, *N*-5-(adamantane-1-yl-methoxy)pentyl-DNJ (**22**) is a very powerful inhibitor of the non-lysosomal ($IC_{50} = 1.7$ nM) and membrane ($IC_{50} =$ 48 nM) glucosylceramidases [65]. Recently, Compain, Martin *et al.* synthesized a series of DNJ derivatives bearing an alkyl chain at the pseudo-anomeric position (C-1) in order to assess GCase inhibitory activity [64, 66–68]. Naturally occurring α-1-*C*-methyl-DNJ (**23**) showed weak inhibition toward GCase, with an IC_{50} value of 150 μM. Introduction of the butyl group at the C-1α position to give α-1-*C*-butyl-DNJ (**24**) slightly improved its inhibition toward GCase, whereas that to the C-1β position to give β-1-*C*-butyl-DNJ (**25**) abolished the inhibitory activity. Further elongation of an alkyl chain at C-1α, as seen in the *N*-alkyl-DNJ derivatives, remarkably enhanced the GCase inhibitory activity with increasing chain length [64]. In particular, α-1-*C*-octyl-DNJ (**26**) and α-1-*C*-nonyl-DNJ (**27**) showed 460-fold and 890-fold stronger inhibition toward GCase than DNJ, with IC_{50} values of 0.50 and 0.27 μM, respectively.

A number of polyhydroxylated nortropane alkaloids have been isolated from the families Solanaceae and Convolvulaceae [1, 4]. They were designated as calystegines. Calystegines possess three structural features in common: a nortropane ring system; two to four secondary hydroxyl groups varying in position and stereochemistry; and, a novel aminoketal functionality, which generates a tertiary hydroxyl group at the bicyclic ring bridgehead. The known members of calystegines have been subdivided into three groups on the basis of the number of the hydroxyl groups present, namely calystegines A with three OH groups, B with four OH groups and C with five OH groups. Among calystegines, calystegines A_3 (**28**), B_1 (**29**), B_2 (**30**) and C_1 (**31**) are inhibitors of almond

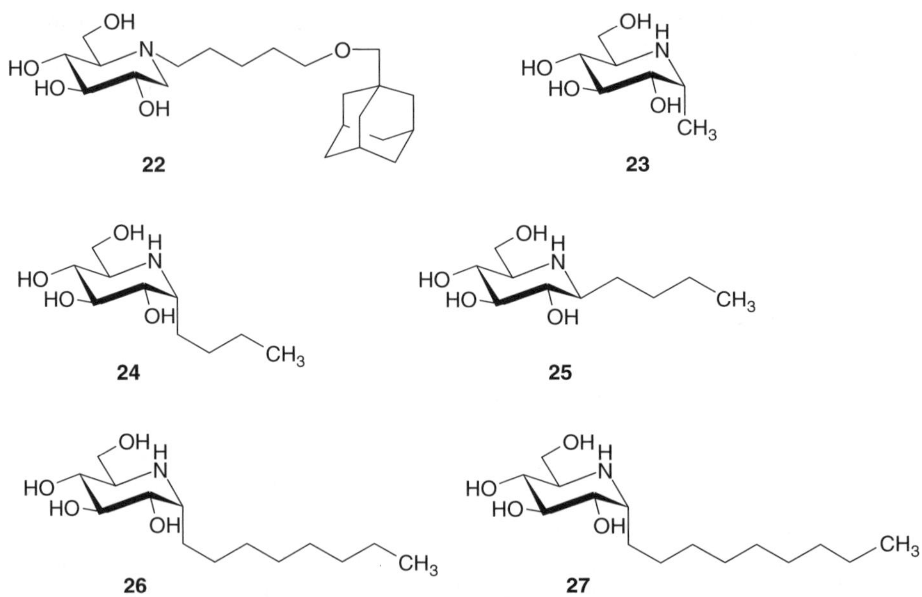

Figure 2.4 Structures of β-glucosidase-inhibiting *N*-alkyl- and *C*-alkyl-DNJ derivatives

β-glucosidase [69] and also potent inhibitors of GCase, with IC_{50} values of 3.1, 2.5, 1.0 and 2.5 μM, respectively [70]. Calystegines can be viewed as derivatives of 1,5-dideoxy-1,5-iminoxylitol (DIX) (**32**), with an ethano bridge across the 1,5-positions. DIX is a very specific inhibitor of GCase, with an IC_{50} value of 1.9 μM [70]. The fact that the inhibitory potency of α-1-*C*-alkyl-DNJ derivatives increases with the length of the alkyl chain suggests that introduction of an alkyl chain to the pseudo-anomeric position of DIX may lead to highly potent and selective inhibitors of GCase. Incorporation of a nonyl chain into DIX to give α-1-*C*-nonyl-DIX (**33**) dramatically improved its inhibitory potency, with an IC_{50} value of 6.8 nM [71]. In addition, α-1-*C*-nonyl-DIX inhibited GCase in a competitive manner, with a K_i value of 2.2 nM. Although *N*-nonyl-DIX (**34**) was prepared for comparison, this compound showed only slight improvement in inhibitory activity toward GCase. Very interestingly, the NMR data indicated that the introduction of a nonyl chain into the C-1 position of DIX caused a piperidine ring inversion from $^4C_1$ to $^1C_4$ conformation [71]. Replacement of the hydroxyl group at C-2 of DIX with a hydroxymethyl group gives isofagomine IFG (**17**), which enhanced its inhibitory potency by 60-fold, with an IC_{50} value of 40 nM [70]. The *N*-alkylation of IFG markedly lowered its activity, suggesting that an intact imino moiety at that position is essential for maintaining potent inhibition of GCase. On the other hand, incorporation of an alkyl chain longer than a butyl group to the C-6α position of IFG remarkably enhanced inhibitory activity toward GCase [72]. The most potent inhibitor α-6-*C*-nonyl-IFG (**35**) displayed a remarkable IC_{50} value of 0.6 nM, which is 93-fold more potent relative to IFG. Unlike **33**, compound **35** exists exclusively in the $^4C_1$ conformation with all substituents equatorial. Interestingly, the inhibition mode for IFG

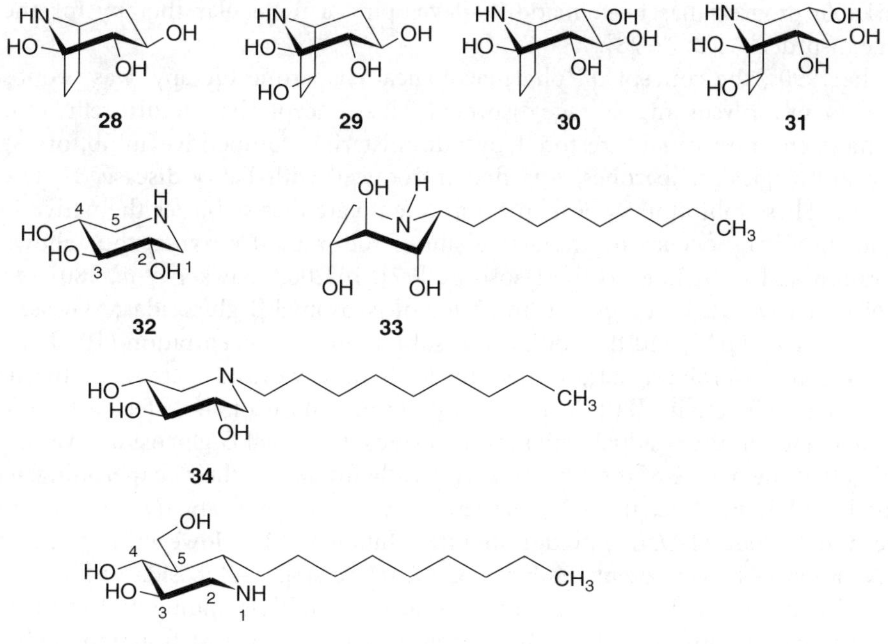

28 29 30 31

32 33

34

35

Figure 2.5 Structures of β-glucosidase-inhibiting iminoxylitol analogues and their alkyl derivatives

and α-6-*C*-butyl-IFG is competitive, whereas that for the α-6-*C*-alkyl-IFG derivatives with a longer chain is a mixed type inhibition [72].

2.3.2 Applications

Recent experimental data show that some human genetic diseases are due to mutations in proteins that influence their folding and lead to the retention of mutant proteins in the ER and to successive degradation [73, 74]. Lysosomes are membrane-bound cytoplasmic organelles that serve as a major degradative compartment in eukaryotic cells. The degradative function of lysosomes is carried out by more than 50 acid-dependent hydrolases contained within the lumen [75]. The glycosphingolipid (GSL) storage diseases are genetic disorders in which a mutation of one of GSL glycohydrolases blocks GSL degradation, leading to lysosomal accumulation of undegraded GSL [76]. Possible strategies for the treatment of these lysosomal storage diseases include enzyme replacement therapy, gene therapy, substrate deprivation and bone marrow transplantation. The successful treatment for such diseases to date is the enzyme replacement therapy for patients with type 1 Gaucher disease and Fabry disease. However, this enzyme replacement therapy is useful only in diseases in the absence of neuropathology since enzymes do not cross the blood-brain barrier. Another problem in this therapy is the cost, which prevents many patients from obtaining this treatment. In recent years, remarkable progress has been made in developing a molecular therapy for the GSL storage disorders [2, 3, 77, 78].

In the 1990s, the concept of 'pharmacological chaperone therapy' was proposed in the treatment for lysosomal storage disorders. This concept, that an intracellular activity of mutant enzymes can be restored by administering competitive inhibitors serving as pharmacological chaperones, was first introduced with Fabry disease, as described later [79]. These inhibitors appear to act as a template that stabilizes the native folding state in the ER by occupying the active site of the mutant enzyme, thus allowing its maturation and trafficking to the lysosome [77]. In 2002, Sawker *et al.* [80] reported that *N*-nonyl-DNJ (**21**) is a potent inhibitor of lysosomal β-glucosidase (GCase), with an IC_{50} value of 1 μM, and the addition of subinhibitory concentration (10 μM) of this compound to a fibroblast culture medium leads to a two-fold increase in the mutant (N370S) enzyme activity [80]. Very recently, examination of the effects of a series of DNJ analogues on the residual activities of various lysosomal β-glucosidase variants has revealed that the nature of the alkyl moiety greatly influences their chaperoning activity: *N*-butyl-DNJ is inactive, the DNJ derivatives with *N*-nonyl and *N*-decyl chains are active, and *N*-dodecyl-DNJ is predominantly inhibitory [81]. However, it is also known that *N*-nonyl-DNJ is a potent inhibitor of ER processing α-glucosidases like *N*-butyl-DNJ and hence has potential as an antiviral agent to inhibit folding and trafficking of viral envelope glycoproteins [62, 63]. Inhibitors targeting a host function such as ER processing α-glucosidases must be carefully considered in terms of side effects since they may inhibit folding, secretion and trafficking of other glycoproteins in a patient's cells or may inhibit directly lysosomal α-glucosidase, after being taken up into cells. In fact, addition of *N*-nonyl-DNJ at 10 μM lowered the cellular lysosomal α-glucosidase activity by 50 per cent throughout the assay period (10 days) in spite of the excellent chaperoning activity for the mutant β-glucosidase [64]. The inhibition of lysosomal α-glucosidase as a side effect may induce storage of glycogen in the lysosomes, as observed in Pompe disease. On the other hand, α-1-*C*-octyl-DNJ (**26**), with a K_i value of 0.28 μM, showed a novel chaperoning activity for N370S Gaucher variants, minimizing the potential for undesirable side effects such as lysosomal α-glucosidase inhibition [64]. Similarly, α-1-*C*-nonyl-DIX, which is a very specific inhibitor of GCase but not an inhibitor of lysosomal α-glucosidase, acts as a pharmacological chaperone and is able to double the residual cellular activity of GCase in N370S fibroblasts from Gaucher patients at 10 nM [71].

Very recently, Fan *et al.* have assessed the efficacy of pharmacological chaperoning activity on calystegines A_3, B_1, B_2, C_1, on IFG and on DIX [70]. Among them, IFG was the most effective pharmacological chaperone capable of increasing residual GCase activity in N370S Gaucher fibroblasts. Intracellular GCase activity increased approximately two-fold when cells were incubated with IFG. The effective concentration for IFG was determined to be 10–50 μM, whereas those of calystegines and DIX were 100–200 μM. This two-fold increase in the residual enzyme activity is expected to have a significant impact in the disease development among the patients with the N370S mutation because this mutation results in mild clinical symptoms and some of the patients having the homozygous N370S mutations are even asymptomatic [82]. Further details on this topic are presented in Chapter 10 of this volume.

2.4 α- and β-Galactosidase inhibitors

2.4.1 Structures and *in vitro* inhibitory activity

In 1988, galactostatin (galactonojirimycin) (**36**) was isolated from the fermentation broth of *Streptomyces lydicus* PA-5726 [83]. Galactostatin as well as NJ (**1**) can be converted to 1-deoxygalactonojirimycin (DGJ) (**37**) by catalytic hydrogenation with platinum catalyst or chemical reduction with $NaBH_4$ [83]. Galactostatin is known to be a powerful inhibitor of bacterial and fungal β-galactosidases and also to be an extremely potent inhibitor of plant and bacterial α-galactosidases, with K_i values ranging from 100–0.7 nM [84, 85]. Although DGJ has not yet been found from a natural source, β-1-*C*-butyl-DGJ (**38**) has been isolated from *Adenophora* sp. (Campanulaceae) [86]. α-Homogalactonojirimycin (α-HGJ) (**39**) bearing a hydroxymethyl group at the C-1α position of DGJ has been chemically synthesized [87] and is a five-fold weaker inhibitor of human lysosomal α-galactosidase A (α-Gal A) than DGJ [88]. The β-isomer of α-HGJ exhibited no significant inhibition toward β-galactosidases and still retained fairly potent inhibitory activity toward coffee bean α-galactosidase [89]. A naturally occurring iminosugar, α-homoallonojirimycin (α-HAJ) (**40**) is also a fairly potent inhibitor of α-Gal A because it can be regarded as the 2-epimer of α-HGJ [88]. Five-membered iminosugars with galactose-like configuration are also potent inhibitors of galactosidases. In 1985, Fleet *et al.* synthesized 1,4-dideoxy-1,4-imino-D-lyxitol (DIL) (**41**) from D-mannose and found it to be a strong inhibitor of coffee bean α-galactosidase, with an IC_{50} value of 0.2 μM [90]. Introduction of a hydroxymethyl group at the C-1β of DIL to give symmetrical 2,5-dideoxy-2,5-imino-D-galactitol (**42**) enhanced inhibitory potency toward this enzyme, with an IC_{50} value of 0.05 μM [91].

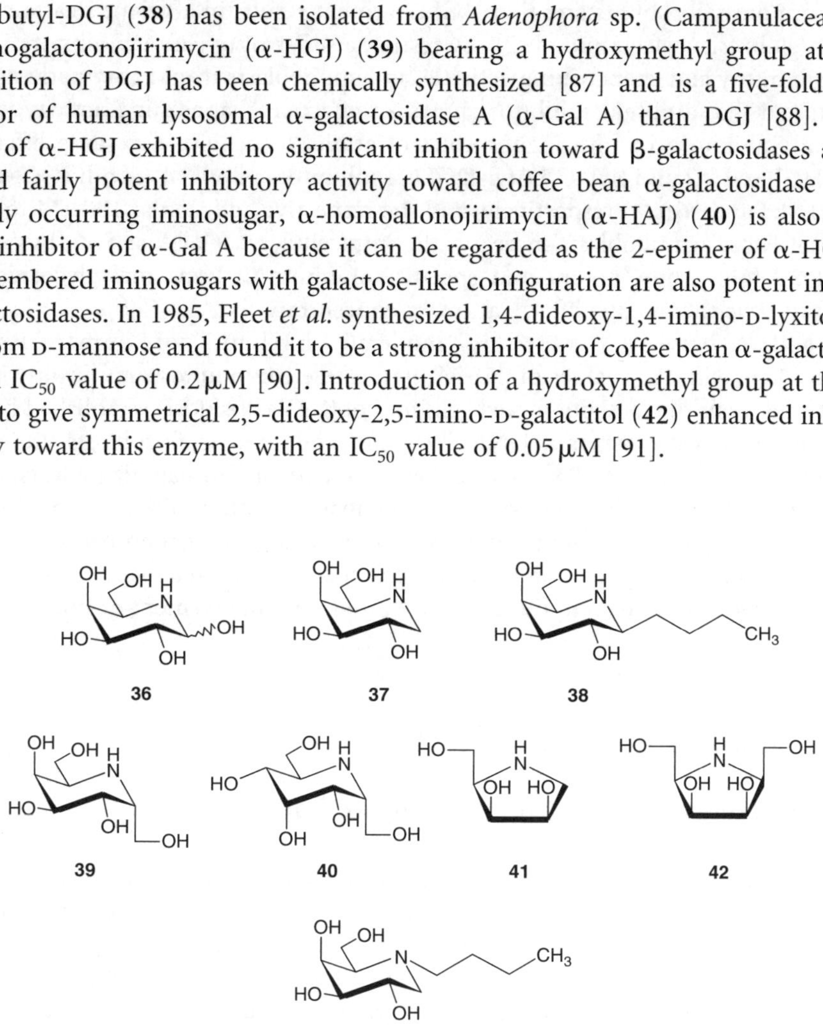

Figure 2.6 Structures of α- and β-galactosidase inhibitors

2.4.2 Applications

In 1999, Fan *et al.* reported that DGJ, a powerful competitive inhibitor of α-Gal A with a K_i value of 40 nM, effectively enhanced the mutant enzyme activity in lymphoblasts established from Fabry patients with the R301Q or Q279E mutation [79]. Furthermore, oral administration of DGJ to transgenic mice expressing a human mutant α-Gal A (R301Q) substantially elevated the enzyme activity in the major organs. DGJ acts as a pharmacological chaperone for the mutant protein to promote the successful escape from the ER quality-control mechanisms, resulting in the successful transport from the ER to the Golgi apparatus and its correct targeting to the lysosomes. In 2005, Yam *et al.* showed that DGJ induces trafficking of ER-retained R301Q α-Gal A to the lysosomes of transgenic mouse fibroblasts and that DGJ treatment results in efficient clearance of the substrate, globotriaosylceramide [92]. By testing a series of α-Gal A inhibitors for both *in vitro* inhibitory and chaperoning activities in lymphoblasts from Fabry patients, it was demonstrated that a potent inhibitor shows an effective chaperoning activity, whereas less potent inhibitors require higher concentrations to achieve the same effect [88]. DGJ, α-HGJ, α-HAJ and β-1-*C*-butyl-DGJ are inhibitors of α-Gal A with IC_{50} values of 0.04, 0.21, 4.3 and 16 μM, respectively, and the respective addition at 100 μM to culture medium of Fabry lymphoblasts increases the intracellular α-Gal A activity by 14-, 5.2-, 2.4- and 2.3-fold. Thus, potent and specific inhibitors of lysosomal glycosidases are expected to have therapeutic effects at lower concentrations.

DGJ and *N*-butyl-DGJ (**43**) are moderate inhibitors of human lysosomal β-galactosidase, with IC_{50} values of 25 μM [93]. These compounds significantly restored cellular mutant enzyme activity model cell lines of β-galactosidosis established by introducing mutant β-galactosidase cDNA into knockout mouse fibroblasts and human β-galactosidosis fibroblasts [93]. The enzyme activity in human fibroblasts with an R201C mutation increased seven-fold after incubation with 1 mM DGJ for four days. However, a survey of much more effective pharmacological chaperones is needed for clinical application of this method to β-galactosidosis because relatively high concentrations (0.5–1 mM) of these compounds are required for enhancement of mutant enzyme activity.

2.5 α-Mannosidase inhibitors

2.5.1 Structures and *in vitro* inhibitory activity

Certain poisonous plants often cause serious livestock losses. Australian legumes, *Swainsona*, are known as 'poison peas' and sheep eating them develop a syndrome called 'pea struck' [94, 95]. There is also the livestock poisoning by the closely related *Astragalus* and *Oxytropis* species, which are found throughout most of the world, and intoxication of livestock by certain of those species known as locoweeds in the western US is called 'locoism' [95, 96]. The poisoning is characterized by cytoplasmic vacuolation of neuronal cells due to accumulation of mannose-rich oligosaccharides in lysosomes [97]. The trihydroxyindolizidine alkaloid swainsonine (**44**) occurs in these legumes and

has been identified as a causative agent in locoism [96, 98]. Swainsonine is a potent inhibitor of lysosomal α-mannosidase [99] and Golgi α-mannosidase II [100]. *Astragalus lentiginosus* contains another indolizidine alkaloid 2-*epi*-lentiginosine (**45**) [101], which is probably an intermediate in the biosynthesis of swainsonine in *Rhizoctonia leguminicola* because the experimental refeeding of [^{3}H]-2-*epi*-lentiginosine resulted in a very high level (45 per cent) of incorporation of radioactivity into swainsonine [102]. Swainsonine has been isolated from convolvulaceous species other than legumes. *Ipomoea carnea* is found in most regions of Brazil and is an evergreen plant blooming throughout the year. Animals eat this plant especially in drought periods since it is one of the few plants that stay green. In experimental studies in which the plant was given to adult goats, all animals showed disorders of behaviours and consciousness as well as abnormalities of gait, ability to stand and posture, and one goat died [103]. *I. carnea* produces swainsonine and 2-*epi*-lentiginosine, which are potent inhibitors of human lysosomal α-mannosidase, with IC_{50} values of 0.04 and 5.0 μM, respectively [104, 105].

There are two classes of processing α-mannosidases in the ER [106, 107]. Class I α1,2-mannosidase (Man I) cleave only α–1,2-mannose residues and ER-resident Man I trims a single mannose in the middle branch of $Man_9(GlcNAc)_2$ to form $Man_8(GlcNAc)_2$ isomer B [108]. This class of enzyme is inhibited by iminosugars with a six-membered ring such as 1-deoxymannojirimycin (DMJ) (**46**) and kifunensine (**47**) [109]. DMJ was first isolated from the seeds of *Lonchocarpus sericeus* (Leguminosae), a native to the West Indies and tropical America [110], and later shown to be present in many disparate species of plants and microorganisms [111]. Kifunensine was initially isolated as a weak inhibitor of Jack Bean α-mannosidase ($IC_{50} = 100$ μM) from the culture broth of the actinomycete, *Kitasatosporia kifunense* 9482 [112]. In contrast, class II α-mannosidase (Man II) including Golgi α-mannosidase II and lysosomal α-mannosidase are not so specific and cleave α–1,2-, α–1,3-, and α–1,6-linked mannose residues [113–115].

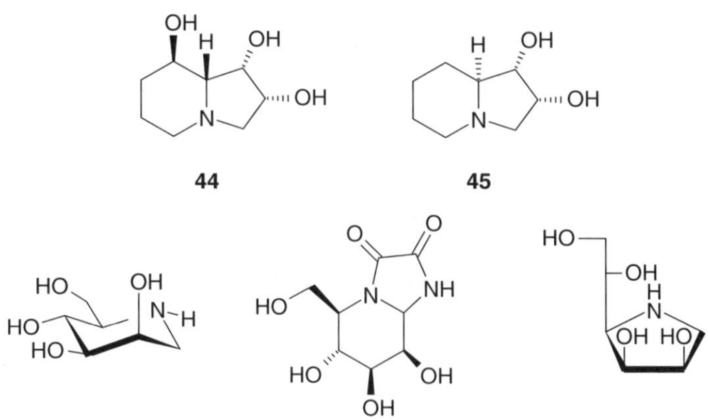

44 **45** **46** **47** **48**

Figure 2.7 Structures of α-mannosidase inhibitors

Man II is inhibited by swainsonine and a synthetic iminosugar 1,4-dideoxy-1,4-imino-D-mannitol (DIM) (**48**) [107, 109].

2.5.2 Applications

Swainsonine is a toxic indolizidine alkaloid found in species of *Astragalus* and *Oxytropis* responsible for causing locoism, as described above. However, this alkaloid appears to have clinical benefit as well. Although a number of alkaloids have been reported to show anticancer activity, research has focused on developing swainsonine as a drug candidate for the management of human malignancies [111]. By inhibiting Golgi α-mannosidase II activity, swainsonine effectively shuts down the mannose trimming pathway after the initiation of β–1,6-GlcNAc-linked chain, which inhibits tumour cell metastasis, decreases solid tumour growth in mice and enhances the cellular immune response [116, 117]. Nevertheless, serious inhibition of lysosomal α-mannosidase makes its clinical use less desirable [118–120].

Recent studies have demonstrated that ER-resident Man I plays a key role in the degradation of misfolded glycoproteins [121–123]. This enzyme creates a signal that targets misfolded glycoproteins for translocation out of the ER and degradation by the proteasome, which is known as ER-associated degradation (ERAD) [124, 125]. The inhibitors of ER-resident Man I such as kifunensine and DMJ are found to suppress ERAD of misfolded glycoproteins [126–129]. These α-mannosidase inhibitors as well as pharmacological chaperones might be candidates for a new molecular therapy of human genetic disorders.

2.6 Concluding remarks and future prospects

From the success of α-glucosidase inhibitors as antidiabetic agents and neuraminidase inhibitors as anti-influenza drugs, the practical use of glycosidase inhibitors appears to be limited to diabetes and viral infection. However, since glycosidases are involved in a wide range of anabolic and catabolic processes of carbohydrates, their inhibitors could have many kinds of beneficial effects as therapeutic agents. The pharmacological chaperone therapy for lysosomal storage disorders is a quite new application of glycosidase inhibitors. Although the safety and effectiveness of enzyme replacement therapy for such diseases were demonstrated in type 1 Gaucher disease and Fabry disease, the application is only for a nonneuronopathic types of diseases since enzyme proteins do not cross the blood-brain barrier. The pharmacological chaperone therapy and substrate reduction therapy with small molecules are attracting considerable interest, particularly for the neuronopathic lysosomal storage disorders. Orally administered pharmacological chaperones that treat lysosomal storage diseases might replace intravenous enzyme replacement therapy. Many inhibitors of lysosomal glycosidases are waiting for the evaluation of pharmacological chaperone therapy for such diseases and some of them are in preclinical or phase I /II clinical trials (Amicus Therapeutics Inc., Cranbury, NJ, USA).

References

1. Asano, N., Nash, R. J., Molyneux, R. J. and Fleet, G. W. J., 2000 *Tetrahedron: Asymmetry*, **11**, 1–36.
2. Asano, N., 2003 *Glycobiology*, **13**, 93R–104R.
3. Butters, T. D., Dwek, R. A. and Platt, F. M., 2005 *Glycobiology*, **15**, 43R–52R.
4. Asano, N., 2003 *Curr. Top. Med. Chem.*, **3**, 471–484.
5. Inoue, S., Tsuruoka, T. and Niida, T., 1966 *J. Antibiot.*, **19**, 288–292.
6. Inoue, S., Tsuruoka, T., Ito, T. and Niida, T. 1968 *Tetrahedron*, **24**, 2125–2144.
7. Yagi, M., Kouno, T., Aoyagi, Y. and Murai, H., 1976 *Nippon Nogei Kagaku Kaishi*, **50**, 571–572.
8. Asano, N., Tomioka, E., Kizu, H. and Matsui, K., 1994 *Carbohydr. Res.*, **253**, 235–245.
9. Asano, N., Oseki, K., Tomioka, E., Kizu, H. and Matsui, K., 1994 *Carbohydr. Res.*, **259**, 243–255.
10. Asano, N., Yamashita, T., Yasuda, K., Ikeda, K., Kizu, H., Kameda, Y., Kato, A., Nash, R. J., Lee, H.-S. and Ryu, K.-S. 2001 *J. Agric. Food Chem.*, **49**, 4208–4213.
11. Asano, N., Oseki, K., Kizu, H. and Matsui, K., 1994 *J. Med. Chem.*, **37**, 3701–3706.
12. Elbein, A. D., 1987 *Ann. Rev. Biochem.*, **56**, 497–534.
13. Hettkamp, H., Legler, G. and Bause, E., 1984 *Eur. J. Biochem.*, **142**, 85–90.
14. Schweden, J., Borgmann, C., Legler, G. and Bause, E., 1986 *Arch. Biochem. Biophys.*, **248**, 335–340.
15. Nash, R. J., Bell, E. A. and Williams, J. M., 1985 *Phytochemistry*, **24**, 1620–1622.
16. Fleet, G. W. J. and Smith, P. W., 1986 *Tetrahedron*, **42**, 5685–5692.
17. Kite, G. C., Fellows, L. E., Fleet, G. W. J., Liu, P. S., Scofield, A. M. and Smith, N. G., 1988 *Tetrahedron Lett.*, **29**, 6483–6486.
18. Liu, P. S., 1987 *J. Org. Chem.*, **52**, 4717–4721.
19. Rhinehart, B. L., Robinson, K. M., Liu, P. S., Payne, A. J., Wheatley, M. E. and Wagner, S. R., 1987 *J. Pharmacol. Exp. Ther.*, **241**, 915–920.
20. Kite, G. C., Horn, J. M., Romeo, J. T., Fellows, L. E., Lees, D. C., Scofield, A. M. and Smith, N. G., 1990 *Phytochemistry*, **29**, 103–105.
21. Asano, N., Kato, A., Miyauchi, M., Kizu, H., Kameda, Y., Watson, A. A., Nash, R. J. and Fleet, G. W. J., 1998 *J. Nat. Prod.*, **61**, 625–628.
22. Asano, N., Nishida, M., Kato, A., Kizu, H., Matsui, K., Shimada, Y., Itoh, T., Baba, M., Watson, A. A., Nash, R. J., de Q. Lilley, P. M., Watkin, D. J. and Fleet, G. W. J., 1998 *J. Med. Chem.*, **41**, 2565–2571.
23. Zeng, Y., Pan, Y. T., Asano, N., Nash, R. J. and Elbein, A. D., 1997 *Glycobiology*, **7**, 297–304.
24. Hohenschutz, L. D., Bell, E. A., Jewess, P. J., Leworthy, D. P., Pryce, R. J., Arnold, E. and Clardy, J., 1981 *Phytochemistry*, **20**, 811–814.
25. Molyneux, R. J., Roitman, J. N., Dunnheim, G., Szumilo, T. and Elbein, A. D., 1986 *Arch. Biochem. Biophys.*, **251**, 450–457.
26. Saul, R., Ghidoni, J., Molyneux, R. J. and Elbein, A. D., 1985 *Proc. Natl. Acad. Sci. USA*, **82**, 93–97.
27. Pan, Y. T., Hori, H., Saul, R., Sanford, B. A., Molyneux, R. J. and Elbein, A. D., 1983, *Biochemistry*, **22**, 3975–3984.
28. Sasak, V. W., Ordovas, J. M. and Elbein, A. D., 1985 *Biochem. J.*, **232**, 759–766.
29. Palamarczyk, G. and Elbein, A. D., 1985 *Biochem. J.*, **232**, 795–804.
30. Molyneux, R. J., Benson, M., Wong, R. Y., Tropea, J. E. and Elbein, A. D., 1988 *J. Nat. Prod.*, **51**, 1198–1206.
31. Nash, R. J., Fellows, L. E., Dring, J. V., Fleet, G. W. J., Girdhar, A., Ramsden, N. G., Peach, J. M., Hegarty, M. P. and Scofield, A. M., 1990 *Phytochemistry*, **29**, 111–114.

32. Kato, A., Kano, E., Adachi, I., Molyneux, R. J., Watson, A. A., Nash, R. J., Fleet, G. W. J., Wormald, M. R., Kizu, H., Ikeda, K. and Asano, N., 2003 *Tetrahedron: Asymmetry*, **14**, 325–331.
33. Nash, R. J., Thomas, P. I., Waigh, R. D., Fleet, G. W. J., Wormald, M. R., de Lilley, P. M. and Watkin, D. J., 1994 *Tetrahedron Lett.*, **35**, 7849–7852.
34. Wormald, M. R., Nash, R. J., Watson, A. A., Bhadoria, B. K., Langford, R., Sims, M. and Fleet, G. W. J., 1996 *Carbohydr. Lett.*, **2**, 169–174.
35. Welter, A., Jaot, J., Dardenne, G., Marlier, M. and Casimir, J., 1976 *Phytochemistry*, **15**, 747–749.
36. Fleet, G. W. J., Nicholas, S. J., Smith, P. W., Evans, S. V., Fellows, L. E. and Nash, R. J., *Tetrahedron Lett.*, **26**, 3127–3130.
37. Scofield, A. M., Fellows, L. E., Nash, R. J. and Fleet, G. W. J., 1986 *Life Sci.*, **39**, 645–650.
38. Yu, C.-Y., Asano, N., Ikeda, K., Wang, M.-X., Butters, T. D., Wormald, M. R., Dwek, R. A., Winters, A. L., Nash, R. J. and Fleet, G. W. J., 2004 *Chem. Commun.* 1936–1937.
39. Asano, N., Ikeda, K., Yu, L., Kato, A., Takebayashi, K., Adachi, I., Kato, I., Ouchi, H., Takahata, H. and Fleet, G. W. J., 2005 *Tetrahedron: Asymmetry*, **16**, 223–229.
40. Kato, A., Kato, N., Kano, E., Adachi, I., Ikeda, K., Yu, L., Okamoto, T., Banba, Y., Ouchi, H., Takahata, H. and Asano, N., 2005 *J. Med. Chem.*, **48**, 2036–2044.
41. Schmidt, D. D., Frommer, W., Müller, L. and Truscheit, E., 1979 *Naturwissenschaften*, **66**, 584–585.
42. Murao, S. and Miyata, S., 1980 *Agric. Biol. Chem.*, **44**, 219–221.
43. Ezure, Y., Murao, S., Miyazaki, K. and Kawamata, M., 1985 *Agric. Biol. Chem.*, **49**, 1119–1125.
44. Junge, B., Matzke, M. and Stltefuss, J., 1996 *Handbook of Experimental Pharmacology* (eds J. Kuhlmann and W. Plus), Springer-Verlag, New York, Vol. 119, pp. 411–482.
45. Joubert, P. H., Foukaridis, G. N. and Bopape, M. L., 1987 *Eur. J. Clin. Pharmacol.*, **31**, 723–724.
46. Joubert, P. H., Venter, H. L. and Foukaridis, G. N., 1990 *Br. J. Clin. Pharmacol.*, **30**, 391–396.
47. Franz, S., Calvillo, L., Tillmanns, J., Elbing, I., Dienesch, C., Bischop, H. and Ertl, G., 2005 *FASEB J.*, **19**, 591–593.
48. Defronzo, R. A., Bonadonna, R. C. and Ferrannini, E., 1992 *Diabetes Care*, **15**, 318–368.
49. Martin, J. L., Veluraja, K., Ross, K., Johnson, L. N., Fleet, G. W. J., Ramsden, N. G., Bruce, I., Orchard, M. G., Oikonomakos, N. G., Papageorgiou, A. C., Leonidas, D. D. and Tsitoura, H. S., 1991 *Biochemistry*, **30**, 10101–10116.
50. Fosgerau, K., Westergaard, N., Quistorff, B., Grunner, N., Kristiansen, M. and Lundgren, K., 2000 *Arch. Biochem. Biophys.*, **380**, 274–284.
51. Andersen, B., Rassov, A., Westergaard, N. and Lundgren, K., 1999 *Biochem. J.*, **342**, 545–550.
52. Jakobsen, P., Lundbeck, J. M., Kristiansen, M., Breinholt, J., Demuth, H., Pawlas, J., Torres Candela, M. P., Andersen, B., Westergaard, N., Lundgren, K. and Asano, N., 2001 *Bioorg. Med. Chem.*, **9**, 733–744.
53. Karpas, A., Fleet, G. W. J., Dwek, R. A., Petursson, S., Namgoong, S. K., Ramsden, N. G., Jacob, G. S. and Rademacher, T. W., 1988 *Proc. Natl. Acad. Sci. USA*, **85**, 9229–9233.
54. Fu, Y.-K, Hart, T. K., Jonak, Z. L. and Bugelski, P. J., 1993 *J. Virol.*, **67**, 3818–3825.
55. Ficher, P. B., Collin, M., Karlsson, G. B., Lames, W., Butters, T. D., Davis, S. J., Gordon, S., Dwek, R. A. and Platt, F. M., 2000 *J. Virol.*, **69**, 5791–5797.
56. Taylor, D. L., Sunkara, P. S., Liu, P. S., Kang, M. S., Bowlin, T. L. and Tyms, A. S., 1991 *AIDS*, **5**, 693–698.
57. Sunkara, P. S., Taylor, D. L., Kang, M. S., Bowlin, T. L., Liu, P. S. and Tyms, A. S., 1998 *Lancet*, 1206.
58. Cook, C. S., Karabatsos, P. J., Schoenhard, G. L. and Karim, A., 1995 *Pharm. Res.*, **12**, 1158–1164.
59. Khan, K. N., Snook, S. S., Semler, D. E., Baron, D. A. and Alden, C. L., 1996 *Toxicol. Pathol.*, **24**, 531–538.

60. Mehta, A., Zitzmann, N., Rudd, P. M., Bock, T. M. and Dwek, R. A., 1998 *FEBS Lett.*, **430**, 17–22.
61. Block, T. M., Lu, X., Platt, F. M., Foster, G. R., Gerlich, W. H., Blumberg, B. S. and Dwek, R. A., 1994 *Proc. Natl. Acad. Sci. USA*, **91**, 2235–2239.
62. Block, T. M., Lu, X., Mehta, A. S., Blumberg, B. S., Tennant, B., Ebling, M., Korba, B., Lansky, D. M., Jacob, G. S. and Dwek, R. A., 1998, *Nat. Med.*, **4**, 610–614.
63. Zitzmann, N., Mehta, A. S., Carrouée, S., Butters, T. D., Platt, F. M., McCauley, J., Blumberg, B. S., Dwek, R. A. and Block, T. M., 1999 *Proc. Natl. Acad. Sci. USA*, **96**, 11878–11882.
64. Yu, L., Ikeda, K., Kato, A., Adachi, I., Godin, G., Compain, P., Martin, O. R. and Asano, N., 2006 *Bioorg. Med. Chem.*, **14**, 7736–7744.
65. Overkleeft, H. S., Renkema, G. H., Neele, J., Vianello, P., Hung, I. O., Strijland, A., van der Burg, A. M., Koomen, G.-J., Pandit, U. K. and Aerts, J. M. F. G., 1998 *J. Biol. Chem.*, **273**, 26522–26527.
66. Godin, G., Compain, P., Masson, G. and Martin, O. R., 2002 *J. Org. Chem.*, **67**, 6960–6970.
67. Godin, G., Compain, P. and Martin, O. R., 2003 *Org Lett.*, **5**, 3269–3272.
68. Godin, G., Compain, P., Masson, G., Martin, O. R., Ikeda, K., Yu, L. and Asano, N., 2004 *Bioorg. Med. Chem. Lett.*, **14**, 5991–5995.
69. Asano, N., Kato, A., Oseki, K., Kizu, H. and Matsui, K., 1995 *Eur. J. Biochem.*, **229**, 369–376.
70. Chang, H.-H., Asano, N., Horii, S., Ichikawa, Y. and Fan, J.-Q., 2006 *FEBS J.*, **273**, 4082–4092.
71. Compain, P., Martin, O. R., Boucheron, C., Godin, G., Yu, L., Ikeda, K. and Asano, N., 2006 *ChemBioChem.*, **7**, 1356–1359.
72. Zhu, X., Sheth, K. A., Li, S., Chan, H.-H. and Fan, J.-Q., 2005 *Angew. Chem. Int. Ed.*, **44**, 7450–7453.
73. Bychkova, V. E. and Ptitsyn, O. B., 1995 *FEBS Lett.*, **359**, 6–8.
74. Welch, W. J. and Brown, C. R., 1996 *Cell Stress Chap.*, **1**, 109–115.
75. de Duve, C., 1963 *Lysosomes* (eds A. V. S. de Reuck and M. P. Cameron) Churchill, London, pp. 1–35.
76. Kornfeld, S. and Mellman, I., 1989 *Annu. Rev. Cell Biol.*, **5**, 483–525.
77. Fan, J.-Q., 2003 *TREND Pharmacol. Sci.*, **24**, 355–360.
78. Cohen, F. E. and Kelly, J. W., 2003 *Nature*, **426**, 905–909.
79. Fan, J.-Q., Ishii, S., Asano, N. and Suzuki, Y., 1999 *Nat. Med.*, **5**, 112–115.
80. Sawker, A. R., Cheng, W.-C., Beutler, E., Wong, C.-H., Balch, W. E. and Kelly, J. W., 2002 *Proc. Natl. Acad. Sci. USA*, **99**, 15428–15433.
81. Sawkar, A. R., Adamsky-Werner, S. L., Chen, W.-C., Wong, C.-H., Beutler, E., Zimmer, K.-P. and Kelly, J. W., 2005 *Chem. Biol.*, **12**, 1235–1244.
82. Elstein, D., Abrahamov, A., Dweck, A., Hadas-Halpern, I. and Zimran, A., 2002 *Paediatr. Drugs*, **4**, 417–426.
83. Miyake, Y. and Ebata, M., 1988 *Agric. Biol. Chem.*, **52**, 661–666.
84. Miyake, Y. and Ebata, M., 1988 *Agric. Biol. Chem.*, **52**, 1649–1654.
85. Legler, G. and Pohl, S., 1986 *Carbohydr. Res.*, **155**, 119–129.
86. Ikeda, K., Takahashi, M., Nishida, M., Miyauchi, M., Kizu, H., Kameda, Y., Arisawa, M., Watson, A. A., Nash, R. J., Fleet, G. W. J. and Asano, N., 2000 *Carbohydr. Res.*, **323**, 73–80.
87. Martin, O. R., Xie, F. and Liu, L., 1995 *Tetrahedron Lett.*, **36**, 4027–4030.
88. Asano, N., Ishii, S., Kizu, H., Ikeda, K., Yasuda, K., Kato, A., Martin, O. R. and Fan, J.-Q., 2000 *Eur. J. Biochem.*, **267**, 4179–4186.
89. Martin, O. R., Saavedra, O. M., Xie, F., Liu, L., Picasso, S., Vogel, P., Kizu, H. and Asano, N., 2001 *Bioorg. Med. Chem.*, **9**, 1269–1278.
90. Fleet, G. W. J., Nicholas, S. J., Smith, P. W., Evans, S. V., Fellows, L. E and Nash R. J., 1985 *Tetrahedron Lett.*, **26**, 3127–3130.
91. Wang, Y.-F., Takaoka, Y. and Wong, C.-H., 1994 *Angew. Chem. Int. Ed. Engl.*, **314**, 1242–1244.

92. Yam, G. H.-F., Zuber, C. and Roth, J., 2005 *FASEB J.*, **19**, 12–18.
93. Tominaga, L., Ogawa, Y., Taniguchi, M., Ohno, K., Matsuda, J., Oshima, A., Suzuki, Y. and Nanba, E., 2001 *Brain Dev.*, **23**, 284–287.
94. James, L. F., van Kampen, K. R. and Hartley, W. J., 1970 *Pathol. Vet.* **7**, 116–125.
95. Hartley, W. J., Baker, D. C. and James, L. F., 1989 *Swainsona and Related Glycosidase Inhibitors* (eds L. F. James, Elbein, A. D., Molyneux, R. J. and Warren, C. D.) Iowa State University Press, Ames, IA, pp. 50–56.
96. Molyneux, R. J. and James, L. F.,1982 *Science*, **216**, 190–191.
97. Dorling, P. R., Huxtable, C. R. and Vogel, P., 1978 *Neuropathol. Appl. Neurobiol.*, **4**, 285–295.
98. Colegate, S. M., Dorling, P. R. and Huxtable, C. R., 1979 *Aust. J. Chem.*, **32**, 2257–2264.
99. Dorling, P. R. and Huxtable, C. R., 1980 *Biochem. J.*, **191**, 649–651.
100. Fuhrmann, U., Bause, E., Legler, G. and Ploegh, H., 1984 *Nature*, **307**, 755–758.
101. Pastuszak, I., Molyneux, R. J., James, L. F. and Elbein, A. D., 1990 *Biochemistry*, **29**, 1886–1891.
102. Harris, T. M., Harris, C. M., Hill, J. E., Ungemach, F. S., Broquist, H. P. and Wickwire, B. M., 1987 *J. Org. Chem.*, **52**, 3094–3098.
103. Tokarnia, C. H., Dobereiner, J. and Peixoto, P. V., 2002 *Toxicon*, **40**, 1635–1660.
104. Haraguchi, M., Gorniak, S. L., Ikeda, K., Minami, Y., Kato, A., Watson, A. A., Nash, R. J. and Molyneux, R. J., 2003 *J. Agric. Food Chem.*, **51**, 4995–5000.
105. Ikeda, K., Kato, A., Adachi, I., Haraguchi, M. and Asano, N., 2003 *J. Agric. Food Chem.*, **51**, 7642–7646.
106. Moremen, K. W., Trimble, A. and Herscovics, A., 1994 *Glycobiology*, **4**, 113–125.
107. Daniel, P. F., Winchester, B. and Warren, C. D., 1994 *Glycobiology*, **4**, 551–566.
108. Helenius, A. and Aebi, M., 2001 *Science*, **291**, 2364–2369.
109. Elbein, A. D., 1991 *FESEB J.*, **5**, 3055–3063.
110. Fellows, L. E., Bell, A., Lynn, D. G., Pilkiewicz, F., Miura, I. and Nakanishi, K., 1979 *J. Chem. Soc. Chem. Commun.*, 977–978.
111. Watson, A. A., Fleet, G. W. J., Asano, N., Molyneux, R. J. and Nash, R. J., 2001 *Phytochemistry*, **56**, 265–295.
112. Kayakiri, H., Kasahara, C., Oku, T. and Hashimoto, M., 1990 *Tetrahedron Lett.*, **31**, 225–226.
113. Moremen, K. W. and Robbins, P. W., 1991 *J. Cell Biol.*, **115**, 1521–1534.
114. Nebes, V. L. and Schmidt, M. C., 1994 *Biochem Biophys. Res. Commun.*, **200**, 239–245.
115. Liao, Y. F., Lal, A. and Moremen, K. W., 1996 *J. Biol. Chem.*, **271**, 28348–28358.
116. Dennis, J. W., 1986 *Cancer Res.*, **46**, 5131–5136.
117. Humphries, M. J., Matsumoto, K., White, S. L. and Olden, K., 1986 *Proc. Natl. Acad. Sci. USA*, **83**, 1752–1756.
118. Tulsiani, D. R., Harris, T. M. and Touster, O., 1982 *J. Biol. Chem.*, **257**, 7936–7939.
119. De Gasperi, R., Daniel, P. F. and Warren, C. D., 1992 *J. Biol. Chem.*, **267**, 9706–9712.
120. Goss, P. E., Reid, C. L., Bailey, D. and Dennis, J. W., 1997 *Clin. Cancer Res.*, **3**, 1077–1086.
121. Helenius, A., 1994 *Mol. Biol. Cell*, **5**, 253–265.
122. Liu, Y., Choudhury, P., Cabral, C. M. and Sifers, R. N., 1999 *J. Biol. Chem.*, **274**, 5861–5867.
123. Knop, M., Hauser, N. and Wolf, D. H., 1996 *Yeast*, **12**, 1229–1238.
124. Jakob, C. A., Burda, P., Roth, J. and Aebi, M., 1998 *J. Cell Biol.*, **142**, 1223–1233.
125. Ellgaard, L., Molinari, M. and Helenius, A., 1999 *Science*, **286**, 1882–1888.
126. Liu, Y., Choudhury, P., Cabral, C. M. and Sifers, R. N., 1997 *J. Biol. Chem.*, **272**, 7946–7951.
127. Marcus, N. Y. and Perlmutter, D. H., 2000 *J. Biol. Chem.*, **275**, 1987–1992.
128. Vallée, F., Karaveg, K., Herscovics, A., Moremen, K. W. and Howell, P. L., 2000 *J. Biol. Chem.*, **275**, 41287–41298.
129. Tokunaga, F., Brostrom, C., Koide, T. and Arvan, P., 2000 *J. Biol. Chem.*, **275**, 40757–40764.

3

General strategies for the synthesis of iminosugars and new approaches towards iminosugar libraries

Barbara La Ferla, Laura Cipolla and **Francesco Nicotra**

3.1 Introduction

The numerous potential therapeutic applications of iminosugars, reviewed in other chapters of this book, have greatly contributed to keeping a high level of interest among synthetic chemists for these compounds. This has stimulated the search for more efficient approaches for the preparation of compounds of already proven activity and, at the same time, it has contributed to the research of novel structures and to the exploration of new and original synthetic strategies. General methods for iminosugar synthesis have already been described in many review articles and will not be further discussed in this occasion. The principal approach to iminosugars adopts carbohydrates as the starting materials and introduces the ring nitrogen through their manipulation. This approach quite often leads to long and difficult synthetic sequences, so recently chemists have adopted new strategies that allow the formation of libraries of compounds by a common synthetic pathway, rather than a parallel synthesis of individual target molecules. This chapter will focus on the methods devoted to the preparation of such iminosugar libraries. Particular attention will also be given to the most innovative synthetic approaches that appeared in the literature mostly from 2002. These innovative approaches can be found mainly in total syntheses proposed from noncarbohydrate starting materials to access both monocyclic piperidine, pyrrolidine and 1-*N*-iminosugar, and bicyclic pyrrolizidine and indolizidine-based compounds. In order to increase potency and/or specificity in the biological activity, other new bicyclic compounds with novel natural or unnatural structures have been designed recently; among them are tetrahydropyridoimidazole, isourea-type indolizidines, spirocyclopropyl and polyhydroxylated azaazulenes iminosugars. Since carbohydrates are usually found conjugated to other biomolecules,

Iminosugars: From synthesis to therapeutic applications Edited by Philippe Compain and Olivier R. Martin
 ISBN 978-0-470-03391-3

such as lipids, proteins, nucleosides or phosphate groups, iminosugars can also be integrated into glycoconjugate mimetics. Among them iminosugar-based peptidomimetics which combine hydrolytic stability with inhibitory properties are particularly interesting, and the synthesis of examples of iminosugar α-amino acids have been reported. Among other conjugates, a few synthetic methods for iminosugar phosphonates and nucleotides are presented. Libraries of iminosugar-conjugates have also been prepared as a new class of glycosidase and glycosyltransferase inhibitors, and as novel biological probes and diagnostics.

The synthesis of compounds such as *C*-glycoside, imino-*C*-disaccharides and *gem*-diamine 1-*N*-iminosugars has been omitted since other chapters of this volume focus on such compounds.

3.2 Monocyclic compounds

Since the discovery of nojirimycin, the first iminosugar isolated and found to be a potent inhibitor of β-glycosidases, many efforts have been devoted to the synthesis of monocyclic polyhydroxylated piperidines and pyrrolidines, in search of compounds with a high selectivity towards specific carbohydrate processing enzymes. A great number of synthetic approaches have been proposed starting mainly from carbohydrate precursors, and these have been very well described in previous review articles [1, 2]. This chapter is not intended to be an overview of syntheses from carbohydrates, which generally focus on the preparation of single target molecules, although this approach is still very popular, as evidenced by the many papers published in recent years. For this reason many efforts have recently been made towards the methods which allow the synthesis of a library of compounds by a common synthetic pathway, or by different decoration of a common scaffold compound. The generation of libraries of molecules is particularly useful in the search for powerful and specific inhibitors of a selected target enzyme, since the rational design of inhibitors is often extremely difficult due to the limited information regarding the structure of enzyme active sites. The idea of a 'scaffold decoration strategy' was first proposed for the generation of inhibitors of fucosidase and fucosyltransferase [3] and was realized a few years later in the preparation of a library of fuconojirimycin derivatives [4] by the condensation of the amino group of fuconojirimycin scaffold 1 (Scheme 1) with a sub-library of 60 aliphatic and aromatic carboxylic acids in a microtiter plate.

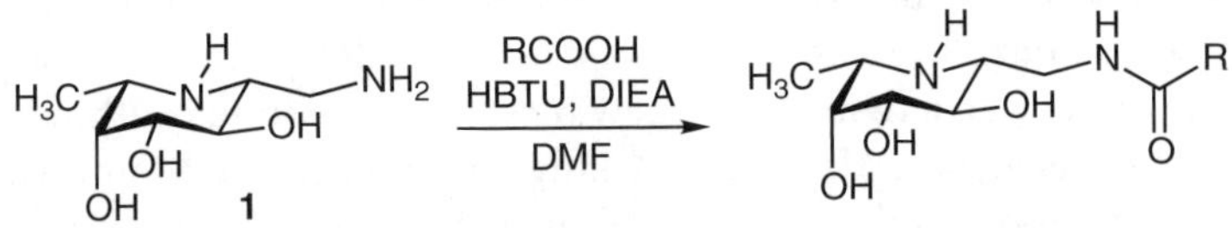

Scheme 1 Generation of a library of fuconojirimycin derivatives

Using a similar strategy a combinatorial library of iminosugars with galacto/manno configuration has also been prepared [5]. The five member ring carrying hydroxyl

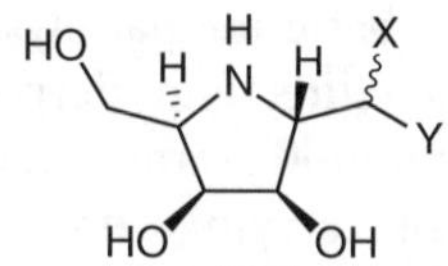

2 (X,Y)

X = BuNH, $(CH_3)_2N(CH_2)_2NH$, $CH_3(CH_2)_9NH$,
$HOCH_2CH_2NH$, $CH_3O(CH_2)_3NH$, $PhCH_2CH_2NH$,

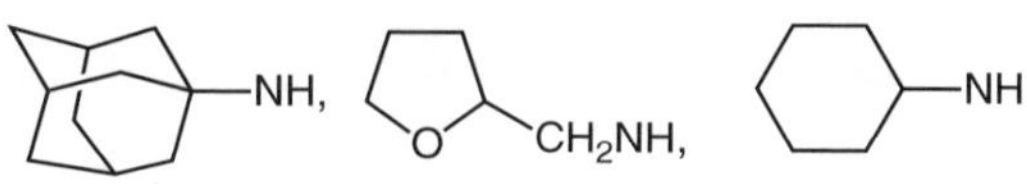

Y = H, CH_2NH_2, $CONH_2$

Figure 3.1 Five-membered ring iminocyclitols library

groups in a specific orientation was chosen to mimic the cation-like transition state of glycosidases and glycosyltransferases. At position C1' diversity was introduced in a combinatorial fashion by derivatization with two series of substituents affording a small library of 27 compounds (Figure 3.1).

The strategy that allows the synthesis of a library of compounds by a common pathway consists of the introduction of possible derivatizations during the synthesis of the iminosugar core structure. As for the previous strategy, the most commonly used starting materials are commercially available monosaccharides and, in this case, a common synthetic scheme is used for the introduction of the ring nitrogen and, at the same time, to generate variously derivatized compounds. This methodology was developed for the synthesis of imino-D-galactitol derivative **7** from protected heptonolactone **3** (Scheme 2), in the search of potential inhibitors of mycobacterial cell wall biosynthesis [6].

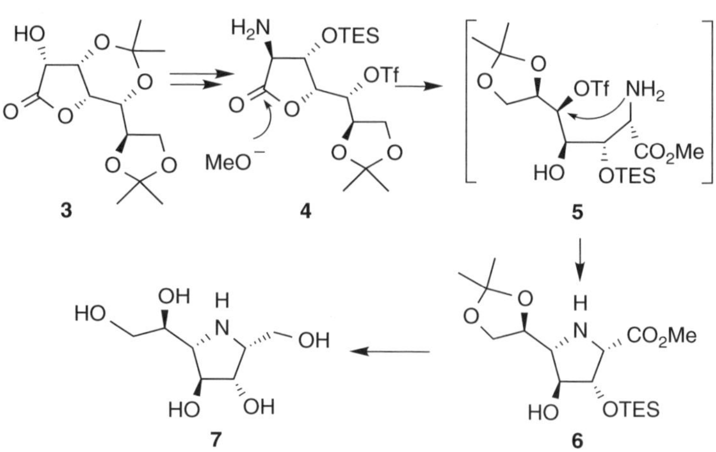

Scheme 2 Synthesis of the core structure of α-imino-D-galactofuranose derivative

The key step in the synthesis is the basic methanolysis of the lactone ring of amino triflate **4** to afford intermediate **5** which directly undergoes ring closure with the formation of the iminosugar ring. This methodology was extended to generate a small library of imino-D-galactitol derivatives **12–14**, carrying different substituted amino groups by realizing the lactone-ring opening with various primary amines in place of methoxide (Scheme 3).

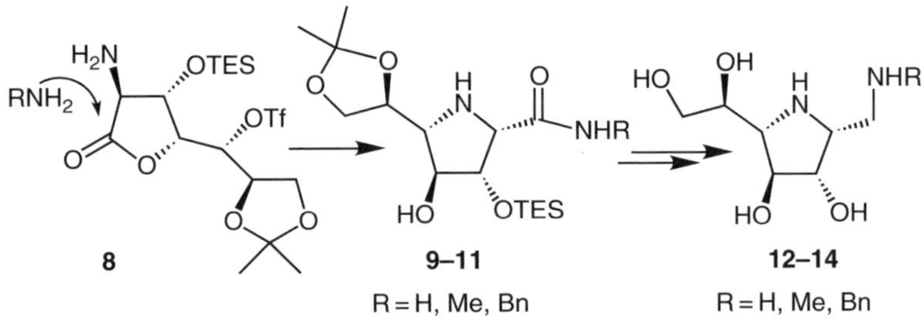

Scheme 3 Synthesis of substituted α-imino-D-galactofuranose library

Recently, we proposed a strategy for the preparation of a small library of iminosugars which combines both approaches described so far [7]. Starting from a commercially available protected glucopyranose derivative and exploiting a common synthetic scheme, a small library of eight iminosugar scaffolds was first realized; then each scaffold within the library was selectively derivatized at different orthogonally protected positions. The types of diversity introduced during the synthesis are the variable configuration of the ring carbons next to the nitrogen atom, the substituents at the nitrogen and carboxyl-derived functional group **16** (Scheme 4). This diversity was obtained by reacting aldehyde

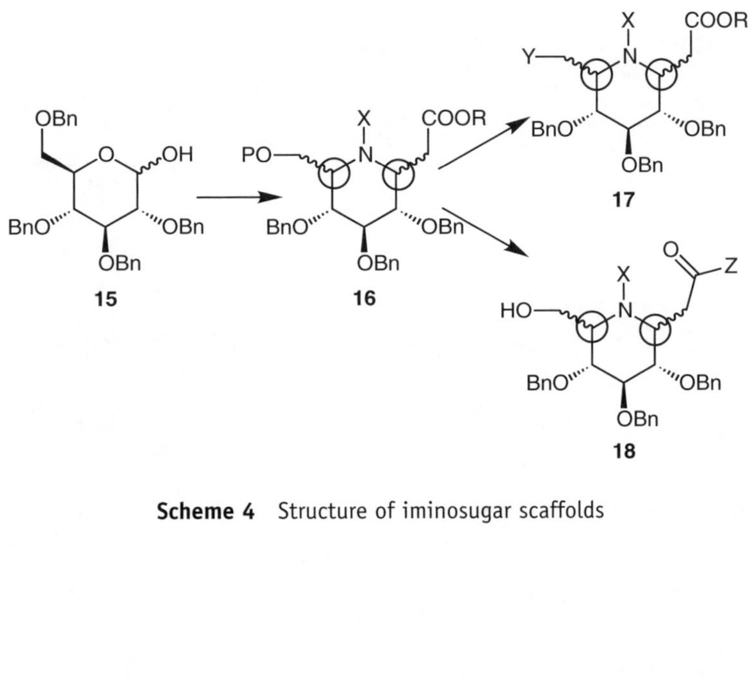

Scheme 4 Structure of iminosugar scaffolds

20 (Scheme 5) with different ylids, thus affording different α,β unsaturated esters 21. Conjugated addition of various primary amines afforded a series of *N*-substituted compounds 22 with concomitant generation of a new stereocentre, and the final reductive amination generated the second stereocentre next to the ring nitrogen, thus affording eight iminosugar scaffolds **23a–h**.

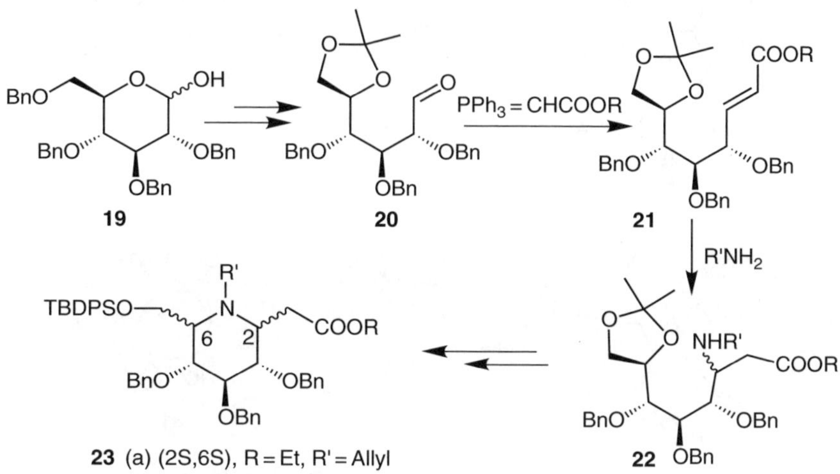

23 (a) (2S,6S), R = Et, R' = Allyl
(b) (2S,6S), R = Et, R' = Bu
(c) (2S,6S), R = Bn, R' = Allyl
(d) (2S,6R), R = Et, R' = Bn
(e) (2R,6R), R = Et,R' = Allyl
(f) (2R,6R), R = Et, R' = Bu
(g) (2R,6R), R = Bn, R' = Allyl
(h) (2R,6R), R = Et, R' = Bn

Scheme 5 Small library of eight iminosugar scaffolds

Each scaffold can then be further derivatized at the primary hydroxyl group and at the carboxylic function; examples of such derivatizations include the introduction of an azido group and the coupling of the carboxylic acid with different amino acids.

Along with classical synthesis of iminosugars from carbohydrates, innovation in organic synthesis has prompted investigation on total synthesis from noncarbohydrate, chiral and nonchiral, starting materials. A first example is represented by the synthesis of (±)-deoxymannojirimycin [8] reported in Scheme 6. The key step of this synthesis is the chemo- and stereoselective palladium-catalysed amination of silylated butendiol dicarbonates **24** that leads to the formation of the intermediate **25**. This was converted to the epoxyaldehyde **26** which afforded through a highly stereoselective intramolecular aldolization the piperidine ring **27** as the major compound. Regioselective epoxide ring opening and Tamao–Flemming oxidation of the C-Si bond led to the protected (±)-deoxymannojirimycin **29** which was then completely deprotected.

Chiral synthons other than carbohydrates are very useful tools in total synthesis. It is the case of an enantiomerically pure 2-(1'-amino-2'-hydroxyalkyl)furans [9]. In this

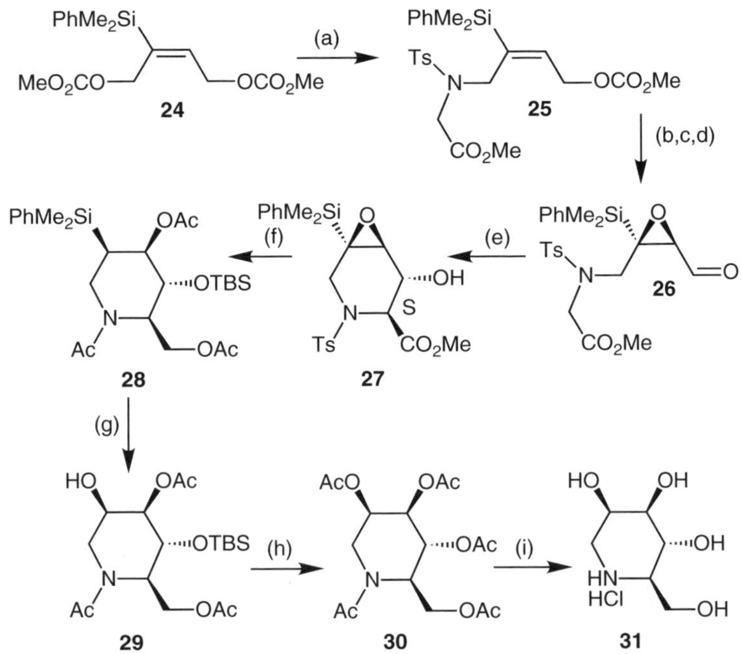

Scheme 6 *Reagents and conditions*: (a) TsGlyMe, NEt_3, *i*-PrOH, 55°C, $Pd(OAc)_2$ (2 mol%), dppe (4 mol%) 71%; (b) MeOH, K_2CO_3 (20 mol%) (quant.); (c) m-CPBA (75%); (d) IBX, DMSO (70%); (e) DBU (92%; 85/15 S/R); (f) $LiAlH_4$ (4 equiv.), Et_2O then Ac_2O, DMAP, NEt_3, 48 h (78%); (g) $Hg(OAc)_2$ (1.5 equiv.), AcOOH/AcOH 18 h (78%); (h) TBAF then Ac_2O (quant.); (i) HCl 6N reflux, 16 h (quant.)

paper the authors developed an efficient route to such compounds from L- or D-serine and then one of the compounds obtained, (1*R*,2*R*)-1-phenyl-2-amino-2-(furan-2'-yl)-ethanol, was used as the starting material for the synthesis of a piperidine derivative **38** (Scheme 7). Thus, oxidation of furan **32** followed by hemiaminal trapping afforded ketone **34** which was stereoselectively reduced to compound **35**. Reductive removal of the ethoxy group provided allylic alcohol **36** which was stereoselectively hydroxylated and deprotected to the final piperidine derivative **38**, whose structure was confirmed by NOESY experiments.

A readily available olefin has been successfully used as the starting material for the synthesis of both piperidinols [10] and pyrrolidinols [11]. As shown in the retrosynthesis in Scheme 8, a regioselective asymmetric aminohydroxylation reaction is shared by both pathways. This reaction leads to the formation of a syn-aminoalcohol with excellent stereoselectivity (>20:1). Then, for the preparation of the piperidine derivatives, the aminoalcohol **40** is converted to the intermediate **41** which, through an RCM reaction, afforded the key cyclic olefin **42** which was dihydroxylated and converted to a series of 1-deoxyiminosugars. For the preparation of the pyrrolidines, intermediate **40** was converted to allylic alcohol **44** which afforded the five-membered ring iminosugar **46** through an epoxidation-intramolecular amination cascade.

Another interesting concise *de novo* synthesis of six membered-ring iminosugars exploits as key reaction a one-pot, tandem organocatalytic asymmetric Mannich–Wittig

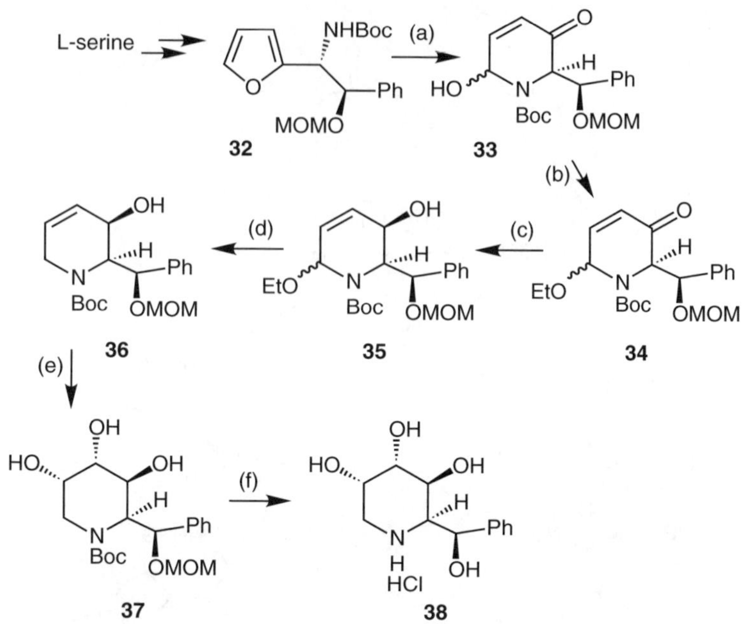

Scheme 7 *Reagents and conditions*: (a) mCPBA (80%); (b) $HC(OEt)_3$, BF_3OEt_2, 4 Å MS, THF, 0°C (97%); (c) $NaBH_4$, $CeCl_3/H_2O$, MeOH, −78°C (97%); (d) $NaBH_4$, HCO_2H, 0°C (80%); (e) OsO_4 (cat.), NMO, acetone/H_2O 9:1 (89%); (f) 3 N HCl/MeOH (96%)

olefination followed by a diastereoselective dihydroxylation (Scheme 9) [12]. In these two steps, the newly formed stereocentres are created with high stereocontrol. Compound **50** is the precursor of the final δ-lactam **51**. Other Wittig olefination reagents and conditions ($Ph_3P{=}CH\text{-}COOMe$, DMF, r.t.) could provide the *Z* olefin as the major product (4:1 *Z*:*E* ratio). This isomer was first cyclized and then dihydroxylated affording the corresponding gulonolactam.

The versatility of total synthesis has been exploited for the preparation of novel structures containing a difluoromethylene group within the piperidine ring. These structures, namely *gem*-difluoromethylenated polyhydroxylated piperidine analogues, are able to modify the binding characteristics by changing the pKa of the nitrogen. When the CF_2 group was introduced in place of a CHOH group important for binding, a decrease in binding affinities was observed [13a]. Instead, when the OH is not important, the fluorinated derivative showed an increase in binding affinities. In Scheme 10, as an example, the total synthesis of 1,4-dideoxy-4,4-difluoro-D-mannonojirimycin, from (R)-glyceraldehyde acetonide **52** and 3-bromo-3,3-difluoro propene **53**, is reported.

A different approach was chosen for the creation of a library of pyrrolidine derivatives. In this case, an iminosugar scaffold providing the possibility of introducing a wide range of substituents was generated [14, 15]. In particular, the authors investigated hydrophobic substituents which, in some cases, were shown to increase glycosidase inhibition and bioavailability. This approach led to the rapid development of iminosugar libraries through parallel synthesis methods. The scaffold was

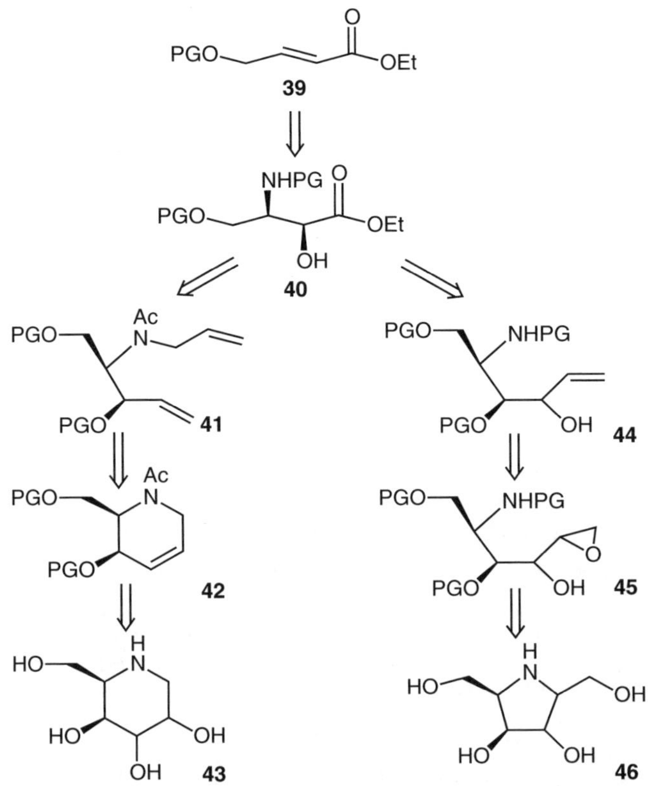

Scheme 8 Retrosynthetic approach to iminosugars from olefin (**39**)

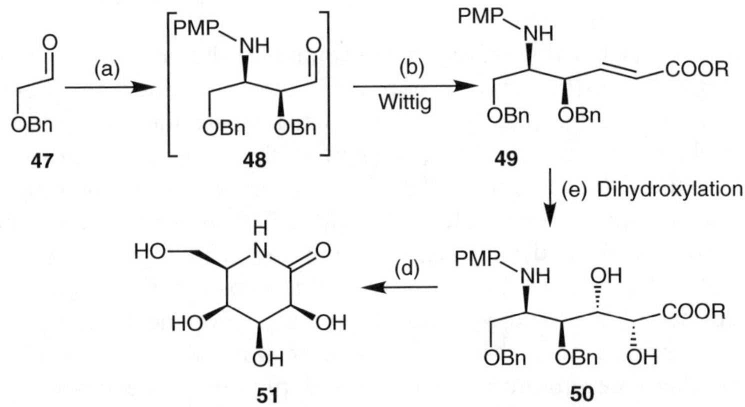

Scheme 9 *Reagents and conditions*: (a) *p*-anisidine, (*S*)-proline, DBU; (b) $(EtO)_2P(O)CH_2COOR$, LiCl (64%); (c) OsO_4 cat., NMO, Acetone/H_2O (67%); (d) AcOH/MeOH, reflux (74%)

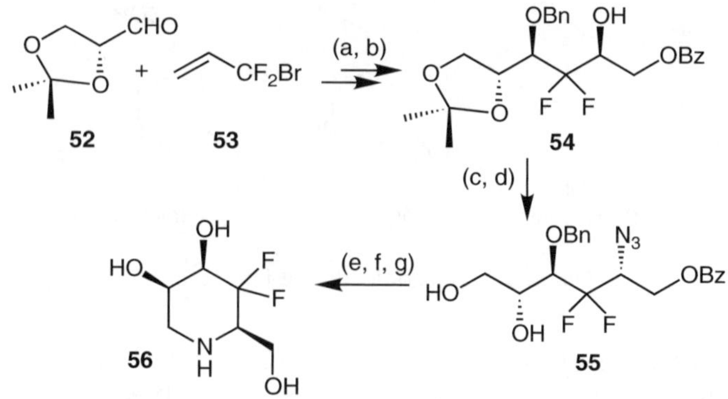

Scheme 10 *Reagents and conditions*: (a) See [13b]; (b) BzCl, pyridine, DCM, −78°C, 2 h; (c) (i) Tf_2O, pyridine, DCM, −25°C, 3 h; (ii) NaN_3, DMF, 10 h; (d) 75% AcOH, 50°C 2 h; (e) MsCl, collidine, DCM, 0°C, 12 h; (f) (i) PPh_3, THF, 20 h; (ii) saturated $NaHCO_3$, 65°C, 12 h; (iii) CbzCl, 3 h; (g) (i) H_2, $Pd(OH)_2$, MeOH, 1 atm, 10 h; (ii) saturated NH_3/MeOH, 36 h

synthesized by regioselective reactions and hydrophobic groups were introduced by way of highly diastereoselective organometallic additions. Imine **59** was obtained in two ways, according to Scheme 11: first, from 3-pyrroline **57**, and secondly via the manipulation of the hydroxyl group of the chiral starting material, tartaric acid (**58**). The methodology of addition of organometallics to cyclic imines offers potential access to interesting iminosugar derivatives, with systematic variations in hydrophobic substituent and stereochemistry, in order to facilitate structure-activity relationship studies.

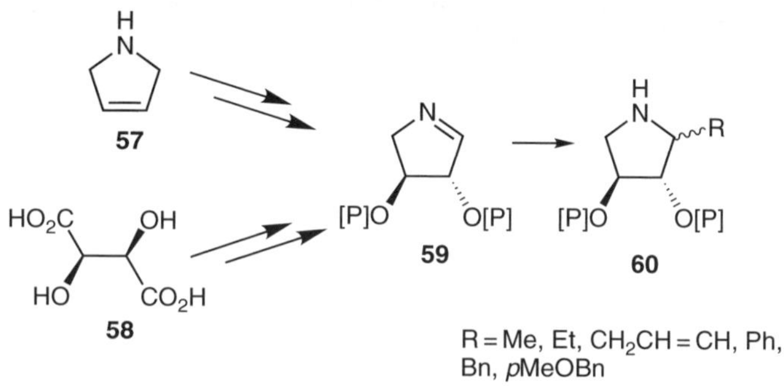

Scheme 11 Library of pyrrolidine-based iminosugars

3.3 1-*N*-Iminosugars

Among the monocyclic polyhydroxylated piperidine and pyrrolidine compounds, 1-*N*-Iminosugars are those carbohydrate mimics in which the anomeric carbon is replaced by a nitrogen atom, and the ring oxygen replaced by a methylene group (Figure 3.2).

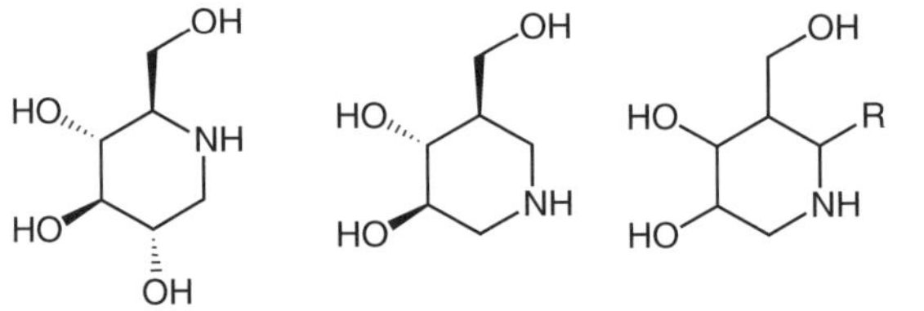

Figure 3.2 Piperidine based iminosugars

As for other iminosugars there is no general synthesis of these compounds. Moreover, the greatest problem in the preparation of these compounds is the introduction of the hydroxymethyl group next to a stereocentre. Most of the synthetic pathways proposed so far start from noncarbohydrate starting materials, although carbohydrates are still used as starting materials for example in the synthesis of isofagomine lactams [16].

Since the utilization of chiral pool compounds means changing starting material for each synthesis, it is more convenient to design a common precursor for the general synthesis of several derivatives. This is the case of compound **61** (Scheme 12) initially designed for the synthesis of isofagomine [17] and later used for the preparation of other 1-*N*-iminosugar derivatives (Scheme 13). Compound **61** was made from D-(−)-tartaric acid according to retrosynthetic analysis shown in Scheme 12, the key step being a photoinduced cyclization of intermediate **62**.

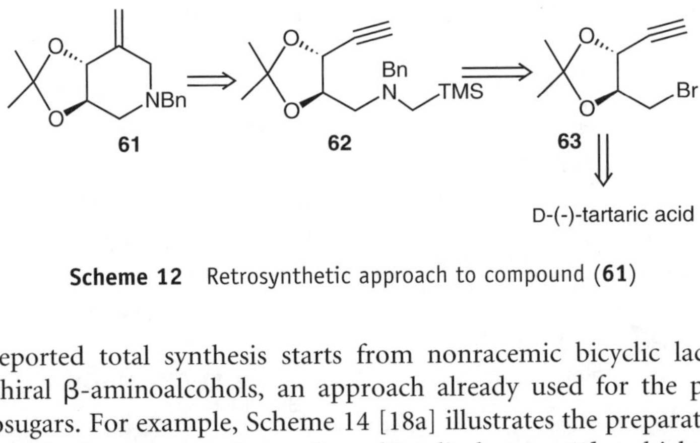

Scheme 12 Retrosynthetic approach to compound (**61**)

Another reported total synthesis starts from nonracemic bicyclic lactams derived from homochiral β-aminoalcohols, an approach already used for the preparation of several iminosugars. For example, Scheme 14 [18a] illustrates the preparation of a novel C-6 substituted isofagomine analogue from bicyclic lactam **68b**, which was obtained stereoselectively from (S)-phenylglycinol. 1-*N*-Iminosugars of pyrrolidine structure have also been synthesized from noncarbohydrate starting materials. The two diastereoisomers **73** and **74** have been designed according to the retrosynthetic analysis proposed in Scheme 15. From this analysis, the hydroxyl group on the ring was introduced by hydration of azacyclopentene **75** which in turn was obtained from the acyclic precursor **76** using a ring closing metathesis (RCM) reaction [19].

The RCM reaction has also been the key reaction for the formation of unsaturated piperidine **82**, a scaffold for the preparation of a small library of 1-*N*-iminosugars. The

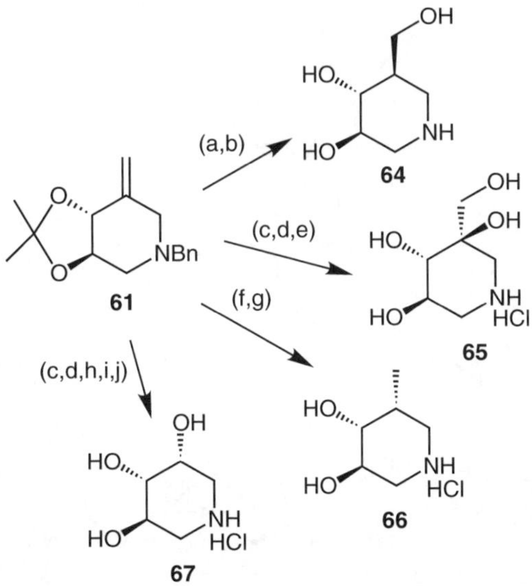

Scheme 13 *Reagents and conditions*: (a) 9-BBN, THF, 0°C to rt, 20 h, then NaOH, H_2O_2, 0°C to rt, 4 h, (45%); (b) (i) HCl, MeOH, 1 h, then NH_4OH (quant.); (ii) $Pd(OH)_2$, H_2, 75 psi, EtOH, 10 h (95%); (c) OsO_4 cat., NMO, pyridine, acetone/H_2O (9/1), 0°C to rt, 24 h (95%); (d) $Pd(OH)_2$, H_2, 65 psi, EtOH, 6 h (90%); (e) HCl, MeOH, 4 h (quant.); (f) (i) Pd/C, H_2, MeOH, HCl, 12 h (89%); (ii) $(Boc)_2O$, TEA, DCM, 48 h (75%); (g) HCl, MeOH, 0°C to rt, 4 h (quant.); (h) $NaIO_4$, EtOH/H_2O (4:1), 1 h (80%); (i) $NaBH_4$, MeOH, 40 h, then saturated NaCl 24 h (85%); (j) $Pd(OH)_2$, HCl, MeOH, H_2, 36 h (quant.)

synthesis of the starting material began with the regioselective opening of butadiene monoxide **79** with allylamine (**80**) followed by protection of the secondary amine to afford the metathesis precursor **81**. The Grubbs' catalyst could be used directly on **81**, affording the ring-closing metathesis product **82** (Scheme 16). Among the members of the library thus efficiently synthesized are isofagomine (**83a**, R = CH_2OH) and its derivatives, such as homoisofagomine (**83b**, R = CH_2CH_2OH), 5-deoxyisofagomine (**83c**, R = CH_3), and their enantiomers **84**.

3.4 Bicyclic compounds

Bicyclic iminosugars have attracted a great deal of attention because of their high and specific glycosidase inhibitory activity, which has been attributed, in part, to their rigid structure. Natural products of this class possess a pyrrolizidine or an indolizidine core structure, representative examples being (+)-australine, (+)-castanospermine and (−)-swainsonine (Figure 3.3).

As for monocyclic compounds, the majority of the syntheses reported so far are based on carbohydrate starting materials, thus taking advantage of the stereocentres already present in the sugar moiety [20–22]. More recently, versatile routes based on easily accessible noncarbohydrate synthons have been investigated. This is the case of the total synthesis of the natural pyrrolizidine derivatives hyacinthacine A_1 and

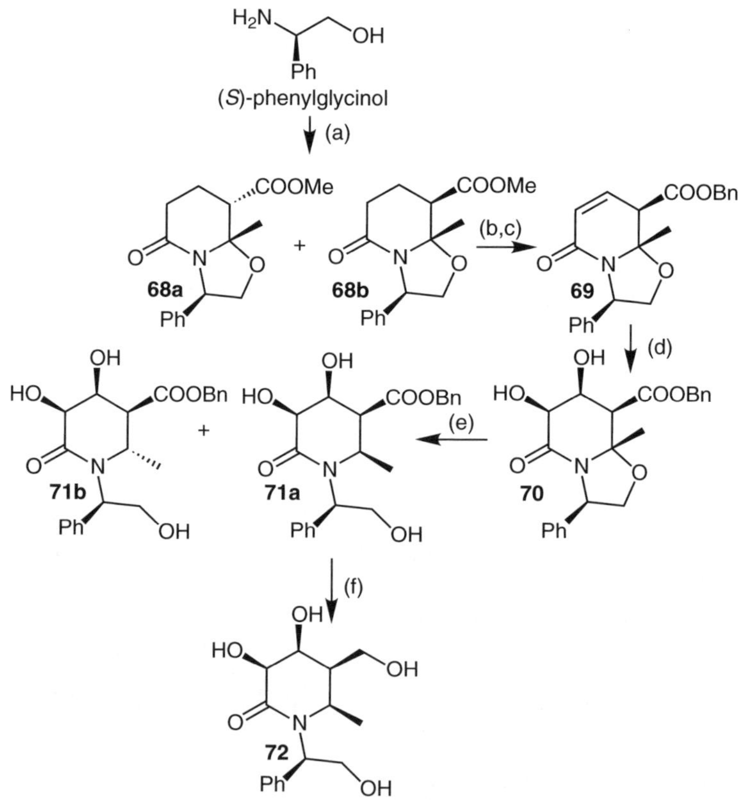

Scheme 14 *Reagents and conditions*: (a) See [18b]; (b) (i) $LiAlH_4$, THF, 0°C to rt, 5 h; (ii) NaH, THF, 0°C, BnBr (38%); (c) (i) LiHMDS, PhSeBr, THF, −78°C; (ii) mCPBA, DCM (44%); (d) OsO_4, $Ba(ClO_3)_2$, THF/H_2O (2/1), 4 d (51%); (e) BH_3DMS, THF, 0°C to rt, 1 d (34% **71a**, 30% **71b**); (f) $Pd(OH)_2$, MeOH, H_2, 4–6 d (79%)

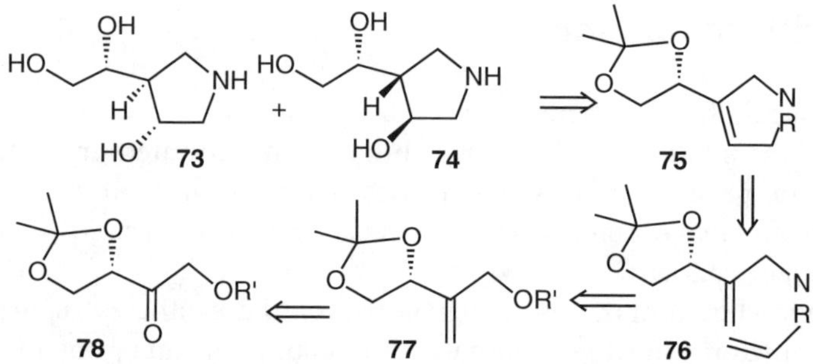

Scheme 15 Retrosynthetic analysis for the synthesis of compounds **73** and **74**

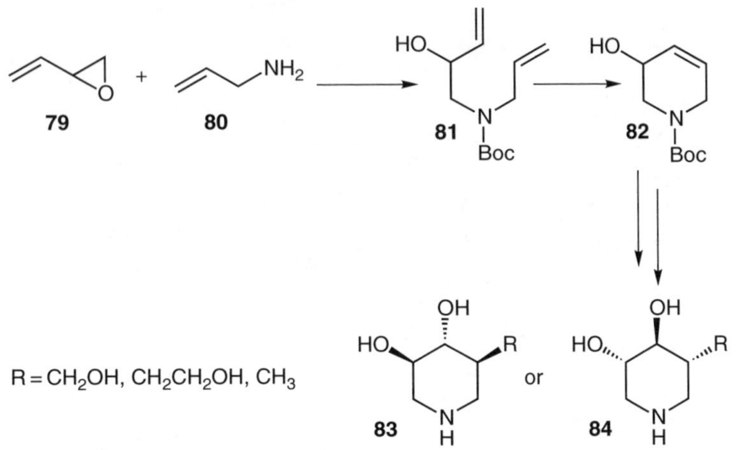

Scheme 16 Synthesis of isofagomine-based library

1-epiaustraline [23], which have been obtained from a commercially available pyrrole derivative **85** (Scheme 17). The retrosynthetic analysis carried out by the authors (Scheme 18) suggested that the second ring should be built on the intermediate compound **89**. This compound, which represents a key intermediate, was successfully obtained from the commercial pyrrole derivative **85** as illustrated in Scheme 19.

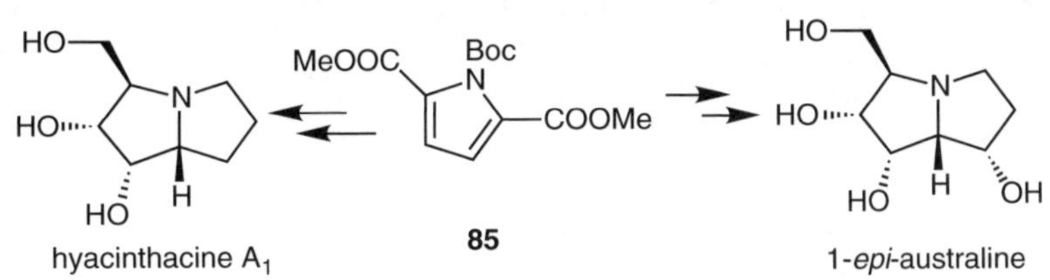

Scheme 17 Hyacinthacine A_1 and 1-epiaustraline from pyrrole derivative **85**

Hyacinthacine A_1 and its epimer in position 3 **93** have been challenging target compound studied by other groups. Another total synthesis [24] builds up the substituted carbon skeleton from a simple starting material such as aldehyde **99** and the

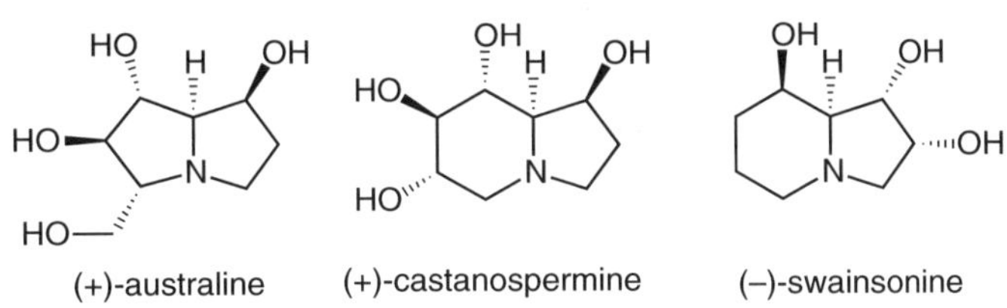

Figure 3.3 Natural polyhydroxylated pyrrolizidine and indolizidines

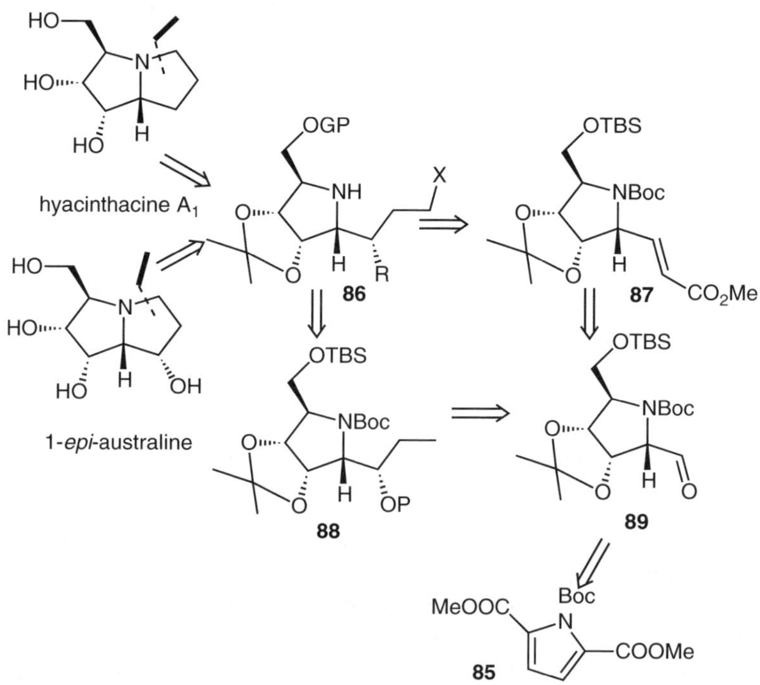

Scheme 18 Retrosynthetic analysis for the synthesis of hyacinthacine A_1 and 1-epiaustraline

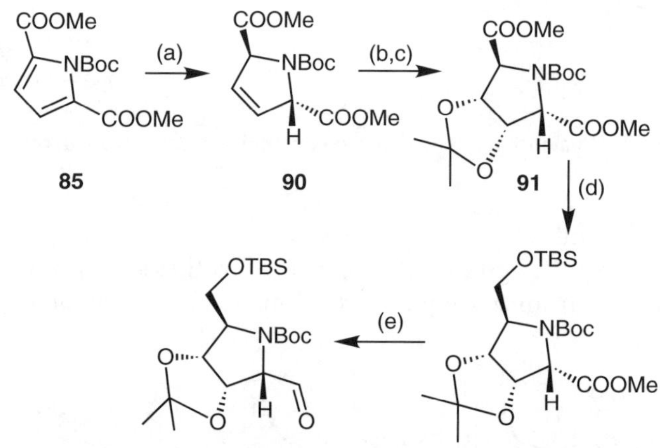

Scheme 19 *Reagents and conditions*: (a) Li, NH_3, −78°C, isoprene, NH_4Cl (73%); (b) cat. OsO_4, DCM, Me_3NO (95%); (c) dimethoxypropane, p-TsOH, acetone (94%); (d) (i) $NaBH_4$, THF/MeOH; (ii) TBSCl, imidazole, DMF (85%); (e) DIBAL-H, DCM, −40°C (97%)

allylsilane **98**, while the nitrogen atom of the fused ring system was introduced through a free radical carboazidation of the allylsilane (Scheme 20).

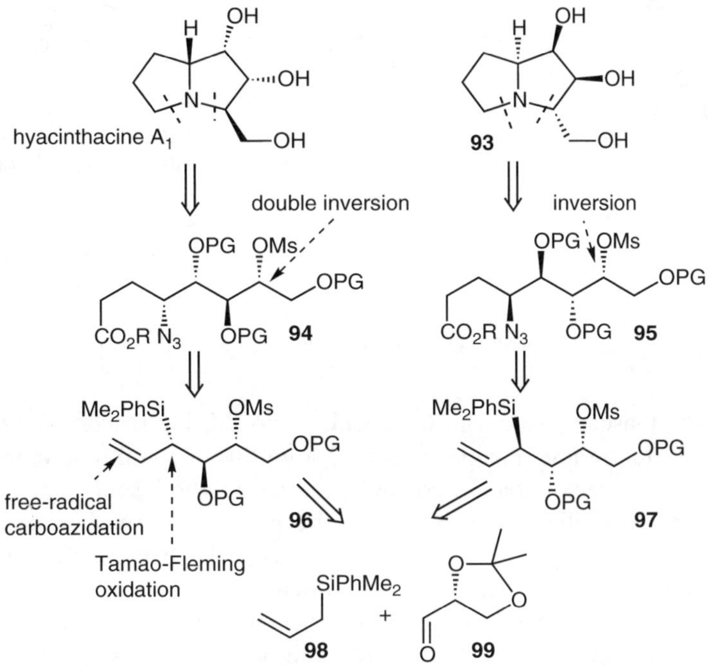

Scheme 20 Retrosynthetic analysis for the synthesis of hyacinthacine A_1 and its epimer **93**

As suggested by the lengthy synthesis of the bicyclic pyrrolizidines shown above, an approach capable of leading to a library of compounds exploiting a common synthetic pathway and common building blocks used in a combinatorial fashion, is of great interest.

The synthesis of highly functionalized pyrrolizidines through cycloaddition reactions has recently been reported [25]. Enantiomerically pure nitrones **100–103** (Figure 3.4) were readily obtained from inexpensive starting materials such as L-malic acid,

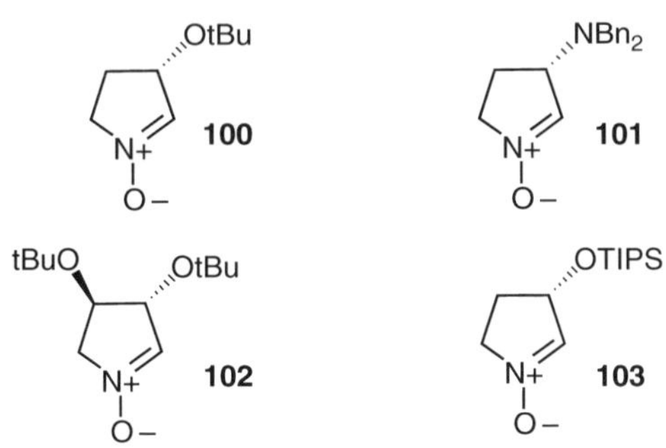

Figure 3.4 Enantiomerically pure nitrones as building blocks for combinatorial synthesis

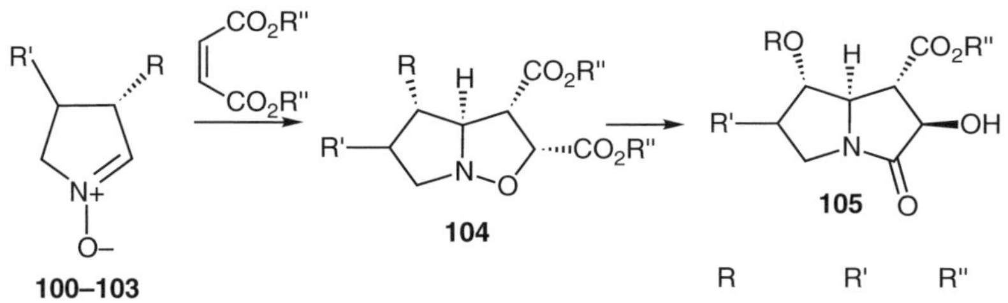

R	R'	R''
OtBu	H	Me
NBn_2	H	Me
OtBu	OtBu	Me
OtBu	H	iPr
OTIPS	H	Me
OTIPS	H	iPr

Scheme 21 Bicyclic iminosugar library

L-diethyl malate, L-aspartic and D-tartaric acid. Cycloaddition reactions of dimethyl and diisopropyl maleate to these nitrones occurred smoothly at room temperature affording a mixture of two to three diastereomeric cycloadducts in good yields. The major adducts were isolated and identified in each case as the products derived from the less sterically hindered *exo-anti* approach. In order to access the pyrrolizidine nucleus, these compounds were subjected to the reductive opening of the isoxazolidine ring in the presence of $Mo(CO)_6$ to give derivatives **105** (Scheme 21). The ready access to highly functionalized, enantiomerically pure pyrrolizidinones by this strategy, together with the high predictability of its stereochemical outcome, suggested that the method could be applicable to the synthesis of hydroxylated pyrrolizidine alkaloids by a simple reduction followed by final deprotection. This strategy was employed to prepare a library of stereochemically differentiated natural necine bases, such as (−)-hastanecine (**107**) and natural analogues (Scheme 22).

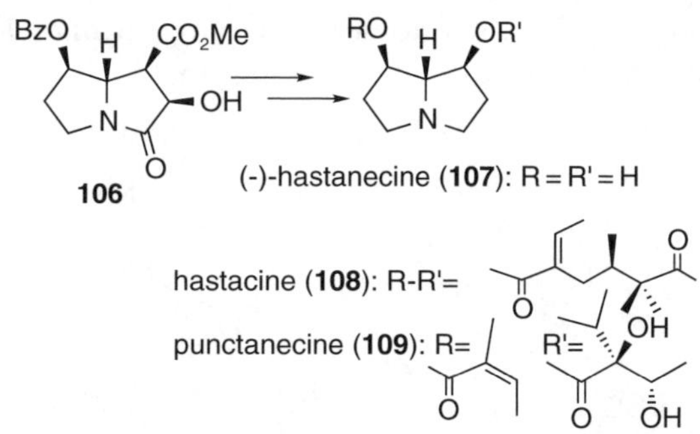

Scheme 22 Library of stereochemically differentiated natural necine bases

Recently, the same procedure was applied for the preparation of rosmarinecine [26] and crotanecine analogues (−)-**111** and (−)-**112** starting from nitrone (−)-**110**

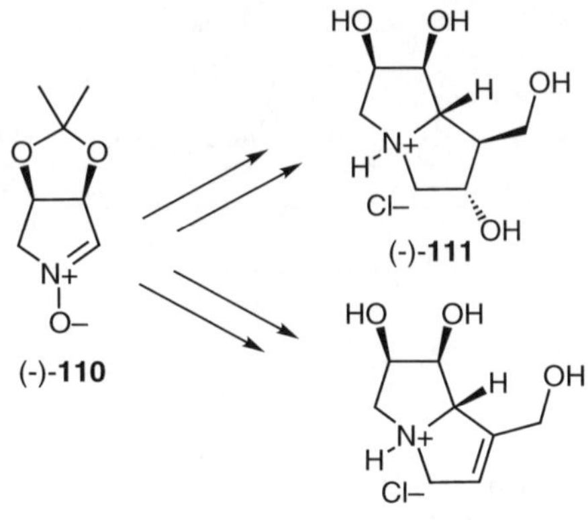

Scheme 23 Crotanecine analogues

(Scheme 23). These compounds were assayed for their inhibitory activities toward 22 commercially available glycosidases, and data showed that **112** is a potent and selective inhibitor of α-mannosidases from jack bean and almonds.

A novel aspect of the reactivity of nitrones has recently been discovered and exploited in the synthesis of hyacinthacine A_2 [27] (Scheme 24). In the presence of SmI_2, nitrone **113** acts as a nucleophile with ethyl acrylate to afford compound **114**. Reduction of the hydroxylamine and lactamization then gave bicyclic compound **115** which was converted to the target product. As for pyrrolizidine derivatives, examples of indolizidine synthesis from noncarbohydrate starting materials have been reported in the literature. Scheme 25 illustrates the total synthesis of (−)-swainsonine and (−)-2-*epi*-swainsonine from pyridine [28]. The first key reaction is the photochemically

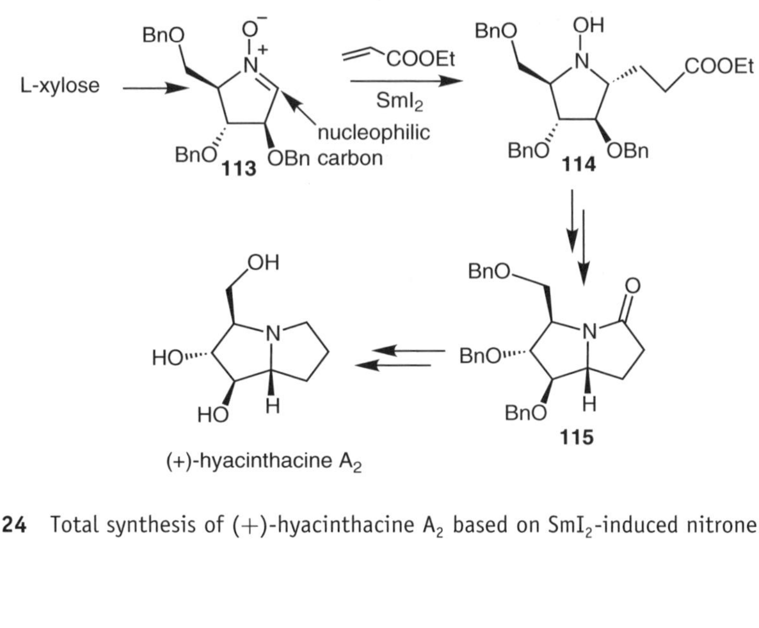

Scheme 24 Total synthesis of (+)-hyacinthacine A_2 based on SmI_2-induced nitrone umpolung

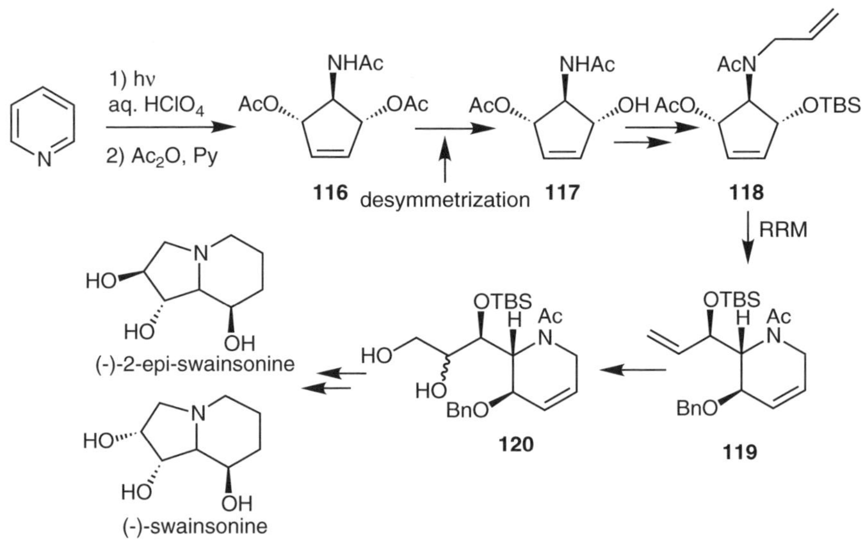

Scheme 25 Synthesis of (−)-swainsonine and (−)-2-*epi*-swainsonine from pyridine

induced transformation of pyridinium perchlorate generated *in situ* into *trans*,*trans*-3, 5-dihydroxy-4-aminocyclopentene which is isolated as triacetylated derivative **116**. This symmetric compound was enzymatically desymmetrized to monoacetate **117**, which was selectively protected and converted to compound **118**. The allyl group and the endocyclic double bond were subjected to a ring rearrangement metathesis (RRM) reaction affording the selectively protected tetrahydropyridine derivative **119**. Regioselective dihydroxylation of the chain double bond afforded the mixture of diastereoisomers **120**, which were separated. An intramolecular Mitsunobu reaction led to the formation of the second ring, then hydrogenation and deprotection provided the final products.

Unnatural indolizidine derivatives **128** and **129** have been recently synthesized from (R)-silyloxy pyrrole **121** (Scheme 26) [29]. The authors developed a new access to bicyclic lactam **122** from the pyrrole derivative **121**. Stereoselective addition of allyltrimethylsilane at C-5 afforded **123** which was manipulated to generate **124**. This substrate was submitted to RCM to form the second cycle. Dihydroxylation led to the mixture of diastereoisomers **126** and **127**, precursors of the final products.

3.5 Other bicyclic compounds

Along with the development of new and more efficient synthetic approaches towards natural/unnatural pyrrolizidines or indolizidines, other bicyclic compounds with novel, unnatural structures have been designed and prepared to improve biological activity. Tetrahydropyridino-imidazole, -triazole and -tetrazole derivatives [30–32] have been designed as transition-state mimetics of enzymatic glycoside hydrolysis. Among them, tetrahydropyridino-imidazoles are the strongest glycosidase inhibitors. The introduction

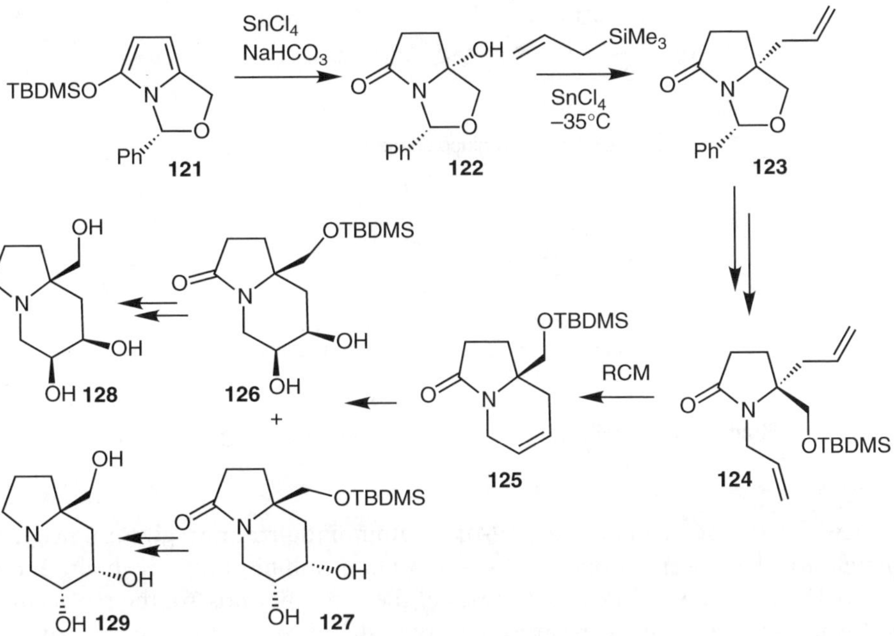

Scheme 26 Synthesis of unnatural indolizidine derivatives (**128**) and (**129**)

of substituents into the imidazole ring, mimicking the aglycon moiety, afforded even stronger inhibitors; a small library of 13 C(2)-substituted tetrahydropyridinoimidazoles has been synthesized from protected imidazole derivative **130** [33] (Scheme 27). As expected, all of these compounds were much stronger inhibitors of β-glucosidases than α-glucosidases and, in particular, derivatives with a hydrophobic flexible substituent, such as **133** and **134** (Figure 3.5), showed K_i values in the nM range.

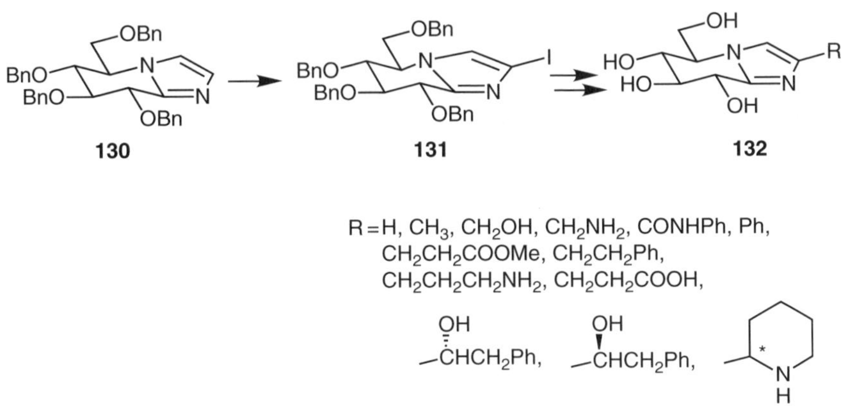

Scheme 27 C(2)-substituted tetrahydroimidazopyridine library

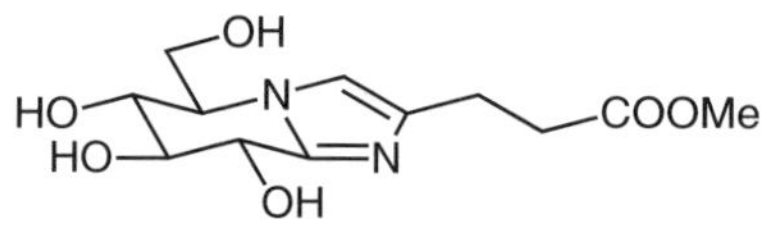

133 $K_i = 9.9$ nM (β-glucosidase from almond)

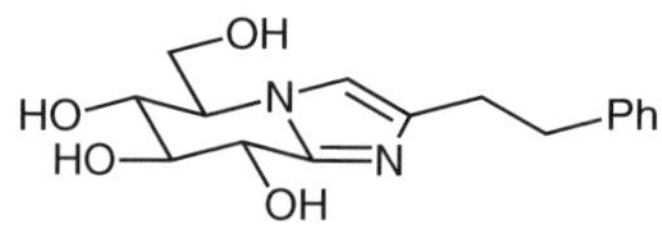

134 $K_i = 1.2$ nM (β-glucosidase from almond)
$K_i = 0.11$ nM (β-glucosidase from *Caldocellum saccharolyticum*)

Figure 3.5 Tetrahydroimidazopyridine inhibitors of β-glucosidases

A new class of compounds, 2-oxa-3-iminoindolizidine glycomimetics **135** (Scheme 28), has been proposed as glycosidase inhibitors with high anomeric selectivity [34, 35]. The main characteristics of these compounds are the replacement of the sp^3 ring nitrogen atom with an isourea-type sp^2 nitrogen, the presence of a stable hemiaminal hydroxyl group anchored in an axial position, and the incorporation of different substituents at the exocyclic nitrogen atom. This last feature was obtained by allowing the azido compound **138** to react with different isothiocyanates. Following this approach, a small library of *N*-glycosylated isourea-containing castanospermine and 6-epicastanospermine analogs **139** and **140** (Figure 3.6) has been synthesized.

Other unnatural bicyclic structures recently proposed as glycosidase inhibitors are the spirocyclopropyl iminosugars **141** and the polyhydroxylated perhydroazaazulenes **142** (Figure 3.7). The spirocyclopropyl derivatives [36] have been prepared from readily available protected aldoses, the key step being the aminocyclopropanation of glyconon-

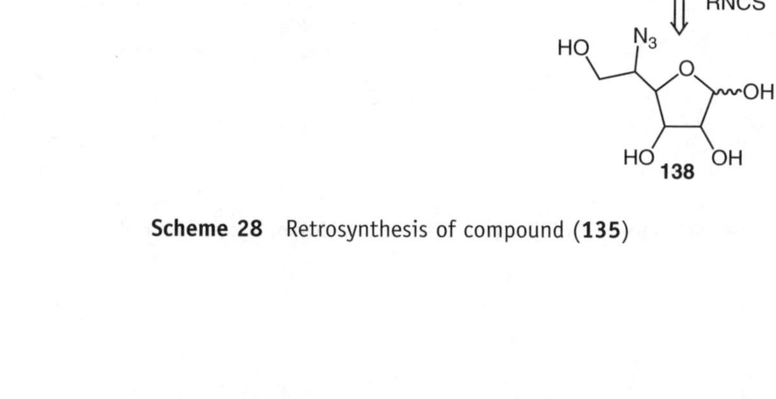

Scheme 28 Retrosynthesis of compound (**135**)

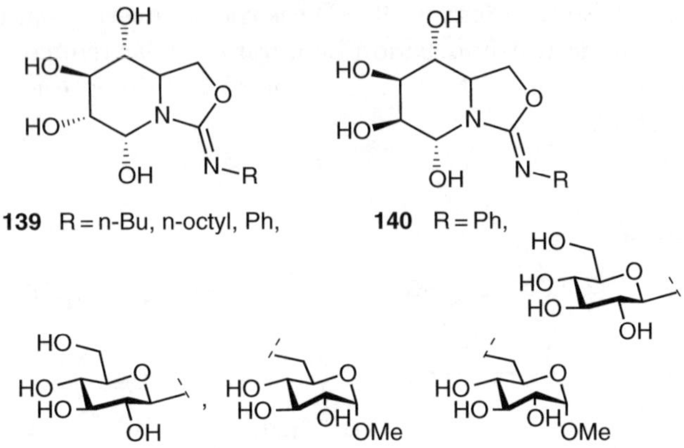

Figure 3.6 Isourea-type glycomimetics (**139**) and (**140**)

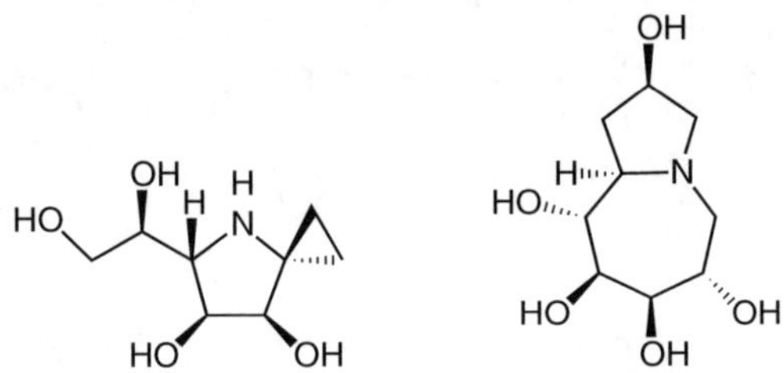

Figure 3.7 Spirocyclopropyl (**141**) and azaazulenes (**142**) iminosugars

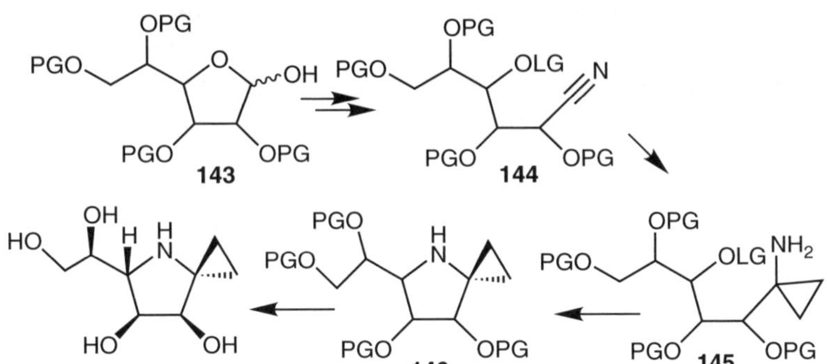

Scheme 29 Synthesis of spirocyclopropyl iminosugars

itriles with subsequent cyclization (Scheme 29). Compound **141** was a potent inhibitor of α-L-fucosidase from kidney ($K_i = 1.6\,\mu M$).

For the synthesis of polyhydroxylated perhydroazaazulenes [37], the first reaction was the 1,3-dipolar cycloaddition of nitrone **147** with ethyl acrylate to afford isoxazolidine

149 as a single diastereoisomer (Scheme 30). The isoxazolidine ring was then transformed into a hydroxypyrrolidine **150**; annellation by intramolecular reductive amination gave rise to the seven member ring. The final compound **142** showed a rather low activity against various glycosidases.

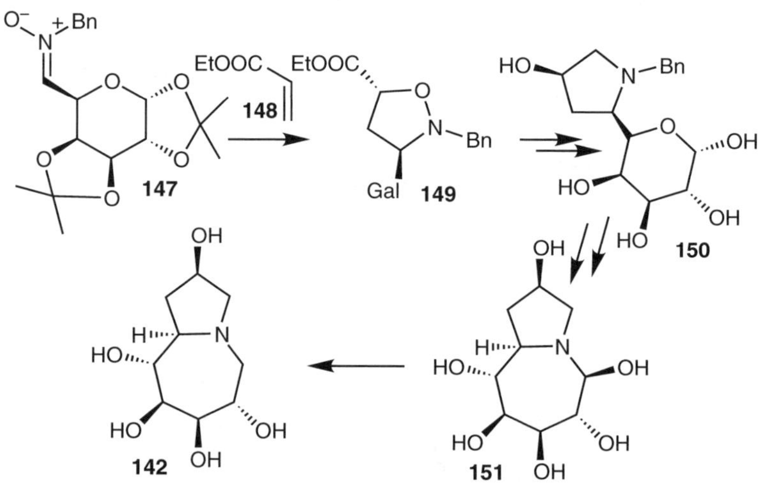

Scheme 30 Synthesis of a polyhydroxylated azaazulene iminosugar

3.6 Iminosugar conjugates

Since carbohydrates are frequently found conjugated to other biomolecules, such as lipids, proteins, nucleosides or phosphate groups, it is of particular interest to consider the synthesis of iminosugar-containing conjugate mimetics. In the following section iminosugar-conjugates will be reviewed.

3.6.1 Glycopeptide mimetics based on iminosugars

Iminosugar-based glycopeptides constitute a new class of unnatural peptides combining hydrolytic stability with inhibitory properties. In general, since the preparation of glycopeptides requires easy access to glycosyl amino acids [38], key precursors to their iminosugar analogues are iminoglycosyl amino acids.

The first example of this type of neoglycoconjugates was described by Schmidt *et al.* [39] who reported the synthesis of nojirimycinyl *C*-L-serine **156**. In this compound the iminosugar moiety is α-linked to an (*S*)-aminobutanoic acid residue which serves as L-serine methylene isostere. The key steps in the synthesis involved the construction of the chiral glycinyl group by a Wittig reaction followed by the hydrogenation of the resulting enamide ester group (Scheme 31). However, this approach

cannot be generalized to the synthesis of various iminoglycosyl amino acids, since the stereochemical outcome of the hydrogenation reaction may vary substantially by changing the iminosugar residue and the length of the side chain [40].

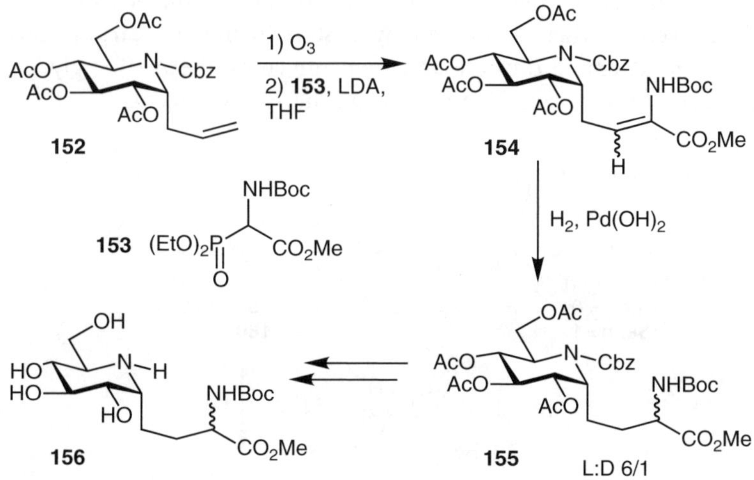

Scheme 31 Synthesis of nojirimycinyl *C*-L-serine

A general synthetic method which is centred on the cross-metathesis (CM) of alkenes **B** and **C** featuring an iminosugar residue and a glycinyl group respectively (Scheme 32) has been reported by Dondoni *et al.* [41,42(a)]. The transformation of the resulting CM product into **A** required the simple reduction of the newly formed carbon-carbon double bond connecting the two functionalized moieties. This method avoids stereochemical concerns involving the two key stereocentres, i.e. the pseudoanomeric carbon of the sugar fragment and the α-carbon of the amino acid moiety.

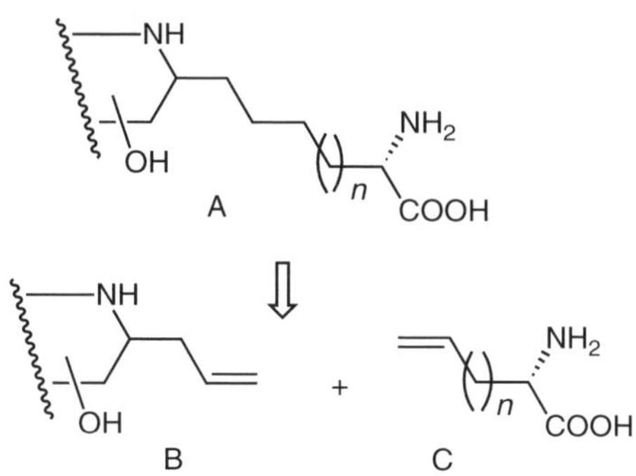

Scheme 32 Retrosynthetic plan to iminosugar α-amino acids

Following this method, a library of novel sugar α-amino acids composed of iminofuranose and iminopyranose residues anomerically linked to the glycinyl group through an alkyl chain was prepared (Scheme 33). A set of eight compounds was prepared by that reaction sequence involving as the initial step the Grubbs Ru-carbene-catalysed cross-metathesis (CM) of various *N*-Cbz-protected *C*-allyl iminoglycosides with *N*-Boc-vinyl-**159** and *N*-Boc-allyloxazolidine **160**. The isolated CM products (mixtures of *E*- and *Z*-alkenes, **162**) were further elaborated by first reducing the carbon–carbon double bond using *in situ* generated diimide and then unveiling the *N*-Boc glycinyl group by oxidative cleavage of the oxazolidine ring using the Jones reagent.

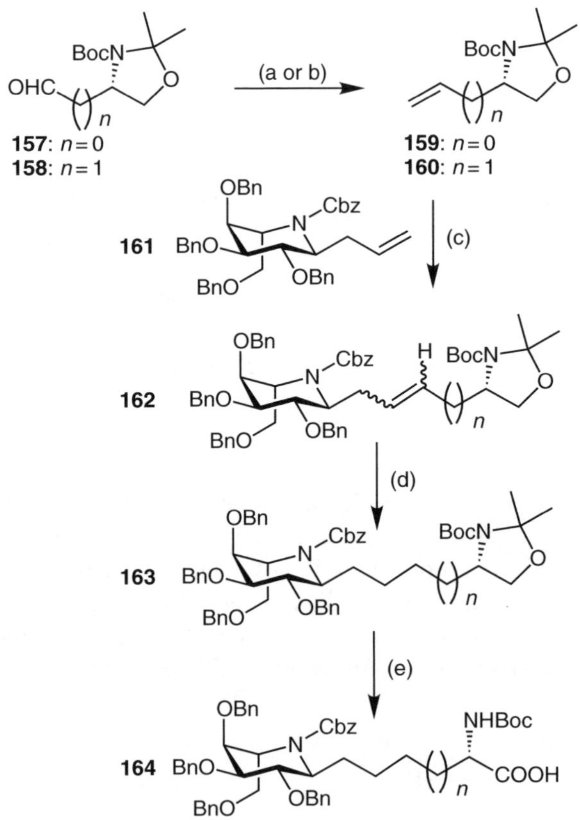

Scheme 33 *Reagents and conditions*: (a) $n = 0$ see [42(b)]; (b) $n = 1$, Ph_3PCH_3Br, KHMDS; (c) Grubbs catalyst 1,3-dimesityl-4,5-dihydroimidazol-2-ylideneruthenium carbene; (d) $TsNHNH_2$, AcONa; (e) Jones reagent

The insertion of a model 5-(*C*-iminoglycosyl)-2-aminopentanoic acid into a tripeptide via sequential carboxylic and amino group coupling with L-phenylalanine derivatives was carried out as a demonstration of the potential of these sugar amino acids in pseudo glycopeptide synthesis. For discovering new peptide-based drugs, many structurally rigid nonpeptidic molecular scaffolds have been designed. Insertion of these moieties

at appropriate sites, a common approach to restrict the conformational degrees of freedom in small peptides, produces the specific three-dimensional structures required for binding to their receptors. A novel structurally constrained iminosugar scaffold consisting of 2,5-dideoxy-2,5-imino-D-idaric acid (**165**) was developed (Figure 3.8) as a building block for peptidomimetic synthesis [43]. The synthesis of the iminosugar scaffold involved the transformation of 1,3:4,6-di-*O*-(*p*-methoxybenzylidene)-D-mannitol **168** into pyrrolidine derivative **169**; a sequence of protection and deprotection steps afforded the corresponding diol, which was oxidized to the diacid **171**(Scheme 34). Bidirectional elongation of the diacid moieties of derivative **171** with identical peptide strands led to the formation of peptides **166** and **167** (Figure 3.8).

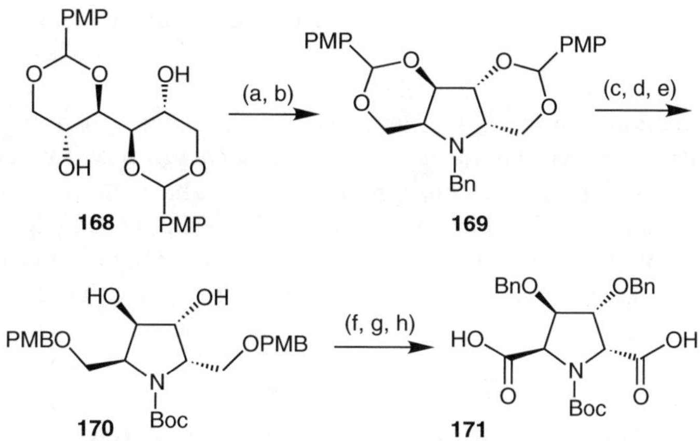

Scheme 34 *Reagents and conditions*: (a) MsCl, Et_3N; (b) $BnNH_2$, reflux; (c) $LiAlH_4$-$AlCl_3$; (d) $Pd(OH)_2$, MeOH; (e) Boc_2O, MeOH; (f) NaH, BnBr, cat. TBAI; (g) DDQ; (h) PDC. In an additional example, deoxymannojirimycin (DMJ) was proposed as iminosugar scaffold to obtain peptidomimetics as new aspartic acid protease inhibitors [60]

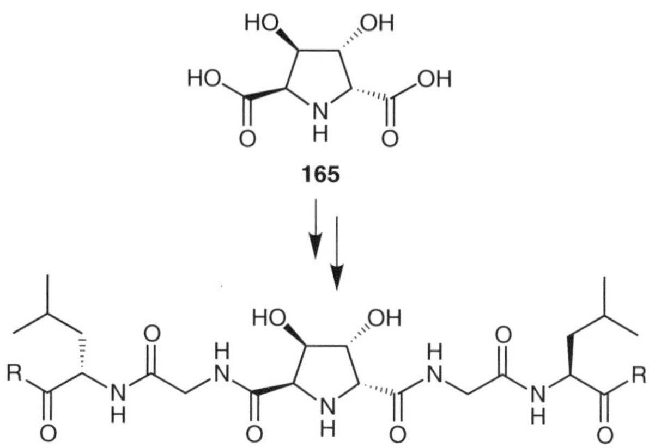

166: R = NHMe; **167**: R = OMe

Figure 3.8 Iminosugar scaffold and peptidomimetics synthesized thereof

3.6.2 Iminosugar phosphonates and iminosugar nucleotides

Due to their relevant role in the biosynthesis of polysaccharides and glycoconjugates, glycosyltransferases have attracted increasing interest as targets for the development of potent inhibitors and therefore for drug discovery [44, 45].

Most glycosyltransferases require glycosyl phosphates or derivatives thereof, such as sugar nucleotides, as glycosyl donors. Glycosyl transfer has been postulated to proceed through transition states exhibiting significant oxocarbenium character, on the basis of the hypothesis that glycosidases and glycosyltransferases have a similar mechanism; the structural studies provide strong encouragement for the design of iminosugar-based inhibitors since they imitate the transition state of the enzymatic reaction, rather than the ground state. Much effort has been devoted to the development of glycosyl phosphate mimics: substitution of the phosphoester oxygen with a carbon atom leads to structurally very similar, yet hydrolytically stable, phosphono analogues. Iminosugar phosphonates combine the functional properties of glycosyl phosphonates and iminosugars. A few examples of iminosugar phosphonates have been reported in recent years [46–49].

A new class of stable glycosyl phosphate mimetics that combines an iminosugar moiety and a nonisosteric phosphate analogue has been proposed by Martin *et al.* [50] (Scheme 35). Compound **175** was designed to display a strong affinity toward certain carbohydrate processing enzymes: glycosyl phosphate mimetics that consist of a phosphonate directly bound to the pseudo anomeric carbon were found to have a polarity similar to that of the natural sugar 1-phosphates [51]. Diethyl phosphite was converted *in situ* to its trimethylsilyloxy P(III) derivative [$TMSOP(OEt)_2$] in the presence of TMSCl and then reacted with imine **173** to afford the α-amino phosphonate **174** as a single diastereoisomer (6*S*). Imine **173** was efficiently prepared from commercially available 2,3;4,6-di-*O*-isopropylidene-α-L-sorbofuranose **172** by a protecting group strategy that allowed regioselective deprotection of the hydroxyl group at C-6, followed by oxidation.

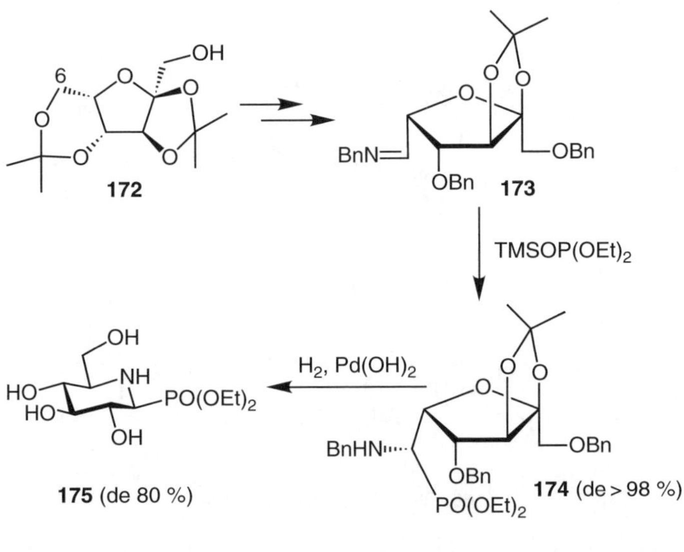

Scheme 35 Synthesis of a nonisosteric iminosugar phosphonate

A new synthetic approach to iminosugar phosphonates was recently proposed [52]; 1-phosphonomethylene-1-deoxy-L-idonojirimycin (**189**) has been synthesized starting from commercially available tetra-*O*-benzyl-D-glucose (Scheme 36). The synthetic strategy involved the introduction of the nitrogen atom at C-5 as an azido group, the introduction of the phosphonate group by stereoselective reaction of the aldehyde group at C-1 with dimethyl methylenephosphonate anion, conversion of the azide into an amino group and finally cyclization to the corresponding 1-phosphonomethylene-1-deoxy-L-idonojirimycin.

Iminosugar nucleotides have also been proposed as unreactive sugar nucleotide analogues in the search for novel glycosyltransferases inhibitors [45]. The first example of a sugar nucleotide analogue having a saturated nitrogen heterocycle as a glycosyl donor mimic was described by Berliner and Wong in 1975 [53]. Since then several examples have been reported in the last few years [45]. A general approach to iminosugar-*C*-glycosides via organometallic addition to *N*-glycosylhydroxylamines was also applied to the synthesis of sugar nucleotide analogues [54, 55]. For example, stereoselective addition of organometallic nucleophiles, such as 2-lithiothiazole, to *N*-benzyl-*N*-ribosylhydroxylamine (**191,** Scheme 37) afforded iminosugar nucleoside **193** after reductive dehydroxylation using zinc copper (II) acetate and treatment with methanesulphonyl chloride.

Pyrrolidine analogues of 2-deoxy-D-ribofuranose having nitrogen in place of the anomeric carbon and a methylene group instead of the ring oxygen have been reported recently [56, 57]. This type of iminosugar has been used in the synthesis of transition-state analogue inhibitors of purine nucleoside phosphorylase (see Chapter 8 of this volume) [58, 59] and for the synthesis of derivatives with a substituted aryl group on the endocyclic nitrogen atom [60a,b]. In particular, a compound carrying a 1-pyrenyl group was introduced on the ring nitrogen atom as described; this compound can be considered as an iminosugar nucleoside analogue of the pyrene deoxynucleoside prepared by Kool and coworkers who used it for DNA synthesis [61, 62]. The key step for obtaining such *N*-aryliminosugars (Scheme 38) was a double aminoalkylation reaction of 1-aminopyrene with the branched-chain sugar dialdehyde masked in the form of 3-deoxy-3-*C*-formyl-1, 2; 5, 6-di-*O*-isopropylidene-α-D-allofuranose (**195**). The iminosugar **196** was obtained in a one-pot reaction by condensing the aldehyde **195** with 1-aminopyrene followed by reductive amination, deprotection and subsequent intramolecular reductive amination reaction of the aldehyde function previously masked in the furanoside. The iminosugar **196** was an advanced intermediate for the preparation of the corresponding lower homologs **197** and **198**, which were obtained by sodium periodate oxidation followed by reduction with sodium borohydride.

Imino-*C*-nucleosides were also synthesized from carbohydrate-based cyclic imines **199** [63] by nucleophilic addition of suitable aryllithium derivatives (Scheme 39). In order to access iminosugar nucleoside libraries, a common synthetic scheme was exploited for the introduction of the ring nitrogen and, at the same time, to generate differently functionalized compounds. One of the first examples of such a strategy was described by Yokoyama *et al.* [64] in the generation of mimetics of iminosugar nucleosides (Scheme 40). Starting from protected 2-deoxyribose **202**, different aromatic substituents, mimicking the nucleoside residues, were first introduced at C1 by reaction

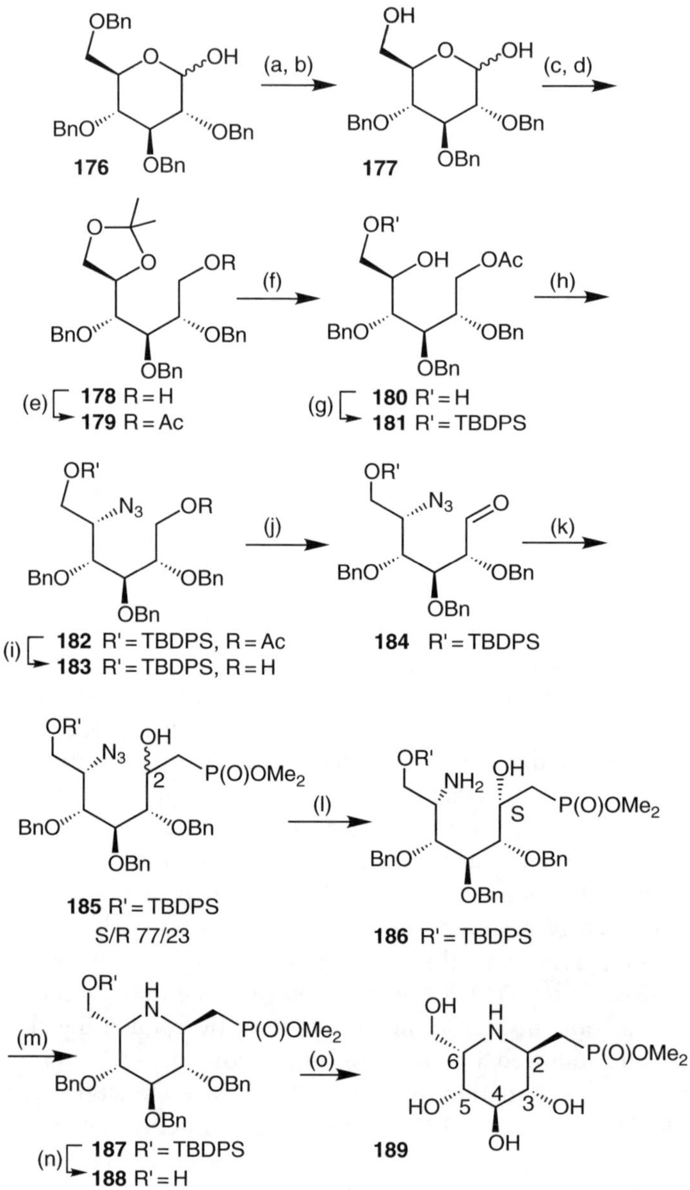

Scheme 36 *Reagents and conditions*: (a) Ac_2O/TFA 4/1 then NaOH 4M; (b) Na, MeOH; (c) $NaBH_4$, EtOH; (d) 2,2-dimethoxypropane, camphor-10-sulfonic acid, CH_3CN; (e) Ac_2O, pyridine, DMAP; (f) H_2O, CSA; (g) TBDPSCl, imidazole; (h) $(PhO)_2P(O)N_3$, PPh_3, DIAD; (i) MeONa, MeOH; (j) Dess–Martin periodinane; (k) $CH_3PO(OMe)_2$, BuLi, −78°C; (l) PPh_3, H_2O, THF, 60°C; (m) PPh_3, DIAD; (n) TBAF; (o) $Pd(OH)_2$, H_2, MeOH/AcOH

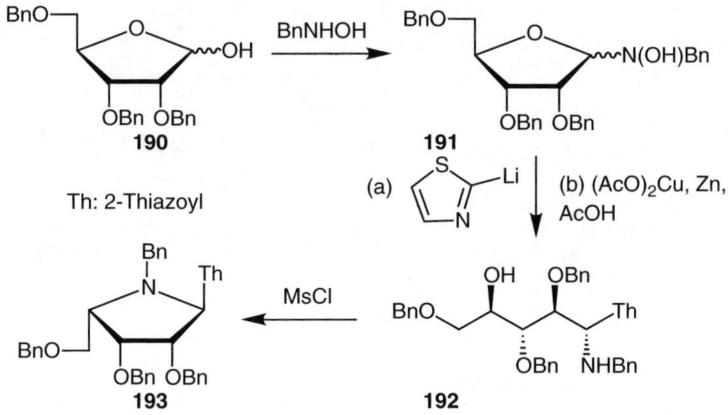

Scheme 37 Synthesis of an iminosugar nucleoside

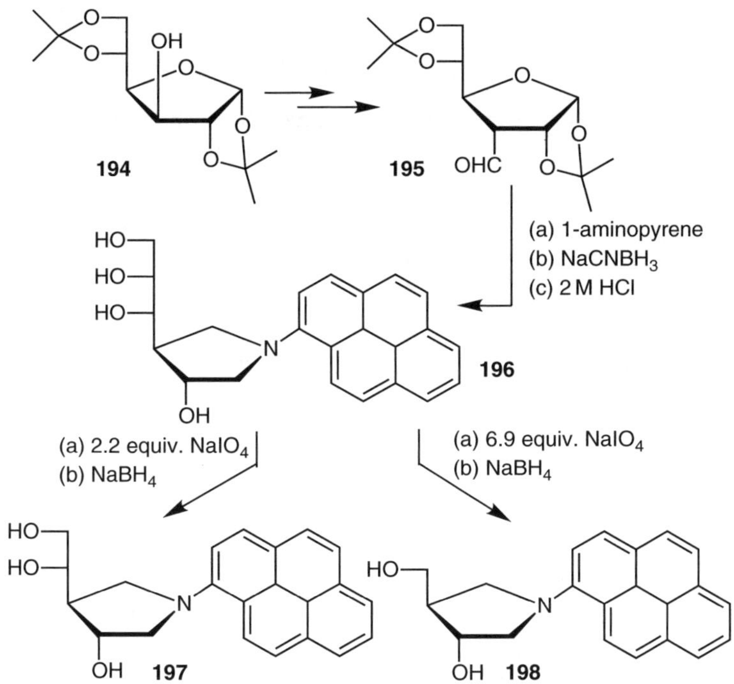

Scheme 38 Synthesis of *N*-aryl-1-N-iminosugars

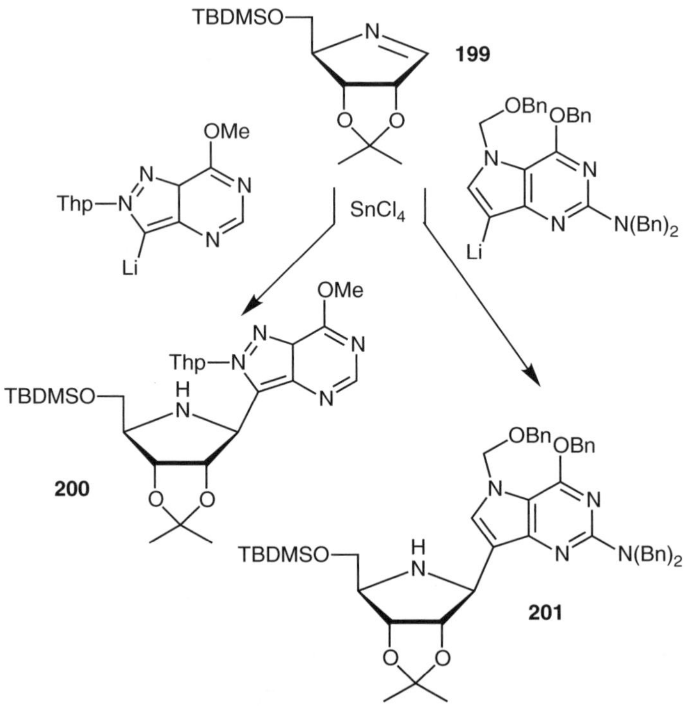

Scheme 39 Synthesis of imino-*C*-nucleosides with aryllithium derivatives

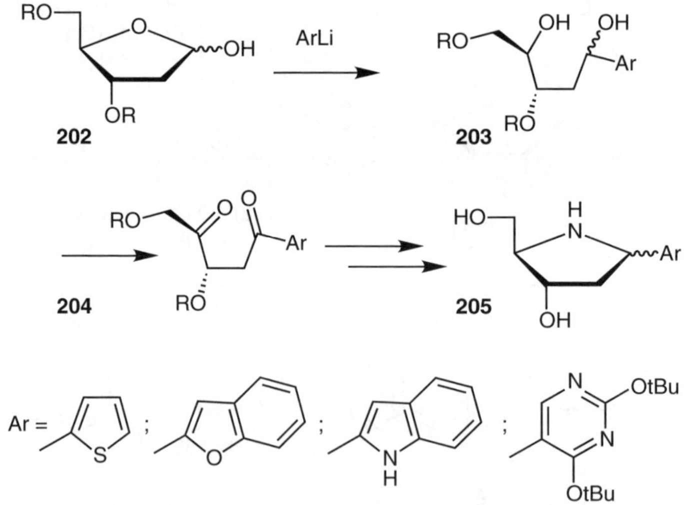

Scheme 40 Synthesis of an iminosugar nucleoside library

with different organolithium nucleophiles, then the intermediates **203** were cyclized to the iminosugar **205** by a double oxidation followed by reductive amination.

3.6.3 Miscellaneous

Iminosugar–oligoarabinofuranoside hybrids have been recently synthesized and proposed as a new class of mycobacterial arabinosyltransferase inhibitors [65]. These compounds were designed on the basis that the association of an iminosugar with a potential ligand of the aglycon-binding site of the glycosyltransferase should lead to more potent and selective inhibitors than simple iminosugars. The mycobacterial cell wall is a promising target for the development of new antituberculosis drugs. These hybrids were built from various chiral polyhydroxylated pyrrolidines linked to small oligoarabinofuranosides (Figures 3.9

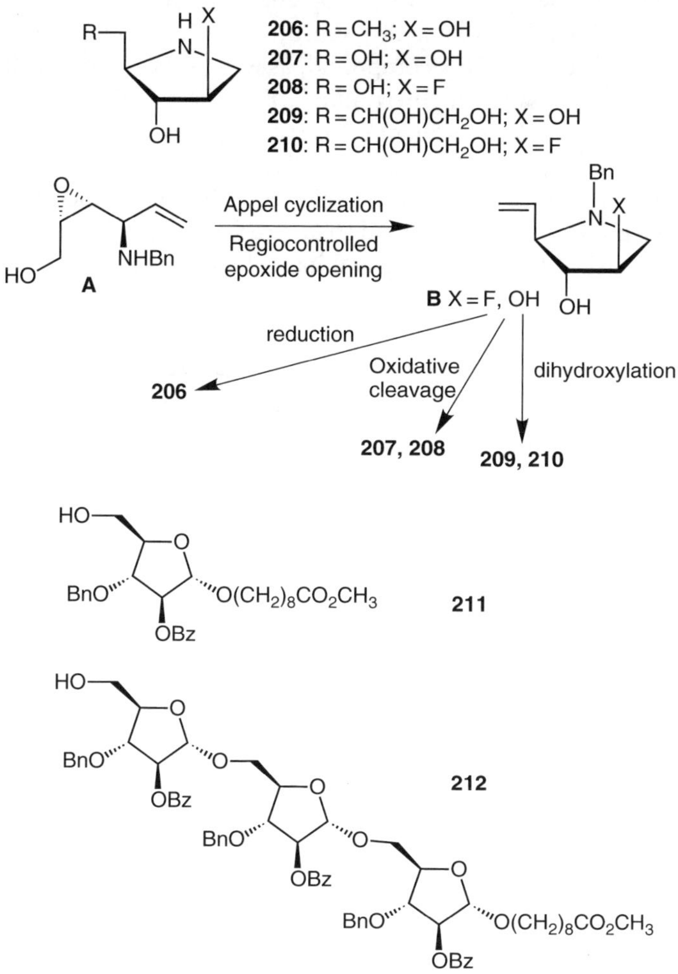

Figure 3.9 Building blocks used for the synthesis of iminosugar-oligoarabinofuranoside hybrids

and 3.10). The coupling reaction was a reductive amination of a suitable oligoarabinofuranoside aldehyde with an unprotected amine counterpart. Good *in vitro* arabinosyltransferase inhibitions were observed for some of these hybrids and, more importantly, their inhibitory activity was modulated by the length of the oligosaccharide moiety. The iminosugar core was 1,4-dideoxy-1,4-imino-D-arabinitol as a mimic of the transferred arabinosyl residue, on which different substituents have been appended, in order to generate structural diversity. The polyhydroxylated pyrrolidines (**206–210**, Figure 3.9) used in this study share this core structure. Compounds **206–210** were synthesized starting from a versatile chiral *cis*-α,β-epoxyamine **A**, readily accessible from the corresponding epoxy aldehyde (Figure 3.9). Appel cyclization on the activated primary alcohol, followed by the regioselective opening of the epoxide formed the hydroxylated or fluorinated vinylpyrrolidine intermediates **B** possessing the required configuration. Catalytic hydrogenation of this vinylpyrrolidine (X = OH) gave ethyl-substituted pyrrolidine **206**, while oxidative cleavage of the double bond of both pyrrolidines **B** (X = F or OH) gave rise to the targeted iminosugar **207** or its 2-deoxy-2-fluoro analog **208**. 1,4-Dideoxy-1,4-imino-D-galactitol **209**, bearing an additional hydroxymethyl substituent, was also used for mimicking the transferred arabinosyl residue. This compound, as well as its 2-deoxy-2-fluoro analogue **210**, was prepared from the same vinylpyrrolidines **B**, by diastereoselective dihydroxylation of the olefin. Finally, a simpler member of this family, (*S*)-prolinol was also included for the library generation.

Two arabinofuranosides (**211–212**) were selected as building blocks: monoarabinofuranoside **211**, which possesses the minimal structure for recognition, and arabinofuranotrimer **212** as a model of the elongating arabinan chain (Figure 3.9).

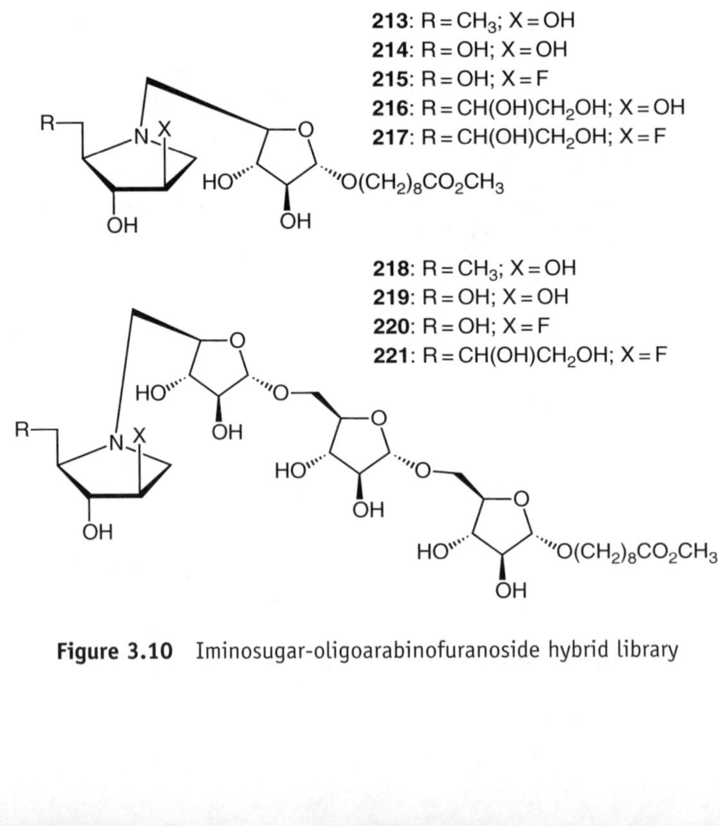

Figure 3.10 Iminosugar-oligoarabinofuranoside hybrid library

Iminosugar–oligoarabinofuranoside hybrids were obtained by coupling the pyrrolidines by reductive amination with the arabinofuranoside-derived aldehydes. Deprotection steps afforded compounds 213–221 (Figure 3.10).

Novel iminoalditol glycosidase inhibitors were designed as novel biological probes and diagnostics and synthesized by appending fluorescently tagged groups to the nitrogen atom [66]. 1,5-Dideoxy-1,5-imino-D-glucitol 222, the corresponding D-manno 225, and L-ido epimers 226, as well as the powerful β-glucosidase inhibitor isofagomine 227, were *N*-alkylated with di-, tri-, as well as tetra-ethylene glycol-derived straight chain spacers by a set of standard procedures. The terminal functional groups of the spacer arms, primary amines, were employed to introduce fluorescent dansyl moieties (Scheme 41).

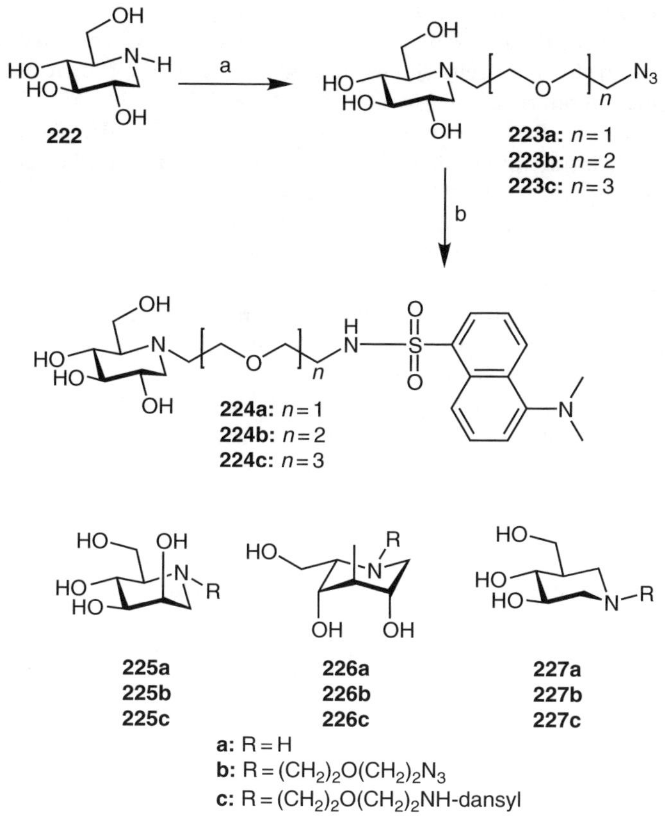

Scheme 41 *Reagents and conditions*: (a) Na_2CO_3, $TsOCH_2(CH_2OCH_2)_nN_3$, $n = 1, 2, 3$; (b) (i): Pd/C, H_2; (ii): dansylCl

3.7 Conclusions

Iminosugars continue to be of great synthetic interest, due to the intrinsic pharmacological potential of many of them and the need for more active and selective compounds. Because of the difficulties encountered in their synthesis, several novel synthetic approaches have been investigated and various examples of small libraries of iminosugars have been reported in the last few years. New classes of iminosugar compounds and conjugates have also been proposed recently for more specific and newer biological applications.

References

1. Cipolla, L., La Ferla, B., Nicotra, F., 2003. General methods for iminosugar synthesis. *Curr. Topics Med. Chem.*, **3**, 485–511.
2. La Ferla, B. and Nicotra, F., 1998. *Synthetic Methods for the Preparation of Iminosugars, from Iminosugars as Glycosidase Inhibitors* Wiley-VCH Verlagsgesellschaft GmbH, pp. 68–92.
3. Wischnat, R., Martin, R., Takayama, S., Wong, C.-H., 1998. Chemoenzymatic synthesis of iminocyclitol derivatives: a useful library strategy for the development of selective fucosyltransfer enzyme inhibitors. *Biorg. Med. Chem. Lett.*, **8**, 3353–3358.
4. Wu, C.-Y., Chang, C.-F., Chen, J. S.-Y., Wong, C.-H., Lin, C.-H., 2003. Rapid diversity-oriented synthesis in microtiter plates for *in situ* screening: discovery of potent and selective α-fucosidase inhibitors. *Angew. Chem. Int. Ed.*, **42**, 4661–4664.
5. Saotome, C., Wong, C.-H. and Kanie, O., 2001. Combinatorial library of five-membered iminocyclitol and the inhibitory activities against glyco-enzymes. *Chem. Biol.*, **8**, 1061–1070.
6. Lee, R. E., Smith, M. D., Pickering, L., Fleet, G. W. J., 1999. An approach to combinatorial library generation of galactofuranose mimics as potential inhibitors of mycobacterial cell wall biosynthesis: synthesis of a peptidomimetic of uridine 5′-diphosphogalactofuranose (UDP-Galf). *Tetrahedron Lett.*, **40**, 8689–8692.
7. La Ferla, B., Bugada, P., Cipolla, L., Peri, P., Nicotra, F., 2004. Synthesis of iminosugar scaffolds for the generation of glycosidase inhibitor libraries. *Eur. J. Org. Chem.*, 2451–2470.
8. Boglio, C., Stahlke, S., Thorimbert, S., Malacria, M., 2005. A stereoselective route towards polyhydroxylated piperidines. A total synthesis of (±)-deoxymannojirimycin. *Org. Lett.*, 7(22), 4851–4854.
9. Cong, X., Liu, K.-G., Liao, Q.-J., Yao, Z.-J., 2005. Preparation of enantiomerically pure 2-(1’-aminomethyl)furan derivatives and synthesis of an unnatural polyhydroxylated piperidine. *Tetrahedron Lett.*, **46**, 8567–8571.
10. Singh, O. V. and Han, H., 2003. A general methodology for the asymmetric synthesis of 1-deoxyiminosugars. *Tetrahedron Lett.*, **44**, 2387–2391.
11. Singh, O. V. and Han, H. A., 2004. Stereodivergent total asymmetric synthesis of polyhydroxylated pyrrolidines via tandem allylic epoxidation and intramolecular cyclization reaction. *Tetrahedron Lett.*, **45**, 6348–6352.
12. Liao, W.-W., Ibrahem, I. and Còrdova, A., 2006. A concise enantioselective synthesis of iminosugar derivatives. *Chem. Commun.*, 674–676.
13. (a) Wang, R.-W., Qiu, X.-L., Bols, M., Ortega-Caballero, F., Qing, F.-L., 2006. Synthesis and biological evaluation of glycosidase inhibitors: *gem*-difluoromethylenated nojirimycin analogues. *J. Med. Chem.*, **49**, 2989–2997. (b) Zhang, X., Xia, H., Dong, X., Jin, J., Meng, W.-D., Qing, F.-L. 2003. 3-Deoxy-3,3-difluoro-D-arabinofuranose: first stereoselective synthesis and application in preparation of *gem*-difluorinated sugar nucleosides. *J. Org. Chem.*, **68**, 9026-9033.
14. Davis, B. G., Maughan, M. A. T., Chapman, T. M., Villard, R., Courtney, S., 2002. Novel cyclic sugar imines: carbohydrate mimics and easily elaborated scaffolds for aza-sugars. *Org. Lett.*, **4**, 103–106.

15. Chapman, T. M., Courtney, S., Hay, P., Davis, B. G., 2003. Highly diastereoselective additions to polyhydroxylated pyrrolidine cyclic imines: ready elaboration of aza-sugar scaffolds to create diverse carbohydrate-processing enzyme probes. *Chem. Eur. J.*, **9**, 3397–3414.
16. Lillelund, V. H., Liu, H., Liang, X., Søhoel, H., Bols, M., 2003. Isofagomine lactams, synthesis and enzyme inhibition. *Org. Biomol. Chem.*, 1, 282–287.
17. Pandey, G. and Kapur, M., 2000. A general strategy towards the synthesis of 1-*N*-iminosugar type glycosidase inhibitors: demonstration by the synthesis of D- as well as L-glucose type iminosugars (isofagomine). *Tetrahedron Lett.*, **41**, 8821–8824.
18. (a) Xie, J., Güveli, T., Hebbe, S., Dechoux, L., 2004. Synthesis of novel 1-*N*-iminosugars from chiral nonracemic bicyclic lactams. *Tetrahedron Lett.*, **45**, 4903–4906. (b)Agami, C., Dechoux, L. and Hebbe, S., 2002. Asymmetric synthesis of nitrogen heterocycles by reaction of chiral β-enaminocarbonyl substrates with acrylate derivatives. *Tetrahedron Lett.*, **43**, 2521-2523.
19. Cren, S., Wilson, C. and Thomas, N. R., 2005. A rapid synthesis of hexofuranose-like iminosgars using ring-closing metathesis. *Org. Lett.*, **7**, 3521–3523.
20. Carmona, A. T., Fuentes, J., Robina, I., Rodriguez Garcia, E., Demange, R., Vogel, P., 2002. Stereoselective synthesis of 1,4-dideoxy-1,4-imino-octitols and novel tetrahydroxyindolizidines. *J. Org. Chem.*, **68**, 3874–3883.
21. Sletten, E. and Liotta, L. J., 2006. A flexible stereospecific synthesis of polyhydroxylated pyrrolizidines from commercially available pyranosides. *J. Org. Chem.*, **71**, 1335–1343.
22. Karanjule, N. S., Markad, S. D., Shinde, V. S., Dhavale, D. D., 2006. Intramolecular 5-endo-trig aminomercuration of β-hydroxy-γ-alkenylamines: efficient route to a pyrrolidine ring and its application for the synthesis of (+)-castanospermine and analogues. *J. Org. Chem.*, **71**, 4667–4670.
23. Donohoe, T. J., Sintim, H. O. and Hollinshead, J., 2005. A non-carbohydrate based approach to polyhydroxylated pyrrolizidines: total syntheses of the natural products hyacinthacine A_1 and 1-epiaustraline. *J. Org. Chem.*, **70**, 7297–7304.
24. Chabaud, L., Landais, Y. and Renaud, P., 2005. Total synthesis of hyacinthacine A_1 and 3-*epi*-hyacinthacine A_1. *Org. Lett.*, **17**, 2587–2590.
25. Goti, A., Cicchi, S., Cacciarini, M., Cardona, F., Fedi, V., Brandi, A., 2000. Straightforward access to enantiomerically pure, highly functionalized pyrrolizidines by cycloaddition of maleic acid esters to pyrroline *N*-oxides derived from tartaric, malic and aspartic acids - synthesis of (−)-hastanecine, 7-*epi*-croalbinecine and (−)-croalbinecine. *Eur. J. Org. Chem.*, 3633–3645.
26. Cicchi, S., Marradi, M., Vogel, P., Goti, A., 2006. One-pot synthesis of cyclic nitrones and their conversion to pyrrolizidines: 7a-*epi*-crotanecine inhibits α-mannosidases. *J. Org. Chem.*, **71**, 1614–1619.
27. Desvergnes, S., Py, S. and Vallée, Y., 2004. Total synthesis of (+)-hyacinthacine A_2 based on SmI_2-induced nitrone umpolung. *J. Org. Chem.*, **70**, 1459–1462.
28. Song, L., Duesler, E. N. and Mariano, P. S., 2004. Stereoselective synthesis of polyhydroxylated indolizidines based on pyridinium salt photochemistry and ring rearrangement metathesis. *J. Org. Chem.*, **69**, 7284–7293.
29. Langlois, N., Le Nguyen, B. K., Retailleau, P., Tarnus, C., Salomon, E., 2006. Creation of quaternary stereocentres: synthesis of new polyhydroxylated indolizidines. *Tetrahedron Asymm.*, **17**, 53–60.
30. Tatsuta, K., Miura, S., Ohta, S., Gunji, H., 1995. Syntheses and glycosidase inhibiting activities of nagstatin analogs. *J. Antibiot.*, **48**, 286–288.
31. Granier, T., Panday, N. and Vasella, A., 1997. Structure activity relations for imidazo-pyridine-type inhibitors of β-D-glucosidases. *Helv. Chim. Acta*, **80**, 979–987.

32. Aoyagi, T., Suda, H., Uotani, K., Kojima, F., Aoyama, T., Horiguchi, K., Hamada, M., Takeuchi, T. 1992. Nagstatin, a new inhibitor of *N*-acetyl-*β*-D-glucosaminidase, produced by Streptomyces amakusaensis MG846-fF3. Taxonomy, production, isolation, physico-chemical properties and biological activities. *J. Antibiot.*, **45**, 1404–1408.
33. Panday, N., Canac, Y. and Vasella, A., 2000. Very strong inhibition of glucosidases by C(2)-substituted tetrahydroimidazopyridines. *Helv. Chim. Acta*, **83**, 58–79.
34. García-Moreno, M. I., Díaz Pérez, P., Ortiz Mellet, C., García-Fernández, J. M., 2002. Castanospermine–trehazolin hybrids: a new family of glycomimetics with tuneable glycosidase inhibitory properties. *Chem. Commun.*, 848–849.
35. García-Moreno, M. I., Díaz Pérez, P., Ortiz Mellet, C., García-Fernández, J. M., 2003. Synthesis and evaluation of isourea-type glycomimetics related to the indolizidine and trehazoline glycosidase inhibitor families. *J. Org. Chem.*, **68**, 8890–8901.
36. Laroche, C., Behr, J.-B., Szymoniak, J., Bertus, P., Schütz, C., Vogel, P., Plantier-Royon, R., 2006. Spirocyclopropyl pyrrolidines as a new series of α-L-fucosidase inhibitors. *Bioorg. Med. Chem.*, **14**, 4047–4054.
37. Torres-Sánchez, M. I., Borrachero, P., Cabrera-Escribano, F., Gómez-Guillén, M., Angulo-Alvarez, M., Mª Diánez, J., Mª Estrada, D., López-Castro, A., Pérez-Garrido, S., 2005. A short and highly stereoselective route to polyhydroxy-perhydroazaazulenes via a *C*-(D-*galacto*-pentopyranos-5-yl)isoxazolidine. *Tetrahedron: Asymm.*, **16**, 3897–3907.
38. Davis, B. G., 2002. Synthesis of glycoproteins. *Chem. Rev.*, **102**, 579–601.
39. Fuchss, T. and Schmidt, R. R., 2000. Synthesis of nojirimycinyl *C*-(L)-serine. *Synthesis*, 259–264.
40. Fuchss, T. and Schmidt, R. R., 1998. Synthesis of the C-analog of 2-acetylamino-2-deoxy-β-D-glucopyranosyl-L- and -D-serine. *Synthesis*, 753–758.
41. Dondoni, A., Giovannini, P. P. and Marra, A., 2001. A concise *C*-glycosyl amino acid synthesis by alkenyl *C*-glycoside–vinyloxazolidine cross-metathesis. Synthesis of glycosyl serine, asparagine and hydroxynorvaline isosteres. *J. Chem. Soc., Perkin Trans.*, **1**, 2380–2388.
42. (a) Dondoni, A., Giovannini, P. P. and Perrone, D., 2005. Cross-metathesis of *C*-allyl iminosugars with alkenyl oxazolidines as a key step in the synthesis of *C*-iminoglycosyl α-amino acids. A route to iminosugar containing *C*-glycopeptides. *J. Org. Chem.*, **70**, 5508–5518. (b) McKillop, A., Taylor, R. J. K., Watson, R. J., Lewis, N. 1994. An improved procedure for the preparation of the Garner aldehyde and its use for the synthesis of *N*-protected 1-halo-2-(R)-amino-3-butenes. *Synthesis*, 31–33.
43. Chakraborty, T. K., Srinivasu, P., Kiran Kumar, S., Kunwar, A. C., 2002. Studies on novel peptidomimetics having bi-directional dispositions of hydroxylated D-Pro-Gly motifs anchored on a C2-symmetric iminosugar-based foundation. *J. Org. Chem.*, **67**, 2093–2100.
44. Compain, P. and Martin, O. R., 2001. Carbohydrate mimetics-based glycosyltransferase inhibitors. *Bioorg. Med. Chem.*, **9**, 3077–3092.
45. Compain, P. and Martin, O. R., 2003. Design, synthesis and biological evaluation of iminosugar-based glycosyltransferase inhibitors. *Curr. Top. Med. Chem.*, **3**, 541–560.
46. Bosco, M., Bisseret, P., Bouix-Peter, C., Eustache, J., 2001. A new concise synthesis of nectrisine and its facile conversion to phosphonoazasugars. *Tetrahedron Lett.*, **42**, 7949–7952.
47. Schuster, M., He, W.-F. and Blechert, S., 2001. Chemical-enzymatic synthesis of azasugar phosphonic acids as glycosyl phosphate surrogates. *Tetrahedron Lett.*, **42**, 2289–2291.
48. Gautier-Lefebvre, I., Behr, J.-B., Guillerm, G., Ryder, N. S., 2000. Synthesis of new (difluoromethylphosphono)azadisaccharides designed as bisubstrate analogue inhibitors for GlcNAc-1,4 glycosyltransferases. *Bioorg. Med. Chem. Lett.*, **10**, 1483–1486.
49. Kajimoto, T., Chen, L., Liu, K. K.-C., Wong, C.-H., 1991. Palladium-mediated stereocontrolled reductive amination of azido sugars prepared from enzymic aldol condensation: a general approach to the synthesis of deoxy aza sugars. *J. Am. Chem. Soc.*, **113**, 6678–6680.

50. Godin, G., Compain, P., Masson, G., Martin, O. R., 2002. A general strategy for the practical synthesis of nojirimycin *C*-glycosides and analogues. Extension to the first reported example of an iminosugar 1-phosphonate. *J. Org. Chem.*, **67**, 6960–6970.
51. Briner, K. and Vasella, A., 1987. Glycosylphosphonates of 2-amino-2-deoxyaldoses. Synthesis of a phosphonate analogue of lipid X. *Helv. Chim. Acta*, **70**, 1341–1356.
52. La Ferla, B., Bugada, P. and Nicotra, F., 2006. Synthesis of the dimethyl ester of 1-deoxy-L-idonojirimycin-1-methylenephosphonate: a new approach to iminosugar phosphonates. *J. Carbohydr. Chem.*, **25**, 151–162.
53. Berliner, L. J. and Wong, S. S., 1975. Manganese(II) and spinlabeled uridine 5'-diphosphate binding to bovine galactosyltransferase. *Anal. Biochem.*, **36**, 43–61.
54. Dondoni, A. and Perrone, D., 2003. A convenient synthesis of iminosugar-*C*-glycosides via organometallic addition to *N*-benzyl-*N*-glycosylhydroxylamines. *Tetrahedron*, **59**, 4261–4273.
55. Dondoni, A., Giovannini, P. and Perrone, D., 2002. New synthesis of pyrrolidine homoazasugars via aminohomologation of furanoses and their use for the stereoselective synthesis of aza-*C*-disaccharides. *J. Org. Chem.*, **67**, 7203–7214.
56. Filichev, V. V., Brandt, M. and Pedersen, E. B., 2001. Synthesis of an aza analogue of 2-deoxy-D-ribofuranose and its homologues. *Carbohydr. Res.*, **333**, 115–122.
57. Filichev, V. V. and Pedersen, E. B., 2001. Synthesis of 1′-aza-*C*-nucleosides from (3R,4R)-4-(hydroxymethyl)pyrrolidin-3-ol. *Tetrahedron*, **57**, 9163–9168.
58. Lewandowicz, A., Tyler, P. C., Evans, G. B., Furneaux, R. H., Schramm, V. L., 2003. Achieving the ultimate physiological goal in transition state analogue inhibitors for purine nucleoside phosphorylase. *J. Biol. Chem.*, **278**, 31465–31468.
59. Lewandowicz, A., Shi, W., Evans, G. B., Tyler, P. C., Furneaux, R. H., Basso, L. A., Santos, D. S., Almo, S. C., Schramm, V. L., 2003. Over-the-barrier transition state analogues and crystal structure with *Mycobacterium tuberculosis* purine nucleoside phosphorylase. *Biochemistry*, **42**, 6057–6066.
60. (a) Chery, F., Cronin, L., O'Brien, J. L., Murphy, P. V., 2004. Facile route for the synthesis of the iminosugar nucleoside (3R,4R)-1-(pyren-1-yl)-4-(hydroxymethyl)pyrrolidin-3-ol. *Carbohydr. Res.*, **339**, 1565–1568. (b) Chery, F., Cronin, L., O'Brien, J. L., Murphy, P. V., 2004. Synthesis of peptidomimetics based on iminosugar and β-D-glucopyranoside scaffolds and inhibiton of HIV-protease. *Tetrahedron*, **60**, 6597–6608.
61. Guckian, K. M., Schweitzer, B. A., Ren, R. X.-F., Sheils, C. J., Paris, P. L., Kool, E. T., 1996. Experimental measurement of aromatic stacking affinities in the context of duplex DNA. *J. Am. Chem. Soc.*, **118**, 8182–8183.
62. Smirnov, S., Matray, T. J., Kool, E. T., de los Santos, C., 2002. Integrity of duplex structures without hydrogen bonding: DNA with pyrene paired at abasic sites. *Nucl. Acids Res.*, **30**, 5561–5569.
63. Evans, G. B., Furneaux, R. H., Hausler, H., Larsen, J. S., Tyler, P. C., 2004. Imino-*C*-nucleoside synthesis: heteroaryl lithium carbanion additions to a carbohydrate cyclic imine and nitrone. *J. Org. Chem.*, **69**, 2217–2220.
64. Yokoyama, M., Ikeue, T., Ochiai, Y., Momotake, A., Yamaguchi, K., Togo, H., 1998. Synthesis and properties of *C*-aza-2-deoxy-L-lyxonucleosides. *J. Chem. Soc. Perkin Trans.*, **1**, 2185–2192.
65. Marotte, K., Ayad, T., Génisson, Y., Besra, G. S., Baltas, M., Prandi, J., 2003. Synthesis and biological evaluation of imino sugar-oligoarabinofuranoside hybrids, a new class of mycobacterial arabinofuranosyltransferase inhibitors. *Eur. J. Org. Chem.*, 2557–2565.
66. Lundt, I., Steiner, A. J., Stuetz, A. E., Tarling, C. A., Ully, S., Withers, S. G., Wrodnigg, T. M., 2006. Fluorescently tagged iminoalditol glycosidase inhibitors as novel biological probes and diagnostics. *Bioorg. Med. Chem.*, **14**, 1737–1742.

4

Iminosugar *C*-glycosides: synthesis and biological activity

Philippe Compain

4.1 Introduction

Since the commercialization of the first iminosugar-based drug in 1996 (Glyset™, 1) [1], the rate of discoveries in the field of sugar mimetics with nitrogen replacing the ring oxygen has increased dramatically (Figure 4.1). The scope of biological activity of iminosugars, historically known as potent inhibitors of glycosidases [2], has been extended to the inhibition of a number of enzymes of medicinal interest such as glycosyltransferases [3], metalloproteinases [4], nucleoside-processing enzymes [5] or a sugar nucleotide mutase [6]. Since these enzymes are involved in numerous fundamental biological processes, iminosugars provide leads for the development of new therapeutic agents for a wide range of diseases [7, 8]. However, due to their synthetic accessibility, most iminosugar derivatives reported are simple iminoalditols related to 1-deoxynojirimycin (2) and do not carry a substituant at the 'anomeric' position. Thus precious aglycone-specific information is lost when such iminosugars are used to mimic oligosaccharides or glycoconjugates as carbohydrate-processing enzyme inhibitors.

A fundamental limitation of imino-analogs of glycosides is indeed their instability caused by the lability of the *N*,*O*-acetal function which prevents their use as biological probes or drug candidates. The replacement of the oxygen atom of the *N*,*O*-acetal function by a methylene group to form imino-*C*-glycosides has frequently been used as a tactic to generate stable analogs of glycoconjugates [9, 10]. Since the first synthesis of α-homonojirimycin **3** [11], one of the simplest examples of this class of compounds, many synthetic efforts have been devoted to develop efficient and stereocontrolled routes to iminosugar *C*-glycosides (Figures 4.2. and 4.3) [9, 10, 12].

Since most glycoconjugates contain carbohydrate residues in the pyranoid form, and since our goal is to present iminosugar-based inhibitors or ligands exhibiting high

Iminosugars: From synthesis to therapeutic applications Edited by Philippe Compain and Olivier R. Martin
 ISBN 978-0-470-03391-3

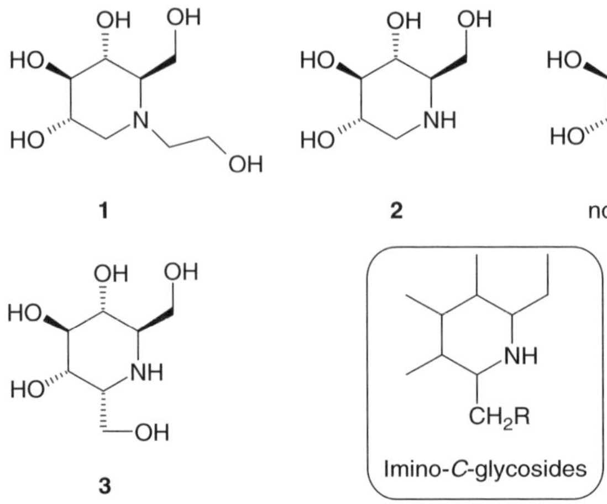

Figure 4.1 Iminoalditols, iminosugars and imino-*C*-glycosides

selectivity, this chapter will focus on the synthesis and biological activity of iminosugar *C*-glycosides in the piperidine series having a functionalized or relatively complex aglycon part [13]. Our purpose is therefore not to provide a comprehensive and exhaustive review on the subject but rather to present an overall view of the field by highlighting some representative synthetic strategies and key biological data. Based on relevant examples, fundamental questions will be addressed such as: what is the advantage of having an aglycone part positioned on the 'anomeric' position instead of the endocyclic nitrogen in terms of ligand binding selectivity? What is the impact of modern synthetic methodologies, such as metathesis, on strategies developed in recent years for the preparation of iminosugar *C*-glycosides? Finally, we hope that this review will be a useful source of information and, more importantly, a source of fruitful ideas for the reader.

4.2 Synthesis of iminosugar *C*-glycosides

The structure of iminosugar *C*-glycosides presents a number of challenges that may be enumerated as follows: (a) at least four contiguous stereogenic centres must be obtained with high stereochemical control, (b) the piperidine ring must be generated efficiently and (c) due to the high density of functional groups, protecting groups must be selected judiciously. The main synthetic challenge remains the design of short and general routes to iminosugar *C*-glycosides of predictable configuration and the generation of structural diversity from advanced precursors to accelerate the discovery of biologically relevant compounds.

The retrosynthetic analysis of iminosugar-*C*-glycosides may be based on three key disconnections of the following critical C-C or C-N bonds: C5-N, C1-N, C1-CH_2R (Figures 4.2 and 4.3). General synthetic strategies reported in the literature may be divided into two main categories: (a) those which construct the *C*-glycosidic structure by way of intramolecular cyclization (C5-N and/or C1-N key disconnection) and (b) those

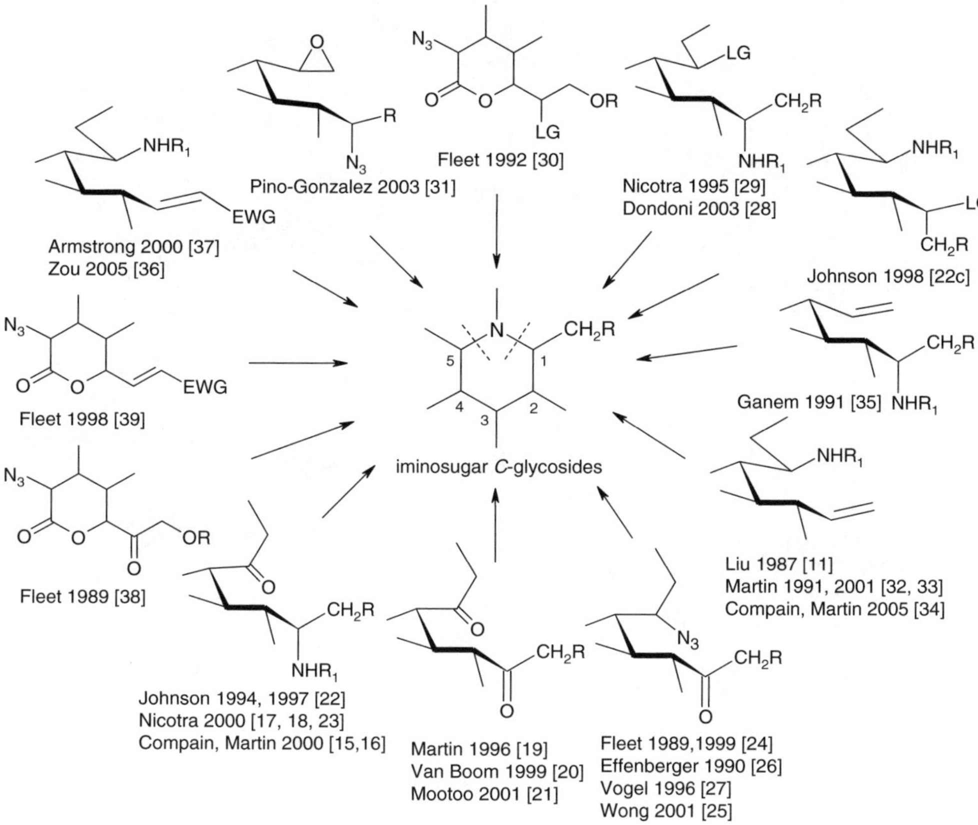

Figure 4.2 The intramolecular cyclization approach

which make use of an electrophilic iminosugar donor (C1-CH_2R key disconnection). This latter intermolecular approach, which is inspired from well-documented strategies developed for the synthesis of classical *C*-glycosides [14], has been less exploited than the intramolecular approach. Because of their availability and their structural relationship with iminosugars, carbohydrates have largely been used as the starting material.

4.2.1 Intramolecular cyclization: C5-N and/or C1-N key disconnection

Reductive amination [15–27]

One of the reactions used most often for the formation of the C5-N and/or C1-N bond has been reductive or double reductive amination. Chemists have thus taken advantage of the reactivity of amino groups to develop specific synthetic strategies that cannot be applicable to classical *C*-glycosides.

A versatile strategy for the synthesis of nojirimycin *C*-glycosides and related compounds with full stereocontrol has been reported recently [15]. The first key step

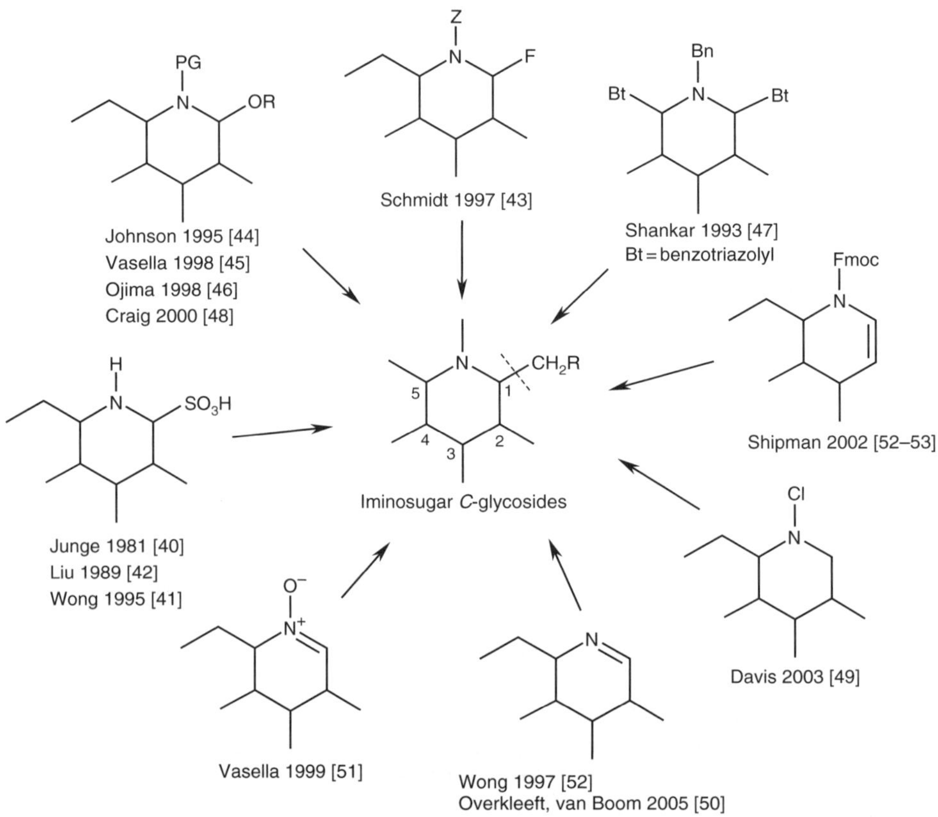

Figure 4.3 The electrophilic iminosugar donor approach

of the process is the chain extension of imine **4** (Scheme 4.1). This addition step, which controls the α- vs. β-configuration at the pseudo-anomeric centre in the final product, is highly diastereoselective (*re*-face addition) and the stereoselectivity can be effectively inverted by adding an external monodentate Lewis acid (*si*-face addition). Structural diversity may be introduced at the 'anomeric' position by using the wide library of organometallic nucleophiles available. The last key steps of the synthetic plan are the intramolecular reductive amination of the latent keto function of the aminosorbofuranose derivative. The complete synthesis could be achieved in only 10 steps from commercially available 2,3;4,6-di-*O*-isopropylidene-α-L-sorbofuranose and provided α- or β-1-*C*-substituted 1-deoxynojirimycin derivatives in 27 to 52 per cent overall yield. A variation of this strategy using *N*-NAP or *N*-allyl protecting group has been developed to access to β-*C*-glycosides derivatives **8** having an unsaturated R group [16].

Access to α-1-*C*-allyl 1-deoxynojirimycin derivatives **10** by way of intramolecular reductive amination with $NaBH(OAc)_3$ is presented in Scheme 4.2 [17]. The amino group is introduced by the reaction of tetra-*O*-benzyl-D-glucose with benzylamine in the presence of toluene-*p*-sulfonic acid. Reaction with allylmagnesium bromide provided the open chain amino alcohol **9** stereoselectively after protection of the amino group

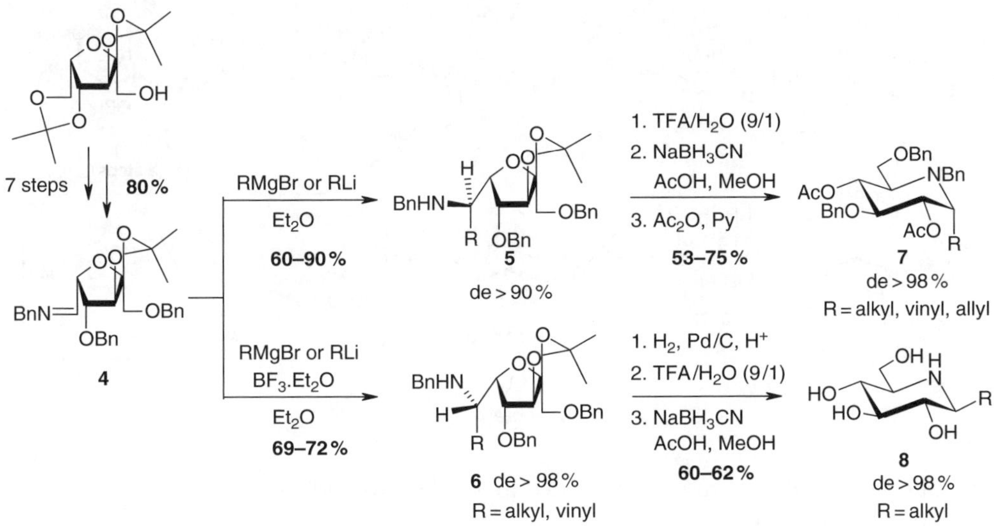

Scheme 4.1

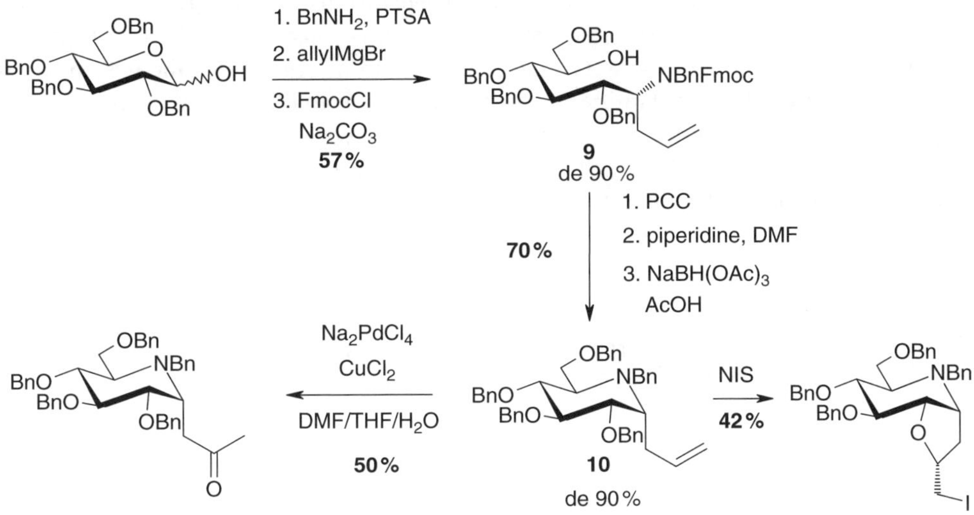

Scheme 4.2

by a Fmoc group. Oxidation of the free hydroxyl group and removal of the Fmoc group followed by subsequent reductive amination afforded *C*-glycosides **10**. Chemical manipulations of the allylic appendage have also been performed [17, 18].

Double reductive amination seems to be a method of choice for the synthesis of iminosugar *C*-glycosides since C5-N and C1-N bonds can be formed stereoselectively in one step. The examples reported in the literature indicated that reductive

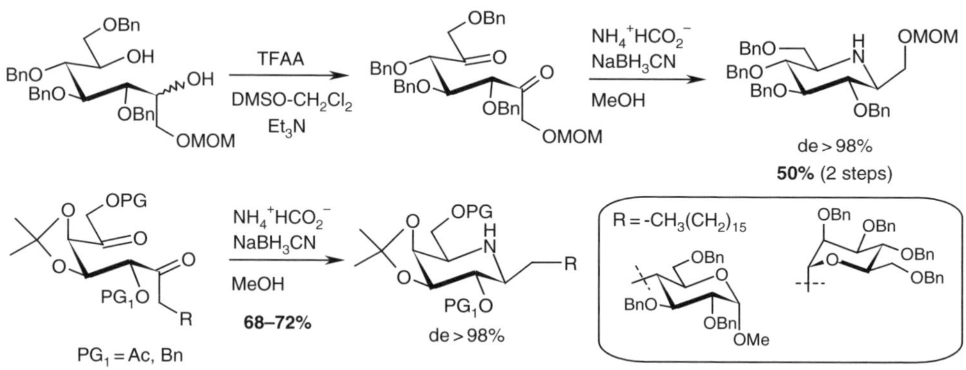

Scheme 4.3

amination of dicarbonyl sugars proceeded in good yields and with complete β-selectivity (Scheme 4.3) [19–21].

Intramolecular S_N2 reaction [22c, 28–31]

In 2003, Dondoni and Perrone reported the preparation of iminosugar *C*-glycosides from tetra-*O*-benzyl-D-glucose and tetra-*O*-benzyl-D-galactose (Scheme 4.4). The synthetic plan was based on the stereoselective addition of organometallic reagents onto *N*-benzyl-*N*-glycosylhydroxylamines **11** followed by intramolecular displacement of a mesylate group by the amino group [28]. This latter step, which involves an inversion of configuration at C-5, afforded the fully protected iminosugar-*C*-glycosides having a pseudo L-*ido* or L-*altro* configuration [29].

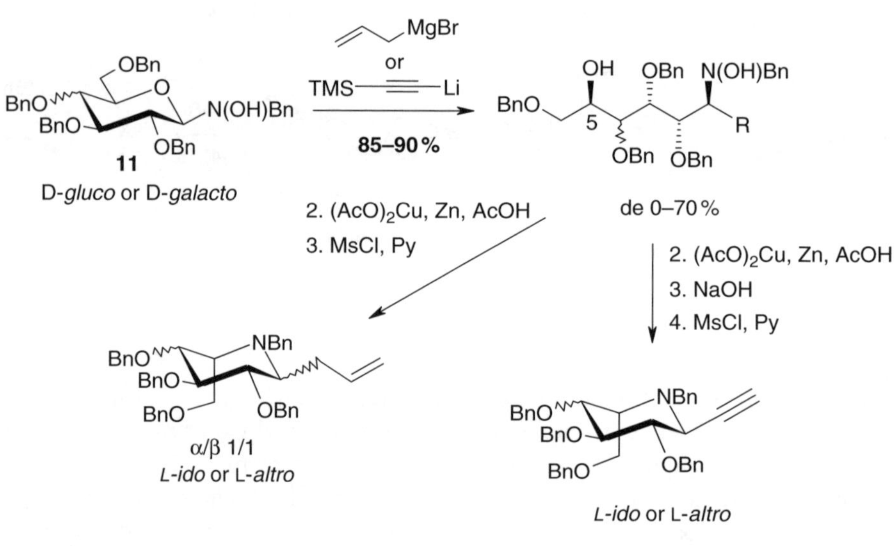

Scheme 4.4

Electrophile-induced cyclization of aminoalkenes [11, 32–35]

Cyclization of unsaturated amido- or aminoalditols have usually been promoted by mercury(II) salts or NIS as the source of electrophile [11, 32–35]. The main advantage of this methodology is to form in one step a piperidine ring and an organomercurial or iodo derivative that may be further functionalized. The reaction generally proceeds in good yield and with high diastereoselectivity. Electrophile-induced cyclization of D-*gluco*, L-*altro*, D-*galacto* amino- or amidoheptenitol afforded the corresponding iminosugar *C*-glycoside having the 2,3-*cis* configuration as dictated by the configuration at C3 (allylic carbon) of the starting material (Scheme 4.5) [11, 32, 33].

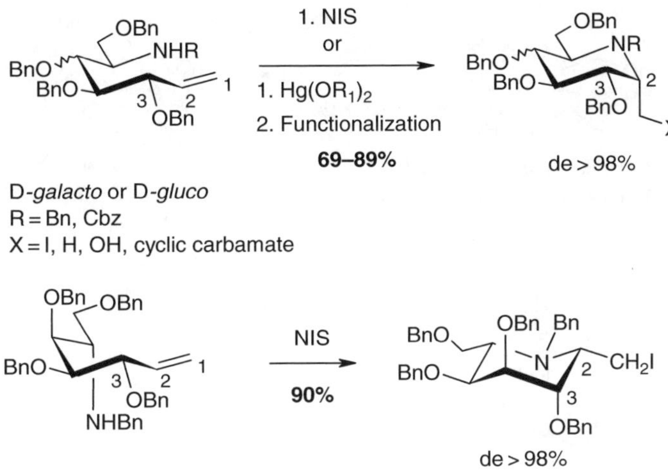

Scheme 4.5

It is worth noting that the reaction performed from L-*ido* amidoheptenitol **12** leads to 2,3-*trans* epimer **13**, because steric interactions between *syn*-diaxial susbtituents probably destabilize the intermediate leading to the 2,3-*cis* isomer (Scheme 4.6) [32]. A loss of

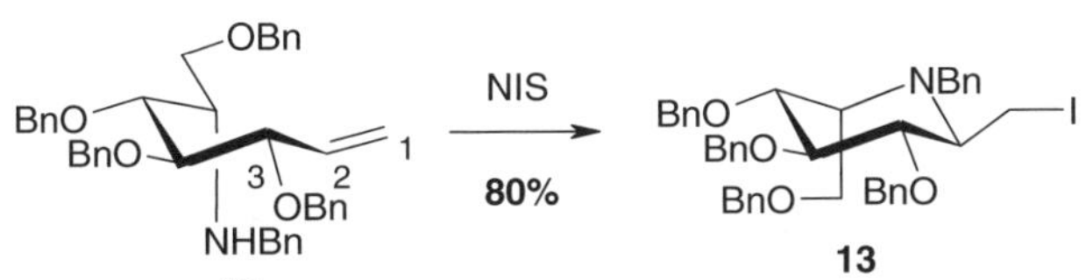

12 **13**

de > 98%

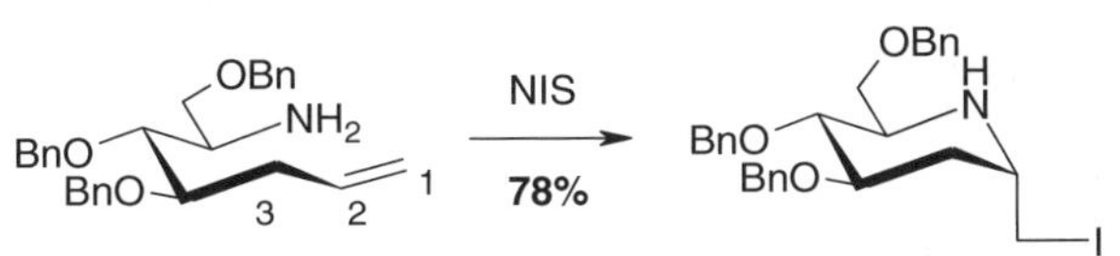

14

de 70%

Scheme 4.6

diastereoselectivity is observed in the absence of an alkoxy group at C-3 as judged by the cyclization of 3-deoxy heptenitol **14** [34].

Liotta, Lee and Ganem utilized a cyclization promoted by mercury(II) salts to form the C5-N bond in iminosugar *C*-glycoside **16** and **17** [35]. The degree of stereoselectivity was found to be strongly dependent on the absolute configuration at C1 of the starting material. The reaction is indeed highly controlled by the incipient axial hexenyl side chain so that only the equatorial organomercurial **16** was obtained from **15a**. In contrast, cyclization performed from **15b**, the epimer of **15a**, afforded a 1:1 mixture of axial and equatorial organomercurials **17** (Scheme 4.7).

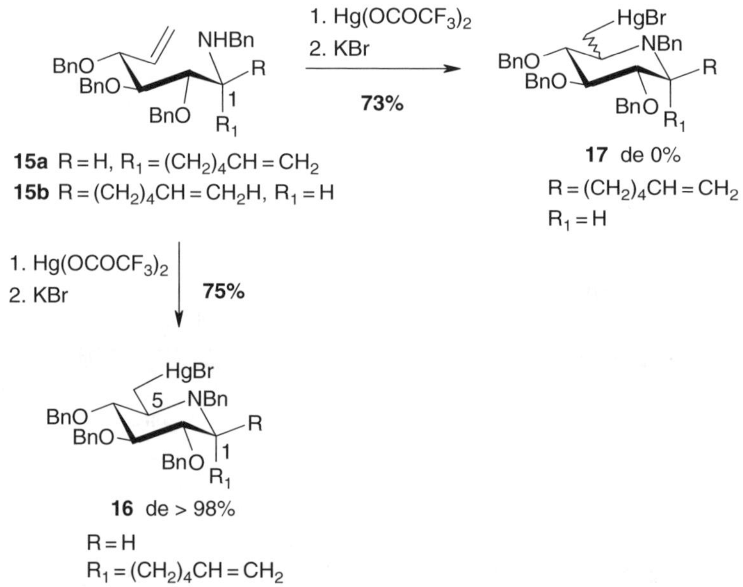

Scheme 4.7

Miscellaneous reactions [36–39]

Iminosugar *C*-glycosides have been prepared by way of intramolecular hetero-Michael addition to form a C1-N bond [36, 37]. Depending on the structure of the starting material, the reaction was found to be not [36] or completely diastereoselective [37]. 5-azido-5-deoxy-glycofuranoside **18** was used as latent substrate for intramolecular Michael addition to provide a 1:1 mixture of α- and β-epimer of iminosugar *C*-glycoside **19** (Scheme 4.8) [36].

Fleet *et al.* reported synthetic strategies which make use of 1,3-dipolar cycloaddition of open-chain azido-enoates or intramolecular aza-Wittig reaction of ketoazides as the key step [38, 39].

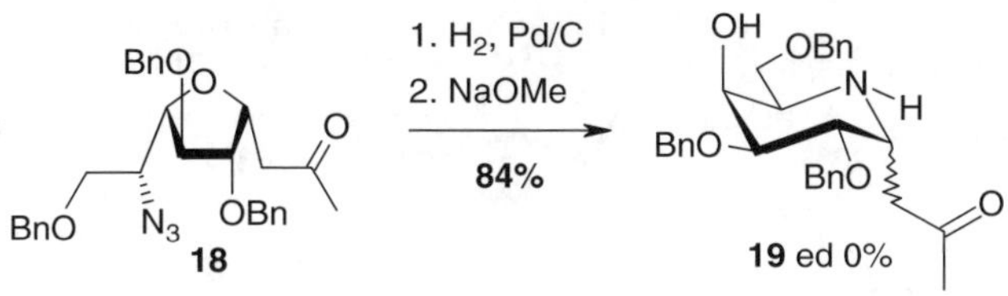

Scheme 4.8

4.2.2 Electrophilic iminosugar donors: C1-CH_2R key disconnection

Piperidinosyl donors [40–48]

One of the most widely used methods for the synthesis of classical *C*-glycosides involves the Lewis-acid catalysed reaction of carbon nucleophiles with activated sugar donors bearing a leaving group at the anomeric position [14]. In the case of iminosugars, the direct application of this strategy is not obvious because of the relative instability of the corresponding iminoglycosides. However, a few examples of such methodologies have been published in the literature. The first general access to iminosugar *C*-glycosides reported in 1981 was based on sulphonate/cyanide and cyanide/organomagnesium reagent exchange reactions [40] (Scheme 4.9). The process afforded 1-α-*C*-substituted-1-deoxynojirimycin derivatives with modest to good stereoselectivity as judged by the diastereomeric excess reported. The sulphonate/cyanide exchange reaction was also applied to synthesize α-homoaminoiminosugars and α-homoiminosugars [41, 42]. The latter method has the advantage of being applicable to unprotected piperidinose derivatives.

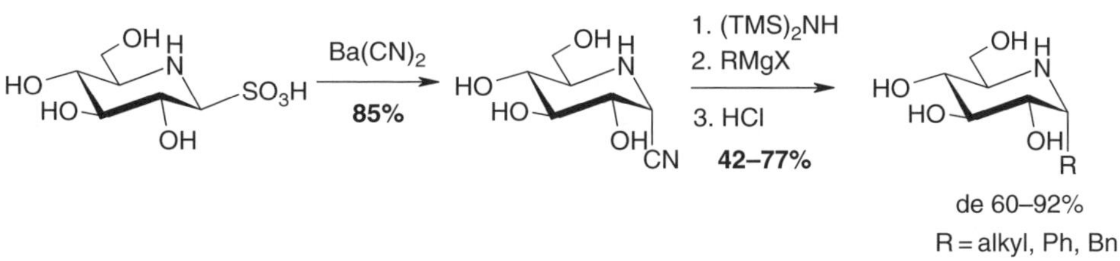

Scheme 4.9

The groups of Johnson, Schmidt and Vasella reported that *N*-alkoxycarbonyl-protected piperidinosyl donors could be key intermediates in the synthesis of iminosugar *C*-glycosides [43–45]. Reaction of iminoglycosyl fluoride **20** with various silylated carbon nucleophiles in the presence of $BF_3.OEt_2$ proceeded in good yields and with modest to high diastereoselectivity in favour of the α-epimer (Scheme 4.10) [43]. The best stereoselectivity was achieved with allyl- and propynyltrimethylsilane.

Exchange of pseudo anomeric alkoxy groups has also been performed to generate key piperidinose-derived acyliminium intermediates. Reactions with TMSCN or

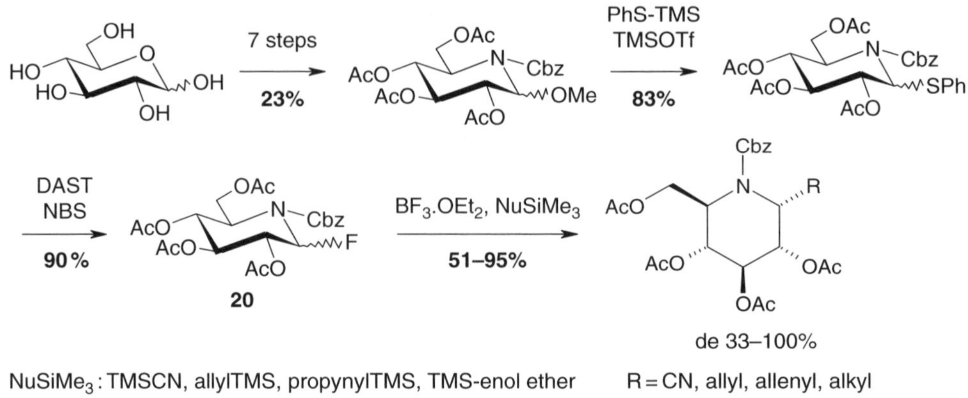

Scheme 4.10

allyltrimethylsilane afforded the expected iminosugar *C*-glycosides with complete diastereoselectivity (Scheme 4.11) [44–46]. Neighbouring-group participation of the acetamido group at C-2 controls the stereochemistry of the bond forming step to generate 1-β-cyano-1-deoxynojirimycin derivative.

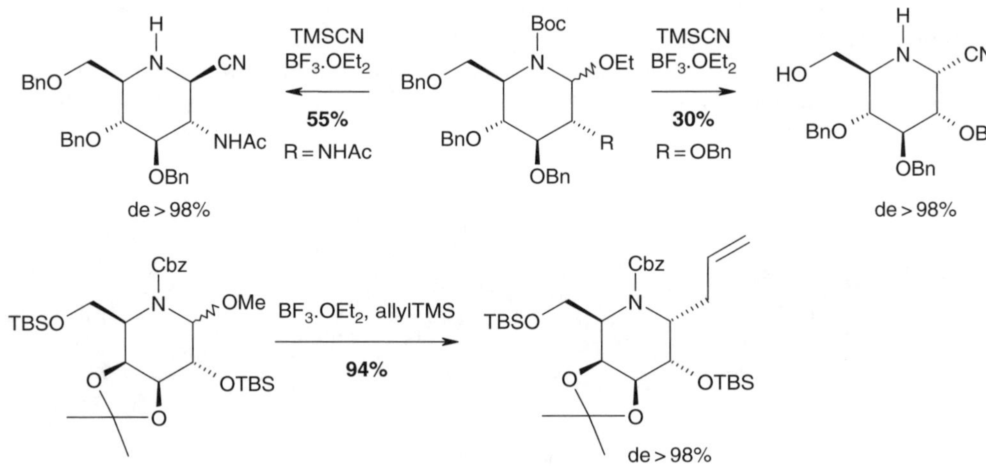

Scheme 4.11

Shankar *et al.* reported a related strategy from polyhydroxylated 2,6 bis(benzotriazolyl) piperidines which allows the one-step formation of two key C-C bonds (Scheme 4.12) [47]. The principal advantages of this straightforward approach are that no hydroxyl protection is required and the displacement reaction can be performed without Lewis acid. However, the double benzotriazolyl/carbon nucleophile exchange, which provides the 1,5-*trans* isomer as the major product, proceeds with modest diastereoselectivity.

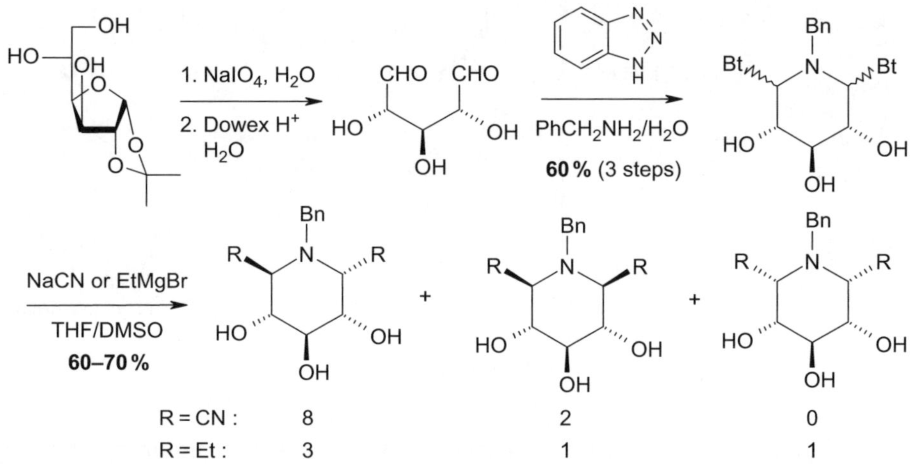

Scheme 4.12

Addition to endocyclic C=N bond [49–52]

In 2003, Davis *et al.* reported the synthesis of adenophorine **21** by way of the nucleophilic addition to a cyclic imine [49]. Starting from 1-deoxynojirimycin derivative **22**, the aldimine **23** was obtained by treatment with *N*-chlorosuccinimide followed by subsequent regioselective elimination of HCl (Scheme 4.13). Addition of EtMgBr afforded protected 1-*epi*-adenophorine **24** as a single diastereoisomer. Epimerization at C-1 was performed using a similar strategy: regioselective formation of ethyl ketimine **25** followed by LAH reduction to give adenophorine after deprotection.

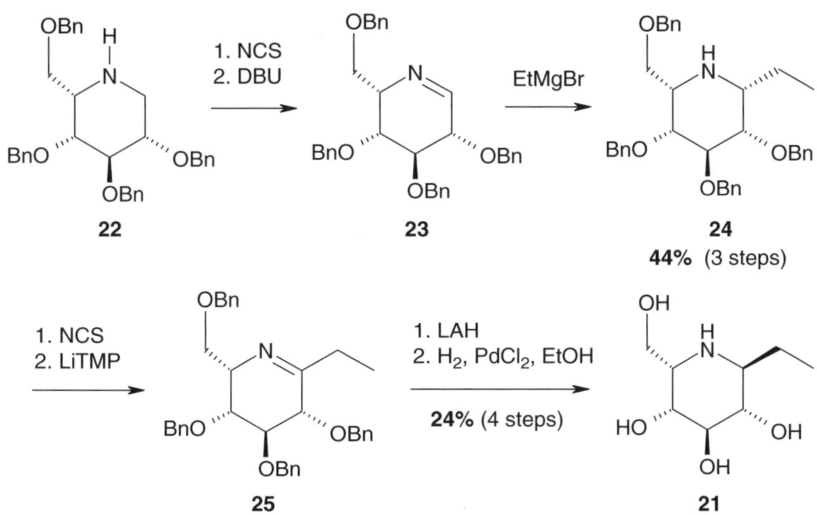

Scheme 4.13

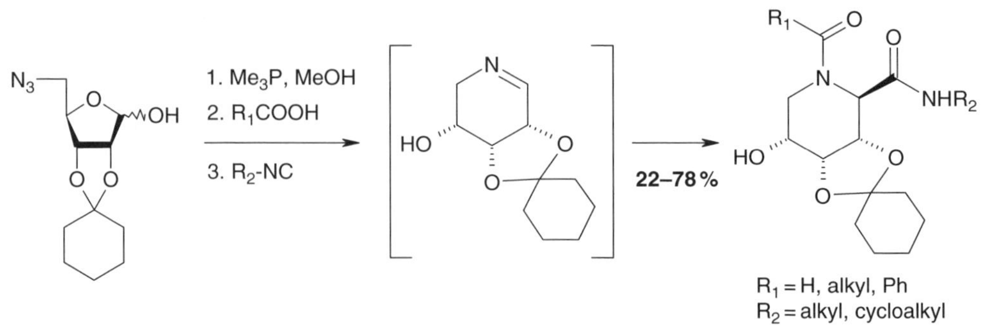

Scheme 4.14

A small library of iminosugar *C*-glycosides was obtained by combining Staudinger-aza-Wittig mediated synthesis of cyclic imine with the Ugi reaction (Scheme 4.14) [50]. The overall process was found to be highly diastereoselective.

Diastereoselective addition of trimethylsilyl cyanide to a cyclic iminosugar-derived nitrone was also reported by Peer and Vasella [51].

Imino glucals [53, 54]

By analogy with classical glycal chemistry, Shipman *et al.* have shown that imino glucal **26**, obtained in 11 steps from D-glucal, could undergo Lewis-acid mediated

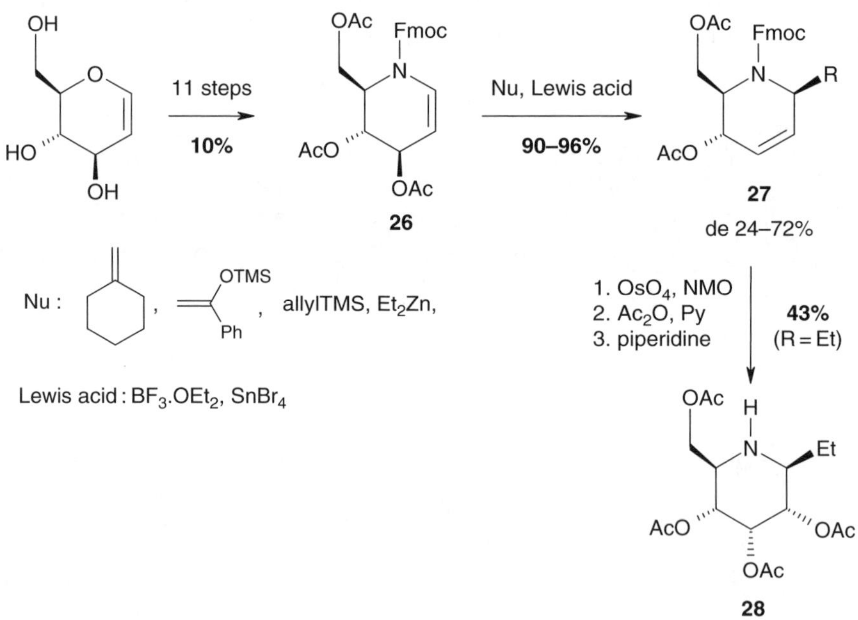

Scheme 4.15

carbon-carbon bond forming reactions by allylic displacement of the C-3 acetate group [53, 54]. The reaction favoured the formation of the β-anomer in high yields but with modest diastereoselectivity (Scheme 4.15). Stereospecific dihydroxylation of the resulting endocyclic double bond afforded the more highly oxygenated iminosugar *C*-glycoside **28** having a pseudo D-*allo* configuration.

4.2.3 The cycloaddition approach

Wightman and Vasella have independently reported 1,3-dipolar cycloaddition reactions between cyclic iminosugar-derived nitrones and methyl acrylate [51] or sugar alkenes [55]. Aza-Diels-Alder reaction was used as a key step to provide polysubstituted piperidines from noncarbohydrate starting precursors [56, 57]. A highly stereocontrolled tandem aza[4+2]/allylboration process has been developed recently by Hall and Tailor (Scheme 4.16) [57]. An asymmetric version of this multicomponent reaction has been achieved using a chiral auxiliary approach.

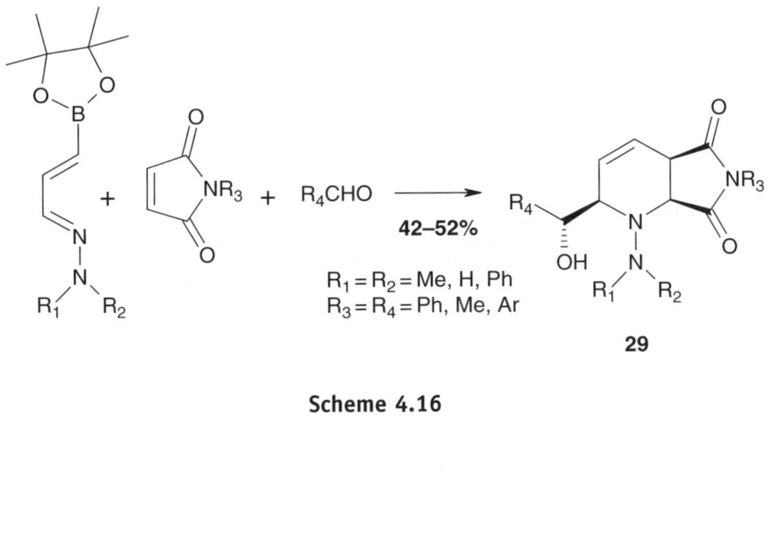

29

Scheme 4.16

4.2.4 Iminosugar C-glycoside building blocks

Iminosugar *C*-glycosides bearing key structural functionalities have been used as advanced intermediates for the convergent synthesis of imino-*C*-glycosides with a great degree of structural diversity in the aglycone. In 2003, it was demonstrated that cross-metathesis was the method of choice to access in one step various glycoconjugate mimic precursors (Scheme 4.17) [16a]. The reaction proceeded with excellent *E*/*Z* selectivity and good to excellent yields. In 2005, Dondoni, Giovannini and Perrone used the same chemistry for the synthesis of iminosugar containing *C*-glycopeptides [58].

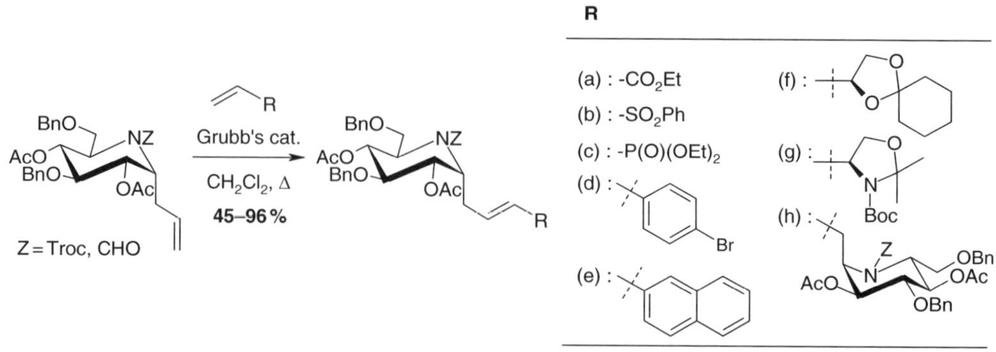

Scheme 4.17

Iminosugar-based aziridine **30** was found to be a versatile intermediate for the synthesis of fagomine *C*-glycoside derivatives [34]. Ring-opening of bicyclic aziridine **30** with various heteroatomic nucleophiles including thiol, amine, alcohol, carboxylate and phosphate occurred in modest to very good yields and was found to be completely regioselective (Scheme 4.18).

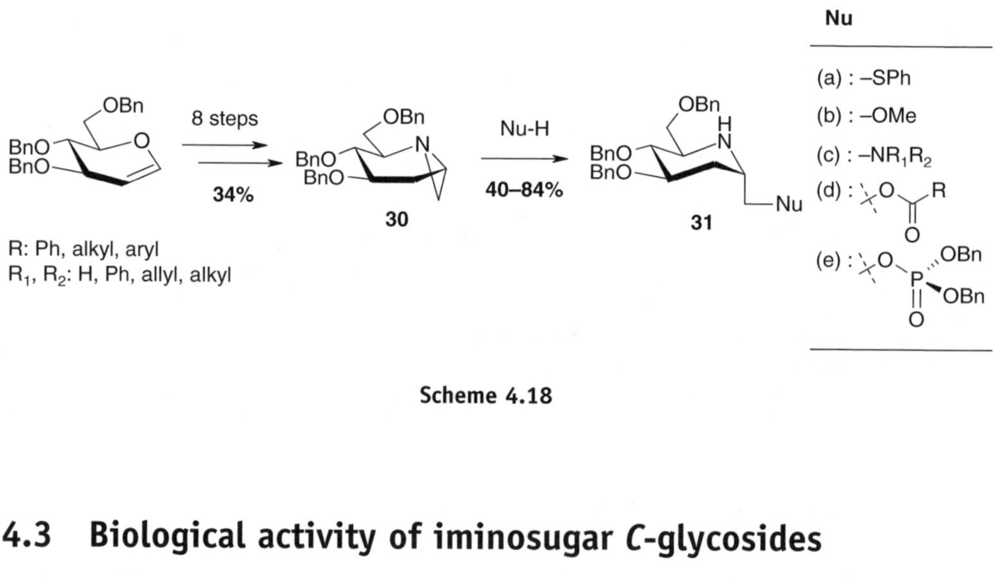

Scheme 4.18

4.3 Biological activity of iminosugar *C*-glycosides

Iminosugar *C*-glycosides constitute an important class of iminosugars. Not surprisingly, most of the chapters of this book will display examples of glycoenzyme inhibitors or potential therapeutic agents based on this type of structure (Scheme 4.19) [2, 3, 7, 10].

Through selected case studies, our aim will mainly be to highlight what are the advantages of iminosugar *C*-glycosides in terms of potency and selectivity compared with simpler iminoalditols including *N*-substituted 1-deoxyiminosugars. It is important to note that these latter compounds are indeed much more readily available and can be

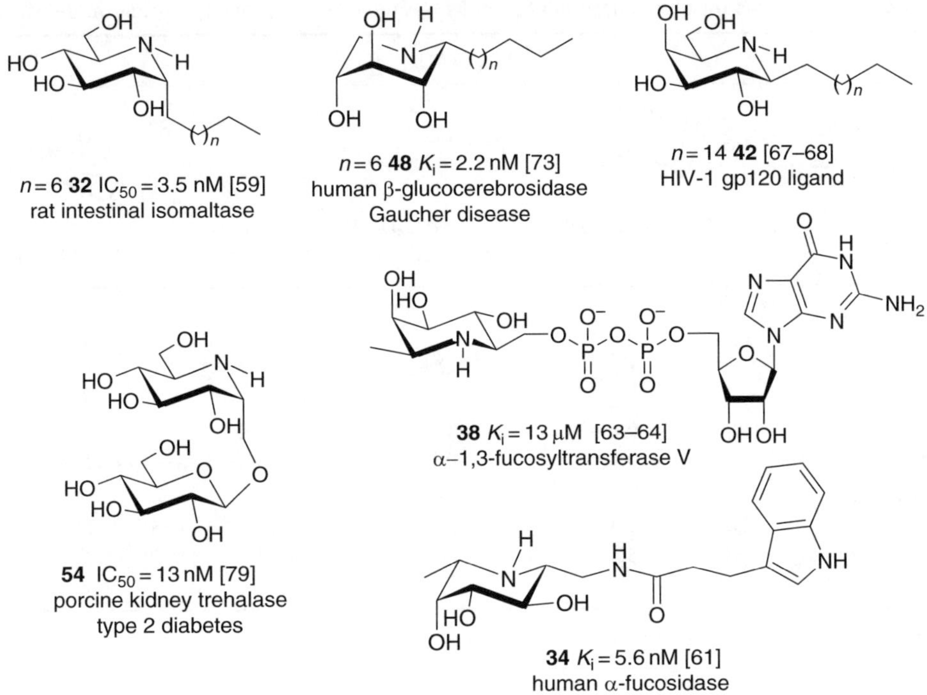

Scheme 4.19 Selected iminosugar *C*-glycosides of biological and therapeutical interest

obtained in a few steps by reductive amination or alkylation reactions from commercially available 1-deoxyiminosugars such as DNJ (2). Consequently, to date, there are five times more examples of *N*-substituted piperidinols (around 1000) than piperidinose *C*-glycosides (around 200) reported in the *Chemical Abstracts Service*. Based on various significant examples, it will be shown that iminosugar *C*-glycosides offer attractive potential in the field of antiviral agents and for the discovery of selective inhibitors of therapeutically relevant carbohydrate-processing enzymes.

4.3.1 Intestinal α-glucosidase inhibitors

Several lipophilic derivatives of 1-deoxynojirimycin bearing an alkyl chain attached to the 'anomeric' position were recently evaluated as intestinal glycosidase inhibitors since these enzymes represent valuable targets for the management of type 2 diabetes [59]. Biological assays showed a marked dependence of the selectivity and potency upon the position of the alkyl chain (i. e., α-1-*C*-, β-1-*C*- or *N*-alkyl derivatives). The results, reported in Table 4.1, indicate clearly that a simple 1,2-shift of the alkyl chain from the endocyclic nitrogen to the 'anomeric' carbon having a β-configuration is detrimental to the inhibition towards α-glycosidases. In contrast, α-1-*C*-alkyl-1-deoxynojirimycins were found to be more efficient and much more selective as isomaltase inhibitors than

Table 4.1 IC_{50} values for glycosidase inhibition by *N*- and 1-*C*-alkyl 1-deoxynojirimycin derivatives

	IC_{50} (μM)				
enzyme	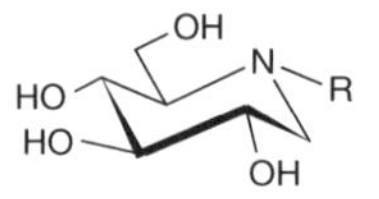 R = *n*-C_4H_9	R = *n*-C_9H_{19} **33**	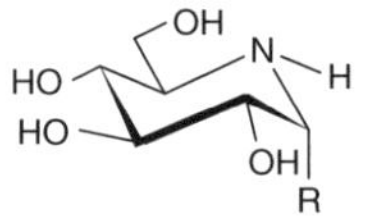 R = *n*-C_4H_9	R = *n*-C_9H_{19} **32**	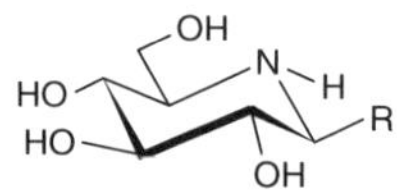 R = *n*-C_4H_9
α-glucosidase					
rice	0.42	0.08	11	1.5	61
yeast	NI[b]	NI	465	56	NI
maltase[a]	2.1	1.3	12	4.0	210
sucrase[a]	58	0.66	3.6	2.5	155
isomaltase[a]	2.7	0.23	0.15	0.0035	115
β-glucosidase					
sweet almond	NI	150	NI	150	NI
Caldocellum saccharolyticum	NI	80	NI	80	NI

[a]From rat intestine. [b]NI: less than 50% inhibition at 1000 μM.

the parent *N*-alkyl derivatives. α-1-*C*-Nonyl-1-deoxynojirimycin (**32**) with an IC_{50} of 3.5 nM is the most potent and selective inhibitor of intestinal isomaltase reported to date. Remarkably, it is 65-fold more potent than the parent *N*-nonyl derivative **33**.

4.3.2 Fucosidase inhibitors

In 2003, Lin, Wong *et al.* reported the discovery of picomolar slow tight-binding inhibitors of the α-fucosidase from *Corynebacterium sp.* [60]. Compound **34** with a K_i of 0.32 nM and a K_i^* of 0.46 pM represents the most potent glycosidase inhibitor reported to date (Scheme 4.19). The simple screening method used was based on the amide-forming reaction of a *C*-glycosylated L-fuconojirimycin core **35** with various carboxylic acids in a microtiter plate, followed by a direct inhibition assay without product isolation (Scheme 4.20). The β-aminomethyl group attached to the C-1 position of 1-deoxyfuconojirimycin thus allowed the identification of an additional high-affinity binding component through amide-bond formation. The optimal aglycones uncovered were based on hydrophobic aromatic heterocycles.

Very recently, in a similar study, Ho *et al.* screened various fuconojirimycin cores, including **36** and **37**, the *N*- and *O*-2 substituted analogs of **35**, bearing an aminopropyl group to avoid the formation of a labile *N*,*N*- or *N*,*O*-acetal function (Scheme 4.21) [61]. β-1-*C*-Aminomethyl-1-deoxynojirimycin (**35**) and its *N*-alkylated analogs **36** were found to display comparable inhibition towards *T. maritima* and human α-fucosidase.

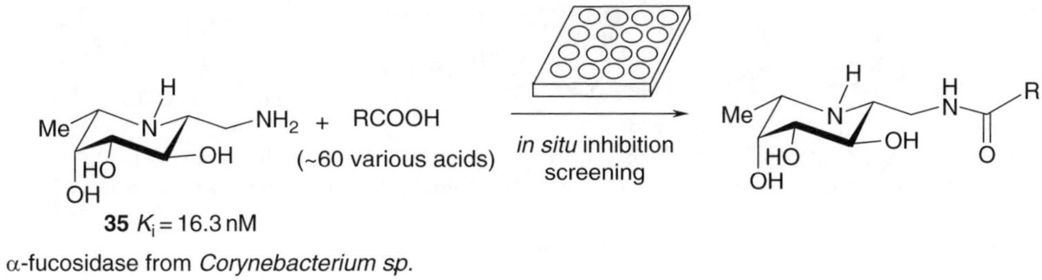

Scheme 4.20

However, if more than 80 per cent of the coupling products of **35** exhibited enhanced inhibition towards both fucosidases, only a few coupling reactions of **36** and **37** resulted in an increase in inhibitory potency.

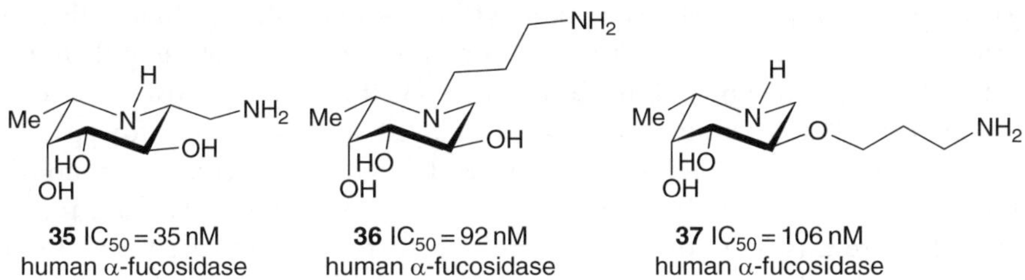

Scheme 4.21

4.3.3 Fucosyltransferase inhibitors

Fucosyltransferases and their regulation have been the subject of intensive investigations due to the correlation of fucosylation patterns in oligosaccharides with adhesion events and various developmental or pathological processes such as cancer and apoptosis [3]. For example, human α-1,3-fucosyltransferase V (FucT V), which catalyses the transfer of the L-fucose moiety from GDP-fucose to the 3-OH group of *N*-acetylglucosamine in sialyllactosamine to produce sialyl Lewis X, has been a valid target for the development of potent inhibitors. Such compounds may have potential medicinal applications as anti-inflammatory or antitumour agents. In 2001 and 2002, the groups of Schuster and Wong reported the first examples of stable iminosugar nucleotide conjugates which were designed to mimic the transition state of the GDP-fucose moiety of fucosyltransferase-catalysed reaction [62–64]. It was shown that covalently linking a six-membered instead of a five-membered iminosugar with GDP provided a slightly better inhibitor with a K_i of 13 μM (Scheme 4.22). It is worth noting that the fluorinated GDP-fucose analogue **40** is one of the most potent inhibitor of α-1,3-fucosyltransferases known to date, a result consistent with a charged sp^2-hybridized transition-state structure [65, 66].

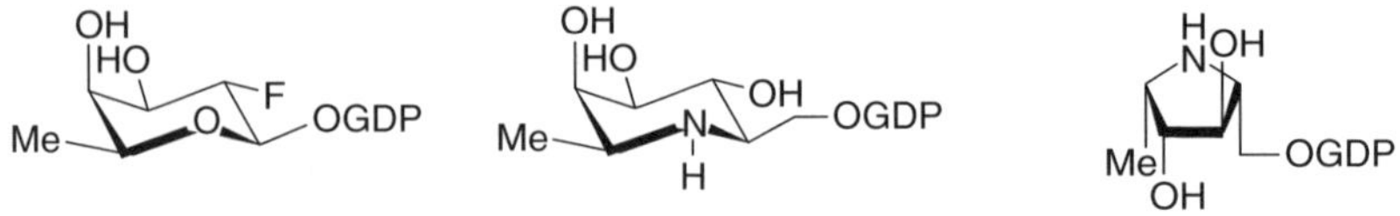

40 $K_i = 4\ \mu M$ [65] α–1,3-fucosyltransferase V

38 $K_i = 13\ \mu M$ [63–64] α–1,3-fucosyltransferase V

39 $IC_{50} = 45\ \mu M$ [62] α–1,3-fucosyltransferase V

Scheme 4.22

4.3.4 Anti-HIV agents

It has been shown that interaction between gp120 and the galactosphingolipid GalCer **41** is important for HIV infectivity in CD4 negative cells [67, 68]. In view of potential therapeutic applications, stable GalCer analogues have been designed as probes for the development of new HIV entry inhibitors. Fantini *et al.* synthesized and evaluated GalCer mimics based on *C*-glycosides, imino-*C*-glycosides or *N*-alkylated iminosugars bearing a simple stearyl chain (Scheme 4.23) [67, 68]. β-1-*C*-heptadecyl-1-deoxygalactonojirimycin (**42**) showed potent affinity for gp120, even higher than GalCer **41**, in an assay based on change of surface pressure when the glycolipid monolayers were exposed to solutions of gp120. Compound **43**, the β-*C*-galactoside analog of **42**, and *N*-heptadecyl-1-deoxynojirimycin (**44**) displayed specific affinity for gp120, equal to that of **41**, whereas, quite unexpectedly, *N*-heptadecyl-1-deoxygalactonojirimycin (**45**) showed less affinity for the viral glycoprotein. Even if more data are required to rationalize these results, the potent affinity observed for **42** compared with the parent *N*-substituted and β-*C*-galactoside analogs further highlights the biological interest of iminosugar *C*-glycosides.

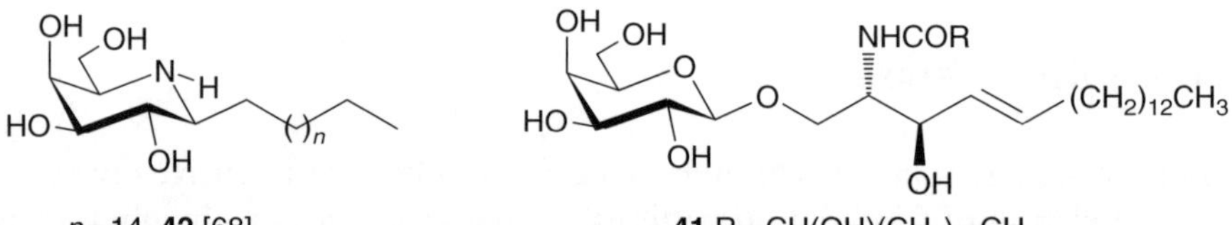

$n = 14$ **42** [68]

41 $R = CH(OH)(CH_2)_{21}CH_3$

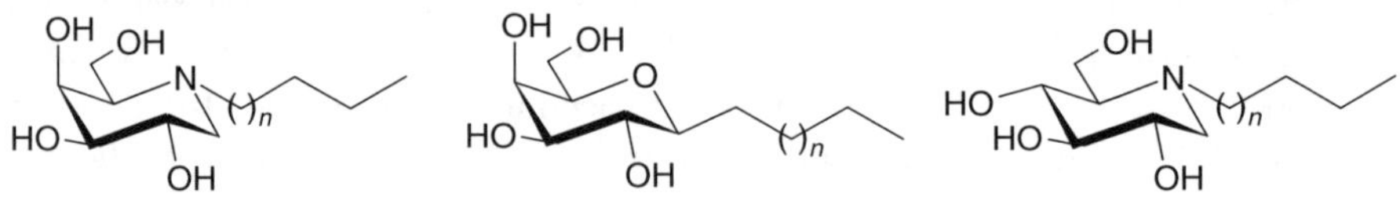

$n = 14$ **45** [67]

$n = 14$ **43** [68]

$n = 14$ **44** [67]

Scheme 4.23

4.3.5 Gaucher disease

Gaucher disease, the most common glycolipid storage disease, is a relatively rare hereditary disorder due the deficiency in a specific β-glucosidase (β-glucocerebrosidase)

involved in the catabolism of glycosphingolipids in lysosomes [69]. Defects in the catalytic activity of this enzyme lead to the accumulation of undegraded glucosylceramide (GlcCer) in macrophages and to severe symptoms. The first-line treatment for Gaucher disease is based on the administration of Cerezyme®, a recombinant form of β-glucocerebrosidase (GCase), to supplement the defective hydrolytic enzyme [70]. The second strategy uses a small molecule, *N*-butyl-1-deoxynojirimycin (**46**) (Zavesca®), to inhibit the biosynthesis of GlcCer (Substrate Reduction Therapy, SRT) [69]. A third strategy is based on the capacity of certain competitive inhibitors of the abnormal enzyme to influence positively its folding state and thus prevent its degradation by the quality control mechanism of the endoplasmic reticulum before its transfer to lysosomes [71].

Chemical chaperone therapy

In the search for a chemical chaperone therapy for Gaucher disease, Sawkar *et al.* disclosed that the addition of sub-inhibitory concentrations (10 μM) of *N*-nonyl-1-deoxynojirimycin (**33**) to a fibroblast culture medium leads to a two-fold increase in the activity of N370S GCase, the most common mutation causing Gaucher disease [72]. α-1-*C*-Alkyl-1-deoxynojirimycin derivatives were found to display slightly better inhibition towards GCase and chaperoning activity comparable to that of the parent *N*-substituted iminosugars (Table 4.2.) [73, 74].

The addition of 20 μM α-1-*C*-octyl-1-deoxynojirimycin (**47**) or 10 μM *N*-nonyl-1-deoxynojirimycin **33** to the N370S GCase culture medium for 10 days led to 1.9-fold and 2.3-fold increases in the enzyme activity, respectively. The major advantage of the *C*-glycoside derivative **47** was that it had no effect on the intracellular lysosomal α-glucosidase activity for 10 days whereas *N*-nonyl-1-deoxynojirimycin (**33**) decreased the activity by 50 per cent throughout the assay period. Removal of the hydroxymethyl groups in α-1-*C*-nonyl-1-deoxynojirimycin **32** to give the corresponding iminoxylitol derivative **48** significantly enhanced inhibitory potency towards GCase ($K_i = 2.2$ nM, $IC_{50} = 6.8$ nM) and abolished inhibition towards various α-glucosidases including rat intestinal sucrase, maltase and isomaltase. These results may be due to a piperidine ring inversion from a classical $^4C_1$(D) to a $^1C_4$(D) conformation. NMR data clearly indicated that compound **48** existed predominantly in an inverted $^1C_4$(D) conformation in which all hydroxyl groups are axial and the alkyl chain equatorial whereas the conformation of its *N*-nonyl analog **49** is $^4C_1$(D). This observation may explain the large difference in biological activity between the two regioisomers **48** and **49**. Such a difference is indeed not observed between *N*-nonyl-1-deoxynojirimycin (**33**) and α-1-*C*-nonyl-1-deoxynojirimycin (**32**) which both have the same $^4C_1$(D) conformation. Iminoxylitol derivative **48** was found to double the residual activity of GCase in fibroblasts from Gaucher patients at nanomolar sub-inhibitory concentrations (10 nM). This compound is therefore a promising candidate for the development of small-molecule drugs for the treatment of Gaucher disease devoid of side effects associated with α-glucosidase inhibition.

In the search for a chemical chaperone for the treatment of Fabry disease, a lysosomal storage disorder caused by deficiency of α-galactosidase A, it was found that β-1-butyl-1-deoxygalactonojirimycin (**50**) displayed better inhibition towards the

Table 4.2 IC_{50} Values for lysosomal glycosidases and effect of iminosugar derivatives on lysosomal GCase activity in N370S Gaucher fibroblasts (GM00372)

Inhibitors ($n = 6$)	GCase [μM]	α-glucosidase [μM]	Enhancement of GCase activity
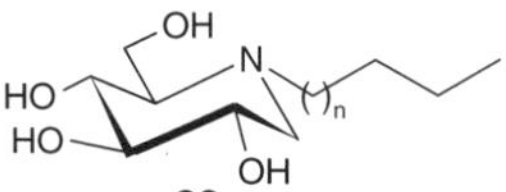 **33**	0.66	1.5	×2.4 at 10 μM
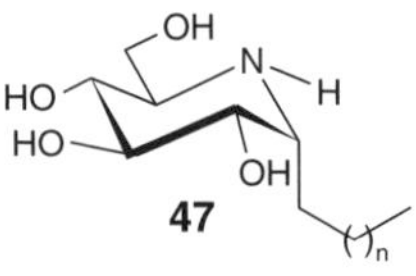 **47**	0.50	5.0	×1.7 at 20 μM
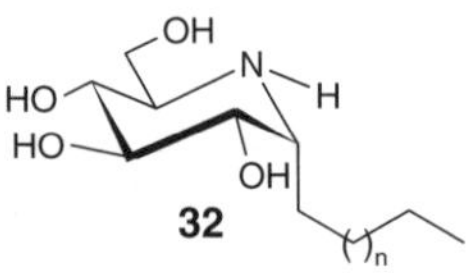 **32**	0.27	4.8	×1.7 at 2.5 μM
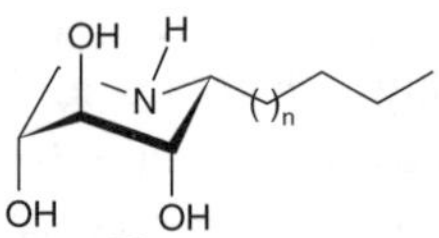 **48**	0.0068	NI[a]	×1.8 at 0.01 μM
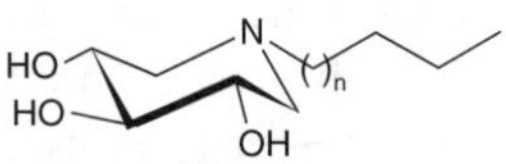 **49**	1.5	NI	ND[b]

[a]NI: less than 50% inhibition at 1000 μM. [b]ND: Not determined.

enzyme than its *N*-butyl analog **51** [75]. However, in this case the best inhibitor reported, 1-deoxygalactonojirimycin (**52**), was not substituted by an alkyl chain (Scheme 4.24).

Substrate reduction therapy

N-butyl-1-deoxynojirimycin (**46**), the first orally administered treatment for lysosomal diseases, acts as an inhibitor of glucosylceramide synthase, the glucosyltransferase involved in the biosynthesis of GlcCer that accumulates pathologically in macrophages of patients with Gaucher disease. Molecular modelling and kinetic experiments have suggested that **46** is a competitive inhibitor that mimics the ceramide acceptor and not

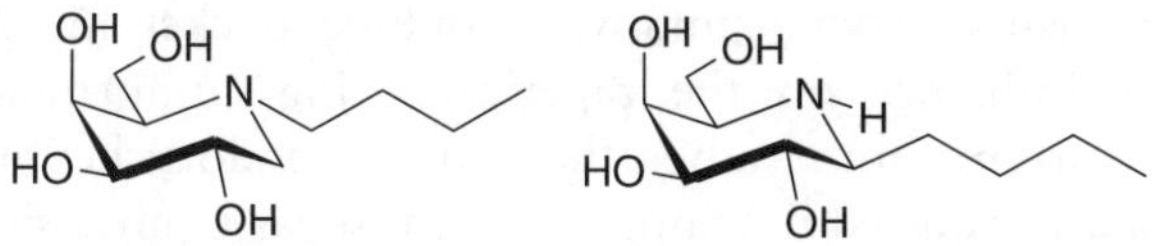

51 $IC_{50} = 300\ \mu M$ [75]
α–galactosidase A
Fabry disease

50 $IC_{50} = 24\ \mu M$ [75]
α–galactosidase A
Fabry disease

52 $IC_{50} = 0.04\ \mu M$ [75]
α–galactosidase A
Fabry disease

Scheme 4.24

the donor substrate (UDP-glucose) in the glucosylceramide synthase-catalysed process. As iminosugar **46** leads to serious side effects arising mainly from the inhibition of digestive glucosidases, various iminoglycolipids bearing one or two alkyl chains were designed as better ceramide mimics to obtain more potent and selective new therapeutic agents [76]. However, it was found that the addition of a second alkyl chain actually led to less potent inhibitors. In addition, α-1-*C*-nonyl-1-deoxynojirimycin (**32**) was 170-fold less potent as glucosylceramide synthase inhibitors than its *N*-alkylated analog **33** (Scheme 4.25) [77].

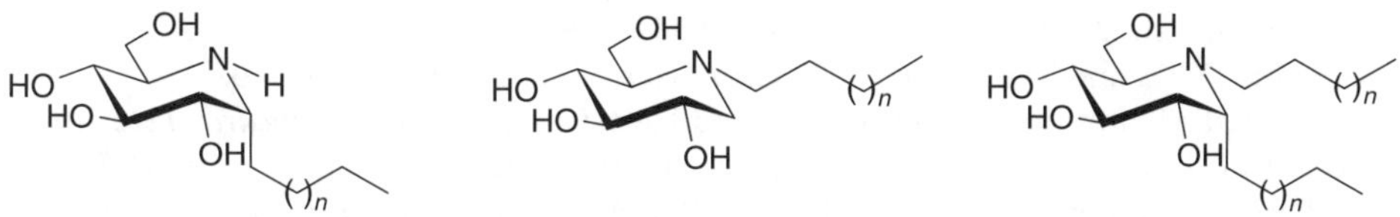

$n = 6$ **32** $IC_{50} = 276\ \mu M$ [77]
glucosylceramide synthase

$n = 1$ **46** $IC_{50} = 20\ \mu M$ [76]
$n = 6$ **33** $IC_{50} = 1.6\ \mu M$ [76]
glucosylceramide synthase

$n = 5$ **53** $IC_{50} = 174\ \mu M$ [76]
glucosylceramide synthase

Scheme 4.25

4.4 Conclusion

Many synthetic efforts have been devoted to develop efficient and stereocontrolled routes to iminosugar *C*-glycosides since the 1990s. The diversity of the strategies designed further highlights the creativity of organic chemists and how manifold solutions can be imagined to solve a given synthetic challenge. Many other approaches are certainly yet to come. By analogy with methods developed for the synthesis of classical *C*-glycosides, we can expect in the near future new strategies based on radical chemistry or using intramolecular aglycon delivery. One may also envision the extension of the ingenious ring-closing metathesis route developed by Postema *et al.* for the preparation of β-*C*-glycoconjugates and β-*C*-disaccharides [78].

Iminosugar *C*-glycosides are well positioned as stable glycoconjugate mimetics of biological and therapeutic interest. Various comparative studies on simple glycolipid analogs have already demonstrated a marked dependence of the selectivity and/or potency of the inhibitors/ligands upon the position of the alkyl chain (1-*C*- or *N*-alkyl derivatives). Concerning glycosidase inhibitors, improved results obtained with iminosugar *C*-glycosides may be partly explained by a better position of the alkyl chain that allows suitable interactions with enzyme putative lipophilic pocket [59, 73, 74]. Another point is the stronger influence on the piperidine ring conformation of substituent linked to the C-1 carbon atom rather than to the endocyclic nitrogen atom [73]. More biological data are required to confirm if iminosugar *C*-glycosides will display decisive advantages compared with their *N*-substituted analogs or with other glycoconjugate mimetics. Much remains to be done and the field is wide open for further exciting studies that may lead to the discovery of the first therapeutic agents based on this important class of iminosugars.

References

1. (a) Mitrakou, A., Tountas, N., Raptis, A. E., Bauer, R. J., Schulz, H. and Raptis, S. A., 1998 *Diab. Med.*, **15**, 657–660. (b) Scott, L. J. and Spencer, C. M., 2000 *Drugs*, **59**, 521–549.
2. (a) Stütz, A. E. (ed.), 1999 *Iminosugars as Glycosidase Inhibitors: Nojirimycin and Beyond.* Wiley-VCH, New York. (b) Asano, N., Nash, R. J., Molyneux, R. J. and Fleet, G. W. J., 2000 *Tetrahedron: Asymmetry*, **11**, 1645–1680.
3. (a) Compain, P. and Martin, O. R., 2001 *Bioorg. Med. Chem.*, **9**, 3077–3092. (b) Compain, P. and Martin, O. R., 2003 *Curr. Top. Med. Chem.*, **3**, 541–560.
4. Moriyama, H., Tsukida, T., Inoue, Y., Yokota, K., Yoshino, K., Kondo, H., Miura, N. and Nishimura, S.-I., 2004 *J. Med. Chem.*, **47**, 1930–1938.
5. Schramm, V. L. and Tyler, P. C. 2003 *Curr. Top. Med. Chem.*, **3**, 525–540.
6. (a) Lee, R. E., Smith, M. D., Pickering, L. and Fleet, G. W. J., 1999 *Tetrahedron Lett.*, **40**, 8689–8692. (b) Lee, R. E., Smith, M. D., Nash, R. J., Griffiths, R. C., McNeil, M., Grewal, R. K., Yan, W., Besra, G. S., Brennan, P. J. and Fleet, G. W. J., 1997 *Tetrahedron Lett.*, **38**, 6733–6736.
7. Martin, O. R. and Compain, P. (eds), 2003 Iminosugars: recent insights into their bioactivity and potential as therapeutic agents, in *Curr. Top. Med. Chem.*, Bentham, Hilversum, Netherlands, Volume 3, issue 5, pp. 471–591.
8. Alper, J., 2001 *Science*, **291**, 2338–2343.
9. Martin, O. R., 1997 In *Carbohydrate Mimics: Concepts and Methods* (Y. Chapleur (ed.), Wiley-VCH, pp. 259–282.
10. Zou, W., 2005 *Curr. Top. Med. Chem.*, **5**, 1363–1391.
11. Liu, P. S., 1987 *J. Org. Chem.*, **52**, 4717–4721.
12. For recent reviews devoted to the synthesis of iminosugars: (a) Cipolla, L., La Ferla, B. and Nicotra, F., 2003 *Curr. Top. Med. Chem.*, **3**, 485–511. (b) Pearson, M. S. M., Mathé-Allainmat, M., Fargeas, V. and Lebreton, J., 2005 *Eur. J. Org. Chem.*, **11**, 2159–2191. (c) Dhavale, D. D. and Matin, M. M., 2005 *Arkivoc*, 110–132. (d) Afarinkia, K. and Bahar, A., 2005 *Tetrahedron: Asymmetry*, **16**, 1239–1287. (e) La Ferla, B. and Nicotra, F., 1999 In *Iminosugars as Glycosidase Inhibitors: Nojirimycin and Beyond*, (A. E. Stütz, ed.). Wiley-VCH, New York, pp. 68–92. (f) Robina, I. and Vogel, P., 2005 *Synthesis*, 675–702.
13. Two chapters devoted to the general synthesis of iminosugars and imino-*C*-disaccharides are included in this book: Chapters 3 and 5.
14. (a) Postema, M. H. D., 1995 *C-Glycoside Synthesis*. CRC Press: Boca Raton. (b) Du, Y., Linhardt, R. J. and Vlahov, I. R., 1998 *Tetrahedron*, **54**, 9913–9959. (c) Xie, J., 1999 *Recent Res. Devel. Organic Chem.*, **3**, 505–523. (d) Lee, D. Y. W. and He, M., 2005 *Curr. Top. Med. Chem.*, **5**, 1333–1350.
15. (a) Godin, G., Compain, P., Masson, G. and Martin, O. R., 2002 *J. Org. Chem.*, **67**, 6960–6970. (b) Masson, G., Compain, P. and Martin, O. R., 2000 *Org. Lett.*, **2**, 2971–2974.
16. (a) Godin, G., Compain, P. and Martin, O. R., 2003 *Org. Lett.*, **5**, 3269–3272. (b) Godin, G., Compain, P. and Martin, O. R., 2003 *Synlett*, 2065–2067.

17. Cipolla, L., La Ferla, B., Peri, F. and Nicotra, F., 2000 *Chem. Commun.*, 1289–1290.
18. Cipolla, L., Palma, A., La Ferla, B. and Nicotra, F., 2002 *J. Chem. Soc., Perkin Trans I*, 2161–2165.
19. Saavedra, O. M. and Martin, O. R., 1996 *J. Org. Chem.*, **61**, 6987–6993.
20. Leeuwenburgh, M. A., Picasso, S., Overkleeft, H. S., Van der Marel, G. A., Vogel, P. and Van Boom, J. H., 1999 *Eur. J. Org. Chem.*, 1185–1189.
21. Cheng, X., Kumaran, G. and Mootoo, D. R., 2001 *Chem. Commun.*, 811–812.
22. (a) Johnson, C. R., Miller, M. W., Golebiowski, A., Sundram, H. and Ksebati, M. B., 1994 *Tetrahedron Lett.*, **35**, 8991–8994. (b) Johns, B. A., Pan, Y. T., Elbein, A. D. and Johnson, C. R., 1997 *J. Am. Chem. Soc.*, **119**, 4856–4865. (c) Johns, B. A. and Johnson, C. R., 1998 *Tetrahedron Lett.*, **39**, 749–752.
23. La Ferla, B., Bugada, P., Cipolla, L., Peri, F. and Nicotra, F., 2004 *Eur. J. Org. Chem.*, 2451–2470.
24. (a) Fleet, G. W. J., Namgoong, S. K., Barker, C., Baines, S., Jacob, G. S. and Winchester, B., 1989 *Tetrahedron Lett.*, **30**, 4439–4442. (b) Shilvock, J. P., Nash, R. J., Watson, A. A., Winters, A. L., Butters, T. D., Dwek, R. A., Winkler, D. A. and Fleet, G. W. J., 1999 *J. Chem. Soc., Perkin Trans I*, 2747–2754.
25. Mitchell, M., Qaio, L. and Wong, C.-H., 2001 *Adv. Synth. Catal.*, **343**, 596–599.
26. Straub, A., Effenberger, F. and Fischer, P., 1990 *J. Org. Chem.*, **55**, 3926–3932.
27. Baudat, A., Picasso, S. and Vogel, P., 1996 *Carbohydrate Res.*, **281**, 277–284.
28. Dondoni, A. and Perrone, D., 2003 *Tetrahedron,* **59**, 4261–4273.
29. For a related synthetic strategy see: Cipolla, L., Lay, L., Nicotra, F., Pangrazio, C. and Panza, L., 1995 *Tetrahedron,* **51**, 4679–4690.
30. Bruce, I., Fleet, G. W. J., Cenci di Bello, I. and Winchester, B., 1992 *Tetrahedron,* **48**, 10191–10200.
31. Pino-Gonzalez, M. S., Assiego, C. and Lopez-Herrera, F. J., 2003 *Tetrahedron Lett.*, **44**, 8353–8356.
32. Martin, O. R., Liu, L. and Yang, F., 1996 *Tetrahedron Lett.*, **37**, 1991–1994.
33. Martin, O. R., Saavedra, O. M., Xie, F., Liu, L., Picasso, S., Vogel, P., Kizu, H. and Asano, N., 2001 *Bioorg. Med. Chem.*, **9**, 1269–1278.
34. Goujon, J.-Y., Gueyrard, D., Compain, P., Martin, O. R., Ikeda, K., Kato, A. and Asano, N., 2005 *Bioorg. Med. Chem.*, **13**, 2313–2324.
35. Liotta, L. J., Lee, J. and Ganem, B., 1991 *Tetrahedron,* **47**, 2433–2447.
36. Yi, T., Wu, A.-A., S.-H and Zou, W., 2005 *Tetrahedron,* **61**, 11716–11722.
37. McAlpine, I. J. and Armstrong, R. W., 2000 *Tetrahedron Lett.*, **41**, 1849–1853.
38. Bruce, I., Fleet, G. W. J., Cenci di Bello, I. and Winchester, B., 1989 *Tetrahedron Lett.*, **30**, 7257–7260.
39. Shilvock, J. P., Hsia, K. Y., Nash, R. J., Lloyd, J. D., Winters, A. L., Asano, N. and Fleet, G. W. J., 1998 *Tetrahedron: Asymmetry*, **9**, 4157–4164.
40. Böshagen, H., Geiger, W. and Junge, B., 1981 *Angew. Chem. Int. Ed. Engl.*, **20**, 806–807.
41. Wong, C.-H., Provencher, L., Porco, J. A., Jung, S.-H., Wang, Y.-F., Chen, L., Wang, R. and Steensma, D. H., 1995 *J. Org. Chem.*, **60**, 1492–1501.
42. Anzeveno, P. B., Creemer, L. J., Daniel, J. K., King, C.-H. R. and Liu, P. S., 1989 *J. Org. Chem.*, **54**, 2539–2542.
43. Fuchss, T., Streicher, H. and Schmidt, R. R., 1997 *Liebigs Ann./Recueil,* 1315–1321.
44. Johnson, C. R., Golebiowski, A., Sundram, H., Miller, M. W. and Dwaihy, R. L., 1995 *Tetrahedron Lett.*, **36**, 653–654.
45. Granier, T. and Vasella, A., 1998 *Helv. Chim. Acta*, **81**, 865–880.
46. Ojima, I. and Vidal, E. S., 1998 *J. Org. Chem.*, **63**, 7999–8003.
47. Shankar, B. B., Kirkup, M. P., McCombie, S. W. and Ganguly, A. K., 1993 *Tetrahedron Lett.*, **34**, 7171–7174.
48. Adelbrecht, J.-C., Craig, D., Dymock, B. W. and Thorimbert, S., 2000 *Synlett*, 467–470.
49. Maughan, M. A. T., Davies, I. G., Claridge, T. D. W., Courtney, S., Hay, P. and Davis, B. G., 2003 *Angew. Chem. Int. Ed. Engl.*, **42**, 3788–3792.

50. Timmer, M. S. M., Risseeuw, M. D. P., Verdoes, M., Filippov, D. V., Plaisier, J. R., Van der Marel, G. A., Overkleeft, H. S. and Van Boom, J. H., 2005 *Tetrahedron: Asymmetry*, **16**, 177–185.
51. Peer, A. and Vasella, A., 1999 *Helv. Chim. Acta*, **82**, 1044–1065.
52. Takayama, S., Martin, R., Wu, J., Laslo, K., Siuzdak, G. and Wong, C.-H., 1997 *J. Am. Chem. Soc.*, **119**, 8146–8151.
53. Dransfield, P. J., Gore, P. M., Shipman, M. and Slawin, A. M. Z., 2002 *Chem. Commun.*, 150–151.
54. Dransfield, P. J., Gore, P. M., Prokes, I., Shipman, M. and Slawin, A. M. Z., 2003 *Org. Biomol. Chem.*, **1**, 2723–2733.
55. Duff, F. J., Vivien, V. and Wightman, R. H., 2000 *Chem. Commun.*, 2127–2128.
56. Tailor, J. and Hall, D. G., 2000 *Org Lett.*, **2**, 3715–3718.
57. Shürer, S. C. and Blechert, S., 1999 *Tetrahedron Lett.*, **40**, 1877–1880.
58. Dondoni, A., Giovannini, P. P. and Perrone, D., 2005 *J. Org. Chem.*, **70**, 5508–5518.
59. Godin, G., Compain, P., Martin, O. R., Ikeda, K., Yu, L. and Asano, N., 2004 *Bioorg. Med. Chem. Lett.*, **14**, 5991–5995.
60. (a) Chang, C.-F., Ho, C.-W., Wu, C.-Y., Chao, T.-A., Wong, C.-H. and Lin, C.-H., 2004 *Chem. Biol.*, **11**, 1301–1306. (b) Wu, C.-Y., Chang, C.-F., Chen, J. S.-Y., Wong, C.-H. and Lin, C.-H., 2003 *Angew. Chem. Int. Ed. Engl.*, **42**, 4661–4664.
61. Ho, C.-W., Lin, Y.-N., Chang, C.-F., Li, S.-T., Wu, Y.-T., Wu, C.-Y., Chang, C.-F., Liu, S.-W., Li, Y.-K. and Lin, C.-H., 2006 *Biochemistry*, **45**, 5695–5702.
62. Schuster, M. and Blechert, S., 2001 *Bioorg. Med. Chem. Lett.*, **11**, 1809–1811.
63. Mitchell, M. L., Tian, F., Lee, L. V. and Wong, C.-H., 2002 *Angew. Chem. Int. Ed. Engl.*, **41**, 3041–3044.
64. Mitchell, M. L., Lee, L. V. and Wong, C.-H., 2002 *Tetrahedron Lett.*, **43**, 5691–5693.
65. Murray, B. W., Wittmann, V., Burkart, M. D., Hung, S.-S. and Wong, C.-H., 1997 *Biochemistry*, **36**, 823–831.
66. Lee, L. V., Mitchell, M. L., Huang, S.-J., Fokin, V. V., Sharpless, B. and Wong, C.-H., 2003 *J. Am. Chem. Soc.*, **125**, 9588–9589.
67. Weber, K. T., Hammache, D., Fantini, J. and Ganem B., 2000 *Bioorg. Med. Chem. Lett.*, **10**, 1011–1014.
68. Augustin, L. A., Fantini, J. and Mootoo, D. R., 2006 *Bioorg. Med. Chem.*, **10**, 1182–1188.
69. Butters, T. D., Dwek, R. A. and Platt, F. M., 2003 *Curr. Top. Med. Chem.*, **3**, 561–574.
70. Futerman, A. H., Sussman, J. L., Horowitz, M., Silman, I. and Zimran, A., 2004 *Trends Pharm. Sci.*, **25**, 147–151.
71. Fan, J.-Q., 2003 *Trends Pharm. Sci.*, **24**, 355–360.
72. Sawkar, A. R., Cheng, W.-C., Beutler, E., Wong, C.-H., Balch, W. E. and Kelly, J. W., 2002 *Proc. Natl. Acad. Sci USA*, **99**, 15428–15433.
73. Compain, P., Martin, O. R., Boucheron, C., Godin, G., Yu, L., Ikeda, K. and Asano, N., 2006 *ChemBioChem*, **7**, 1356–1359.
74. Yu, L., Ikeda, K., Kato, A., Adachi, I., Godin, G., Compain, P., Martin, O. R. Asano, N., 2006 *Bioorg. Med. Chem.*, **14**, 7736–7744.
75. Asano, N., Ishii, S., Kizu, H., Yasuda, K., Martin, O. R. and Fan J.-Q., 2000 *Eur. J. Biochem.*, **267**, 4179–4186.
76. Boucheron, C., Desvergnes, V., Compain, P., Martin, O. R., Lavi, A., Mackeen, M., Wormald, M. R., Dwek, R. A. and Butters, T. D., 2005 *Tetrahedron: Asymmetry*, **16**, 1747–1756.
77. Godin, G., Compain, P., Martin, O. R., Lavi, A. and Butters, T. D., unpublished results.
78. See, for example,: (a) Liu, L. and Postema, M. H. D., 2001 *J. Am. Chem. Soc.*, **123**, 8602–8603. (b) Postema, M. H. D. and Piper, J. L., 2003 *Org. Lett.*, **5**, 1721–1723.
79. Ikeda, K., Takahashi, M., Nishida, M., Miyauchi, M., Kizu, H., Kameda, Y., Arisawa, M., Watson, A. A., Nash, R. J., Fleet, G. W. J. and Asano, N., 2000 *Carbohydr. Res.*, **323**, 73–80.

5

Imino-*C*-disaccharides and analogues: synthesis and biological activity

Pierre Vogel, Sandrine Gerber-Lemaire and **Lucienne Juillerat-Jeanneret**

5.1 Introduction

The biological importance of oligosaccharides was first recognized for their role in metabolism and energy storage. More recently, it became clear that complex oligosaccharides also regulate a large number of biological processes. Most important are the oligosaccharides formed on the surfaces of cells and their role on protein and glycoprotein conformation. The cell-surface oligosaccharides are the 'words' used by cells to communicate with the outer world [1–4]. They guide their social behaviour such as cell/cell interactions, and cell/invader interactions, including HIV/cell penetration [5]. Glycosidation and deglycosidation of proteins contribute to their folding [6], their fate and cell localization [7]. The most frequent glycoconjugates are grouped in four main classes: N-linked glycoproteins (oligosaccharides attached to the $CONH_2$ moiety of asparagine units) [8], O-linked glycoproteins (oligosaccharides attached to the hydroxyl group of serine or threonine units) [9], glycosyl phosphatidylinositols (GPIs: glycolipids of type Manα(1$\rightarrow$4)GlcNAcα(1$\rightarrow$6)myo-inositol-P-lipid anchored to the cell membrane and connected to proteins via phosphoethanolamine linker) [10], and glycolipids (classified into glycosyllipids, glycoglycerolipids and others including lipid A of lipopolysaccharides) [11]. In general, glycoproteins are more stable than their corresponding unglycosylated counterparts, even when there is no major change in conformation associated with glycosylation [12]. Targeting glycosyltransferases and glycosidases, the enzymes intervening in the biosynthesis of the glycoproteins, has led to a better understanding of cell functions and now represents a new chemotherapeutic approach [13]. In this review we shall concentrate on glycosidase inhibitors and on their potential as anticancer drugs. Carbohydrate mimetics are often potent and selective

Iminosugars: From synthesis to therapeutic applications Edited by Philippe Compain and Olivier R. Martin
 ISBN 978-0-470-03391-3

glycosidase inhibitors [14]. They have found applications in the treatment of diabetes of type II (e.g. miglitol (**1**) [15] and acarbose (**2**) [16]) and obesity [17], lysosomal diseases such as Gaucher's disease [18], or influenza [19] (e.g. zanamivir (**3**) (Relenza®) [20], oseltamivir phosphate (Tamiflu®) (**4**) [21]).

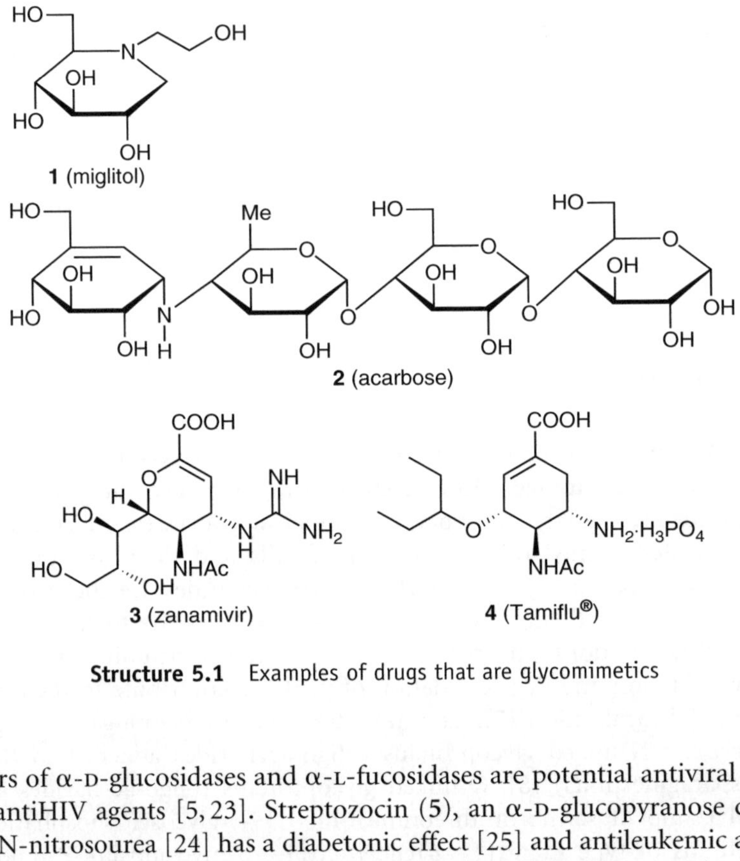

Structure 5.1 Examples of drugs that are glycomimetics

Inhibitors of α-D-glucosidases and α-L-fucosidases are potential antiviral agents [22], including antiHIV agents [5, 23]. Streptozocin (**5**), an α-D-glucopyranose derivative of N-methyl-N-nitrosourea [24] has a diabetonic effect [25] and antileukemic activity [26]. It is now approved for the treatment of carcinoid and pancreatic islet tumours in humans [27]. Tunicamycin, which belongs to a family of nucleoside antibiotics (**6**) [28], has been shown to interfere with glycoprotein synthesis as it blocks the first step in the biosynthesis of N-glycans in cells [29]. It potentiates drug cytotoxicity of vincristine [30], of cisplatin and carboplatin [31], and of staurosporine [32]. Castanospermine (**7**), a polyhydroxyindolizidine isolated from the seeds of the Australian leguminous tree, *Castanospermun australe* [33], is an α-glucosidase I inhibitor [34]. It inhibits platelets aggregation [35] and tumour colonization [36], angiogenesis and tumour growth [37] (see also Chapter 12). Other α-glucosidase inhibitors such as 1,6-*epi*-cyclophellitol (**8**) also inhibit the metastatic process [38]. N-Methyl-1-deoxynojirimycin (**9**) reduced endothelial cell migration [37]. Inhibitors of α-glucosidase that are not glucose mimetics might also show cytotoxicity. This is the case with sintenin (**10**) [39] and its synthetic analogues [40].

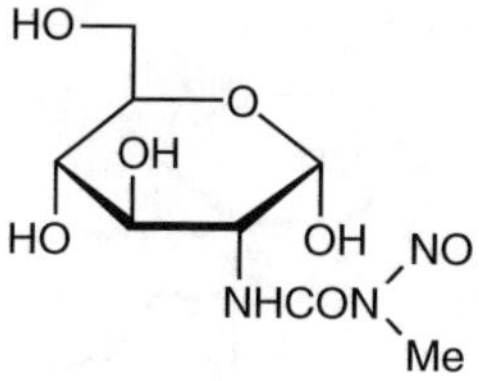

5 (streptozocin)

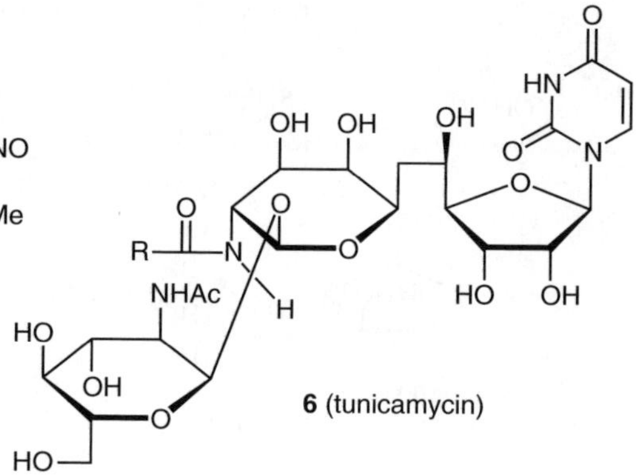

6 (tunicamycin)

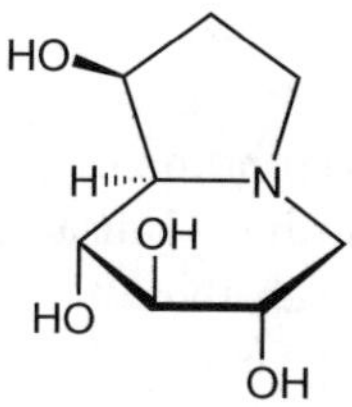

7 ((+)-castanospermine)

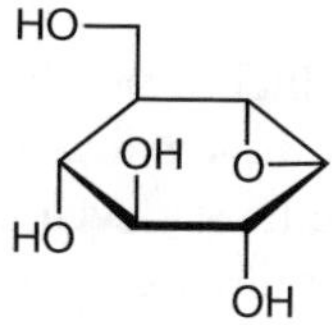

8

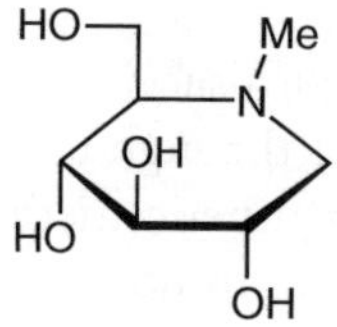

9

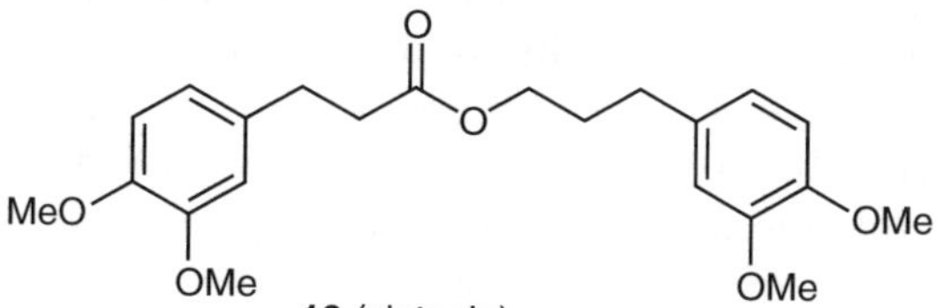

10 (sintenin)

Structure 5.2 Glycosidase inhibitors as antitumour agents

Swainsonine (**11**), isolated in 1979 from the Australian plant *Swainsonia canescens* [41], is a potent inhibitor of lysosomal α-mannosidase [42] and Golgi α-mannosidase II [43, 44]. In 1986, Humphries *et al.* [45] reported that B16-F10 murine melanoma cells treated with swainsonine in growth medium and then injected intravenously into syngeneic C57BL/6 mice resulted in dramatic inhibition of colonization. This was confirmed with further experimentation [46]. Swainsonine increases the cytotoxicity of human large granular lymphocytes, a phenomenon also seen to some extent with 1-deoxymannonojirimycin (**12**), an α-mannosidase I inhibitor [47]. Both **11** and **12** inhibit human melanoma cell invasion [48]. The potentials of **11** and **12** as anticancer agents [49–54] will be discussed in Chapter 12.

In our research group we have searched for alternative inhibitors of α-mannosidases with the hope of uncovering a new lead in anticancer drugs. We reasoned that disaccharides mimetics might be more potent and more selective than monosaccharide mimetics such as **7**, **8**, **9**, **11** or **12** because they would include functional and stereochemical information of the hexose to which α-mannose is attached. As disaccharides are readily hydrolysed *in vivo*, nonhydrolysable mimetics were sought and we have concentrated our efforts on the preparation of C-linked disaccharides [55], and more specifically

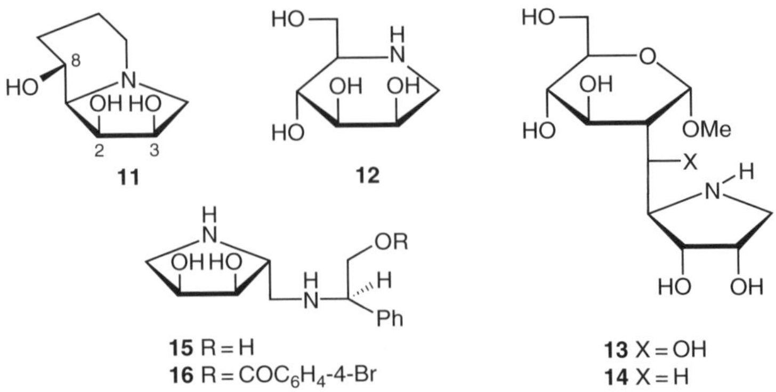

Structure 5.3 Examples of α-mannosidase inhibitors

on imino-*C*-linked disaccharides in which a polyhydroxypyrrolidine or polyhydroxypiperidine is linked through a methano linker to a hexose (imino-*C*-disaccharides). We shall review the synthetic methods we have developed to reach these targets. Unfortunately, none of the imino-*C*-disaccharides we have made so far has shown interesting inhibitory activity towards various glycosidases. The disaccharide mimics **13** [56] and **14** [57] with a (2*R*,3*R*,4*S*)-3,4-dihydroxypyrrolidin-2-yl moiety (imitates α-mannosides and swainsonine) attached at C-2 of methyl α-D-glucopyranoside through a hydroxymethano and methano linker, respectively, are poor inhibitors of α-mannosidase from Jack bean. In parallel with this we have searched for simpler disaccharide mimetics containing dihydroxypyrrolidine moieties [58, 59] and have found that **15** is a potent and highly selective inhibitor of α-mannosidases. Although **15** did not affect cancer cells, its parabromobenzoate **16** was found to inhibit the growth of human glioblastoma and melanoma cells [60] more efficiently than swainsonine (**11**).

5.2 Synthesis of imino-*C*-disaccharides

There are two main classes of 'imino-*C*-disaccharides': the linear (1→1)-*C*- and (1→ω)-*C*-linked disaccharide mimetics on the one hand, and the branched (1→n)-*C*-linked disaccharides (Figure 5.1), on the other hand. The first example of a linear 'imino-*C*-disaccharide' (1,5-dideoxy-1,5-imino-D-mannitol linked at C-6 of D-galactose through a CH_2 linker) was prepared by Johnson and coworkers [61] in 1994. Other linear 'imino-*C*-disaccharides' were then reported by the groups of Martin [62], van Boom [63], Dondoni [64], Mootoo [65], Wightman [66] and Nicotra [67]. In 1996, our group presented the synthesis of the first example of a branched 'imino-*C*-disaccharide' (β-D-(1→3)-CH(OH)-linked 1,5,6-trideoxy-1,5-iminogalactoside of a D-altrofuranouronic derivative) [68]. Shortly after, further examples of branched 'imino-*C*-disaccharides' were presented by Johnson [69], and more recently by Mootoo and Argyropoulos [70]. Pseudo-'imino-*C*-disaccharides' (no linker between the two monosaccharide analogues: X = no atom) have been obtained by Cardona *et al.* [71–74]. In our group, Cardona realized the syntheses of homo-(1→3)-*C*-linked iminodisaccharides, with X = CH_2-CH= [75] and X = -CH_2-CH_2- [76] (Figure 5.1).

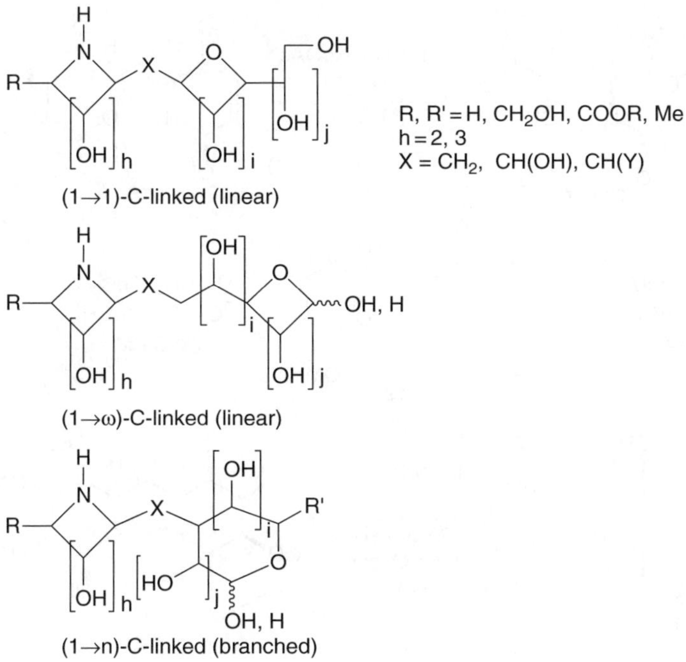

Figure 5.1 Classes of imino-*C*-disaccharides

5.2.1 Synthesis of linear 'imino-*C*-disaccharides'

These systems have been obtained applying the Miyaura–Suzuki coupling [61, 69], alkylsamarium diiodide addition to sugar-derived aldehydes [62], acetylide addition of lithium acetylide to aldonolactones [63, 65], Wittig olefination [64, 67, 77] and dipolar cycloadditions of cyclic nitrones [66]. A recent review described these syntheses [78] and thus they will not be discussed here. Recently, Nelson and coworkers [79] have proposed a general, two-directional approach to imino-*C*-(1→1)-linked disaccharide mimetics. The 1,3-amino alcohol derivative *syn*-**17** [80] was converted into a 7:3 mixture of pyran anomers **18** by treatment with NBS/AcOH and subsequent reaction with methyl orthoformate (Scheme 5.1). The differential reactivity of the pyran and piperidine rings allows the two-directional synthetic approaches to be interchanged. For instance, allylation of **18** with allyltrimethylsilane and BF_3 gave **19**. Luche reduction [81] of **19** gave **20**. Dihydroxylation of **20** and subsequent acetylation provided a 1:1 mixture of imino-*C*-disaccharide precursors **21** in 61 per cent yield. The same reaction sequence applied to *anti*-**17** gave a 1:1 mixture of diastereoisomeric **22**.

When *syn*-**17** was treated with Et_3SiH, BF_3 and then with $NaBH_4/CeCl_3$ **23** was obtained. Its dihydroxylation under Upjohn [82] (NMO, cat. OsO_4, acetone/H_2O) and Donohoe's (OsO_4/TMEDA, CH_2Cl_2, −78°C) conditions [83] gave after acetylation the imino-*C*-disaccharide precursors **24** and **25**, respectively.

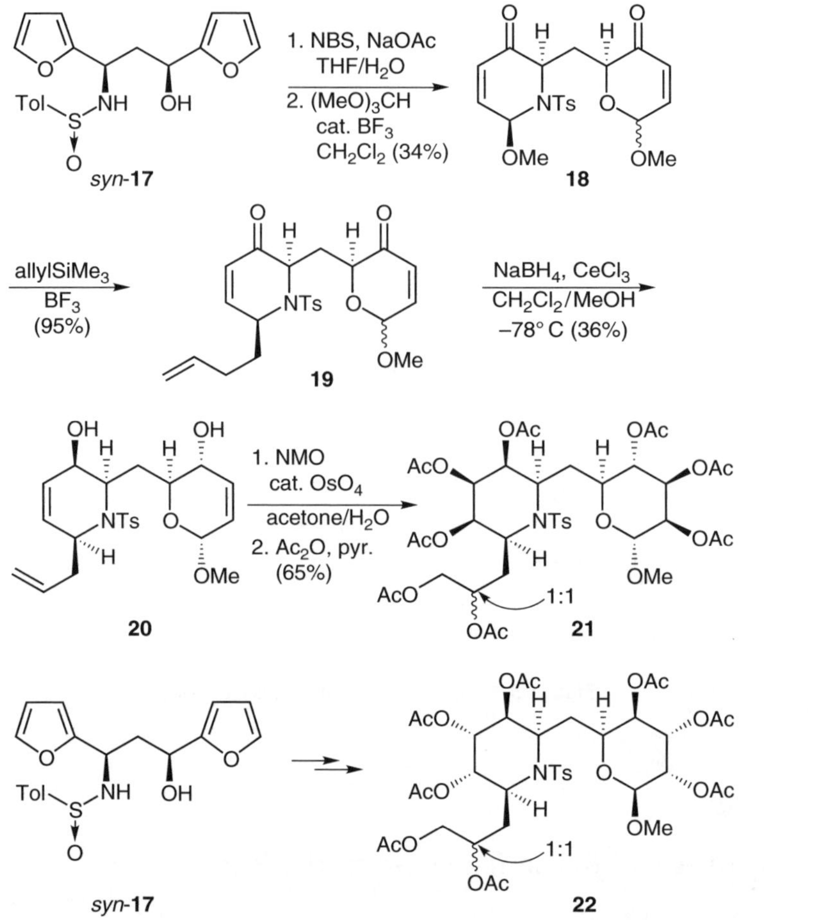

Scheme 5.1 Two-directional approach to imino-*C*-(1→1)-linked disaccharide mimetics

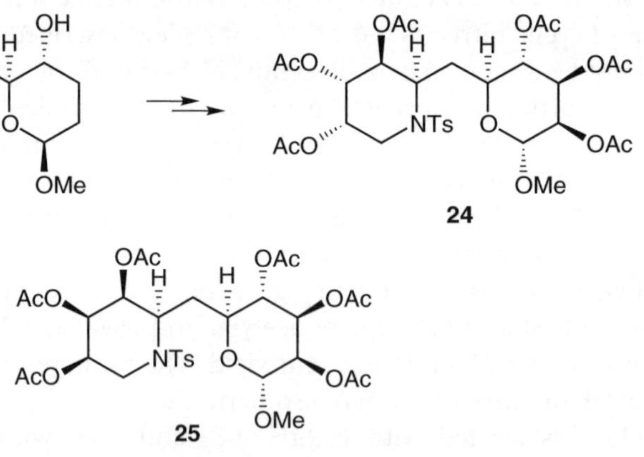

Structure 5.4 Examples of imino-(1→1)-linked disaccharide precursors

5.2.2 Synthesis of branched imino-*C*-(1→3)-disaccharides via aldol reaction of 7-oxanorbornanone derivatives

The first example of 'branched imino-*C*-disaccharides' was prepared by Baudat and Vogel [68] in 1996. It attaches 6-deoxygalactonojirimycin at position C-3 of D-altrose through a hydroxymethano linker (Scheme 5.2(a)). The same synthetic approach has also allowed the preparation of 6-deoxygalactonojirimycin β-*C*-(1→3) linked with D-galactose (Scheme 5.2(b)) [84]. The method relies on the stereoselective cross-aldol addition of a 2,6,7-trideoxy-2,6-imino-D-*glycero*-L-*manno*-heptose and a 7-oxabicyclo[2.2.1]heptan-2-one derivative ('naked sugar methodology') [85]. The commercially available D-*glycero*-D-*gulo*-heptono-1,4-lactone (**26**, two steps from D-glucose) is converted into aldehyde **27**. The enantiomerically pure (-)-(1*S*,4*S*)-7-oxabicyclo[2.2.1]hept-5-en-2-one (**28**, an example of 'naked sugars') adds electrophiles such as PhSeCl with high *exo* stereoselectivity and regioselectivity giving adduct **29**. The high regioselectivity of the addition is attributed to the electron-releasing ability of the homoconjugated ketone (frangomeric effect: $n(C{=}O)/\sigma C(1,2)/2pC^+(6)$ hyperconjugative interaction) [85–87]. Ketone (+)-**30** and aldehyde **27** undergo highly diastereoselective cross-aldol reaction (Zimmerman–Traxler transition state [88, 89]) giving aldol **31** and the debenzylated derivative **32**.

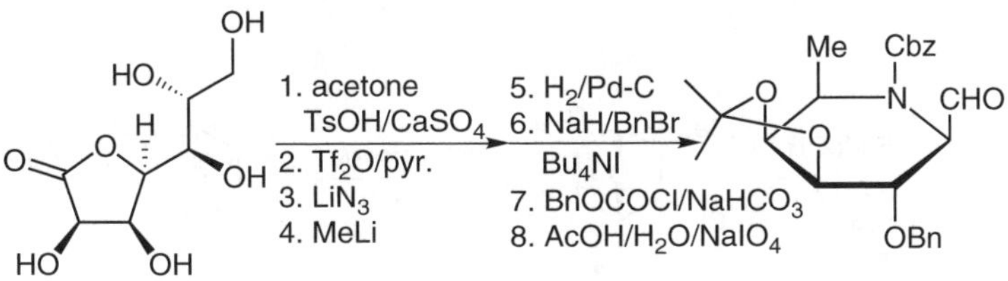

26 **27**

Structure 5.5 Conversion of an heptonolactone into an iminosugar carbaldehyde

Highly *exo* face selective reduction of the bicyclic ketone moiety with $NaBH_4$ and subsequent acetylation produces **33**. Oxidative removal of the benzeneselenyl group with mCPBA gives chloroalkene **34**. Its dihydroxylation, then acetylation and highly regioselective Baeyer–Villiger oxidation of the intermediate α-acetoxyketone (regioselectivity controlled by the 7-oxa ethereal group) with mCPBA provides uronolactone **35**. Acidic methanolysis of **35** furnishes uronic ester **36**. Its reduction with $LiBH_4$, and subsequent polyacetylation, hydrogenolysis of the benzylcarbonate and ammonolysis in MeOH gives the imino-*C*-disaccharide **37** (Scheme 5.2(a)). Cross-aldol reaction of aldehyde **27** with (-)-**28**, the enantiomer of (+)-**28**, generates aldol **31B** in 70 per cent yield. Then uronolactone **35B** can be obtained (α-hydroxyl group protected as a brosylate rather than as an acetate). On treatment of **35B** with methanol in DMF and in the presence of K_2CO_3, the intermediate methyl uronic ester **35C** undergoes intramolecular S_N2 displacement (which does not occur with the corresponding α-acetate) giving anhydrosugars **36B**. Its reduction with $LiBH_4$ and subsequent acidic treatment with CF_3COOH in 1:8 H_2O/THF

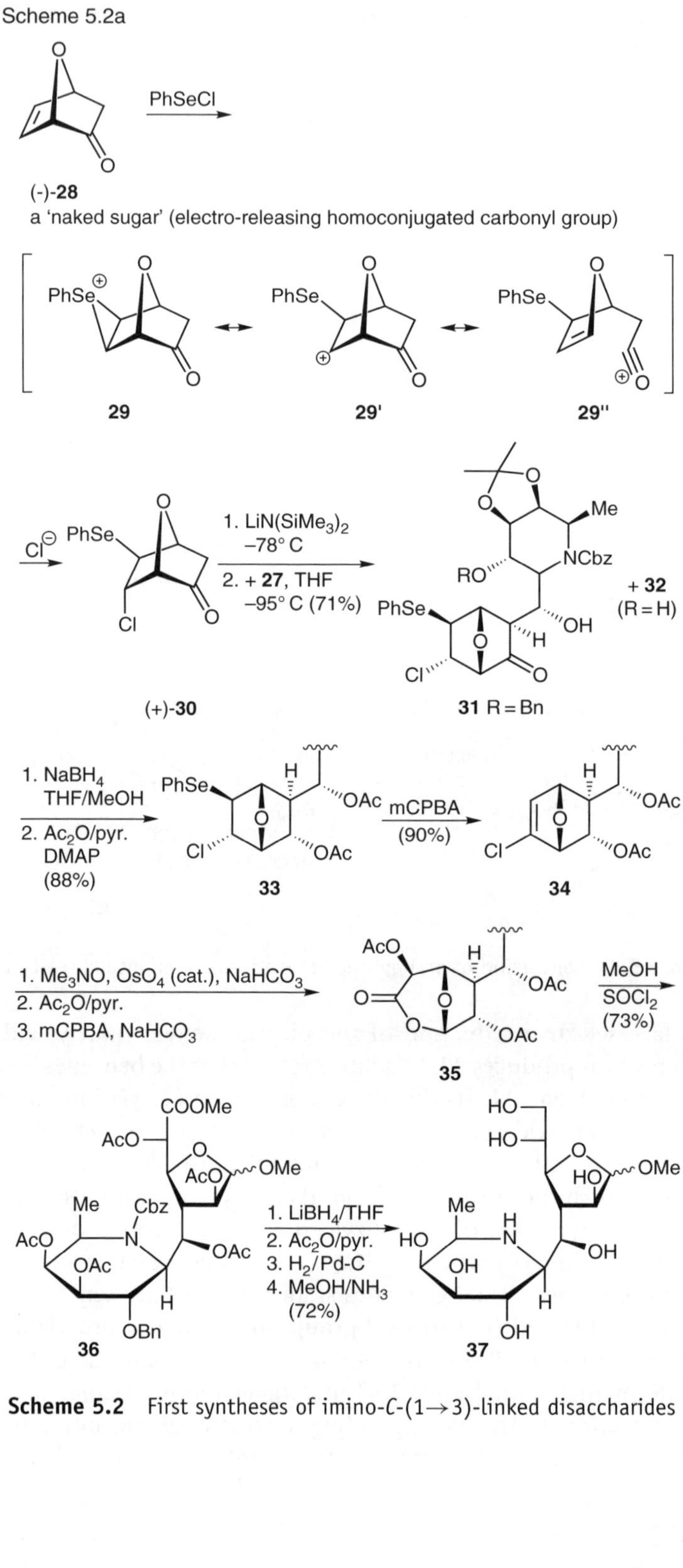

Scheme 5.2 First syntheses of imino-C-(1→3)-linked disaccharides

Scheme 5.2b

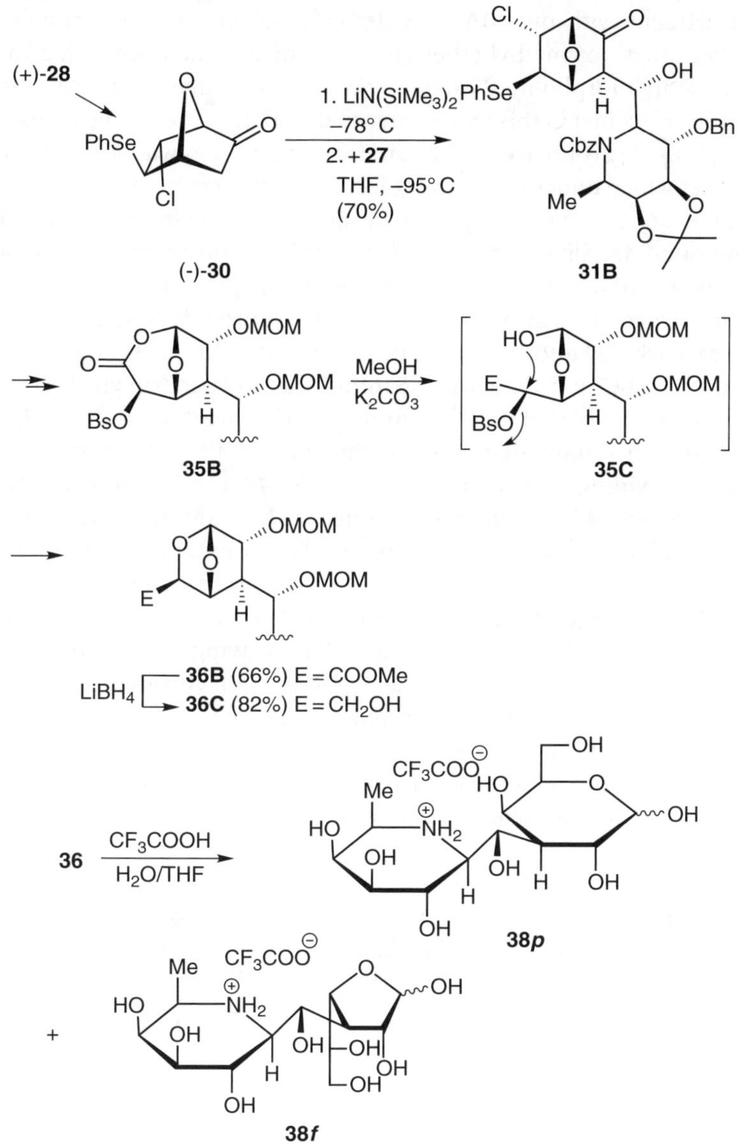

Scheme 5.2 (Continued)

at 37°C leads to a 68:32 mixture of the galactopyranose **38*p*** and galactofuranose **38*f*** (Scheme 5.2(b)).

The cross-aldol reaction of furfuraldehyde with racemic 7-oxabicyclo[2.2.1]heptan-2-one derivatives [90] has been applied to generate the racemic 'imino-*C*-disaccharides' in which 1,4-dideoxy-1,4-imino-DL-erythritol is linked to C(3) of DL-galactose through a hydroxymethano linker (Scheme 5.3) [90]. The lithium enolate of enone (±)-**30** reacts with furan-2-carbaldehyde at −78°C to yield a single aldol **39** (Zimmerman–Traxler transition state [88]). It is protected as silyl ether **40**. Reduction of the ketone with $NaBH_4$ and subsequent treatment with mCPBA provides chloroalkene **41**. Protection of the *endo*-alcohol moiety as methoxymethyl ether gives **42** which is then submitted to dihydroxylation with N-methylmorpholine N-oxide (NMO) and a catalytic amount of OsO_4. The α-hydroxyketone obtained in this way is esterified as a brosylate giving **43**. Baeyer–Villiger oxidation gives a single uronolactone **44**. Careful methanolysis with MeOH in DMF generates a uronic ester which undergoes intramolecular displacement of the brosylate (see reaction **35B**→**35C**→**36B**, Scheme 5.2) producing anhydro uronate **45** (major), together with epimeric compound **46**. Single crystallization of this mixture provides pure **45** in 50 per cent yield. Reduction of the methyl ester with $LiAlH_4$ and protection of the primary alcohol as a methoxymethyl ether furnishes **47**. A multigram procedure has been developed which transforms ketone (±)-**30** into lactone **44** in 67 per cent yield (32.5 g, eight steps). Conversion of **44** into **47** can be carried out on a 25 mmol scale (47 per cent yield, three steps). The Clauson–Kaas oxidation [92] of the furan moiety of **47** generates a mixture of isomeric 1,5-dimethoxy-1,5-dihydrofurans that is hydrogenated into a γ-ketoaldehyde reacting with primary amines providing a small library of pyrroles **48**. Oxidation of **47** with dimethyldioxirane provides pure **49** quantitatively. Under Luche's conditions [81] **49** was reduced into enediol **51** (major) and an isomeric compound (9:1 mixture). Pure **51** was esterified as di(methanesulphonate) **52**, which was not isolated but directly submitted to reaction with LiN_3 in DMF. Selective S_N2 displacement of the primary mesylate moiety gives azide **53**, which is not isolated but submitted to dihydroxylation giving a 2:1 mixture of diols **54** and **55** (55 per cent, three steps), that can be separated by flash chromatography on silica gel. Diols **54** and **55** were converted into diacetonides **56** and **57**. Reduction of the azido moiety of **56** with $HCOONH_4$ in the presence of Pd/C yields pyrrolidine **58**. Desilylation and subsequent acidic hydrolysis provides the unprotected racemic imino-*C*-disaccharide **59** · HCl, as a 44:44:12 mixture of α-DL-pyranose, α-DL-furanose, and β-DL-furanose. Similarly, **57** is converted into **60** · HCl.

When the desilylation of **56** is followed by treatment of the hydroxymesylate with DBU, epoxide **61** is generated. Hydrogenation of the azido moiety of **61** gives pyrrolidine **62** which is then converted into **63** · HCl. Similarly, **57** is transformed into **64** · HCl (Scheme 5.3). If diols **54** and **55** are protected as di(methoxymethyl) diethers, the reaction sequences described in Scheme 5.3 lead to protected imino-*C*-disaccharides, whose acidic methanolysis allows preparing racemic imino-*C*-disaccharides in which the galactose moiety is under the form of methyl galactopyranosides and/or methyl galactofuranosides [91].

5.2.3 Synthesis of an imino-*C*-(1→3)-disaccharide via Michael addition of a 'naked sugar'-derived enone

Enantiomerically pure ketone (+)-**30** is converted into enone (−)-**65** by reaction of its potassium enolate with Eschenmoser's salt ($CH_2{=}NMe_2I$) [55a]. Epoxidation of 'naked sugar' (+)-**66** followed by treatment with 70 per cent aqueous $HClO_4$ in $CF_3CH(OH)CF_3$ provides **67** in 60 per cent overall yield. Protection of **67** as MOM ether, then treatment with MeOH/DBU, liberates the chiral auxiliary (R*OH) and an *endo* alcohol that is

Scheme 5.3a

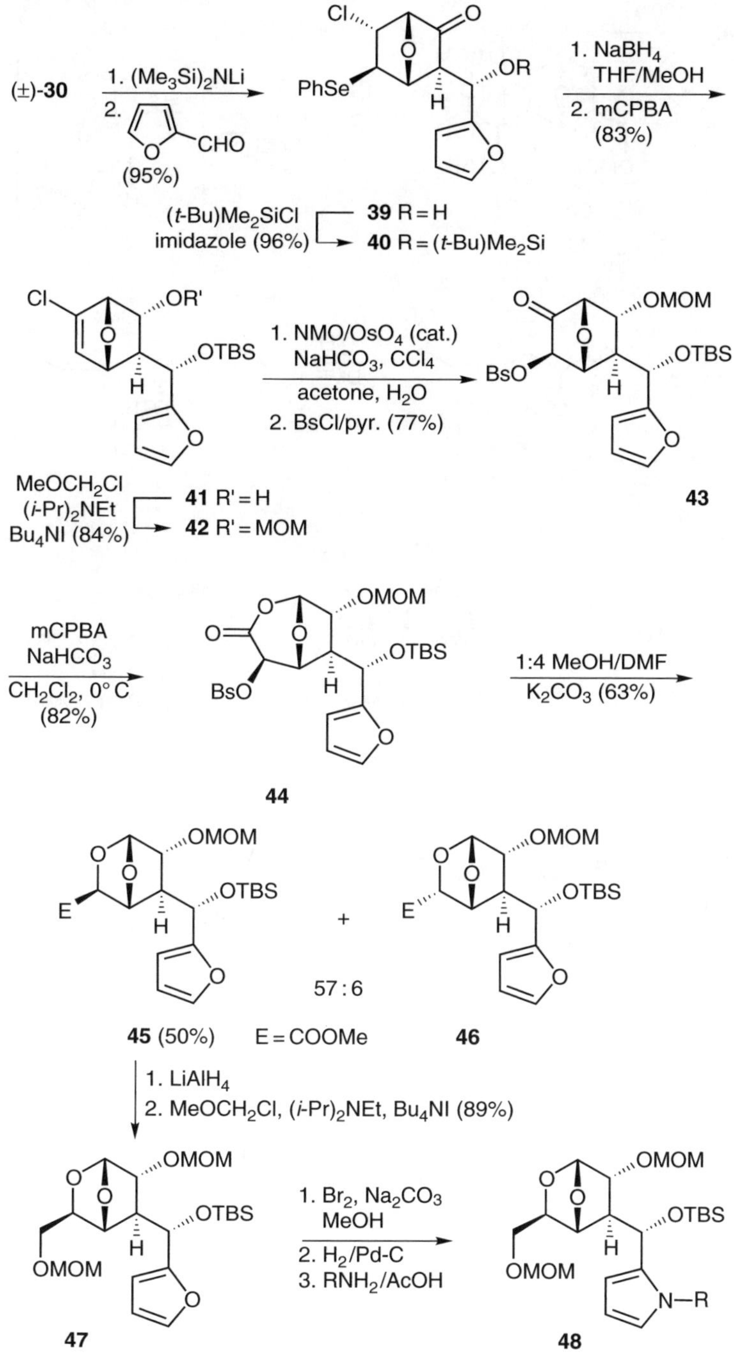

Scheme 5.3 Syntheses of racemic imino-C-(1→3)-disaccharide attaching a 2,3-dihydroxypyrrolidine to D-galactose

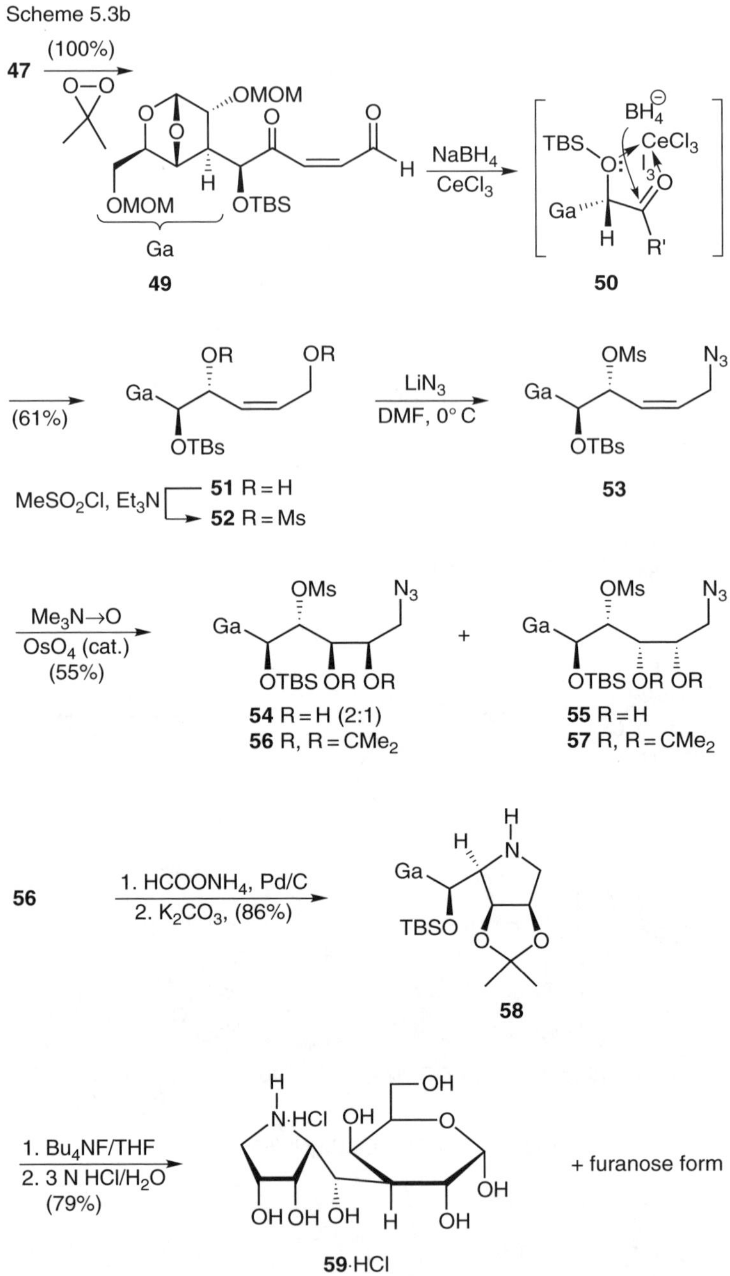

Scheme 5.3 (Continued)

Scheme 5.3c

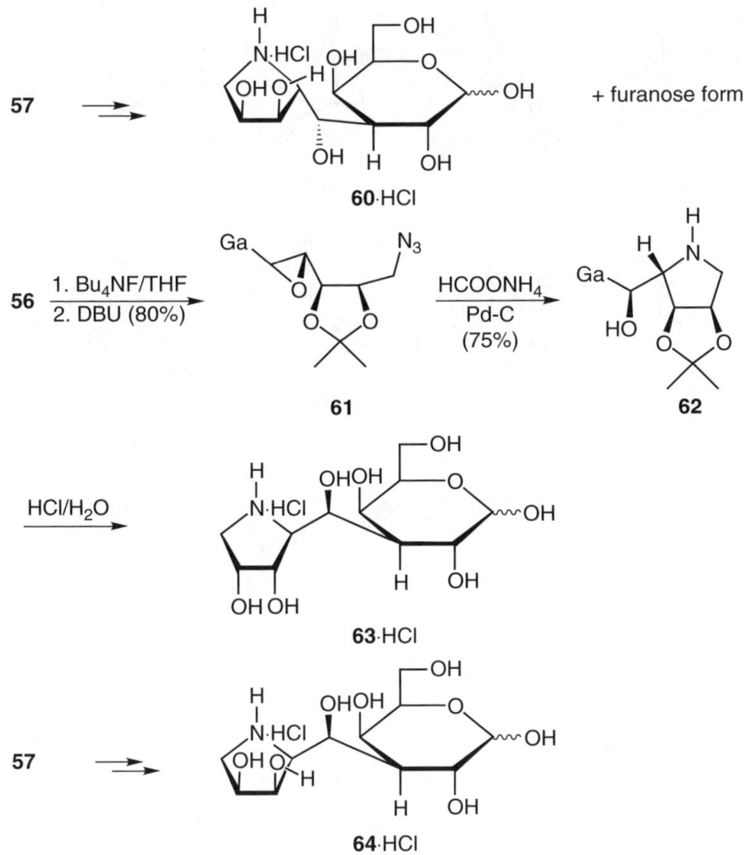

Scheme 5.3 (Continued)

protected as MOM ether. This gives **68** in 80 per cent overall yield. Baeyer–Villiger oxidation of ketone **68** is also highly regioselective (control by the 7-oxa ethereal group) and produces uronolactone **69** as a single product. The lithium enolate of **69** adds to enone (−)-**65** giving a single aldol **71** after cautious work-up with anhydrous AcOH/MeOH in THF at low temperature. Under these conditions the intermediate enolate **70** is protonated exclusively onto its *exo* face (steric factor) and the resulting product **71** is not epimerized. Ketone reduction, oxidative removal of the selenyl group and alcohol protection as MOM ether provides chloroalkene **72**. The lactone moiety of **72** adds BnOLi generating a mixture of α, β-furanoses that are silylated into **73**. Dihydroxylation of chloroalkene **73**, followed by acetylation of the resulting α-hydroxyketone furnishes **74**. Ketone **74** undergoes a highly regioselective Baeyer–Villiger oxidation producing uronolactone **75**. Liberation of the carboxylic acid upon hydrogenolysis of benzyl ester **75** generates **76** which is converted into the corresponding acyl azide. The latter undergoes *in situ* Curtius rearrangement, the intermediate isocyanate being quenched with benzyl alcohol to provide benzylcarbamate **77** (85 per cent yield, 3:1 mixture of α- and

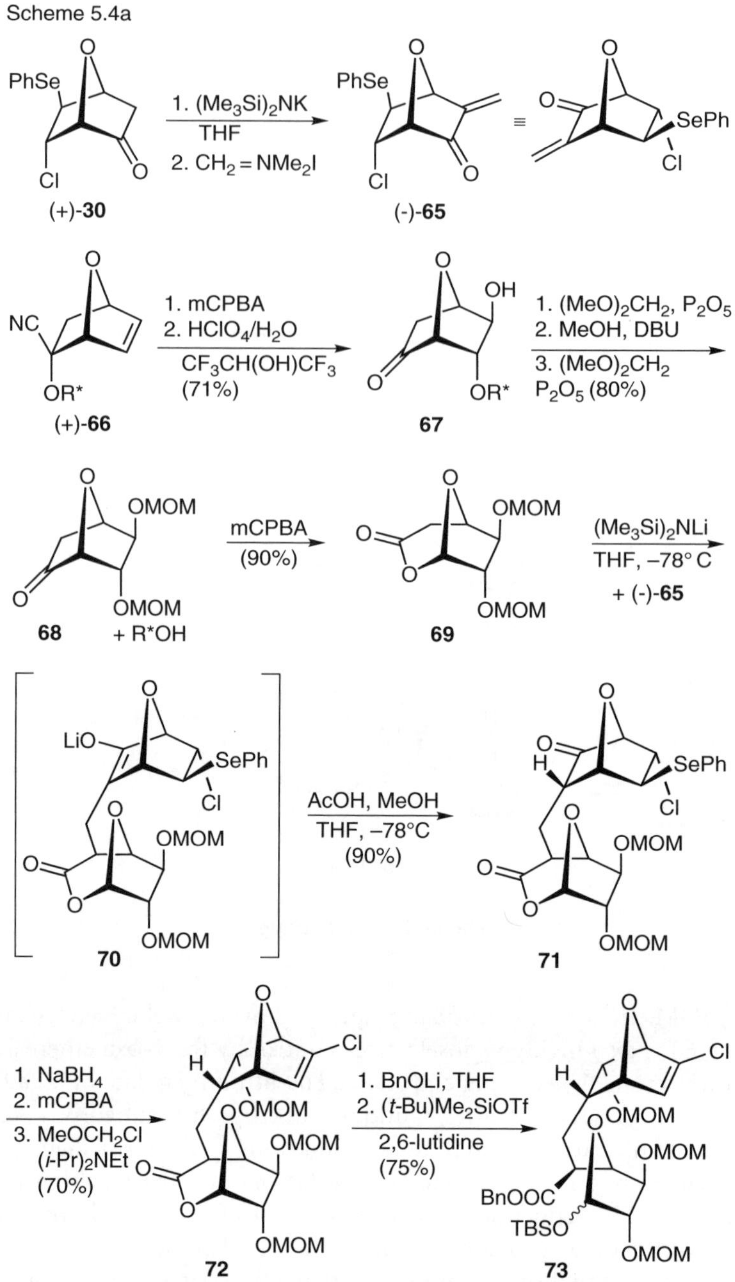

Scheme 5.4 Total asymmetric synthesis of methyl 3-deoxy-3-(1′, 2′, 6′-trideoxy-2′,6′-imino-D-galactit-1-yl)-α-D-mannopyranoside

Scheme 5.4b

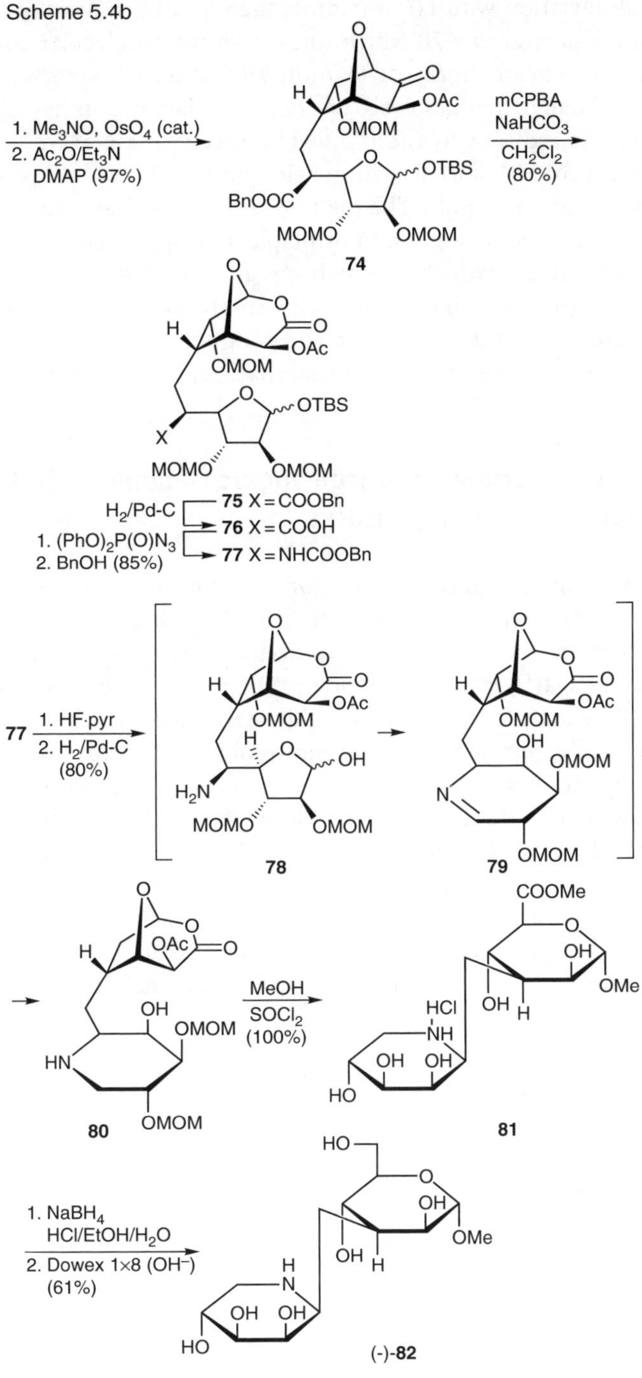

Scheme 5.4 (Continued)

β-furanosides). Desilylation with HF · pyridine, then hydrogenolysis of the benzylcarbamate generates a primary amine **78** which undergoes intramolecular condensation with the aldose moiety to form an intermediate imine **79** that is hydrogenated *in situ* affording **80** in high yield. Finally, methanolysis of the uronolactone under acidic conditions produces **81** which is reduced to the imino-*C*-disaccharide (−)-**82** (61 per cent, two steps). In this approach [93], a total asymmetric synthesis is realized which does not use carbohydrates as starting materials. The method is thus applicable to generate the enantiomer of (−)-**82** and other analogues. In principle, this approach allows the preparation of imino-*C*-disaccharides in which the hydroxy groups of the two sugar units can be exchanged for other groups such as amino, alkoxy, chloro and fluoro substituent, since monosaccharide containing these groups can be obtained readily from the 'naked sugars'. Furthermore, all possible stereomeric monosaccharides have been prepared [85, 94, 95].

5.2.4 Cross-aldol reactions with isolevoglucosenone derivatives and 2,5-dideoxy-2,5-iminopentoses

Although the methods presented above for the synthesis of branched 'imino-*C*-disaccharides' are highly stereoselective and lead to a high diversity (stereoisomers, functionalities), they are too lengthy. As enantiomerically pure 7-oxabicyclo[2.2.1]hept-5-en-2-one ('naked sugar'), levoglucosenone (**83**) and isolevoglucosenone (**84**) [96] are bicyclic systems allowing predictable and highly stereoselective chemical modifications. These enantiomerically pure enones are readily obtained by pyrolysis of cotton [97] and from D-glucose (four steps) [98], respectively. Isolevoglucosenone (**84**) adds nucleophiles [96, 98, 99] and radicals [100] exclusively on its less hindered face (*syn* with respect to the oxa bridge). Oshima and coworkers [101] have shown that conjugate addition of Me_2AlSPh to simple enones, followed by the reaction of the aluminum enolates with aldehydes allows the preparation of the corresponding aldols in one-pot procedures. Applying this reaction, we have developed a one-pot, three-component approach to the synthesis of *C*-linked disaccharides and analogues (Scheme 5.5) [102].

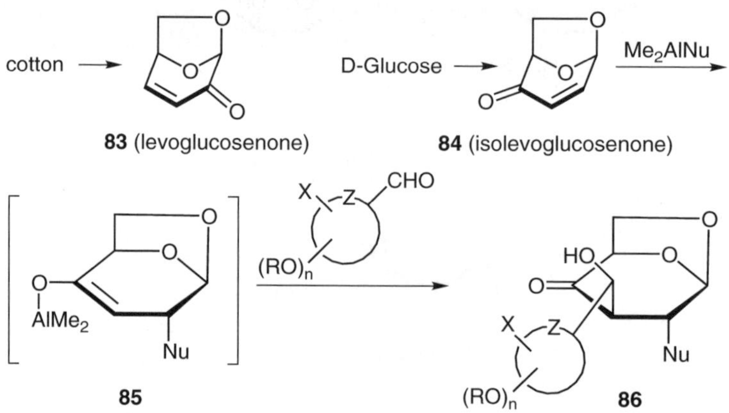

Scheme 5.5 Use of isolevoglucosenone as template

When the nucleophile is an alkoxy group, the one-step procedure fails because the high affinity of oxygen for aluminum makes the reaction **84** + $Me_2AlOR \rightarrow$ **85** not sufficiently exothermic. Therefore, a two-step process is required, as illustrated in Scheme 5.6 for the synthesis of imino-*C*-(1→3)disaccharide **93** [103]. Isolevoglucosenone (**84**) adds benzyl alcohol giving **87** [104]. The enolates of ketone **87** do not induce β-elimination at low temperatures. The chlorozinc enolate, generated by deprotonation of **87** with $(Me_3Si)_2NK$ followed by treatment with anhydrous $ZnCl_2$, added to aldehyde **88** giving an unstable zinc aldol that is reduced *in situ* at −78°C with L-Selectride ($LiB[CH(Me)C_2H_5]_3H$) into the D-galactose derivative **90** (53 per cent based on **87**). Desilylation of **90** and subsequent debenzylation and acidic methanolysis gives a 3:2 mixture of methyl α- and β-D-galactopyranoside **91**α, β (82 per cent yield).

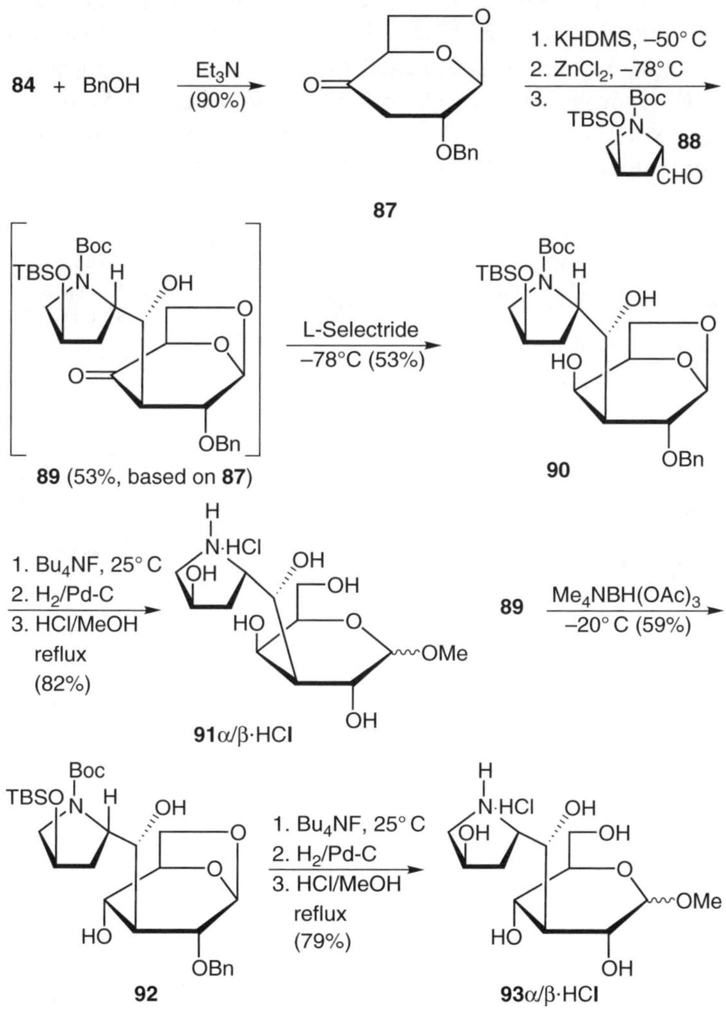

Scheme 5.6 Convergent synthesis of imino-*C*-(1→3)-disaccharides from isolevoglucosenone

Interestingly, reduction of ketone **89** with $Me_4NB(OAc)_3H$ [105] produces the D-glucose derivative **92** which, after deprotection and acidic methanolysis, is converted into imino-*C*-disaccharide **93** (7:3 mixture of α- and β-D-glucopyranosides).

5.2.5 Takai–Hiyama–Nozaki–Kishi couplings

Applying the Takai–Hiyama–Nozaki–Kishi coupling of hydroxyproline-derived carbaldehyde **88** with the enol triflate **94** derived from ketone **87**, a quick and convergent access to imino-*C*-(1→4)-disaccharides has been opened (Scheme 5.7). Adduct **87** of benzyl alcohol to isolevoglucosenone is enolized on treatment with $(Me_3Si)_2NLi$ in 95:5 THF/HMPA at −78°C. Quenching of the corresponding lithium enolate with 2-[bis(trifluoromethanesulphonyl)amino]-5-chloropyridine [106] provides enol triflate **94** in 87 per cent yield. Takai–Hiyama–Nozaki–Kishi coupling [107] of **94** with aldehyde **88** generates alcohols **95** and **96** isolated in 62 and 17 per cent yield, respectively. Interestingly, the reaction was accelerated by ultrasound and O_2 (5 mol per cent with respect to $CrCl_2$). Moreover, O_2 suppressed the formation of products of reduction of the triflate [108]. Hydroboration of **95** with $BH_3 \cdot Me_2S$ in THF followed by oxidative work-up (H_2O_2/NaOH) furnishes the D-*gluco* derivative **97**. Deprotection and acidic methanolysis provide imino-*C*-(1→4)-disaccharide **98** · HCl.

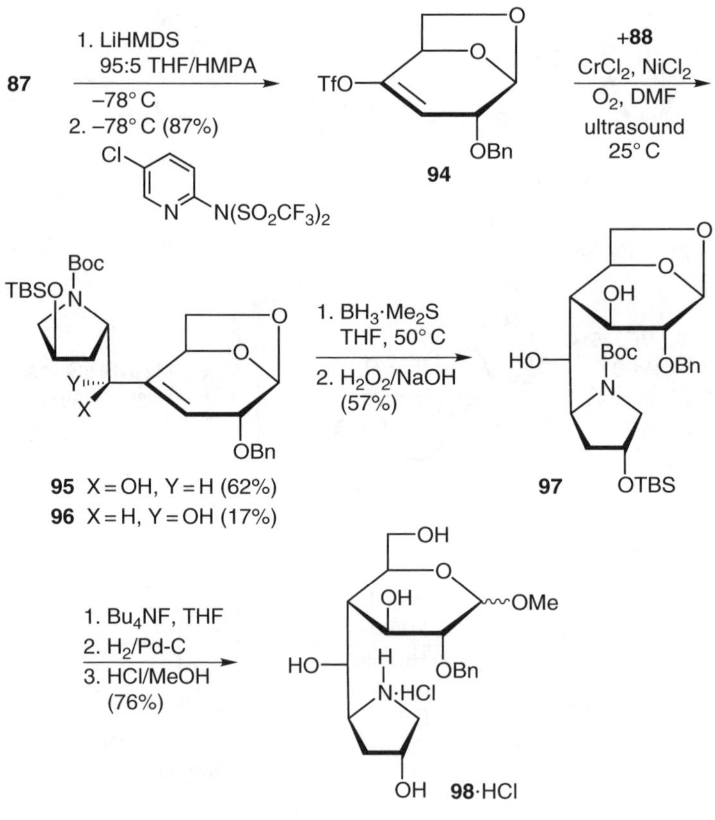

Scheme 5.7 Synthesis of an imino-*C*-(1→4)-disaccharide from isolevoglucosenone

Johnson and coworkers [69] have obtained the branched imino-*C*-disaccharide D-iminoManβ-CH_2(1→4)-D-Talo-α-OMe applying the Miyaura–Suzuki coupling between a polyhydoxylated 1-bromocyclohexene and a methylidene-substituted hexose derivative [77]. Intramolecular oxyallylation of a sugar-derived enol ether has been used by Mootoo and coworkers [65] to prepare D-azaGal*p*-CH_2(1→4)-D-Glc-α-OMe.

5.2.6 Imino-*C*-(1→2)-disaccharides via Takai–Hiyama–Nozaki–Kishi coupling

Starting from levoglucosenone (**83**), its adduct **99** with benzyl alcohol is converted into triflate **100**, as above. In the Takai–Hiyama–Nozaki–Kishi coupling reaction, **100** is less reactive than isomer **94** (Scheme 5.7), thus explaining the modest yield (53 per cent) of its reaction with aldehyde **88** giving allylic alcohols **101** and **102**. In this case the product of reduction **103** is formed even using ultrasound and O_2 (5 mol per cent). Acidic

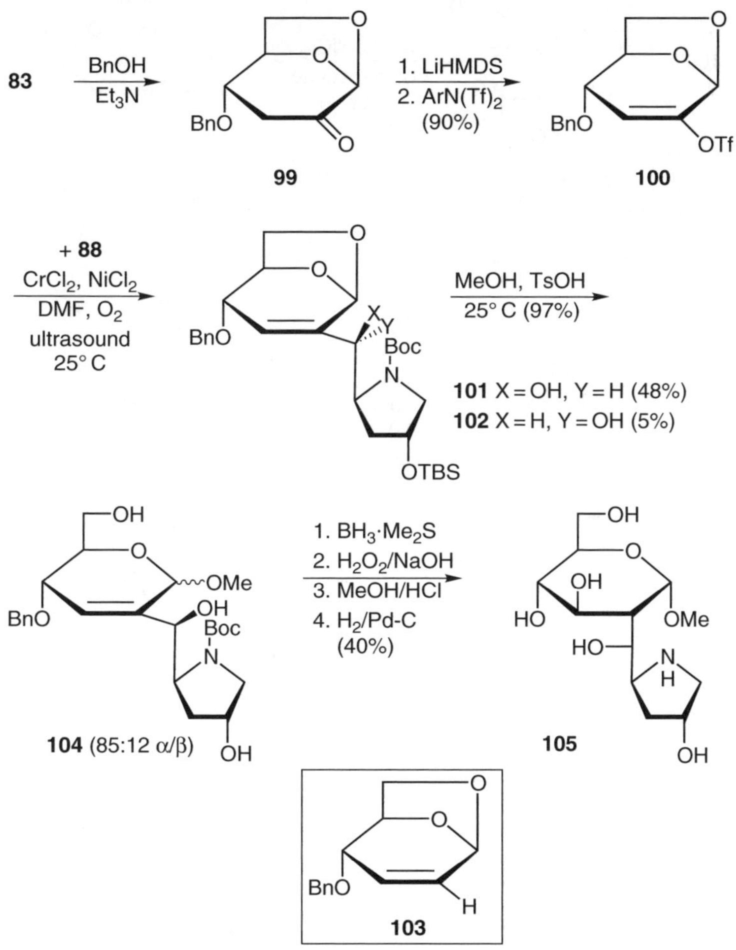

Scheme 5.8 Synthesis of an imino-*C*-(1→2)-disaccharide from levoglucosenone

methanolysis of **101** gives **104**. Hydroboration of **104** and subsequent deprotection produces imino-*C*-(1→2)-disaccharide **105** [108].

The procedure presented in Scheme 5.8 has been applied to prepare the imino-*C*-(1→2) disaccharide **13** which combines β-L-iminoerythrose at C-2 of methyl α-D-glucopyranoside (L-iminoerythro-β-CH(OH)-(1→2)-D-Glc-α-OMe) through a (1'*S*)-hydroxymethano linker. In this case aldehyde **106** is coupled with bicyclic triflate **100**. Probably for steric reasons, the yield of this Takai–Hiyama–Nozaki–Kishi coupling remains mediocre. Acidic methanolysis of triflate **100** generates the monocyclic triflate **108**. Its primary alcohol is protected as a benzyloxymethyl ether **109**. The latter is coupled with **106** with a better yield (48 per cent) giving allylic alcohol **110**. Subsequent hydroboration and deprotection provides imino-*C*-(1→2)-disaccharide **13** (Scheme 5.9) [56]. The syntheses described in Schemes 5.6–5.9 demonstrate the utility of levoglucosenone and isolevoglucosenone templates as they give rapid access to

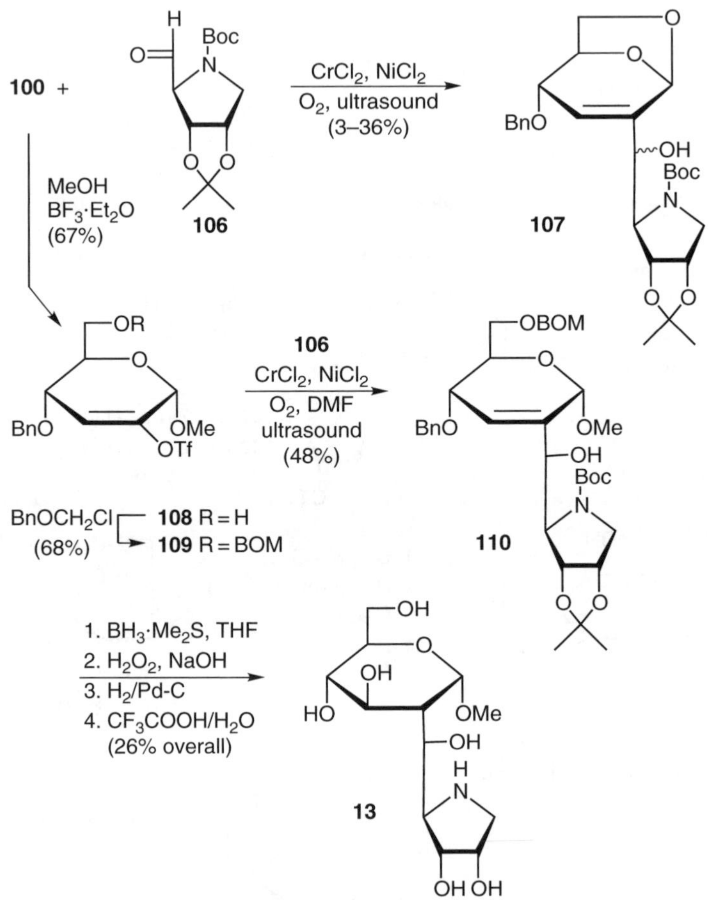

Scheme 5.9 Synthesis of methyl 2-deoxy-2-[(1*S*)-2,5-dideoxy-2,5-imino-L-ribit-1-yl]-α-D-glucopyranoside

imino-*C*-(1→2), imino-*C*-(1→3) and imino-*C*-(1→4)-disaccharides, using the same iminosugar-derived carbaldehydes.

5.2.7 Imino-*C*-(1→2)-disaccharides via aldol condensation

Methyl 4,6-*O*-benzylidene-2-deoxy-α-D-*erythro*-hexopyranosid-3-ulose (+)-**111** is readily available from D-mannose [109]. The groups of Fraser-Reid [110] and Chapleur [111] have established that β-elimination of methoxide anion from enolates of **111** is a relatively slow reaction permitting the condensation of the latter with various electrophiles, without subsequent elimination. We thus envisaged a synthetic approach to imino-*C*-disaccharides applying the cross-aldol condensations of ketone **111** with protected 2,5-dideoxy-2,5-imino-L-ribose-derived carbaldehydes. The preparations of imino-*C*-disaccharides **13**, **14** and **120** linking 4-deoxy-4-amino-β-L-*erythro*furanose at C-2 of D-glucose and D-allose are outlined in Scheme 5.10 [57].

Aldol condensation of aldehyde **113** (obtained by periodic oxidation of diol **112** [112]) with **111** gave **114** in 81 per cent yield. Its reduction with $NaBH_4$ in EtOH furnished diol **115**. Subsequent hydrogenolysis of the benzylamine and benzylidene acetal and acidic work-up produced **13**. DAST-promoted elimination of water from aldol **114** gave a 34:49 mixture of enones **116** and **117**. Under Luche's conditions [81] **116** was reduced stereoselectively into allylic alcohol **118** (87 per cent). Subsequent hydrogenolysis catalysed first by Pd on charcoal, and then by $PtO_2 \cdot H_2O$ under acidic conditions provided imino-C-disaccharide **14**. Following a similar route, transformation of **117** gave a low yield of several products. However, conjugate addition of thiophenol to **117** in the presence of a catalytic amount of Et_3N furnished a single adduct **119**. Its reduction with $NaBH_4$ and subsequent hydrogenolysis with 10 per cent $PtO_2 \cdot H_2O$, H_2SO_4 as catalyst, followed by successive treatment with an excess of Raney nickel in MeOH, and then with CF_3COOH/H_2O, provided imino-*C*-disaccharide **120**.

5.2.8 Synthesis of homo-imino-*C*-(1→3)-disaccharides

The general method presented in Scheme 5.5 for the preparation of advanced precursors of imino-*C*-(1→3)-disaccharides has been applied to the synthesis of homo-imino-*C*-(1→3)-disaccharides (Scheme 5.11). Cross-aldol condensation of **123** with 1,6-anhydro-2-*O*-benzyl-3-deoxy-β-D-*erythro*-hexopyrano-4-ulose (the adduct of benzyl alcohol to isolevoglucosenone **84**) generates, after water elimination, a single enone **124**. It is then reduced selectively into allylic alcohol **125**, the deprotection of which affords **126** and **127**.

Interestingly, benzyl ether hydrogenolysis of **125** did not reduce its alkene moiety. In methanol and under acidic conditions, **126** and **127** are equilibrating [75]. Debenzylation of **124** (H_2/10 per cent $Pd(OH)_2$/C in MeOH) followed by hydrogenation in the presence of PtO_2 gives the D-galactose derivative **128**. The ring opening of the anhydrogalactose moiety of **128** is a difficult operation. Complex mixtures are formed on treatment of **128** under standard acidic conditions such as CF_3COOH/MeOH or HCl/MeOH.

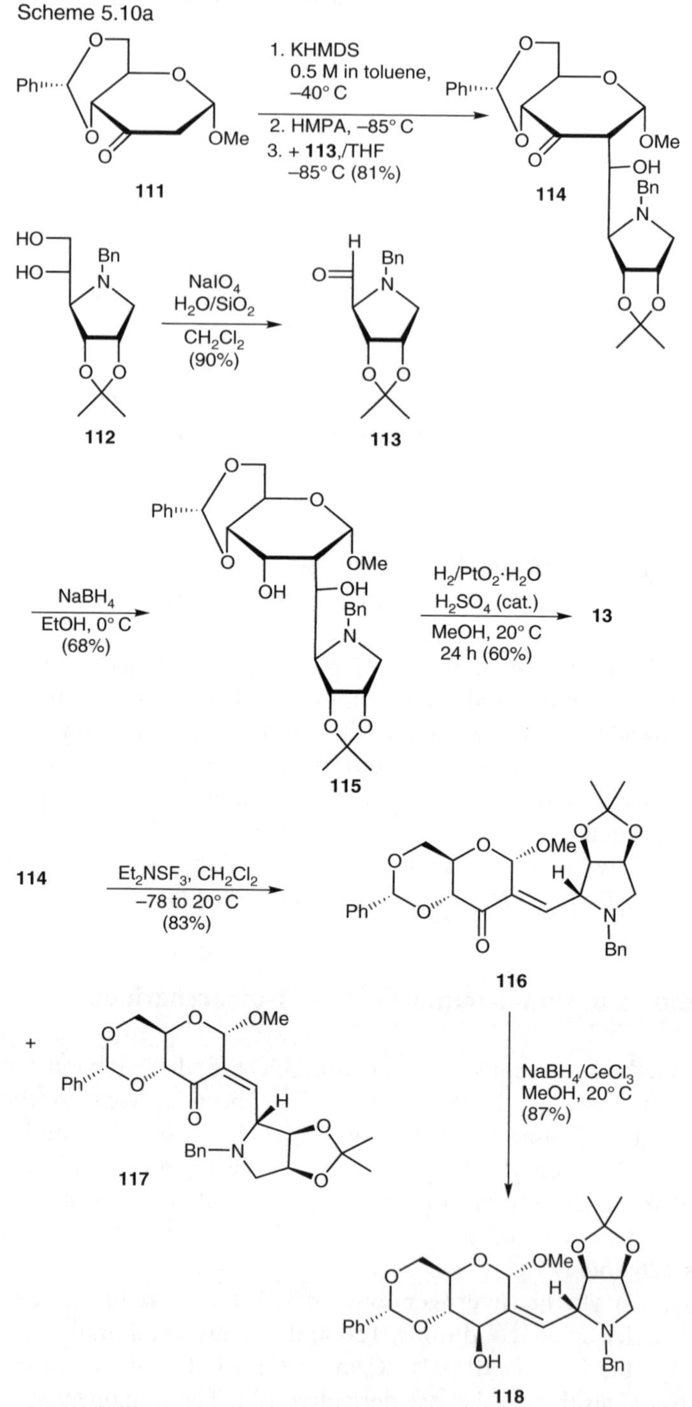

Scheme 5.10 Imino-C-(1→2)-disaccharide via aldol condensation

Scheme 5.10b

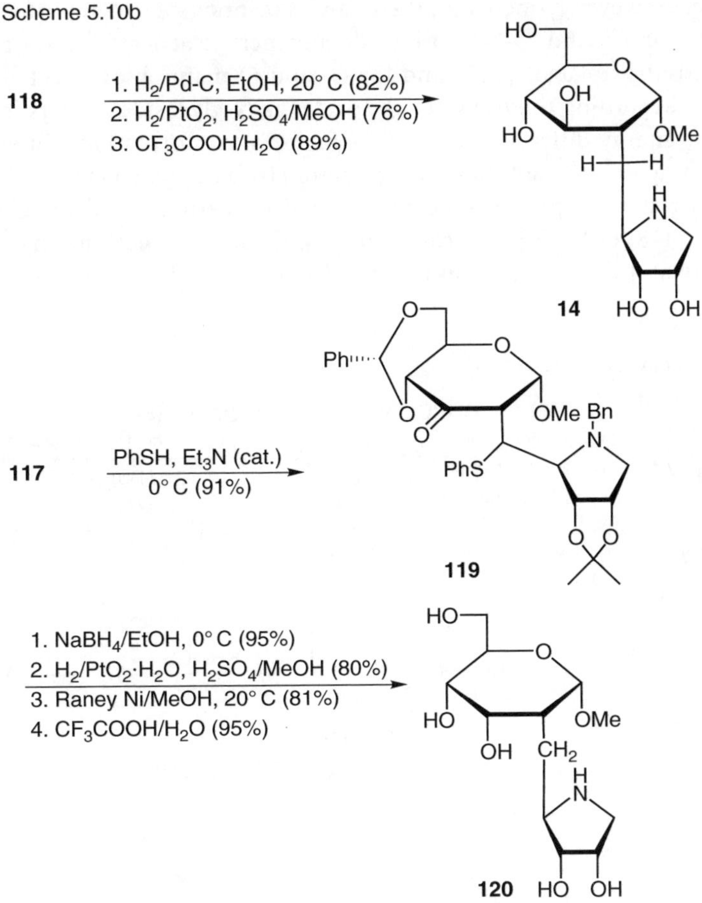

Scheme 5.10 (Continued)

The methodology proposed by Witczak and coworkers [113] for the hydrolysis of anhydropyranoses had to be applied. Exchange of the Boc protection group of **128** as trifluoroacetamide is carried out with $(CF_3CO)_2O$ in pyridine, followed by methanolysis in the presence of a catalytical amount of ammonia [114]. This produces **129** which is esterified to give **130**. Acetolysis of **130** with Ac_2O and CF_3COOH furnishes **131**. Final deprotection with $MeOH/NH_3$ gives the homo-imino-*C*-(1→3)-disaccharide **132** [76].

5.3 Conformations of imino-*C*-disaccharides

Kishi and coworkers have proposed that *C*- and *O*-glycosides share the same conformational characteristics in solution [115, 116]. Moreover, the conformations of C-lactose bound to peanut agglutinin is basically identical to the conformation of its parent O-lactose bound to the same protein. This has led to the claim that the conformational similarity between *O*- and *C*-glycosides is a general phenomenon [117].

On their side, however, Jimenez-Barbero and coworkers have reported that similar conformations for *C*- and *O*- glycosides do not persist at least for *C*- and *O*-lactose ($\beta(1\rightarrow4)$-glycosidic linkage) [118] and for *C*- and *O*-mannobiose ($\alpha(1\rightarrow2)$-glycosidic linkage) [119]. Regarding the use of *C*-glycosides as *O*-glycoside isosters it appears that, due to the low-energy difference among conformers, conformations different from the major one existing in water solution may be bound by biopolymers without major energy conflicts. This fact is of prime importance for drug design based on *C*-disaccharides and analogues [118, 119]. Topological features of the biopolymer binding site restricting ligand mobility and demanding conformer selection, as well as changes in the inherent

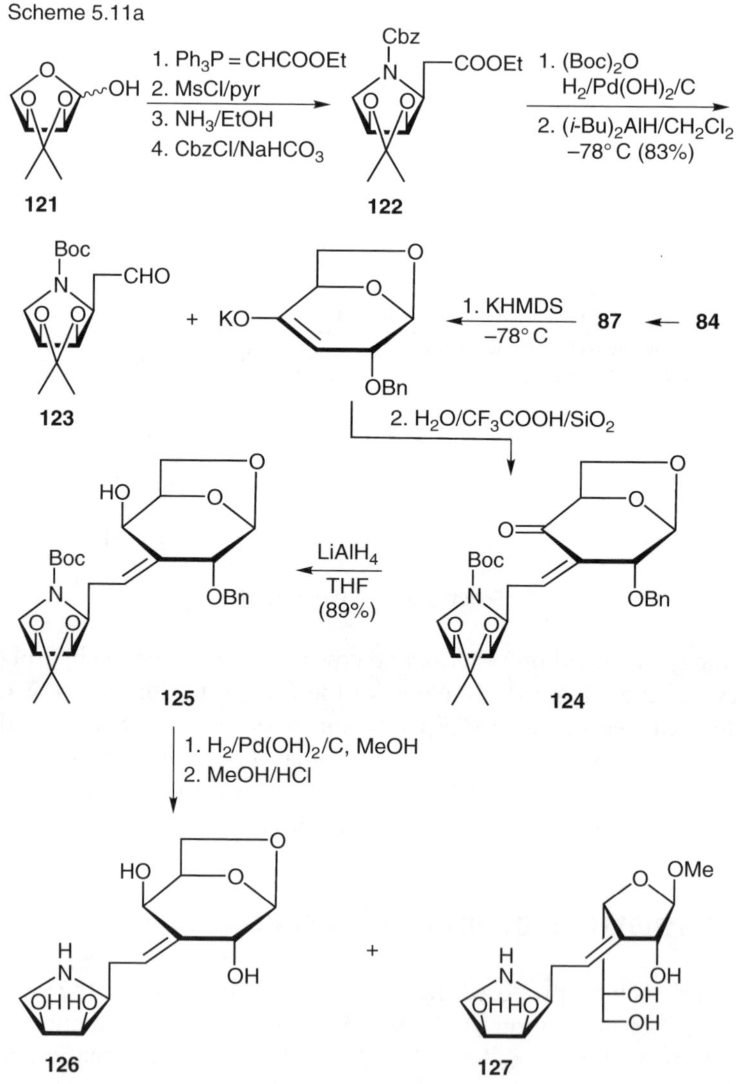

Scheme 5.11 Syntheses of homo-imino-*C*-(1→3)-disaccharides starting from isolevoglucosenone

Scheme 5.11b

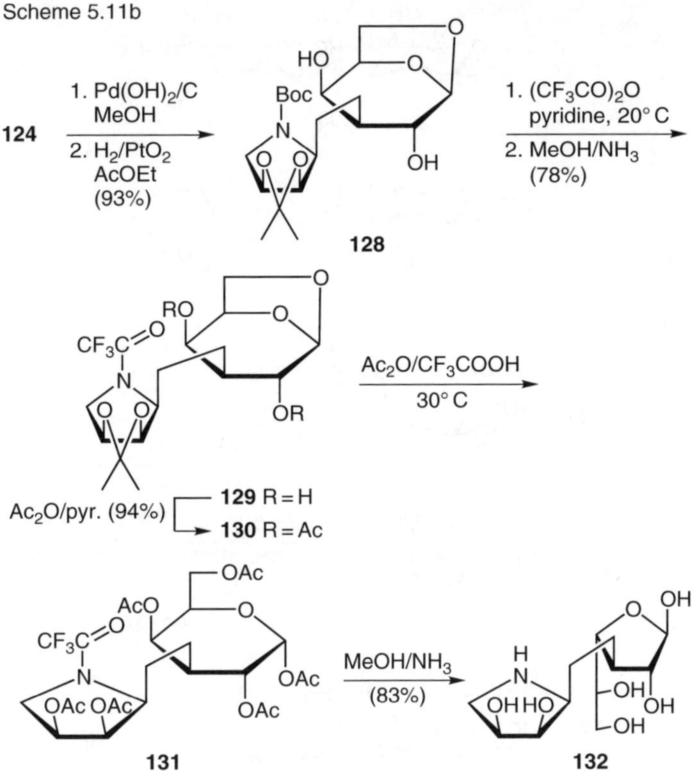

Scheme 5.11 (Continued)

dynamic equilibrium of the flexible *C*-disaccharide can affect the binding constant. When a lectin, for instance, imposes a constraint by establishing interactions with the two homosaccharide units, then the mobility decreases and only a limiting number of favoured conformers will fit into the binding site. Alternatively, the possibility exists that the intramolecular mobility is maintained, if the entropy penalty exceeds the enthalpic gain by weak ligand/protein interactions.

For the imino-*C*-(1→1)-disaccharides **133** (D-iminoMan-β-CH_2(1→1)-β-D-Glc) and **134** (D-iminoGlc-β-CH_2(1→1)-β-D-Glc) in water (Figure 5.2), by a combination of NMR studies (vicinal H/H coupling constants, NOE's) and time-averaged restrained molecular dynamics calculations, Jimenez-Barbero and coworkers obtained the population distributions of conformers about their C(1')-CH_2 and C(1)-CH_2 bonds shown in Figure 5.2 ^{1}H-NMR spectra confirm the $^4C_1$ chair conformations of both monosaccharide units, both in acidic and basic media [120]. All possible conformers are accessible within a free energy difference of 1.5 kcal/mol and the major ones are *exo-syn*/*exo-syn* for both imino-*C*-disaccharides. There is one small change in conformer populations between basic and acidic medium. The results obtained with the imino-*C*-(1→6)-disaccharides (D-iminoMan-β-CH_2(1→6)-D-Man-α-OMe), (D-iminoGlc-β-CH_2(1→6)-D-Glc-α-OMe) are similar to those reported for **133** and **134** [118]. Since these compounds have three bonds in the interresidue linkages, no important polar and/or steric interactions between the

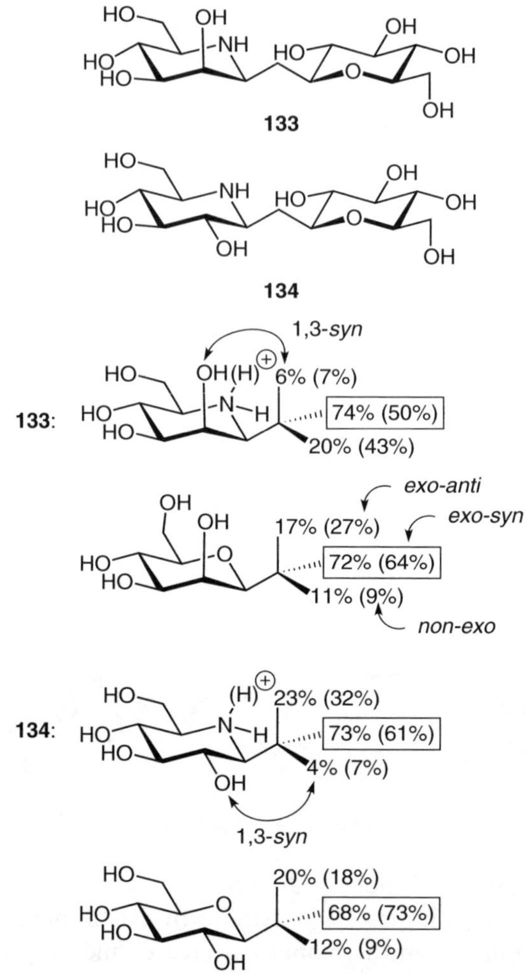

Figure 5.2 Population of conformers for imino-C-(1→1)-disaccharides **133** and **134** in aqueous base and comparison with corresponding C-disaccharides. Values in parentheses are for aqueous acid solutions

residues are likely to occur. It is established that *O*-(1→6)-linked oligosaccharides are rather flexible systems [121]. ^{1}H-NMR data of imino-*C*-(1→3)-disaccharide (−)-**82** (D-iminoLyx*p*-β-CH_2(1→3)-D-Man*p*-α-OMe) demonstrates that the *non-exo* conformer **B**′ is slightly more abundant than the *exo-anti* conformer **C**′ (conformers about the CH_2-C(1′) bond) and both **B**′ and **C**′ conformers are more stable than the *exo-anti* conformer **A**′ (gauche interactions between the R group and the piperidine ring: Figure 5.3). Similarly, conformers **B** and **C** (rotamers about the CH_2-C(3) bond) have similar stability and are more populated than conformer **A** (destabilizing *gauche* interactions). Temperature dependence of the vicinal coupling constants CH_2-HC(1′) of (-)-**82** gave equilibrium constant K(**B**′ ⇄ **C**′) as a function of temperature. Van't Hoff plots lead to $\Delta\bar{H}_r(\mathbf{B}' \rightleftarrows \mathbf{C}') = +0.7 \pm 0.1$ kcal/mol and $\Delta\bar{S}_r(\mathbf{B}' \rightleftarrows \mathbf{C}') \cong 0$ cal/mol · K [93].

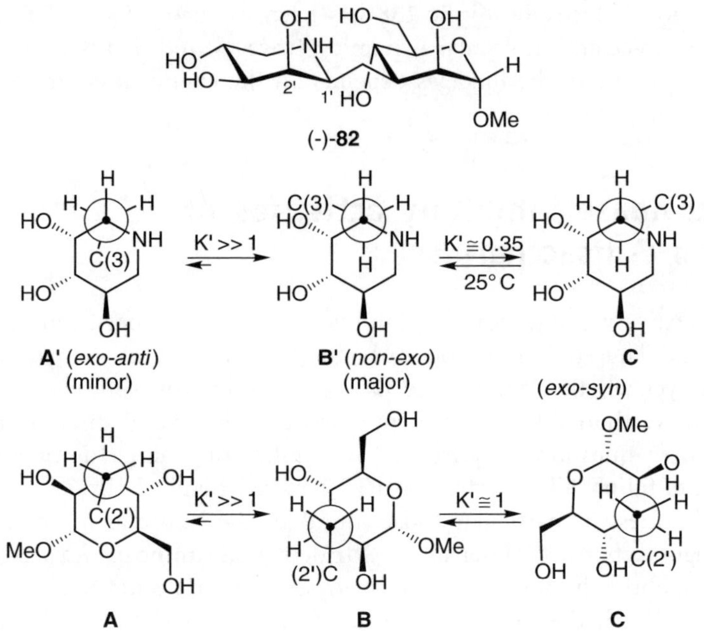

Figure 5.3 Conformer population in imino-*C*-(1→3)disaccharide (−)-**82** as determined by ^{1}H-NMR in D_2O

Recent studies [121] that combined molecular mechanics calculations and NMR experiments concluded that lactose mimetic **135** resembles lactose (**136**) in its conformational behaviour, with somewhat higher flexibility. Importantly, all the conformations sampled by **136** are also accessible to **135**. Transfer NOE experiments for **135** in the presence of mistletoe lectin have demonstrated that the glycosidic angles of **135** can access a larger range of conformational space than those of lactose (**136**). This suggests that details of the chemical nature of the disaccharide mimetic can affect properties of the bound state, here flexibility. Interestingly, the carba-analogue **137** of lactose (**136**) which lacks the *exo*-anomeric effect, resembles both the natural counterpart **136** and the imino-*C*-disaccharide mimetic **135** in terms of conformational behaviour. The conformational behaviour of β-O-D-Gal-(1→1)-α-D-Man (**139**) and the *C*-glycoside

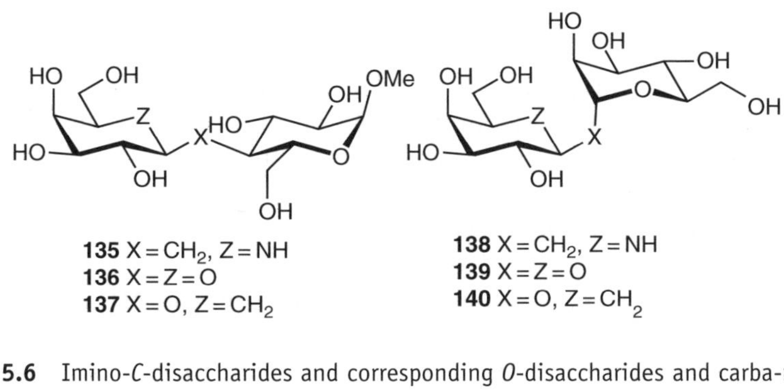

135 X = CH_2, Z = NH
136 X = Z = O
137 X = O, Z = CH_2

138 X = CH_2, Z = NH
139 X = Z = O
140 X = O, Z = CH_2

Structure 5.6 Imino-*C*-disaccharides and corresponding *O*-disaccharides and carba-mannosides

(**140**) and imino-*C*-glycoside mimic **138** has been evaluated by J/NOE NMR studies, molecular mechanics and molecular dynamics. It was found that the population distribution around the glycosidic linkages depends on the chemical nature of the acetal or pseudoacetal residue [122].

5.4 Glycosidase inhibitory activities of imino-*C*-disaccharides

Up to now, relatively few glycosidase inhibitory assays have been reported for imino-*C*-disaccharides. Therefore it is too early to present structure-activity relationships. It is not clear yet whether these disaccharide mimetics will show better potency and enzyme selectivity than other simpler monosaccharide or/and disaccharide mimetics. Answers to these questions require the availability of a large number of imino-*C*-disaccharides fulfilling all the structural and stereochemical diversity of natural *O*-linked disaccharides, in particular those found in the nonreducing end of biological significant oligosaccharides. Whereas 3,7,8-trideoxy-3,7-imino-D-*threo*-L-*galacto*-octitol (**141**) is a competitive inhibitor of β-glucosidases from almonds and from *Caldocellum saccharolyticum* [123], the more complicated imino-*C*-galactosides **37** and **38** showed neither significant inhibition of these enzymes, nor did they inhibit other commercially available glycosidases [84].

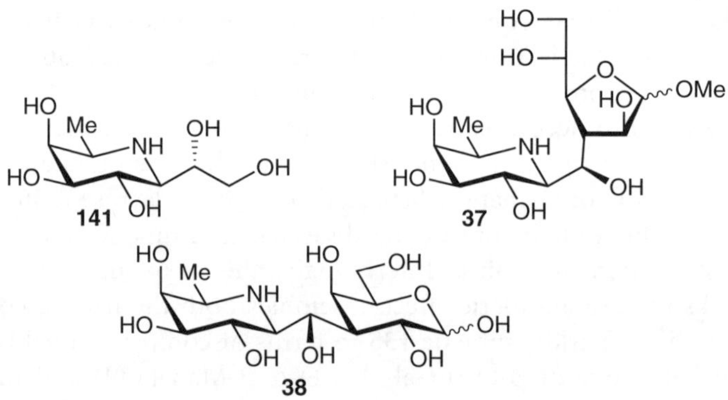

Structure 5.7 Examples of imino-*C*-disaccharides

Among the racemic imino-*C*-disaccharides **59**, **60**, **63** and **64** (Scheme 5.3) in which a 2,3-dihydroxypyrrolidine unit is linked to position C(3) of a galactose moiety through a hydroxymethano group, only **60** that mimics the mannopyranosyl cation intermediate during the hydrolysis of an α-mannopyranoside has a weak, but specific (no other glycosidase being inhibited) α-mannosidase inhibitory activity (Table 5.1) [91]. Interestingly, the D-iminomannosides **142**, **143**, **144** and **145** inhibit neither α-mannosidases, nor β-mannosidase from *Helix promatia*. It was surprising that mimetics of β-D-mannopyranosides **142**, **143**, **144**, **145** and **146** (Table 5.2) showed significant inhibitory activities toward amyloglucosidase from *Aspergillus niger* [69].

Table 5.1 D-iminomannose and analogues *C*-linked to monosaccharides: inhibitory activities toward α-mannosidase from almond **(A)** and from Jack bean (**B**) at 1 mM concentration, optimal pH, 35°C

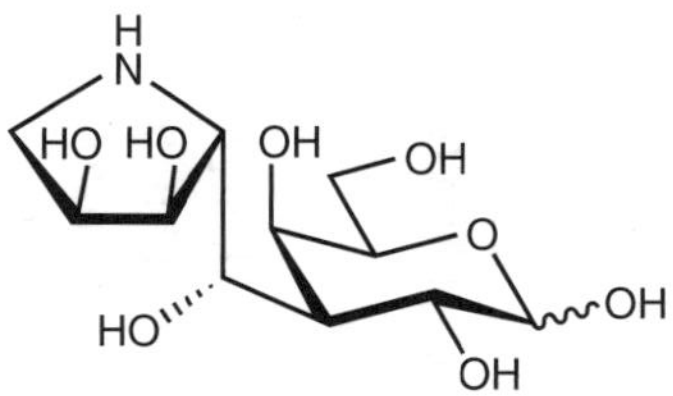 **60** [89]	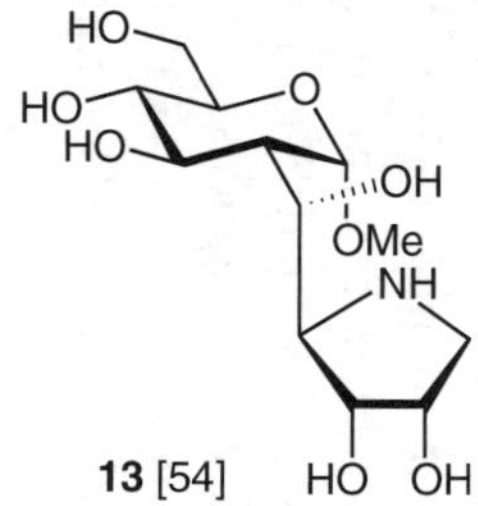 **13** [54]	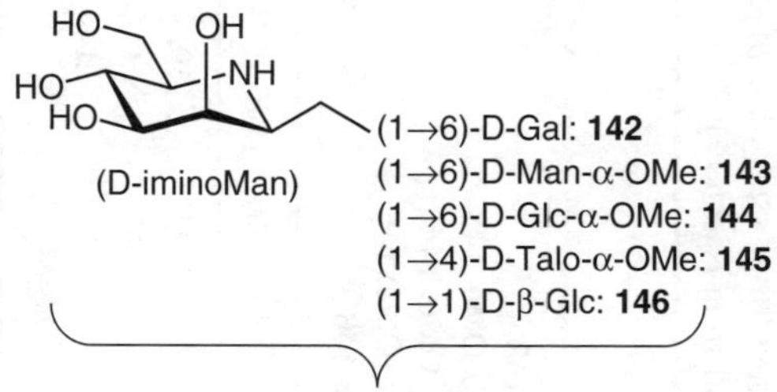
(A): 27% (B): 43%	26% 39%	No inhibition "

Table 5.2 Inhibitors of amyloglucosidase from *Aspergillus niger* (**C**), of α-galactosidase from coffee beans (**D**), from *Aspergillus niger* (**E**) and from *E. coli* (**F**)[a,b]

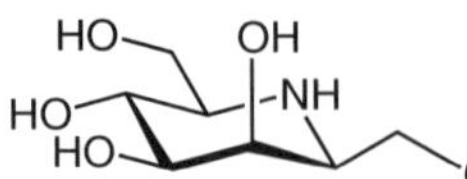

(D-iminoMan)

(1→6)-D-Gal: **142**
(1→6)-D-Man-α-OMe: **143**
(1→6)-D-Glc-α-OMe: **144**
(1→4)-D-Talo-α-OMe: **145**
(1→1)-D-β-Glc: **146**

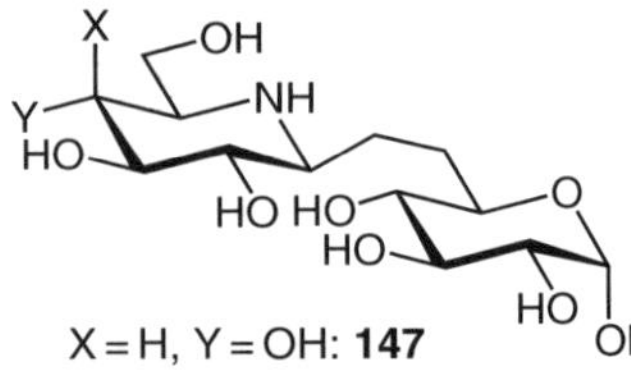

X = H, Y = OH: **147**
X = OH, Y = H: **148**

	142 [69]	**143** [69]	**144** [69]	**145** [69]	**146** [69]	**147** [63]	**148** [63]
C: IC_{50} (μM):	N.I.	25	12(21)	150	26	5(5)	N.I
D: IC_{50} (μM):	N.I	N.I	N.I	N.I	N.I	N.I	0.77(0.09)
E: IC_{50} (μM):	N.I	N.I	N.I	N.I	N.I	N.I	10(2)
F: IC_{50} (μM):	N.I	N.I	228(630)	N.I	N.I	310(200)	0.2(0.2)

[a] IC_{50} values are given in μM (K_i in μM in parenthesis).
[b] N.I. = no inhibition at 1 mM concentration.

The D-iminoglucopyranoside **147** is a better inhibitor of the amylase, although it is not an α-D-glucopyranoside but a β-D-glucopyranoside! As expected, the β-D-iminogalactoside **148** is only recognized by α-galactosidases (Table 5.2) [69]. The imino-*C*-disaccharide (−)-**82** which mimics an α-D-mannopyranoside (1→3)-linked to an α-D-mannopyranoside is neither recognized by α-mannosidases nor by β-mannosidases, α- and β-glucosidases, α- and β-galactosidases, amylases, isomaltases, α − *N*-acetylgalactoaminidases, α-*N*-acetylglucosaminidases at 1 mM concentration. However, α-L-fucosidase from bovine liver (K_i = 100 μM) and α-L-fucosidase from human placenta (58 per cent inhibition at 1 μM) are inhibited by (−)-**82**.

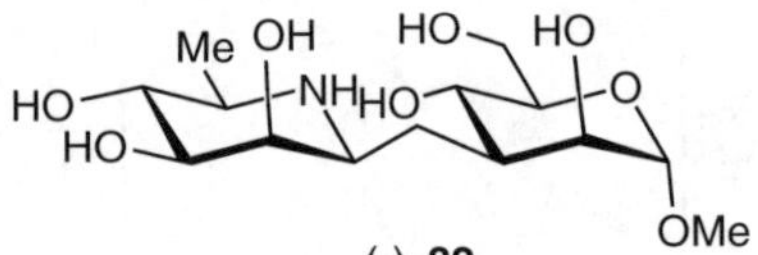

(-)-**82**

Structure 5.8 An α-L-fucosidase inhibitor

5.5 Efficient combinatorial method for the discovery of glycosidase inhibitors

As none of the imino-*C*-disaccharides evaluated thus far did present potent inhibitory activities toward glycosidases (too weak inhibitors, lack of selectivity towards various types of glycosidases), we turned our attention to a more classical approach which consists of generating a large collection of disaccharide mimetics that are much simpler

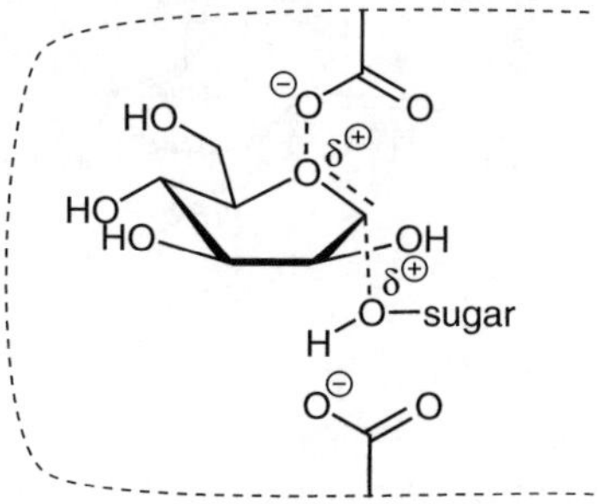

149

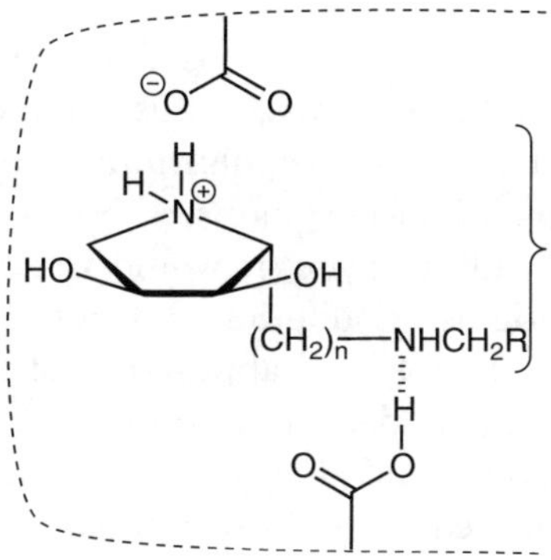

150

Figure 5.4 Design of mimetics of a transition state or intermediate structure of an α-mannosidase-catalysed hydrolysis of an α-D-mannopyranoside

in terms of structure and polyfunctionality. Furthermore, to become a drug, a good inhibitor must satisfy a number of conditions [124] such as membrane permeability, which often requires the presence of lipophilic groups. As we are interested in finding potent and selective α-mannosidase inhibitors, we envisioned that our synthetic targets should be (2*R*,4*R*,4*S*)-2-(aminoalkyl)-3,4-dihydroxypyrrolidines **150** in which the dihydroxypyrrolidine mimics the mannosyl cation intermediate formed during the hydrolytic process (transition state **149**, Figure 5.4) and the R groups attached to the aminoalkyl moiety would mimic the oligosaccharide to which the α-mannopyranoside is attached. The length of the aminoalkyl group ($n = 1, 2, 3$) must be chosen to give the best fit for **150** with the two carboxyl groups of the active site of the α-mannosidase, the latter interacting with the two amine moieties of **150** (see, however, [125]).

We envisioned also that imines of type **151** could be used as equivalents of diamines **150** for their ability to interact with α-mannosidases and to inhibit them. The latter could in principle be formed in equilibrium with diamines of type **152** and aldehydes **153**. If this mixture could be done in the presence of the α-mannosidases, a very efficient method would be found for the quick discovery of α-mannosidase inhibitors. The inhibition assays would thus be carried out on these mixtures directly. Such assays performed in the wells of multiwell plates would allow the screening of a large number of diamines and aldehydes using small amounts as each assay needs less than 0.2 ml of solution of imines in the mM range.

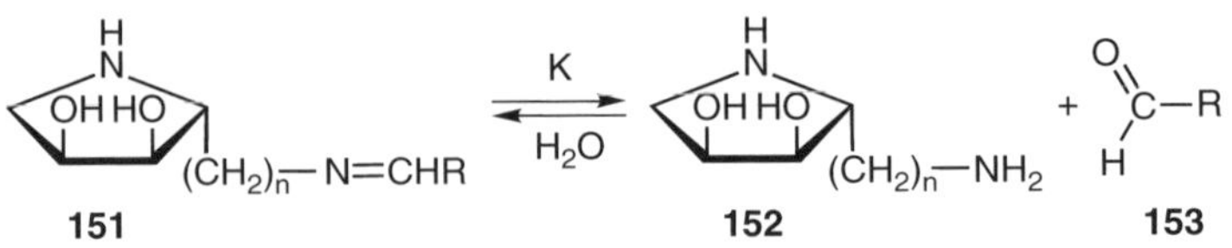

$$K = \frac{[\mathbf{151}][H_2O]}{[\mathbf{152}][\mathbf{153}]} \quad \text{(eq. (1))}$$

Structure 5.9 Dynamic libraries of imines

One of the difficulties with this plan [126] is that the equilibrium constants for the formation of imines might be too small. Thus, in the reaction of amines with aldehydes to form imines and water, the equilibrium would have to be obtained at relatively high amine and aldehyde concentrations. Although this method would require large amounts of each amine and aldehyde, it would not be a problem if, under the pH conditions of the enzymatic assay, dilution did not lead to fast re-equilibration with the consequence of nearly complete disappearance of the imines. Alternatively, if equilibrium was established rapidly under dilute conditions, the mixtures of imines that result from a combination of amine and aldehyde sublibraries could be incubated with the enzyme. The enzyme is expected to bind preferentially to the imine that is the best inhibitor (the solution forms a dynamic library [126] of imines) and thus a rapid assay of a large number of imines should be possible. We have observed that the unprotected diamine **154** reacts rapidly with all kinds of aldehydes **153** at pH 8 and these reagents equilibrate with the corresponding imines **151**. For instance, a solution of **154** (5 mM) in water generates imine **155** in less than 1 hour at 25°C. ^{1}H-NMR spectroscopy showed that less than 5 per cent of the solution consists of each of the reagents **154** and PhCHO, which suggests a minimum equilibrium constant $K > 4.2 \times 10^6$ (equation (1)) [58].

Under similar conditions, a solution of diamine **156** (2 mM) and benzaldehyde (8 mM) in water equilibrates with imine **157** after 24 hours at 25°C (Scheme 5.12). The ^{1}H-NMR spectrum of this solution shows less than 10 per cent compound **156** to be present and thus suggests a minimum equilibrium constant of $K > 8 \times 10^4$. In contrast, Boc-protected pyrrolidines derived from **154** and **156** did not react with aldehydes in solution at pH 8. As expected [127], the unprotected secondary amine β-amino and γ-amino moieties in diamines **154** and **156**, respectively, autocatalyse their addition to aldehydes to give first a α-hydroxyamine (e.g. **159**) and then the products by elimination of water from these intermediates (Scheme 5.12). We also examined the reactivity of pyrrol-2-carbaldehyde towards all kinds of primary amines and found that it equilibrates with the corresponding imines in D_2O at 25°C. ^{1}H-NMR spectroscopy after 24 hours gave $K > 2.2 \times 10^4$ pyrrol-2-carbaldehyde and benzylamine.

A 0.25 mM solution of diamine **154** leads to 58 per cent inhibition of α-mannosidase form Jack beans at the optimal pH value (pH = 5; 0.02 U ml^{-1} enzyme; substrate is *p*-nitrophenyl α-mannopyrannoside). This enzyme was chosen as it is a useful model for mammalian α-mannosidases such as Golgi α-mannosidase II [128]. A mixture of **154** (0.25 mM) and ethanal, propanal, butanal, pentanal and hexanal (0.25 mM each)

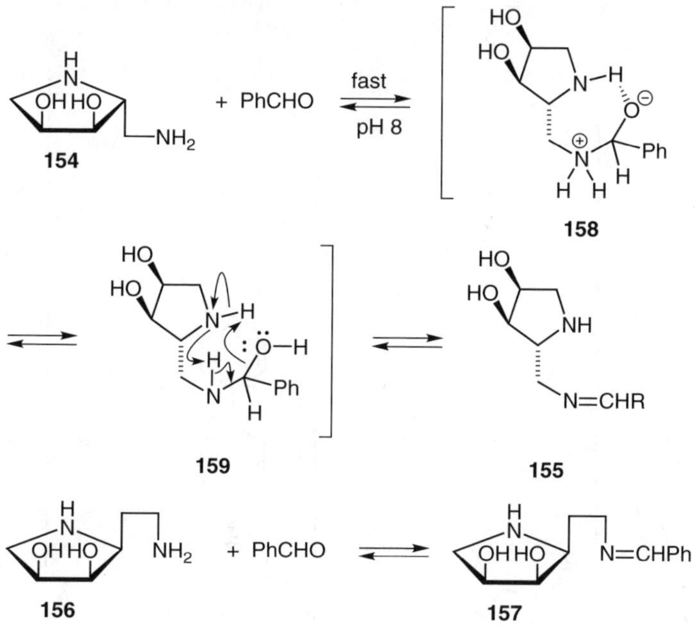

Scheme 5.12 Interpretation of the assistance of neighbouring secondary amine to imine formation

was left at 25°C for 24 hours (pH 8) and then buffered (pH 5). The inhibitory activity of this mixture (about 65 per cent) was not significantly better than for pure **154**. When mixtures containing aromatic aldehydes (0.25 mM) were prepared, significantly higher inhibitory activity was detected. The method described above allowed one to find that diamine **154** and aromatic or heteroaromatic aldehydes gave imines with the best inhibitory activity. For the best imines we prepared the corresponding diamines following Scheme 5.13. The diamines **160R′** obtained in this way had inhibitory activities toward α-mannosidases form Jack bean and from almonds [129] that paralleled the inhibitory activities observed for the amino-imines [58]. Examples of the best inhibitors obtained in this way are given in Table 5.3.

N-Benzylation of the primary amine moiety of **154** increases not only its inhibitory activity but also its selectivity toward α-mannosidases [59b]. This was also observed

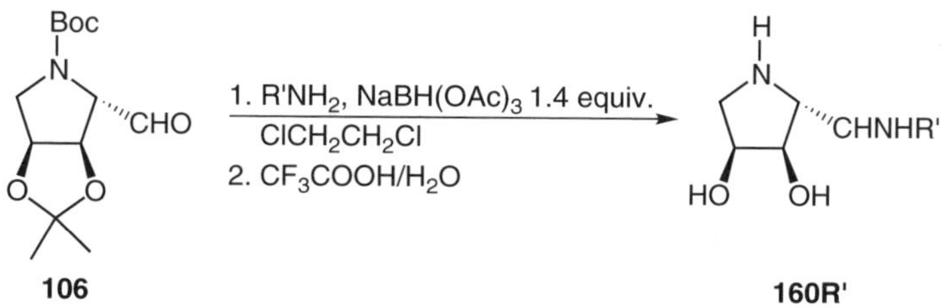

Scheme 5.13 Preparation of α-mannosidase inhibitors

Table 5.3 Examples of α-mannosidase inhibitors (α-mannosidase from Jack bean, optimal pH, all competitive, IC_{50} and K_i values in μM)

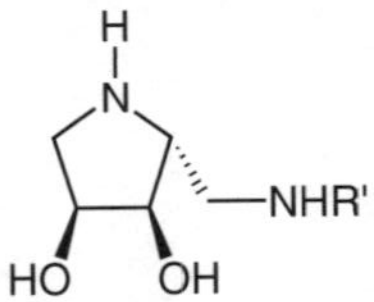

160R'

R':	H	CH_2Ph	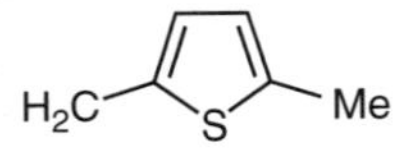	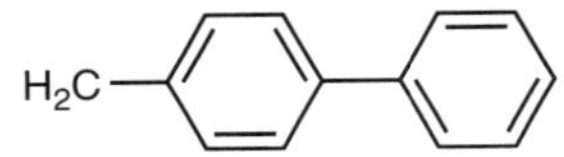
IC_{50}:	170	60	10	5
K_i:	53	7.4	3.5	2.3

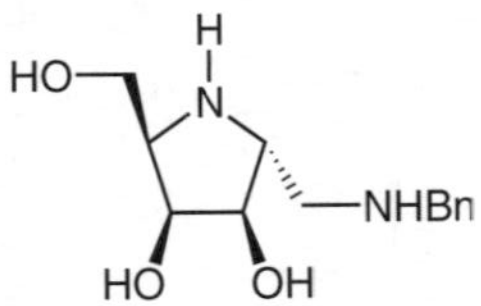

161

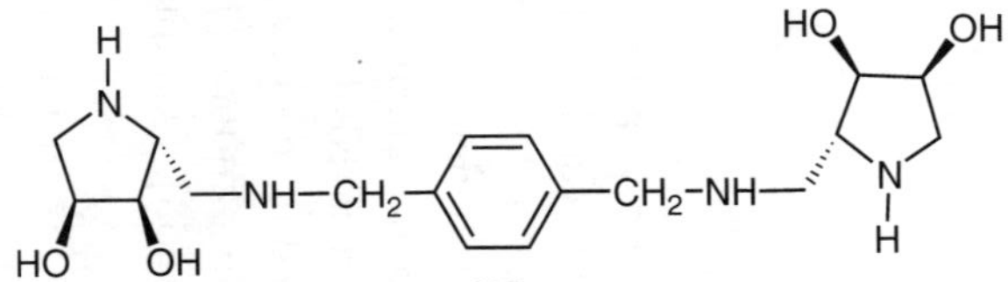

162

	161	162
IC_{50}:	6.2	50
K_i:	1.2	12

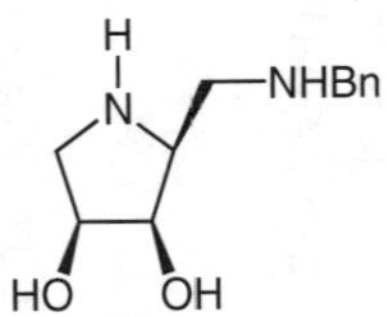

163

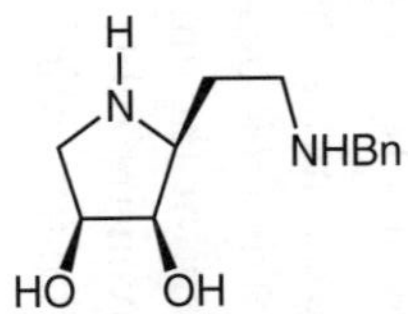

164

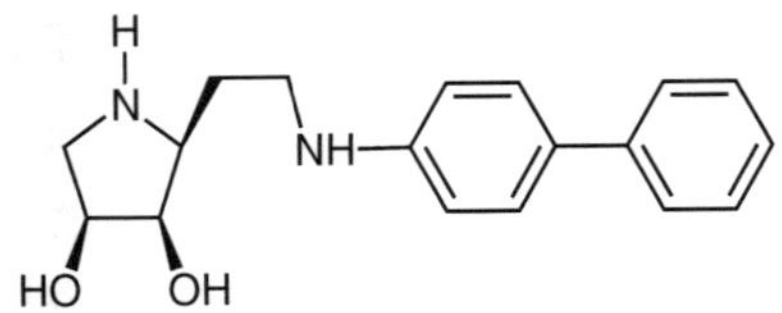

165

	163	164	165
IC_{50}:	360	620	360
K_i:	8.7	360	102

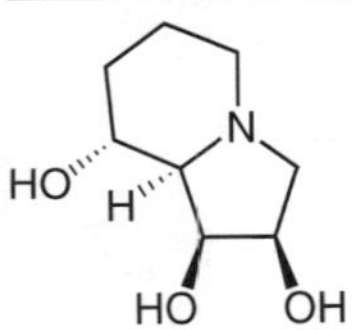

11

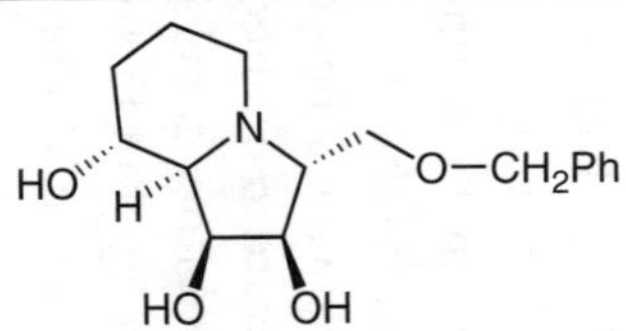

166

	11	166
IC_{50}:	0.4	0.05

Table 5.4 Substituted pyrrolidine-3,4-diol derivatives made form 1-phenyl- and 1,2-diphenyl-glycinols and their inhibitory activities toward α-mannosidase from Jack bean (IC_{50} and K_i values in μM)

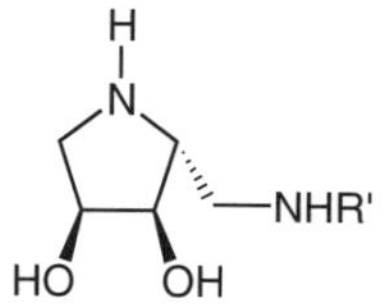

R':	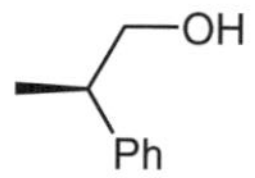	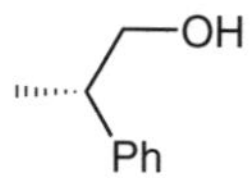	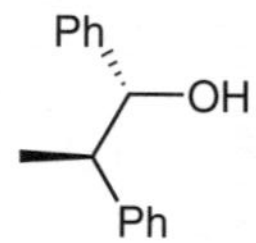	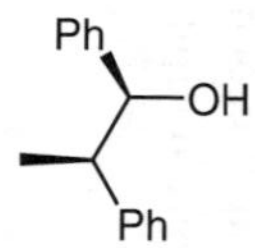	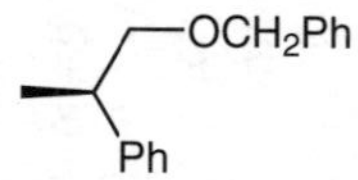	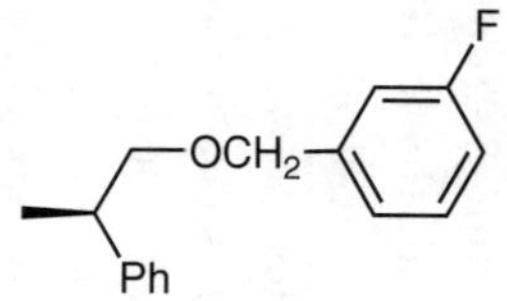
	15	**167**	**168**	**169**	**170**	**171**
IC:50	0.7	100	110	128	58	30
K_i:	0.135				16	7

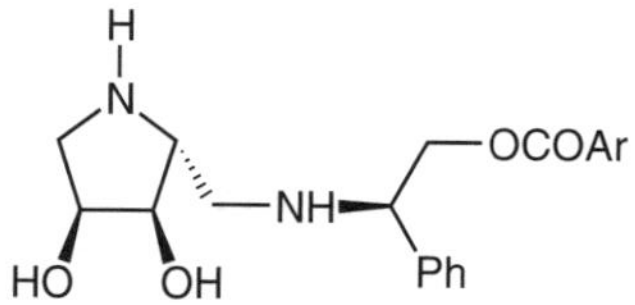

Ar:	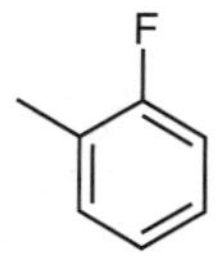	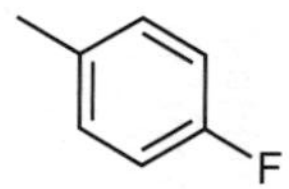	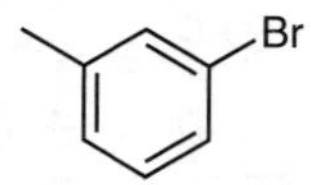	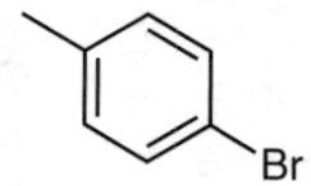
	172	**173**	**174**	**16**
IC:50	63	60	ca. 80	55
K_i:	24	19		

	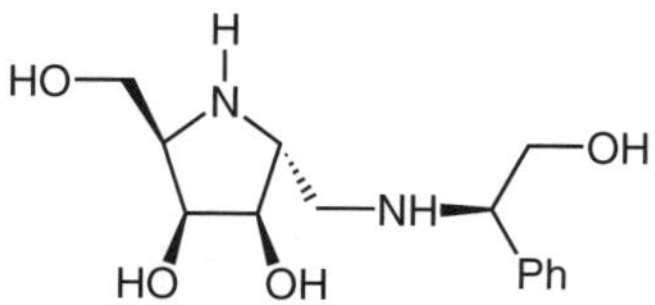	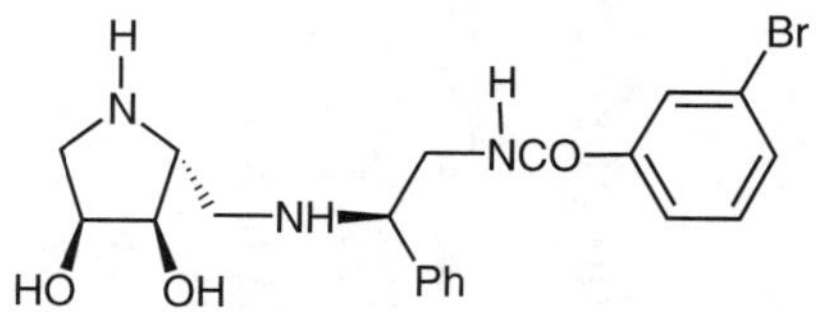
	175	**176**
IC:50	4.2	89
K_i:	1.3	

for (2*R*,3*R*,4*S*,5*R*)-(2-aminomethyl)-5-(hydroxymethyl)pyrrolidine-3,4-diol (see **161** [59g]). Compound **162** prepared via reductive amination of aldehyde **106** with 1,4-di(aminomethyl)benzene was found to be a better inhibitor of α-mannosidases than **154**(= **160R′** with R′ = H) [130]). The 2-epimer of **160R'** are less active as shown with **163**. Analogues of **160R'** with longer aminoalkyl chain are also less active, as shown with **164** and **165**. These data must be compared with those obtained for swainsonine (**11**) [4a] and derivative **166** [131].

Applying the method of Scheme 5.13 to phenylglycinols a new series of α-mannosidase inhibitors was made (Table 5.4). Interestingly, triol **15** was a much better inhibitor than **160R′** with R′ = Bn and diastereomer **167**. Etherification of the primary hydroxyl group of **15**, giving derivatives **170** and **171**, or introduction of a supplementary phenyl substituent (**168, 169**) reduced the inhibitory activity. Benzoic esters **172–174** and **16** were also less active than **15**. Furthermore, contrary to the pair **160R′** with R′ = Bn and **161** (Table 5.3) which showed that the (5*R*)-5-hydroxymethyl substituent improves the α-mannosidase inhibitory activity, the data for the pair of compounds **15** and **175** showed a decrease of activity when substituting **15** with (5*R*)-5-hydroxymethyl group. It appears, therefore, that the hydroxymethyl group attached at the benzylic centre of **160R′** with R′ = Bn is much more important than a 5-hydroxymethyl group for the α-mannosidase inhibitory activity.

5.6 Antitumour activity of new α-mannosidase inhibitors

Inhibitors **15, 167, 171-174, 16** and **176** were evaluated for their effects on human LN18 and LNZ308 glioblastoma cells, melanoma Me275 and Me237 cells and human fibroblasts (model for nontumoural cells). Whereas swainsonine (**11**) did not affect the growth of these tumour cells, these compounds showed a time- and concentration-dependent inhibition of growth of the tumour cells but much less of fibroblasts. The most active compound was ester **16**. As **15** was much less active than **16**, it is possible that cellular esterases are able to release the more hydrophilic triol **15**. The evaluation of the incorporation of [^{3}H]-leucine and [^{3}H]-thymidine following exposure of the cells to **16** demonstrated that it inhibits thymidine incorporation (DNA synthesis) at a slightly lower concentration and to a higher extent than inhibiting leucine incorporation (protein synthesis). These results suggest that **16** acts initially by inhibiting DNA synthesis, then the rate of protein synthesis decreases resulting in diminished cell survival [60].

5.7 Conclusion

Efficient synthetic methods to prepare disaccharide mimetics that link iminosugars (polyhydroxylated pyrrolidines or piperidines) to a monosaccharide unit through methano or hydroxymethano linker have been developed. These methods should now be applied to generate the molecular and stereochemical diversity required for finding potent and specific inhibitors of glycosidases. Studies on the conformer populations of imino-*C*-disaccharides have shown that they can adopt conformations similar to those available to the natural *O*-linked disaccharides. The few imino-*C*-disaccharides made thus far and that are supposed to mimic transition states or intermediates of the hydrolytic process catalysed by α-mannosidases have not yet led to interesting inhibitory activities. A higher molecular diversity is probably required for attaining more interesting inhibitors. Nevertheless, simpler compounds than imino-*C*-disaccharides such as derivatives of 2-(aminomethyl)pyrrolidine-3,4-diol are potent and selective inhibitors of α-mannosidases. For instance pyrrolidine **15** has an inhibition constant $K_i = 135\,nM$ for its competition of α-mannosidase from Jack bean. Its ester (2*S*)-2-({[(2*R*,3*S*,4*S*)-3,4-dihydroxypyrrolidin-2-yl]methyl}amino)-2-phenylethyl] 4-bromobenzoate, which is more lipophilic, showed increased bioavailability with improved growth inhibitory activity for human glioblastoma (brain cancer) and melanoma (skin cancer) cells, whereas fibroblasts growth was much less inhibited. Swainsonine had a much weaker tumour cell inhibitory effect than these new α-mannosidase inhibitors which can be considered as 'simplified' imino-*C*-disaccharides.

References

1. Varki, A. (1993) *Glycobiology*,**3**, 97–130.
2. Roth, J. (2002) *Chem. Rev.*, **102**, 285–303.
3. Chrispeels, M., Hindsgaul, O., Paulson, J. C., Lowe, J., Manzi, A., Powell, L. and van Halbeek, H. (1999) *Essentials of Glycobiology* (Varki, A., Cummings, R., Esko, J., Freeze, H., Hart, G. G. and Marth, J. eds), Cold Spring Harbor Laboratory Press, Cold Spring Harbor, New York.
4. Puri, A., Rawat, S. S., Lin, H-M. J., Finnegan, C. M., Mikovits, J., Ruscetti, F. W. and Blumenthal, R. (2004) *AIDS*, **18**, 849–858.
5. See, for example: Goffard, A. and Dubuisson, J. (2003) *Biochimie*, **85**, 295–301; Greimel, P., Spreitz, J., Stütz, A. E. and Wrodnigg, T. M. (2003) *Curr. Top. Med. Chem.*, **3**, 513–523.
6. See, for example: Imperiali, B. and O'Connor, S. E. (1998) *Pure Appl. Chem.*, **70**, 33–40; Stevens, F. J. and Argon, Y. (1999) *Sem. Cell. Develop. Biology*, **10**, 443–454; Parodi, A. J. (2000) *Biochem. J.*, **348**, 1–13; Saxon, E. and Bertozzi, C. R. (2001) *Ann. Rev. Cell. Develop. Biol.*, **17**, 1–23; Roth, J., Ziak, M. and Zuber, C. (2003) *Biochimie*, **85**, 287–294; von Anken, E. and Braakman, I. (2005) *Crit. Rev. Biochem. Molec. Biol.*, **40**, 191–228; Ceriotti, A., Duranti, M. and Bollini, R. (1998) *J. Experiment. Botany*, **49**, 1091–1103; Imperiali, B. and O'Connor, S. E. (1999) *Curr. Opin. Chem. Biol.*, **3**, 643–649; Dejgaard, S., Nicolay, J., Taneri, M., Thomas, D. Y. and Bergeron, J. J. M. (2004) *Curr. Issues Mol. Biol.*, **6**, 29–42; Bosques, C. J., Tschampel, S. M., Woods, R. J. and Imperiali, B. (2004) *J. Am. Chem. Soc.*, **126**, 8421–8425; Zhang, K. and Kaufman, R. J. (2006) *Neurology*, **66**, S102–S109; Mitra, N., Sinha, S., Ramya, T. N. C. and Surolia, A. (2006) *Trends Biochem. Sciences*, **31**, 156–163.
7. See, for example: Rudd, P. M., Wormald, M. R. and Dwek, R. A. (1999) *Trends Glycoscience Glycotechnol.*, **11**, 1–21; Rudd, P. M., Merry, A. H., Wormald, M. R. and Dwek, R. A. (2002) *Curr. Opin. Struct. Biol.*, **12**, 578–586; Arnold, S. M. and Kaufman, R. J. (2003) *New Compreh. Biochem.*, **38**, 411–432; Alder, N. N. and Johnson, A. E. J. (2004) *Biol. Chem.*, **279**, 22787–22790.
8. Roth, J. (2002) *Chem. Rev.*, **102**, 285–303; Betenbaugh, M. J., Tomiya, N., Narang, S., Hsu, J. T. A. and Lee, Y. C. (2004) *Curr. Opin. Struct. Biol.*, **14**, 601–606; Trombetta, E. S. (2003) *Glycobiology*, **13**, 77R–91R.

9. Schachter, H. and Brockhausen, I. (1992) *Glycoconjugates*, 263–332; Dennis, J. W. and Zhuang, P. (1993) *Trends Glycoscience Glycotechnology*, **5**, 287–295; Brockhausen, I. (1995) *New Comprehensive Biochem.*, **29a**, 201–259; Marth, J. D. (1999) *Essentials of Glycobiology* (Varki, A., ed.) Cold Spring Harber Lab. Press, Cold Spring Harbor, New York, USA, pp.101–113; Brockhausen, I. (2000) *Carbohydrates in Chemistry and Biology* (Ernst, B., Hart, G. W. and Sinaÿ, P., eds), Wiley-VCH Verlag GmbH, Weinheim, Germany, Vol. 3, pp.313–328; Van den Steen, P., Rudd, P. M., Dwek, R. A. and Opdenakker, G. (1998) *Crit. Rev. Biochemistry Mol. Biology*, **33**, 151–208; Endo, T. (2004) *Glycoconj. J.*, **21**, 3–7; Wopereis, S., Lefeber, D. J., Morava, E. and Wevers, R. A. (2006) *Clin. Chem.*, **52**, 574–600.
10. Ferguson, M. A., Homans, S. W., Dwek, R. A. and Rademacher, J. W. (1988) *Science*, **239**, 753–759; Kojima, N., Shiota, M., Sadahira, Y., Handa, K. and Hakamori, S. (1992) *J. Biol. Chem.*, **267**, 17264–17270; Baumann, N. A., Menon, A. K. and Rancour, D. M. (2000) *Carbohydrates in Chemistry and Biology* (Ernst, B., Hart, G. W. and Sinaÿ, P., eds) Wiley-VCH Verlag GmbH, Weinheim, Germany, Vol. 4, pp. 757–769; McConville, M. J. and Menon, A. K. (2000) *Mol. Membrane Biol.*, **17**, 1–16; Low, M. G. (2000) *Frontiers Mol. Biol.*, **27**, 210–238; Campbell, A. S. (2001) *Glycoscience* (Fraser-Reid, B. O., Tatsuta, K. and Thiem, J., eds), Springer-Verlag, Berlin, Germany, Vol. 2, pp.1695–1777; Menon, A. K. (2004) *Encyclopedia of Biological Chemistry* (Lennarz, W. J. and Lane, M. D., eds) Elsevier, Ed. Oxford, UK, Vol. 2, pp.308–311.
11. Ishida, H. (2000) *Carbohydrates in Chemistry and Biology* (Ernst, B., Hart, G. W. and Sinaÿ, P., eds), Wiley-VCH Verlag GmbH, Weinheim, Germany, Vol. 1, pp. 305–317; Lloyd, K. O. and Furukawa, K. (1998) *Glycoconj. J.*, **15**, 627–636; Smith, G. L., Dodson, K., Waksman, G. and Hultgren, S. J. (2005) *Struct. Biol. Bacterial Pathogenesis*, 37–48; Abe, Y. and Harata, K. (2005) *Polysaccharides*, 2nd edition (Dumitriu, S., ed.), Marcel Dekker, Inc., New York, USA, pp.743–771; Hakomori, S. (2004) *Encyclopedia of Biological Chemistry* (Lennarz, W. J. and Lane, M. D., eds) Elsevier Ltd., Oxford, UK, Vol. 2, pp. 261–265.
12. See, for example: Rudd, P. M. and Dwek, R. A. (1997) *Crit. Rev. Biochem. Mol. Biol.*, **32**, 1–100; Welinder, K. G. and Tams, J. W. (1995) *Progr. Biotechnol.*, **10**, 205–210; Wormald, M. R. and Dwek, R. A. (1999) *Structure (London)*, **7**, R155–R160; Khan, R. H., Rasheedi, S. and Haq, S. K. (2003) *J. Biosciences (Bangalore, India)*, **28**, 709–714; Yusa, A., Kitajima, K. and Habuchi, O. (2005) *Biochem. J.*, **388**, 115–121; Zhu, L., Jang, G.-F., Jastrzebska, B., Filipek, S., Pearce-Kelling, S. E., Aguirre, G. D., Stenkamp, R. E., Acland, G. M. and Palczewski, K. (2004) *J. Biol. Chem.*, **279**, 53828–53839; Wujek, P., Kida, E., Walus, M., Wisniewski, K. E. and Golabek, A. A. (2004) *J. Biol. Chem.*, **279**, 12827–12839; Watanabe, I., Zhu, J., Recio-Pinto, E. and Thornhill, W. B. (2004) *J. Biol. Chem.*, **279**, 8879–8885; Sola, R. J. and Griebenow, K. (2006) *FEBS letters*, **580**, 1685–1690.
13. Dwek, R. A., Butters, T. D., Platt, F. M. and Zitzmann, N. (2002) *Nature Reviews Drug Discovery*, **1**, 65–75; Elbein, A. D. (2004) *Encyclopedia of Biological Chemistry* (Lennarz, W. J. and Lane, M. D., eds), Elsevier Ltd., Oxford, UK, Vol. 3, pp. 500–503.
14. Stütz, A. E. (ed.) (1999) *Iminosugars as Glycosidase Inhibitors, Nojirimycin and Beyond*, Wiley-VCH-Verlag GmbH, Weinheim, Germany; Ernst, B., Kolb, H. C. and Schwardt, O. (2006) *Organic Chemistry of Sugars* (Levy, D. E. and Fügedi, P., eds), CRC Press LLC, Boca Raton, Florida, USA, pp. 803–861; Postema, M. H. D., Piper, J. L. and Betts, R. L. (2005) *Synlett*, 1345–1358.
15. (a) Mooradian, A. D. and Thurman, J. E. (1999) *Drugs*, **57**, 19–29; (b) Scott, L. J. and Spencer, C. M. (2000) *Drugs*, **59**, 521–549.
16. (a) Truscheit, E., Frommer, W., Junge, B., Müller, L., Schmidt, D. D. and Wingender, W. (1981) *Angew. Chem. Int. Ed. Engl.*, **20**, 744–761; (b) Clissold, S. P. and Edwards, C. (1988) *Drugs*, **35**, 214–243; (c) Chen, X., Fan, Y., Zheng, Y. and Shen, Y. (2003) *Chem. Rev.*, **103**, 1955–1977.

17. (a) Segal, P., Eliahou, H. E., Petzinna, D., Neuser, D., Brückner, A. and Spengler, M. (2005) *Clinical Drug Investigation*, **25**, 589–595; (b) Neuser, D., Benson, A., Brückner, A., Goldberg, R. B., Hoogwerf, B. J. and Petzinna, D. (2005) *Ibid.*, **25**, 579–587; (c) Moyers, S. B. (2005) *J. Am. Dietetic Association*, **105**, 948–959.
18. (a) Platt, F. M., Neises, G. R., Reinkensmeier, G., Townsend, M. J., Perry, V. H., Proia, R. L., Winchester, B., Dwek, R. A. and Butters, T. D. (1997) *Science*, **276**, 428–431; (b) Cox, T., Lachmann, R., Hollak, C., Aerts, J., van Weely, S., Hrebicek, M., Platt, F. M., Butters, T. D., Dwek, R., Moyses, C., Gow, I., Elstein, D. and Zimran, A. (2000) *Lancet*, **255**, 1481–1485; (c) Platt, F. M., Jeyakumar, M., Andersson, U., Priestmann, D. A., Dwek, R. A., Butters, T. D., Cox, T. M., Lachmann, R. H., Hollak, C., Aerts, J. M., van Weely, S., Hrebicek, M., Moyses, L., Gow, I., Elstein, D. and Zimran, A. (2001) *J. Inherit Metab. Dis.*, **24**, 275–290.
19. (a) Garman, E. and Laver, G. (2004) *Curr. Drug Targets*, **5**, 119–136; (b) Schmidt, A. C. (2004) *Drugs*, **64**, 2031–2046.
20. (a) von Itzstein, M., Wu, W. Y., Kok, G. B., Pegg, M. S., Dyason, J. C., Jin, B., Phan, T.-V., Smythe, M. L., White, H. F. and Oliver, S. W. (1993) *Nature*, **363**, 418–423; (b) von Itzstein, M., Wu, W. Y. and Jin, B. (1994) *Carbohydr. Res.*, **259**, 301–305; (c) von Itzstein, M. and Taylor, N. R. (1994) *J. Med. Chem.*, **37**, 616–624; (d) Ison, M. G., Gnann, J. W. Jr., Nagy-Agren, S., Treanor, J., Paya, C., Steigbigel, R., Elliott, M., Weiss, H. L. and Hayden, F. G. (2003) *Antiviral Ther.*, **8**, 183–190.
21. (a) Kim, C. U., Lew, W., Williams, M. A., Wu, H., Zhang, L., Chen, X., Escarpe, P. A., Mendel, D. B., Laver, W. G. and Stevens, R. C. (1998) *J. Med. Chem.*, **41**, 2451–2459; (b) Lew, W., Chen, X. and Kim, C. U. (2000) *Curr. Med. Chem.*, **7**, 663–672; (c) Karpf, M. and Trussardi, R. (2001) *J. Org. Chem.*, **66**, 2044–2051; (d) Abrecht, S., Harrington, P., Iding, H., Karpf, M., Trussardi, R., Wirz, B. and Zutter, U. (2004) *Chimia*, **58**, 621–629; (e) Ward, P., Small, I., Smith, J., Suter, P. and Dutkowski, R. (2005) *J. Antimicrob. Chemother.*, **55** (Suppl. 1), i5–i21.
22. (a) Sunkara, P. S., Bowlin, T. L., Liu, P. S. and Sjoerdsma, A. (1987) *Biochem. Biophys. Res. Commun.*, **148**, 206–210; (b) El Ashry, E. S. H., Rashed, N. and Shobier, A. H. S. (2000) *Pharmazie*, **55**, 251–262; (c) Asano, N. (2003) *Glycobiology*, **13**, 93R–104R; (d) Houston, T. A. and Blanchfield, J. T. (2003) *Mini-Reviews in Med. Chem.*, **3**, 669–678; (e) Sorbera, L. A., Castaner, J. and García-Capdevila, L. (2005) *Drugs of the Future*, **30**, 545–552; (f) Sunkara, P. S., Kang, M. S., Bowlin, T., Liu, P. S., Tyms, A. S. and Sjoerdsma, A. (1990) *Ann. New York Acad. Science*, **616**, 90–96.
23. (a) Gruters, R. A., Neefjes, J. J., Tersmette, M., de Goede, R. E., Tulp, A., Huisman, H. G., Miedema, F. and Ploegh, H. L. (1987) *Nature*, **330**, 74–77; (b) Papandreou, M. J., Barbouche, R., Guieu, R., Kieny, M.-P. and Fenouillet, E. (2002) *Mol. Pharmacol.*, **61**, 186–193; (c) Robina, I,; Moreno-Vargas, A. J., Carmona, A. T. and Vogel, P. (2004) *Curr. Drug. Metab.*, **5**, 329–361, and references cited therein.
24. Herr, R. R., Jahnke, H. K. and Argoudelis, A. D. (1967) *J. Am. Chem. Soc.*, **89**, 4808–4809.
25. Rerup, C. and Tarding, F. (1969) *Eur. J. Clin. Pharmacol.*, **7**, 89–96; Like, A. A. and Rossini, A. A. (1976) *Science*, **193**, 415–417.
26. Bhuyan, B. K., Fraser, T. J., Buskirk, H. H. and Neil, G. L. (1972) *Cancer Chemotherapy Reports. Part 1*, **56**, 709–720.
27. Wiley, P. F. (1980) *Anticancer Agents Based on Natural Products Models* (Cassidy, J. M. and Douros, J. D., eds), Academic Press, New York, USA, pp. 167–200; Yoshimori, M., Tajiri, H., Nakamura, K. and Ozaki, H. (1985) *Cancer & Chemotherapy*, **12**, 227–234.
28. Takatsuki, A. and Tamura, G. (1971) *J. Antibiot.*, **24**, 215–223.
29. Takatsuki, A., Kohno, K. and Tamura, G. (1975) *Agr. Biol. Chem.*, **39**, 2089–2091; Kuo, S. and Lampen, J. O. (1976) *Arch. Biochem. Biophy.*, **172**, 574–581; King, I. A. and Tabiowo, A. (1981) *Biochem. J.*, **198**, 331–338.

30. Hiss, D., Gabriels, G., Jacobs, P. and Folb, P. (1996) *Eur. J. Cancer*, **32A**, 2164–2172.
31. Yamada, M., Tomida, A., Yun, J., Cai, B., Yoshikawa, H., Taketani, Y. and Tsuruo, T. (1999) *Cancer Chemother. Pharmacol.*, **44**, 59–64; Noda, I., Fujieda, S., Seki, M., Tanaka, N., Sunaga, H., Ohtsubo, T., Tsuzuki, H., Fan, G.-K. and Saito, H. (1999) *Internat. J. Cancer*, **80**, 279–284.
32. Cusinato, F., Pighin, I., Luciani, S. and Trevisi, L. (2006) *Biochem. Pharmacol.*, **71**, 1562–1569.
33. Hohenschutz, L. D., Bell, E. A., Jewess, P. J., Leworthy, D. P., Pryce, R. J., Arnold, E. and Clardy, J. (1981) *Phytochemistry*, **20**, 811–814; Nash, R. D., Fellows, L. E., Dring, J. V., Stirton, C. H., Carter, D., Hegarty, M. P. and Bell, E. A. (1988) *Phytochemistry*, **27**, 1403–1404.
34. Molyneux, R. J., Roitman, J. N., Dunnheim, G., Szumilo, T. and Elbein, A. D. (1986) *Arch. Biochem. Biophys.*, **251**, 450–457.
35. Hu, L. H., Zou, H. B., Gong, J. X., Li, H. B., Yang, L. X., Cheng, W., Zhou, C. X., Bai, H., Guéritte, F. and Zhao, Y. (2005) *J. Nat. Prod.*, **68**, 342–348; Spearman, M. A., Ballon, B. C., Gerrard, J. M., Greenberg, A. H. and Wright, J. A. (1991) *Cancer Lett.*, **60**, 185–191.
36. (a) Humphries, M. J., Matsumoto, K., White, S. L. and Olden, K. (1986) *Cancer Res.*, **46**, 5215–5222; Ostrander, G. K., Scribner, N. K. and Rohrschneider, L. R. (1988) *Cancer Res.*, **48**, 1091–1094; (b) Yee, C. S., Schwab, E. D., Lehr, J. E., Quigley, M. and Pienta, K. J. (1997) *Anticancer Res.*, **17**, 3659–3663.
37. Pili, R., Chang, J., Partis, R. A., Mueller, R. A., Chrest, F. J. and Passaniti, A. (1995) *Cancer Res.*, **55**, 2920–2926.
38. Atsumi, S., Nosaka, C., Ochi, Y., Iinuma, H. and Umezawa, K. (1993) *Cancer Res.*, **53**, 4896–4899.
39. Chen, J. J., Duh, C. Y., Huang, H. Y. and Chen, I. S. (2003) *Helv. Chim. Acta*, **86**, 2058–2064.
40. Hu, L. H., Zou, H. B., Gong, J. X., Hai, B., Yang, L. X., Zhou, C. X., Bai, H., Wu, X. M., Cheng, W. and Zhao, Y. (2005) *Chin. Chem. Lett.*, **16**, 997–1000; Sharma, S. D., Sharma, M. L. and Rathee, R. (2005) *J. Indian Chem. Soc.*, **82**, 938–939.
41. Colegate, S. M., Dorling, P. R. and Huxtable, C. R. (1979) *Austr. J. Chem.*, **32**, 2257–2264; see also: Perrone, G. G., Barrow, K. D. and McFarlane, I. J. (1999) *Bioorg. Med. Chem.*, **7**, 831–835.
42. Dorling, P. R., Huxtable, C. R. and Colegate, S. M. (1980) *Biochem. J.*, **191**, 649–651; Elbein, A. D., Solf, R., Dorling P. R. and Vosbeck, K. (1981) *Proc. Natl. Acad. Sciences USA*, **78**, 7393–7397; Elbein, A. D. (1984) *CRC Crit. Rev. Biochem.*, **16**, 21–49; Cenci di Bello, I., Fleet, G., Namgoong, S. K., Tadano, K. and Winchester, B. (1989) *Biochem. J.*, **259**, 855–861.
43. Tulsiani, D. R., Harris, T. M. and Touster, O. (1982) *J. Biol. Chem.*, **257**, 7936–7939; Tulsiani, D. R., Broquist, H. P. and Touster, O. (1985) *Arch. Biochem. Biophys.*, **236**, 427–434.
44. Nakabayashi, S., Warren, C. D. and Jeanloz, R. W. (1988) *Carbohydr. Res.*, **174**, 279–289.
45. Humphries, M. J., Matsumoto, K., White, S. L. and Olden, K. (1986) *Proc. Natl. Aca. Sciences USA*, **83**, 1752–1756; Roberts, J. D., Klein, J. L., Palmantier, R., Dhume, S. T., George, M. D. and Olden, K.(1998) *Cancer Detect. Prevent.*, **22**, 455–462.
46. Dennis, J. W. (1986) *Cancer Res.*, **46**, 5131–5136; Dennis, J. W. and Laferte, S. (1987) *Cancer Metastasis Rev.*, **5**, 185–204.
47. (a) Yagita, M. and Saksela, E. (1990) *Scand. J. Immunol.*, **31**, 275–282; (b) Yagita, M., Noda, I., Maehara, M., Fujieda, S., Inoue, Y., Hashino, T. and Saksela, E. (1992) *Int. J. Cancer*, **52**, 664–672.
48. Seftor, R. E. B., Seftor, E. A., Gimes, W. J., Liotta, L. A., Stetler-Stevenson, W. G., Welch, D. R. and Hendrix, M. J. C. (1991) *Melanoma Res.*, **1**, 43–54; Fernandes, B., Sagman, M., Auger, M., Demetrio, M. and Dennis, J. W. 1991) *Cancer Res.*, **51**, 718–723; Olden, K., Breton, P., Grzegorzewski, K., Yasuda, Y., Gause, B. L., Oredipe, O. A., Newton, S. A. and White, S. L. (1991) *Pharmacol. Therapeutics*, **50**, 285–290.

49. Goss, P. E., Baptiste, J., Fenandes, B., Baker, M. and Dennis, J. W. (1994) *Cancer Res.*, **54**, 1450–1457; Goss, P. E., Baker, M. A., Carver, J. P. and Dennis, J. W.(1995) *Clin. Cancer Res.*, **1**, 935–944.
50. Mohla, S., White, S., Grzegorzewski, K., Nielsen, D., Dunston, G., Dickson, L., Cha, J. K., Asseffa, A. and Olden, K. (1990) *Anticancer Res.*, **10**, 1515–1522.
51. Goss, P. E., Reid, C. L., Bailey, D. and Dennis, J. W. (1997) *Clin. Cancer Res.*, **3**, 1077–1086.
52. Dennis, J. W., White, S. L., Freer, A. M. and Dime, D. (1993) *Biochem. Pharmacol.*, **46**, 1459–1466.
53. Shaheen, P. E., Stadler, W., Elson, P., Knox, J., Winquist, E. and Bukowski, R. M. (2005) *Invest. New Drugs*, **23**, 577–581.
54. Oredipe, O. A., Furbert-Harris, P. M., Laniyan, I., Green, W. R., Griffin, W. M. and Sridhar, R. (2003) *Cell. Mol. Biol.*, **49**, 1037–1048; see also: Klein, J.-L. D., Roberts, J. D., George, M. D., Kurtzberg, J., Breton, P., Chermann, J.-C. and Olden, K. (1999) *Brit. J. Cancer*, **80**, 87–95.
55. (a) Bimwala, R. M. and Vogel, P. (1992) *J. Org. Chem.*, **57**, 2076–2083; (b) Ferritto, R. and Vogel, P. (1994) *Tetrahedron:Asymmetry*, **5**, 2077–2092; (c) Zhu, Y.-H., Demange, R. and Vogel, P. (2000) *Tetrahedron:Asymmetry*, **11**, 263–282; (d) Pasquarello, C., Picasso, S., Demange, R., Malissard, M., Berger, E. G. and Vogel, P. (2000) *J. Org. Chem.*, **65**, 4251–4260; (e) Viodé, C. and Vogel, P. (2001) *J. Carbohydr. Chem.*, **20**, 733–746; (f) Jiménez-Barbero, J., Demange, R., Schenk, K. and Vogel, P. (2001) *J. Org. Chem.*, **66**, 5132–5138; (g) Demange, R., Bühlmann, C. and Vogel, P. (2003) *Helv. Chim. Acta*, **86**, 361–376; (h) Demange, R., Awad, L. and Vogel, P. (2004) *Tetrahedron:Asymmetry*, **15**, 3575–3585; (i) Steunenberg, P., Jeanneret, V., Zhu, Y.-H. and Vogel, P. (2005) *Tetrahedron:Asymmetry*, **16**, 337–346; (j) Awad, L., Riedner, J. and Vogel, P. (2005) *Chem. Eur. J.*, **11**, 3565–3573.
56. Navarro, I. and Vogel, P. (2002) *Helv. Chim. Acta*, **85**, 152–160.
57. Rodríguez-García, E., Brimble, M. A. and Vogel, P. (2006) *Eur. J. Org. Chem.*, 3845–3855.
58. Gerber-Lemaire, S., Popowycz, F., Rodriguez-García, E., Carmona-Asenjo, A. T., Robina, I. and Vogel, P. (2002) *ChemBioChem*, **3**, 466–470.
59. (a) Moreno-Vargas, A. J., Demange, R., Fuentes, J., Robina, I. and Vogel, P. (2002) *Bioorg. Med. Chem. Lett.*, **12**, 2335–2339; (b) Popowycz, F., Gerber-Lemaire, S., Rodríguez-García, E., Schütz, C. and Vogel, P. (2003) *Helv. Chim. Acta*, **86**, 1914–1948; (c) Moreno-Vargas, A. J., Schütz, C., Scopelliti, R. and Vogel, P. (2003) *J. Org. Chem.*, **68**, 5632–5640; (d) Moreno-Vargas, A. J., Robina, I., Demange, R. and Vogel, P. (2003) *Helv. Chim. Acta*, **86**, 1894–1913; (e) Carmona-Asenjo, A. T., Popowycz, F., Gerber-Lemaire, S., Rodríguez-García, E., Robina, I. and Vogel, P. (2003) *Bioorg. Med. Chem.*, **11**, 4897–4911; (f) Popowycz, F., Gerber-Lemaire, S., Schütz, C. and Vogel, P. (2004) *Helv. Chim. Acta*, **87**, 800–810; (g) Lysek, R. and Vogel, P. (2004) *Helv. Chim. Acta*, **87**, 3167–3181; (h) Moreno-Vargas, A. J., Carmona, A. T., Mora, F., Vogel, P. and Robina, I. (2005) *Chem. Commun.*, 4949–4951; (i) Cardona, F., Moreno, G., Guarna, F., Vogel, P., Schütz, C., Merino, P. and Goti, A. (2005) *J. Org. Chem.*, **70**, 6552–6555.
60. Fiaux, H., Popowycz, F., Favre, S., Schütz, C., Vogel, P., Gerber-Lemaire, S. and Juillerat-Jeanneret, L. (2005) *J. Med. Chem.*, **48**, 4237–4246.
61. Johnson, C. R., Miller, M. W., Golebiowski, A., Sundram, H. and Ksebati, M. B. (1994) *Tetrahedron* Lett., **35**, 8991–8994.
62. Martin, O. R., Liu, L. and Yang, F. (1996) *Tetrahedron Lett.*, **37**, 1991–1994.
63. Leeuwenburgh, M. A., Picasso, S., Overkleeft, H. S., van der Marel, G. A., Vogel, P. and van Boom, J. H. (1999) *Eur. J. Org. Chem.*, 1185–1189.
64. Dondoni, A., Giovannini, P. P. and Marra, A. (2000) *Tetrahedron Lett.*, **41**, 6195–6199.
65. Cheng, X., Kumaran, G. and Mootoo, D. R. (2001) *Chem. Commun.*, 811–812.
66. Duff, F. J., Vivien, V. and Wightman, R. H. (2000) *Chem. Commun.*, 2127–2128.

67. La Ferla, B., Cipolla, L., Peri, F. and Nicotra, F. (2001) *J. Carbohydr. Chem.*, **20**, 667–680.
68. Baudat, A. and Vogel, P. (1996) *Tetrahedron Lett.*, **37**, 483–484.
69. Johns, B. A., Pan, Y. T., Elbein, A. D. and Johnson, C. R. (1997) *J. Am. Chem. Soc.*, **119**, 4856–4865.
70. Argyropoulos, N. G. and Sarli, V. C. (2004) *Tetrahedron Lett.*, **45**, 4237–4240.
71. Cardona, F., Valenza, S., Picasso, S., Goti, A. and Brandi, A. (1998) *J. Org. Chem.*, **63**, 7311–7318.
72. Cardona, F., Salanski, P., Chmielewski, M., Valenza, S., Goti, A. and Brandi, A. (1998) *Synlett*, 1444–1446.
73. Cardona, F., Valenza, S., Goti, A. and Brandi, A. (1999) *Eur. J. Org. Chem.*, 1319–1323.
74. Cardona, F., Goti, A. and Brandi, A. (2003) *Org. Lett.*, **5**, 1475–1478.
75. Cardona, F., Robina, I. and Vogel, P. (2000) *J. Carbohydr. Chem.*, **19**, 555–571.
76. Marquis, C., Cardona, F., Robina, I., Wurth, G. and Vogel, P. (2002) *Heterocycles*, **56**, 181–208.
77. Dondoni, A., Giovannini, P. P. and Perrone, D. (2002) *J. Org. Chem.*, **67**, 7203–7214.
78. Robina, I. and Vogel, P. (2005) *Synthesis*, 675–702.
79. Kennedy, A., Nelson, A. and Perry, A. (2005) *Chem. Commun.*, 1646–1648.
80. Kennedy, A., Nelson, A. and Perry, A. (2004) *Synlett*, **6**, 967–970.
81. Luche, J.-L. (1978) *J. Am. Chem. Soc.*, **100**, 2226–2227; Gemal, A. L. and Luche, J.-L. (1981) *J. Am. Chem. Soc.*, **103**, 5454–5459.
82. Cha, J. K., Christ, W. J. and Kishi, Y. (1984) *Tetrahedron*, **40**, 2247–2255.
83. Donohoe, T. J., Blades, K., Moore, P. R., Waring, M. J., Winter, J. J. G., Helliwell, M., Newcombe, N. J. and Stemp, G. (2002) *J. Org. Chem.*, **67**, 7946–7956.
84. Baudat, A. and Vogel, P. (1997) *J. Org. Chem.*, **62**, 6252–6260.
85. (a) Black, K. A. and Vogel, P. (1986) *J. Org. Chem.*, **51**, 5341–5348; (b) Gasparini, F. and Vogel, P. (1990) *J. Org. Chem.*, **55**, 2451–2457; (c) Vogel, P., Fattori, D., Gasparini, F. and Le Drian, C. (1990) *Synlett*, 173–185; (d) Reymond, J.-L. and Vogel, P. (1990) *Tetrahedron:Asymmetry*, **1**, 729–736; (e) Forster, A., Kovac, T., Mosimann, H., Renaud, P. and Vogel, P. (1999) *Tetrahedron: Asymmetry*, **10**, 567–571.
86. Carrupt, P.-A. and Vogel, P. (1988) *J. Phys. Org. Chem.*, **1**, 287–298.
87. Vogel, P. (1997) *Chimica Oggi*, **25**, 37–43.
88. Zimmerman, H. E. and Traxler, M. D. (1957) *J. Am. Chem. Soc.*, **79**, 1920–1923.
89. Anh, N. T., El Kaïm, L., Thank, B. T., Maurel, F. and Flament, J. P. (1992) *Bull. Soc. Chim. Fr.*, **129**, 468–477.
90. Kraehenbuehl, K. and Vogel, P. (1995) *Tetrahedron Lett.*, **36**, 8595–8598.
91. (a) Kraehenbuehl, K., Picasso, S. and Vogel, P. (1997) *Bioorg. Med. Chem. Lett.*, **7**, 893–896; (b) Kraehenbuehl, K., Picasso, S. and Vogel, P. (1998) *Helv. Chim. Acta*, **81**, 1439–1479.
92. (a) Clauson-Kaas, N. (1947) *Acta Chem. Scand.*, **1**, 379–381; (b) Clauson-Kaas, N. and Tyle, Z. (1952) *Acta Chem. Scand.*, **6**, 667–670.
93. Marquis, C., Picasso, S. and Vogel, P. (1999) *Synthesis*, 1441–1452.
94. Allemann, S. and Vogel, P. (1994) *Tetrahedron*, **50**, 2469–2478.
95. Vogel, P. (2000) *Curr. Org. Chem.*, **4**, 455–480.
96. Horton, D., Roski, J. P. and Norris, P. (1996) *J. Org. Chem.*, **61**, 3783–3793; see also: (a) Witczak, Z. J., Chen, H. andKaplon, P. (2000) *Tetrahedron:Asymmetry*, **11**, 519–532; (b) Kadota, K., ElAzab, A. S., Taniguchi, T. and Ogasawara, K. (2000) *Synthesis*, 1372–1374; (c) Witczak, Z. J., Kaplon, P. and Kolodziej, M. (2002) *J. Carbohydr. Chem.*, **21**, 143–148.
97. (a) Witczak, Z. J. (1994) *Levoglucosenone and Levoglucosans: Chemistry and Applications*, ATL Press, Inc., Science Publishers, Mount Prospect, Illinois, USA. (b) Witczak, Z. L. (1994) *Stud. Nat. Prod. Chem.*, **14**, 267–281; (c) Tsuchiya, Y. and Sumi, K. (1970) *J. Appl. Polym. Sci.*, **14**, 2003–2013; (d) Shafizadeh, F. and Chin, P. P. S. (1976) *Carbohydr. Res.*, **46**, 149–154;

(e) Shafizadeh, F., Furneaux, R. H. and Stevenson, T. T. (1979) *Carbohydr. Res.*, **71**, 169–191; (f) Shibagaki, M., Takahashi, K., Kuno, H., Honda, I. and Matsushita, H. (1990) *Chem. Lett.*, 307–310.

98. Furneaux, R. H., Gainsford, G. J., Shafizadeh, F. and Stevenson, T. T. (1986) *Carbohydr. Res.*, **146**, 113–128.
99. Taniguchi, T., Nakamura, K. and Ogasawara, K. (1996) *Synlett*, 971–972.
100. Venkateswara Rao B., Chan, J. B., Moskowitz, N. and Fraser-Reid, B. (1993) *Bull. Soc. Chim. Fr.*, **130**, 428–432.
101. Itoh, A., Ozawa, S., Oshima, K. and Nozaki, H. (1980) *Tetrahedron Lett.*, **21**, 361–364.
102. Zhu, Y-H. and Vogel, P. (1998) *Tetrahedron Lett.*, **39**, 31–34.
103. Zhu, Y.-H. and Vogel, P. (1999) *J. Org. Chem.*, **64**, 666–669.
104. Mori, S., Ohno, T., Harada, H., Aoyama, T. and Shioiri, T. (1991) *Tetrahedron*, **47**, 5051–5070.
105. Evans, D. A., Chapman, K. T. and Carreira, E. M. (1988) *J. Am. Chem. Soc.*, **110**, 3560–3578.
106. Comins, D. L. and Dehghani, A. (1992) *Tetrahedron Lett.*, **33**, 6299–6302.
107. (a) Takai, K., Tagashira, M., Kuroda, T., Oshima, K., Utimoto, K. and Nozaki, H. (1986) *J. Am. Chem. Soc.*, **108**,6048–6050; (b) Cintas, P. (1992) *Synthesis*, 248–257; (c) Fürstner, A. (1999) *Chem. Rev.*, **99**, 991–1045; (d) Chen, D.-W., Takai, K. and Ochiai, Y. (1997) *Tetrahedron Lett.*, **38**, 8211–8214; (e) Kishi, Y. (1992) *Pure Appl. Chem.*, **64**, 343–350.
108. Zhu, Y.-H. and Vogel, P. (1999) *Chem. Commun.*, 1873–1874.
109. (a) Klemer, A. and Rodemeyer, G. (1974) *Chem. Ber.*, **107**, 2612–2614; (b) Horton, D. and Weckerle, W. (1975) *Carbohydr. Res.*, **44**, 227–240; (c) Jones, D. N., Taylor, M. G. and Wood, W. W. (1992) *Synth. Commun.*, **22**, 1687–1690.
110. (a) Tsang, R. and Fraser-Reid, B. (1984) *J. Chem. Soc., Chem. Commun.*, 60–62; (b) Handa, S., Tsang, R., McPhail, A. T. and Fraser-Reid, B. (1987) *J. Org. Chem.*, **52**, 3489–3491; (c) Yu, K.-L. and Fraser-Reid, B. (1989) *J. Chem. Soc., Chem. Commun.*, 1442–1445; (d) Yu, K.-L., Handa, S., Tsang, R. and Fraser-Reid, B. (1991) *Tetrahedron*, **47**, 189–204.
111. (a) Chapleur, Y. and Euvard, M. N. (1987) *J. Chem. Soc., Chem. Commun.*, 884–885; (b) Chapleur, Y., Longchambon, F. and Gillier, H. (1988) *J. Chem. Soc., Chem. Commun.*, 564–566.
112. Fleet, G. W. J. and Son, J. C. (1988) *Tetrahedron*, **44**, 2637–2647.
113. Witczak, Z. J., Chabra, R., Chen, H. and Xie, X.-Q. (1997) *Carbohydr. Res.*, **301**, 167–175.
114. Chen, X.-Y., Link, T. R. and Schramm, V. L. (1996) *J. Am. Chem. Soc.*, **118**, 3067–3068.
115. (a) Wei, A., Boy, K. M. and Kishi, Y. (1995) *J. Am. Chem. Soc.*, **117**, 9432–9436; (b) Wang, Y., Goekjian, P. G., Ryckman, D. V., Miller, W. H., Babirad, S. A. and Kishi, Y. (1992) *J. Org. Chem.*, **57**, 482–489; (c) Wu, T., Goekjian, P. G. and Kishi, Y. (1987) *J. Org. Chem.*, **52**, 4819–4829; (d) Wei, A. and Kishi, Y. (1994) *J. Org. Chem.*, **59**, 88–96 and references cited therein.
116. (a) Ravishankar, R., Surolia, A., Vijayan, M., Lim, S. and Kishi, Y. (1998) *J. Am. Chem. Soc.*, **120**, 11297–11303; (b) Espinosa, J. F., Bruix, M., Jarreton, O., Skrydstrup, T., Beau, J.-M. and Jiménez-Barbero, J. (1999) *Chem. Eur. J.*, **5**,442–448.
117. (a) Espinosa, J. F., Cañada, F. J., Asensio, J. L., Martin-Pastor, M.,. Dietrich, H., Martin-Lomas, M., Schmidt, R. R. and Jiménez-Barbero, J. (1996) *J. Am. Chem. Soc.*, **118**, 10862–10871; (b) Rubinstenn, G., Sinaÿ, P. and Berthault, P. (1997) *J. Phys. Chem. A.*, **101**, 2536–2540.
118. (a) Espinosa, J. F., Cañada, F. J., Asensio, J. L., Dietrich, H,; Martin-Lomas, M., Schmidt, R. R. and Jiménez-Barbero, J. (1996) *Angew. Chem. Int. Ed. Engl.*, **35**, 303–306; (b) Espinosa, J. F., Montero, E., García, J. L., Dietrich, H., Martin-Lomas, M., Schmidt, R. R., Imberty, A., Canada, F. J. and Jiménez-Barbero, J. (1998) *J. Am. Chem. Soc.*, **120**, 1309–1318.
119. Asensio, J. L., Cañada, F. J., García-Herrero, A., Murillo, M. T., Fernández-Mayoralas, A., Johns, B. A., Kozak, J., Zhu, Z., Johnson, C. R. and Jimenez-Barbero, J. (1999) *J. Am. Chem. Soc.*, **121**, 11318–11329.

120. Dowd, M. K., Reilly, O. J. and French, A. D. (1994) *Biopolymers*, **34**, 625–638.
121. Fernández-Alonso, M. C., Cañada, F. J., Solís, D., Chery, X., Kumaran, G., André, S., Siebert, H.-C., Mootoo, D. R., Gabius, H.-C. and Jiménez-Barbero, J. (2004) *Eur. J. Org. Chem.*, 1604–1613.
122. García-Aparicio, V., Fernández-Alonso, M. d. C., Angulo, J., Asensio, J. L., Cañada, F. J., Jiménez-Barbero, J., Mootoo, D. R. and Cheng, X. (2005) *Tetrahedron:Asymmetry*, **16**, 519–527.
123. Baudat, A., Picasso, S. and Vogel, P. (1996) *Carbohydr. Res.*, **281**, 277–284.
124. See, for example: Dubravko, J., Mesar, V., Basic, I., Nadramija, D. and Verbanac, D. (2003) *PharmaChem.*, **2**, 64–67; Andrews, C. W., Bennett, L. and Yu, L. X. (2000) *Pharmaceutical Research*, **17**, 639–644; Lipinski, C. A., Lombardo, F., Doniny, B. W. and Feeney, P. J. (1997) *Adv. Drug Delivery Rev.*, **23**, 3–25.
125. Kawatkar, S. P., Kuntz, D. A., Woods, R. J., Rose, D. R. and Boons, G.-J. (2006) *J. Am. Chem. Soc.*, **128**, 8310–8319.
126. Ganesan, A. (1998) *Angew. Chem.*, **110**, 2989–2992; Ganesan, A. (1998) *Angew. Chem. Int. Ed.*, **37**, 2828–2831; Lehn, J.-M. (1999) *Chem. Eur. J.*, **5**, 2455–2463; Polyakov, V. A., Nelen, M. I., Nazarpack-Kandlousy, N., Ryabov, A. D. and Eliseev, A. V. (1999) *J. Phys. Org. Chem.*, **12**, 357–363; Eliseev, A. V. and Lehn, J.-M. (1999) *Curr. Top. Microbiol. Immunol.*, **243**, 159–172; Karan, C. and Miller, B. L. (2000) *Drug Discovery Today*, **5**, 67–75; Lehn, J.-M. and Eliseev, A. V. (2001) *Science*, **291**, 2331–2332 and references cited therein; Brady, P. A., Bonar-Law, R. P., Rowan, S. J., Suckling C. J. and Sanders, J. K. (1996) *Chem. Commun.*, 319–320; Rowan, S. J., Brady, P. A. and Sanders, J. K. M. (1996) *Angew. Chem.*, **108**, 2283–2285; Rowan, S. J., Brady, P. A. and Sanders, J. K. M. (1996) *Angew. Chem. Int. Ed. Engl.*, **35**, 2143–2145.
127. See, for example: Hine, J., Cholod, M. S. and Chess, W. K., Jr. (1973) *J. Am. Chem. Soc.*, **95**, 4270–4276.
128. Howard, S., He, S. and Withers, S. G. (1998) *J. Biol. Chem.*, **273**, 2067–2072.
129. Popowycz, F., Gerber-Lemaire, S., Demange, R., Rodriguez-Garcia, E., Carmona-Asenjo, A. T., Robina, I. and Vogel, P. (2001) *Bioorg. Med. Chem. Lett.*, **11**, 2489–2493.
130. Gerber-Lemaire, S., Popowycz, F., Rodriguez-García, E., Schütz, C., Carmona-Asenjo, A. T., Robina, I. and Vogel, P. (2002) *ARKIVOC*, 23–37.
131. Pearson, W. H. and Guo, L. (2001) *Tetrahedron Lett.*, **42**, 8267–8271; Pearson, W. H. and Hembre, E. J. (2001) *Tetrahedron Lett.*, **42**, 8273–8276.

6

Isofagomine, noeuromycin and other 1-azasugars, iminosugar-related glycosidase inhibitors

Oscar Lopez and **Mikael Bols**

6.1 Introduction

1-Azasugars are a series of synthetic nitrogen-containing sugar analogues consisting of a monosaccharide structure where the anomeric carbon has been replaced by a nitrogen atom (Figure 6.1). Interest in these compounds is based on the discovery that many of them are potent glycosidase inhibitors. The name 1-azasugar is used here rather than the name iminosugar in accordance with IUPAC nomenclature that recommends that carbohydrates in which the ring-oxygen has been substituted with N be called iminosugars, and those where a carbon has been substituted with N be called azasugars [1]. It should be noted though that this distinction is strictly based on the comparison of these compounds with sugars, as both iminosugars and 1-azasugars are hydroxylated piperidines, and as such are very similar. Since coverage of hydroxylated piperidines as glycosidase inhibitors would be too broad and not very meaningful, the present review will take this distinction very seriously and only cover work with compounds that were made or investigated as sugar-mimics with N in the anomeric position. The review will focus on work from our laboratory but every attempt will be made to cover the work of others exhaustively.

Iminosugars: From synthesis to therapeutic applications Edited by Philippe Compain and Olivier R. Martin
 ISBN 978-0-470-03391-3

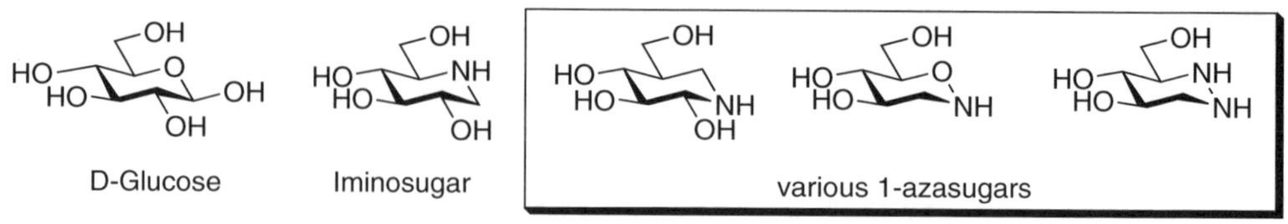

Figure 6.1 Glucopyranose mimics: (from left) glucose, 1-deoxynojirimycin, noeuromycin, glucooxazine and 1-azafagomine

6.2 1-Azasugars that are piperidines (isofagomine, noeuromycin, etc.)

6.2.1 Glucose analogues

The idea that 1-azasugars might be glycosidase inhibitors came to us from work by Reymond *et al.* on catalytic antibodies. These scientists had produced an antibody capable of catalysing the hydrolysis of the aryl tetrahydropyranyl ether **1** by using piperidine derivative **2** as part of the antigen (Figure 6.2) [2]. The rationale behind eliciting catalytic antibodies is that a transition state analogue is used as antigen; hence the success of this project implies that **2** is indeed a transition state analogue for the hydrolysis of **1** (it also inhibits 14D9-catalysed hydrolysis). The structure is unusual in that it has a positively charged nitrogen atom in the position equivalent to C1 in a sugar and led us to test the idea that having basic nitrogen in the 1-position of a true sugar structure would lead to glycosidase inhibitors. Thus we designed isofagomine **4**, an isomer of the natural product fagomine, as a potential transition state-like analogue inhibitor of the enzyme-catalysed hydrolysis of **3**. As it happened this hypothesis turned out to be true as **4** is a very potent inhibitor of some of the enzymes that catalyse the above transformation; most remarkable is the observation that **4** inhibits β-glucosidase with a K_i of 110 nM [3] and glycogen phosphorylase with a K_i of 0.7 μM [4] whereas 1-deoxynojirimycin (Figure 6.1) has a much weaker inhibition (K_i of 47 μM and 55 mM, respectively, for these two enzymes). On the other hand, 1-deoxynojirimycin is a more potent α-glucosidase inhibitor than isofagomine (**4**) (Table 6.1).

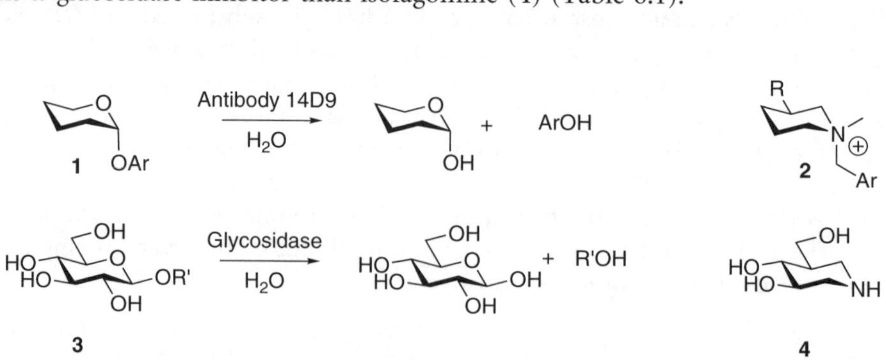

Figure 6.2 Ar = -$C_6H_4CONHCH_2CH_2OH$, R = CH_2NH_2, R′ = carbohydrate

Table 6.1 K_i values in μM of *gluco*-configured 1-azasugars towards various glycosidases. 1-deoxynojirimycin is shown for comparison. α-glu = α-glucosidase from yeast, β-glu = β-glucosidase from almonds, Isomalt. = isomaltase from yeast, GA = glucoamylase from *A. Awamori*, α-man = α-mannosidase from jack bean, GP = glycogen phosphorylase from pig liver

	α-glu	β-glu	Isomalt.	GA	α-man	GP
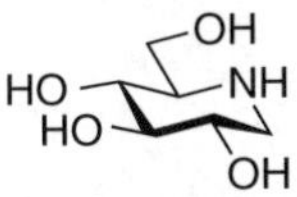	25	47	11	9.8	270	55000
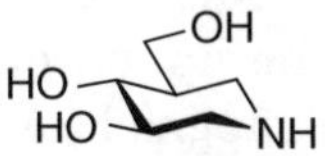 **4**	86	0.11	7.2	3.7	770	0.7
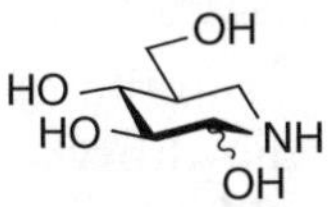 **7**	0.022	0.069	0.025			
 **30**	6.9	0.32	0.27		3306	13.5
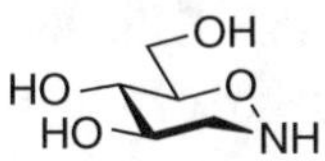 **35**	>1000	60				
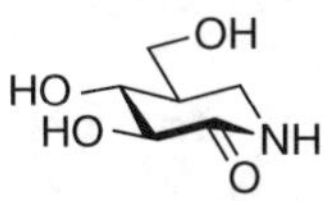 **37**	>250	29	>250		>250	
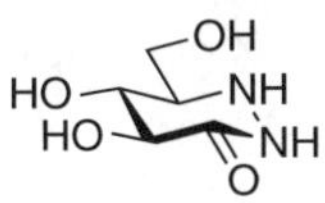 **40**	>1000	13			25	

Isofagomine (4) has been synthesized from a number of different starting materials. The syntheses fall into several categories, some starting from carbohydrates such as levoglucosan [5], D-lyxose [6], D-arabinose [7] and other chiral starting materials [8,9] such as D-glyceraldehyde [10] and D-tartrate [11–13], and those starting from achiral starting materials [14–18]. Most of the syntheses involve some problems of diastereoisomer or enantiomer separation.

In the chiral pool syntheses, the secondary hydroxyl groups of isofagomine come directly from C-2 and C-3 of D-arabinose, L-xylose or D-tartrate, or from C-3 and C-4 of D-glucose or D-lyxose. In general these syntheses enjoy the comfort of 100 per cent enantiopurity, and flexibility in modification around nitrogen and C-2, C-3, C-5 and C-6, but provide little flexibility in modifications of stereochemistry. It should be noted though that many of these starting materials can be obtained in both enantiomeric forms at nearly the same price. One of the syntheses, the one from D-arabinose, is shown in Figure 6.3. D-Arabinose is converted into the benzyl glycoside and then oxidized selectively at the 4-position using dibutyltin oxide/bromine [19] to give ketone 5. This compound is reacted with nitromethane in a Henry reaction which adds the extracyclic

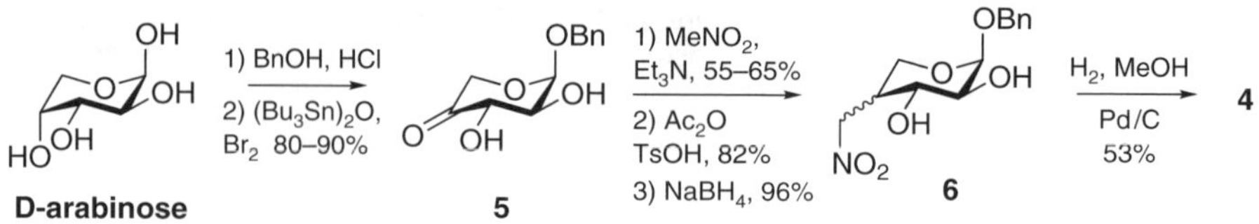

Figure 6.3 Synthesis of **4** from D-arabinose

carbon atom, followed by peracetylation, and elimination and reductive addition to give **6** as a mixture of diastereomers favouring the desired isomer. Hydrogenolysis of **6** directly leads to **4**, which is obtained in 53 per cent yield after chromatographical removal of the epimer.

Most of the syntheses from achiral starting materials prepare a 5-substituted 1,2,5,6-tetrahydropyridine, and then introduce the hydroxyl groups of **4** by dihydroxylation, thus giving the synthesis flexibility in terms of stereochemistry and modification at C-3 and C-4. While attractive, this strategy has, so far, been hampered by the significant and somewhat surprising problem that the 1,2,5,6-tetrahydropyridine cannot be epoxidized or even dihydroxylated with diastereoselectivity. These syntheses also fight the significant problem of introducing asymmetric control in order to effectively compete with the chiral pool synthesis, and this problem has been nicely solved in the recent synthesis by Ouchi *et al.* [17].

Apart from the stereoisomers, which are mentioned below, a number of structural analogues of **4** have been investigated. The most important of these are the 2-substituted derivatives, such as **7** (Figure 6.4). This 2-OH-analogue, noeuromycin (**7**), is generally more potent than **4**, in some cases up to 4000 times (Table 6.1) [20]. The reason for better binding is obviously the presence of the 2-OH group similar to the substrate, though in some cases the electron-withdrawing effect of the OH group on the amine may actually compromise the binding. Mutarotation at C-2 allows the compound to bind to proteins in both *gluco* and *manno* configuration showing that the OH group is certainly not superfluous. The hemiaminal at C2 results in the compound decomposition at neutral pH [21, 22] which is similar to what has been found for nojirimycin. Replacing the 2-OH group with hydroxymethyl, which avoids this problem, does not result in very good inhibition [23]. A phosphate of this hydroxymethyl group has also been reported as a potential inhibitor [24].

The disaccharide analogues of **4**, compounds **8** and **9**, which resemble isomaltose [25] and maltose [26] respectively, have been prepared but were found to be weaker inhibitors

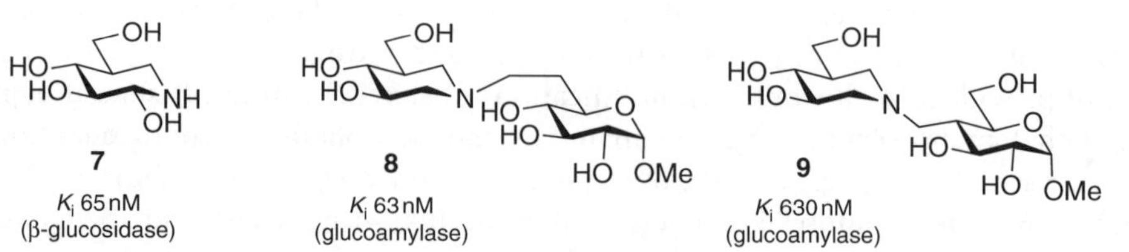

Figure 6.4 The 2-hydroxy analogue of **4**, noeuromycin (**7**), and some disaccharide azasugars

10 ($n = 1$)
11 ($n = 2$)
12 ($n = 3$)

13
IC_{50} 0.6 nM
(GCase)

Figure 6.5 Cellulose analogues of isofagomine (**10–12**) and the pseudo ring-oxygen substituted derivative **13**

than **4** with the single exception that **8** was much more potent than **4** versus *A. niger* glucoamylase.

A more successful type of disaccharide analogue than those discussed above were, however, obtained by glycosylation of **4** (Figure 6.5). Stick and collaborators have prepared the series of oligosaccharide analogues **10–12** by enzymatic glycosylation of the 4-OH group of **4** and showed that these compounds are potent cellulase inhibitors [27]. The inhibition of endo-glycanase Cex from *Cellulomonas fimi* increased from **10** to **12**, which is consistent with the substrate specificity of this enzyme. An X-ray structure of **10** bound to the active site of endocellulase Cel5A from *Bacillus agaradhaerens* showed that the nitrogen was protonated and that the enzyme was in the dicarboxylate form [28].

Several groups have investigated the influence of substitution at nitrogen with simple alkyl groups because it is easy to carry out and may improve bioavailability. However, it was generally found that binding was decreased by substitution at N [6, 29, 30]. Researchers at Novo-Nordisk have investigated a large series of these derivatives in an attempt to obtain better glycogen phosphorylase inhibitors. They also found that binding decreased when nitrogen was substituted with small groups, but the loss could be partially regained when the chain was extended. The propylphenyl derivative gave an IC_{50} of 1 μM where **4** gave 0.7 μM towards glycogen phosphorylase [31]. These findings are in disappointing contrast to what has been found for iminosugars, like 1-deoxynojirimycin, where *N*-substitution frequently improves binding, as shown in other chapters of this volume (see, for example, Chapters 4, 9, 10 and 11). Clearly these results, together with the findings with **8** and **9**, show that there is little space for substitution around the anomeric centre.

This problem has been elegantly solved by Fan's group who incorporated alkyl groups at the pseudo ring-oxygen position rather than at the anomeric position (Figure 6.5) [32]. Compounds with alkyl groups of different length were investigated as inhibitors of β-glucocerebrosidase and the nonyl derivative **13** was found to be the most potent compound with sub-nanomolar inhibition. This compound had an IC_{50} of 0.6 nM, while **4** had an IC_{50} of 56 nM. The synthesis was carried out as outlined in Figure 6.6. The benzyl glycoside of L-xylose **14** was selectively converted into the 2,3-ketal **15** and the 4-OH group, subsequently transformed into a leaving group and substituted with cyanide ion to give **16**. A Grignard reaction followed by reduction with sodium borohydride gave the amine **17** as a single stereoisomer. Finally,

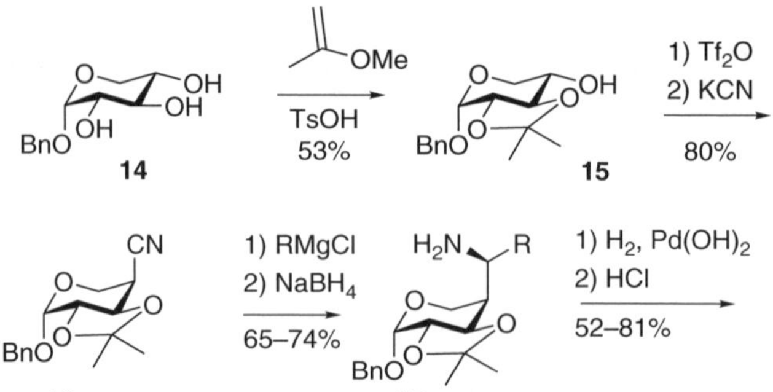

Figure 6.6 Synthesis of the pseudo ring-oxygen substituted derivative **13** (R = n-nonyl)

internal reductive amination and deprotection led to **13**. An alternative synthesis of isofagomine derivatives with an alkyl substituent in this position has recently been disclosed [33].

Isofagomine analogues deoxy or with other substituents at positions 3, 4, 5 or 6 have also been prepared and investigated, but changes at these positions generally decrease inhibitory strength a great deal [34–37]. The K_i values of some of these derivatives are summarized in Table 6.2 for the inhibition of β-glucosidase, which was the enzyme that best tolerated these modifications. It is seen that the 3-OH group is far more important than the 4-OH group and that only when the stereochemistry is changed can an amino or hydroxymethyl group compensate for the loss in binding.

Introducing an extra OH at C5 (in **4**) has also been investigated by several groups but the loss in activity by that modification is considerable [6, 38].

Table 6.2 K_i values in μM of 3- and 4-substituted isofagomines against β-glucosidase from almonds

R	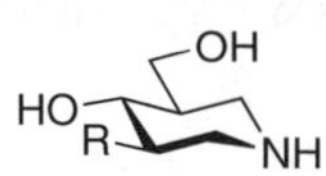	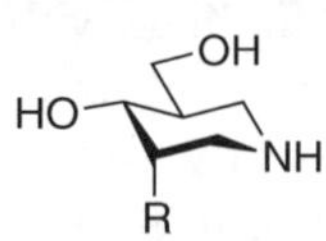	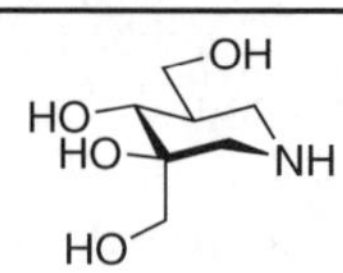 7.0
H	2200		
OH	0.11		
NH_2	209	26	
CH_2OH	>1000	47	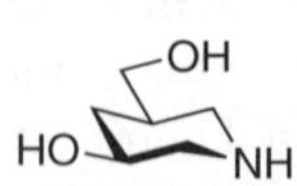
COOH	>1000	>1000	5.6

6.2.2 Galactose analogues

Independently of the work described above, Ichikawa *et al.* designed and prepared the *galacto* analogue of isofagomine (**18**) [39]. The design was based on the hypothesis that the transition state of retaining β-glycosidases stabilize charge at anomeric carbon rather than at ring-oxygen as conventionally assumed [6]. Indeed, **18** is a very potent β-galactosidase inhibitor ($K_i = 4\,nM$), while the inhibition of α-galactosidase is much more modest ($K_i = 50\,\mu M$). The inhibition of various glycosidases by **18** is shown in Table 6.3, where it is compared with 1-deoxygalactonojirimycin (top row) [6, 40]. It is seen that the contrast between the two compounds is remarkable, the former being extremely potent against β-glycosidases, and the latter being extremely potent against α-galactosidases. Ichikawa *et al.* also made the 5-hydroxy and *N*-butyl analogues, but these compounds were poor inhibitors, which is analogous to what was found for isofagomine (see above). The 5-hydroxy group may reduce the basicity of the nitrogen besides any possible negative influence of an axial polar group, while the *N*-butyl apparently is too bulky at the anomeric position.

Table 6.3 K_i values in μM of *galacto*-configured 1-azasugars against glycosidases. 1- Deoxy-galactonojirimycin is shown for comparison (top row). α-gal = α-galactosidase from green coffee bean, β-gal *A.O.* = β-galactosidase from *A. Oryzae*, β-gal *E.C.* = β-galactosidase from *E. Coli*, β-gal *S.F.* = β-galactosidase from *S. Fragilis*, β-glu = β-glucosidase from almonds

	α-gal	β-gal *A.O.*	β-gal *E.C.*	β-gal *S.F.*	β-glu
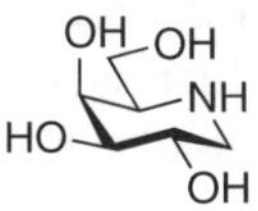	0.0016		12.5	81	540
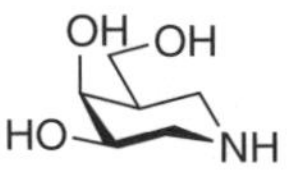 **18**	50	0.004	0.2	0.33	0.097
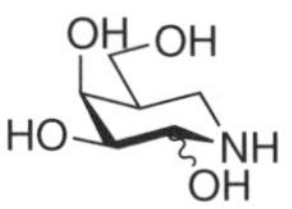 **19**	0.74	0.035	0.40	0.091	
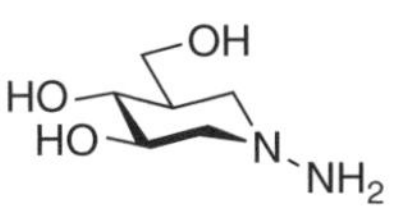 **20**	73	4.5			410
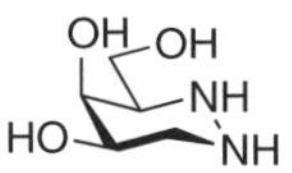 **31**	0.28	0.04	0.30	7.8	0.13
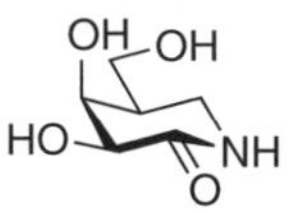 **38**	5.2	0.067		9	3

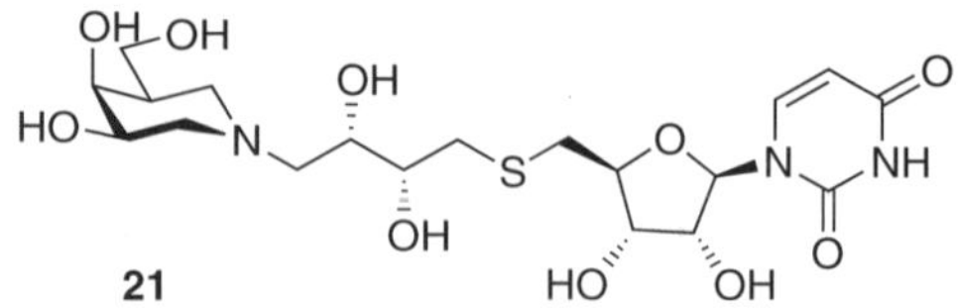

21

K_{is} = 4.4 μM (α-1,3-galactosyl transferase)

Figure 6.7 A 1-azasugar galactosyl transferase inhibitor

We have made **19**, the 2-hydroxy analogue of **18** (noeuromycin type), and found that the inclusion of the 2-OH does not improve inhibition of β-galactosidases [20]. The inhibition of α-galactosidase does increase significantly, but not as much as 1-deoxygalactonojirimycin, so the position of the nitrogen appears to be more important than the presence of a 2-OH group for α-galactosidase inhibition.

A 1-*N*-amino analogue of **18**, the hydrazine derivative **20**, has also been investigated but it is a considerably weaker inhibitor than **18** (Table 6.3) [41]. The most likely explanation is that the amino group is too bulky and this is similar to the findings with *N*-butyl derivatives.

Interestingly, the azasugar **18** has been substituted at nitrogen with a nucleotide mimic in order to create a galactosyl transferase inhibitor (Figure 6.7). The UDP sugar-like **21** was prepared from **18**, tartaric acid and uridine and fulfilled the expectations: it is a low micromolar range inhibitor of porcine α-1,3-galactosyl transferase with a potency about double that of **18** [42]. A compound that contained merely the azasugar and the tartaric acid-based linker was considerably weaker, which again seems to confirm that substitution at N with small substituents is unfavourable, but as the substituents size increases the loss in inhibition may be regained. Still the N-substitution is less successful for 1-azasugars than with 1-deoxy nojirimycin-like iminosugars.

6.2.3 Fucose analogues

The L-fucose analogue of isofagomine **22** has been reported both by Ichikawa's group and by ours (Table 6.4), and was synthesized from either D-ribose [43] or L-arabinose [44]. The compound is a reasonably good inhibitor of α-fucosidase with K_i values in the low micromolar range, but it is nevertheless much weaker than the fucose version of 1-deoxynojirimycin which binds 1000-fold better. Again this supports the notion that iminosugars are better α-glycosidase inhibitors than 1-azasugars. However, when a 2-OH group is included in the 1-azasugar, as in the noeuromycin analogue **23** [20], the α-fucosidase inhibition increases to a level that is essentially equal to the iminosugar. Therefore it is clear that the presence of the 2-OH group is more important than the placement of the nitrogen for strong inhibition of these enzymes.

Table 6.4 K_i values in μM of L-*galacto*-configured 1-azasugars against fucosidases. 1-Deoxyfuconojirimycin is shown for comparison (top row). α-fuc B. K. = α-fucosidase from bovine kidney, α-fuc H. P. = α-fucosidase from human placenta

	α-fuc B. K.	α-fuc H.P.
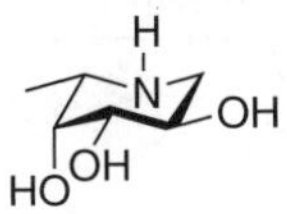	0.029	0.0013
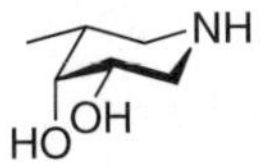 22	4	6.4
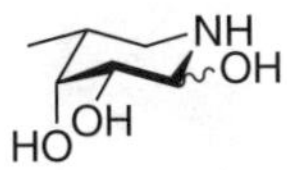 23	0.0047	0.0032
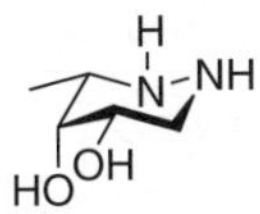 32	0.81	0.63
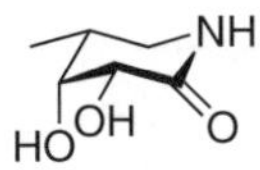 39	26	18

6.2.4 Glucuronic acid analogues

The glucuronic acid analogue of isofagomine, 24 was reported by Ichikawa's group (Table 6.5) [45]. The compound was synthesized from D-arabinose in 11 steps as a mixture of isomers requiring separation, but the synthesis has subsequently been improved considerably [10]. It is a very potent glucuronidase inhibitor, and in fact more than 1000-fold stronger than the iminosugar analogue (top row).

Nishimura *et al.* have reported the 2-trifluoroacetamido analogue of 24, 25, and the compound is indeed a very potent inhibitor of β-glucuronidase. Compound 25 is an analogue of Siastatin B, which is the 2-acetamidogalacturonic acid analogue. Many different analogues and isomers have been made and investigated as has been described in detail by Professor Nishimura himself in this volume (Chapter 12) and in reviews [46–48]. Generally the trifluoroacetamido derivatives are the most potent compounds, being much stronger inhibitors than the acetamido derivatives. They are, however, also unstable at neutral or slightly acidic pH and decompose to 2-OH derivatives (noeuromycins) and eventually to the 3-ketone, which is inactive [21]. The high activity is therefore presumably due to the conversion to the 2-OH compound.

Table 6.5 K_i values in μM of 1-azasugars that are analogues of glucuronic acid towards glucuronidases. The glucuronic acid analogue of 1-deoxynojirimycin is shown for comparison (top row). β-gluU B.L. = β-glucuronidase from bovine liver, β-gluU H. L. = β-glucuronidase from human liver

	β-gluU B.L.	β-gluU H.L.
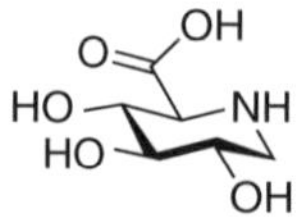	>560	80
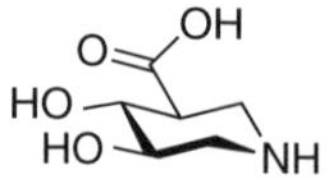 **24**	0.079	
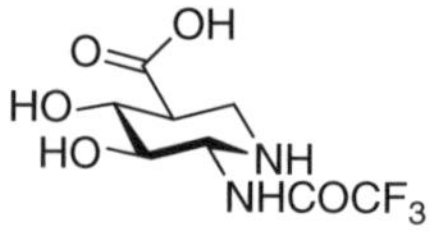 **25**	0.065	
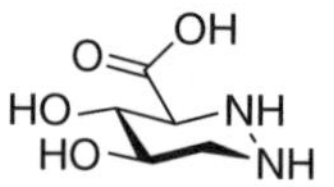 **33**	1.0	

6.2.5 Ribose analogues

The observation that isofagomine was a glycogen phosphorylase inhibitor gave us the idea that a ribosyl cation mimic might be an inhibitor of purine nucleoside phosphorylase. Compound **26** was the first 1-azasugar analogue of ribofuranose that was investigated and the compound had a K_i of 180 μM [49]. Subsequently the pyrrolidine without the unnecessary 4-OH has been prepared, but it is not more potent against the enzyme [50, 51]. Several syntheses of **27** and nucleoside analogues of it have appeared [52–59]. In the present context it is important to note that the *N*-substituted analogue **28** has been found to be a very strong inhibitor of purine nucleoside phophorylase displaying a K_i of 0.16 nM [60–63]. More details on this family of compounds can be found in Chapter 8 in this volume.

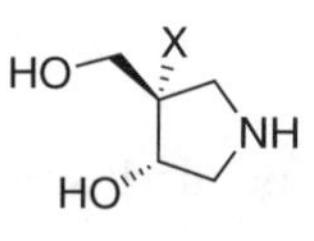

26 X = OH
27 X = H

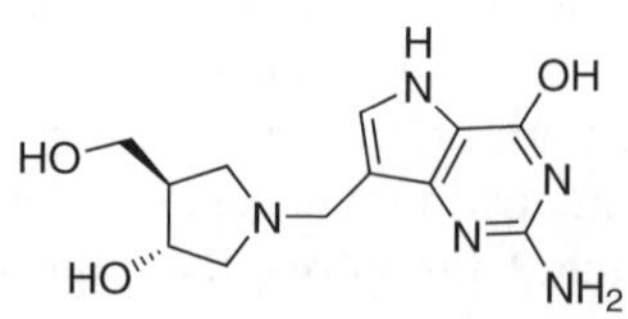

28

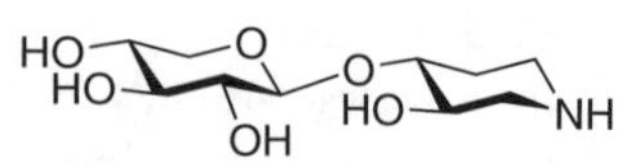

29

Figure 6.8 Pentose-based 1-azasugars

6.2.6 Xylose analogues

Withers' group has synthesized a xylobiose-like 1-azasugar **29** (Figure 6.8) and compared its inhibition of xylobiases with those of the corresponding iminosugar [64]. They found that the isofagomine analogue was up to 40 times more potent in its inhibition of *Cellulomonas fimi* β-xylosidase than the iminosugar, while a *Bacillus circulans* β-xylosidase was poorly inhibited by both compounds. This supports again the notion that 1-azasugars are better β-glycosidase inhibitors than iminosugars, but not as strongly as some of the examples above. An X-ray crystal structure of the complex of **29** bound to *Cellulomonas fimi* β-xylosidase has been reported.

6.2.7 Glucosamine analogues

Nishimura's group has made the 2-acetamido and 2-trifluoroacetmido analogues of **4** [65]. These compounds are submicromolar α- and β-glucosidase inhibitors, but disappointingly poor β-glucosaminidase inhibitors. Nishimura suggests that these 2-amido derivates are hydrolysing to **7** in the medium [21]. Incorporation of an extra CH_2 in the molecule by making the 2-acetamidomethyl derivative of **4** gives a stable compound and this idea has been tested [66]. The compound has a K_i of 2.4 μM versus lysosomal β-hexosaminidase.

6.2.8 Other stereoisomers

A diastereomer of **27** corresponding to xylo- or lyxofuranose has been reported [67]. The compound was not a potent glycosidase inhibitor. An analogue of **9** in which the azasugar has *ido/gulo* configuration has been made in an attempt to mimic a distorted substrate, but the inhibition of an endo-cellulase was modest ($K_i = 200$ μM) [68]. An acetylxylosamine derivative has been prepared and this compound inhibits *N*-acetyl-β-hexosaminidase from *Streptomyces plicatus* with a K_i of 21 μM [69]. A GalNAc analogue has also been reported and this compound had a K_i of 2.7 μM [70].

6.3 1-Azasugars that are hydrazines

Azafagomines are monosaccharide analogues where a hydrazine has been introduced in the ring. Azafagomine (**30**) resembles a natural product, fagomine, but has a nitrogen atom in place of the anomeric carbon (Table 6.1) [71].

A wide range of stereoisomeric azafagomines have been prepared mimicking D-glucose (**30**) [72], D-xylose [36], D-galactose (**31**) [40], D-glucuronic acid (**33**) and L-fucose (**32**) [36]. All of them are potent inhibitors of both the corresponding α- and β-glycosidases (Tables 6.1, 6.3 and 6.4) with K_i values in the low micro or nanomolar range. The L-*gluco* isomer has also been prepared and is a very poor inhibitor of glucosidases [73]. The azafagomines are generally more potent towards α-glycosidases than the isofagomines and less potent towards β-glycosidases. Thus the glucoisomer **30** displays a K_i of 0.32 μM towards β-glucosidase and 6.9 μM towards α-glucosidase (Table 6.1), while the galactoisomer **31** displays a K_i of 0.04 μM towards β-galactosidase and 0.28 μM towards α-galactosidase (Table 6.3). This inhibition profile can be explained by the higher versatility of the inhibitor allowing it to become charged at either nitrogen and resembling both nojirimycins and isofagomines, since nojirimycins generally are the best α-glycosidase inhibitors, while isofagomines are the best β-glycosidase inhibitors. An azafagomine analogue of castanospermine, **34**, has also been made (Figure 6.9), but this compound is a disappointing inhibitor compared with castanospermine [74]. The weaker inhibition, in some cases, of the azafagomine compared with the corresponding amine inhibitor is undoubtedly due to the lower base strength of the former. Azafagomine (**30**) has a pK_a of 5.3, while **4** has a pK_a of 8.4 and 1-deoxynojirimycin has a pK_a of 6.7, which could make the inhibitor's interaction with the acid groups in the glycosidase active site weaker. Noteworthy is also that **32** is a considerably stronger inhibitor than the isofagomine analogue. As has been observed for **4**, azafagomines display slow onset inhibition of some glycosidases, where the binding step is the slow step [75, 76]. This is probably due to either inhibitor and/or enzyme binding in little populated protonation states.

Depending on their stereochemistry, these inhibitors are best prepared from chiral pool starting material or asymmetric synthesis. The *gluco* isomer, azafagomine (**30**), has been synthesized from both achiral [77] and chiral [72] starting materials. The most efficient route to the enantiopure compound is the synthesis from L-xylose which is

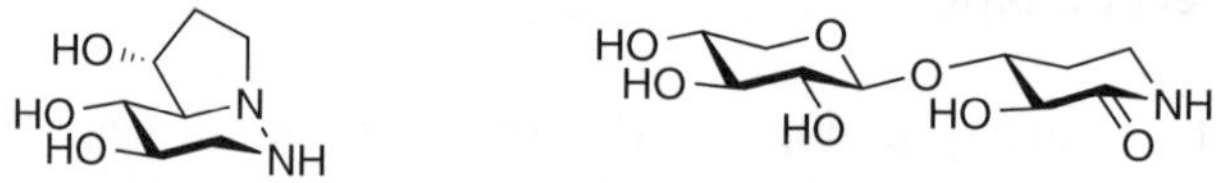

34

36
β-Xylanase
$K_i = 0.34\ \mu M$

Figure 6.9 Azafagomines

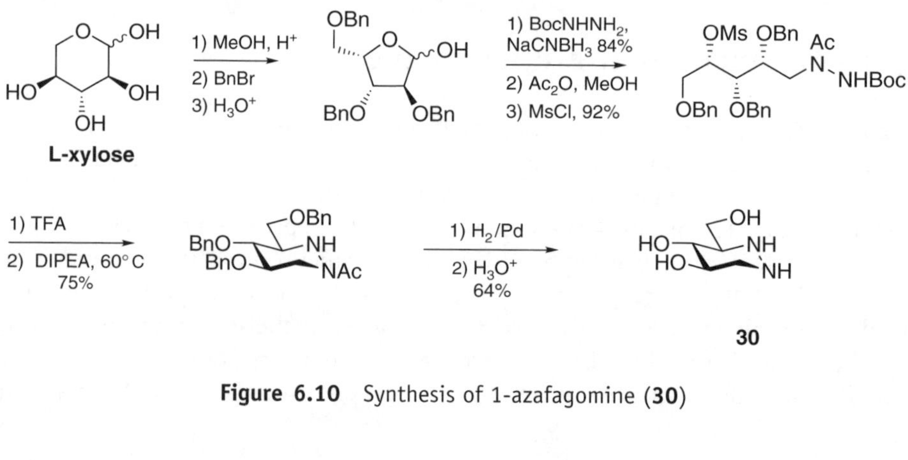

30

Figure 6.10 Synthesis of 1-azafagomine (**30**)

shown in Figure 6.10. L-xylose is converted into the 2,3,5-tri-*O*-benzylfuranoside, which is subjected to reductive amination with Boc hydrazine, *N*-acetylation and mesylation of the 4-OH group. Removal of the Boc allows the molecule to cyclize and deprotection yielded **30**.

6.4 1-Azasugars that are oxazines

The glucooxazine **35** is similar to **4** except that the ring-oxygen normally present in monosaccharides has been retained [78, 79]. So while these two molecules resemble each other chemically, they are functionally distinct in that one of them, **35**, may be considered a better transition state or intermediate analogue than isofagomine (**4**).

The glycooxazine **35** was proposed quite early [78], later synthesized in racemic form [79] and eventually made enantiomerically pure [8]. It has also been converted into oligosaccharide analogues by substitution at the 4-position with glucosyl residues giving cellulase inhibitors [27]. While the compound at neutral pH was found to be a relatively poor inhibitor ($K_i = 60\,\mu M$), it has subsequently been found that at low pH it becomes up to 100-fold better [80].

6.5 1-Azasugars that are piperidones

Withers' group [81] made the remarkable discovery that isofagomine lactams could be inhibitors. They designed and prepared xylobiose isofagomine lactam **36** (Figure 6.9), which was a potent inhibitor ($K_i = 0.34\,\mu M$) of Cex-xylanase. The affinity of **36** was studied by X-ray crystallography of its complex with the enzyme (xylanase Xyn10A) to clarify the mode of binding; it was found that the lactam bound to the enzyme as the amide tautomer [82]. Other isofagomine lactams were subsequently investigated by us, such as the *gluco*/*manno*-, *galacto*- and L-*fuco*-isofagomine-lactams (Tables 6.1, 6.3 and 6.4) [83, 84]. The *gluco*/*manno* (**37**) and *galacto* isofagomine lactam (**38**) were synthesized from D-arabinose in nine and 11 steps, respectively and the L-*fuco*-isofagomine lactam (**39**) was synthesized from L-arabinose in 12 steps. These lactams were also found to inhibit glycosidase in micro- to nanomolar range, but generally much less than the isofagomines. Compound **37** was found to inhibit β-glucosidase with a K_i of $29\,\mu M$ and β-mannosidase with a K_i of $9\,\mu M$. This 'dual' inhibition of β-glucosidase and β-mannosidase was studied by Davies and collaborators [85] by X-ray crystallography of three-dimensional structures of enzyme (*Tm*GH1 and *Cm*Man5) complex with **37**. In the case of *Tm*GH1, **37** adopted a $^4H_3$ conformation, whereas with *Cm*Man5 a $B_{2,5}$ conformation was adopted (Figure 6.11). They proposed that this 'dual' inhibition was related to the structure of the transition-state ($^4H_3$ and $B_{2,5}$) more than an interaction with the catalytic acid.

Lactam **38** was found to be a remarkably potent inhibitor of β-galactosidase, having a K_i of 67 nM (Table 6.3). Nevertheless it is about 17-fold weaker than **18** which is remarkably potent against this enzyme. The inhibition of other galactosidase and

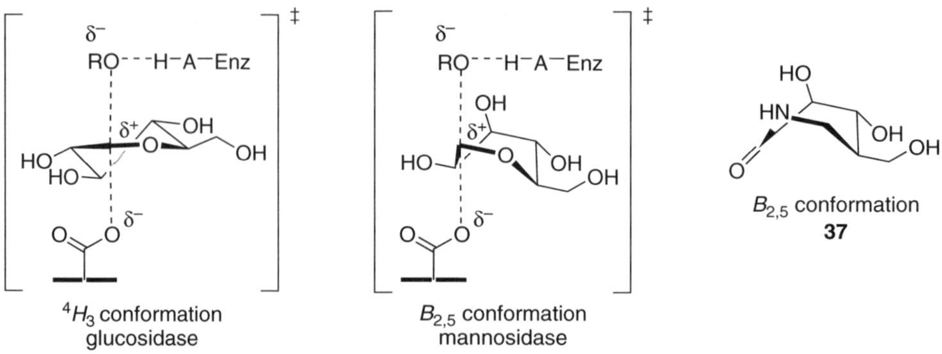

Figure 6.11 Binding mode of **37** according to X-ray crystallography (right), and how that may mimic the transition state in *Cm*Man5 (middle), but not in *Tm*GH1 (left)

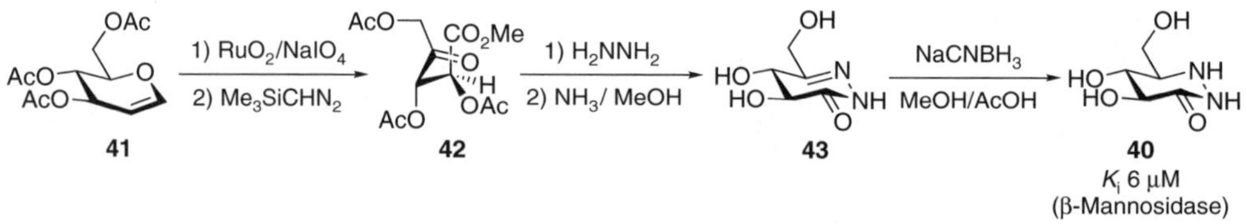

Figure 6.12 Synthesis of **40**

β-glucosidase is comparatively better than the *gluco* isomer 37. The L-*fuco* lactam **39** was found to be a moderate α-fucosidase inhibitor (Table 6.4).

Vasella's group has reported the synthesis of **40** (Table 6.1), an azafagomine-type lactam [86]. The synthesis was carried out starting from triacetyl-D-glucal **41** by oxidative cleavage with $RuO_2/NaIO_4$ followed by esterification with Me_3SiCHN_2 to give oxo ester **42** in a 63 per cent yield. This ester was treated with an excess of hydrazine followed by deacetylation with ammonia in methanol to give the dihydropyridazinone (**43**) in a 78 per cent yield. Finally, reduction of **43** with $NaBH_3CN$ in methanol in the presence of acetic acid, gave the tetrahydropyridazinone (**40**) in a 75 per cent yield as a diastereoisomeric mixture (Figure 6.12). Compound **40** was found to be a good inhibitor of β-glucosidase (K_i 13 μM), α-mannosidase (K_i 25 μM) and β-mannosidase (K_i 6 μM). It is a better β-glucosidase and α-mannosidase inhibitor than the isofagomine lactam 37, but a surprisingly poor α-glucosidase inhibitor bearing in mind that it has a basic site at the ring-oxygen position.

6.6 Sulphur-containing analogues of 1-azasugars

The observation that charge at the anomeric position appeared to be so advantageous led to the idea of investigating incorporation of charge by other means than nitrogen. Therefore a series of sulphur analogues, the sulphide **44**, the sulphoxide **47**, the sulphone **48** and the sulphonium salts **45** and **46**, of the fucose-configured isofagomine were prepared

Table 6.6 K_i values in μM of sulphur analogues of 1-azasugars against fucosidases. α-fuc B. K. = α-fucosidase from bovine kidney, α-fuc H. P. = α-fucosidase from human placenta

	α-fuc B. K.	α-fuc H.P.
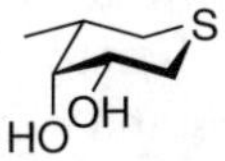 **44**	73.000	>100.000
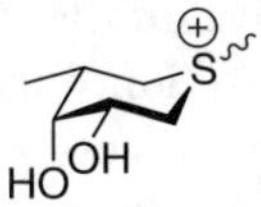 **45**	102	303
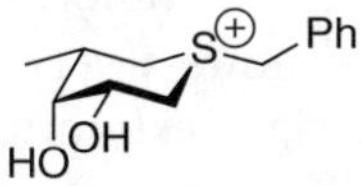 **46**	161	309
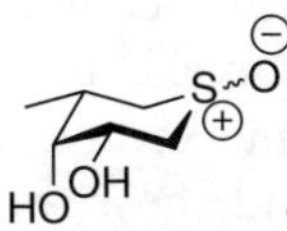 **47**	740	>1000
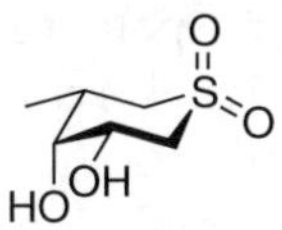 **48**	2000	>100.000

(Table 6.6) [87]. All were competitive inhibitors or not inhibitors and clearly weaker than the nitrogen derivatives but mechanistically interesting. The sulphide derivative **44** was an extremely poor inhibitor of both enzymes. This is not surprising since neither in terms of charge nor geometry does it mimic the transition state. However, as soon as the sulphur atom becomes charged, as in the methyl sulphonium analogue **45**, the inhibition is increased 700 times. This very clearly demonstrates the advantage of having a positive charge in this position. The increased binding may be the result of a salt bridge between the catalytic nucleophile in the enzyme and the charged sulphur. Compound **45** is 25–50 times weaker than isofagomine analogue **22**. This may be explained to some extent by poor fit of the methyl group in the enzyme active site as *N*-alkylated analogues of isofagomines are generally much weaker inhibitors than the secondary amines themselves, presumably because of unfavourable steric interactions of the methyl group in that area. The *S*-benzyl sulphonium salt **46** has an inhibitory activity much like **45**. This suggests that the size of the *S*-alkyl group is of minor importance. Interestingly, the sulphoxide **47** is a weaker inhibitor than **45** although still a 100 times stronger inhibitor than **44**. This suggests that **47** can benefit from some electrostatic interaction between the enzyme nucleophile and the charged sulphur atom. However, the negative oxygen atom appears to be unfavourable in the electronegative surroundings around the anomeric carbon. This situation is intensified in the sulphone **48**, which is a weaker inhibitor. This may, however, also be due to limited space in the active site to accommodate the sulphone.

6.7 Slow inhibition and thermodynamics of binding

Several 1-azasugars have been found to display slow onset inhibition of glycosidases, i.e. the inhibition is noninstantaneous, but requires an appreciable amount of time to reach maximum inhibition. Alternatively, if the azasugar is incubated with the enzyme and the substrate is added subsequently a slow displacement of the inhibitor by the substrate is observed. This phenomenon, which is known from iminosugars and other glycosidase inhibitors too, is dependent on the structure of the inhibitor and on the specific enzyme being inhibited, and is difficult to predict. By following the reaction during slow-onset inhibition it was possible to obtain an insight into the inhibition process. For 1-azasugars it was found that the slow-onset inhibition was not due to a slow conformational change in the enzyme [76]. Rather the binding of the inhibitor follows a very simple kinetic scheme with one step which is slow. This seems at first sight extraordinary, because binding is commonly believed to take place nearly at diffusion-controlled rates, and indeed older work suggested that most slow inhibition phenomena are caused by slow conformational changes in the protein. The kinetic constants involved, k_{on} and k_{off}, ($K_i = k_{off}/k_{on}$) were measured for a number of inhibitors and show some remarkable trends (Table 6.7). First of all k_{on} is remarkably small even for a potent inhibitor like **4** and **30**. Furthermore the release rate from almond β-glucosidase is remarkably similar even for inhibitors with widely different K_i values.

To gain a further insight into this remarkable behaviour, the inhibition of 1-azasugars was also studied at different temperatures to obtain thermodynamic parameters. For

Table 6.7 Dissociation constants and rate constants for 1-azasugar inhibitors of almond β-glucosidase at temperature 25°C and pH 6.8. k_{on} is the rate constant for the binding of the inhibitor to the enzyme, while k_{off} is the rate constant for release of the inhibitor from the enzyme. K_i has been calculated as k_{off}/k_{on}

Inhibitor	K_i (μM)	k_{on} ($M^{-1}s^{-1}$)	k_{off} (s^{-1})
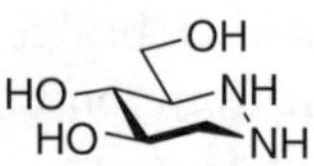 **30**	0.33 ± 0.06	$(3.3 \pm 0.2) \times 10^4$	$(1.1 \pm 0.2) \times 10^{-2}$
	56 ± 6	$(1.50 \pm 0.09) \times 10^2$	$(8.4 \pm 0.7) \times 10^{-3}$
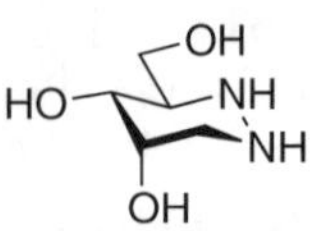	128 ± 17	$(1.4 \pm 0.4) \times 10^2$	$(1.8 \pm 0.4) \times 10^{-2}$
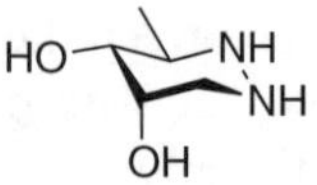	28 ± 6	$(2.4 \pm 0.5) \times 10^3$	$(6.7 \pm 0.1) \times 10^{-2}$
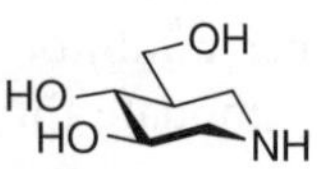 **4**	0.22 ± 0.06	$(7.3 \pm 0.3) \times 10^3$	$(1.6 \pm 0.4) \times 10^{-3}$

Table 6.8 Dissociation constants ($K1_i$) and rate constants (k_{on}/k_{off}) for inhibition of almond β- glucosidase by **4** at five temperatures and pH 6.8. k_{on} is the rate constant for the binding of the inhibitor to the enzyme, while k_{off} is the rate constant for release of the inhibitor from the enzyme. K_i has been calculated as k_{off}/k_{on}

Temperature (°C)	K_i (μM)	k_{on} ($M^{-1}s^{-1}$)	k_{off} (s^{-1})
0	1.90 ± 0.3	$(1.03 \pm 0.07) \times 10^3$	$(1.75 \pm 0.6) \times 10^{-3}$
15	0.80 ± 0.1	$(2.5 \pm 0.15) \times 10^3$	$(2.3 \pm 0.2) \times 10^{-3}$
25	0.27 ± 0.2	$(7.3 \pm 0.3) \times 10^3$	$(1.6 \pm 0.4) \times 10^{-3}$
35	0.12 ± 0.1	$(1.7 \pm 0.2) \times 10^4$	$(1.7 \pm 0.4) \times 10^{-3}$
45	0.05 ± 0.006	$(3.6 \pm 0.3) \times 10^4$	$(3.6 \pm 0.6) \times 10^{-3}$

isofagomine (4) k_{off} is virtually independent of temperature while the k_{on} increase with temperature, resulting in a decrease in K_i with temperature (Table 6.8) [75]. This is in contrast to **30**, for which the K_i does not change much with temperature, while the K_i of 1-deoxynojirimycin was found to decrease with increasing temperature. The consequence of this behaviour in a thermodynamic sense is that the binding of **4** must have a positive standard enthalpy associated with binding, while it is negative for 1-deoxynojirimycin and essentially zero for **30** [88].

This result, which has been regarded as controversial, was questioned by Davies and collaborators who measured binding enthalpies of **4** and 1-deoxynojirimycin using microcalorimetry [89]. They found that both compounds gave negative enthalpy of binding in the calorimetry experiment and called it a reassessment of the above results. Apparently these authors were not aware that the enthalpy measured in the calorimeter is not a standard value and therefore cannot be relied on to be identical with ΔH standard. In fact their results confirm that **4** is a stronger inhibitor at higher temperatures meaning that it must have positive ΔH standard.

6.8 Are 1-azasugars (and iminosugars) transition state analogues?

It is a widely recognized idea that the most potent inhibitors are those that mimic the transition state of an enzymatic reaction. The classification of inhibitors as to whether they are transition state analogues or not, is not a totally academic question, since transition state analogues should be a better starting point for the design of new inhibitors. 1-Azasugars and iminosugars, like 1-deoxynojirimycin (Figure 6.1) have, in their protonated forms, an obvious similarity with the transition state of glycoside hydrolysis, though certain imperfections like the absence of sp^2 geometry in the molecule are also evident.

It has been argued by Withers [90], and later Davies [89], that the fact that these nitrogen-containing compounds have a structural resemblance with the glycosidase transition state and are strong glycosidase inhibitors, does not necessarily prove that they are transition state analogues. Withers has encouraged the use of so-called 'rigorous' criteria to classify whether glycosidase inhibitors are transition state analogues or what he has termed as fortuitous binders (supposedly meaning inhibitors by chance). Some of the criteria termed 'rigorous' are :

1. The inhibitor should be most potent at the pH optimum of the enzyme. Isofagomine's (**4**) inhibition of β-glucosidase has a pH basic optimum in a more basic range than the enzyme catalysis. Azafagomine (**30**) and glucooxazine (**35**) have inhibition optima that more closely resemble the enzyme pH optimum and should therefore better qualify according to this criterion.
2. The inhibitor should bind to the catalytically active form of the enzyme. Davies has found (see above) that **10**, when it is bound to endocellulase Cel5A, is protonated while the enzyme is di-deprotonated, and argues that, since the di-deprotonated form is not catalytically active, **10** cannot be a transition state analogue. The complex formation could, however, occur by binding of unprotonated **10** to the catalytically active form of the enzyme with the proton being transferred subsequently.
3. A satisfactory free-energy relationship must exist between inhibition and catalysis when the inhibitor and the substrate are analogously modified. This means that the same structural modification, such as an epimerization, in the inhibitor and the substrate should have the same effect on inhibition and catalysis. Such a plot has actually not been made for 1-azasugars, but Withers has shown that iminosugars do not give a good free-energy correlation with α-glucosidase activity. There are several indications that 1-azasugars will not give a good correlation, such as the observation that *galacto*-isofagomin (**18**) is a stronger inhibitor of β-glucosidase than **4** even though glucosides are better substrates than galactosides.

Thus according to several if not all of these 'rigorous' criteria, 1-azasugars fail to be transition state analogues. Since it has only been investigated for a few cases, and since each enzyme and each substrate will have a new transition state, the inhibitors may very well be transition state analogues in cases other than those investigated; nevertheless the conclusion that isofagomines are not transition state analogue inhibitors of the enzymes against which they are most potent is remarkable, and it follows logically that we should not use these compounds as starting points for further improvement of inhibitor design. The latter approach would obviously not be very judicious, since much of the most remarkable progress within the glycosidase inhibitor field in recent years has been made with iminosugars and azasugars. It is interesting to observe that even Withers and Davies do not believe so firmly in the 'rigorous' criteria that it prevents them from working very intensely with these compounds.

So what is the problem with the 'rigorous' criteria for determining transition state analogy in the present instance? First of all, it is necessary to realize that perfect transition state analogues do not exist, and that all will have imperfections that may or may not be revealed by such analysis. Further the value of using criteria 1 and 2 for inhibitors that change ionization state with pH is debatable. It is of concern that rule 1 points to azafagomine (**30**) as being a better transition state analogue than **4** simply because its pK_a matches the enzyme pH optimum better, even though it is a weaker inhibitor. Also the use of rule 2 on the various ionized forms of the enzyme is questionable; while the substrate definitely interacts with the monoprotonated form (the catalytic form), in the transition state the enzyme has essentially delivered a proton and may actually resemble more the unprotonated (noncatalytic) form. It is also important that the significance of the linear free relationship plots (rule 3) should not be not overinterpreted. The results of such an analysis only apply to the hydrolysis of a specific substrate, which naturally contains an aglycon as well, and since the studied azasugars/iminosugars only mimic the glycon (except for **8** and **9**) they cannot be more than partial transition state analogues in any case.

References

1. McNaught, A. D. (1996) Nomenclature of carbohydrates. *Pure & Appl. Chem.* **68**, 1919–2008.
2. Reymond, J. –L., Janda, K. D. and Lerner, R. A. (1991) *Angew. Chem. Int. Ed. Engl.*, **30**, 1711–1713.
3. Jespersen, T. M., Dong, W., Skrydstrup, T., Sierks, M.R., Lundt, I. and Bols, M. (1994) *Angew. Chem. Int. Ed. Engl.*, **33**, 1778–1779.
4. Lundgren, K., Rassov, A. and Bols, M. (1996) *Diabetes*, **45**, 521.
5. Jespersen, T. M., Bols, M., Sierks, M. R. and Skrydstrup, T. (1994) *Tetrahedron*, **50**, 13449–13460.
6. Ichikawa, Y., Igarashi, Y., Ichikawa, M. and Suhara, Y. (1998) *J. Am. Chem. Soc.*, **120**, 3007–3018.
7. Andersch, J. and Bols, M. (2001) *Chem Eur. J.*, **7**, 3744–3747.
8. Best, W. M., MacDonald, J. M., Skelton, B. W., Stick, R. V., Tilbrook, D. M. G. and White, A. H. (2002) *Can. J. Chem.*, **80**, 857–865.
9. Pandey, G. and Kapur, M. (2002) *Org. Lett.*, **4**, 3883–3886.
10. Kim, Y. J., Ichikawa, M. and Ichikawa, Y. (2000) *J. Org. Chem.*, **65**, 2599–2602.
11. Pandey, G. and Kapur, M. (2000) *Tetrahedron Lett.*, **41**, 8821–8824.
12. Pandey, G., Kapur, M., Khan, M. I. and Gaikwad, S. M. (2003) *Org. Biomol. Chem.*, **1**, 3321–3326.
13. Pandey, G. and Kapur, M. (2001) *Synthesis*, **8**, 1263–1267.
14. Zhao, G., Deo, U. C. and Ganem, B. (2001) *Org. Lett.*, **3**, 201–203.
15. Hansen, S. U. and Bols, M. (2000) *J. Chem. Soc. Perkin Trans 1*, **6**, 911–915.
16. Banfi, L., Guanti, G., Paravidino, M. and Riva, R. (2005) *Org. Biomol. Chem.*, **3**, 1729–1737.
17. Ouchi, H., Mihara, Y., Watanabe, H. and Takahata, H. (2004) *Tetrahedron Lett.*, **45**, 7053–7056.
18. Ouchi, H., Mihara, Y. and Takahata, H. (2005) *J. Org. Chem.*, **70**, 5207–5214.
19. Tsuda, Y., Matsuhira, N. and Kanemitsu, K. (1985) *Chem. Pharm. Bull.*, **33**, 4095–4097.
20. Liu, H., Liang, X., Søhoel, H., Bülow, A. and Bols M. (2001) *J. Am. Chem. Soc.*, **123**, 5116–5117.
21. Kondo, K. -I., Adachi, H., Shitara, E., Kojima, F. and Nishimura, Y. (2001) *Bioorg. Med. Chem.*, **9**, 1091–1095.
22. Liu, H., Lillelund, V. H., Andersch, J., Liang, X. and Bols M. (2004) *J. Carbohydr. Chem.*, **23**, 223–238.
23. Schuster, M. (1999) *Bioorg. Med. Chem. Lett.*, **9**, 615–618.
24. Schuster, M. and Blechert, S. (1999) *Tetrahedron: Asymmetry*, **10**, 3139–3145.
25. Dong, W., Jespersen, T. M., Skrydstrup,T., Bols, M. and Sierks, M.R. (1996) *Biochemistry*, **35**, 2788–2795.

26. Andreassen, V., Svensson, B. and Bols, M. (2001) *Synthesis*, 339–342.
27. MacDonald, J. M., Stick, R. V., Tilbrook, D. M. G. and Withers, S. G. (2002) *Aust. J. Chem.*, **55**, 747–752.
28. Varrot, A., Tarling, C. A., MacDonald, J. M., Stick, R. V., Zechel, D. L., Withers, S. G. and Davies, G. J. (2003) *J. Am. Chem. Soc.*, **125**, 7496–7497.
29. Lundt, I., Steiner, A. J., Stütz, A. E., Tarling, C. A., Ully, S., Withers, S. G. and Wrodnigg, T. M. (2006) *Bioorg. Med. Chem.*, **14**, 1737–1742.
30. Greimel, P., Häusler, H., Lundt, I., Rupitz, K., Stütz, A. E., Tarling, C. A., Withers, S. G. and Wrodnigg, T. M. (2006) *Bioorg. Med. Chem. Lett.*, **16**, 2067–2070.
31. Jakobsen, P., Lundbeck, J. M., Kristiansen, M., Breinholt, J., Demuth, H., Pawlas, J., Candela, M. P. T., Andersen, B., Westergaard, N., Lundgren, K. and Asano, N. (2001) *Bioorg. Med. Chem,.* **9**, 733–744.
32. Zhu, X., Sheth, K. A., Shihong, L., Hui-Hwa, C. and Fan, J. –Q. (2005) *Angew. Chem. Int. Ed.*, **44**, 7450–7453.
33. Xie, J., Güveli, T., Hebbe, S. and Dechoux, L. (2004) *Tetrahedron Lett.*, **45**, 4903–4906.
34. Lohse, A., Jensen, H. H., Bach, P. and Bols, M. (2000) *J. Chem. Soc. Perkin Trans 1*, 659–665.
35. Jensen, H. H., Lohse, A., Petersen, B. O., Duus, J. Ø and Bols, M. (2000) *J. Chem. Soc. Perkin Trans 1*, 667–670.
36. Jensen, H. H., Jensen, A., Hazell, R. and Bols, M. (2002) *J. Chem. Soc. Perkin Trans 1*, 1190–1198.
37. Lohse, A. and Bols, M. (1997) *Tetrahedron*, **53**, 6917–6924.
38. Matin, M. M., Sharma, T., Sabharwal, S. G. and Dhavale, D. D. (2005) *Org. Biomol. Chem.*, **3**, 1702–1707.
39. Ichikawa, Y. and Igarashi, Y. (1995) *Tetrahedron Lett.*, **36**, 4585–4586.
40. Jensen, H. H. and Bols, M. (2001) *J. Chem. Soc. Perkin Trans 1*, 905–909.
41. Søhoel, H., Liang, X. and Bols, M. (2000) *Synlett.*, **7**, 347–348.
42. Kim, Y. J., Ichikawa, M. and Ichikawa, Y. (1999) *J. Am. Chem. Soc.*, **121**, 5829–5830.
43. Igarashi, Y., Ichikawa, M. and Ichikawa, Y. (1996) *Bioorg. Med. Chem. Lett.*, **6**, 553–558.
44. Hansen, A., Tagmose, T. M. and Bols, M. (1996) *J. Chem Soc. Chem. Commun.*, 2649–2650.
45. Igarashi, Y., Ichikawa, M. and Ichikawa, Y. (1996) *Tetrahedron Lett.*, **37**, 2707–2708.
46. Nishimura, Y. (1995) *Studies in Natural Products Chemistry*, **16**, 75–121.
47. Nishimura, Y. (1992) *Studies in Natural Products Chemistry*, **10**, 495–583.
48. Lillelund, V., Liang, X., Jensen, H. H. and Bols, M. (2002) *Chem. Rev.*, **102**, 515–553.
49. Bols, M., Persson, M. P., Butt, W. M., Jørgensen, M., Christensen, P. and Hansen, L. T. (1996) *Tetrahedron Lett.*, **37**, 2097–2100.
50. Hansen, S. U. and Bols, M. (1998) *Acta Chem. Scand.*, **52**, 1214–1222.
51. Makino, K. and Ichikawa, Y. (1998) *Tetrahedron Lett.*, **39**, 8245–8248.
52. Sørensen, M. D., Khalifa, N. G. and Pedersen, E. B. (1999) *Synthesis*, **11**, 1937–1943.
53. Filichev, V. V., Brandt, M. and Pedersen, E. B. (2001) *Carbohydr. Res.*, **333**, 115–122.
54. Filichev, V. V. and Pedersen, E. B. (2001) *Tetrahedron*, **57**, 9163–9168.
55. Karlsson, S. and Högberg, H. –E. (2001) *Tetrahedron: Asymmetry*, **12**, 1977–1982.
56. Filichev, V. V. and Pedersen, E. B. (2003) *Org. Biomol. Chem.*, **1**, 100–103.
57. Galeazzi, R., Martelli, G., Mobbili, G., Orena, M. and Rinaldi, S. (2004) *Tetrahedron: Asymmetry*, **15**, 3249–3256.
58. Hassan, A. A., Jørgensen, P. T., Stein, P. C., Abdel Fattah, M. E., El Gawad, I. I. A. and Pedersen, E. B. (2004) *Carbohydr. Res.*, **339**, 1565–1568.
59. Zhou, G. –C., Parikh, S. L., Tyler, P. C., Evans, G. B., Furneaux, R. H., Zubkowa, O. V., Benjes, P. A. and Schramm, V. L. (2004) *J. Am. Chem. Soc.*, **126**, 5690–5698.

60. Evans, G. B., Furneaux, R. H., Lewandowicz, A., Schramm, V. L. and Tyler, P. C. (2003) *J. Med. Chem.*, **46**, 5271–5276.
61. Evans, G. B., Furneaux, R. H., Tyler, P. C. and Schramm, V. L. (2003) *Org. Lett.*, **5**, 3639–3640.
62. Evans, G. B., Furneaux, R. H., Lenz, D. H., Painter, G. F., Schramm, V. L., Singh, V. and Tyler, P. C. (2005) *J. Med. Chem.*, **48**, 4679–4689.
63. Kotian, P. L. and Chand, P. (2005) *Tetrahedron Lett.*, **46**, 3327–3330.
64. Williams, S. J., Hoos, R. and Withers, S. G. (2000) *J. Am. Chem. Soc.*, **122**, 2223–2235.
65. Shitara, E., Nishimura, Y., Kojima, F. and Takeuchi, T. (1999) *Bioorg. Med. Chem,.* **7**, 1241–1246.
66. van den Berg, R. J. B. H. Donker-Koopman, Wi., van Boom, J. H., Aerts, H. M. F. G. and Noort, D. (2004) *Bioorg. Med. Chem.*, **12**, 891–902.
67. Godskesen, M. and Lundt, I. (1998) *Tetrahedron Lett.*, **39**, 5841–5844.
68. Fort, S., Coutinho, P. M., Schülein, M., Nardin, R., Cottaz, S. and Driguez, H. (2001) *Tetrahedron Lett.*, **42**, 3443–3446.
69. Knapp, S., Yang, C., Pabbaraja, S., Rempel, B., Reid, S. and Withers, S. G. (2005) *J. Org. Chem.*, **70**, 7715–7720.
70. Mark, B. L., Vocadlo, D. J., Zhao, D., Knapp, S., Withers, S. G. and James, M. N. G. (2001) *J. Biol. Chem.*, **276**, 42131–42137.
71. Bols, M., Hazell, R. and Thomsen, I. B. (1997) *Chem. Eur. J.*, **3**, 940–947.
72. Ernholt, B. V., Thomsen, I. B., Lohse, A., Plesner, I., Jensen, K. B., Hazell, R. G., Liang, X., Jacobsen, A. and Bols, M. (2000) *Chem. Eur. J.*, **6**, 278–287.
73. Ernholt, B. V., Thomsen, I. B., Jensen, K. B. and Bols, M. (1999) *Synlett.*, **6**, 701–704.
74. Søndergaard, K., Liang, X. and Bols, M. (2001) *Chem. Eur. J.*, **7**, 2324–2331.
75. Bülow, A., Plesner, I. and Bols M. (2001) *Biochem. Biophys. Acta*, **1545**, 207–215.
76. Lohse, A., Hardlei, T., Jensen, A., Plesner, I. and Bols, M. (2000) *Biochem. J.*, **349**, 211–215.
77. Liang, X. and Bols, M. (1999) *J. Org. Chem.*, **64**, 8485–8488.
78. Stick, R. V. (1997) *Top. Curr. Chem.*, **187**, 187–213.
79. Bach, P. and Bols, M. (1999) *Tetrahedron Lett.*, **40**, 3461–3464.
80. Gloster, T. M., MacDonald, J. M., Tarling, C. A., Stick, R. V., Withers, S. G. and Davies, G. J. (2004) *J. Biol. Chem.*, **279**, 49236–49242.
81. Williams, S. J., Notenboom, V., Wicki, J., Rose, D. R. and Withers, S. G. (2000) *J. Am. Chem. Soc.*, **122**, 4229–4230.
82. Gloster, T., Williams, S. J., Tarling, C. A., Roberts, S., Dupont, C., Jodoin, P., Shareck, F., Withers, S. G. and Davies, G. J. (2003) *Chem. Commun.*, 944–945.
83. Søhoel, H., Liang, X. and Bols, M. (2001) *J. Chem. Soc., Perkin Trans. 1*, 1584–1585.
84. Lillelund, V., Liu, H., Liang, X., Søhoel, H. and Bols, M. (2003) *Org. Biomol. Chem.*, **1**, 282–287.
85. Vincent, F., Gloster, T. M., Macdonald, J., Morland, C., Stick, R. V., Dias, F. M. V., Prates, J. A. M., Fontes, C. M. G. A., Gilbert, H. J. and Davies, G. J. (2004) *ChemBioChem.*, **5**, 1596–1599.
86. Ramana, C. V. and Vasella, A. (2000) *Helv. Chim. Acta*, **83**, 1599–1610.
87. Ulgar, V., Fernandez, J. M. and Bols, M. (2002) *J. Chem. Soc. Perkin Trans 1*, 1242–1246.
88. Bülow, A., Plesner I. and Bols M. (2000) *J. Am. Chem. Soc.*, **122**, 8567–8568.
89. Zechel, D. H., Baraston, A. B., Gloster, T. M., Boraston, C. M., Macdonald, J. M., Tilbrook, D. M. G. and Stick, R. V. (2003) *J. Am. Chem. Soc.*, **125**, 14313–14323.
90. Withers, S. G., Namchuk, M. and Mosi, R. (1999) In *Iminosugars as Glycosidase Inhibitors* (Stütz, A., ed.), Wiley-VCH, Weinheim, Germany.

7

Iminosugar-based glycosyltransferase inhibitors

Lisa J. Whalen, William A. Greenberg, Michael L. Mitchell and **Chi-Huey Wong**

7.1 Biological role and structural features of glycosyltransferases

7.1.1 Biological role of glycosyltransferases

Enzymatic glycosylations are among the most important and abundant reactions in living systems [1, 2]. Glycosyltransferases (GTs, E.C. 2.4.x.y) are involved in many details of cellular biochemistry, such as protein and lipid glycosylation, as well as in the synthesis of polysaccharides such as cellulose, the main component of biomass [3]. The role of cell surface oligosaccharides in processes such as intercellular recognition, cancer cell metastasis and the immune response to viral and bacterial infection, has generated interest in the inhibition of GTs involved in the biosynthesis of oligosaccharides and glycoconjugates, which may lead to the discovery of novel drug therapies [4–7].

Glycosyltransferases are divided into two mechanistic classes: those of the Leloir pathway [8–10] or of the non-Leloir pathway. The GTs of the Leloir pathway utilize nucleoside mono- or diphosphate sugars as monosaccharide donors for the synthesis of oligosaccharides and glycoconjugates [11]. The non-Leloir pathway GTs typically use glycosyl phosphates, sucrose or glycolipid phosphates as activated donors [12, 13].

The Leloir GTs catalyse the transfer of a sugar moiety from an activated nucleotide sugar to the hydroxyl group of an acceptor, which may be an oligosaccharide, a lipid or a protein [14]. Depending on the GT involved, the reaction can proceed with inversion or retention of the anomeric configuration of the nucleotide sugar donor. For inverting GTs, the acceptor nucleophile attacks C1 of the donor in an S_N2-like manner, leading to inversion. Retaining GTs are believed to proceed through two inversions, first generating

Iminosugars: From synthesis to therapeutic applications Edited by Philippe Compain and Olivier R. Martin
 ISBN 978-0-470-03391-3

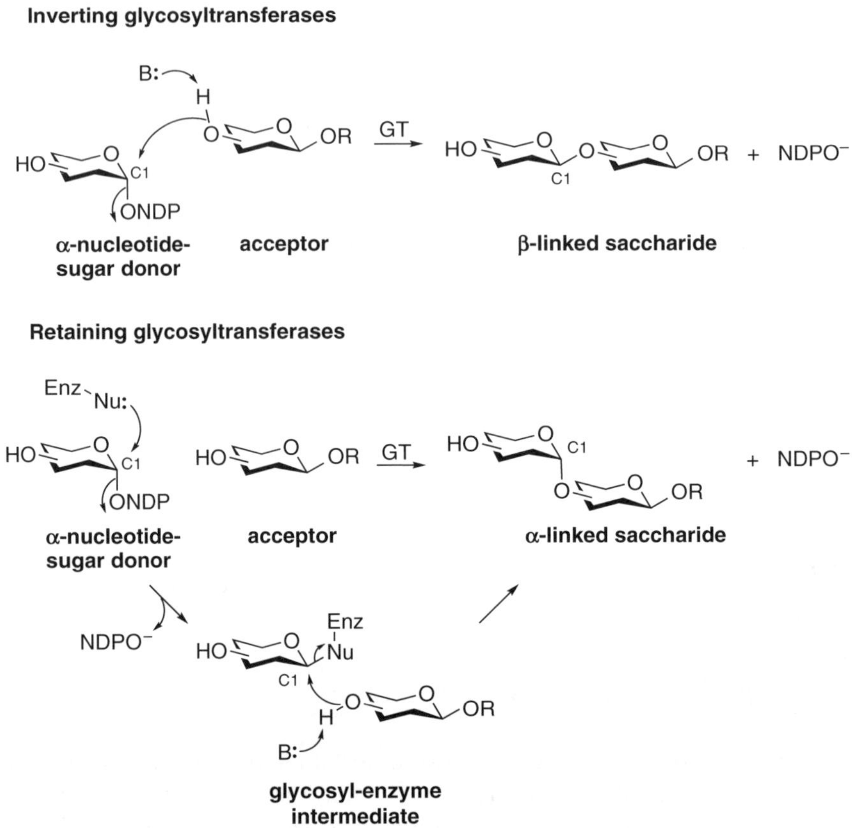

Figure 7.1 The proposed mechanisms for GT-catalysed reactions

a glycosyl-enzyme intermediate and then proceeding to the reaction product in the second step (Figure 7.1).

There are nine commonly found monosaccharide donors in mammals: UDP-glucose (UDP-Glc, **1**), UDP-*N*-acetylglucosamine (UDP-GlcNAc, **2**), UDP-galactose (UDP-Gal, **3**), UDP-*N*-acetylgalactosamine (UDP-GalNAc, **4**), UDP-xylose (UDP-Xyl, **5**), GDP-mannose (GDP-Man, **6**), GDP-fucose (GDP-Fuc, **7**), UDP-glucuronic acid (UDP-GlcUA, **8**) and CMP-*N*-acetylneuraminic acid (CMP-NeuAc, **9**) (Figure 7.2) [15]. Though other monosaccharides are found in mammalian systems, they are usually the result of modifications after incorporation into an oligosaccharide [16, 17]. Glycosyltransferases are typically classified according to the sugar nucleotide donor used; for example, a GT that uses UDP-galactose as the donor is described as a galactosyltransferase [14, 18].

7.1.2 Structural features of glycosyltransferases

To date, classification of over 10 000 known and putative GTs based on amino acid sequence distributes them into 85 families (*http://afmb.cnrs-mrs.fr/CAZY)* [19]. The 228

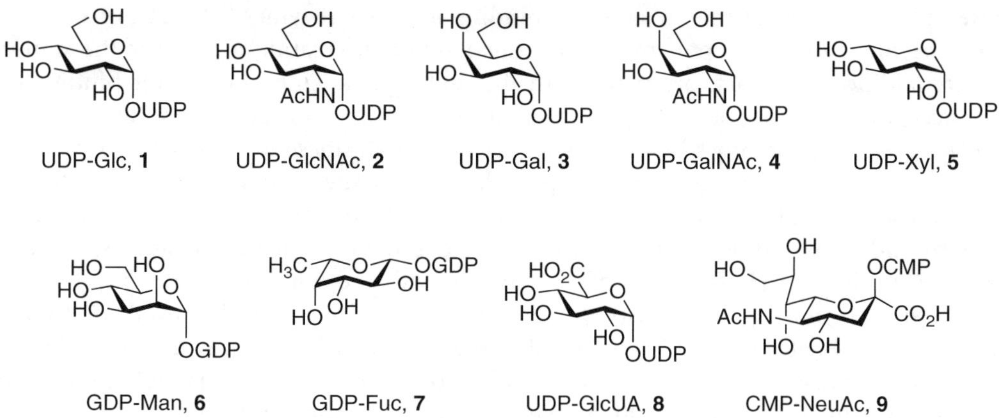

Figure 7.2 The nine common mammalian sugar nucleotide donors

human GT genes divide into 42 separate families, reflecting the large number of acceptor molecules used by GTs. Further complicating matters, many examples of closely related sequences exist which possess different catalytic activity [20], making sequence-based prediction of enzyme mechanism and function difficult.

Since the X-ray crystal structure of the DNA-modifying enzyme bacteriophage T4 β-glucosyltransferase (BGT) was solved [21], over 100 crystal structures of GTs have been reported (http://www.cermav.cnrs.fr/glyco3d). These structures represent 38 distinct GTs, covering 22 different families. Not surprisingly, the nucleotide-binding GTs possess folds similar to the classic Rossmann fold [22], with α/β/α sandwiches that form β-sheets for the binding of nucleotides. Two structural superfamilies, GT-A and GT-B, are well studied and were first described in the crystal structures of *Bacillus subtilis* SpsA [23] and BGT. A third structural superfamily has recently emerged from studies of the crystal structure of the bacterial sialyltransferase Cst II [24].

The GT-A fold comprises an α/β/α sandwich, which contains a seven-stranded β-sheet in which strand 6 is antiparallel to the remaining strands [19]. A smaller β-sheet flanks the central β-sheet, and association of both sheets creates the active site. Most GT-A family members also possess another characteristic structural feature, the DxD (or ExD or equivalent) motif (Asp-x-Asp). This motif is involved in the coordination of a divalent metal cation, which is required for activity, in the binding of the nucleotide-sugar [25]. However, an example exists of a GT-A family member that does not possess a DxD motif: it is a metal-independent enzyme [26]. The DxD motif is always found at the same location, in a short loop connecting one β-strand of the central β-sheet to a smaller β-strand. Although not always observed in the crystal structures, the divalent cation is believed to coordinate the oxygen atoms of the α and β phosphates and may bridge the protein surface and the nucleotide-sugar using the DxD motif [27, 28]. Interestingly, the DxD motif interacts differently with the divalent metal cation in inverting and retaining GT-A enzymes. In inverting enzymes, only the last Asp residue interacts with the cation, whereas both Asp residues interact with the cation in retaining enzymes [29, 30]. In addition to the DxD motif's importance, flexible loops often play a significant role in substrate binding in GT-A enzymes [31]. The disordered loop becomes ordered upon nucleotide sugar binding, creating a lid over the donor where additional contacts are made with the diphosphate and a pocket for binding of the acceptor. This may serve to prevent abortive hydrolysis of the nucleotide sugar.

The GT-A superfamily's binding mode for both manganese ion and UDP was elucidated from the crystal structure of SpsA, leading to speculation about the catalytic mechanism [23]. Three invariant residues in the N-terminal nucleotide-binding domain were found to have direct contact with the nucleotide diphosphate: Tyr 11, Asp 39 and Asp 99. While Tyr 11 was involved in stacking with the nucleotide base, Asp 39 was involved with binding UDP. Asp 99 was positioned adjacent to the distal phosphate where it coordinates the Mn^{2+} ion, ideally suited to play a role in leaving group departure. Another residue, Asp 191, was located in a position to function as the general base in this inverting GT, although three other residues were also candidates.

The GT-B fold comprises two similar domains reminiscent of a Rossmann fold, with a connecting linker region and a catalytic site located between the domains. While this C-terminal nucleotide-binding domain is well conserved amongst the members of the GT-B class, variations are widely found in the N-terminal domains where loops and helices point towards the active site. This allows for the accommodation of different acceptors. The interaction of a glutamate residue and glycine-rich loops with the ribose and phosphate of the sugar-nucleotide donor, respectively, have been reported [32]. No bound metal ions have been found even though divalent metals may be required for full activity.

The first example of a structure of a member of the GT-B superfamily, bacteriophage T4 β-glucosyltransferase, was described by Freemont and coworkers [21]. Upon binding of UDP-Glc by bacteriophage T4 β-glucosyltransferase, a conformational change occurs to produce a closed conformation with a channel through which the acceptor hydroxyl can attack C1 of glucose. The reaction proceeds with inversion of configuration at C1 to give glucosylated 5-hydroxymethyl cytosine in the DNA. The UDP portion of the donor UDP-Glc was fully resolved while no significant electron density was observed for the glucose moiety.

A third superfamily has recently been described with the crystal structure of Cst II, a sialyltransferase from *Campylobacter jejuni* [24]. As the first structure of a GT that utilizes a nucleotide monophosphosugar, CMP-NeuAc, new structural elements were anticipated. Indeed, this enzyme possesses a unique seven-stranded parallel β-sheet and has no DxD motif. As with the GT-A superfamily, flexible loops play a significant role in substrate binding.

Truncated Cst II was cocrystallized with CMP-NeuAc and sialyllactose, the wild type donor and acceptor, as well as with the nonhydrolysable donor substrate analogue CMP-3-fluoro-NeuAc for comparison. In the case of CMP-NeuAc, only CMP was observed to bind in a deep cleft in the nucleotide-binding domain. It was assumed that the NeuAc moiety had been transferred to the acceptor, or the CMP-NeuAc was hydrolysed within the crystal. Aromatic stacking with Tyr 156 and hydrogen bonds between the cytidine carbonyl O2 with Asp 154 and Phe155 or cytidine N4 with Ser 161 helped secure the nucleotide base. For CMP-3-fluoro-NeuAc, little deviation was observed in the overall structure as compared with CMP-NeuAc. However, a large change in the order of residues 175–187 created an effective lid that closed over the donor sugar and formed the acceptor binding site.

The high-resolution X-ray crystal structures of rabbit *N*-acetylglucosaminyltransferase 1 (GnT 1) [33] and human β-1,3-glucuronyltransferase (GlcAT-I) [34] provided further insight into the mechanisms of GTs. The X-ray crystal structure of GnT 1 was the first structure to reveal the complete sugar nucleotide donor, which was not observed in any of the other structures. Negishi and coworkers determined the first crystal structure of a GT (GlcAT-I) in the presence of hydrolysed donor substrate (UDP), the catalytic Mn^{2+} ion and an acceptor substrate analog (Galβ1-3Galβ1-4Xyl). This structure implicated a glutamate carboxyl residue as the catalytic base responsible for deprotonating the incoming hydroxyl group of the acceptor.

The crystal structure of galactosyltransferase LgtC from *Neisseria meningitidis* in complex with manganese and UDP-2-deoxy-2-fluorogalactose (a donor sugar analogue) both in the presence and absence of 4'-deoxylactose (acceptor analogue) provides some insight into the catalytic mechanism of retaining GTs [29]. The donor analog is buried deeper within the enzyme than in the inverting GTs, reflecting the greater need to exclude water from the active site to limit hydrolysis of a reactive glycosyl-enzyme intermediate. The sugar nucleotide donor adopts a folded conformation (i.e., the UDP moiety is bound in the extended conformation while the galactose tucks back under the phosphates) which contrasts with the fully extended conformation seen in GnT 1 structure [33]. Upon binding of the sugar nucleotide, a conformational change creates the binding pocket for the acceptor sugar. The structure also revealed a Gln residue which may attack the anomeric centre to form an imidic ester intermediate, stabilized by a nearby (4.0 Å) Asp residue. The reaction may be similar to the *N*-acetylhexosaminidase-catalysed hydrolysis of *N*-acetylglucosylpolysaccharides, which proceeds through an oxazolinium ion intermediate resulting from the attack of the carbonyl of the 2-acetamide on C1. This Gln residue is contained within an invariant motif found in all members of the same family of retaining GTs [35], although definitive evidence of the catalytic role of the Gln residue remains elusive. Alternatively, the reaction may proceed via a front-side attack. The approach of the reactive hydroxyl of the acceptor toward the reaction centre would occur from the same side from which UDP would depart. This mechanism would most likely proceed through a highly dissociative (oxocarbenium ion-like) transition state [29].

7.1.3 Mechanistic insights

Much of the insight on the mechanism of inverting glycosyltransferases comes from biochemical studies of α-1,3-fucosyltransferase V (FucT V). Fucosylated oligosaccharide structures are critical to numerous cell–cell interactions and have generated interest as therapeutic targets for treatment of inflammatory diseases and cancer. Understanding the mechanism of the fucosyltransferase reaction would help in the design of inhibitors for use as anti-inflammatory or cancer therapeutics.

α-1,3-Fucosyltransferase V catalyses the transfer of L-fucose from GDP-Fuc to the 3-hydroxyl of the acceptor, with inversion of configuration at the anomeric centre of L-fucose. The transition state of glycosyltransferase reactions is thought to be similar

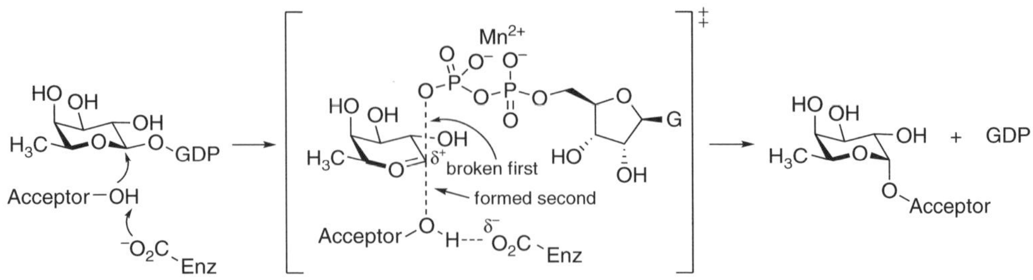

Figure 7.3 The proposed mechanism of α-1,3-fucosyltransferase V [27]

to those of glycosidase reactions [36, 37], exhibiting a flattened half-chair conformation with substantial oxocarbenium ion character at the anomeric position. Product inhibition studies established that FucT V has an ordered, sequential, 'bi-bi' mechanism with GDP-Fuc bound first and GDP released last [38]. FucT V has a catalytic residue with a pK_a of 4.1, presumably an active-site carboxylate residue responsible for shuffling a proton from the acceptor hydroxyl group to the departing nucleotide [39]. A solvent kinetic isotope effect was observed ($D_V = 2.9$, $D_{V/K} = 2.1$) and exploited in a proton inventory study to show a single proton transfer in the transition state [39]. A secondary isotope effect on the fucosyltransfer reaction with GDP-[1-^{2}H]-β-L-fucose was determined to be $D_v = 1.32$ and $D_{V/K} = 1.27$ [27]. The observed secondary isotope effect and the inhibition of the enzymatic reaction by GDP-2-fluorofucose are consistent with a charged, sp^2-hybridized transition state [27]. Based on the above data, a mechanism for FucT V was proposed in which the glycosidic cleavage occurs prior to the nucleophilic attack, a process between S_N1 and S_N2 mechanisms (Figure 7.3). Mechanistic insights gained in these studies have helped in the design of inhibitors of fucosyltransferases and other glycosyltransferases.

7.2 Development of inhibitors of glycosyltransferases

The glycosyltransferases that have received the most attention as targets for inhibitors are those that append terminal sugars to glycoconjugates; these enzymes play a major role in intercellular recognition and signalling events. These targets include fucosyltransferases, galactosyltransferases and sialyltransferases. Other prominent GT targets are involved in the biosynthesis of structural components of the fungal or bacterial cell wall (chitin synthase or MurG, respectively). Several different strategies have been used to identify potent inhibitors of GTs [40]. The most common inhibitor designs are analogues of the acceptor, the donor (i.e. unreactive sugar nucleotide) or the transition state, including bi- and tri-substrate analogues. Analogues which mimic the transition state are advantageous as they are expected to bind the enzyme more tightly than the natural substrate or analogues in the ground state [41]. Bisubstrate analogues, containing the glycosyl donor and acceptor in an arrangement simulating the transition state complex, are of particular interest because incorporating aspects of the acceptor saccharide leads to higher specificity for a particular glycosyltransferase [42]. Recent reviews [43–45] have summarized a broad variety of general approaches to inhibition of glycosyltransferases. In this chapter we focus on efforts that incorporate iminosugars into their design, and concentrate on recent developments and trends that have emerged since the latest reviews were published.

Carbohydrate mimetics simulate the shape and functionalities of the natural substrates in the ground state or in the transition state. One well-studied structural modification is the replacement of the endocyclic oxygen atom with a nitrogen atom, which produces an iminosugar (also improperly known as 'azasugar' or 'iminocyclitol') [46]. Under physiological conditions, the amino group of iminosugars is protonated and can interact with anionic groups in the enzyme active site, making them attractive targets as potential transition state analogues (Figure 7.4) [47]. Six-membered piperidinols exhibit hydroxyl group configurations similar to natural substrates, while the five-membered pyrrolidinols mimic the flattened, half-chair conformation of the glycosyl residue in the proposed transition state. Iminosugars are well known to be potent inhibitors of glycosidases [46], and the mechanistic similarities between glycosidases and GTs [36, 37] have led to the investigation of iminosugars as inhibitors of GTs.

General methods for the synthesis of iminosugars are described in other chapters of this volume. In this chapter, we highlight the methods that have been used in our laboratory to prepare a variety of these compounds, and illustrate the methods with several examples. In the following section we summarize the inhibitory activities of iminosugars against glycosyltransferases.

A powerful method for the synthesis of iminosugars and analogues involves the use of aldolases, a class of enzymes capable of forming carbon–carbon bonds in a stereoselective fashion by catalysing the aldol reaction between an aldehyde or ketone donor and another aldehyde or ketone acceptor. Many aldolases are commercially available, and directed evolution and structure-based mutagenesis have been used to modify or expand the substrate selectivity of these enzymes [48, 49]. The nitrogen atom of the iminosugar is typically introduced as an azide. Following the aldol reaction, the stereochemistry of which depends on the type of aldolase used, and enzymatic dephosphorylation if necessary, the azide is reduced by hydrogenation, followed by spontaneous formation of a cyclic imine and *in situ* reduction to the final iminosugar.

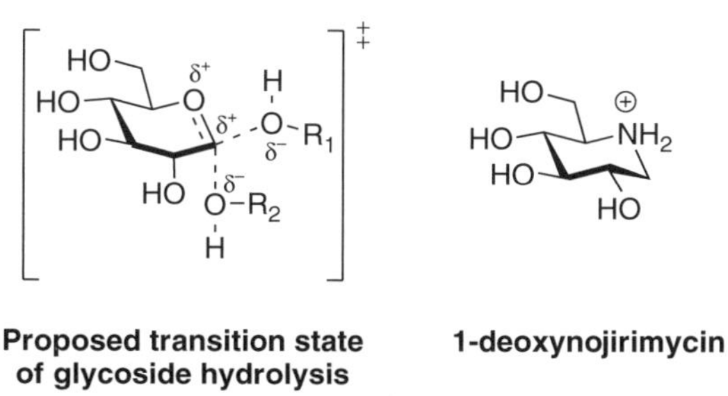

Figure 7.4 Comparison of the transition state of glycosidase hydrolysis and the iminosugar 1- deoxynojirimycin

The reduction is typically diastereoselective, with the stereochemistry of the new bond induced by the stereochemistry of the hydroxyl groups set with the aldolase. If the azido group is α- to the aldehyde in the acceptor, a five-membered ring is formed; if the azide is β, a six-membered ring results (Figures 7.5 and 7.6). By using a wide range of donor

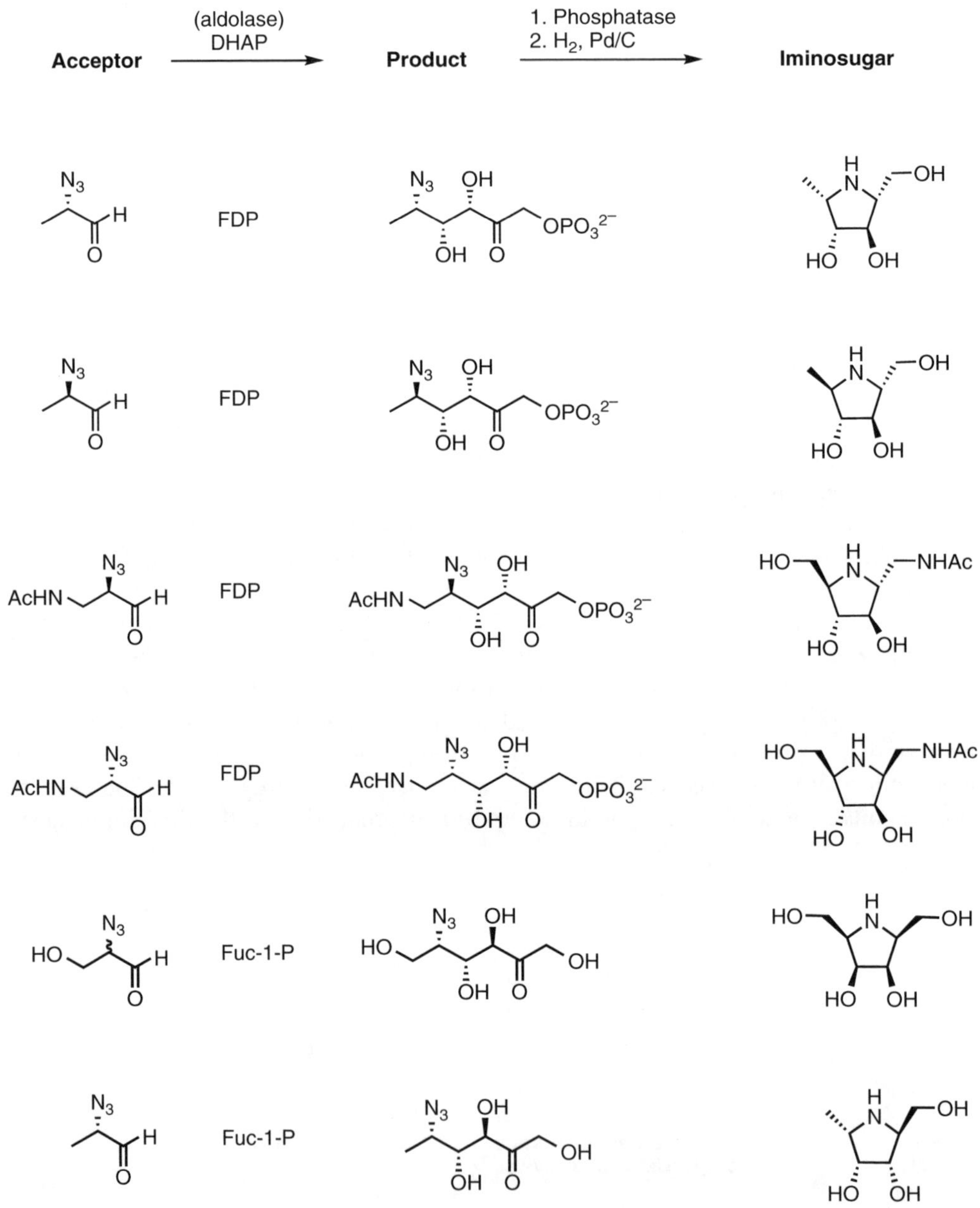

Figure 7.5 Examples of synthesis of five-membered iminosugars using aldolases [49]

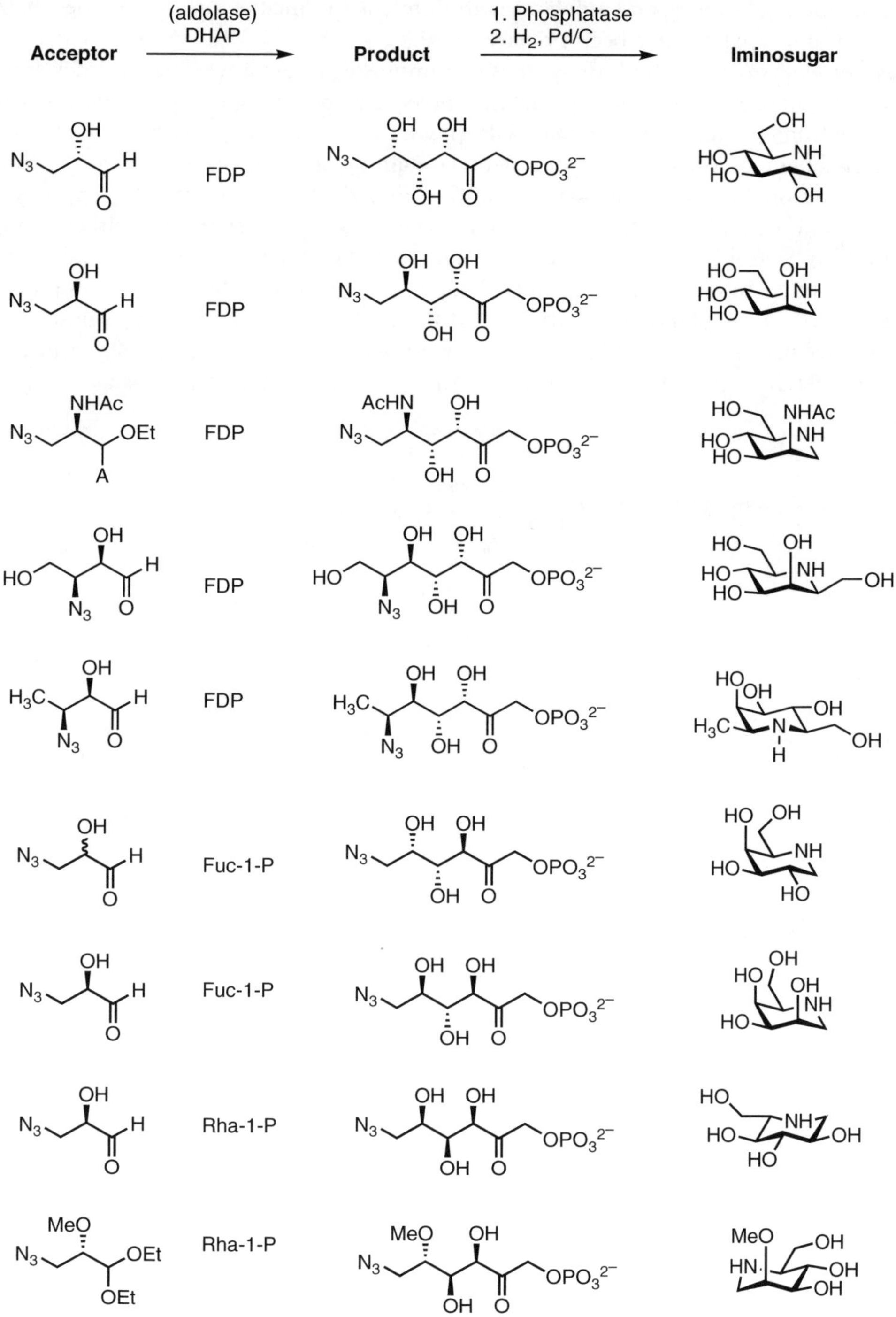

Figure 7.6 Examples of synthesis of six-membered iminosugars using aldolases [49]

and acceptor molecules, and aldolases with different specificities and selectivities, many different iminosugars have been prepared [49].

A non-enzymatic method for synthesis of iminosugar derivatives that was developed in our laboratory takes advantage of intramolecular epoxide opening of intermediates derived from pentose starting materials (Figure 7.7) [50]. The Wittig reaction of a protected pentose leads to an acyclic olefin containing one unprotected hydroxyl group. Depending on the choice of pentose starting material, iminosugars with a variety of configurations can be produced. The olefin is converted to an epoxide of either configuration by Sharpless asymmetric epoxidation, and the free hydroxyl group is inverted and converted to the azide by standard methods, through a chloromethanesulfonate ester intermediate. Staudinger reduction of the azide is accompanied by regioselective intramolecular epoxide opening, with complete selectivity for the *5-exo-tet* cyclization over the Baldwin-disfavoured *6-endo-tet*, to produce the protected iminosugar. Further

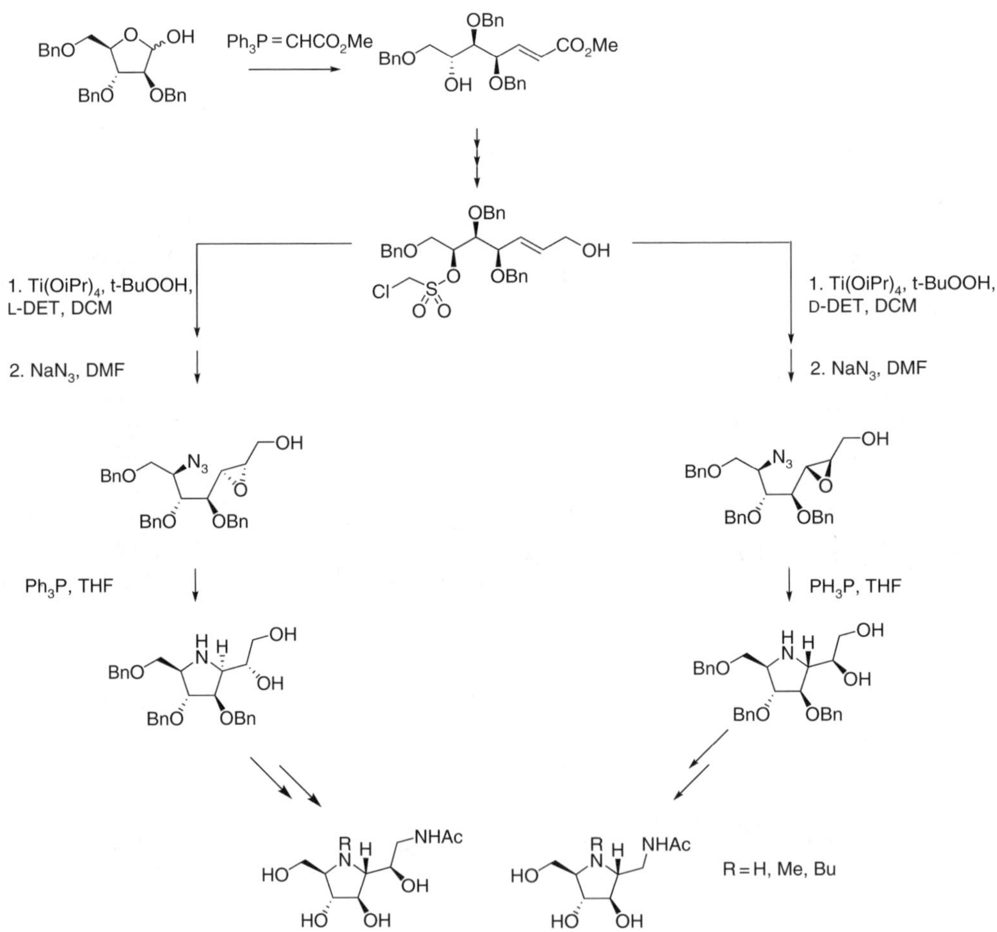

Figure 7.7 Chemical synthesis of iminosugars by epoxide-opening cyclization

transformations on these intermediates can lead to a variety of new derivatives, as shown in Figure 7.7.

7.2.1 Fucosyltransferase inhibitors

As discussed briefly at the beginning of this chapter, the fucosyltransferases (FucTs) play key roles in the biosynthesis of many important fucose-containing oligosaccharides such as sialyl Lewis^x (sLe^x), a determinant in numerous cell–cell interactions including inflammation [51, 52] and tumour metastasis [53–55]. The terminal step in the biosynthetic pathway of these fucose-containing saccharides is the addition of the L-fucose moiety by α-1,3-fucosyltransferase (α-1,3-FucT) [56–58]. Since inhibitors of fucosyltransferase may disrupt the biosynthesis of these saccharides, they have potential medicinal applications as antiinflammatory or antitumour agents.

Results for designed α-1,3-FucT inhibitors are illustrative of the general trends observed for inhibition of other glycosyltransferase inhibitors. Although simple monocyclic iminosugars can be potent inhibitors of glycosidases, with IC_{50} values in the low micromolar and nanomolar range, they are not generally very potent inhibitors of glycosyltransferases, only in the millimolar range. Clearly, inhibition of glycosyltransferases is much more complex than inhibition of glycosidases. In fact, one of the most potent inhibitors of α-1,3-FucT discovered thus far is GDP, the nucleotide by-product of the glycosyltranferase reaction, with an IC_{50} of 67 μM [38]. This product inhibition effect was exploited by us and others, first in the observation of a synergistic inhibitory effect when iminosugars **10**, **11**, **14** or **15** and GDP were added in combination, and then in the design of iminosugars that covalently incorporated the GDP moiety (**12** and **13**) (Figures 7.8 and 7.9) [38, 59–63].

Another approach to increasing inhibitor potency has been the design of bisubstrate analogues that incorporate both the iminosugar fucose mimetic and an acceptor sugar to provide additional contacts in the enzyme active site [38, 64]. Again, improved inhibitors were discovered, albeit at the cost of synthesizing quite complex molecules. In the case of bisubstrate analogs **17** and **20**, again a synergistic effect upon addition of GDP was observed, resulting in low micromolar inhibitors (Figure 7.10). The iminosugar-containing GDP-Fuc analogues **21** and **22** compared favourably with the bisubstrate inhibitors against both FucT V and FucT VI [65]. Despite the effort that has been expended towards the development of iminosugar-based inhibitors of fucosyltransferases, greater success has been realized with other types of inhibitors, such as the non-hydrolysable 2-fluoro substrate analogue [27].

7.2.2 Chitin synthetase inhibitors

Chitin synthetase (CS) catalyses the polymerization of *N*-acetyl-D-glucosamine starting from UDP-GlcNAc (Figure 7.11). CS is an essential enzyme for fungal cell wall biosynthesis but is absent in humans, making it an ideal drug target for antifungal agents.

	Reference	IC_{50}	IC_{50}, + 30 μM GDP
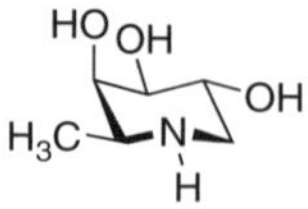 **10**	[38]	73 mM	3.55 mM
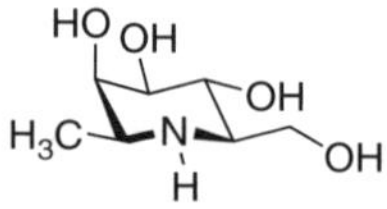 **11**	[38]	71.5 mM	1.54 mM
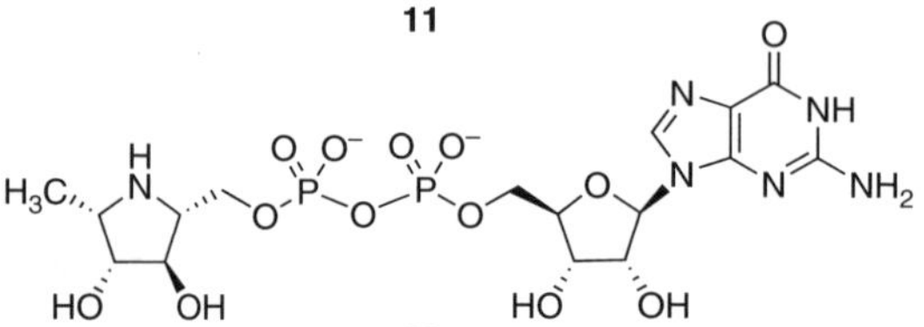 **12**	[60]	45 μM	
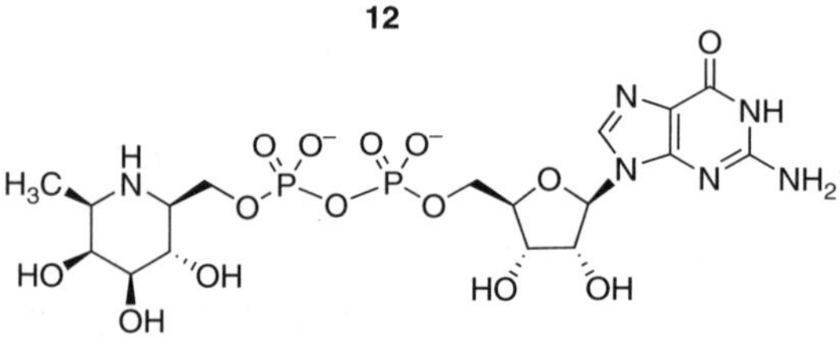 **13**	[59]	6 μM	

Figure 7.8 Improved fucosyltransferase inhibition by addition of GDP

	Reference	% Inhibition	% Inhibition, + 50 μM GDP
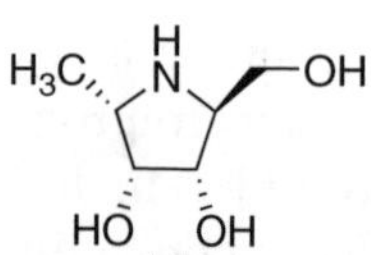 **14**	[61–63]	50% at 80 mM	80% at 80 mM
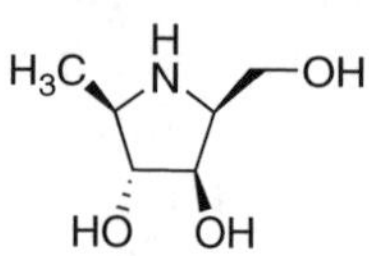 **15**	[61–63]	50% at 34 mM (K_i = 19 mM)	90% at 34 mM
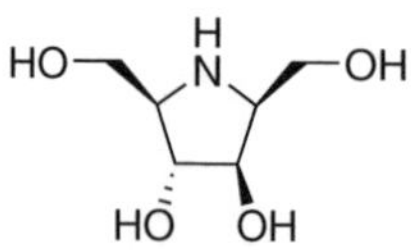 **16**	[63]	50% at 52 mM	

Figure 7.9 Improved fucosyltransferase inhibition by addition of GDP

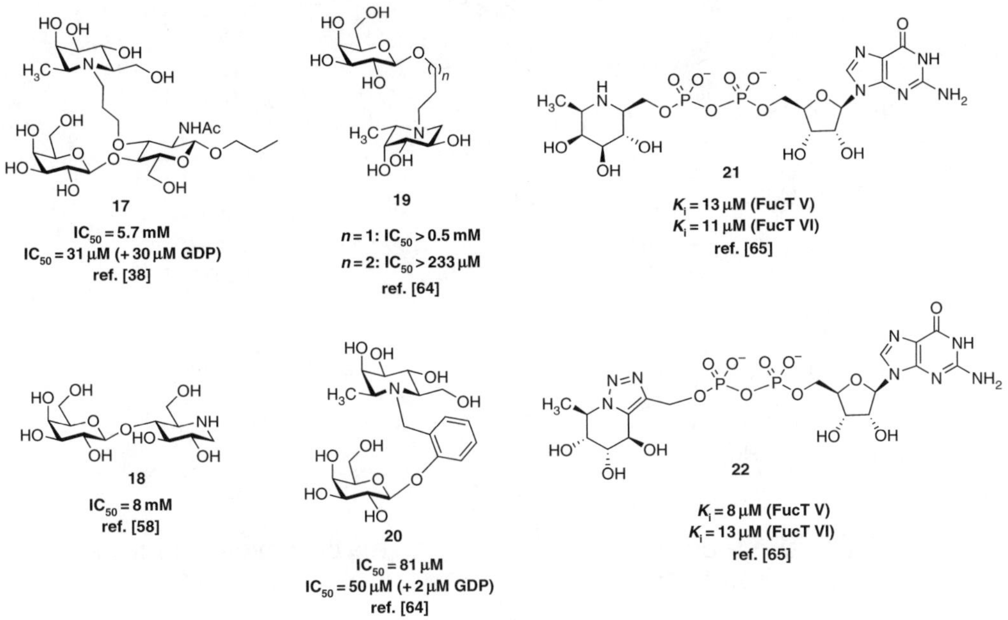

Figure 7.10 Bisubstrate and other inhibitors of fucosyltransferase

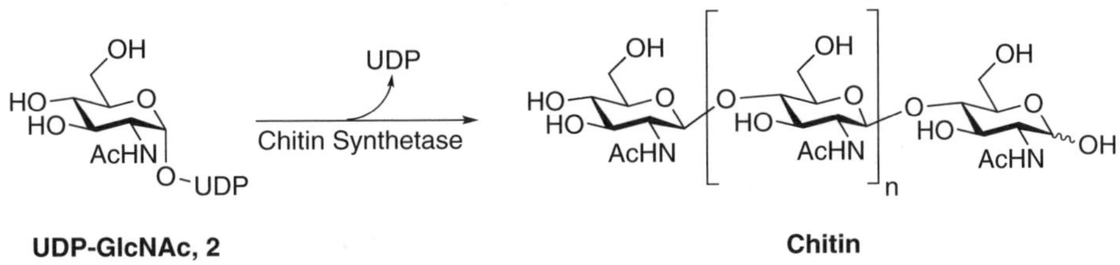

Figure 7.11 Biosynthesis of chitin by chitin synthetase

Selected inhibitors of chitin synthetase are illustrated in Figure 7.12. The sugar nucleotide analogue **23** features a five-membered iminosugar coupled with a malonate diester as a UDP mimic [66]. Interestingly, the benzyl protective groups were not removed due to instability, but the compound still exhibited modest activity against chitin synthetase although minimal antifungal activity was observed. The pyrrolidinol 6-deoxy-homoDMDP **24** was found to inhibit CS uncompetitively with a low micromolar K_i [67]. Furthermore, a similar synergistic inhibition pattern was observed with UDP and 6-deoxy-homoDMDP as was observed for GDP and the fucosyltransferase inhibitors (see above). Seeking to elaborate on the design of 6-deoxy-homoDMDP and find stronger inhibitors, a series of C2-substituted pyrrolidinols **25** and **26** was prepared and tested for inhibition, but without success [68]. No significant antifungal activity was observed for any of the 6-deoxy-homoDMDP derivatives.

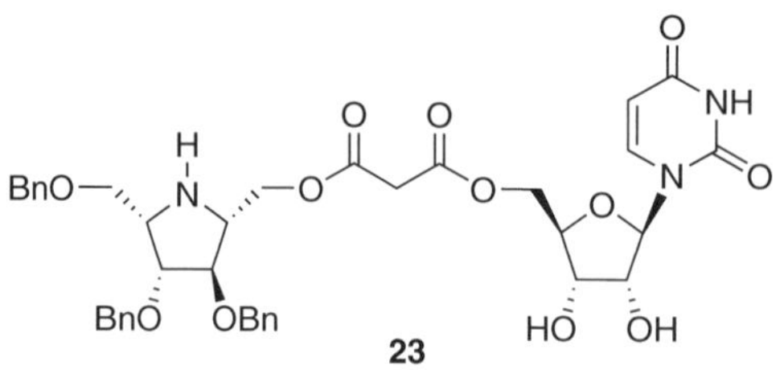

23
$K_i = 30\,\mu M$
ref. [66]

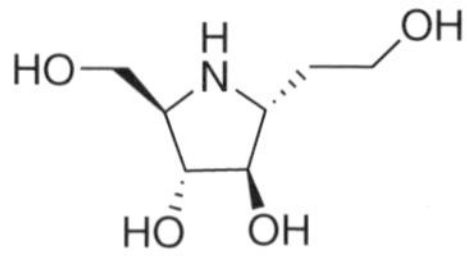

24
$K_i = 38\,\mu M$
ref. [67]

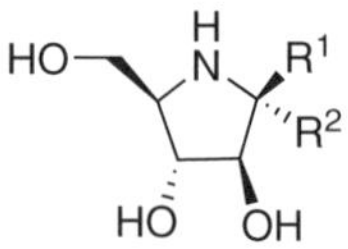

25

$R^1 = CH_2OH$, $R^2 = H$
$IC_{50} = 5.7$ mM

$R^1 = H$, $R^2 = CF_2P(O)(OH)(OEt)$
$IC_{50} = 38$ mM
ref. [68]

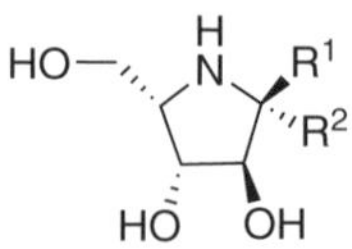

26

$R^1 = CF_2P(O)(OH)(OEt)$, $R^2 = H$
$IC_{50} = 4.0$ mM

$R^1 = H$, $R^2 = CF_2P(O)(OH)(OEt)$
$IC_{50} = 1.6$ mM
ref. [68]

Figure 7.12 Selected inhibitors of chitin synthetase

7.2.3 Ceramide glucosyltransferase inhibitors

Glycosphingolipids (GSLs) are ubiquitous components of the cellular membranes of eukaryotic cells [69]. GSLs are composed of at least one monosaccharide residue linked to a hydrophobic ceramide or sphingoid long-chain aliphatic amino alcohol that resides in the lipid bilayer. These molecules enrich the outer surface of the plasma membrane in a layer of carbohydrate that protects the cell membrane from chemical and mechanical damage [70]. However, insufficient catabolism of GSLs results in their accumulation, which is the physiological basis of Gaucher disease. Therefore, partial inhibition of the biosynthesis of GSLs may be an effective therapeutic strategy [70, 71]. Indeed, the iminosugar derivative *N*-butyl-1-deoxynojirimycin (NB-DNJ, Zavesca®) has become the first orally administered treatment for lysosomal storage disorders such as Gaucher disease. This small molecule inhibits ceramide glucosyltransferase, but it also inhibits glucosidase I, leading to a need for more selective therapy. Ceramide glucosyltransferase catalyses the transfer of glucose from UDP-Glc to the acceptor normal fatty acid ceramide *in vivo* and has been the focus of many inhibition studies (Figure 7.13).

Alternative ceramide glucosyltransferase inhibitors such as *N*-alkylated- deoxynojirimycin compounds 27–32 were also identified (Figure 7.14). The presence of an *N*-alkyl chain is obligatory for the inhibition as deoxynojirimycin (DNJ) is not an inhibitor of ceramide glucosyltransferase. There is a direct relationship between increasing the

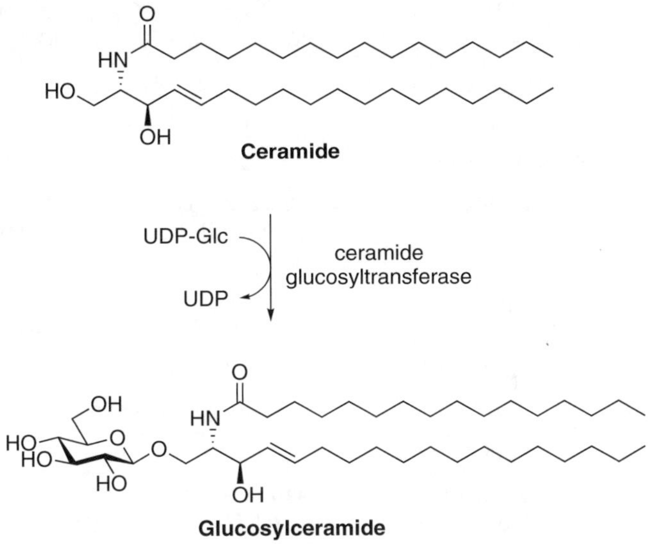

Figure 7.13 The reaction catalysed by ceramide glucosyltransferase *in vivo*

length of the alkyl chain and inhibitory potency, perhaps reflecting the better mimicry of the ceramide alkyl chains [72]. Five-membered iminosugars **33–37** were also investigated and displayed promising *in vitro* inhibition but only one possessed activity in tissue cells comparable to NB-DNJ [72].

Given the strong inhibition displayed by NB-DNJ, a library of derivatives (compounds **38, 39, 42–44,** Figure 7.15) based on the addition of a second alkyl chain to the NB-DNJ scaffold at either O-2, C-1 or O-4 was synthesized [73]. Compared with *N*- nonyl DNJ or NB-DNJ [74] the new iminosugars possessed less affinity for ceramide glucosyltransferase, but were still within an order of magnitude of the parent compounds.

7.2.4 β-1,4-Galactosyltransferase inhibitors

One of the most studied glycosyltransferases, β-1,4-galactosyltransferase (β-1,4-GalT), catalyses the transfer of galactose from UDP-Gal to the 4-hydroxyl of *N*- acetylglucosamine residues (Figure 7.16). Other galactosyltransferases have roles in the biosynthesis of many cell surface oligosaccharides such as blood group antigens and sialyl Lewis^X, a ligand for the selectin family of proteins involved in tumour metastasis and inflammatory response [56–58].

Using a novel mass spectrometry screening method, cyclic imines **43** and **44** were identified as weak inhibitors of β-1, 4-galactosyltransferase (Figure 7.17) [75]. Five-membered iminosugar **45** with the pseudo *galacto*-configuration was also investigated

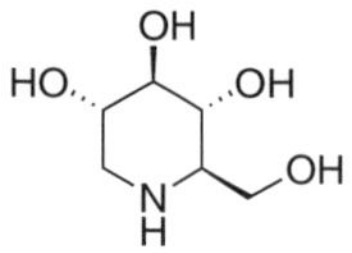

DNJ

NI
ref. [72]

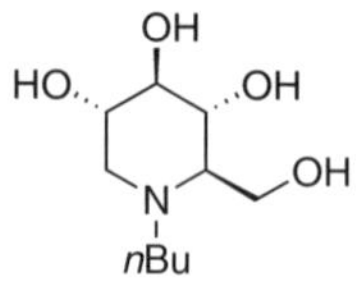

NB-DNJ (Zavesca®)

87% at 200 μM
ref. [71]

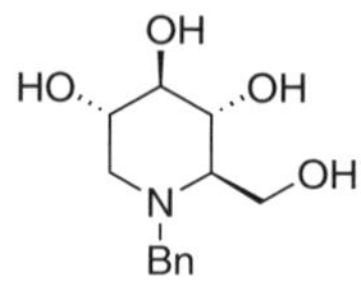

27

73% at 200 μM
ref. [72]

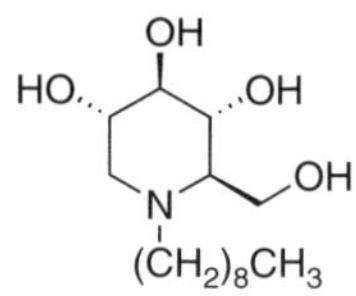

28

96% at 200 μM
ref. [72]

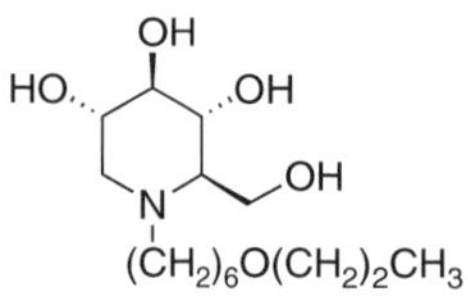

29

97% at 200 μM
ref. [72]

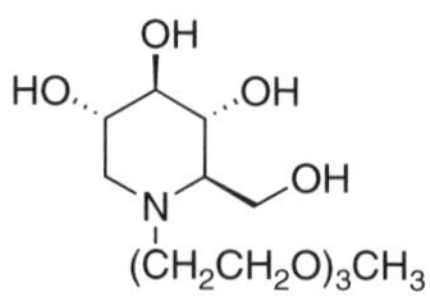

30

20% at 200 μM
ref. [72]

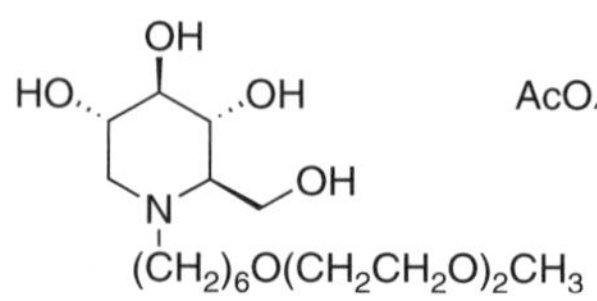

31

93% at 200 μM
ref. [72]

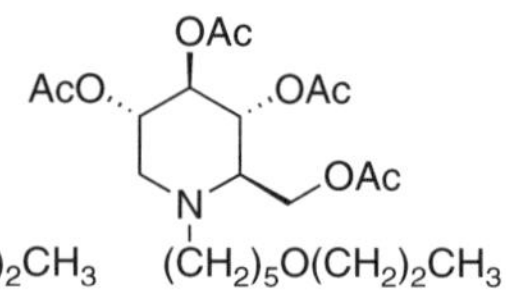

32

18% at 200 μM
ref. [72]

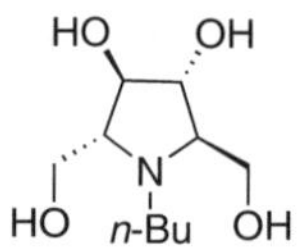

33

86% at 200 μM
ref. [72]

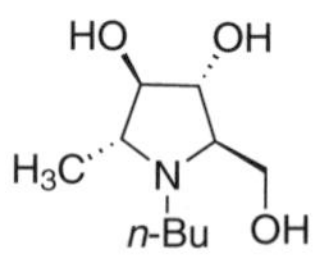

34

11% at 200 μM
ref. [72]

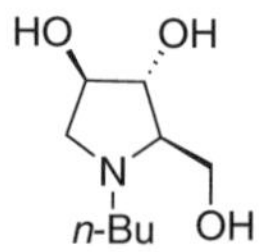

35

21% at 200 μM
ref. [72]

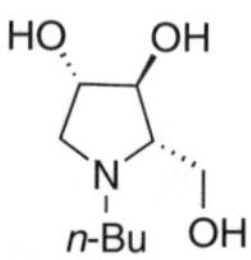

36

25% at 200 μM
ref. [72]

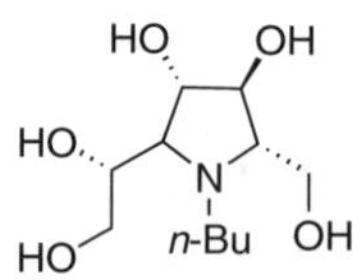

37

6% at 200 μM
ref. [72]

Figure 7.14 Iminosugar-based inhibitors of ceramide glucosyltransferase (percentage inhibition at 200 μM *in vitro*); NI = no inhibition

and showed weak inhibitory activity against β-1, 4-GalT [76]. In a separate study, malonic, tartaric and monosaccharide moieties were used as replacements of the pyrophosphate moiety (**47**, Figure 7.17) [77]. The polyhydroxylated pyrrolidine moiety ($K_i = 1$ mM) may act as a transition state analogue leading to inhibition.

7.2.5 α-1,3-Galactosyltransferase inhibitors

Size, availability and other factors have led to investigations into using porcine organs as replacements for those failing in humans, a process known as xenotransplantation.

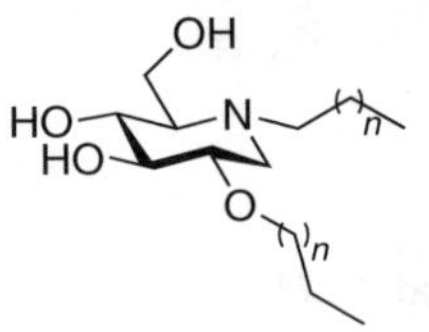

38
n = 6, IC_{50} = 164 μM
ref. [73]

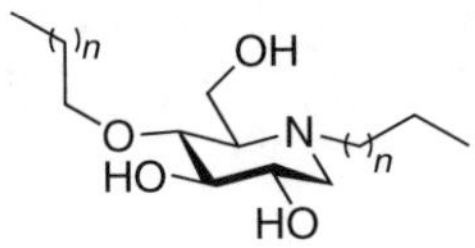

39
n = 6, IC_{50} = 134 μM
ref. [73]

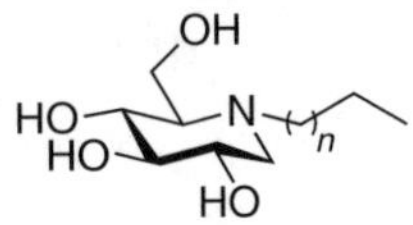

NB-DNJ
n = 2, IC_{50} = 20 μM
***N*-nonyl DNJ**
n = 7, IC_{50} = 1.6 μM
ref. [74]

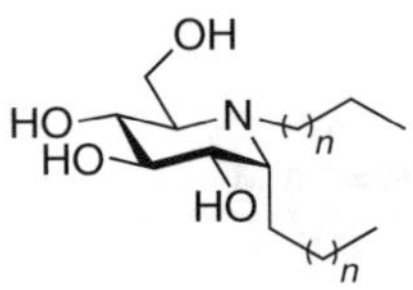

40
n = 2, IC_{50} = 609 μM
41
n = 6, IC_{50} = 174 μM
ref. [73]

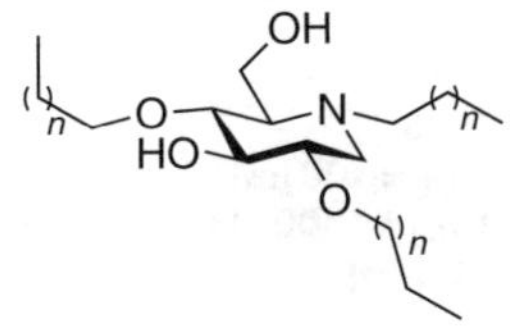

42
n = 2, 29% inhibition at 1 mM
ref. [73]

Figure 7.15 Inhibitory data for *N*- and *O*-alkylated DNJ derivatives

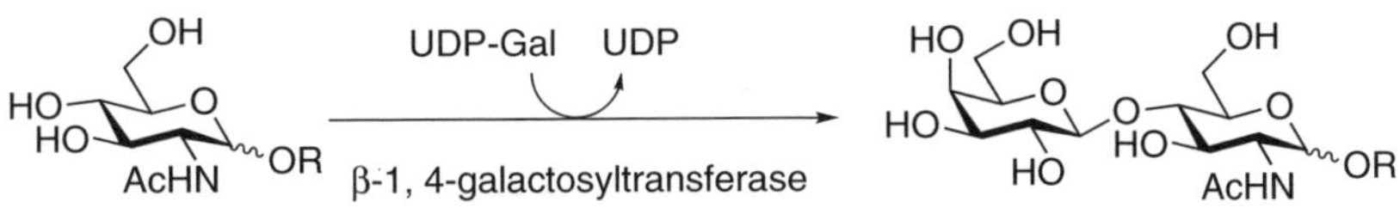

Figure 7.16 The reaction catalysed by β-1,4-galactosyltransferase

The α-Gal epitope (Figure 7.18) is expressed on the surface of pig vascular endothelial cells, and humans produce more anti-Gal antibodies than any other kind of antibody [78]. The binding of human anti-Gal antibody to the porcine α-Gal epitope results in hyperacute rejection of the organ. Porcine α-1,3-galactosyltansferase (α-1,3-GalT) appends the terminal galactose residue to form α-Gal epitopes. As a result, eradication of this epitope by inhibiting α-1,3-GalT has attracted attention as a therapy for hyperacute rejection during xenotransplantation.

The 1-*N*-Iminosugar **48a** was identified as a good inhibitor of α-1,3-GalT (IC_{50} = 15 μM) [75], and gave rise to derivatives **49** and **50** (Figure 7.19) [79]. These derivatives are believed to form a strongly associated complex in the active site via favourable electrostatic interactions. A simple diol moiety was used to mimic the pyrophosphate and the resulting conjugate showed improved inhibitory potency against α-1,3-GalT and suppressed undesired β-galactosidase inhibition.

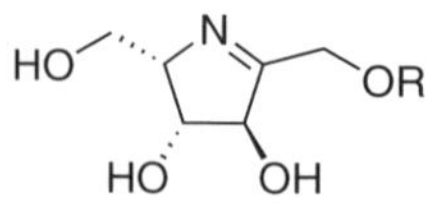

43
R = H, 30% at 2 mM
R = butyrate, 30% at 2 mM
ref. [75]

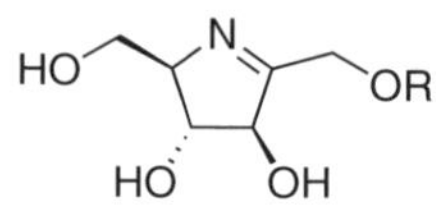

44
R = H, 30% at 2 mM
R = butyrate, 30% at 2 mM
ref. [75]

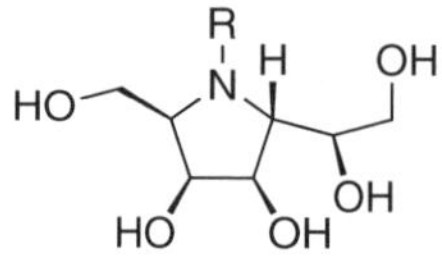

45
R = Me, IC_{50} > 10 mM
ref. [76]

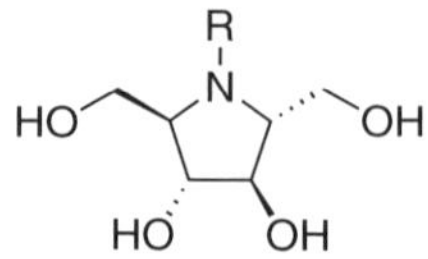

46
R = H, IC_{50} = 100 μM
R = Me, IC_{50} = 1500 μM
ref. [76]

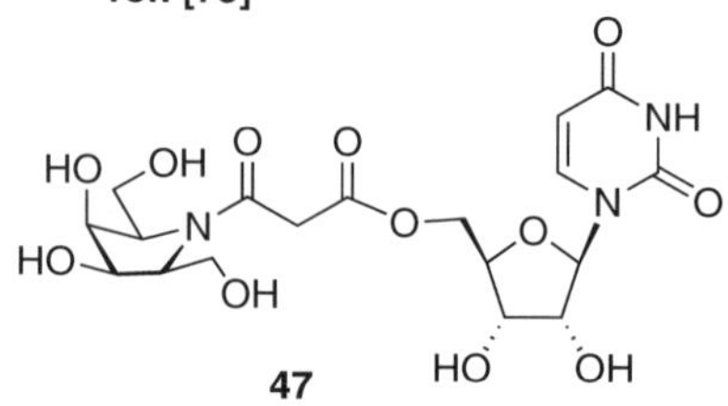

47
K_i = 1 mM
ref. [77]

Figure 7.17 Inhibitors of β-1,4-galactosyltransferase

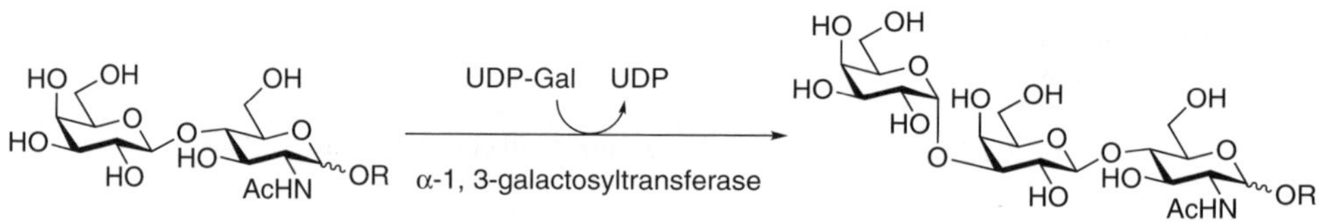

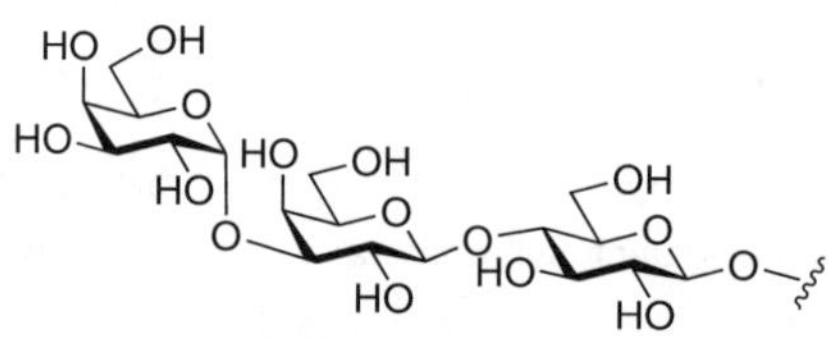

α-Gal epitope
(Galactose α-1, 3-lactose)

Figure 7.18 The reaction catalysed by α-1,3-galactosyltransferase

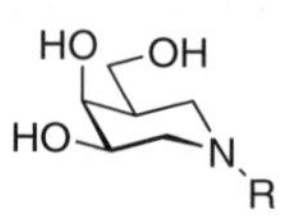

48a R = H, IC_{50} = 15 μM
48b R = *n*Bu, 45% inh. at 5 mM
ref. [75], [79]

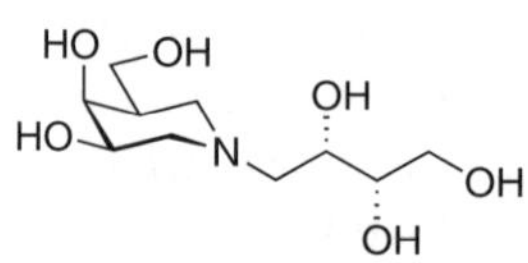

49
IC_{50} = 70 μM
ref. [75], [79]

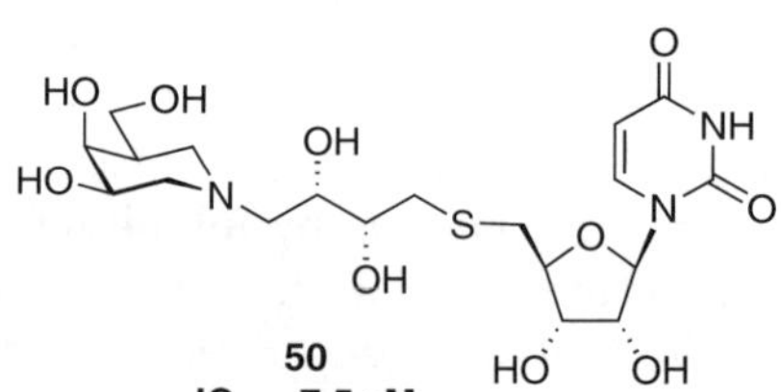

50
IC_{50} = 7.5 μM
ref. [75], [79]

Figure 7.19 Inhibitors of α-1,3-GalT

7.2.6 Arabinofuranosyl transferase inhibitors

Two major mycobacterial human pathogens are *Mycobacterium leprae* (the agent that causes leprosy) and *Mycobacterium tuberculosis* (responsible for tuberculosis). Both of these pathogens utilize the sugar D-arabinose in their cell wall biosynthetic pathways, in the form of the arabinofuranoside oligomers arabinogalactan and lipoglycan [80, 81]. However, D-arabinose is not used in human biosynthetic pathways, and as such any enzymes utilizing D-arabinose are promising therapeutic targets against mycobacterial infections. A series of iminosugar-oligoarabinofuranoside hybrids **54–61** (Figure 7.20) was prepared with the goal of inhibiting the mycobacterial arabinosyltransferases [82]. The simplest iminosugars were low to moderate inhibitors, but linkage of the iminosugar

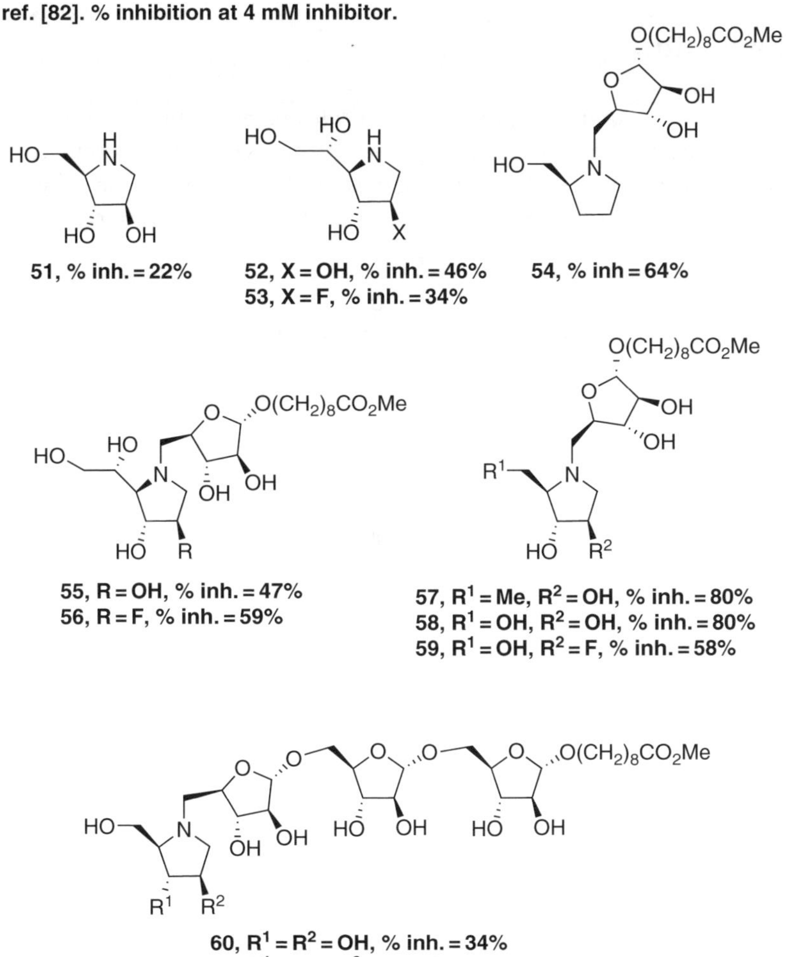

Figure 7.20 Inhibitors of mycobacterial arabinofuranosyltransferase

to a single arabinofuranoside moiety increased the activity. With the addition of three arabinosyl units a decrease in activity was observed.

7.2.7 Galactofuranosyltransferase inhibitors

A second mycobacterial target, utilized by *M. tuberculosis* for processing of its cell wall, is UDP-galactofuranosyl transferase (UDP-Gal*f* transferase). This enzyme catalyses the transfer of galactofuranose to a growing oligofuranose chain (Figure 7.21). It is capable of catalysing the synthesis of both β-1,5 and β-1,6 linkages [83]. Two iminosugar inhibitors, **62** and **63**, were synthesized and tested for inhibition of UDP-Gal*f* transferase (Figure 7.22) [84]. The iminosugars were found to be moderate inhibitors of the enzyme, with IC_{50} values in the millimolar range.

7.3 Conclusion

Despite the increasing number of potential glycosyltransferase inhibitors, very few exhibit potent activity. The rational design of glycosyltransferase inhibitors remains a difficult task due to intrinsic aspects of these enzymes. The transition state of glycosyltransferase reactions is inherently complex as it includes the sugar donor, acceptor glycan,

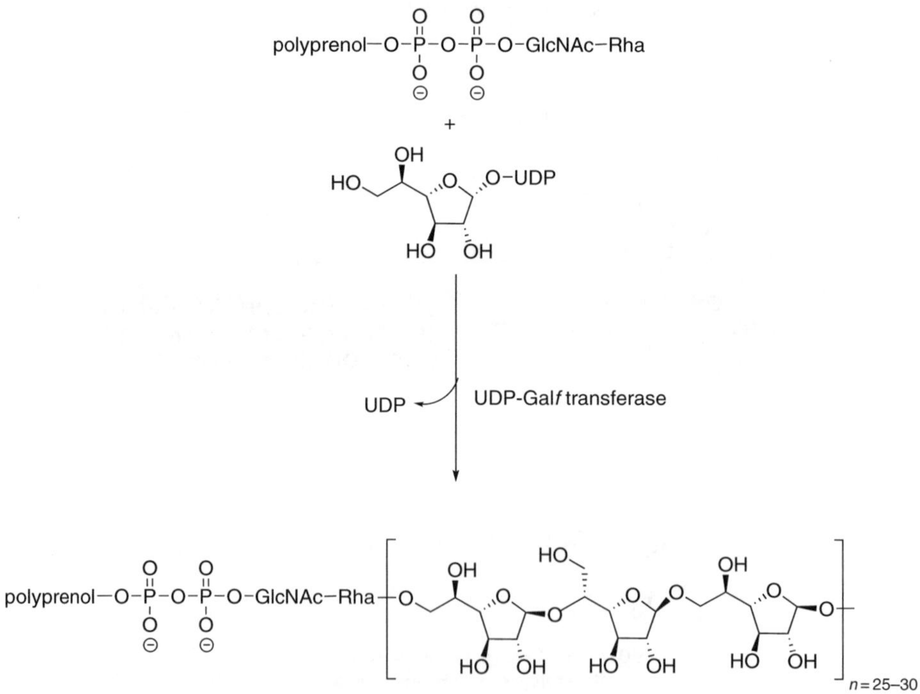

Figure 7.21 The reaction catalysed by UDP-galactofuranosyl transferase

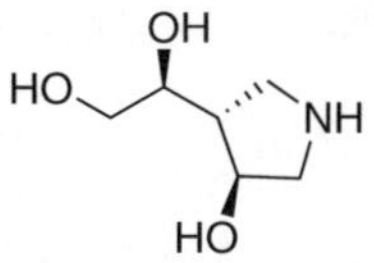
62
% inh. = 40% at 8 mM

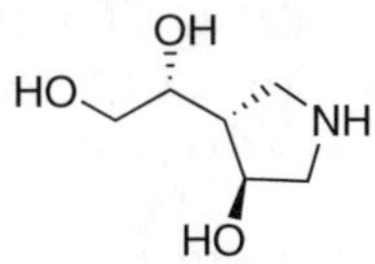
63
IC_{50} = 4.8 mM

ref. [84]

Figure 7.22 Inhibitors of UDP-Gal*f* transferase

divalent metal and nucleotide. The weak binding interactions of the enzymes with their natural substrates also make designing high-affinity inhibitors difficult. Exacerbating these difficulties is the limited amount of structural and mechanistic data on these enzymes. Iminosugar-based transition state analogues offer one of the more promising approaches toward inhibiting glycosyltransferases, and advances in enzymatic and chemical methods for their synthesis makes a wide variety of these compounds available. Bisubstrate analogues that contain a transition state mimic and acceptor in similar steric arrangement to the transition state complex are of particular interest, as such compounds are expected to exhibit higher affinity for the active site and higher specificity due in part to acceptor selectivity in glycosyltransferases. Due to the critical role glycosyltransferases play in many biological events, further investigation is warranted, not only to increase our understanding of their mechanism, but also because of the potential for therapeutic intervention.

References

1. Hu, Y. N. and Walker, S. (2002) *Chem. Biol.*, **9**, 1287–1296.
2. Breton, C. and Imberty, A. (1999) *Curr. Opin. Struct. Biol.*, **9**, 563–571.
3. Taniguchi, N., Honke, K. and Fukuda, M. (2002) *Handbook of Glycosyltransferases and Related Genes.*, Springer, Tokyo, Japan.
4. Boons, G. J. (1996) *Tetrahedron*, **52**, 1095–1121.
5. Reichel, F. and Boons, G. J. (1998) *Chemistry in Britain*, **34**, 43–46.
6. Varki, A. (1993) *Glycobiology*, **3**, 97–130.
7. Dwek, R. A. (1996) *Chem. Rev.*, **96**, 683–720.
8. Leloir, L. F. (1971) *Science*, **172**, 1299–1303.
9. Sadler, J. E., Beyer, T. A., Oppenheimer, C. L., Paulson, J. C., Prieels, J. P., Rearick, J. I. and Hill, R. L. (1982) *Methods Enzymol.*, **83**, 458–514.
10. Ginsburg, V. (1964) *Adv. Enzymol.*, **26**, 35–88.
11. Kornfeld, R. and Kornfeld, S. (1985) *Annu. Rev. Biochem.*, **54**, 631–664.
12. Wong, C.-H., Halcomb, R. L., Ichikawa, Y. and Kajimoto, T. (1995) *Angew. Chem. Int. Ed.*, **34**, 521–546.
13. Wong, C.-H., Halcomb, R. L., Ichikawa, Y. and Kajimoto, T. (1995) *Angew. Chem. Int. Ed.*, **34**, 412–432.
14. Paulson, J. C. and Colley, K. J. (1989) *J. Biol. Chem.*, **264**, 17615–17618.

15. Wong, C.-H. and Whitesides, G. M. (1994) *Enzymes in Synthetic Organic Chemistry*; Elsevier Science, Tarrytown, New York, USA, Vol. 12.
16. Ferrier, R. J. (ed.) (2003) *Carbohydrate Chemistry (Parts I and II)*, Royal Society of Chemistry, London, UK.
17. Aspinall, G. O. (ed.) (1982–1985) *The Polysaccharides*, Academic Press, New York, USA, Vol. I-III.
18. Yamamoto, F., Marken, J., Tsuji, T., White, T., Clausen, H. and Hakomori, S. (1990) *J. Biol. Chem.*, **265**, 1146–1151.
19. Breton, C., Snajdrova, L., Jeanneau, C., Koca, J. and Imberty, A. (2006) *Glycobiology*, **16**, 29R-37R.
20. Yamamoto, F., Clausen, H., White, T., Marken, J. and Hakomori, S. (1990) *Nature*, **345**, 229–233.
21. Vrielink, A., Ruger, W., Driessen, H. P. C. and Freemont, P. S. (1994) *EMBO J.*, **13**, 3413–3422.
22. Lesk, A. M. (1995) *Curr. Opin. Struct. Biol.*, **5**, 775–783.
23. Charnock, S. J. and Davies, G. J. (1999) *Biochemistry*, **38**, 6380–6385.
24. Chiu, C. P., Watts, A. G., Lairson, L. L., Gilbert, M., Lim, D., Wakarchuk, W. W., Withers, S. G. and Strynadka, N. C. (2004) *Nat. Struct. Mol. Biol.*, **11**, 163–170.
25. Busch, C., Hofmann, F., Selzer, J., Munro, S., Jeckel, D. and Aktories, K. (1998) *J. Biol. Chem.*, **273**, 19566–19572.
26. Pak, J. E., Arnoux, P., Zhou, S. H., Sivarajah, P., Satkunarajah, M., Xing, X. K. and Rini, J. M. (2006) *J. Biol. Chem.*, **281**, 26693–26701.
27. Murray, B. W., Wittmann, V., Burkart, M. D., Hung, S. C. and Wong, C.-H. (1997) *Biochemistry*, **36**, 823–831.
28. Wiggins, C. A. R. and Munro, S. (1998) *Proc. Nat. Acad. Sci. U.S.A.*, **95**, 7945–7950.
29. Persson, K., Ly, H. D., Dieckelmann, M., Wakarchuk, W. W., Withers, S. G. and Strynadka, N. C. J. (2001) *Nat. Struct. Biol.*, **8**, 166–175.
30. Tarbouriech, N., Charnock, S. J. and Davies, G. J. (2001) *J. Mol. Biol.*, **314**, 655–661.
31. Qasba, P. K., Ramakrishnan, B. and Boeggeman, E. (2005) *Trends Biochem. Sci.*, **30**, 53–62.
32. Wrabl, J. and Grishin, N. (2001) *J. Mol. Biol.*, **314**, 365–374.
33. Unligil, U. M., Zhou, S. H., Yuwaraj, S., Sarkar, M., Schachter, H. and Rini, J. M. (2000) *EMBO J.*, **19**, 5269–5280.
34. Pedersen, L. C., Tsuchida, K., Kitagawa, H., Sugahara, K., Darden, T. A. and Negishi, M. (2000) *J. Biol. Chem.*, **275**, 34580–34585.
35. Campbell, J. A., Davies, G. J., Bulone, V. and Henrissat, B. (1997) *Biochem. J.*, **326**, 929–939.
36. Rye, C. S. and Withers, S. G. (2000) *Curr. Opin. Chem. Biol.*, **4**, 573–580.
37. Sinnott, M. L. (1990) *Chem. Rev.*, **90**, 1171–1202.
38. Qiao, L., Murray, B. W., Shimazaki, M., Schultz, J. and Wong, C.-H. (1996) *J. Am. Chem. Soc.*, **118**, 7653–7662.
39. Murray, B. W., Takayama, S., Schultz, J. and Wong, C.-H. (1996) *Biochemistry*, **35**, 11183–11195.
40. Khan, S. H. and Matta, K. L. (1992) *Glycoconjugates*, 361–378.
41. Kraut, J. (1988) *Science*, **242**, 533–540.
42. Waldscheck, B., Streiff, M., Notz, W., Kinzy, W. and Schmidt, R. R. (2001) *Angew. Chem. Int. Ed.*, **40**, 4007–4011.
43. Compain, P. and Martin, O. R. (2003) *Curr. Top. Med. Chem.*, **3**, 541–560.
44. Compain, P. and Martin, O. R. (2001) *Bioorg. Med. Chem.*, **9**, 3077–3092.
45. Jung, K.-H. and Schmidt, R. R. (2003) In *Carbohydrate-based Drug Discovery*, Wiley-VCH, Weinheim, Germany, Vol. 2, pp. 609–659.
46. Stutz, A. E. (ed.) (1999) *Iminosugars as Glycosidase Inhibitors: Nojirimycin and Beyond*, Wiley-VCH, Weinheim, Germany.

47. Legler, G. (1990) *Adv. Carbohydr. Chem. Biochem.*, **48**, 319–384.
48. Silvestri, M. G., Desantis, G., Mitchell, M. and Wong, C.-H. (2003) *Asymmetric Aldol Reactions Using Aldolases*, Wiley, Hoboken, New York, USA, Vol. 23.
49. Whalen, L. J. and Wong, C.-H. (2006) *Aldrichim. Acta*, **39**, 63–71.
50. Takebayashi, M., Hiranuma, S., Kanie, Y., Kajimoto, T., Kanie, O. and Wong, C.-H. (1999) *J. Org. Chem.*, **64**, 5280–5291.
51. McEver, R. P., Moore, K. L. and Cummings, R. D. (1995) *J. Biol. Chem.*, **270**, 11025–11028.
52. Springer, T. A. (1990) *Nature*, **346**, 425–434.
53. Muramatsu, T. (1993) *Glycobiology*, **3**, 294–296.
54. Sakamoto, J., Furukawa, K., Cordoncardo, C., Yin, B. W. T., Rettig, W. J., Oettgen, H. F., Old, L. J. and Lloyd, K. O. (1986) *Cancer Res.*, **46**, 1553–1561.
55. Itzkowitz, S. H., Yuan, M., Fukushi, Y., Palekar, A., Phelps, P. C., Shamsuddin, A. M., Trump, B. F., Hakomori, S. and Kim, Y. S. (1986) *Cancer Res.*, **46**, 2627–2632.
56. Lowe, J. B., Stoolman, L. M., Nair, R. P., Larsen, R. D., Berhend, T. L. and Marks, R. M. (1990) *Cell*, **63**, 475–484.
57. Walz, G., Aruffo, A., Kolanus, W., Bevilacqua, M. and Seed, B. (1990) *Science*, **250**, 1132–1135.
58. Phillips, M. L., Nudelman, E., Gaeta, F. C. A., Perez, M., Singhal, A. K., Hakomori, S. I. and Paulson, J. C. (1990) *Science*, **250**, 1130–1132.
59. Mitchell, M. L., Lee, L. V. and Wong, C.-H. (2002) *Tetrahedron Lett.*, **43**, 5691–5693.
60. Schuster, M. and Blechert, S. (2001) *Bioorg. Med. Chem. Lett.*, **11**, 1809–1811.
61. Wang, Y. F., Dumas, D. P. and Wong, C.-H. (1993) *Tetrahedron Lett.*, **34**, 403–406.
62. Wong, C.-H., Dumas, D. P., Ichikawa, Y., Koseki, K., Danishefsky, S. J., Weston, B. W. and Lowe, J. B. (1992) *J. Am. Chem. Soc.*, **114**, 7321–7322.
63. Ichikawa, Y., Lin, Y. C., Dumas, D. P., Shen, G. J., Garciajunceda, E., Williams, M. A., Bayer, R., Ketcham, C., Walker, L. E., Paulson, J. C. and Wong, C.-H. (1992) *J. Am. Chem. Soc.*, **114**, 9283–9298.
64. Jefferies, I. and Bowen, B. R. (1997) *Bioorg. Med. Chem. Lett.*, **7**, 1171–1174.
65. Mitchell, M. L., Tian, F., Lee, L. V. and Wong, C.-H. (2002) *Angew. Chem. Int. Ed.*, **41**, 3041–3044.
66. Behr, J.-B., Gautier-Lefebvre, I., Mvondo-Evina, C., Guillerm, G. and Ryder, N. S. (2001) *J. Enz. Inhib.*, **16**, 107–112.
67. Djebaili, M. and Behr, J. -B. (2005) *J. Enzym. Inhib. Med. Chem.*, **20**, 123–127.
68. Gautier-Lefebvre, I., Behr, J.-B., Guillerm, G. and Muzard, M. (2005) *Eur. J. Med. Chem.*, **40**, 1255–1261.
69. Chester, M. A. (1999) *Glycoconjugate J.*, **16**, 1–6.
70. Butters, T. D., Dwek, R. A. and Platt, F. M. (2000) *Chem. Rev.*, **100**, 4683–4696.
71. Cox, T., Lachmann, R., Hollak, C., Aerts, J., van Weely, S., Hrebicek, M., Platt, F., Butters, T., Dwek, R., Moyses, C., Gow, I., Elstein, D. and Zimran, A. (2000) *Lancet*, **355**, 1481–1485.
72. Butters, T. D., van den Broek, L., Fleet, G. W. J., Krulle, T. M., Wormald, M. R., Dwek, R. A. and Platt, F. M. (2000) *Tetrahedron: Asymmetry*, **11**, 113–124.
73. Boucheron, C., Desvergnes, V., Compain, P., Martin, O. R., Lavi, A., Mackeen, M., Wormald, M., Dwek, R. and Butters, T. D. (2005) *Tetrahedron: Asymmetry*, **16**, 1747–1756.
74. Butters, T. D., Dwek, R. A. and Platt, F. M. (2003) *Curr. Top. Med. Chem.*, **3**, 561–574.
75. Takayama, S., Martin, R., Wu, J. Y., Laslo, K., Siuzdak, G. and Wong, C.-H. (1997) *J. Am. Chem. Soc.*, **119**, 8146–8151.
76. Saotome, C., Kanie, Y., Kanie, O. and Wong, C.-H. (2000) *Bioorg. Med. Chem.*, **8**, 2249–2261.
77. Wang, R., Steensma, D. H., Takaoka, Y., Yun, J. W., Kajimoto, T. and Wong, C.-H. (1997) *Bioorg. Med. Chem.*, **5**, 661–672.

78. Xu, Y., Lorf, T., Sablinski, T., Gianello, P., Bailin, M., Monroy, R., Kozlowski, T., Awwad, M., Cooper, D. K. C. and Sachs, D. H. (1998) *Transplantation*, **65**, 172–179.
79. Kim, Y. J., Ichikawa, M. and Ichikawa, Y. (1999) *J. Am. Chem. Soc.*, **121**, 5829–5830.
80. Daffe, M. and Draper, P. (1998) *Adv. Microb. Physiol*, **39**, 131–203.
81. Daffe, M., Brennan, P. J. and McNeil, M. (1990) *J. Biol. Chem.*, **265**, 6734–6743.
82. Marotte, K., Ayad, T., Genisson, Y., Besra, G. S., Baltas, M. and Prandi, J. (2003) *Eur. J. Org. Chem.*, 2557–2565.
83. Kremer, L., Dover, L. G., Morehouse, C., Hitchin, P., Everett, M., Morris, H. R., Dell, A., Brennan, P. J., McNeil, M. R., Flaherty, C., Duncan, K. and Besra, G. S. (2001) *J. Biol. Chem.*, **276**, 26430–26440.
84. Cren, S., Gurcha, S. S., Blake, A. J., Besra, G. S. and Thomas, N. R. (2004) *Org. Biomol. Chem.*, **2**, 2418–2420.

8

Transition state analogue inhibitors of *N*-ribosyltransferases

Vern L. Schramm and **Peter C. Tyler**

8.1 Introduction

8.1.1 Transition states and isotope effects

All chemical reactions reach a point in chemical bond breaking where the reactants have equal probability of going forward to products or reverting to reactants. That point is the transition state. In transition state theory, this state has a short lifetime, usually considered to be less than a single bond vibration, or the time for the restoring mode for the bond of interest to be converted to the translational mode that will transform the transition state to products or reactants. Since bond vibrations are fast (10^{-13} to $10^{-12}\,s^{-1}$) at temperatures of biological interest, direct observations of transition state bond vibrational properties have only been accessible in the gas phase [1]. For solution chemistry, physical organic chemists have relied on the indirect method of kinetic isotope effects (KIE) to probe the nature of transition states. A KIE is the difference in reaction rate as a result of a specific isotopic substitution being made in the reactant molecule. The difference in reaction rates of the isotopically substituted and unsubstituted molecules reports on the nature of the transition state [2–4]. This theory has been developed over the past 60 years, and continues to be developed as a result of increased computational power and algorithms for estimating bond configuration and electron distributions at reactant and transition states in reactions of ever increasing complexity, including enzymatic reactions [5–7].

8.1.2 Enzymatic transition states and inhibitor design

Enzymes are remarkable in their ability to increase reaction rates, typically 10^{10}- to 10^{15}-fold, relative to the same reaction conditions without the enzyme catalyst. For

Iminosugars: From synthesis to therapeutic applications Edited by Philippe Compain and Olivier R. Martin
 ISBN 978-0-470-03391-3

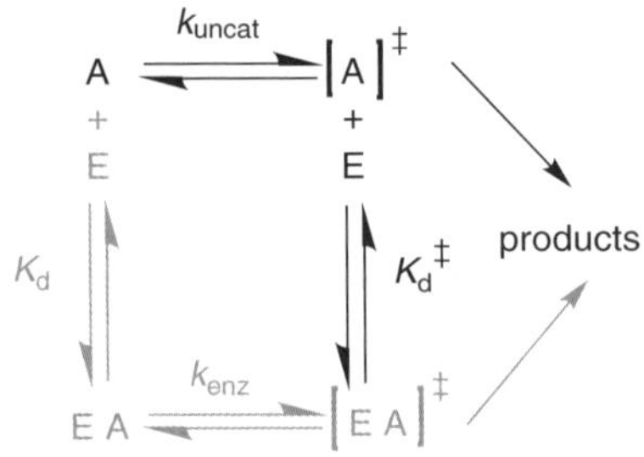

$$K_d^{\ddagger} = K_d \frac{k_{uncat}}{k_{enz}}$$

Figure 8.1 Thermodynamic box for the reaction rate, $k_{(uncat)}$ of an uncatalysed (solvent only) reaction compared with the same reaction rate $k_{(enz)}$ for an enzyme-catalysed reaction. K_d is the binding of enzyme to the reactant (A) and $K_d^{\ddagger}$ is the binding of the transition state species $[A]^{\ddagger}$ to the enzyme. Note that the binding of the transition state species is tighter than A by the catalytic rate enhancement imposed by the enzyme. Since typical enzymes increase reaction rates by 10^{10}–10^{15}, potential binding affinity for analogues of the transition state is large. Adapted from Schramm, V. L. (1998) *Ann. Rev. Biochem.*, **67**, 693–720 (A colour reproduction of this figure can be found in the colour section towards the centre of the book)

decades, investigators recognized that enzymes could increase the reaction rate by increasing the concentration of the transition state species, which occurs rarely in uncatalysed reactions [8–10]. This relationship was interpreted as transition state binding by Wolfenden [11] and others [12–14], and expressed as shown in Figure 8.1. In this construct, enzymes accelerate reactions by binding more tightly to the transition state than the Michaelis complex by the factor of the rate enhancement. Therefore, a typical enzyme is proposed to bind 10^{10}- to 10^{15}-fold more tightly to the transition state than to the reactants. Chemically stable analogues of the transition state are therefore expected to bind much tighter than the reactants and, being nonreactive, will inhibit the enzyme and be released slowly. More recently, it has been recognized that tight-binding of the transition state complex is not required for catalysis or to explain tight-binding of analogues, but this does not change the validity of the thermodynamic argument of Figure 8.1.

8.1.3 Design of transition state analogues

Based on transition state theory applied to enzymatic reactions, a protocol has been developed to establish the transition state structure of enzyme-catalysed reactions [15–17]. This information is then used for the design and synthesis of transition state analogues. The problem is to convert KIE information into stable chemical structures related to the transition state. Steps used in this process are:

1. Selection of an enzyme that is amenable to KIE analysis and is a physiological target.
2. Synthesis of reactants with specific isotope labels near the bond of interest.

3. Measurement of intrinsic KIE (KIE from transition state chemistry without influence of non-chemical steps).
4. Calculation of a quantum chemical model of the transition state that generates the intrinsic KIEs.
5. Conversion of the transition state electronic structure into a molecular electrostatic potential map.
6. Design of stable chemical analogues with geometry and molecular electrostatic van der Waals surface closely related to that of the transition state.

Design of transition state analogues based on experimentally determined transition state structures is in its infancy and has been most developed with the *N*-ribosyltransferases described here. The transition state structures of these enzymes will not be emphasized here, as they have been published elsewhere [18–26]. In this chapter, our intent is to provide a summary of selected inhibitors that explore the nature of transition state binding energy. That is, how well can we match the theoretical energy of the transition state and what are the chemical features that are essential to provide the tight binding that characterizes transition state analogues?

8.2 Nucleoside hydrolases

8.2.1 Inosine-uridine nucleoside hydrolase transition state analogue considerations

The transition state structure of the inosine-uridine nucleoside hydrolase (IU-NH) from *Crithidia fasciculata*, a non-pathogenic protozoan parasite, revealed a highly dissociative (S_N1) and nearly symmetrical transition state [18]. At the transition state, inosine demonstrated approximately 0.07 Pauling bond order between N9 and C1' in the reaction coordinate and a protonated or strongly hydrogen bonded hypoxanthine leaving group. There is approximately 0.05 bond order between C1' and the attacking water nucleophile (Figure 8.2). This creates a transition state with carbocation character in the ribosyl group. The weakly bound attacking water nucleophile is activated by a Ca^{2+} ion that is tightly bound in the catalytic site. Since the bond order is low to the water nucleophile, a transition state analogue would not require a water mimic, as the catalytic site nucleophile will be formed from solvent water, also a feature of the actual transition state.

Since IU-NH uses both inosine and uridine as substrates, high specificity for the leaving group is not required. The specificity of the catalytic site for leaving groups was investigated by the catalytic action of IU-NH on *p*-NO_2-phenyl-O-β-D-riboside (pNPR) [27]. The enzyme uses inosine with a k_{cat}/K_m of $7.6 \times 10^4\ M^{-1}\ s^{-1}$ but its catalytic efficiency is 54-fold greater with *pNPR* to give a k_{cat}/K_m of $4.1 \times 10^6\ M^{-1}\ s^{-1}$ [28]. Furthermore, ionizable groups at the catalytic site of IU-NH required for acid–base catalysis of inosine are not required for *pNPR*. Transition state stabilization is therefore proposed to occur through activation of the ribosyl group toward the oxacarbenium ion. Thermodynamic analysis of the ΔΔG barrier for the transition state using different

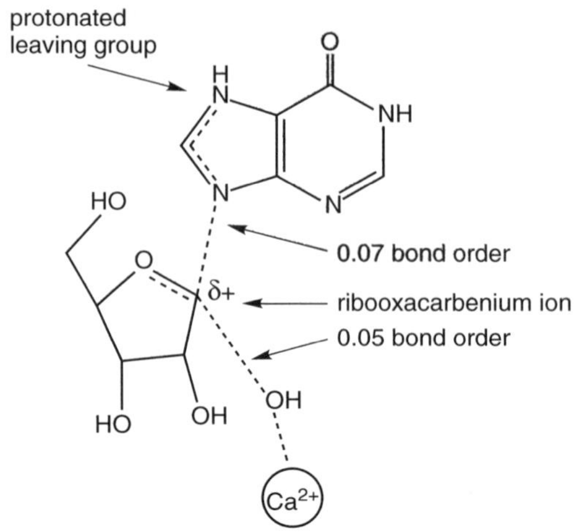

Figure 8.2 Transition state structure for IU-NH showing the bond order to the leaving group and the attacking nucleophile at the moment of the transition state. The electron deficiency caused by the delayed attack of the water nucleophile generates a ribooxacarbenium ion at the transition state

reactants, indicated that IU-NH lowers the energy of activation for hydrolysis of inosine by 17.7 kcal/mol with 13.1 kcal/mol from formation of the ribooxacarbenium ion and only 4.6 kcal/mol from leaving group interactions [29].

To summarize, structural characteristics expected to provide good transition state analogues of IU-NH should contain features of the ribooxacarbenium ion transition state. However, leaving groups can be pyrimidines, purines or activated phenols as in *pNPR*. The attacking water nucleophile is not needed in the stable mimic of the transition state since at the transition state it is more closely bound to the catalytic site Ca^{2+} than to C1'.

8.2.2 Inosine-adenosine-guanosine nucleoside hydrolase transition state analogue considerations

The transition state structure of the inosine-adenosine-guanosine nucleoside hydrolase (IAG-NH) from *Trypanosoma brucei brucei*, a pathogenic protozoan parasite, revealed an early dissociative (S_N1) highly asymmetric transition state [19]. At the transition state, inosine as the reactant demonstrated approximately 0.19 Pauling bond order between N9 and C1' in the reaction coordinate and a protonated or strongly hydrogen bonded hypoxanthine leaving group. There is no significant (0.05) bond order to the attacking water nucleophile and these can be compared with the IU-NH (Figure 8.2). As for IU-NH, this creates a transition state with carbocation character in the ribosyl group. The weakly bound attacking water nucleophile is activated by the catalytic site Ca^{2+} ion. Since the bond order is not significant to the water nucleophile, a complex between IAG-NH and an analogue of the transition state would not require a water mimic, as the catalytic site nucleophile will be formed from solvent water, also a feature of the actual transition state.

Since IAG-NH is specific for purine nucleosides as substrates, specific leaving group structure is implicated. The specificity of the catalytic site for leaving groups was again investigated by the catalytic action on *pNPR*. IAG-NH uses inosine with a k_{cat}/K_m of $1.9 \times 10^6\,M^{-1}\,s^{-1}$ and its catalytic efficiency decreased to give a k_{cat}/K_m of $1.5 \times 10^3\,M^{-1}s^{-1}$ with *pNPR*, only 10^{-4} of the rate with inosine [28]. Transition state stabilization is therefore proposed to occur through a combination of activation of the ribosyl group toward the oxacarbenium ion and leaving group activation. Quantitation of these effects revealed that of the 17.7 kcal/mol $\Delta\Delta G$ catalytic rate enhancement imposed by IAG-NH, 8.8 kcal/mol comes from leaving group activation and 8.9 kcal/mol from ribosyl activation to the ribooxacarbenium ion [29].

Inhibitors expected to provide transition state interactions at the catalytic site of IAG-NH should contain features both of the ribooxacarbenium ion and the purine leaving group with an elevated pKa, two major features of the transition state. As with IU-NH, the attacking water nucleophile is not a covalent part of the inhibitor design, since its bond order is insignificant at the transition state and it is more closely associated with the catalytic site calcium ion than with C1'.

8.2.3 Inhibitors for IU-NH and IAG-NH

A synthetic scheme for inhibitors of the nucleoside hydrolases is shown in Scheme 1. Table 8.1 (with references to Tables 8.2 and 8.4) gives the structures and dissociation constants for potential transition state analogue inhibitors with these enzymes. The inhibition constant values listed are equilibrium dissociation constants for the enzyme–inhibitor complexes. Dissociation constants for simple competitive inhibitors are indicated as K_i values. K_i^* values are also equilibrium dissociation constants, but in these cases the inhibition occurs in two steps: a rapid inhibition followed by the time-dependent slow-onset of a second, higher affinity inhibition. The dissociation constant for this tighter-binding slow-onset inhibition is K_i^*. Thermodynamically, both K_i and K_i^* are binding constants reached at equilibrium.

8.2.4 Inhibitor specificity for IU-NH

The six tightest-binding inhibitors for IU-NH are [**2.22**], [**4.17**], [**1.3**], [**1.1**], [**1.7**] and [**2.1**] with dissociation constants from 3 to 42 nM. These values are to be compared with the K_m for inosine binding of 380 μM. Thus, at 3 nM, [**2.22**] binds 127 000 times more tightly than the substrate (K_m is known to be a dissociation constant for IU-NH). Note that all of these compounds contain the iminoribitol sugar and thus mimic the important ribooxacarbenium ion of the IU-NH transition state. The iminoribitol group is known to have a pK_a of 6.9 in [**2.1**] [30]. Analysis of kinetic constants as a function of pH has shown that the neutral forms of [**1.1**], [**1.10**] and [**1.12**] iminoribitol-based inhibitors bind to IU-NH in the initial encounter. It was proposed that binding the neutral form of the iminoribitol inhibitors mimics the binding of the neutral substrate, but that these inhibitors become protonated after binding to mimic the cationic transition state [29].

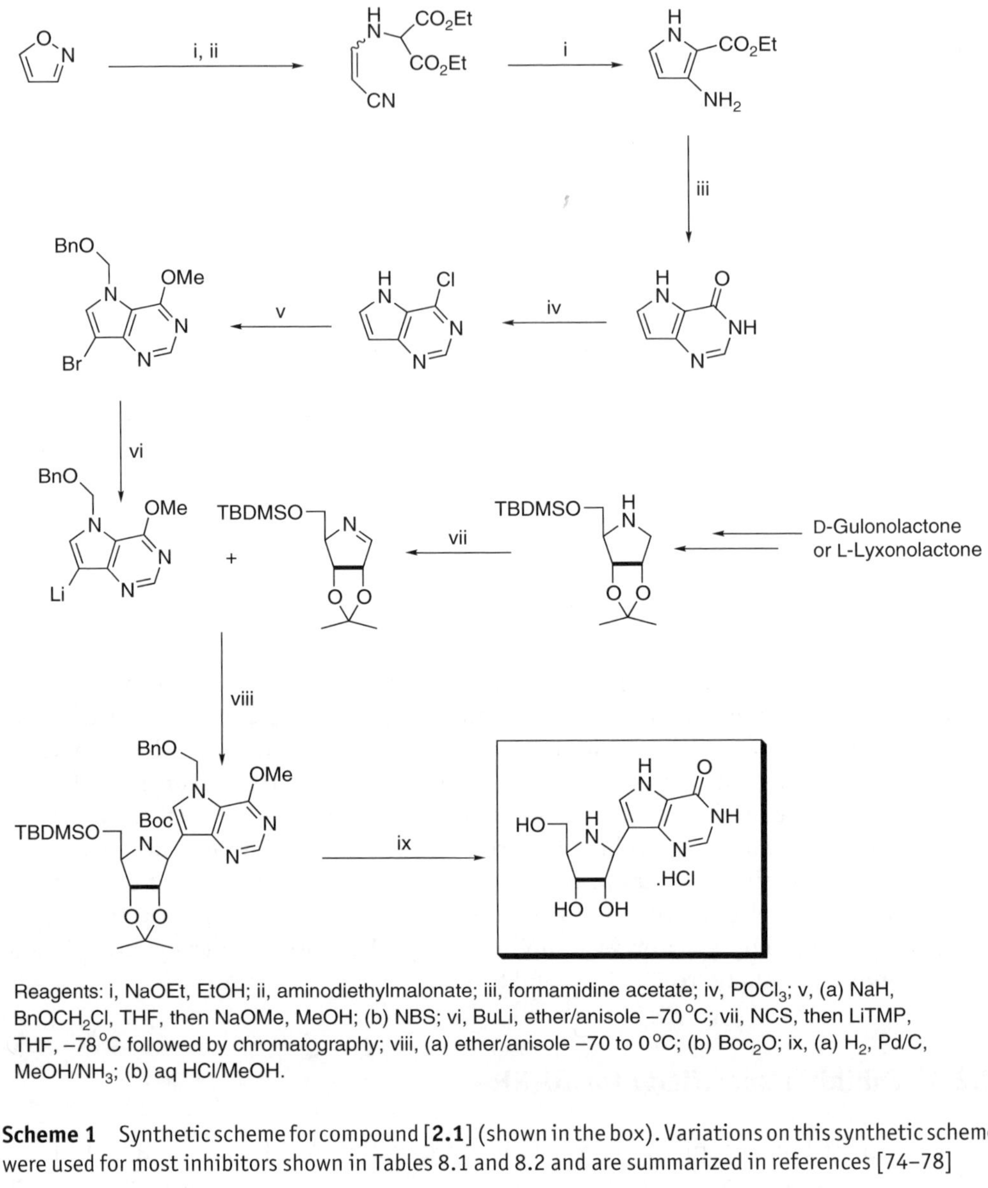

Reagents: i, NaOEt, EtOH; ii, aminodiethylmalonate; iii, formamidine acetate; iv, $POCl_3$; v, (a) NaH, $BnOCH_2Cl$, THF, then NaOMe, MeOH; (b) NBS; vi, BuLi, ether/anisole –70 °C; vii, NCS, then LiTMP, THF, –78 °C followed by chromatography; viii, (a) ether/anisole –70 to 0 °C; (b) Boc_2O; ix, (a) H_2, Pd/C, $MeOH/NH_3$; (b) aq HCl/MeOH.

Scheme 1 Synthetic scheme for compound [**2.1**] (shown in the box). Variations on this synthetic scheme were used for most inhibitors shown in Tables 8.1 and 8.2 and are summarized in references [74–78]

The ionic state of bound iminoribitols has not been experimentally confirmed for IU-NH, but is cationic for the *N*-ribosyltransferases human PNP and hypoxanthine-guanine-phosphoribosyltransferase [30] and is discussed below.

8.2.5 Inhibitor specificity for IAG-NH

IAG-NH differs from IU-NH in its leaving group specificity, and by deriving more of its transition state energy from leaving group interactions. Thus, it shows greater

Table 8.1 Iminoribitol analogues for the transition states of leaving-group nonspecific nucleoside hydrolase (IU-NH) and purine-specific nucleoside hydrolase (IAG-NH). Dissociation constants were determined from the assumption that the inhibitors are competitive with inosine or *p*-nitrophenyl-O-β-D-ribofuranoside as substrates. K_i and K_i^* values are dissociation constants for the enzyme-inhibitor complex. K_i^* values result from slow-onset tight-binding inhibition and usually occur only when the value shown in the table is below 10 nM. Structures of the compounds [**2.x**] and [**4.x**] can be found in Tables 8.2 and 8.4, respectively

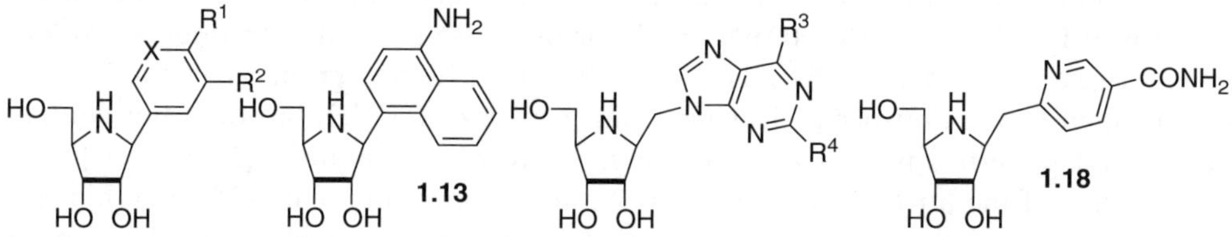

1.1–1.12 **1.13** **1.14–1.17** **1.18**

	X	R^1	R^2	R^3	R^4	IU-NH from *C. fasciculata* K_i (or K_i^*) nM	IAG-NH from *T. brucei brucei* K_i (or K_i^*) nM
1.1	CH	Cl	H			30	190 000
1.2	CH	F	H			57	205 000
1.3	CH	Br	H			28	113 000
1.4	CH	OH	H			75	35 000
1.5	CH	NO_2	H			1100	>360 000
1.6	CH	COOH	H			96	>480 000
1.7	CH	NH_2	H			30	12 000
1.8	N	H	H			7900	>240 000
1.9	CH	H	NO_2			7500	>360 000
1.10	CH	H	NH_2			51	38 000
1.11	N	NH_2	NH_2			770	>30 000
1.12	CH	H	H			300	180 000
1.13						430	>50 000
1.14				OH	H	5400	3600
1.15				OH	NH_2	22 000	5000
1.16				NH_2	H	21 000	13 000
1.17				H	H	>50 000	>50 000
1.18						5200	2300
2.1						42	24
2.2						84	110
2.4						>50 000	660
2.5						>50 000	690
2.9						>50 000	>50 000
2.11						970	190
2.22						3	23
4.17						7	0.9

dependence of inhibition on the nature of the leaving group. The six most powerful inhibitors for IAG-NH are [**4.17**], [**2.22**], [**2.1**], [**2.2**], [**2.11**] and [**1.18**] (Table 8.1). The most powerful inhibitor is 0.9 nM, but only five of the inhibitors are in the nanomolar range with [**1.18**] binding with a K_i value of 2.3 μM, similar to the K_m for the reactant inosine with a value of 18 μM [31]. As predicted, the three best inhibitors for IAG-NH all have the full component of H-bond acceptors/donors found in purine leaving groups. Remarkable specificities are found in Table 8.1 considering that the catalytic sites of these enzymes both accept inosine with similar catalytic efficiencies. Thus, [**1.1**] is a 30 nM inhibitor of IU-NH but a 190 000 nM inhibitor of IAG-NH, demonstrating a 6300 fold preference in affinity for the IU-NH. The *p*-Cl-phenyl leaving group analogue in [**1.1**] is well suited to the nonspecific leaving group nature of IU-NH, but does not contain the H-bond pattern found in a purine ring. However, compounds [**2.22**] and [**4.17**] with features of the purine ring in the leaving group position are 0.9–23 nM inhibitors of both enzymes since IU-NH accepts purine leaving groups while IAG-NH requires them. It is instructive that specificity for the IAG-NH relative to IU-NH can be achieved by altering the iminoribitol common to the inhibitors of Table 8.1. Since the primary transition state interaction of IU-NH is for the ribosyl group, while IAG-NH attains a large fraction of its catalytic power from the leaving group, the 2'-deoxy of [**4.17**] binds with a dissociation constant of 700 nM to IAG-NH and shows no binding to IU-NH even at 100 μM [32]. Manipulation of these features provides inhibitors powerful for both N-ribosyl nucleosidases like [**4.17**], specific for IU-NH like [**1.1**] or those specific for IAG-NH like the 2'-deoxy of [**4.17**].

8.2.6 Biological significance of nucleoside hydrolases and their inhibitors

Nucleoside hydrolases are common in certain pathogenic but not all protozoan parasites. These organisms are purine auxotrophs and rely on purine salvage for RNA and DNA synthesis. The original hypothesis for investigation of the nucleoside hydrolases was that these organisms depend on the activity of purine nucleoside salvage for nucleic acid synthesis [33, 34]. Many of the compounds of Table 8.1 were screened for action in a protozoan parasite assay operated by Professor Simon Croft, London School of Hygiene and Tropical Medicine. None of the compounds showed strong growth inhibition of protozoan parasites cultured *in vitro* (unpublished). However, more physiologically relevant assays in animal models have not yet, to our knowledge, been initiated.

8.3 Purine nucleoside phosphorylases (PNPs)

8.3.1 Relevance of PNP for human disease

The discovery by Eloise Gibblet and coworkers in 1975 that the human genetic deficiency of PNP causes a specific T-cell immunodeficiency [35] triggered extensive research to find inhibitors of PNP [36–39]. Subsequent studies revealed that only activated, rapidly dividing T-cells were killed by the presence of 2'-deoxyguanosine, the metabolite that accumulated as a consequence of PNP inhibition. The idea for physiological intervention by PNP inhibitors was that controlled whole-body inhibition of human PNP would also cause 2'-deoxyguanosine accumulation and would kill rapidly dividing T-cells in T-cell cancers and in inappropriate division of T-cells as in autoimmune diseases or tissue transplant rejection [40, 41]. Several nanomolar inhibitors of human PNP were taken into phase III clinical trials, but failed to cause sufficient inhibition of PNP to cause accumulation of 2'-deoxyguanosine or inhibition of T-cell proliferation [42, 43]. Studies of human genetic deficiency of PNP established that T-cell deficiency does not develop if the mutant PNP has 1 per cent or more of the native enzymatic activity [44]. Therefore, greater than 99 per cent inhibition of whole body PNP catalytic activity is required to generate the T-cell killing phenotype. This is a difficult goal since PNP is an abundant enzyme in mammals, present at micromolar concentrations in erythrocytes, intestine, liver and spleen [45].

8.3.2 Relevance of PNP for malaria

The most deadly malaria parasite is *Plasmodium falciparum*, which has long been known to be a purine auxotroph [46]. With the advent of the *P. falciparum* genome sequencing and annotation [47] together with biochemical studies, it has recently been established that PNP is an essential step in the conversion of purine nucleotides from the adenylate pool of erythrocytes to the nucleic acids of the parasites [48, 49]. These findings suggested that powerful inhibitors acting on the *P. falciparum* PNP may be effective as antimalarials.

8.3.3 Transition state structures of bovine, human and malarial PNPs

Mammalian PNPs are homotrimers and the bovine and human share 87 per cent sequence identity, with 100 per cent conservation in the catalytic site. *P. falciparum* resembles the bacterial PNPs and is a homohexamer with less than 20 per cent identity to the known mammalian PNPs. Kinetic isotope effects and computational methods have been used to solve transition states for bovine, human and malarial PNPs [20]. The phosphorolysis of inosine by these enzymes did not give intrinsic KIE and the transition states were solved using arsenate as a chemically irreversible analogue of phosphate. The transition state for bovine PNP is an early S_N1 transition state with 0.38 Pauling bond order between C1' and N9 of the hypoxanthine leaving group and very weak participation of the attacking arsenate nucleophile [20, 21]. Transition states for the human and *P. falciparum* PNPs are remarkably different with fully dissociated S_N1 transition states with a fully developed ribooxacarbenium ion at the transition state with no significant bond order either to the leaving group or to the attacking arsenate nucleophile (Figure 8.3). The bovine and human PNP transition states are sufficiently different to predict that inhibitors can be designed with specificity for each enzyme.

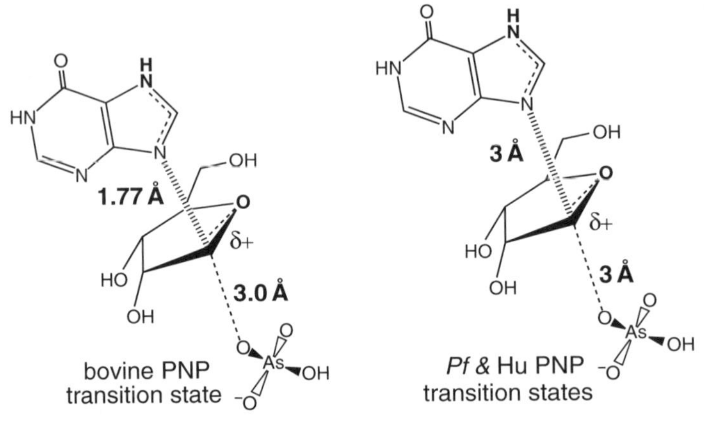

Figure 8.3 Transition state structures for bovine, human and *P. falciparum* PNPs showing the bond length to the leaving group and the attacking nucleophile at the moment of the transition states. Although the physiological nucleophile is phosphate, transition state analysis used arsenate for determination of these transition states. The electron deficiency caused by the delayed attack of the nucleophilic oxygen generates a ribooxacarbenium ion at the transition state. The ribosyl cation is more developed in human and *P. falciparum* transition states than in the bovine enzyme (A colour reproduction of this figure can be found in the colour section towards the centre of the book)

The transition state with inosine as reactant for the *P. falciparum* PNP is indistinguishable from that of the human enzyme; however, the human and parasite PNP have different properties. *P. falciparum* PNP uses 5'-methylthioinosine as a good substrate, but the human enzyme does not. The intrinsic catalytic efficiency of human and *P. falciparum* PNPs is also different with the k_{cat} values being greater for the human enzyme by a factor of 20 [50]. These features suggested that it might be possible to design specific features into transition state analogue inhibitors to distinguish between bovine, human and *P. falciparum* PNPs.

8.3.4 Inhibitors for bovine, human and *P. falciparum* PNPs

Synthetic schemes for inhibitors of the PNPs and references are shown in Schemes 1 and 2. Tables 8.2 and 8.3 list some representative inhibitors for the three PNPs exemplified here. Based on the transition state structure of bovine PNP, it was predicted that iminoribitol linked to 9-deazahypoxanthine or 9-deazaguanine would be powerful inhibitors since inosine and guanosine are good substrates [51]. The transition state for bovine PNP has a C1'-N9 bond length of 1.77 Å due to the 0.38 residual bond order in this early transition state but the attacking arsenate oxygen nucleophile is 3.01 Å from C1'. Immucillin-H and Immucillin-G, compounds [**2.1**] and [**2.2**], were first synthesized to match the transition state of bovine PNP and related *N*-ribosyltransferases and are 23 pM and 30 pM slow-onset, tight-binding inhibitors of the bovine enzyme [51]. Mammalian PNPs also use 2'-deoxypurine nucleosides and [**2.4**] and [**2.5**] are the 2'-deoxy analogues of [**2.1**] and [**2.2**], respectively, to give dissociation constants of

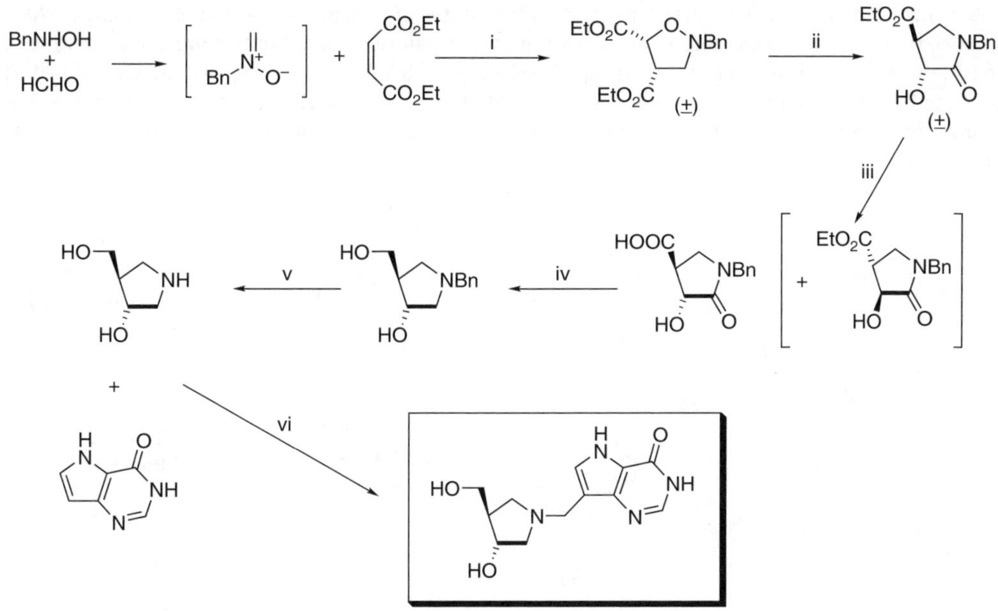

Reagents: i, EtOH, reflux; ii, Zn, HOAc; iii, Novozyme 435, aq acetone pH 7.5; iv, $NaBH_4$, $BF_3.OEt_2$, THF; v, H_2, Pd/C; vi, HCHO, aq dioxane, 80 °C.

Scheme 2 Synthetic scheme for compound [**3.1**] (shown in the box). Variations on this synthetic scheme were used for most inhibitors shown in Table 8.3 and are summarized in references [79, 80]

100 pM and 190 pM for bovine PNP. Other notable inhibitors for bovine PNP include [**2.3**], [**2.7**] and [**2.21**] which are 90 pM, 140 pM and 40 pM inhibitors, respectively. These are the 8-methyl, 8-F and 8-N analogues of [**2.1**] designed to perturb the pK_a of the nitrogen in the five-membered ring. Each change decreases the binding affinity from the 23 pM of [**2.1**] for bovine PNP, suggesting that the pK_a of [**2.1**] is near optimal for binding of this analogue. Integrity of the 9-deazapurine ring is not required for tight binding: thus, [**2.22**] and [**2.23**] are missing C2 of the purine ring system. Remarkably, [**2.23**] with a dissociation constant of 60 pM, binds 200-fold more tightly than [**2.22**] making it one of the best of the bovine PNP inhibitors. Changes in binding energy for transition state analogues of bovine PNP, as a function of atomic substitutions in [**2.1**], are shown in Figure 8.4 [52].

Comparison of inhibition constants for bovine and human PNPs gave the surprising result that [**2.1**] binds 2.4 times more tightly to the bovine enzyme. Given that catalytic site residues are 100 per cent conserved, and the proteins are 87 per cent identical overall, it was assumed that catalytic features, including the transition state would also be indentical, but the difference in binding a transition state analogue suggested that the transition states might be different. Kinetic isotope effects for the arsenolysis reaction of the human PNP demonstrated that the transition state for human PNP is later than for bovine PNP as shown in Figure 8.3. Based on this transition state, it was necessary to synthesize a new family of inhibitors to mimic the features of a fully-dissociated

Table 8.2 Iminoribitol analogues for the transition states of human, bovine and *P. falciparum* PNPs. Dissociation constants were determined from the assumption that the inhibitors are competitive with inosine in the presence of near-saturating phosphate as substrates. K_i and K_i^* values are defined as indicated in the legend to Table 8.1. The chemical structure of **[4.17]** can be found in Table 8.4. Blanks indicate that dissociation constants have not been published for this combination of enzyme and inhibitor

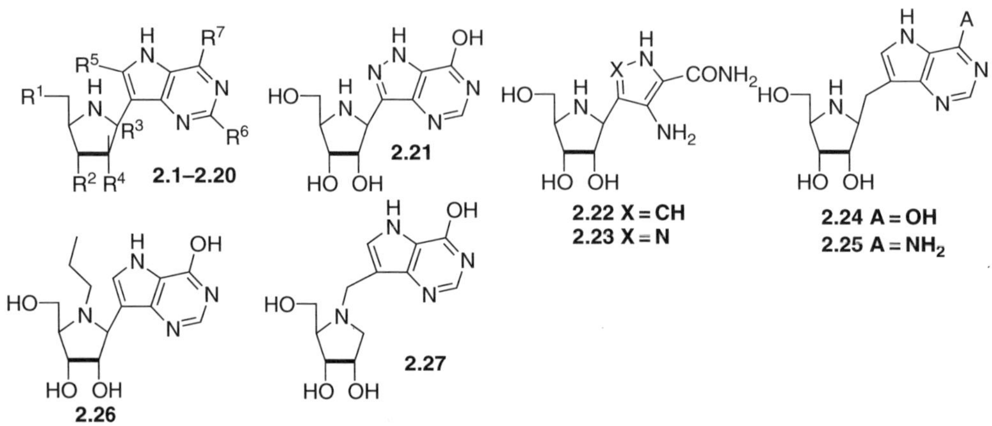

	R^1	R^2	R^3	R^4	R^5	R^6	R^7	Human PNP K_i (or K_i^*) nM	Bovine PNP K_i (or K_i^*) nM	*P. falciparum* PNP K_i (or K_i^*) nM
2.1	OH	OH	H	OH	H	H	OH	0.056	0.023	0.86
2.2	OH	OH	H	OH	H	NH_2	OH	0.042	0.03	0.90
2.3	OH	OH	H	OH	CH_3	H	OH	20	0.090	>100 000
2.4	OH	OH	H	H	H	H	OH	0.14	0.10	2.2
2.5	OH	OH	H	H	H	NH_2	OH	0.18	0.19	8.5
2.6	OH	H	H	OH	H	H	OH	7.5	1.0	42
2.7	OH	OH	H	OH	F	H	OH	0.39	0.14	9.1
2.8	OH	OH	OH	H	H	H	OH	3.5	4.4	9.1
2.9	H	OH	H	OH	H	H	OH	25	15	53
2.10	H	OH	H	OH	H	NH_2	OH		37	
2.11	F	OH	H	OH	H	H	OH	6.8		60
2.12	OH	OH	H	OH	H	OH	OH		310	
2.13	OH	OH	H	OH	H	H	SH	25	1900	58
2.14	OH	OH	H	OMe	H	H	OH	6		840
2.15	OH	OH	F	F	H	H	OH	1.4		>15 000
2.16	MeS	OH	H	OH	H	H	OH	300		2.7
2.17	PhS	OH	H	OH	H	H	OH	250		150
2.18	OH	OH	H	OH	H	H	OMe	4.7		26
2.19	$CONH_2$	OH	H	OH	H	H	OH	>120 000		>120 000
2.20	COOH	OH	H	OH	H	H	OH	>190 000		>190 000
2.21								0.18	0.040	13
2.22									12	
2.23								0.096	0.060	

2.24	250	91	800 000
2.25	440		150 000
2.26	410		>300 000
2.27	2.8		530
4.17		2600	

Table 8.3 Hydroxypyrrolidine and *tert*-amino substituents of 9-deazapurines as inhibitors of human, bovine and *P. falciparum* PNPs. Dissociation constants were determined as indicated in the legends to Tables 8.1 and 8.2. Blanks indicate that dissociation constants have not been published for this combination of enzyme and inhibitor. Abbreviations are: ⁿPrS, *n*-propylthio; MeS, methylthio

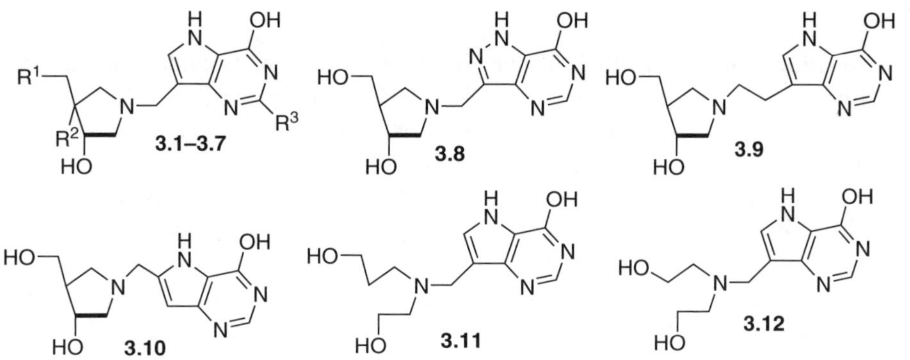

	R^1	R^2	R^3	Human PNP K_i(or K_i*) nM	Bovine PNP K_i(or K_i*) nM	*P. falciparum* PNP K_i(or K_i*) nM
3.1	OH	H	H	0.016	0.11	0.50
3.2	OH	H	NH_2	0.007	0.023	0.89
3.3	Et	H	H	0.36		1.6
3.4	ⁿPrS	H	H	0.041		160
3.5	MeS	H	H	0.071		0.90
3.6	OH	OH	H	14		4300
3.7	MeS	OH	H	430		45 000
3.8				2.0		5.5
3.9				0.46		610
3.10				6600		130 000
3.11				120		>3000
3.12				1.3		170

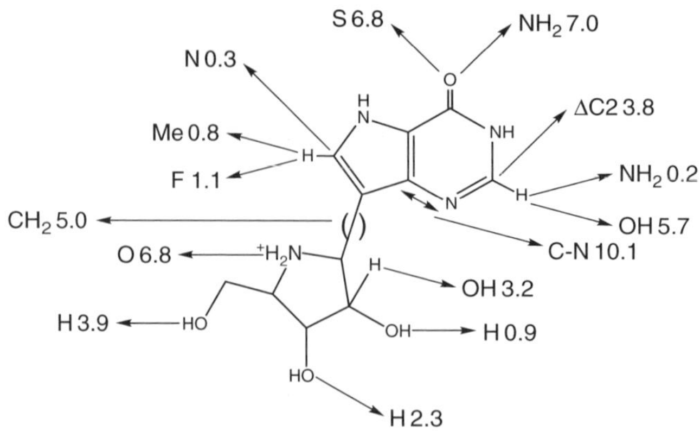

Figure 8.4 Energetics of transition state analogue binding to bovine PNP relative to [**2.1**]. Atomic substitutions are shown for specific positions together with the $\Delta\Delta G$ in kcal/mol for the indicated substitution relative to [**2.1**]. Reprinted with permission from [52].

S_N1 transition state where the ribosyl group exists as a metastable ribooxacarbenium ion without significant bond order to either the attacking arsenate oxygen or the N9 of the hypoxanthine leaving group. This transition state places the leaving group N9 approximately 3.0 Å from C1' at the transition state.

First-generation inhibitors designed for bovine PNP such as [**2.1**] and [**2.2**] have these groups separated by the 1.5 Å of the carbon–carbon covalent bond. To extend this distance to mimic the human PNP transition state, a methylene bridge was placed between the iminosugar and the 9-deazahypoxanthine leaving group analogue. In addition, the carbocation charge in the human PNP transition state is no longer shared between the ribosyl and leaving groups, but is centred primarily on C1'. Therefore, the nitrogen of the iminosugar replaced C1' to locate the nitrogen cation at the same position as the carbocation in the actual transition state. The 1'N in these compounds requires that the 2'C do not carry a hydroxyl group to provide chemical stability. Thus the second generation compounds including [**3.1**] through [**3.8**] are all 2'-deoxy (Table 8.3). The parent compound of the Table 8.3 compounds is [**3.1**] with the trivial name of DADMe-Immucillin-H (4'-Deaza-1'-Aza-2'-Deoxy-1',9-Methylene-Immucillin-H). [**3.1**] is a slow-onset, tight-binding transition state analogue of human PNP with a dissociation constant of 16 pM [53]. Its deazaguanine analogue [**3.2**] binds even tighter to human PNP with a dissociation constant of 7 pM. The closer structural match of the DADMe compounds to the human PNP transition state relative to that from bovine PNP is supported by a comparison of the dissociation constants. Human PNP binds [**3.1**] sevenfold and [**3.2**] threefold more tightly than does bovine PNP, whereas the Immucillin family compounds [**2.1**] and [**2.2**] bind better to the bovine PNP (Table 8.2).

Transition state analysis of *P. falciparum* PNP gives a fully-dissociated S_N1 transition state closely related to human PNP (21). However, the catalytic efficiency of *P. falciparum* PNP is 20-fold less than the human enzyme. According to the thermodynamic interpretation of transition state binding theory (Figure 8.1), transition state analogues designed to match a fully-dissociated transition state should bind 20-fold less well to the *P. falciparum* PNP than to human PNP. Compounds [**2.1**], [**2.2**], [**2.4**], [**2.5**], [**2.6**], [**2.7**] and [**2.8**] give tighter binding to human PNP by factors of 15-, 21-, 16-, 47-, six-, 23- and threefold respectively in reasonable agreement with the theory of Figure 8.1. Notable exceptions to this relationship occur when inhibitor structure incorporates novel aspects of catalytic site interactions. Thus, the *P. falciparum* PNP uses 5'-methylthioinosine as a good substrate while human PNP uses it only 50 times less efficiently than inosine [49]. Accordingly, [**2.16**] incorporates a 5'-methylthiogroup to replace the 5'-hydroxyl group of [**2.1**] (Table 8.2). This change reduces the affinity of binding for human PNP by a factor of 5400 but makes only a threefold reduction in binding to the *P. falciparum* PNP relative to [**2.1**]. This simple change endows a 112-fold specificity for *P. falciparum* PNP in [**2.16**].

Other analogues for the PNPs show surprising results that are not yet readily interpretable. [**2.3**] is the 8-methyl analogue of [**2.1**] and is a 90 pM inhibitor of bovine PNP (Table 8.2). Yet, for the human enzyme it binds with a constant of 20 nM and shows no significant binding to the PNP from *P. falciparum*. The results suggest different steric contacts to the eight-position of the purine ring in the three enzymes, but there is no structural or other inhibitor data to resolve this unusual feature of the catalytic site specificities. Likewise, [**2.15**] the 2'-difluoro analogue of [**2.1**] would be expected to have similar geometry and electronic properties. However, it binds to human PNP at least 10 000 times more tightly than to *P. falciparum* PNP.

The transition states of human and *P. falciparum* PNPs have greater than 3 Å separation between the ribose and the leaving group. In the inhibitor family of [**2.1**], this distance is 1.5 Å and in the inhibitor family of [**3.1**] the distance is 2.5 Å. To increase the distance to 3.5 Å, and introduce additional rotational freedom, the connecting bonds between the imino sugar and the leaving group were increased with an ethyl group [**3.9**]. Instead of increasing binding affinity, the increased separation apparently introduces sufficient increased entropic energy to cause decreased binding to both human and *P. falciparum* PNPs by factors of 29 and 1220 respectively when compared with [**3.1**]. Despite this decreased affinity, it should be noted that [**3.9**] is still a powerful inhibitor of human PNP with a dissociation constant of 0.46 nM.

In compounds [**2.22**] and [**2.23**] much of the binding affinity is retained when the purine ring is opened, provided that the H-bonding elements are retained (Table 8.2). In contrast, opening the iminosugar ring causes large losses of binding energy. Analogue [**3.11**] contains all of the ionic and H-bonding elements of the imino sugar, but the ring is opened between carbons 2' and 3' (Table 8.3). Remarkably, human PNP experiences a 7500-fold loss of binding energy compared with [**3.1**] while the *P. falciparum* PNP experiences more than a 6000-fold loss of binding energy. In another demonstration of the effect of geometry, [**3.12**], with only a single carbon distance changed from [**3.11**], binds approximately 100 times better than [**3.11**] to both human and *P. falciparum* PNPs.

The tight binding of [**2.1**] and [**3.1**] to PNPs suggested that it might be possible to combine their features with a methylene bridge separating the 9-deazahypoxanthine and imino sugar of [**2.1**]. This compound, [**2.24**] binds 2500-fold less well to human PNP than [**2.1**], while it binds 930 000 less well to *P. falciparum* PNP. An important lesson from these comparisons is that molecular features that establish transition state binding interactions are cooperative. Destruction of any single interaction that is a characteristic of the transition state has more effect than the loss of a single H-bond interaction. Thus the interactions that characterize the transition state must be accurately aligned and assist in forming neighbouring interactions. The energetic interactions of Immucillins and DADMe-Immucillins are compared for human and *P. falciparum* PNPs in Figure 8.5. In the upper panel, the $\Delta\Delta G$ values are with respect to [**2.1**] and in the lower panel, with respect to [**3.1**].

8.3.5 Physiological significance of the human PNP inhibitors

The medical relevance of human and *P. falciparum* PNP inhibitors was introduced above. The PNP inhibitors produced by the pharmaceutical industry originated from a structure-based drug design program and exhibited dissociation constants in the nanomolar range. One of the compounds, the 3-pyridinylmethyl-9-deazaguanine analogue of guanosine (BCX-34) was used in clinical trials. Topical and oral formulations of BCX-34 were studied in patients with psoriasis or cutaneous T-cell lymphomas. The results of these clinical trials showed no significant clinical improvements, and development of these drugs was cancelled [42, 43].

The availability of PNP inhibitors with picomolar affinity from transition state design raised the possibility that sustained inhibition of human PNP might occur with these agents. Administration of a single oral dose of Immucillin-H [**2.1**] to mice gave rapid onset ($t^1/_2 = 15$ min) of blood PNP inhibition followed by the slow regain of PNP activity showing a $t^1/_2 = 100$ hour [53]. The same experiment using [**3.1**] also showed a rapid onset of blood PNP inhibition following oral administration ($t^1/_2 = 10$ min) followed by an even slower regain ($t^1/_2 = 11.5$ days) of PNP activity [53]. The time to regain 50 per cent of the original PNP activity of blood is equivalent to the time for regeneration of erythrocytes in the mouse. Therefore, in the mouse, [**3.1**] has been termed 'the ultimate physiological inhibitor', meaning that a single oral dose goes onto the target enzyme and stays there for the lifetime of the enzyme.

A physiological demonstration of whole body inhibition of PNP would be the accumulation of 2'-deoxyguanosine in the blood. In primates, intravenous or oral administration of [**2.1**] caused elevated plasma 2'-deoxyguanosine, and oral administration was more effective at sustaining 2'-deoxyguanosine in the blood than intravenous administration [54]. These studies and the low toxicity of [**2.1**] in animal toxicology studies led to a phase I/II clinical trial for relapsed T-cell leukaemia. Low toxicity in the phase I/II trial led to a phase II trial where significant response rates were found in patients with pretreated and relapsed T-cell leukaemia. Clinical trials with [**2.1**] are also underway with an oral therapy for cutaneous T-cell lymphoma (CTCL), a phase II trial in chronic lymphocytic leukaemia (CLL) and a phase I/II trial in B-cell acute lymphoblastic leukaemia (B-ALL). Orphan drug status has been granted for [**2.1**] by the FDA for three indications: T-cell non-Hodgkin's lymphoma, including CTCL; CLL and related leukaemias including T-cell prolymphocytic leukaemia, adult T-cell leukaemia and hairy cell leukaemia; and for treatment of B-ALL [55].

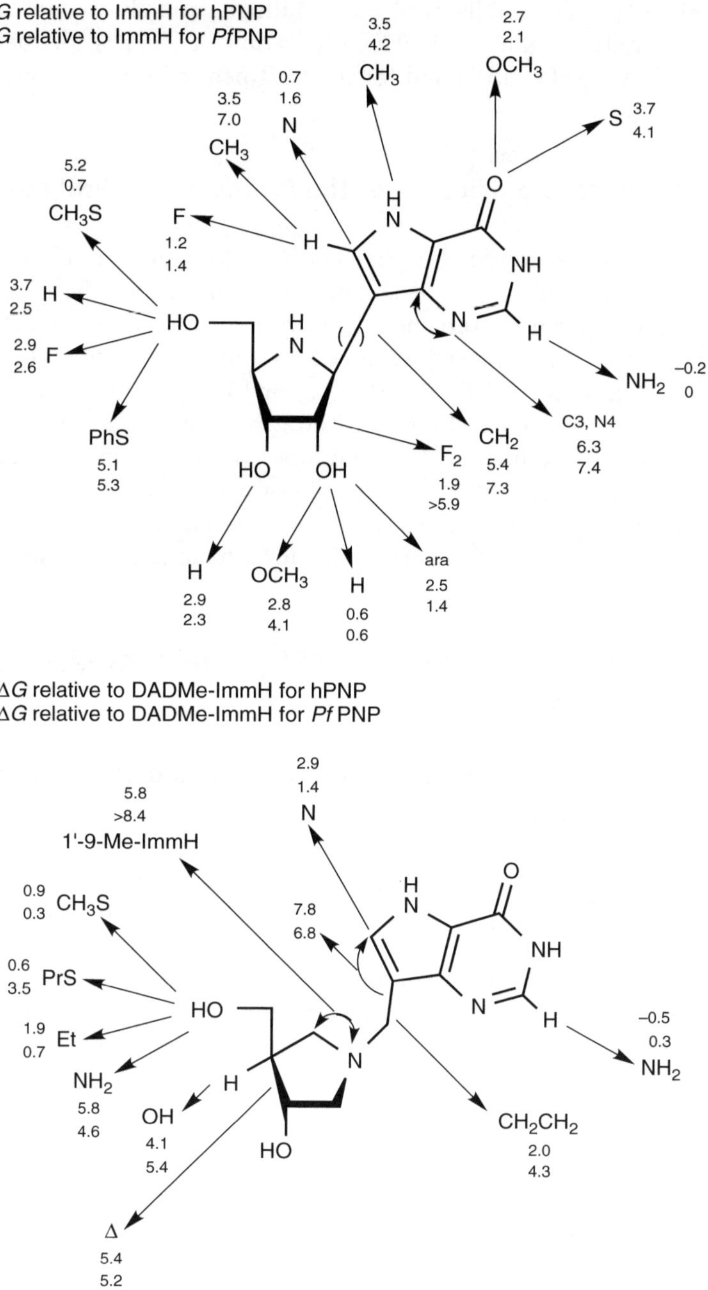

Figure 8.5 Energetics of transition state analogue binding to human and *P. falciparum* PNPs relative to [**2.1**] (upper panel) or to [**3.1**] (lower panel). Atomic substitutions are shown for specific positions together with the ΔΔ*G* in kcal/mol for the indicated substitution relative to the parent compound. Reproduced from Lewandowicz et al. (2005) *J. Biol. Chem.*, **280**, 30320–30328 with permission from the American Society for Biochemistry and Molecular Biology (A colour reproduction of this figure can be found in the colour section towards the centre of the book)

Compound [**3.1**] is also orally available in humans and shows low animal toxicity. Phase I clinical trials are underway. The application of [**3.1**] is for the prevention of acute rejection in transplantation and for the treatment of autoimmune diseases [56].

8.3.6 Physiological significance of the *P. falciparum* PNP inhibitors

Plasmodium falciparum cultured in human erythrocytes require purines in the medium, usually added as hypoxanthine. In the absence of added purines, the parasites are unable to salvage sufficient purines from the adenylate pool of the erythrocytes for optimal growth, although they can achieve several cell division cycles before growth arrest [57]. Addition of PNP inhibitors [**2.1**], [**2.16**], [**3.1**] or [**3.2**] to cultures of *P. falciparum* in human erythrocytes, washed to be free of exogenous purines, causes rapid death by purine starvation [48]. Addition of hypoxanthine but not inosine rescues cells from the purine-starvation death caused by PNP inhibitors. Thus, inhibitors for *P. falciparum* PNP may find application in the treatment of malaria. At the time of writing, there have been no clinical trials of the Immucillins for a malaria indication.

8.4 5′-Methylthioadenosine (MTA) nucleosidases and phosphorylases

8.4.1 Functions of 5′-methylthioadenosine nucleosidases (MTANs) and phosphorylases (MTAPs)

The pathways of polyamine biosynthesis are widely but not universally distributed among prokaryotes and eukaryotes. The synthetic pathway involves donation of the propylamino group from decarboxylated S-adenosylmethionine to putrescine or spermidine to generate spermidine or spermine (Figure 8.6) [58]. Each reaction generates 5'-methylthioadenosine (MTA) as the product, to yield two moles of MTA produced for each mole of spermine synthesized. The MTA metabolite has no known use in prokaryotic or eukaryotic metabolism except to be converted back to active precursor pools. The metabolic roles of MTAN and MTAP are to release adenine from MTA for conversion into AMP by adenine phosphoribosyltransferase, and hence back to the adenylate pool. The 5'-methylthioribosyl group is converted to methionine, and conversion of the resulting methionine and ATP regenerates S-adenosylmethionine, the precursor for the polyamine pool as well as the donor in methyl transfer. In these reactions, the polyamine pathway recycles both the methylthio group of methionine and the adenine of MTA back to S-adenosylmethionine.

In many bacteria, quorum sensing molecules as well as polyamines are synthesized from pathways leading from S-adenosylmethionine [59]. MTAN is an N-ribosyl hydrolase that uses both MTA and S-adenosylhomocysteine as substrates. MTAN (abbreviated MTA/SAHase (Pfs) in Figure 8.7) functions at two sites in the linked pathways between S-adenosylmethionine (SAM) and quorum sensing molecules. It is the only pathway

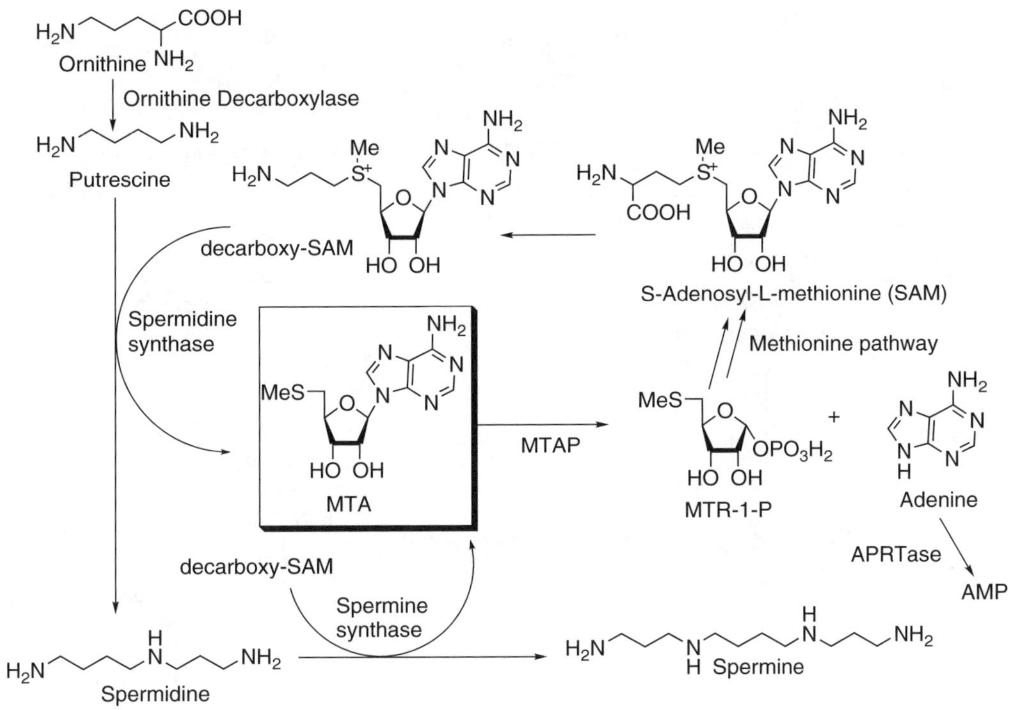

Figure 8.6 The biosynthetic pathway for polyamine synthesis. Reprinted with permission from [58] © 2004 American Chemical Society

to produce ribosyl-homocysteine, a direct precursor for autoinducer-2 molecules such as the furanosyl boron diester. Autoinducer-1 molecules are formed directly from S-adenosylmethionine (SAM), and blocking MTAN prevents the recycling of MTA to adenine nucleotides and the methylthio group to methionine (Figure 8.7). Autoinducers-1 and -2 are important in pathogenic bacteria to signal production of biofilms, toxins and antibiotic resistance factors. This has led to the proposal that MTAN inhibitors may function as antibacterial agents [60].

8.4.2 Inhibitors of 5′-methylthioadenosine phosphorylases (MTAPs)

The polyamine pathway produces cations involved in charge neutralization of nucleic acids and is therefore important for rapidly dividing tissues, including cancers. The rate-limiting step for this pathway is ornithine decarboxylase (ODC) and interventions in the polyamine pathway have used α-difluoromethylornithine (DFMO), a covalent inhibitor of ODC. Despite favourable responses in cancer clinical trials, the rapid cellular turnover of ODC and the ototoxicity of DFMO have limited the utility of this therapy [61]. Blocking MTAP was proposed to achieve inhibition of the polyamine pathway by causing the accumulation of MTA and thereby cause product

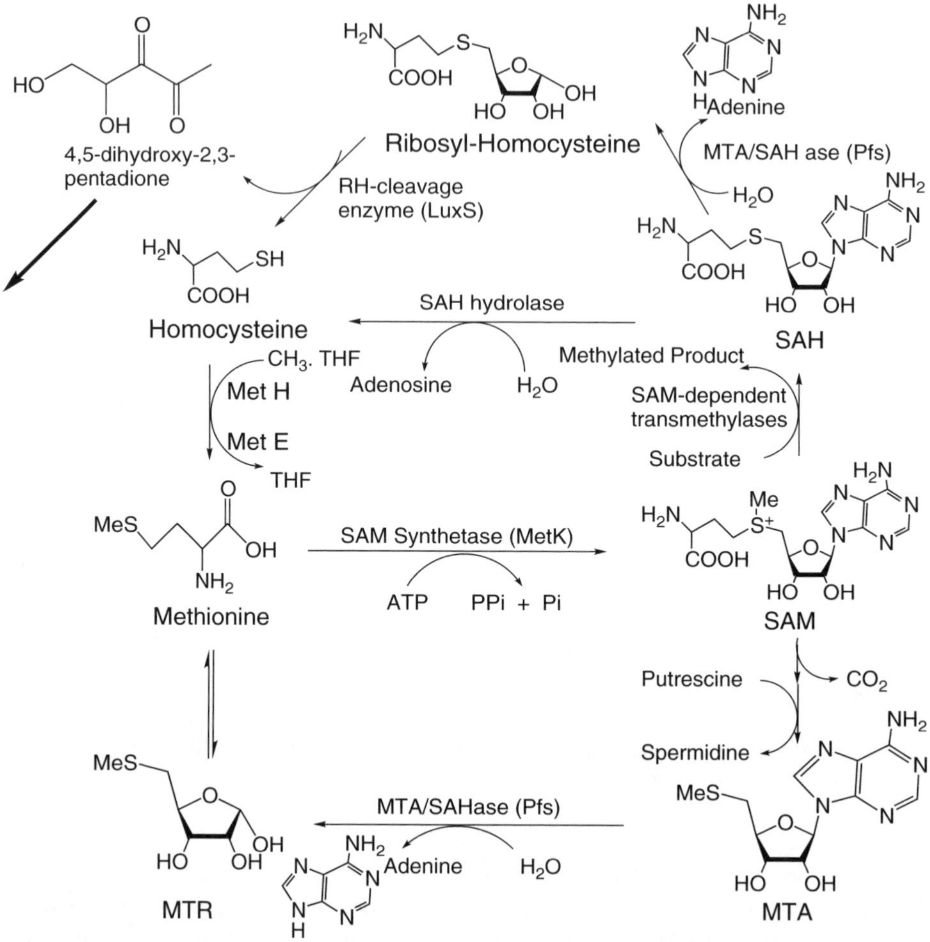

Figure 8.7 The biosynthetic pathway for autoinducer-2 quorum sensing molecules in bacteria

inhibition of both spermidine and spermine synthases in the polyamine pathway (Figure 8.6).

The transition state structure of human MTAP [22] indicated a 5'-methylthioribooxacarbenium ion stabilized by ionization of the 3'-hydroxyl with full dissociation of the adenine leaving group and with low bond order to the attacking nucleophile (arsenate was used for transition state analysis). Adenosine is not a substrate for MTAP. Therefore, transition state analogues are required to include the 5'-methylthio functionality. Synthetic schemes for transition state analogue inhibitors that include the 5'-methylthio functionality are shown in Schemes 3 and 4, together with references to complete synthetic schemes. 9-Deazaadenine linked to the 5'-methylthioiminoribitol group [**4.1**] is a 1 nM inhibitor (Table 8.4). Increasing hydrophobic interactions by replacing the methyl group by ethyl [**4.2**], *n*-propyl

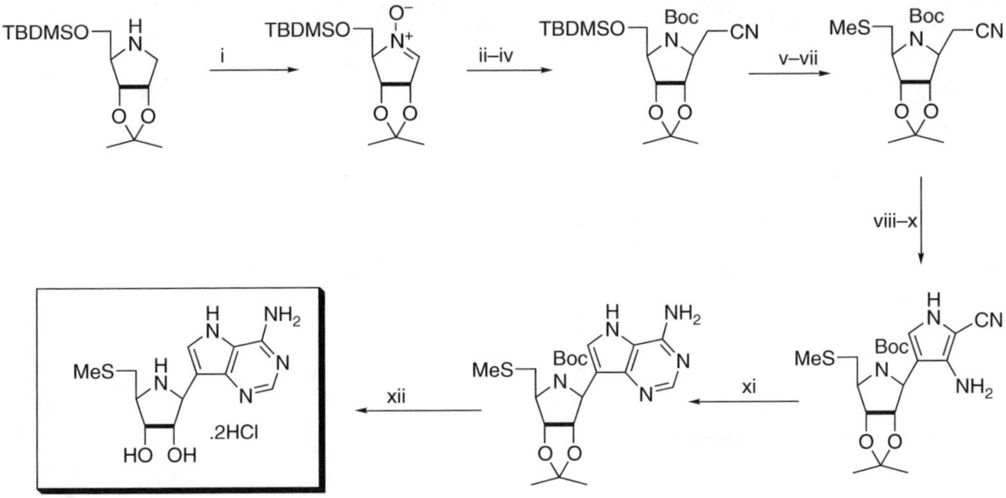

Reagents: i, SeO_2, H_2O_2 aq acetone; ii, BuLi, CH_3CN, THF −70 °C; iii, Zn, HOAc; iv, Boc_2O; v, Bu_4NF; vi, MsCl, Et_3N, CH_2Cl_2; vii, NaSMe, DMF; viii, NaH, EtOCHO, THF; ix, aminoacetonitrile, MeOH; x, MeOCOCl, DBU, CH_2Cl_2, then MeOH; xi, formamidine acetate, EtOH; xii, aq HCl.

Scheme 3 Synthetic scheme for compound **[4.1]** (shown in the box). Variations on this synthetic scheme were used for most inhibitors shown in Table 8.4 and are summarized in reference [81]

[**4.3**], 4-Cl-phenyl [**4.7**], 4-methyl-phenyl [**4.8**] or 3-methylphenyl [**4.10**] enhanced binding to subnanomolar dissociation constants, the best being 166 pM for [**4.7**]. These tight-binding inhibitors are proposed to reflect the cationic charge similarity between the inhibitors and the transition state. However, the compounds of Table 8.4 are nonoptimal mimics of the transition state since the distance between C1' of the cationic ribosyl group and N9 of the leaving group is greater than 3 Å in the actual transition state whereas for the compounds of Table 8.4 it is only 1.5 Å. A methylene bridge increases the separation between leaving group and 5'-methylthioribosyl mimic to 2.5 Å. Together with a positively charged nitrogen to replace the C1' cation of the actual transition state, the compounds of Table 8.5 are superior to those of Table 8.4 for inhibition of human MTAP. Thus, 13 compounds of this series exhibit pM dissociation constants, with [**5.1**]–[**5.13**] giving values down to 10 pM for [**5.10**]. As expected, the closer match of the compounds in Table 8.5 to the transition state of human MTAP gives higher affinity binding than those of Table 8.4. Members of this group are now being explored as agents against cancer indications.

8.4.3 Inhibitors of *E. coli* 5′-methylthioadenosine nucleosidase (MTANs)

The transition state structure has been solved for *E. coli* MTAN. It has a highly dissociated transition state with protonation of the adenine leaving group, presumably at N7, although other sites are possible [23]. Based on transition state structure mimicry, it

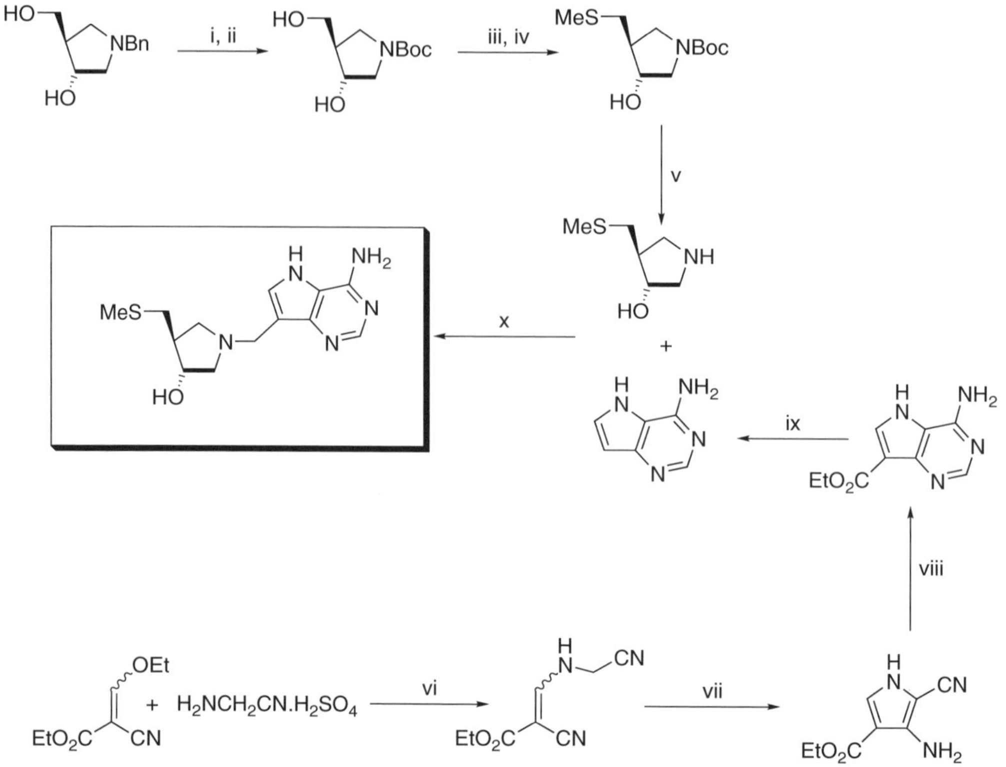

Reagents: i, H_2, Pd/C; ii, Boc_2O; iii, MsCl, Et_3N; iv, NaSMe, DMF; v, aq HCl/MeOH; vi, NaOAc, MeOH; vii, MeOCOCl, Et_3N, then DBU, CH_2Cl_2; viii, formamidine acetate; ix, 10% aq KOH, reflux; x, 37% aq. formaldehyde, NaOAc, dioxane, H_2O.

Scheme 4 Synthetic scheme for compound [**5.1**] (shown in the box). Variations on this synthetic scheme were used for most inhibitors shown in Table 8.5 and are summarized in references [32, 82, 83]

would be expected that the inhibitors of Table 8.5 would be superior to those of Table 8.4 for the *E. coli* enzyme. This is indeed the case. The 47 fM inhibitor [**5.10**] for *E. coli* MTAN is one of the most powerful inhibitors yet described for any enzyme. *E. coli* MTAN is a catalytically highly efficient enzyme with a k_{cat}/K_m of $4.7 \times 10^7\ M^{-1}\ s^{-1}$ [59]. According to transition state theory (Figure 8.1), the more efficient the enzyme, the tighter the transition state analogues should bind. Both the iminoribitol series (Table 8.4) and the methylene-bridged hydroxypyrrolidines (Table 8.5) contain inhibitors with pM dissociation constants but, for the hydroxypyrrolidines of Table 8.5, there are also a group of fM analogues. This affinity supports close mimicry of the transition state for *E. coli* MTAN.

Although the chemistry of the N-ribosyl bond cleavage is similar for *E. coli* MTAN and the phosphorolysis reaction catalysed by human MTAP, it is clear from the results of Tables 8.4 and 8.5 that these analogues are superior inhibitors of the *E. coli* enzyme. One reason for this is that the transition state for human MTAP has an ionized 3'-hydroxyl

Table 8.4 Iminoribitol analogues for the transition states of human MTAP, *E. coli* MTAN and *S. pneumoniae* MTAN. Dissociation constants for MTANs were determined from the assumption that the inhibitors are competitive with 5′-methylthioadenosine (MTA) as substrate. Dissociation constants for MTAP were determined with the assumption that inhibitors are competitive with MTA in the presence of near-saturation phosphate. K_i and K_i^* values are as described in the legends to Tables 8.1, 8.2 and 8.3. The structure of **[2.16]** can be found in Table 8.2. Abbreviations not defined previously are: EtS, ethylthio; PhS, phenylthio; BnS, benzylthio; OMe, methylether; Et, ethyl

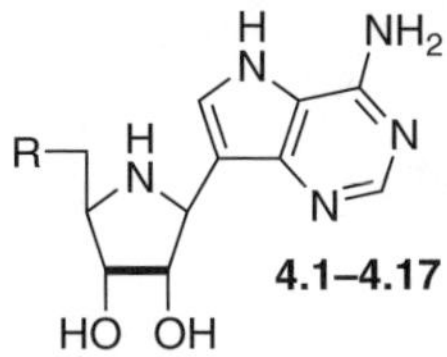

4.1–4.17

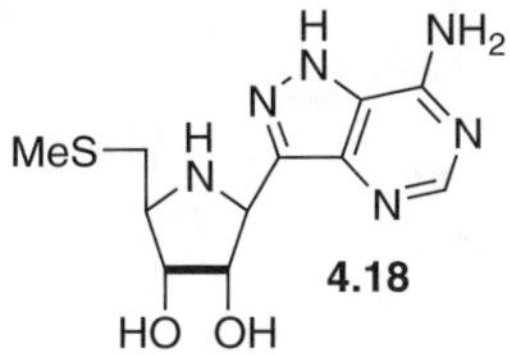

4.18

	R	Human MTAP K_i (or K_i^*) nM	*E coli* MTAN K_i(or K_i^*) nM	*S pneumoniae* MTAN K_i (or K_i^*) nM
4.1	MeS	1.0	0.077	1000
4.2	EtS	0.266	0.027	40
4.3	^{n}PrS	0.214		
4.4	$HO(CH_2)_2S$	14	0.41	
4.5	PhS	1.0	0.032	335
4.6	4-F-PhS	2.0	0.02	360
4.7	4-Cl-PhS	0.166	0.002	193
4.8	4-Me-PhS	0.64	0.008	60
4.9	3-Cl-PhS	6.4	0.02	100
4.10	3-Me-PhS	0.628	0.009	77
4.11	BnS	26	0.012	206
4.12	1-NaphthylS	90	0.75	220
4.13	$F(CH_2)_2S$	3.2	0.03	394
4.14	OMe	134	10	>10 000
4.15	H	720	13	>10 000
4.16	Et	44	0.038	
4.17	OH			30 000
4.18			0.026	1400
2.16			>10 000	>20 000

group, and unprotonated adenine leaving group. These features are not captured in the compounds shown here, and therefore the binding of inhibitors is weaker to human MTAP than to *E. coli* MTAN. This difference is also apparent in the lower catalytic efficiency exhibited by human PNP with a k_{cat}/K_m of $3.2 \times 10^6\,M^{-1}\,s^{-1}$ [22], but this only accounts for one order of magnitude in binding, while most or the transition state analogues shown in Table 8.5 bind 100 to 1000 times better to *E. coli* MTAN than to human MTAP. Despite these differences between enzymes, the inhibitors of Tables 8.4 and 8.5 provide the most powerful inhibitors known for both enzymes.

Table 8.5 Hydroxypyrrolidine substituents of 9-deazapurine and analogues as inhibitors of human MTAP, *E. coli* MTAN and *S. pneumoniae* MTAN. Dissociation constants were determined as indicated in the legend to Table 8.4. Blanks indicate that dissociation constants have not been published for this combination of enzyme and inhibitor. Abbreviations not defined previously are: ⁱPrS, isopropylthio; ⁿBuS, n-butylthio; BnO, benzyl ether; MeS(O), methylsulfoxide; $MeSO_2$, methylsulfone

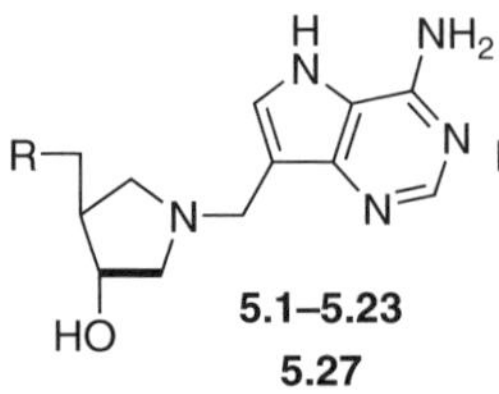

5.1–5.23
5.27

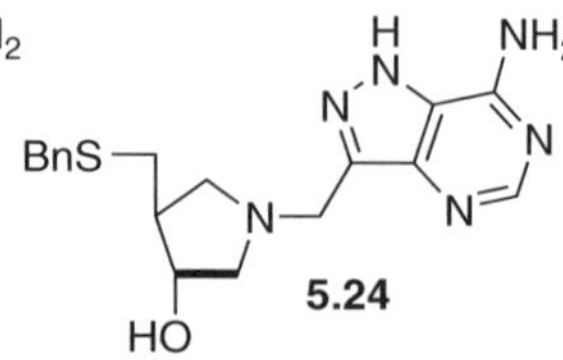

5.24

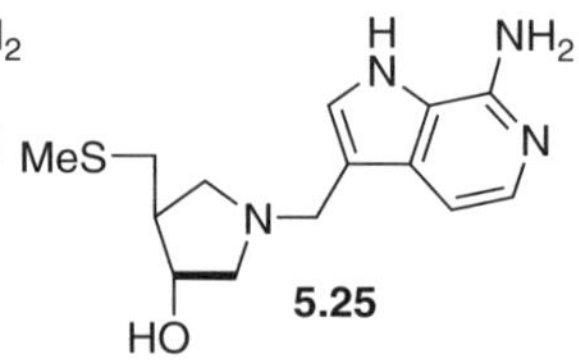

5.25

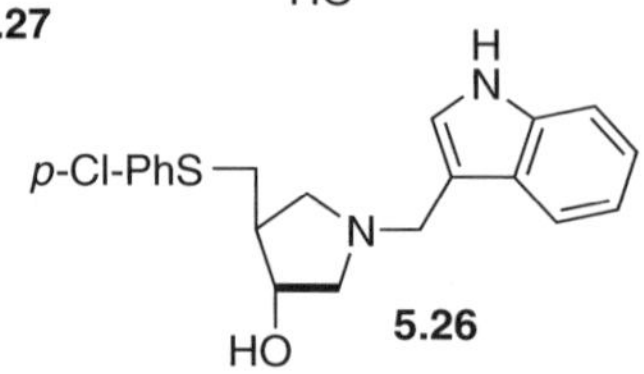

5.26

	R	Human MTAP K_i (or K_i^*) nM [32, 58]	*E coli* MTAN K_i (or K_i^*) nM [59]	*S pneumoniae* MTAN K_i (or K_i^*) nM [84]
5.1	MeS	0.09	0.002	24
5.2	EtS	0.034	0.00095	10
5.3	ⁿPrS	0.12	0.00058	4.0
5.4	ⁱPrS	0.26	0.014	10
5.5	ⁿBuS	0.11	0.0003	2.0
5.6	PhS	0.17	0.002	2.0
5.7	BnS	0.7	0.00046	2.4
5.8	Cyclohexyl-S	0.37	0.00074	0.858
5.9	CyclohexylMe-S		0.00067	
5.10	4-Cl-PhS	0.01	0.000047	0.36
5.11	4-F-PhS	0.16	0.00055	3.5
5.12	4-Pyridyl-S	0.16	0.002	1.4
5.13	3-Cl-PhS	0.27	0.00074	2.5
5.14	Et	3.0	0.006	38
5.15	L-Homocysteinyl		0.006	0.61
5.16	BnO	42	0.009	50
5.17	MeO	8.0	0.062	320
5.18	$PhCH_2$	>2000	1.0	23 000
5.19	Cycloheptyl-S		0.0013	
5.20	Cyclopentyl-S		0.0015	
5.21	Cyclobutyl-S		0.0022	
5.22	MeS(O)		1.0	
5.23	MeSO2		2.0	
5.24		55	0.0004	0.934
5.25			6.5	4500
5.26			0.214	4000
5.27	OH			

8.4.4 Inhibitors of *S. pneumoniae* 5′-methylthioadenosine nucleosidase (MTANs)

The transition state of *S. pneumoniae* MTAN has been solved and has been found to be remarkably different from that of the *E. coli* enzyme. Whereas the *E. coli* enzyme has a protonated leaving group and fully-dissociated transition state, the transition state of *S. pneumoniae* MTAN has an unprotonated, anionic leaving group with an anionic 3'-hydroxyl [24]. These anionic groups sandwich the fully developed cationic C1'-ribosyl cation, giving a transition state with distinct charge differences from that of the *E. coli* enzyme, which is a monocation at C1'. The inhibitors in Tables 8.4 and 8.5 are also monocations, and one notices the superior inhibition ability for the *E. coli* MTAN compared with the *S. pneumoniae* enzyme. By way of example, compounds [**5.1**]–[**5.7**] bind better to *E. coli* MTAN than to *S. pneumoniae* MTAN by factors of 12 000, 10 500, 6900, 700, 6700, 1000 and 5200, respectively.

8.4.5 Physiological significance of bacterial MTAN inhibitors

Genetic experiments in which pathogenic bacteria are made deficient in quorum sensing pathways have revealed reduced pathogenicity or biofilm formation in *Vibrio cholera*, *Streptococcus pneumoniae*, *Neisseria meningitides* and *Pseudomonas aeruginosa* [62–65]. Therefore the hypothesis is that blocking quorum sensing pathways at the level of MTAN will render pathogenic bacterial more vulnerable to the immune system and/or antibiotic treatment. MTAN inhibitors are not expected to block bacterial growth and therefore they are not expected to induce genetic changes that lead to bacterial antibiotic resistance.

8.5 Ricin A-chain

8.5.1 Biological significance of ricin

Ricin is a plant toxin present in castor beans. It attaches to cell surfaces, is translocated, activated and destroys the function of ribosomal protein production [66]. Ricin is composed of A- and B-chains, where the B-chain is a lectin that attaches to cell surfaces. Following translocation, the disulphide bonds that join the A- and B-chains are reduced and the A-chain is released and activated to perform its catalytic function. The 28S ribosomal RNA subunit is the substrate for active ricin A-chain, and a single covalent bond is cleaved to release the adenine from position 4324 of the (rat sequence) rRNA. A specific secondary stem-loop structure is required that includes a GAGA tetraloop. The first adenine in the loop is the substrate for ricin. This adenine is required for efficient binding of eukaryotic elongation factor-2 and without it, ribosome function ceases. The catalytic rate of ricin A-chain is approximately 1800 ribosomes hydrolysed per min, rendering the cell unable to function.

Table 8.6 RNA stem-loop structures as inhibitors of ricin A-chain. The base-paired G-C and C-G bases form the stem. For 4, 6, 10 and 14 base oligomers, this region consists of 0, 1, 3 and 5 GC base pairs, respectively. The G-Inh-G A sequence is the tetraloop with an inhibitor replacing the A at the position of ricin A-chain action. Variation in the first G of the loop also alters the binding affinity of the stem-loop constructs as indicated. Compound **[5.27]** is shown in Table 8.5. The K_i values have the same meaning as defined in the legends to other tables except that stem-loop RNA is the substrate against which these inhibitors compete

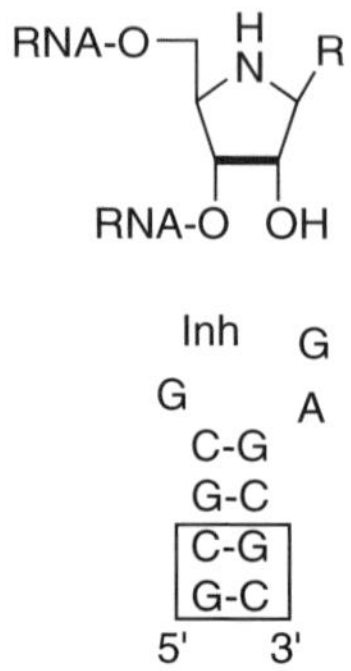

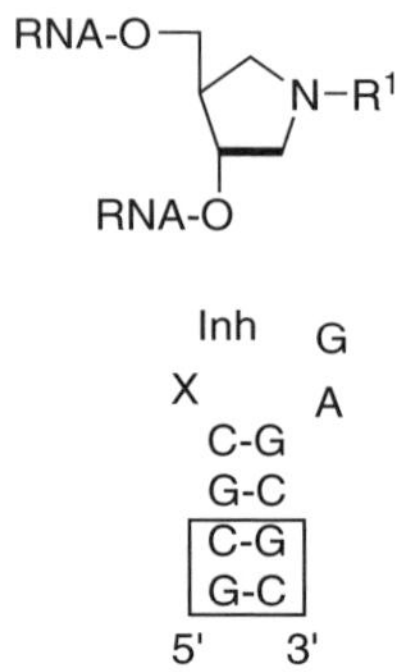

Compound	Oligomer	R	R^1	X	K_i μM
6.1	10	H			1.3
6.2	14				0.18
6.3	10	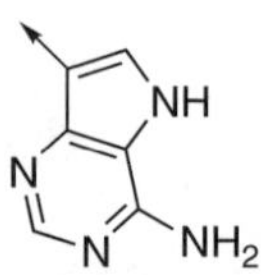			0.57
6.4	14		H	G	0.48
6.5	10		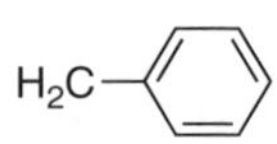	G	0.099
6.6	10		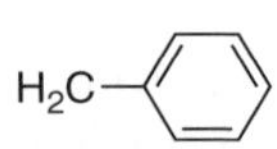	dG	0.026
6.7	4		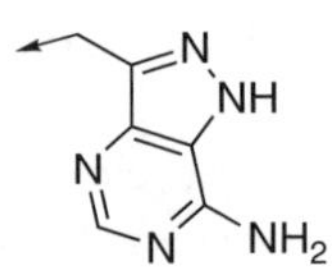	G	22

6.8	6	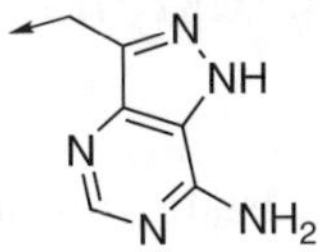	G	43
6.9	10	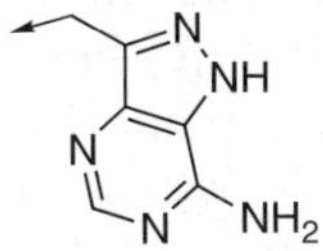	G	0.163
6.10	14	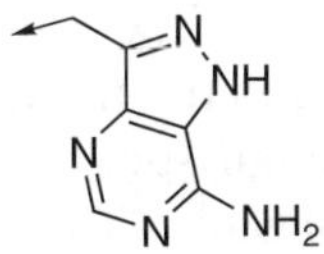	G	0.094
6.11	14	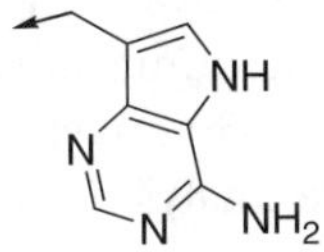	G	0.28
5.27				88

Ricin is one of the most powerful toxins and it is listed as a Class B bioterrorism agent. It gained notoriety in the assassination of Georgi Markov in London in 1978 [67]. Ricin has also been used in immunochemotherapy by linking it to antibodies that recognize cancer cells. In a number of clinical trials with ricin immunotoxins, there have been many remarkable remissions, establishing promise for the method. However, in most studies, the severe vascular leak syndrome limited effective therapy for most patients [68–70]. The purpose for designing transition state analogue inhibitors of ricin A-chain is twofold, to neutralize the toxin in bioterrorism applications and to serve as a rescue agent in ricin immunochemotherapy.

8.5.2 Inhibitors for ricin A-chain

The transition state for ricin A-chain showed a fully-dissociated ribooxacarbenium ion transition state using both stem-loop RNA and DNA as substrates [25, 26]. Small, chemically pure RNA or DNA stem-loops require pH 4 for maximal catalytic activity and the inhibitors have also been characterized at this pH (Table 8.6). Small molecules such as [**5.27**] are poor inhibitors, binding 44-fold less well than small stem-loop substrates which exhibit K_m values near 2 μM [71–73]. When incorporated into stem-loop oligonucleotides of 10 to 14 bases, iminoribitol or hydroxypyrrolidine mimics of the transition state bind with dissociation constants (K_i values) to 26 nM (Table 8.6). Although this is 77-fold tighter binding than substrate, additional chemical design is required to capture the transition state binding energy for the ricin A-chain. Recent advances in small covalently closed, circular nucleic acids may provide a new generation of ricin inhibitors [85].

8.6 Summary and conclusions

Transition state theory, coupled with innovative chemistry, has been successful in producing geometric and electrostatic mimics of enzymatic transition states. These inhibitors are the most powerful catalytic site inhibitors yet produced for every target described here. For bovine PNP, [**2.1**] binds 739 000 times tighter than the substrate inosine. For human PNP, [**3.2**] binds 5 200 000 times more tightly than inosine. For human MTAP, [**5.10**] binds 500 000 times more tightly than MTA. For *E. coli* MTAN, [**5.10**] binds 91 000 000 more tightly than the substrate S-adenosylhomocysteine. Inhibitors for the ricin A-chain need additional chemical development with [**6.6**], the best inhibitor, which binds 77 times tighter than small stem-loop RNAs. The application of transition state theory to design specific mimics is a relatively young science and it remains to be seen how applicable this methodology is for a broader spectrum of enzymes. However, two inhibitors from this research are now progressing in human clinical trials. Five additional compounds are currently in animal preclinical studies for other indications. These results make it likely that transition state analogue design based on experimental transition state analysis will have a significant impact on the future development of pharmaceuticals.

References

1. Zewail A. H. (2000) Femtochemistry: atomic-scale dynamics of the chemical bond using ultrafast lasers (Nobel Lecture). *Angew Chem. Int. Ed. Engl.*, **39**, 2586–2631.
2. Cleland, W.W. (1982) The use of isotope effects to determine transition-state structure for enzymic reactions. *Methods Enzymol.*, **87**, 625–641.
3. Rodgers, J., Femec, D.A. and Schowen, R.L. (1982) Isotopic mapping of transition-state structural features associated with enzymatic catalysis of methyl transfer. *J. Am. Chem. Soc.*, **104,** 3263–3268.
4. Schramm, V.L. (1998) Enzymatic transition states and transition state analogue design. *Ann. Rev. Biochem.*, **67**, 693–720.
5. Gao, J., Ma, S., Major, D. T., Nam, K., Pu, J. and Truhlar, D. G. (2006) Mechanisms and free energies of enzymatic reactions. *Chem. Rev.*, **106**, 3188–3209.
6. Pu, J., Gao, J. and Truhlar, D. G. (2006) Multidimensional tunneling, recrossing, and the transmission coefficient for enzymatic reactions. *Chem. Rev.*, **106**, 3140- 3169.
7. Antoniou, D., Basner, J., Nunez, S. and Schwartz, S. D. (2006) Computational and theoretical methods to explore the relation between enzyme dynamics and catalysis. *Chem. Rev.*, **106**, 3170–3187.
8. Pauling, L. (1948) Chemical achievement and hope for the future. *Amer. Sci.*, **36,** 50–58.
9. Eyring, H. and Johnson, F. H. (1971) The elastomeric rack in biology. *Proc. Natl. Acad. Sci. USA*, **68**, 2341–2344.
10. Jencks, W.P. (1975) Binding energy, specificity, and enzymatic catalysis: the Circe effect. *Adv. Enzymol.*, **43,** 219–410.
11. Wolfenden, R. (1969) Transition state analogues for enzyme catalysis. *Nature*, **223,** 704–705.
12. Leinhard, G. E. (1973) Enzymatic catalysis and transition-state theory. *Science*, **180**, 149–154.

13. Jencks, W. P. (1987) Economics of enzyme catalysis. *Cold Spring Harb. Symp. Quant. Biol.*, **52**, 65–73.
14. Schowen, K. B., Limbach, H. H., Denisov, G. S. and Schowen, R. L. (2000) Hydrogen bonds and proton transfer in general-catalytic transition-state stabilization in enzyme catalysis. *Biochem. Biophys. Acta*, **1458**, 43–62.
15. Schramm, V.L. (1998) Enzymatic transition states and transition state analogue design. *Ann. Rev. Biochem.*, **67**, 693–720.
16. Schramm, V.L. (1999) Enzymatic transition-state analysis and transition-state analogs. *Methods Enzymol.*, **308**, 301–355.
17. Schramm, V. L. (2005) Enzymatic transition states: thermodynamics, dynamics and analogue design. *Arch. Biochem. Biophys.*, **433**, 13–26.
18. Horenstein, B.A., Parkin, D.W., Estupinan, B. and Schramm, V.L. (1991) Transition-state analysis of nucleoside hydrolase from *Crithidia fasciculata. Biochemistry*, **30,** 10788–10795.
19. Schramm, V. L. (2003) Enzymatic transition state poise and transition state analogues. *Acc. Chem. Res.*, **36**, 588–596.
20. Kline, P.C. and Schramm, V.L. (1993) Purine nucleoside phosphorylase. Catalytic mechanism and transition state analysis of the arsenolysis reaction. *Biochemistry*, **32,** 13212–13219.
21. Lewandowicz, A. and Schramm, V. L. (2004) Transition state analysis for human and *Plasmodium falciparum* purine nucleoside phosphorylases. *Biochemistry*, **43**, 1458–1468.
22. Singh, V. and Schramm, V. L. (2006) Transition state structure of human 5'-methylthioadenosine phosphorylase. *J. Am. Chem. Soc.*, **128**, 14691–14696.
23. Singh, V., Lee, J. E., Nunez, S., Howell, P. L. and Schramm, V. L. (2005) Transition state structure of 5'-methylthioadenosine/S-adenosylhomocysteine nucleosidase from *Escherichia coli* and its similarity to transition state analogues. *Biochemistry*, **44**, 11647–11659.
24. Singh, V. and Schramm, V. L. (2007) Transition state structure of *S. pneumoniae* 5'-methylthioadenosine phosphorylase. *J. Am. Chem. Soc.*, **129**, 2783–2795.
25. Chen, X.-Y., Berti, P. J. and Schramm, V. L. (2000) Ricin A-chain: kinetic isotope effects and transition state structures with stem-loop RNA. *J. Am. Chem. Soc.*, **122**, 1609–1617.
26. Chen, X.-Y., Berti, P.J. and Schramm, V. L. (2000) Transition state structure for depurination of DNA by ricin A-chain. *J. Am. Chem. Soc.*, **122,** 6527–6534.
27. Cherian, X. M., Van Arman, S. A. and Czarnik, A. W. (1990) Models for nucleoside glycosylase enzymes. Evidence that the hydrolysis of β-D-Ribofuranosides requires a "backside" preassociated nucleophile. *J. Am. Chem. Soc.*, **112,** 4490–4498.
28. Mazzella, L. J., Parkin, D. W., Tyler, P. C., Furneaux, R. H. and Schramm, V. L. (1996) Mechanistic diagnoses for N-ribohydrolases and purine nucleoside phosphorylase. *J. Am. Chem. Soc.*, **118,** 2111–2112.
29. Parkin, D. W., Limberg, G., Tyler, P. C. Furneaux, R. H., Chen, X.-Y. and Schramm, V.L. (1997) Isozyme-specific transition state inhibitors for the trypanosomal nucleoside hydrolases. *Biochemistry*, **36,** 3528–3534.
30. Sauve, A. A., Cahill, S. M., Zech, S. G., Basso, L.A., Lewandowicz, A., Santos, D. S., Grubmeyer, C., Evans, G. B., Furneaux, R. H., Tyler, P. C., McDermott, A., Girvin, M. E. and Schramm, V. L. (2003) Ionic states of substrates and transition state analogues at the catalytic sites of N-ribosyltransferases. *Biochemistry*, **42**, 5694–5705.
31. Parkin D. W. (1996) Purine-specific nucleoside N-ribohydrolase from *Trypanosoma brucei brucei*. Purification, specificity, and kinetic mechanism. *J. Biol. Chem.*, **271**, 21713–21719.
32. Schramm V. L. and Tyler, P. C. (2003) Imino-sugar-based nucleosides. *Curr. Top. Med. Chem.*, **3**, 525–540.
33. Estupiñán, B. and Schramm, V.L. (1994) Guanosine-inosine preferring nucleoside N-glycohydrolase from *Crithidia fasciculata. J. Biol. Chem.*, **269,** 23068–23073.

34. Horenstein, B.A., Abdulah, D.R., Estupiñán, B. and Schramm, V.L. (1991) Nucleoside hydrolase from *Crithidia fasciculata*. Metabolic role, purification, specificity, and kinetic mechanism. *J. Biol. Chem.*, 266, 20658–20665.
35. Giblett, E. R., Ammann, A. J., Wara, D. W., Sandman, R. and Diamond, L. K. (1975) Nucleoside-phosphorylase deficiency in a child with severely defective T-cell immunity and normal B-cell immunity. *Lancet*, **1**, 1010–1013.
36. Montgomery, J.A., Niwas, S., Rose, J.D., Secrist III, J.A., Babu, Y.S., Bugg, C.E., Erion, M.D., Guida, W.C. and Ealick, S.E. (1993) Structure-based design of inhibitors of purine nucleoside phosphorylase. 1. 9-(Arylmethyl) derivatives of 9-deaza-aguanine. *J. Med. Chem.*, **36**, 55–69.
37. Morris, P.E. Jr, and Montgomery, J.A. (1998) Inhibitors of the enzyme purine nucleoside phosphorylase. *Exp. Opin. Ther. Patents*, **8**, 283–299.
38. Stoeckler, J.D., Ealick, S.E., Bugg, C.E. and Parks, R.E., Jr. (1986) Design of purine nucleoside phosphorylase inhibitors. *Fed. Proc., Fed. Am. Soc. Exp. Biol.*, **45**, 2773–2778.
39. Ealick, S.E., Babu, Y.S., Bugg, C.E., Erion, M.D., Guida, W.C., Montgomery, J.A. and Secrist III, J.A. (1991) Application of crystallographic and modeling methods in the design of purine nucleoside phosphorylase inhibitors. *Proc. Natl. Acad. Sci. USA*, **88**, 11540–11544.
40. Ullman, B., Cohen, A. and Martin, D. W. (1976) Characterization of a cell culture model for the study of adenosine deaminase- and purine nucleoside phosphorylase-deficient immunologic disease. *Cell*, **9**, 205–211.
41. Carson, D.A. and Kaye, J. (1978) Deoxyribonucleoside toxicity in adenosine deaminase and purine nucleoside phosphorylase deficiency: implications for the development of new immunosuppressive agents. *Ciba Found. Symp. 1978*, 115–133.
42. Morris, P.E. Jr. and Omura, G. A. (2000) Inhibitors of the enzyme purine nucleoside phosphorylase as potential therapy for psoriasis. *Curr. Pharm. Des.*, **6**, 943–959.
43. Duvic, M., Olsen, E. A., Omura, G. A., Maize, J. C., Vonderheid, E. C., Elmets, C. A., Shupack, J. L., Demierre, M. F., Kuzel, T. M. and Sanders, D. Y. (2001) A phase III, randomized, double-blind, placebo-controlled study of peldesine (BCX-34) cream as topical therapy for cutaneous T-cell lymphoma. *J. Am. Acad. Dermatol.*, **44**, 940–947.
44. Biggar, W. D., Giblett, E. R., Ozere, R. L. and Grover, B. D. (1978) A new form of nucleoside phosphorylase deficiency in two brothers with defective T-cell function. *J. Pediatr.*, **92**, 354–357.
45. Mohamedali, K.A., Guicherit, O.M., Kellems, R.E. and Rudolph, F.B. (1993) The highest levels of purine catabolic enzymes in mice are present in the proximal small intestine. *J. Biol. Chem.*, **268**, 23728–23733.
46. Trager, W. (1971) Malaria parasites (*Plasmodium lophurae*) developing extracellularly in vitro: incorporation of labeled precursors. *J. Protozool.*, **18**, 392–399.
47. Gardner, M. J., Hall, N., Fung, E., White, O., Berriman, M., Hyman,R.W., Carlton, J. M., Pain, A., Nelson, K. E., Bowman, S. *et al.* (2002) The genome sequence of the human malaria parasite *Plasmodium falciparum*. *Nature*, **419**, 498–511.
48. Kicska, G. A., Tyler, P. C., Evans, G. B., Furneaux, R. H., Schramm, V. L. and Kim, K. (2002) Purine-less death in *Plasmodium falciparum* induced by immucillin-H, a transition state analogue of purine nucleoside phosphorylase. *J. Biol. Chem.*, **277**, 3226–3231.
49. Shi, W., Ting, L. M., Kicska, G. A., Lewandowicz, A., Tyler, P. C., Evans, G. B., Furneaux, R. H., Kim, K., Almo, S. C. and Schramm, V. L. (2004) *Plasmodium falciparum* purine nucleoside phosphorylase: crystal structures, immucillin inhibitors, and dual catalytic function. *J. Biol. Chem.*, **279**, 18103–18106.
50. Lewandowicz, A., Ringia, E. A., Ting, L. M., Kim, K., Tyler, P. C., Evans, G. B., Zubkova, O. V., Mee, S., Painter, G. F., Lenz, D. H., Furneaux, R. H. and Schramm, V. L. (2005) Energetic

mapping of transition state analogue interactions with human and *Plasmodium falciparum* purine nucleoside phosphorylases. *J. Biol. Chem.*, **280**, 30320–30328.

51. Miles, R. W., Tyler, P. C., Furneaux, R. H., Bagdassarian, C. K. and Schramm, V. L. (1998) One-third-the-sites transition state inhibitors for purine nucleoside phosphorylase. *Biochemistry*, **37**, 8615–8621.
52. Kicska, G. A., Tyler, P. C., Evans, G. B., Furneaux, R. H., Shi, W., Fedorov, A., Lewandowicz, A., Cahill, S. M., Almo, S. C. and Schramm, V. L. (2002) Atomic dissection of the hydrogen bond network for transition-state analogue binding to purine nucleoside phosphorylase. *Biochemistry*, **41**, 14489–14498.
53. Lewandowicz, A., Tyler, P. C., Evans, G. B., Furneaux, R. H. and Schramm, V. L. (2003) Achieving the ultimate physiological goal in transition state analogue inhibitors for purine nucleoside phosphorylase. *J. Biol. Chem.*, **278**, 31465–31468.
54. Kilpatrick, J. M., Morris, P. E., Serota, D.G., Jr., Phillips, D., Moore, D. R., Bennett, J. C. and Babu, Y. S. (2003) Intravenous and oral pharmacokinetic study of BCX-1777, a novel purine nucleoside phosphorylase transition-state inhibitor. In vivo effects on blood 2'-deoxyguanosine in primates. *Int. Immunopharmacol.*, **3**, 541–548.
55. http://www.biocryst.com/bcx_1777.htm (29 April 2007)
56. http://www.biocryst.com/bcx_4208.htm (29 April 2007)
57. El Bissati, K., Zufferey, R., Witola, W. H., Carter, N. S., Ullman, B. and Ben Mamoun, C. (2006) The plasma membrane permease PfNT1 is essential for purine salvage in the human malaria parasite *Plasmodium falciparum. Proc. Natl. Acad. Sci. U. S. A.*, **103**, 9286–9291.
58. Singh, V., Shi, W., Evans, G. B., Tyler, P. C., Furneaux, R. H., Almo, S. C. and Schramm, V. L. (2004) Picomolar transition state analogue inhibitors of human 5'-methylthioadenosine phosphorylase and X-ray structure with MT-immucillin-A. *Biochemistry*, **43**, 9–18.
59. Singh, V., Evans, G. B., Lenz, D. H., Mason, J. M., Clinch, K., Mee, S., Painter, G. F., Tyler, P. C., Furneaux, R. H., Lee, J. E., Howell, P. L. and Schramm, V. L. (2005) Femtomolar transition state analogue inhibitors of 5'-methylthioadenosine/S-adenosylhomocysteine nucleosidase from *Escherichia coli. J. Biol. Chem.*, **280**, 18265–18273.
60. Persson, T., Givskov, M. and Nielsen, J. (2005) Quorum sensing inhibition: targeting chemical communication in gram-negative bacteria. *Curr. Med. Chem.*, **12**, 3103–3115.
61. Casero, R. A., Jr., Frydman, B., Stewart, T.M. and Woster, P.M. (2005) Significance of targeting polyamine metabolism as an antineoplastic strategy: unique targets for polyamine analogues. *Proc. West. Pharmacol. Soc.*, **48**, 24–30.
62. Zhu, J. and Mekalanos, J. J. (2003) Quorum sensing-dependent biofilms enhance colonization in *Vibrio cholerae. Dev. Cell.*, **5**, 647–656.
63. Joyce, E. A., Kawale, A., Censini, S., Kim, C. C., Covacci, A. and Falkow, S. (2004) *LuxS* is required for persistent pneumococcal carriage and expression of virulence and biosynthesis genes. *Infect. Immun.*, **72**, 2964–2975.
64. Winzer, K., Sun, Y., Green, A., Delory, M., Blackley, D., Hardie, K. R., Baldwin, T. J. and Tang, C. M. (2002) Role of *Neisseria meningitidis luxS* in cell-to-cell signaling and bacteremic infection. *Infect. Immun.*, **70**, 2245–2248.
65. Mittal, R., Sharma, S., Chhibber, S. and Harjai, K. (2006) Contribution of quorum-sensing systems to virulence of *Pseudomonas aeruginosa* in an experimental pyelonephritis model. *J. Microbiol. Immunol. Infect.*, **39**, 302–309.
66. Lord, J.M., Roberts, L. M. and Robertus, J. D. (1994) Ricin: structure, mode of action, and some current applications. *FASEB J.*, **8**, 201–208.
67. Crompton, R. and Gall, D. (1980) Georgi Markov – death in a pellet. *Med. Leg. J.*, **48**, 51–62.
68. Frankel, A. E. and Kreitman, R. J. (2005) CLL immunotoxins. *Leuk. Res.*, **29**, 985–986.

69. Kreitman, R. J. (2001) Toxin-labeled monoclonal antibodies. *Curr. Pharm. Biotechnol.*, **2**, 313–325.
70. Schnell, R., Borchmann, P., Staak, J. O., Schindler, J., Ghetie, V., Vitetta, E. S. and Engert, A. (2003) Clinical evaluation of ricin A-chain immunotoxins in patients with Hodgkin's lymphoma. *Ann. Oncol.*, **14**, 729–736.
71. Tanaka, K. S. E, Chen, X.-Y, Ichikawa, Y., Tyler, P. C., Furneaux, R. H. and Schramm, V.L. (2001) Ricin A-chain inhibitors resembling the oxacarbenium ion transition state. *Biochemistry*, **40**, 6845–6851.
72. Chen, X.-Y., Link, T. M. and Schramm, V. L. (1998) Ricin A-chain: kinetics, mechanism and RNA-stem-loop inhibitors. *Biochemistry*, **37**, 11605–11613.
73. Roday, S., Amukele, T., Evans, G. B., Tyler, P. C., Furneaux, R. H. and Schramm, V. L. (2004) Inhibition of ricin A-chain with pyrrolidine mimics of the oxacarbenium ion transition state. *Biochemistry*, **43**, 4923–4933.
74. Fleet, G. W. J. and Son, J. C. (1988) Polyhydroxylated pyrrolidines from sugar lactones: Synthesis of 1,4-dideoxy-1,4-imino-D-glucitol from D-galactonolactone and synthesis of 1,4-dideoxy-1,4-imino-D-allitol, 1,4-dideoxy-1,4-imino-D-ribitol and (2*S*,3*R*,4*S*)-3,4-dihydroxyproline from D-gulonolactone. *Tetrahedron*, **44**, 2637–2647.
75. Horenstein, B. A., Zabinski, R. F. and Schramm, V.L. (1993) A new class of C-nucleoside analogues. 1-(*S*)-1,4-dideoxy-1,4-imino-D-ribitols, transition state analogue inhibitors of nucleoside hydrolase. *Tetrahedron Lett.*, **34**, 7213–7216.
76. Furneaux, R. H., Schramm, V. L. and Tyler, P. C. (1999) Transition state analogue inhibitors of protozoa nucleoside hydrolases. *Bioorg. Med. Chem.*, **11**, 2599–2606.
77. Evans, G. B., Furneaux, R. H., Hutchison, T. L., Kezar, H. S., Morris, P.E., Jr., Schramm, V. L. and Tyler, P. C. (2001) Addition of lithiated 9-deazapurine derivatives to a carbohydrate cyclic imine: convergent synthesis of the aza-*C*-nucleoside immucillins. *J. Org. Chem.*, **66**, 5723–5730.
78. Furneaux, R. H. and Tyler, P. C. (1999) Improved syntheses of 3*H*,5*H*-pyrrolo[3,2-d]pyrimidines. *J. Org. Chem.*, **64**, 8411–8412.
79. Evans, G. B., Furneaux, R. H., Tyler, P. C. and Schramm, V. L. (2003) Synthesis of a transition state analogue inhibitor of purine nucleoside phosphorylase via the Mannich reaction. *Org. Lett.*, **5**, 3639–3640.
80. Evans, G. B., Furneaux, R. H., Lewandowicz, A., Schramm, V. L. and Tyler, P. C. (2003) Synthesis of second-generation transition state analogues of human purine nucleoside phosphorylase. *J. Med. Chem.*, **46**, 5271–5276.
81. Evans, G. B., Furneaux, R. H., Schramm, V. L., Singh, V. and Tyler, P. C. (2004) Targeting the polyamine pathway with transition-state analogue inhibitors of 5'-methylthioadenosine phosphorylase. *J. Med. Chem.*, **47**, 3275–3281.
82. Evans, G. B., Furneaux, R. H., Lenz, D. H., Painter, G. F., Schramm, V. L., Singh, V. and Tyler, P. C. (2005) Second generation transition state analogue inhibitors of human 5'-methylthioadenosine phosphorylase. *J. Med. Chem.*, **48**, 4679–4689.
83. Evans, G. B. (2004) The synthesis of N-ribosyl transferase inhibitors based on a transition state blueprint. *Aust. J. Chem.*, **57**, 837–854.
84. Singh, V., Shi, W., Almo, S. C., Evans, G. B., Furneaux, R. H., Tyler, P. C., Painter, G. F., Lenz, D. H., Mee, S., Zheng, R. and Schramm, V. L. (2006) Structure and inhibition of a quorum sensing target from *Streptococcus Pneumoniae. Biochemistry*, **45**, 12929–12941.
85. Sturm, M. B., Roday, S. and Schramm, V. L. (2007) Circular DNA and DNA/RNA hybrid molecules as scaffolds for ricin inhibitor design. *J. Am. Chem. Soc.*, **129**, 5544–5550.

9

Iminosugars as antiviral agents

Pamela A. Norton, Baohua Gu and **Timothy M. Block**

9.1 Introduction

A number of iminosugars, such as DNJ, are glucose mimetics, acting as competitive inhibitors of the N-glycan processing enzymes glucosidase I and glucosidase II; structures of DNJ and some other iminosugars of interest are shown in Figure 9.1. All N-linked glycans, following transfer to acceptor asparagine on nascent glycoproteins from the dolichol donor, are trimmed in the ER, and then in the Golgi, by a series of sequentially active glycoprocessing enzymes [1]. ER α-glucosidases I and II are the first enzymes to function in this pathway (Figure 9.2). The reason why cells remodel glycans immediately after transfer of the oligosaccharide to the polypeptide backbone was not obvious. However, it has become clear that polypeptide trafficking within the cell may depend upon stepwise glycan processing as a means of monitoring protein folding and/or assembly [1–3].

It is now generally accepted that aberrant trafficking and other failures of the biogenesis of N-linked glycoproteins in glucosidase-inhibited cells is a secondary consequence of interference with the calnexin and calnexin-like protein-mediated protein folding (reviewed in [2, 3]). All nascent N-linked glycoproteins contain three terminal glucose residues at the distal termini of their N-glycans. These are removed sequentially, with the external glucose unit removed first by glucosidase I followed by removal of the next two residues by glucosidase II. At this point, if the protein is properly folded, the glycoprotein will continue in transit to the Golgi for further glycan remodelling and other post-translational processing. However, if the glycoprotein is not folded properly, as may typically be the case for either an incompletely synthesized protein and/or a protein destined to be assembled with other subunits, a single glucose residue is re-added by UDP-glucose glycoprotein:glucosyltransferase (UGGT) [3]. UGGT recognizes unfolded proteins, sensing both glycan and protein features [4]. Resident ER protein chaperones calnexin and calreticulin bind to monoglucosylated glycoproteins, produced

Iminosugars: From synthesis to therapeutic applications Edited by Philippe Compain and Olivier R. Martin
 ISBN 978-0-470-03391-3

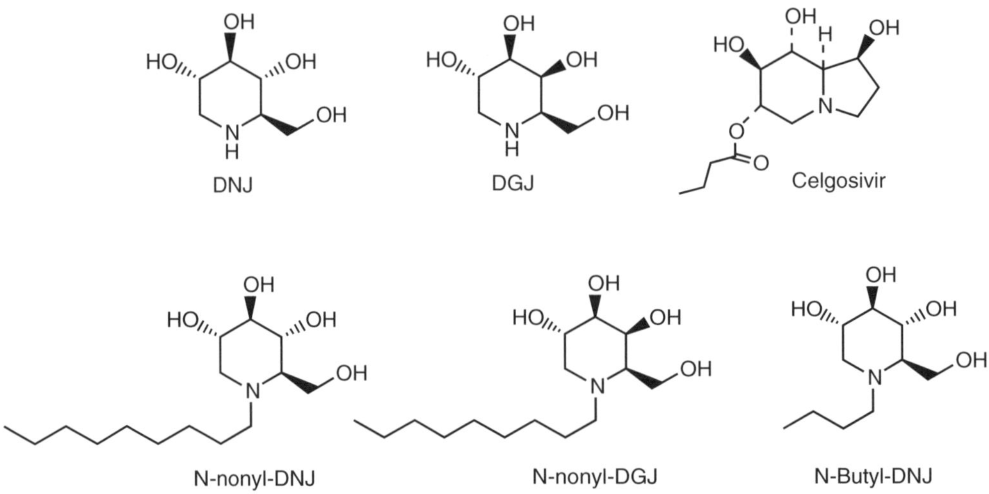

Figure 9.1 Structures of compounds referred to in the text

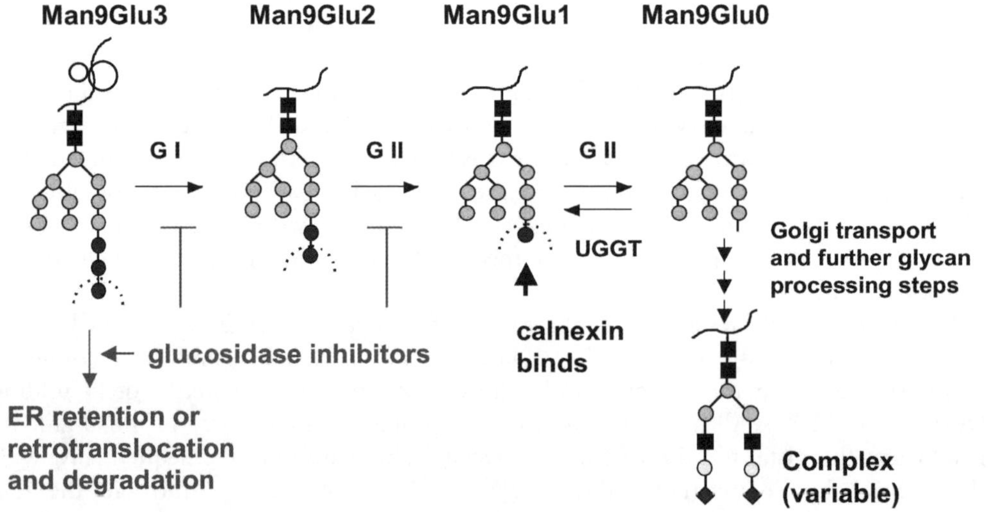

Figure 9.2 The normal pathway of glycoprotein processing and folding, and mechanisms by which misfolded proteins can be detected and eliminated. The structures of the N-linked oligosaccharide side chains are shown schematically. The growing polypeptide chain is indicated by a thin black line. Actions of glucosidases are indicated by a dotted line. Black squares, N-acetyl -D- glucosamine; dark grey circles, mannose; black circles, glucose; pale grey circles, galactose; diamonds, N-acetyl-neuraminic acid (A colour reproduction of this figure can be found in the colour section towards the centre of the book)

either by glucosidase II or UGGT and facilitate proper protein folding [1,2]. These lectin-type chaperones retain the monoglucosylated proteins in the ER. Additional cycles of de-glucosylation/re-glucosylation can occur as long as the protein remains misfolded/unassembled. However, inhibitors of glucosidases I and II abrogate this quality control system by blocking formation of mono-glucosylated glycans, thereby preventing interaction with the lectin-type chaperones (Figure 9.2).

For instance, in the case of the α1-ATZ mutant form of alpha-1-antitrypsin, which is relatively unstable, the result of glucosidase inhibition is an even shorter protein half-life [5]. This effect is reversed by co-administration of the proteasomal inhibitor lactacystin, suggesting increased entry into the ER-associated degradation system. However, glucosidase inhibition may also target the misfolded protein to alternative degradative pathways [6]. Thus, the failure to trim outer glucose residues prevents interaction with lectin chaperones and may serve to facilitate entry into degradative pathways. It is this interaction that prompted the investigation of glucosidase inhibitors as antiviral agents [7–9].

9.2 The relationship between glucosidase inhibition and antiviral action

All enveloped viruses, with a membrane surrounding the viral capsid, contain glycoproteins [10], and are thus potentially susceptible to inhibitors of glucosidases. Indeed, more than 20 years ago it was found that treatment of cells with glucosidase inhibitors could interfere with the maturation of Sindbis virus and the influenza virus fowl plague virus envelope proteins and disrupt the release of infectious particles [11, 12]. We had predicted, in a 'biogenesis' theory that viruses that bud from the endoplasmic reticulum would be sensitive to glucosidase inhibitors [8, 13]. This was not to say that viruses that did not bud from the ER would not be sensitive to glucosidase inhibitors. Moreover, in a refinement of the 'biogenesis' theory, viruses that strongly depend upon the calnexin/calreticulin type pathway for morphogenesis would be expected to be sensitive to glucosidase inhibition.

These predictions, for the most part, seem to be supported by experimental results. For example, hepatitis B and bovine viral diarrhoea virus (BVDV) are two completely different viruses. HBV, with a DNA genome, is a human para-retrovirus that primarily infects hepatocytes. BVDV is a pestivirus with an RNA genome that grows primarily in non-liver cells; it has been used as a surrogate model for the hepacivirus hepatitis C virus. Both, however, acquire their envelope by budding into the endoplasmic reticulum [14–17]. Both viruses, we now know, also depend upon calnexin for the maturation of specific viral glycoproteins [18, 19]. Both viruses are also significantly inhibited by glucosidase inhibitors at concentrations that do not apparently affect cell viability [18, 20–22]. However, as discussed below, the precise mechanisms by which glucosidase inhibitors exert antiviral action might not be identical.

Since other related members of the flaviviridae family bud from the ER, the relationship between calnexin dependency/budding from the ER and sensitivity to glucosidase inhibition can be extended to other members of the family. Among those of medical significance are hepatitis C virus and many of the haemorrhagic fever-causing flaviviruses. Table 9.1 shows that, as predicted, Dengue, West Nile Virus and Japanese Encephalitis Virus are all sensitive to glucosidase inhibitors. These are also viruses of bioterror concern, raising the level of interest in drugs that can be helpful in their

Table 9.1 Envelopment location and sensitivity to glucosidase inhibition

Virus	Organelle in which envelope is acquired	Morphogenesis sensitive?	Infectivity sensitive?	References
Hepatitis B	Endoplasmic reticulum	Yes	Not known	[7]
Bovine viral diarhhoea virus	Endoplasmic reticulum	Yes	Yes	[20, 90]
Hepatitis C	Endoplasmic reticulum	Yes	Not known	[56]
West Nile virus	Endoplasimc reticulum*	Yes	Not known	[70]
Dengue virus	Endoplasmic reticulum	Yes	Yes	[65, 67]
Herpes simplex	Nucleus	No	No	Block (unpublished)
Sindbis	Plasma membrane	Yes	Not known	[11]
Influenza	Plasma membrane	Unclear**	Unclear**	[11, 73]
HIV	Plasma membrane	No	Yes	[23, 76]

* For a variant that may bud from the plasma membrane, see Li *et al.* [71]

** Conflicting results might be due to strain differences, see references cited.

control. Since the model predicts that glucosidase inhibitors would inhibit HCV, clinical trials have been initiated to test this hypothesis, and this is considered below.

Considerable interest followed from the observation that a number of glucosidase inhibitors can inhibit HIV infectivity and cytopathicity [23–25], although HIV buds from the plasma membrane, not from the ER. The HIV envelope glycoproteins are highly glycosylated and this might confer some greater degree of sensitivity. In addition, other viruses such as hepatitis C virus might be rendered more sensitive due to the complexity of glycoprotein folding requiring cotranslational cleavage events [10]. The various factors that confer a degree of sensitivity to glucosidase inhibition have yet to be fully elucidated. However, the special requirements that viruses require higher-order assembly might be sufficient to confer sensitivity to viruses that bud through the ER [26].

9.3 Fate of viral glycoproteins in glucosidase-inhibited cells

The immediate fate of a glycoprotein unable to fold/assemble correctly is that it will be unable to exit the ER via the normal secretory pathway. Eventually, many such proteins will be degraded via a process know as ERAD, for ER-associated degradation [27]. This process requires de-glycosylation, retrotranslocation back to the cytosol followed by degradation by the proteasome. However, in some cases where protein misfolding occurs, proteins can aggregate, causing retention of the protein in the ER. Indeed, this is the case for certain variants of alpha-1-antitrypsin. The mutant α1-ATZ protein contains a single amino acid substitution that results in polymerization within the ER; this contributes to liver disease in some individuals, especially younger children [28]. When cells expressing such a mutant protein are exposed to the glucosidase inhibitor castanospermine, the bulk of the protein synthesized failed to bind to calnexin and was degraded inside the cell in a proteasome-dependent manner, with kinetics similar to degradation in untreated cells. However, enhanced secretion of protein bearing high mannose oligosaccharides was also observed [6], suggesting that at least some material is able to bypass the normal quality control mechanisms. It has been reported that an alternative pathway exists that permits secretion of at least some glycoproteins despite glucosidase inhibition; the Golgi-resident endo-α-D-mannosidase can remove a glucosylated mannose, permitting further glycan maturation and protein secretion [29]. Because the secreted mutant protein retains some degree of enzyme activity, there was interest in glucosidase inhibitors as chemoprophylaxis agents.

The ER retention of misfolded proteins has important implications for viral glycoproteins. Failure of calnexin binding due to lack of glucose trimming can lead to enhanced degradation, both proteasomal and via other mechanisms [30]. However, if misfolded proteins are not degraded (perhaps because they become aggregated or sequestered) they may accumulate in the ER [31]. Whether a given protein undergoes increased or decreased turnover might reflect the ability of the individual protein to interact with other ER chaperones such as BiP [31], and must be evaluated for each individually. Either of these processes may be of limited impact to the infected host in the case of acute viral infection, as any adverse consequences of glucosidase inhibition, such as ER stress due to protein accumulation [32], will be transient, reversing with drug withdrawal. The potential consequences of these processes in the context of a chronic infection will be considered in a later section; here, we can consider the potential ramifications for viral replication and assembly. Glucosidase inhibition is likely to be a more attractive therapeutic strategy for viral proteins that become destabilized with drug treatment. First, by reducing levels of viral glycoproteins (via enhanced protein degradation), glucosidase inhibitors have the potential to interfere with virus assembly and/or reduce virus secretion. Second, by increasing viral glycoprotein entry into the proteasome, glucosidase inhibitors might increase the presentation of viral peptide epitopes by MHC class I. Although the latter is of limited consequence *in vitro*, it is potentially significant *in vivo*, especially in the context of chronic viral infection, as it might lead to T-cell activation. However, it must be noted that MHC class I molecules themselves can be destabilized by glucosidase inhibition [33].

One aspect of glycoprotein degradation by ERAD is that, at some stage, the glycan must be removed from the protein. Peptide:N-glycosidases have been ascribed to both the ER and the cytoplasm [34–36], making it unclear whether de-N-glycosylation occurs prior to, or after, retrotranslocation out of the ER and into the proteasome. Indeed, it has been suggested that the ERAD-associated glycan removal from individual glycoproteins might vary, as there is seemingly contradictory evidence for ER versus cytoplasmic de-N-glycosylation [37]. However, it has been reported that cytosolic N-glycanase can process the triglucosylated side chains that accumulate as a consequence of glucosidase inhibition [38]. One interesting consequence of de-glycosylation prior to protein degradation is that the action of N-glycanase results in deamination of the glycosylated asparagine residues, resulting in aspartic acid. This process can thus produce novel peptides with the potential to be presented to the immune system. The natural occurrence of such nongenetically encoded peptides is described below.

The effects of glucosidase inhibitors on the envelope glycoproteins of HBV have been studied by us and our colleagues. The virus encodes three glycoproteins that share a common C-terminus, but differ in their initiation sites (reviewed in [39]). These proteins are all transmembrane proteins within the ER, but within a few hours, the proteins bud into the ER lumen as lipoprotein particles [17]. These particles, either containing viral capsids or empty 22-nm particles, are then rapidly secreted. Both the large and small glycoproteins, LHBs and HBs, are essential for viral replication, although the role of the middle glycoprotein remains unclear [39].

LHBs and HBs can both be N-glycosylated at a single site within the common domain. The middle protein, MHBs, shares this site, but is also glycosylated near its N-terminus, at amino acid 4. Note that the latter glycosylation sequon is also present in LHBs, but is not used, apparently due to failure of this region to translocate into the ER, rendering it inaccessible to the glycosylation machinery [40]. Glycosylation at amino acid 4 seems to be essential for secretion of viral particles [41]. This glycan at amino acid 4 of MHBs is also responsible for interaction with calnexin [19, 42]. Thus, it is not surprising to find that secretion of MHBs is particularly sensitive to glucosidase inhibition [13, 30, 43]. Recently, it has become clear that secretion of LHBs is also sensitive to glucosidase inhibition, and an example is shown in Figure 9.3, but that small HBs is insensitive [30]. Since LHBs lacks glycosylation at the calnexin sensitive site – topologically, the site apparently is not exposed to the ER environment – this sensitivity was unexpected. The result suggests that LHBs might be rendered sensitive by virtue of close association with MHBs.

Based on the above discussion of HBV glycoproteins, we might predict that in addition to reduced secretion of viral glycoproteins and virion-associated DNA as a consequence of glucosidase treatment, increased intracellular glycoprotein retention may occur. In contrast, recent work has indicated that DNJ treatment results in reduced levels of intracellular viral glycoproteins [30]. Simultaneous treatment with lactacystin

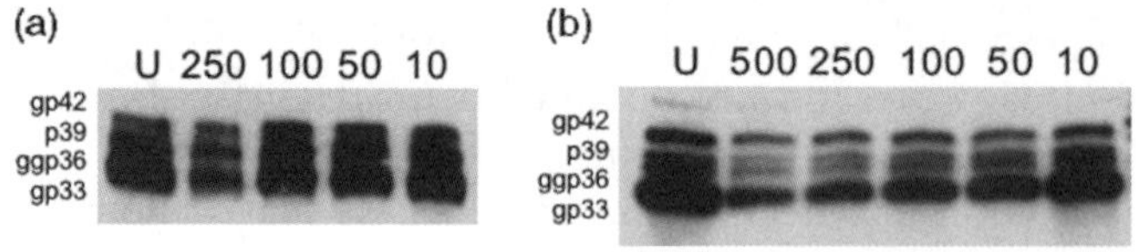

Figure 9.3 Example of cell-specific differences in sensitivity to the same glucosidase inhibitor. (a) HepG2.2.15 cells were grown to confluence in replicate 1.0 cm wells, and then were left untreated (U) or were treated with the indicated concentration (in μg/ml) of N-butyl-DNJ for 1 week. The level of secreted LHBs and MHBs was determined by immunoblotting as described previously [30]. Note that the unglycosylated form of LHBs, p39, does not resolve from the doubly glycosylated form of MHBs, ggp36. (Unglycosylated MHBs is only detected intracellularly.) (b) A similar experiment was performed using Q7-21 cells [98]

results in protein accumulation, implicating the proteasome in the degradation of retained viral proteins. However, in the absence of glucosidase inhibitor treatment, the HBV envelope glycoproteins seem to be relatively resistant to proteasomal degradation [44], which might somewhat offset enhanced turnover. The increased levels of di- and tri-glucosylated free oligosaccharides detected in glucosidase inhibited cells (Figure 9.2) are also consistent with generally increased intracellular turnover of glycoproteins [45], supporting the notion that glucosidase inhibition leads to enhanced proteasome-mediated degradation.

There are circumstances under which HBV glycoproteins are known to accumulate in the ER. The LHBs protein, which normally is the least abundant of the three glycoproteins, is not secreted when expressed either on its own or in excess with respect to the smaller glycoproteins [46–49]. Using transgenic mice that overexpress LHBs, it has become clear that excess LHBs production causes accumulation of HBs as well [50, 51]. This accumulation results in the appearance of so-called ground glass hepatocytes, which are often observed in the livers of individuals with chronic HBV infection. A number of naturally occurring envelope mutants have been associated with ER retention, and the appearance of such mutations tends to correlate with histopathology and disease severity [52–54]. This might also be the case for fibrosing cholestatic hepatitis, which rapidly progresses in immunocompromised individuals. In this case, it appears that accumulation of LHBs results in cell vacuolization followed by rapid apoptotic cell death [55]. It is not yet certain whether specific viral mutations are associated with this particular clinical outcome. Thus, it is clear that HBV glycoprotein retention may be deleterious to infected cells, and it is a concern that glucosidase inhibition could exacerbate the problem. However, it remains possible that, as with the alpha-1-antitrypsin mutant, glucosidase inhibition might have the apparently paradoxical ability to reduce HBV-mediated hepatocyte injury, by enhancing the turnover of retained envelope proteins inside the cell.

There has been considerable interest in the use of glucosidase inhibitors against chronic infection with the hepacivirus hepatitis C virus (HCV) [56–58]. Due to the lack of suitable culture models, much preclinical work has centred on the related pestivirus bovine viral diarrhoea virus (BVDV) as a surrogate model for HCV [13]. BVDV contains two envelope glycoproteins, E1 and E2. Glucosidase inhibitors that disturb the glycan processing and thus folding and trafficking of E1 and E2 were shown to inhibit BVDV infections [59]. It has been reported that glucosidase inhibition reduces the number of viral genomes released from infected cells [58], and virions released from glucosidase-inhibited cells also exhibit reduced infectivity [20]. Consistent with these data, BVDV RNA replication immediately after virus entry into the cell is not reduced by the iminosugars (Gu *et al.*, unpublished results).

Recently iminosugar inhibitors were shown to affect the morphogenesis and binding properties of hepatitis C virus-like particles [56], suggesting possible efficacy against this pathogen. Clinical trials evaluating *in vivo* efficacy of glucosidase inhibitor in patients chronically infected with HCV are in progress. Results to date suggest that 6-O-butanoyl castanospermine (Celgosivir, Figure 9.1) was well-tolerated at doses ranging from 200 to 400 mg once daily or 200 mg twice daily, but exhibited little antiviral efficacy as monotherapy [60]. However, the corporate sponsor, Migenix (Vancouver,

Canada), has reported clinical benefit in a Phase II trial that combined celgosivir with both ribavirin and pegylated interferon (November 6, 2006, news release at www.migenix.com). Although this study was small and fairly short (12 weeks), and has not undergone peer review, the treated population consisted of individuals classed as nonresponders to interferon and ribavirin, suggesting possible benefit for this group. The results of this trial are also in good agreement with results in cell culture which showed that Celgosivir was the most potent of several glucosidase inhibitors tested, and that coadministration of interferon or ribavirin could enhance antiviral efficacy [58].

The flavivirus West Nile virus encodes two viral envelope glycoproteins, prM and E within the viral polyprotein. The proteins prM and E form heterodimers in the ER, where they facilitate virus budding into the ER. The prM ectodomain is then removed and the cleavage of prM enables E to form head-to-tail homodimers which assemblye to cover the virus particle. prE and M contain N-linked glycosylation sites and the glycosylation state of these proteins has been shown to influence the viral particle assemble process and viral infectivity [61]. In addition, the glycosylation of the E protein might also play a significant role in cell surface attachment via the lectin DC-SIGNR on the cell surface, and has been implicated in West Nile virus-related neuro-pathogenesis [62, 63], although the exact mechanisms are not clear.

Although the importance of glycan in the life cycles of flaviviridae is not yet entirely clear, the potential functional importance of glycosylation events might render West Nile and related flaviviruses sensitive to glucosidase inhibition. The glucosidase inhibitor N-nonyl-DNJ has been shown to reduce titers of Kunjin virus, a subtype of West Nile virus, although little reduction of prM secretion or processing was observed [64]. Similarly, N-nonyl-DNJ reduced the titers of Japanese encephalitis virus and protected cells from virus-induced cytopathic effects in tissue culture; some N-nonyl-DNJ treated mice were also protected from lethal challenge with the virus [65]. Dengue virus has also glycosylated prM and E proteins and a similar mechanism of inhibition has been proposed [65–67].

The aforementioned flaviviruses are generally understood to bud into the ER [68–70], although it appears that some variants might bud from the plasma membrane as well [71]. However, the glycoproteins of non-ER budding viruses can also exhibit a requirement for calnexin-mediated folding. One well-studied viral glycoprotein is the influenza virus haemaglutinin (HA), which has been central to our understanding of the importance of monoglucosylated proteins interacting with calnexin [72]. However, HA appears able to interact with the ER chaperone BiP as well as calnexin and calreticulin, which might make it less sensitive to glucosidase inhibitors. Although glucosidase inhibition did not seem to prevent liberation of influenza virus PR-8, the virus released bears high mannose oligosaccharides, suggesting that it is not retained in the ER [11]. However, another study suggests that the appearance of neuraminidase on the cell surface is affected by glucosidase inhibition, and a limited antiviral effect was observed against the reassortant virus NWS-N8 [73]. It is possible that both the viral strain and the host cell modulate the extent of antiviral action induced by glucosidase inhibition. For instance, different cells vary in the extent by which incompletely trimmed glycoproteins can enter the Golgi endomannosidase shunt pathway [74].

For HIV, glucosidase inhibition interferes with proper folding of the envelope glycoprotein, which impairs infectivity, but does not result in reduced secretion [75, 76]. Specifically, membrane fusion is impaired, apparently by altered interaction between the envelope protein and the coreceptor CXCR4 [77]. Thus, non-ER budding viruses can potentially represent targets for antiviral action of glucosidase inhibitors. Interest in the use of glucosidase inhibitors against HIV has diminished somewhat with the development of highly specific inhibitors of viral proteins and their use in highly active antiretroviral combination therapies.

9.4 Specificity of glucosidase inhibition

As can be concluded from the above discussion, glucosidase inhibitors have considerable potential as antiviral agents, but an obvious concern is that inhibiting cellular enzymes such as glucosidases will have deleterious effects on the host. In cell culture assays, reduced glucosidase activity is well-tolerated, as many proteins will fold with adequate efficiency despite the lack of interaction with calnexin/calreticulin, in part due to up-regulation of other classes of ER chaperones [10]. However, a number of viral glycoproteins appear to be unusually dependent upon calnexin for proper biogenesis, especially those in which the virus buds into the ER [10, 24]. As mentioned above, this is thought to be at least in part due to their need for higher order assembly [26]. On the other hand, one advantage of an antiviral agent targeting a cellular enzyme is that it could reduce the emergence of virus resistance [78]. Inhibitors of cellular targets can also be used in combination with direct antiviral drugs to achieve synergistic effects [58, 78], since they differ in their mechanisms of action.

Some DNJ derivatives also competitively inhibit sugar-processing enzymes other than the ER glucosidases. The ability to inhibit ceramide-specific glucosyltransferase and prevent the lysosomal accumulation of glycosphingolipids is, in fact, the basis of its efficacy in the management of Gaucher's disease. DNJs with N-alkyl tails of at least three carbon atoms, such as N-butyl-DNJ (Figure 9.1), have been tested as a so-called substrate deprivation therapy for Gaucher's and other lysosomal storage diseases [79]. The inhibition of intestinal α-glucosidases is apparently the basis of a potential efficacy in the management of type II diabetes as well as the source of sometimes use-limiting gastrointestinally side effects. The latter problem was observed in a Phase I trial when AIDS patients were treated with N-butyl-DNJ at doses starting at 8 mg/kg/day; the lack of any improvement in CD4+ cell number or reduction in HIV-1 p24 levels limited further studies with this drug as a single antiviral agent [80]. However, some control of adverse effects could be managed by an appropriate diet. Another factor limiting the use of N-butyl-DNJ is the high dose needed to achieve pharmacologically relevant levels. In a Phase II trial, N-butyl-DNJ was administered at 1000 mg every 8 hours (in combination with the nucleoside analogue zidovudine) [81]. This dose regimen failed to keep circulating drug levels above the levels needed for antiviral activity *in vitro* (20–200 μM [82, 83]), with trough levels averaging 4 μg/ml (*c.*16 μM).

Thus, a key challenge has been to increase the potency of these compounds without compromising their toxicity profile. One predictor of *in vivo* specificity is the *in vitro* selectivity index. The K_i of DNJ for purified porcine ER glucosidase is 1 μM [84]. However, the compound exhibits considerably lower activity in cells, with an IC_{50} of *c.*200 μM for inhibiting BVDV and HBV [22], with CC_{50} of more than 5000 μM. Therefore, there has been considerable interest in examining other structures that retain glucosidase activity but have improved selectivity, in the hope of identifying compounds that are active at lower concentration. A similar disconnection between *in vitro* enzyme inhibition and in cell activity has been observed for another iminosugar, castanospermine and its derivatives. The derivative 6-O-butanoyl castanospermine (Celgosivir, Figure 9.1) is more active against HIV in culture relative to castanospermine, with IC_{50}s = 20 μM compared with more than 200 μM, but less active against glucosidase I in a cell-free assay (IC_{50} = 1.27 μM compared with 0.12 μM [85]). This difference in activity is attributable, in part, to increased cellular uptake afforded by the addition of the acyl side chain, which is then removed inside the cell [86]. Improved uptake *in vivo* may help ameliorate drug-related side effects.

Other modifications have been evaluated for improved drug selectivity. Block *et al.* showed that a molecule with a DNJ head group and a 9-carbon alkyl side chain, N-nonyl-DNJ was about 50 times more potent than DNJ in inhibiting HBV secretion from tissue cultures [22]. The longer chain compound also exhibited greater antiviral activity against BVDV [22, 58, 59]. Similar effects of chain length were reported by Tan *et al.* [87]. However, the increase in side chain length was also associated with increased toxicity, with CC_{50}s for N-nonyl-DNJ in the vicinity of 100 μM [22, 58, 59]. Introduction of oxygen atom(s) into the chain and addition of hydroxyl groups to alkylated DNJs with straight chains reduced toxicity [88], but can reduce potency (Gu *et al.*, manuscript submitted). Moreover, the increase in antiviral efficacy could not be entirely attributed to improved inhibition of ER glucosidase by N-nonyl DNJ versus short chain N-butyl DNJ [21]. Of course, N-nonyl DNJ uptake by cells in culture might be superior to that of the smaller DNJs. There is some evidence that N-nonyl-DNJ is taken up by the liver preferentially in comparison with DNJs with shorter alkyl side chains, supporting this idea [59]. It is also possible that the side chains affect intracellular localization. For instance, it appears that an octadecyl side chain reduced the accessibility of DNJ to the ER [45].

Thus, although modifications can be made to improve antiviral activities in cell-based assays, certain general considerations related to the specificity of action of glucosidase inhibitors will need to be kept in mind for any potential antiviral application. One level of concern is the viability of the infected cell. As mentioned above, cellular toxicity, whether due to ER stress-related apoptosis induced by accumulation of misfolded proteins or due to off-target activities, is of less concern in the context of acute virus infection, and may be advantageous, helping to clear infected cells. The ability of these drugs to induce ER stress seems to be highly cell-line dependent (Block, unpublished results). However, antiviral activity is clearly observed in cultures in the absence of any cellular toxicity, as described above. Viral proteins may have enhanced sensitivity to the drugs, possibly because even a low percentage of misfolded viral glycoprotein can perturb morphogenesis. Choosing the correct dose will need to balance antiviral efficacy against unwanted side effects, particularly in the context of chronic viral disease. For instance, in the case of HBV or HCV, regular clinical evaluation including tests of liver function will be important, and a liver biopsy may be needed periodically to monitor tissue inflammation and check for distortions of the ER.

9.5 *N*-Alkyl DNJs inhibit virus growth by non-glucosidase inhibitory mechanisms – other potential activities of these compounds

Differential uptake by cells does not entirely explain N-nonyl DNJ's superior ability to inhibit HBV in culture, relative to N-butyl DNJ. Surprisingly, it appeared that N-nonyl DNJ was inhibiting secretion of HBV as well as N-butyl DNJ at concentrations that inhibited very little glycan processing [21]. That is, at very low concentrations of N-nonyl DNJ, the inhibition of HBV could not be accounted for by inhibition of glycan processing. It was hypothesized that N-nonyl-DNJ, in addition to activity as a glucosidase inhibitor, also possessed a second antiviral mechanism. It was further speculated that the second mechanism was related to the alkyl side chain.

This hypothesis seemed to be supported by demonstration of the antiviral activity associated with N-nonyl-1-deoxygalactonojirimycin (N-Nonyl DGJ) [89]. N-Nonyl DGJ was nearly as effective an inhibitor of enveloped HBV DNA secretion from tissue culture as was N-nonyl-DNJ but, with a galactose head group, it is not an inhibitor of glucosidase. Similarly, alkylated DGJs are inhibitory against BVDV [90]. We have called iminosugars with alkyl side chains alkovirs to emphasize the importance of their alkyl side chains [91]. As shown by Durantel *et al.* for BVDV [80], the length of the side chain plays a critical role in affecting the biological activity of the molecule. One consistent observation with alkyl-DNJs and -DGJs is that a side chain length of 8–9 carbons or a cLogP of approximately 2.8–3.0 seems to be the optimal compromise between increased efficacy and acceptable cytotoxicity. We call this the 'rule of nines'.

The surprising nature of the non-glucosidase mediated inhibition of HBV and other viruses by the alkovirs is under investigation. There is evidence that an alkovir called SP231 interferes with the HCV p7 ion channel, which is essential for virus growth [92]. We have proposed that alkylated iminosugar molecules such as N-nonyl DNJ and N-nonyl DGJ are actually glycolipid mimetics and activate glycolipid receptors on cells, stimulating innate host defense pathways [13]. One intriguing possibility is that these glycomimetics stimulate Toll-like receptors, acting in much the same way as α-galactosylceramide. Indeed, we have reported that both α-galactosylceramide and N-nonyl DGJ activate an arm of the interferon pathway, following incubation in tissue culture [93]. Recently, others have reported that *in vivo* activation of Toll-like receptors can reduce replication of HBV in transgenic mice [94]. Thus, this class of molecules might exert antiviral activity via multiple mechanisms.

It was mentioned above that glucosidase inhibition can lead to the generation of nontemplated peptides as a consequence of deglycosylation; there is circumstantial evidence that these are presented to the immune system. Cytotoxic T lymphocytes have been detected in an infected chimpanzee that react against an HCV peptide containing a deaminated aspartic acid, but the same cells poorly recognize the cognate peptide with the genetically encoded asparagine [95]. These authors speculate that certain reactivities that have been identified in HCV-infected humans might also be best explained by the presence of a nontemplated aspartic acid residue. The ability of glucosidase inhibitors to promote proteasome-mediated degradation of viral proteins, thereby producing nontemplated peptides has potentially important implications for the repertoire of antigens presented to the immune system, and is currently under active investigation by our group.

9.6 New directions for improving glucosidase inhibitors as antiviral agents

In addition to alkylation of the DNJ head group with straight carbon side chains, other modifications have been made and tested for antiviral activity. We have been using the BVDV plaque assay, in tissue culture, as our 'sentinel' of antiflavivirus activity and a secretion of HBV surface glycoprotein antigens from a tissue culture line that constitutively produces the virus, as the surrogate for anti-HBV agents. Since BVDV is cytopathic in tissue culture and can be studied under routine laboratory conditions (Biohazard level 2), but has morphogenesis properties similar to HCV and other flaviviruses of bioterror concern, it has been considered to be a reasonable surrogate.

Using these assays, we have tested compounds of three general categories for activity – DNJs with:

1. conformation locking (cyclized) tails of varying length,
2. modified cyclized tails of varying length, or
3. modifications to the ring nitrogen (Gu *et al.*, manuscript submitted).

In this way, we hope to identify compounds that are broadly antiviral while retaining low toxicity.

One challenge that is encountered in assessing antiviral activity is the influence of the host cell. For instance, individual viral glycoproteins may traffic in distinct patterns when different cells are examined [96]. This effect might account for some of the conflicting results mentioned in Table 9.1. We have also observed that iminosugars can have variable activity in different cell types. Against West Nile virus, iminosugars tested are more active when BHK cells are being used for antiviral testing than when Vero cells are being used (Gu *et al.*, manuscript submitted). The exact mechanism is unclear, but a probable cause could be the sensitivity of the glucosidase in these cells to the iminosugar inhibitors, as discussed previously. When tested *in vivo* for sensitivity to

DNJ, total cell glycan processing in both cell lines are sensitive to DNJ incubation, as determined by the appearance of tri-glucosylated species in glycan released from total secreted protein. It remains to be seen whether quantitative differences occur.

Others have reported that flavivirus require substantially differing amounts of drug in two different cell lines to see a similar inhibition of production of infectious virus [67]. An example of differential cell-type susceptibility of an iminosugar toward HBV is shown in Figure 9.3. The short alkyl chain antiviral N-butyl-DNJ was used to treat two different cell lines that stably produce HBV. The HepG2.2.15 cell line was derived from the human hepatoblastoma line HepG2 [97], whereas the Q721 line was derived from a rat McA-RH7777 hepatoma subline [98]. In the former line, NB-DNJ has little activity at doses below 250 μM, as assessed by reduction of secreted envelope glycoproteins. In contrast, some antiviral activity is observed even at 50 μM in the latter cell line. Further work will be needed to establish whether this difference is accounted for by differing extents of glucosidase inhibition, or whether alternative pathways to secretion exist. It is also possible that other antiviral activities are coming into play, although the relatively short side chain of NB-DNJ should not confer alkovir activity. A better understanding of the cellular variability in response to glucosidase inhibitors should help us to develop more specific compounds.

In summary, glucosidase inhibitors have a number of properties that merit continued investigation. In addition to the potential for broad spectrum, if rather low potency, antiviral activity due to enzyme inhibition, they may have other properties as potential immunomodulators. For instance, in the case of HBV, there remains the possibility of using glucosidase inhibitors in combination with current therapies that target the viral polymerase. For HCV, glucosidase inhibitors offer a possible complement to the only currently approved therapy, alpha interferon. Refining the structure–activity relationship should permit the continued development of improved compounds with either broad antiviral activity, or greater potency against single viruses.

Acknowledgements

Work described in this review was supported by grants AI053884 and AI061441 from the NIH and a grant from the Commonwealth of Pennsylvania via the Hepatitis B Foundation.

References

1. L. W. Ruddock and M. Molinari (2006) *J. Cell Sci.*, **119**, 4373–4380.
2. A. Helenius and M. Aebi (2001) *Science*, **291**, 2364–2369.
3. A.J. Parodi (2000) *Annu. Rev. Biochem.*, **69**,69–93.
4. S.C. Taylor, P. Thibault, D.C. Tessier, J.J.M. Bergeron and D.Y. Thomas (2003) *EMBO Reports*, **4**, 405–411.
5. Y. Liu, P. Choudhury, C.M. Cabral and R.N. Sifers (1999) *J. Biol. Chem.*, **274**, 5861–5867.
6. N.Y. Marcus and D.H. Perlmutter (2000) *J. Biol. Chem.*, **275**, 1987–1992.

7. T.M. Block, F. Platt, X. Lu, W. Gerlich, G. Foster, B.S. Blumberg and R.A. Dwek (1994) *Proc. Natl. Acad. Sci. USA*, **91**, 2235–2239.
8. A. Mehta, N. Zitzmann, P.M. Rudd, T.M. Block and R.A. Dwek (1998) *FEBS Lett.*, **430**, 17–22.
9. P. Greimel, J. Spreitz, A.E. Stutz and T.M. Wroodnigg (2003) *Curr. Top. Med. Chem.*, **3**, 513–523.
10. I. Braakman and E. van Anken (2000) *Traffic*, **1**, 533–539.
11. R. Datema, P.A. Romero, R. Rott and R.T. Schwarz (1984) *Arch. Virol.*, **81**, 25–39.
12. W. McDowell, P.A. Romero, R. Datema and R.T. Schwarz (1987) *Virol.*, **161**, 37–44.
13. T.M. Block and R. Jordan (2002) *Antivir. Chem. Chemother.*, **12**, 317–25.
14. M.S. Collett (1992) *Curr. Top. Microbiol. Immunol.*, **15**, 145–155.
15. J. Dubuisson and C.M. Rice (1996) *J. Virol.*, **70**, 778–786.
16. A.P. Huovila, A.M. Eder and S.D. Fuller (1992) *J. Cell Biol.*, **118**, 1305–1320.
17. K. Simon, V.R. Lingappa and D. Ganem (1988) *J. Cell Biol.*, **107**, 2163–2168.
18. N. Branza-Nichita, D. Durantel, S. Carrouee-Durantel, R.A. Dwek and N. Zitzmann (2001) *J. Virol.*, **75**, 3527–3536.
19. M. Werr and R. Prange (1988) *J. Virol.*, **72**, 778–782.
20. R. Jordan, O.V. Nikolaeva, B. Conyers, L. Wang, A. Mehta, R.A. Dwek and T.M. Block (2002) *Virol.*, **295**, 10–19.
21. A. Mehta, S. Carrouee, B. Conyers, R. Jordan, T. Butters, R.A. Dwek and T.M. Block(2001) *Hepatol.*, **33**, 1488–95.
22. A. Mehta, S. Ouzounov, R. Jordan, E. Simsek, X. Lu, R.M. Moriarty, G. Jacob, R.A. Dwek and T.M. Block (2002) *Antivir. Chem. Chemother.*, **13**, 299–304.
23. R.A. Gruters, J.J. Neefjes, M. Tersmette, R.E.Y. De Goede, A. Tulp, H.G. Huisman, F. Miedema and H.L. Ploegh (1987) *Nature*, **330**, 74–77.
24. A. Karpas, G.W. Fleet, R.A. Dwek, S. Petursson, S.K. Namgoong, N.G. Ramsden, G.S. Jacob and T.W. Rademacher (1988) *Proc. Natl. Acad. Sci. USA*, **85**, 9229–9233.
25. A.S. Tyms, E.M. Berrie, T.A. Ryder, R.J. Nash, M.P. Hegarty, D.L. Taylor, M.A. Mobberley, J.M. Davis, E.A. Bell, D.J. Jeffries, D. Taylor-Robinson and L.E. Fellows (1987) *Lancet*, **2**, 1025–1026.
26. P.M. Rudd and R.A. Dwek (1997) *Crit. Rev. Biochem. Molec. Biol.*, **32**, 1–100.
27. B. Meusser, C. Hirsch, E. Jarosch and T. Sommer (2005) *Nature Cell Biol.*, **7**, 766–772.
28. D.A. Rudnick and D.H. Perlmutter (2005) *Hepatol.*, **42**, 514–521.
29. S.E.H. Moore and R.G. Spiro (1990) *J. Biol. Chem.*, **265**, 13104–13112.
30. E. Simsek, A. Mehta, T. Zhou, R.A. Dwek and T. Block (2005) *J. Virol.*, **79**, 12914–12920.
31. E. Ogier-Denis, C. Bauvy, F. Cluzeaud, A. Vandewalle and P. Codogno (2000) *Biochem. J.*, **345**, 459–466.
32. E. Szegezdi, S.E. Logue, A.M. Gorman and A. Samali (2006) *EMBO Rep.*, **7**, 880–885.
33. S.E. Moore and R.G. Spiro (1993) *J. Biol. Chem.*, **268**, 3809–3812.
34. T. Suzuki, K. Kitajima, Y. Emori, Y. Inoue and S. Inoue (1997) *Proc. Natl. Acad. Sci. USA*, **94**, 6244–6249.
35. T. Suzuki, H. Park and W.J. Lennarz (2002) *FASEB J.*, **16**, 635–641.
36. S. Weng and R.G. Spiro (1997) *Biochem. J.*, **322**, 655–661.
37. M.J. Spiro and R.G. Spiro (2001) *Glycobiol.*, **11**, 803–811.
38. C. Hirsch, D. Blom and H.L. Ploegh (2003) *EMBO Journal*, **22**, 1036–1046.
39. V. Bruss (2004) *Virus Res.*, **106**, 199–209.
40. P. Ostapchuk, P. Hearing and D. Ganem (1994) *EMBO Journal*, **13**, 1048–1057.
41. A. Mehta, X. Lu, T.M. Block, B.S. Blumberg and R.A. Dwek (1997) *Proc. Natl. Acad. Sci. USA*, **94**, 1822–1827.
42. R. Prange, M. Werr and H. Loffler-Mary (1999) *Biol. Chem.*, **380**, 305–314.

43. P.A. Norton, B. Conyers, Q. Gong, L.F. Steel, T.M. Block and A.S. Mehta (2005) *J. Virol. Methods*, **124**, 167–172.
44. T.M. Block, A.S. Mehta, B.S. Blumberg and R.A. Dwek (2006) *DNA Cell Biol.*, **25**, 165–170.
45. H.R. Mellor, D.C. Neville, D.J. Harvey, F.M. Platt, R.A. Dwek and T.D. Butters(2004) *Biochem. J.*, **381**, 867–875.
46. V. Bruss and D. Ganem (1991) *Proc. Natl. Acad. Sci. USA*, **88**, 1059–1063.
47. K.L. Molnar-Kimber, V. Jarocki-Witek, S.K. Dheer, S.K. Vernon, A.J. Conley, A.R. Davis and P.P. Hung (1988) *J. Virol.*, **62**, 407–416.
48. J.H. Ou and W.J. Rutter (1987) *J. Virol.*, **61**, 782–786.
49. D.H. Persing, H.E. Varmus and D. Ganem (1986) *Science*, **234**, 1388–1391.
50. F.V. Chisari, P. Filippi, J. Buras, A. McLachlan, H. Popper, C.A. Pinkert, R.D. Palmiter and R.L. Brinster (1987) *Proc. Natl. Acad. Sci. USA*, **84**, 6909–6913.
51. F.V. Chisari, P. Fillipi, A. McLachlan, D.R. Milich, M. Riggs, S. Lee, R.D. Palmiter, C.A. Pinkert and R.L. Brinster (1986) *J. Virol.*, **60**, 880–887.
52. P.K. Chua, R.Y.-L. Wang, M.-H. Lin, T. Masuda, F.-M. Suk and C. Shih (2005) *J. Virol.*, **79**, 13483–13496.
53. T. Kalinina, A. Riu, L. Fischer, T. Sanantonio, H. Will and M. Sterneck (2003) *Gastroenterol.*, **125**, 1077–1084.
54. H.-C. Wang, H.-C. Wu, C.-F. Chen, N. Fausto, H.-Y. Lei and I.-J. Su (2003) *Am. J. Pathol.*, **163**, 2441–2449.
55. N.-C. Foo, B.Y. Ahn, X. Ma, W. Hyun and T.S.B. Yen (2002) *Hepatol.*, **36**, 1400–1407.
56. C. Chapel, C. Garcia, P. Roiingeard, N. Zitzmann, J. Dubuisson, R.A. Dwek, C. Trepo, F. Zoulim and D. Durantel (2006) *J. Gen. Virol.*, **87**, 861–871.
57. L.A. Sorbera, J. Castaner and L. Garcia-Capdevila (2005) *Drugs of the Future*, **30**, 545.
58. K. Whitby, D.L. Taylor, D. Patel, P. Ahmed and A.S. Tyms (2004) *Antivir. Chem. Chemother.*, **15**, 141–151.
59. N. Zitzmann, A.S. Mehta, C. S., T.D. Butters, F.M. Platt, J. McCauley, B.S. Blumberg, R.A. Dwek and T.M. Block (1999) *Proc. Natl. Acad. Sci. USA*, **96**, 11878–11882.
60. E. Yoshida, D. Kunimoto, S.S. Lee, M. Sherman, J. Heathcote and R. Enns (2006) *Gastroenterol.*, **130**, A-784, #S1059.
61. S.L. Hanna, T.C. Pierson, M.D. Sanchez, A.A. Ahmed, M.M. Murtadha and R.W. Doms(2005) *J. Virol.*, **79**, 13262–13274.
62. D.W. Beasley, M.C. Whiteman, S. Zhang, C.Y. Huang, B.S. Schneider, D.R. Smith, G.D. Gromowski, S. Higgs, R.M. Kinney and A.D. Barrett (2005) *J. Virol.*, **79**, 8339–8347.
63. C.W. Davis, H.Y. Nguyen, S.L. Hanna, M.D. Sanchez, R.W. Doms and T.C. Pierson (2006) *J. Virol.*, **80**, 1290–1301.
64. J.M. Mackenzie and E.G. Westaway (2001) *J. Virol.*,75, 10787–10799.
65. S.F. Wu, C.J. Lee, C.L. Liao, R.A. Dwek, N. Zitzmann and Y.L. Lin (2002) *J. Virol.*, **76**, 3596–3604.
66. M.P. Courageot, M.P. Frenkiel, C.D. Dos Santos, V. Deubel and P. Despres (2000) *J. Virol.*, **74**, 564–572.
67. K. Whitby, T.C. Pierson, B. Geiss, K. Lane, M. Engle, Y. Zhou, R.W. Doms and M.S. Diamond (2005) *J. Virol.*, **79**, 8698–8706.
68. J. Dubuisson, F. Penin and D. Moradpour (2002) *Trends Cell Biol.*, **12**, 517–523.
69. C.M. Rice (1996) In *Field's Virology* (B.N. Fields, D.M. Knipe and P.M. Howley, eds) Lippincott-Raven, Philadelphia, USA, pp. 931–959.
70. E.G. Westaway, J.M. Mackenzie and A.A. Khromykh (2002) *Curr. Top. Microbiol. Immunol.*, **267**, 323–351.
71. J. Li, R. Bhuvanakantham, J. Howe and M.-L. Ng (2006) *J. Gen. Virol.*, **87**, 613–622.

72. D.N. Hebert, J.X. Zhang, W. Chen, B. Foellmer and A. Helenius (1997) *J. Cell Biol.*, **139**, 613–623.
73. T. Saito and I. Yamaguchi (2000) *J. Vet. Med. Sci.*, **62**, 575–581.
74. C. Rabouille and R.J. Spiro (1992) *J. Biol. Chem.*, **267**, 11573–11578.
75. P.B. Fischer, G.B. Karlsson, T.D. Butters, R.A. Dwek and F.M. Platt (1996) *J. Virol.*, **70**, 7143–7152.
76. P.B. Fischer, G.B. Karlsson, R.A. Dwek and F.M. Platt (1996) *J. Virol.*, **70**, 7153–7160.
77. M.J. Papandreou, R. Barbouche, R. Guieu, M.P. Kieny and E. Fenouillet (2002) *Mol. Pharmacol.*, **61**, 186–193.
78. S. Ouzounov, A. Mehta, R.A. Dwek, T.M. Block and R. Jordan (2002) *Antiviral Res.*, **55**, 425–435.
79. T.D. Butters, R.A. Dwek and F.M. Platt (2005) *Glycobiol.*, **15**, 43R-52R.
80. M. Tierney, J. Pottage, H. Kessler, M. Fischl, D. Richman, T. Merigan, W. Powderly, S. Smith, A. Karim, J. Sherman and M. Hirsh (1995) *J. Acquir. Immune Defic. Syndr. Hum. Retrovirol.*, **10**, 549–553.
81. M. Fischl, L. Resnick, R. Coombs, A.B. Kremer, J.C. Pottage, R.J. Fass, K.H. Fife, W.G. Powderly, A.C. Collier, R.L. Aspinall, S.L. Smith, K.G. Kowalski and C.B. Wallemark (1995) *J. Acquir. Immune Defic. Syndr. Hum. Retrovirol.*, **10**, 549–553.
82. D. Dedera, N. Vander Heyden and L. Ratner (1990) *AIDS Res. Hum. Retroviruses,* **6**, 785–794.
83. N. Asano, H. Kizu, K. Oseki, E. Tomioka, K. Matsui, M. Okamoto and M. Baba (1995) *J. Med. Chem.*, **38**, 2349–2356.
84. J. Schweden, C. Borgmann, G. Legler and E. Bause (1986) *Arch. Biochem. Biophys.*, **248**, 335–340.
85. D.L. Taylor, M.S. Kang, T.M. Brennan, C.G. Bridges, P.S. Sunkara and A.S. Tyms (1994) *Antimicrob. Agents Chemother.*, **38**, 1780–1787.
86. M.S. Kang (1996) *Glycobiol.*, **6**, 209–216.
87. A. Tan, L. van den Broek, S. van Boeckel, H.L. Ploegh and J. Bolscher (1991) *J. Biol. Chem.*, **266**, 14504–14510.
88. A. Tan, L. van den Broek, J. Bolscher, D.J. Vermass, L. Pastoors, C. van Boeckel and H.L. Ploegh (1994) *Glycobiol.*, **4**, 141–149.
89. X. Lu, T. Tran, E. Simsek and T.M. Block (2003) *J. Virol.*, **77**, 11933–11940.
90. D. Durantel, N. Branza-Nichita, S. Carrouee-Durantel, T.D. Butters, R.A. Dwek and N. Zitzmann (2001) *J. Virol.*, **75**, 8987–8998.
91. A.S. Mehta, B. Conyers, D.L.J. Tyrrell, K.-A. Walters, G.A. Tipples, R.A. Dwek and T.M. Block (2002) *Antivir. Chem. Chemother.*, **46**, 4004–4008.
92. D. Pavlovic, D.C. Neville, O. Argaud, B.S. Blumberg, R.A. Dwek, W.B. Fischer and N. Zitzmann (2003) *Proc. Natl. Acad. Sci. USA*, **100**, 6104–6108.
93. A. Mehta, B. Gu, B. Conyers, S. Ouzounov, L. Wang, R.M. Moriarty, R.A. Dwek and T.M. Block (2004) *Antimicrob. Agents Chemother.*, **48**, 2085–2090.
94. M. Isogawa, M.D. Robek, Y. Furuichi and F.V. Chisari (2005) *J. Virol.*, **79**, 7269–7272.
95. M. Selby, A. Erickson, C. Dong, S. Cooper, P. Parham, M. Houghton and C.M. Walker (1999) *J. Immunol.*, **162**, 669–676.
96. C. Zurzolo, C. Polistina, M. Saini, R. Gentile, L. Aloj, G. Migliaccio, S. Bonatti and L. Nitsch (1992) *J. Cell Biol.*, **117**, 551–564.
97. M.A. Sells, M.L. Chen and G. Acs (1987) *Proc. Natl. Acad. Sci. USA*, **84**, 1005–1009.
98. C. Shih, L.-S. Li, S. Roychoudhary and M.-H. Ho (1989) *Proc. Natl. Acad. Sci. USA*, **86**, 6323–6327.
99. B. Gu, P. Mason, L. Wang, P. Norton, N. Bourne, R. Moriarty, A., Mehta, M. Despande, R. Shah and T. Block (2007) *Antivir. Chem. Chemther.*, **18**, 49–59.

10

Iminosugars as active-site-specific chaperones for the treatment of lysosomal storage disorders

Jian-Qiang Fan

10.1 Introduction

Lysosomal storage disorders are a group of disorders that result from the abnormal metabolism of macro substances such as glycosphingolipids, glycogen, mucopolysaccharides and glycoproteins [1]. More than 50 disorders caused by mutations in metabolic enzymes or activators that are required for the degradation of such substances have been identified and characterized. The majority of them are neuronopathic and involve neurological changes, degeneration and damage, the severity of which is variable. The overall prevalence of lysosomal storage disorders is estimated as 1 in 7700–13 000 live births [2–4]. Gaucher disease is the most common and may be divided into types I, II and III. Both types II and III are neuronopathic with involvement of the central nervous system (CNS). Over the last 20 years, progress in the development of therapies for lysosomal storage disorders has been dramatic. Enzyme replacement therapy (ERT), bone marrow transplantation and substrate reduction therapy (SRT) are currently being used to treat these disorders, and the potential of gene and stem cell therapies is under investigation [5–7]. Since the first ERT for the treatment of Gaucher disease was approved in 1992, ERT was authorized for Fabry disease, mucopolysaccharidoses I and VI and Pompe disease. Approval of ERT for mucopolysaccharidosis II is anticipated in the near future. Bone marrow transplantation in lysosomal storage diseases was first demonstrated in a patient with mucopolysaccharidosis I [8] and has been performed for mucopolysaccharidoses II and VI, Gaucher disease, Krabbe disease, metachromatic dystrophy and several others with varying clinical benefit [5, 9]. SRT has been approved and evaluated for type I Gaucher disease [10, 11], which is the subject of Chapter 11 of this book. For most of lysosomal storage disorders, particularly those with neuronopathic symptoms, supportive management remains the only treatment at the moment.

Iminosugars: From synthesis to therapeutic applications Edited by Philippe Compain and Olivier R. Martin
 ISBN 978-0-470-03391-3

Active-site-specific chaperone (ASSC) therapy, also referred to as pharmacological chaperone therapy, is a novel and emerging therapeutic strategy using small molecules as potential oral drugs for the treatment of lysosomal storage disorders, which are caused by deficiencies in lysosomal enzymes (for recent reviews, see [12, 13]). Certain mutations in the disease-causing enzymes result in the synthesis of improperly folded proteins that are retarded in the endoplasmic reticulum (ER) and degraded by 'quality control' mechanism, named as ER-associated degradation (ERAD). However, these proteins might be enzymatically active if they could be transported properly to lysosomes. At subinhibitory concentrations, potent competitive inhibitors of the mutant enzymes can act as ASSCs, or pharmacological chaperones, that either induce or stabilize the proper conformation of the mutant enzyme. This promotes normal trafficking through the secretory pathway of the ER and, ultimately, increases enzyme activity in lysosomes. ASSCs are attractive therapeutic agents as they can be administered orally and presumably can gain access to most cell types and CNS.

The effect of ASSCs was first demonstrated with Fabry disease which is caused by an α-galactosidase A (α-Gal A) deficiency; in this study, mutant α-Gal A enzymes encoded by missense mutations could be rescued by various α-Gal A inhibitors, including the iminosugar 1-deoxygalactonojirimycin (DGJ) [14, 15]. Subsequently, ASSCs have been identified for Gaucher disease, Tay–Sachs and Sandhoff diseases and G_{M1}-gangliosidosis [16–19]. In addition, clinical proof-of-concept has been demonstrated in a patient with later-onset Fabry disease by infusion of galactose, a weak inhibitor of α-Gal A [20]. These findings have been translated into several FDA-approved clinical trials of ASSC therapy for Fabry disease and Gaucher disease. Phase I clinical trials for Fabry disease have been successfully completed and, remarkably, no single adverse event was reported during the trials (www.amicustherapeutics.com). Phase II clinical trials are now being conducted. Phase I clinical trials for treating Gaucher disease were also completed and Phase II clinical trails are in progress.

Small molecules that rescue misfolded/mistargeted mutant proteins in diseases other than lysosomal disorders have been identified and evaluated for nephrogenic diabetes insipidus [21], retinitis pigmentosa [22], hypogonatropic hypodonadism [23], cystic fibrosis [24] and other diseases. These small molecules are referred to as 'pharmacological chaperones'. (for recent reviews see [25–28]). In this review, chaperone-like small molecules that specifically target the catalytic active-site of an enzyme are referred to as 'active-site-specific chaperones' or 'ASSCs'. This chapter will review the molecular basis of ASSC therapy for lysosomal storage disease, provide a general procedure for the identification of ASSCs and update the current *in vitro* and *in vivo* studies using iminosugars as ASSCs. This therapeutic strategy, namely using small organic molecules as ASSCs, could be applied broadly to other lysosomal storage disorders and genetic metabolic diseases that are caused by misfolding of mutant proteins.

10.2 Degradation of glycosphingolipids

Degradation of glycosphingolipids occurs in lysosomes. Normally, the catabolism of glycosphingolipids is regulated in a stepwise fashion, involving various enzymes

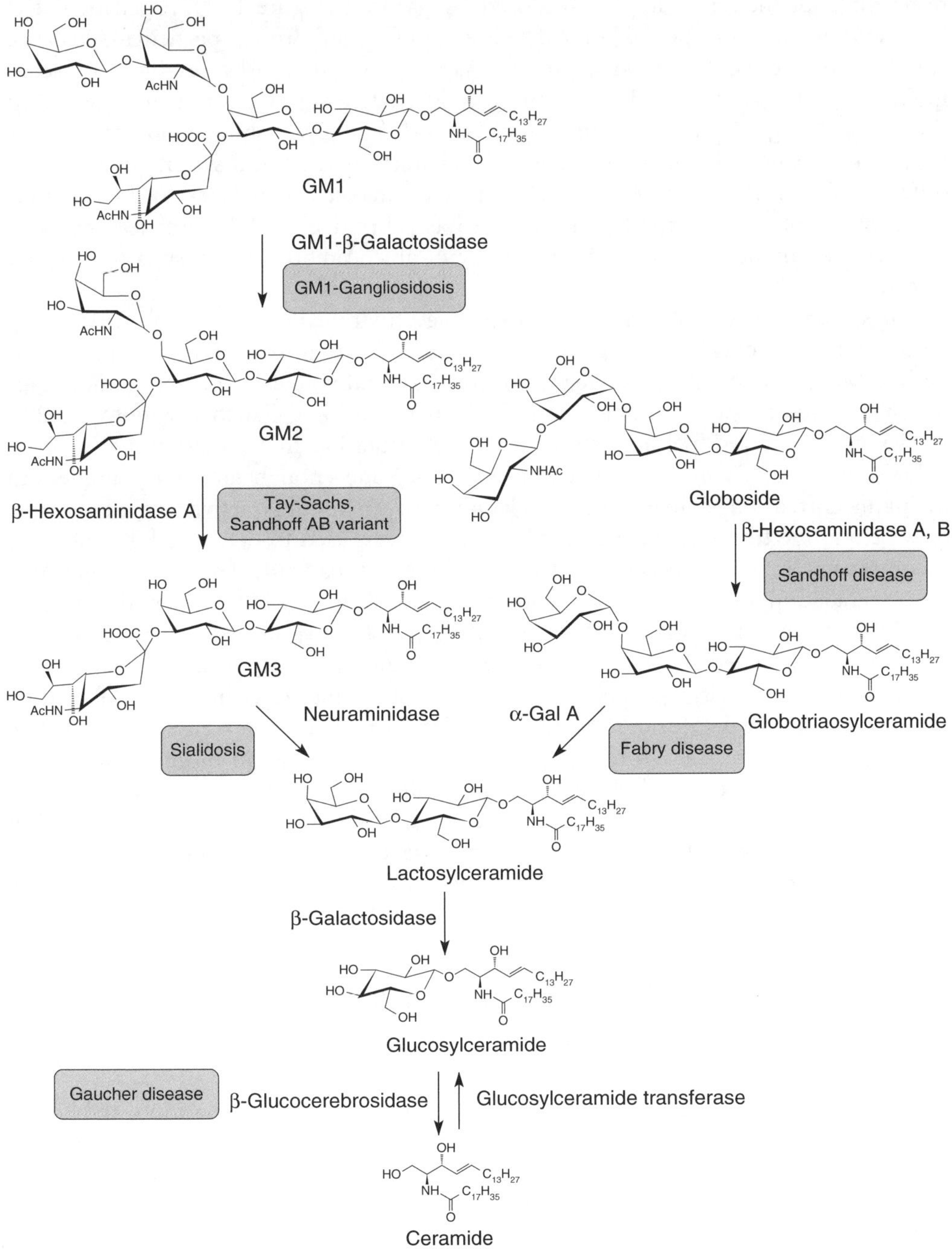

Figure 10.1 Degradation pathway of glycosphingolipids and related glycosphingolipidoses. Glycosphingolipids are degraded by various degradation enzymes stepwise. The deficiency in one enzyme results in the accumulation of the undegraded substrate, leading to a specific lysosomal storage disease

responsible for the removal of each terminal sugar unit (Figure 10.1). A genetic defect in an enzyme that is responsible for the cleavage of a sugar from a particular substance could retard the whole catabolic process. As a result, this causes an abnormal accumulation of the undegraded glycosphingolipid in lysosomes, leading to a so-called lysosomal storage disorder. Accumulation of different glycosphingolipids often leads to different lysosomal storage disorders with characteristic clinical symptoms involving different tissues. For example, the deficient lysosomal α-Gal A leads to the accumulation of globotriaosylceramide, characterized as Fabry disease. Likewise, accumulation of glucosylceramide caused by the deficiency of glucocerebrosidase results in Gaucher disease.

It has been known that a full level of enzyme activity is not required for preventing an accumulation of glycosphingolipids, the cause of the disease. For example, it has been estimated that approximately two-thirds of individuals who are homozygous for the N370S glucocerebrosidase mutation do not exhibit Gaucher symptoms [29], implying that the residual activity in the N370S mutation in a large number of individuals carrying this mutation is sufficient to degrade enough glucosylceramide and maintain normal physiology. The correlation between residual activity of a lysosomal enzyme and the turnover rate of its substrate has been used to determine the substrate accumulation rate in individual cells and whole organs [30]. The results indicated that degradation increased steeply with residual enzyme activity, when the enzyme activity is above a critical threshold. This critical threshold varies with each disease. The threshold is estimated to be relatively low in mucopolysaccharidosis I, since residual α-L-iduronidase activity was below 0.1 per cent of normal level in fibroblasts extract of all three subtypes: Hurler (severe type), Scheie (milder type) and Hurler–Scheie (intermediate type) [31]. On the other hand, rapid accumulation of glucosylceramide is observed in patients with Gaucher disease when glucocerebrosidase activity falls below 11–15 per cent of normal level [32], indicating that this may be the critical threshold for the disease. Generally, an increase of residual enzyme activity above the critical threshold is expected to accomplish a clinical effect for lysosomal storage diseases.

10.3 Lysosomal enzyme biosynthesis and ER-associated degradation (ERAD)

Lysosomal enzymes are synthesized in the cytoplasm and then secreted into the lumen of the ER in a largely unfolded state (Figure 10.2). In general, protein folding is governed by the principle of self-assembly [33]. Newly synthesized polypeptides fold into their native (active) conformation based on their amino acid sequences in a thermodynamic fashion with help from the residual molecular chaperones (e.g. BiP, calnexin, HSPs, etc.). In order to monitor the *in vivo* folding process, the ER has a 'quality control' mechanism, termed ER-associated degradation (ERAD), which uses molecular chaperones to bind and rescue unstable misfolded conformers to facilitate their proper folding and assembly, and to prevent the aggregation of non-native forms [34, 35]. This machinery ensures that only properly folded and assembled proteins are transported to the Golgi

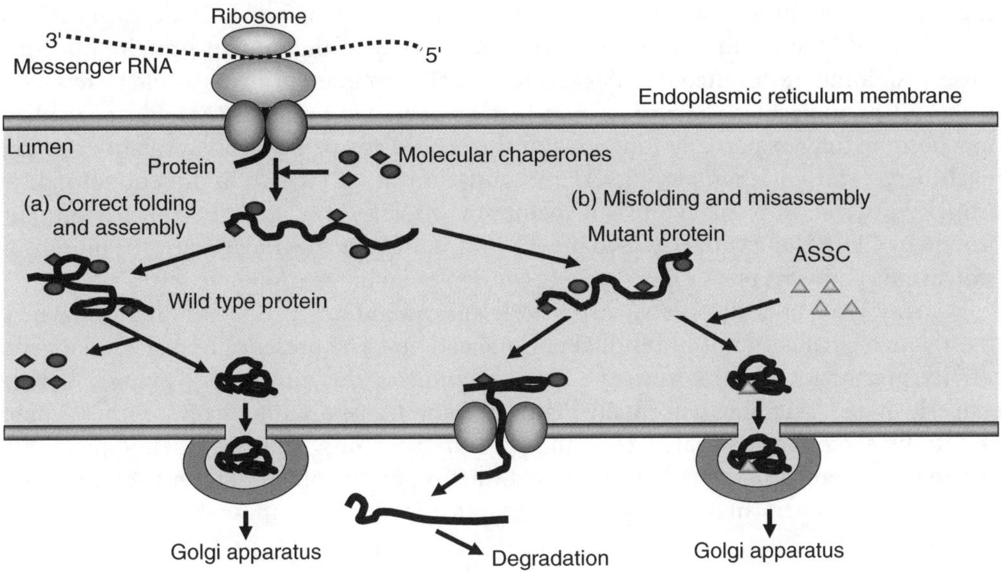

Figure 10.2 Endoplasmic reticulum (ER) quality-control system and active-site-specific chaperone (ASSC) therapy. The quality-control system for proteins in the ER is a 'proofreading' mechanism for newly synthesized proteins and is important for the fidelity of cellular function. This monitoring is enforced by several molecular chaperones and folding-assistant enzymes (ovals and diamonds). (a) Appropriately folded proteins are transported out of the ER, whereas (b) misfolded and unfolded proteins are retained in the ER and are eventually degraded by ER-associated degradation [34]. Wild-type enzymes tend to fold and assemble into their native conformation and are quickly transported to the Golgi apparatus for further maturation. By contrast, mutant proteins that do not fold properly are retained and subsequently degraded. ASSCs (triangles) bind to the active sites of the mutant enzymes and induce their properly folded conformation. This prevents ER-associated degradation and promotes transport of the mutant enzymes to the Golgi apparatus, thereby increasing the concentration of mutant enzymes and residual enzyme activity. Reprinted from Fan, J.-Q. (2003) *Trends Pharmacol. Sci*, **24**, 355, © (2003) with permission from Elsevier (A colour reproduction of this figure can be found in the colour section towards the centre of the book)

complex for further maturation. Improperly folded proteins are retained in the ER and then transported to the cytosol for ubiquitin-mediated degradation within the cytosolic proteasomes. In such a way, misfolded proteins are eliminated to maintain the integrity of the cells [35–38]. It is estimated that misfolding and aggregated proteins occur in up to 30 per cent of wild-type proteins even after interaction with molecular chaperones, and they are rapidly degraded within minutes of their synthesis by the cell's quality control machinery [39]

The consequences of genetic errors that lead to the dysfunction of coding proteins involved in genetic disorders can be various [40]. Nonsense and frame-shift mutations usually result in the premature termination of the biosynthesis, or dramatic alteration in the structure, thereby eliminating the biological activity of the proteins. Splicing mutations cause abnormal protein processing, resulting in the synthesis of functional proteins at a totally or partially diminished level. Missense mutations involving the substitution of critical amino acids often result in the biosynthesis of mutant proteins that are nonfunctional. In other cases, missense mutations or small in-frame deletions/insertions could have little or no impact on the biological activity of the mutant protein, but may cause misfolding and an altered tertiary structure. These mutant proteins would be retained in the ER by molecular chaperones and further degraded by the ERAD, although they may become fully or partially functional if they could escape the cellular quality control machinery [41]. For example, the ΔF508 mutation in CFTR is fully functional if it is properly inserted into the cytoplasm membrane *in vitro* [42]. Retention of the mutant protein by CHIP along with Hsp70 in the ER, and subsequent degradation by the ubiquitin proteasome pathway, plays a direct role in the depletion of the protein *in vivo* [43].

In many lysosomal disorders, certain missense mutations produce mutant enzymes that retain a small amount of residual enzyme activity. The presence of residual enzyme activity presumably results from the small amount of the mutant polypeptide that is properly folded, assembled, normally maturated and transported to the lysosomes. These mutations are often associated with the mild or later-onset disease phenotypes. This provides a therapeutic hope that rescuing a fraction of misfolded mutant proteins from the ERAD machinery may have significant clinical benefits for these diseases.

10.4 Active-site-specific chaperones and iminosugars

An ASSC is a small molecular entity which is capable of specifically binding to the catalytic domain of an enzyme and inducing the proper conformation of misfolded enzymes to prevent the retardation of the mutant proteins within the ER quality control machinery [12] (Figure 10.2). The small molecule serves as a folding template, thereby accelerating the folding process and shifting the folding dynamics in favour of the native-like conformation. In this way, the ASSC prevents the excessive degradation of the mutant enzyme that may retain full or partial activity if it can be properly folded, processed and transported to its normal site of action. Once the mutant enzyme reaches the native-like active conformation, it may maintain its conformational stability after posttranslational maturation and modification, even following dissociation from the ASSC. Theoretically, any compound that interacts with a mutant protein and is capable of inducing its proper folding state could have chaperone activity. A compound that targets the functional site of the protein can be rationally designed, however, since more structural information is generally available for the active site than for other nonfunctional domains. For enzymes, the catalytic site is the preferable domain for ASSC binding. Indeed, the huge quantity of information generated from natural or synthetic substrates, analogues and inhibitors can be readily used to design an ASSC. More importantly, any misfolded enzyme that would be catalytic active after the chaperone rescue is expected to have a preserved catalytic domain with a structure close to that of the native form.

Competitive enzyme inhibitors are expected to be effective ASSCs because of their high affinity to the catalytic domain. The inhibitors are ideal as folding templates for those mutant proteins with a fragile conformation during the protein folding process to induce the correct or proper folding. Once the mutant enzyme/inhibitor complex is secreted out of the ER, the inhibitor at subinhibitory concentrations can be replaced by the highly concentrated substrate to allow the function of the enzyme, since the dynamic exchange of competitive inhibitor and substrate is dependent upon the relative concentrations of each.

Iminosugars mimicking glycoside residue have long been recognized as potent inhibitors for glycosidases. The imino group introduced into the sugar may generate a positive charge that can interact with a transitory state of an enzyme and occupy its active site. Common iminosugar inhibitors of enzymes involved in lysosomal storage disorders are summarized in Table 10.1. Further discussion of the mechanism of inhibition can be found in Chapter 6 of this work. Most iminosugar ASSCs identified for the treatment of lysosomal storage disorders are listed in Table 10.2. Among them, Phase I clinical trials with 1-deoxygalactonojirimycin (DGJ) have been completed and DGJ is now under clinical Phase II evaluation for the treatment of Fabry disease (http://www.amicustherapeutics.com). Similarly, Phase I and II clinical trials for Gaucher disease with isofagomine (IFG) are in progress.

Iminosugars can be effective ASSCs because of their high affinity to the catalytic domain. They can diffuse relatively easily into the cell and bind to folding intermediates of the mutant enzyme protein specifically. Because they are reversible inhibitors, they can be replaced by a high concentration of substrate for metabolic action, once the mutant enzyme reaches the lysosome. In contrast to ERT with exogenous replacement enzymes, small iminosugar molecules may have several advantages such as the ability of crossing the blood–brain barrier, of diffusing through connective tissue matrices and of reaching target sites of pathology which infused lysosomal enzymes cannot reach, or can only when administered at very high doses. To design an effective iminosugar ASSC for lysosomal diseases, the iminosugars listed in Table 10.1 can be used as a convenient starting point.

10.5 Basic requirements for effective ASSCs

The first requirement for an effective ASSC is a high affinity to the biological active site of the target protein. Biological active site can be a catalytic domain of an enzyme or a ligand binding domain of a receptor or lectin. A compound with higher affinity can bind more efficiently to the protein and serves as a better folding template. Asano *et al.* [15] compared the inhibitory action of α-Gal A and the effectiveness in rescuing residual mutant enzyme activity in lymphoblasts of Fabry patients for a series of alkaloids (Figure 10.3). The results indicated a strong correlation whereby a potent enzyme inhibitor is an efficient ASSC. This role can be used as an initial test for the identification of an effective ASSC as detailed in 10.6.

ASSCs rescue mutant enzyme protein by shifting the folding equilibrium toward the native folded state in the ER, allowing formation of its appropriate tertiary conformational structure, export from the ER and trafficking through the Golgi apparatus to its final destination. The second requirement for an effective ASSC would be a better cellular permeability and subcellular distribution. As shown in Figure 10.3, β-1-*C*-butyl-deoxygalactonojirimycin (β-1-*C*-butyl-DGJ, **12**) was a less potent inhibitor of α-Gal A than α-*allo*-homonojirimycin (α-*allo*-HNJ, **5**) ($K_i = 16\,\mu M$ vs $2.6\,\mu M$), but both ASSC

Table 10.1 Summary of possible ASSCs for lysosomal storage disorders

Disorder	Deficient enzyme	Competitive inhibitors	Ref.
Pompe disease	α-glucosidase	1-deoxynojirimycin (DNJ) α-homonojirimycin castanospermine	[71–73]
Gaucher disease	Acid β-glucosidase, or glucocerebrosidase	isofagomine *N*-dodecyl-DNJ calystegines A_3, B_1, B_2, C_1 6-nonyl-isofagomine *N*-octyl-β-valienamine (NOV)	[19, 44, 62]
Fabry disease	α-galactosidase A	1-deoxygalactonojirimycin α-*allo*-homonojirimycin α-*galacto*-homonojirimycin β-1-*C*-butyl-deoxynojirimycin calystegines A_3, B_2 *N*-methyl calystegines A_3, B_2	[14, 15]
G_{M1}-gangliosidosis	Acid β-galactosidase	4-*epi*-isofagomine 1-deoxygalactonojirimycin	[74]
Krabbe disease	Galactocerebrosidase	4-*epi*-isofagomine 1-deoxygalactonojirimycin	[71, 75]
Morquio disease B	Acid β-galactosidase	4-*epi*-isofagomine 1-deoxygalactonojirimycin	[71, 75]
α-Mannosidosis	Acid α-mannosidase	1-deoxymannojirimycin Swainsonine Mannostatin A	[75–77]
β-Mannosidosis	Acid β-mannosidase	2-hydroxy-isofagomine	[75]
Fucosidosis	Acid α-L-fucosidase	1-deoxyfuconojirimycin β-homofuconojirimycin 2,5-imino-1,2,5-trideoxy-L-glucitol 2,5-dideoxy-2,5-imino-D-fucitol 2,5-imino-1,2,5-trideoxy-D-altritol	[78]
Sanfilippo disease B	α-N-Acetylglucosaminidase	1,2-dideoxy-2-acetamido-nojirimycin	[75]
Schindler–Kanzaki disease	α-N-Acetylgalactosaminidase	1,2-dideoxy-2-acetamido-galactonojirimycin	[75]
Tay–Sachs disease	β-Hexosaminidase A	2- acetylamido-isofagomine 1,2-dideoxy-2-acetamido-nojirimycin nagstatin and its derivatives	[75, 78–80]
Sandhoff disease	β-Hexosaminidase B	2- acetamido-isofagomine 1,2-dideoxy-2-acetamido-nojirimycin nagstatin and its derivatives	[75, 78–80]
Hurler–Scheie disease	α-L-Iduronidase	1-deoxy-L-iduronojirimycin 3,4,5-trihydroxypipecolic acid	[75, 81]
Sly disease	β-Glucuronidase	4,5-dihydroxypiperidin-3-carboxylic acid 3,4,5-trihydroxypipecolic acid	[75, 81]
Sialidosis	Sialidase	2,6-dideoxy-2,6-imino-sialic acid Siastatin B	[75, 82]

Table 10.2 ASSCs effective for lysosomal storage disorders

Disorders	ASSCs	References
Fabry disease	1-deoxygalactonojirimycin (DGJ) α-*galacto*-homonojirimycin α-*allo*-homonojirimycin β-1-C-butyl-deoxygalactonojirimycin	[14, 15]
Gaucher disease	*N*-nonyl-deoxynojirimycin (NN-DNJ) *N*-octyl-2,5-anhydro-2,5-imino-D-glucitol *N*-octyl-isofagomine *N*-octyl-β-valienamine(NOV) isofagomine calystegines A_3, B_1, B_2, C_1 1,5-dideoxy-1,5-iminoxylitol (DIX) α-1-*C*-nonyl-DIX α-1-*C*-octyl-1-DNJ	[16, 19, 63–66]
Tay–Sachs and Sandhoff diseases	*N*-acetyl-glucosamine-thiazoline (NGT) 6-acetamido-6-deoxycastanospermine (ACAS)	[17]
G_{M1}-Gangliosidosis	*N*-octyl-4-*epi*-β-valienamine (NOEV)	[18]

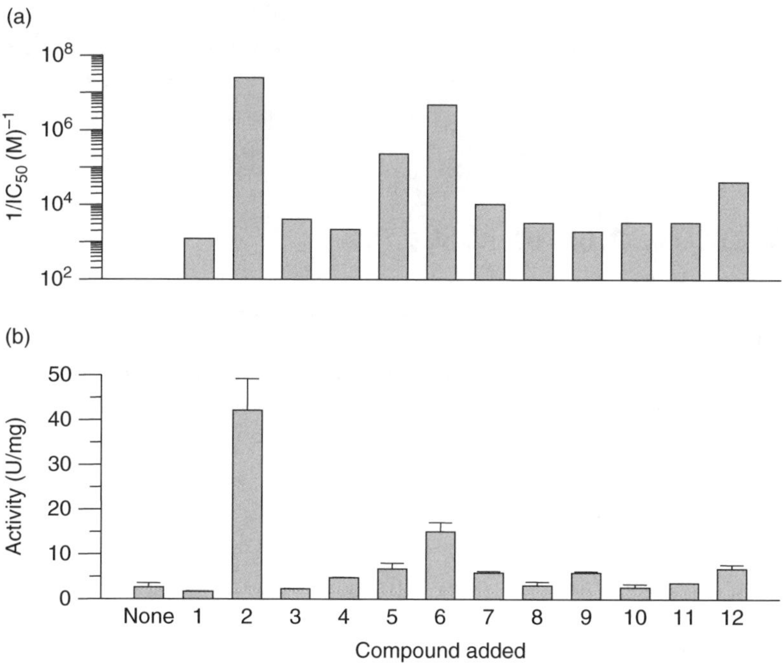

Figure 10.3 *In vitro* inhibition (a) and intracellular enhancement (b) of α-Gal A by inhibitors.(a) $1/IC_{50}$ values against α-Gal A were shown. (b) Residual enzyme activity was determined in cultured R301Q lymphoblasts incubated with each inhibitor at 100 μM for 4 days. Compound **1**, 1-deoxynojirimycin (DNJ); **2**, 1-deoxygalactonojirimycin (DGJ); **3**, 2-deoxy-DGJ; **4**, α-*manno*-HNJ; **5**, α-*allo*-HNJ; **6**, α-*galacto*-HNJ; **7**, *N*-methyl-DGJ; **8**, *N*-ethyl-DGJ; **9**, *N*-propyl-DGJ; **10**, *N*-butyl-DGJ; **11**, *N*-hydroxyethyl-DGJ; **12**, β-1-*C*-butyl-DGJ. Reprinted from Asano *et al.* (2000) *Eur. J. Biochem.*, **267**, 4179 with permission from Blackwell Publishing

effects were the same at 100 μM, and the effect of β-1-*C*-butyl-DGJ was even higher than that of α-*allo*-HNJ at 1000 μM [15]. It is considered that the cellular permeability of β-1-*C*-butyl-DGJ is better than that of α-*allo*-HNJ, because the increased lipophilicity resulting from the *C*-alkylation at C-1 of DGJ may enhance the efficient transport across cell and ER membranes. In ASSCs effective for Gaucher disease, IFG with an IC_{50} value at 40 nM [44] is a much more potent inhibitor for glucocerebrosidase compared with *N*-nonyl-DNJ which has an IC_{50} value at approximate 2 μM [16]. Both compounds showed, however, a maximum ASSC response in cellular glucocerebrosidase activity in Gaucher fibroblasts cultured in the presence of compounds at 10–50 μM. These results clearly indicate that a modest affinity of an ASSC can be compensated greatly by increased cellular permeability and/or ER accessibility.

The third requirement for an effective ASSC is smooth dissociation of the ASSC from the target enzyme. A properly rescued mutant enzyme can return to its physiological function only after the ASSC is replaced by its natural substrate at its native cellular compartment. Zhu *et al.* [44] have designed and synthesized the most potent inhibitor (6-nonyl-IFG) to date for glucocerebrosidase with an IC_{50} value at 0.6 nM. The compound combined two important elements, a sugar mimic domain for the carbohydrate binding site of the enzyme and an effective alkyl chain for a putative hydrophobic binding domain. While binding to both sites are competitive, individually, with both sites acts in a manner to provide a noncompetitive-like inhibitory mechanism for this compound. Although a potent and slow dissociating inhibitor is likely to be effective in rescuing a mutant enzyme protein, however, the rescued enzyme may require a substantial time to eliminate the inhibitor and return to its functioning state, which may not be favourable for a therapeutic agent.

10.6 Identification of ASSCs

Fan and Ishii [45] proposed a three-step screening procedure for identification of ASSCs. The procedure comprises an *in vitro* inhibition assay as the initial first-line screening step, an intracellular cell-based enhancement assay and an *in vivo* animal testing. This procedure was adapted for the screening of effective ASSCs for Gaucher disease and led to the discovery of IFG [19] which is now under FDA-approved clinical evaluations.

10.6.1 *In vitro* inhibition assay

Because effective ASSCs require high affinity to the targeted enzyme and enzyme inhibition assays can be readily done without requiring complicated reagents, this assay is suitable for the screening of a large number of compounds, or using high throughput screening methods. For many lysosomal enzymes, 4-methylumbelliferyl glycosides are often used as substrates for the determination of enzyme activity, and inhibition can be assayed against the wild-type enzyme. Typically, a compound with an IC_{50} value below ~10 μM may enter the subsequent cell-based assay for chaperone enhancement activity. A compound with an IC_{50} value over about 100 μM may not have sufficient affinity to the targeted enzyme, and may not be suitable for the further development as an efficient drug candidate. It should be noted that this *in vitro* step is intended to estimate interaction strength between a compound and a targeted protein. Other factors determining cellular permeability and ER accessibility will be further examined in the following cell-based assay system.

10.6.2 Cell-based chaperone enhancement assay

ASSC activity is derived from a combination of affinity to the targeted protein as well as cellular permeability and ER accessibility, because ASSC is required to cross both cell and ER membranes to be deliverered to the ER where it binds to and rescues its counterpart. Therefore, to evaluate ASSC activity, a cell-based enhancement assay is necessary. Lymphoblasts or fibroblasts derived directly from patients who have mutations known to cause misfolding are preferable for use as cellular models. Transient expression cell lines such as COS-7 cells can also be used as a cellular model for assays, when the patient cells are not available or residual enzyme activity is too low for assaying. In such cases, an expression plasmid encoding a mutant sequence that is known to result in misfolded enzyme protein needs to be prepared. Test compounds can be added at various concentrations to the culture medium of model cells for 4–5 days to allow the accumulation of enough rescued enzyme for assaying the enzyme activity in cell homogenates. Unlike traditional drug screening, in which the potency of a drug is usually correlated to fold-increase of enzyme activity (i.e. a higher fold-increase in enzyme activity translates to an effective drug), the maximum increase in enzyme activity is related to the degree of misfolding conformation of the mutant. In addition, since most compounds for the cell-based assay are potent inhibitors, a decrease in residual enzyme activity would be expected in cells cultured with compounds at higher concentrations, because of overwhelming inhibition. The effective ASSC concentration (C_{ASSC}) [19], defined as the concentrations in which the maximum rescue effect is achieved, is recommended for use in assessing the efficiency of ASSC activity. A lower C_{ASSC} typically translates to a more effective ASSC.

10.6.3 *In vivo* evaluation

Turnover rate of mutant enzyme in tissues could be longer than a cell's lifespan in cultured conditions. Therefore, small amounts of residual protein may well accumulate over time in the tissues of animals, and that can effectively diminish the substrate storage to reverse the course of disease development. Another advantage for evaluating ASSCs *in vivo* is that it is efficient to wash out iminosugar inhibitors to allow functioning of the rescued enzyme, providing that the turnover rate of small molecules is much quicker than that of the mutant protein. In order to perform *in vivo* evaluation of ASSCs, an animal model that expresses misfolded mutant enzyme and presents clear clinical symptoms is ideal. Typically, transgenic animals expressing a mutant enzyme known to be misfolding are suitable for the studies. Matsuda and colleagues introduced a rescuable mutation into β-galactosidase-null mice, and generated a mouse model that expresses the human mutant R201C β-galactosidase but lacking the endogenous mouse β-galactosidase [18]. The mice had about 4 per cent of the wild-type β-galactosidase activity in the brain. Neuropathological examination revealed vacuolated or ballooned neurons. These mice were used for testing chaperone compounds that may be effective for G_{M1}-gangliosidosis. In Fabry disease, Ishii *et al.* generated transgenic mice expressing human R301Q mutant α-Gal A (a rescuable mutation in cellular studies) and then crossbred to α-Gal A knock-out mice [46]. Despite the fact that mice expressed only the human mutant enzyme without endogenous α-Gal A activity, these mice were devoid of typical clinical symptoms for Fabry disease, presumably because of a sufficient level of α-Gal A activity that diminished the accumulation of globotriaosylceramide. The mice, however, are an excellent biochemical model for examination of compounds to accomplish a biochemical proof-of-concept.

10.7 Examples of ASSC therapy for lysosomal storage disorders

10.7.1 Fabry disease

Fabry disease is an X-linked inborn error of glycosphingolipids metabolism that is caused by deficient lysosomal α-Gal A activity [47, 48], resulting in progressive deposition of neutral glycosphingolipids with α-galactosyl residues at the nonreducing terminus, predominantly globotriaosylceramide, in body fluids and tissue lysosomes. Classic Fabry disease, which is associated with either no or low residual enzyme activity, reduces the affected male's life expectancy, and death usually occurs in the fourth or fifth decade as a result of vascular disease that affects the heart, brain and/or kidneys. By contrast, patients with the mild 'cardiac variant' have 5–10 per cent of the normal α-Gal A activity level and present left ventricular hypertrophy. Studies of residual α-Gal A activity in milder cardiac Fabry patients with different mutations [49, 50] revealed that some mutant α-Gal A have kinetic properties similar to those of normal α-Gal A, but they were significantly less stable [51]. Furthermore, mutant proteins with R301Q and Q279E mutations identified in cardiac patients [52, 53] formed aggregates in the ER and were then quickly degraded in transfected COS-1 cells [54]. These findings indicate that deficient activity of the mutant enzyme is caused primarily by retention and subsequent degradation of the misfolded enzyme in the ER.

In an attempt to rescue misfolded mutant enzymes from excessive degradation in the ER and increase the residual α-Gal A activity, Fan and colleagues evaluated a series of enzyme substrate analogues, and demonstrated that DGJ, a potent competitive inhibitor of α-Gal A, effectively increased α-Gal A activity in Fabry lymphoblasts derived from hemizygous Fabry patients with the R301Q or Q279E mutations, when administered at concentrations lower than that usually required for intracellular inhibition of the enzyme [14, 15]. The enzyme activity in R301Q or Q279E lymphoblasts increased 8- or 7-fold after incubation with DGJ for 4 days, respectively, and the increase was dose-dependent at concentrations that were not inhibitory intracellularly (Figure 10.4). DGJ was α-Gal A specific and increased enzyme activity by accelerating transport and

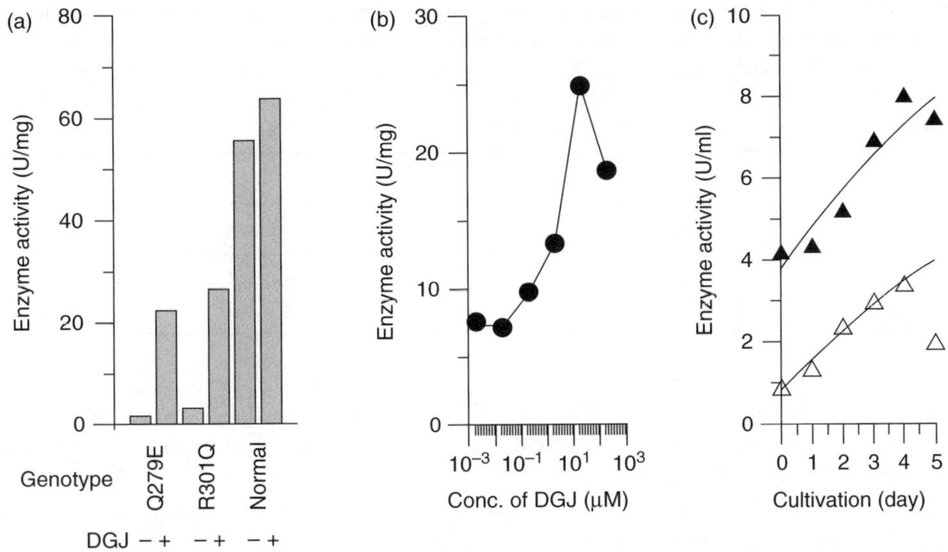

Figure 10.4 Increase of α-Gal A in lymphoblasts from patients with Fabry disease. (a) α-Gal A activity in lymphoblasts (Q279E, R301Q and normal) cultured in the presence (+) or absence (−) of 20 μM DGJ. (b) Effect of DGJ concentration on α-Gal A activity. Lymphoblasts (R301Q) were cultured with DGJ at various concentrations before being collected for enzyme assay. (c) Stabilization of enhanced α-Gal A by DGJ. Lymphoblasts (R301Q) cultured with (▲) or without (▲) 20 μM DGJ were washed and transferred to fresh medium. The cells in 2 ml of the medium were taken each day for enzyme assay. For (a)–(c), each value is an average of triplicate experiments, and standard deviation is less than 10 per cent. Reprinted by permission from Fan *et al.* (1999) *Nature Medicine*, **5**, 112–115 © 1999

maturation of the mutant enzyme. Independent studies in transgenic mouse fibroblasts overexpressing human R301Q α-Gal A confirmed that the mutant enzyme was retained in the ER and not correctly folded, because it formed complexes with BiP [55]. Incubation of the cells with DGJ significantly reduced the association with BiP, indicating that DGJ exerted a chaperone-like effect on the conformation of the enzyme. In human Fabry R301Q and Q357X fibroblasts, DGJ treatment resulted in clearance of lysosomal storage, which was accompanied by the disappearance of multilamellar lysosomal inclusions. Genes involved in cell stress signaling, heat shock response, unfolded protein response and ER-associated degradation show no apparent difference in expression between untreated and DGJ-treated fibroblasts [56]. In addition, oral administration of DGJ to transgenic mice carrying the rescuable R301Q mutation in an α-Gal A-null background caused a dose-dependent increase in α-Gal A activity in the major tissues of the mice [46] (Table 10.3). No apparent toxic effects were observed in the transgenic mice treated with DGJ for 140 days [14], indicating that DGJ may be well tolerated in mice.

The clinical 'proof-of-concept' for ASSC therapy in cardiac Fabry disease has been established by Frustaci and colleagues [20]. Galactose, which is a much less effective inhibitor of α-Gal A than DGJ, was administered to a cardiac Fabry patient by intravenous infusion at 1 g/kg three times weekly. After a 3 month period of treatment, there was remarkable improvement in increase in the left ventricular ejection fraction, and

Table 10.3 Effect of DGJ administration on α-Gal A activity in mice tissues. Transgenic mice expressing human R301Q α-Gal A activity in a knockout background were administered with DGJ in drinking water (0.05 mM) for 2 weeks. Daily dosage was estimated at 3mg/kg body weight/day. The α-Gal A activities in tissue homogenates were determined Control group represents transgenic mice fed with tap water without DGJ. Data are the means ± SD of three mice

	α-Gal A activity (U/mg protein)			
	Heart	Kidney	Spleen	Liver
Control	53.6 ± 12.1	14.7 ± 7.8	26.0 ± 5.6	42.9 ± 7.7
DGJ-treated	740.4 ± 74.2	49.0 ± 5.0	102.4 ± 12.7	110.8 ± 8.5

a moderate reduction in ventricular-wall thickness. The galactose served as an ASSC to rescue the mutant enzyme specifically and thus improved clinical manifestations, suggesting that ASSC therapy may be effective in treating cardiac Fabry disease. Since DGJ is about 120 000-fold more potent as a competitive inhibitor than galactose, it may be an effective orally active drug for Fabry disease.

10.7.2 Gaucher disease

Gaucher disease is the most prevalent lysosomal storage disorder, resulting from the deficient activity of β-glucocerebrosidase [57]. The enzyme is responsible for the degradation of glucosylceramide (glucocerebroside), a natural intermediate in the catabolism of globoside and gangliosides [58]. A deficiency in enzyme activity results in the accumulation of undegraded substrate in the lysosomes of macrophages, leading to various clinical manifestations that include hepatosplenomegaly, secondary hypersplenism, anemia, skeletal abnormality and neurologic dysfunctions. Three clinical phenotypes are generally distinguishable, depending on the extent of primary neurologic involvement and age of onset. The type I or adult form presents clinical symptoms without neuronopathic involvement. The type II (infantile) and the type III (juvenile) forms have clinical manifestations involving the CNS. Type I Gaucher disease is the most prevalent form, and the severity and clinical course of this variant is particularly heterogeneous, ranging from early onset to no clinical manifestations [59]. In contrast, patients with the neurologic forms (types II and III) are rare.

Recent intracellular localization studies of mutant glucocerebrosidases have demonstrated that these mutant enzymes are subject to ER retention and proteasomal degradation, and that the extent of these two events could be important factors in determining disease severity [60, 61]. These studies provide a rational basis for the development of ASSC therapy for Gaucher disease. Sawkar and colleagues were the first to demonstrate that the addition of a competitive inhibitor of β-glucocerebrosidase, *N*-nonyl-deoxynojirimycin (NN-DNJ), at subinhibitory concentrations to the culture medium of Gaucher fibroblasts harbouring the N370S led to an approximately two-fold increase in residual enzyme activity [16]. This indicates that the mutation is amenable to ASSC rescue. Ogawa *et al.* [62] synthesized a potent β-glucocerebrosidase inhibitor,

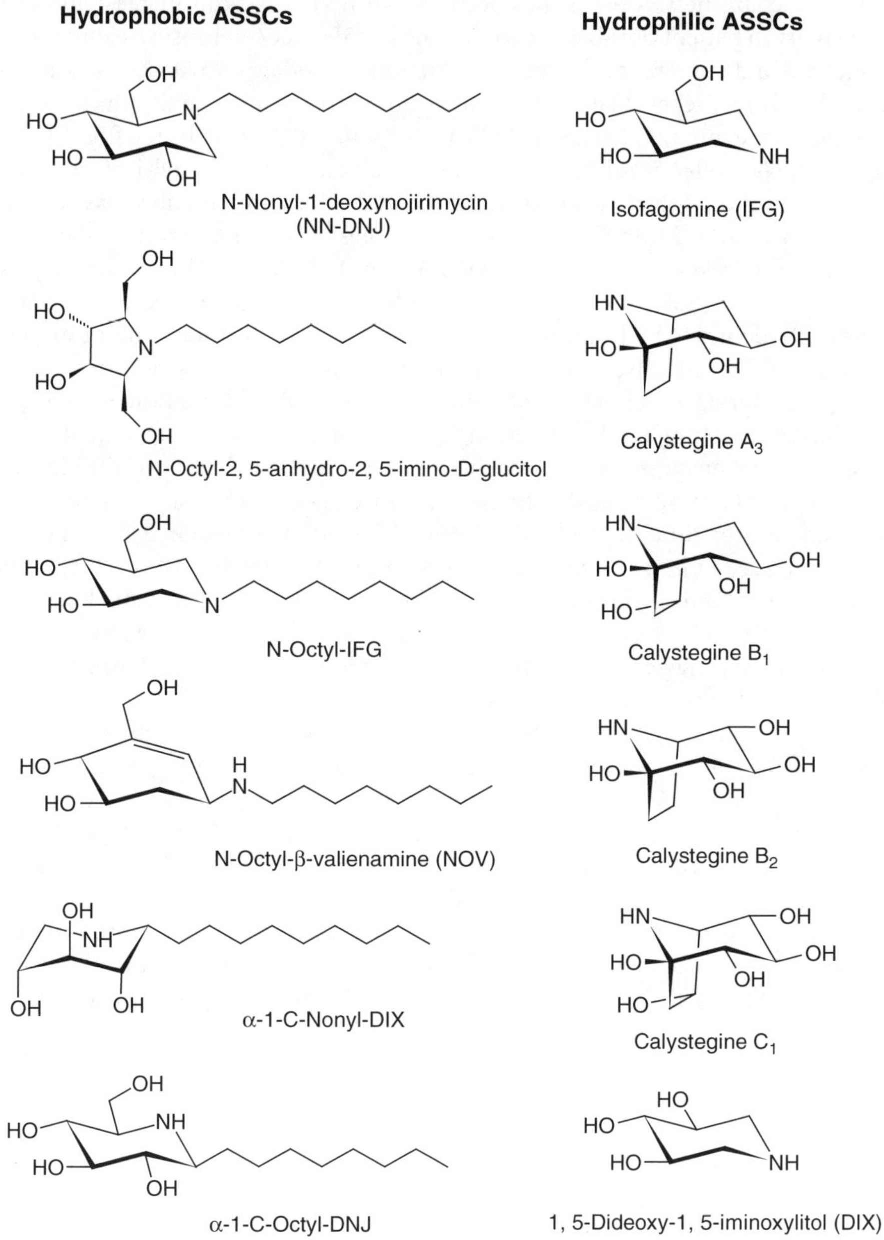

Figure 10.5 Structures of potential ASSCs for Gaucher disease. Hydrophobic ASSCs are *N*-nonyl-DNJ [16], *N*-octyl-2,5-anhydro-2,5-imino-D-glucitol [64], *N*-octyl-IFG [64], *N*-octyl-β-valienamine [62], α-1-C-nonyl-DIX [65] and α-1-C-octyl-DNJ [66]. Hydrophilic ASSCs [19] are IFG, calystegines A_3, B_1, B_2, C_1 and DIX

N-octyl-β-valienamine (NOV). It has been shown that NOV can increase intracellular enzyme activity in patient fibroblasts carrying the F213I glucocerebrosidase mutation [63]. However, NOV did not rescue other mutant glucocerebrosidase (N370S, L444P and 84GG). Sawkar and Kelly further evaluated 34 potential glucocerebrosidase inhibitors to determine if they could rescue mutant enzymes N370S, L444P and G202R in cultured fibroblasts [64]. Among them, *N*-octyl-2,5-anhydro-2,5-imino-D-glucitol and *N*-octyl-isofagomine were able to increase residual enzyme activity in N370S and G202R mutant fibroblasts, but not in L444P fibroblasts. In addition, Compain and colleagues [65], and Asano and colleagues [66] designed and synthesized α-1-*C*-nonyl-DIX and α-1-*C*-octyl-DNJ, respectively, which were both effective ASSCs for N370S mutant glucocerebrosidase. Recently, Chang and colleagues [19] identified hydrophilic iminosugar β-glucocerebrosidase inhibitors, namely isofagomine (IFG) and calystegines, that can effectively increase β-glucocerebrosidase activity in fibroblasts of patients with homozygous N370S mutation through an *in vitro* inhibition screening, and then through intracellular ASSC enhancement assays. In particular, residual enzyme activity was increased approximately by two-fold in the cells cultivated with IFG at 10–50 μM. The increase in glucocerebrosidase activity was both dose-dependent and time-dependent. Western blotting demonstrated that there was a substantial increase in glucocerebrosidase protein in cells after IFG treatment. Immunocytochemistry revealed an improvement in the glucocerebrosidase trafficking pattern, which overlaps that of lysosomal-associated membrane protein 2 in Gaucher fibroblasts cultivated with IFG, suggesting that the transport of mutant glucocerebrosidase is at least partially improved in the presence of IFG.

Zhu *et al.* [44] used synthetic chemicals to demonstrate that β-glucocerebrosidase contains two substrate-binding sites in the catalytic domain: one recognizes the glucosyl residue and the other recognizes the hydrophobic ceramide moiety. The carbohydrate recognition domain is the primary binding site, and the binding is dependent upon strict orientation and configuration of hydroxyl groups, identical to those of β-glucoside. On the other hand, hydrophobic recognition is not specific, and binding depends upon the length of the alkyl chain; a longer chain is better. From a structural point of view, the above ASSC compounds can be classified into two groups: hydrophobic compounds including those containing long alkyl chains or a hydrophobic moiety and hydrophilic compounds including IFG and calystegines (Figure 10.4). N-Dodecyl-DNJ, a potent inhibitor of β-glucocerebrosidase, has been shown to cause cytotoxicity even at lower μM concentrations [16]. C_{tox} values defined as compound concentration at which 50 per cent cell growth retardation occurs, for *N*-nonyl-DNJ were found to be about 50 μM [19]. Although glucocerebrosidase chaperones with alkyl chains have been shown to be effective in cellular studies, nonspecific hydrophobic interactions may cause undesirable side effects in other biological pathways. On the other hand, IFG showed no apparent cytotoxicity at concentrations up to 1 mM, suggesting that it may be an ideal clinical candidate for the treatment of Gaucher disease.

The mechanisms responsible for the intracellular increase of glucocerebrosidase activity by IFG were examined independently in great detail [67]. A major effect of IFG is to facilitate the folding and transport of newly synthesized glucocerebrosidase in the ER, thereby increasing the lysosomal pool of the enzyme. In addition, N370S glucocerebrosidase synthesized in the presence of IFG exhibits a shift in pH optimum from 6.4 to

5.2 and altered sensitivity to SDS. Although IFG fully inhibited β-glucocerebrosidase in the lysosomes in an *in situ* assay, washing of the drug led to partial recovery of glucocerebrosidase activity within 4 hours and full recovery by 24 hours. These findings further provide support for the possible use of IFG for the treatment of some forms of Gaucher disease. As mentioned above, IFG has entered clinical evaluation stages for the treatment of Gaucher disease.

10.7.3 G_{M1}-gangliosidosis

G_{M1}-gangliosidosis and Morquio B disease are autosomal recessive disorders caused by the deficiency of lysosomal acid β-galactosidase [68]. G_{M1}-gangliosidosis is a generalized neurosomatic disease occurring mainly in early infancy, and rarely in childhood or young adults. Morquio B disease is a rare systemic bone disease without CNS involvement.

Ogawa *et al.* synthesized a potent inhibitor of lysosomal β-galactosidase, *N*-octyl-4-*epi*-β-valienamine (NOEV), with an IC_{50} value of 300 nM [69]. Matsuda and colleagues [18] added NOEV to the culture medium of patient fibroblasts and demonstrated a substantial decrease of intracellular substrate storage. Short-term oral administration of NOEV to a mouse model of juvenile G_{M1}-gangliosidosis, expressing a human mutant enzyme protein R201C in a knockout background, resulted in a significant increase in the enzyme activity in the brain and other tissues. Immunohistochemical stain revealed a decrease in the amount of G_{M1} and G_{A1}-gangliosides in neuronal cells in the fronto-temporal cerebral cortex and brainstem. This study demonstrates that small molecular iminosugars can be orally administered and may readily cross the blood–brain barrier, which makes this approach particularly attractive for treating diseases with neuronopathic involvement.

10.7.4 Tay-Sachs and Sandhoff diseases

Tay–Sachs and Sandhoff diseases are lysosomal storage disorders that result from an inherited deficiency of β-hexosaminidase A and B [70]. β-Hexosaminidase A is a heterodimer with α and β subunits encoded by *HEXA* and *HEXB*, respectively, whereas β-hexosaminidase B is a homodimer of β subunit. Mutations in *HEXA* cause the deficiency of β-hexosaminidase A which results in Tay–Sachs disease. Sandhoff disease results from the deficient activity in both isozymes caused by mutations in *HEXB*. Both diseases are autosomal recessive, presenting three major subtypes: the infantile-, juvenile- and adult-onset phenotypes. Because the critical threshold for residual β-hexosaminidase A is relatively low (about 10 per cent of normal), these diseases are good candidates for the chaperone-based therapy.

Tropak and colleagues evaluated various known β-hexosaminidase inihibitors as potential ASSCs for the adult type of Tay–Sachs disease [17]. Among seven compounds tested, *N*-acetylglucosamine-thiazoline (NGT) can increase the level of lysosomal hexoaminidase A activity to 35 per cent of the normal levels in fibroblasts from an adult Tay–Sachs disease patient who is homozygous of G269S mutation in the α-subunit. Western blots documented an increased level of mature (lysosomal) α-subunit in treated cells, and subcellular fractionation of NGT-treated cells resulted in a further 10-fold increase in an ER-depleted lysosomal-enriched fraction. The increased enzyme activity and α-subunit were associated with an increase in the active β-hexosaminidase A by separating the isozymes in treated cells. Similarly, NGT was able to increase the β- hexosaminidase β-subunit P504S mutant enzyme activity in fibroblasts from a patient with adult Sandhoff disease. Thus, these results indicated that ASSC therapy may be an effective approach to treat Tay–Sachs and Sandhoff diseases.

10.8 Future perspectives

Over the last decade, therapeutic developments for lysosomal storage diseases have been dramatic. Enzyme replacement therapy remains the therapy of choice for many diseases including Gaucher and Fabry diseases, MPS I and VI, and Pompe disease at the moment. Substrate reduction therapy has been used for Gaucher disease and, off-label, for other glycosphingolipidoses. Bone marrow transplantation, gene and stem cell therapies are experimental and clinically proposed and tested [1, 5]. Despite all these remarkable advances, significant hurdles still remain. The primary challenge is the need to treat neurodegenerative lysosomal storage disorders. ASSC therapy is an attractive and emerging new strategy that uses small orally active molecules to rescue mutant proteins that are misfolded, mistargeted, and excessively degraded before reaching lysosomes, thereby increasing their activity and providing therapeutic benefit. Since small molecular ASSCs can be designed to cross the blood–brain barrier, ASSC therapy offers the potential to treat diseases with significant CNS involvement. ASSC therapy also offers various advantages such as convenient administration, noninvasive and possibly safer treatment over other therapeutic options. It is also an undeniable advantage that massive amounts of information on enzyme inhibitors accumulated over time can readily provide the initial structure design for ASSCs, thus significantly accelerating the drug development effort, which also shares many similarities with traditional drug discovery and drug development processes. With the basic concept clearly established and encouraging results obtained in independent studies, it is anticipated that a variety of small molecules with ASSC function for various lysosomal and other inborn errors of metabolism will be identified and developed.

It is likely that ASSC therapy alone will have a significant therapeutic effect in treating some forms of Fabry disease, particularly the cardiac variants that are often caused by missense mutations and that have residual enzyme activity, as shown in animal model and patient studies [20, 46]. A combination therapy with a replacement enzyme may provide additional therapeutic effect, since ASSC small molecules may stabilize the replacement enzyme *in situ* and increase its half-life. Similarly, a combination therapy with substrate reduction agents may be beneficial for some glycosphingolipidoses, particularly Gaucher disease. Compounds combining both chaperone and SRT functions are worth designing to test this hypothesis. Rapid drug development has brought two ASSC drugs into clinical development already. There is no question that the chaperone therapy field will be

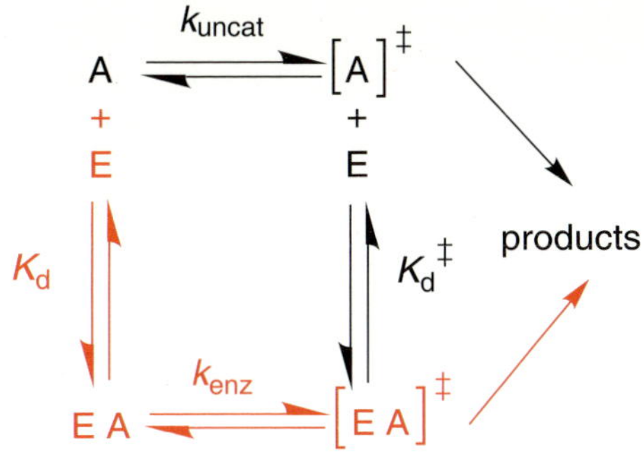

$$K_d^{\ddagger} = K_d \frac{k_{uncat}}{k_{enz}}$$

Figure 8.1 Thermodynamic box for the reaction rate, $k(_{uncat})$ of an uncatalysed (solvent only) reaction compared with the same reaction rate $k(_{enz})$ for an enzyme-catalysed reaction. K_d is the binding of enzyme to the reactant (A) and $K_d^{\ddagger}$ is the binding of the transition state species $[A]^{\ddagger}$ to the enzyme. Note that the binding of the transition state species is tighter than A by the catalytic rate enhancement imposed by the enzyme. Since typical enzymes increase reaction rates by 10^{10}–10^{15}, potential binding affinity for analogues of the transition state is large. Adapted from Schramm, V. L. (1998) *Ann. Rev. Biochem.*, **67**, 693–720

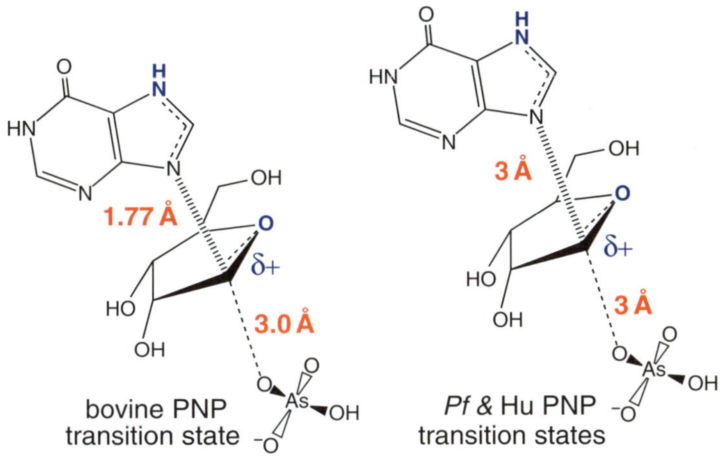

Figure 8.3 Transition state structures for bovine, human and *P. falciparum* PNPs showing the bond length to the leaving group and the attacking nucleophile at the moment of the transition states. Although the physiological nucleophile is phosphate, transition state analysis used arsenate for determination of these transition states. The electron deficiency caused by the delayed attack of the nucleophilic oxygen generates a ribooxacarbenium ion at the transition state. The ribosyl cation is more developed in human and *P. falciparum* transition states than in the bovine enzyme

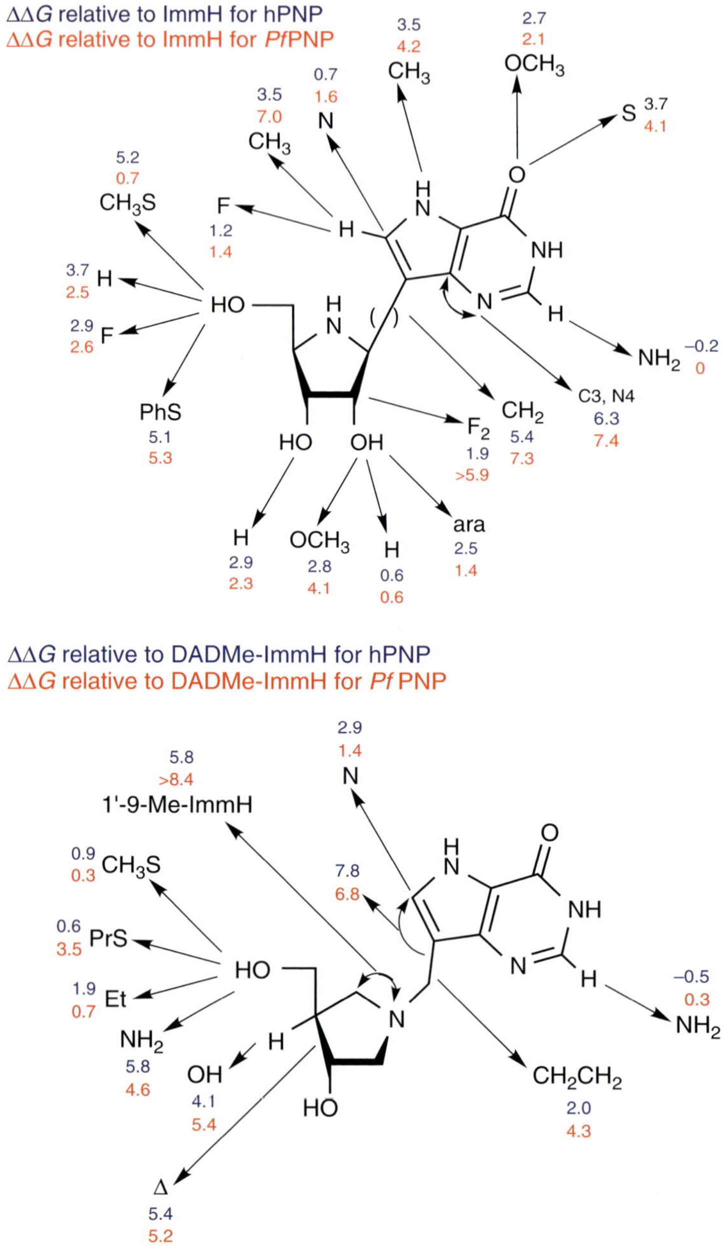

Figure 8.5 Energetics of transition state analogue binding to human and *P. falciparum* PNPs relative to [**2.1**] (upper panel) or to [**3.1**] (lower panel). Atomic substitutions are shown for specific positions together with the $\Delta\Delta G$ in kcal/mol for the indicated substitution relative to the parent compound. Reproduced from Lewandowicz et al. (2005) *J. Biol. Chem.*, **280**, 30320–30328 with permission from the American Society for Biochemistry and Molecular Biology

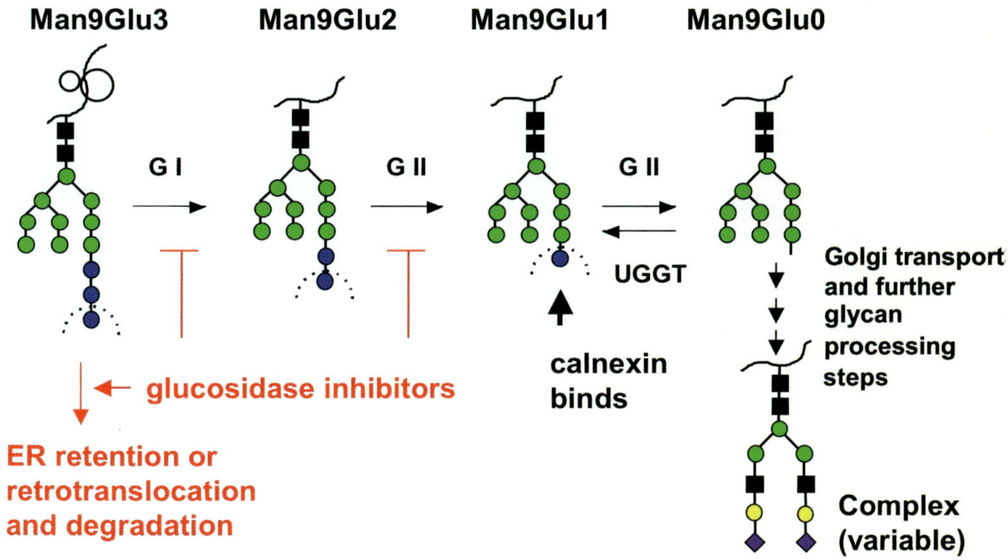

Figure 9.2 The normal pathway of glycoprotein processing and folding, and mechanisms by which misfolded proteins can be detected and eliminated. The structures of the N-linked oligosaccharide side chains are shown schematically. The growing polypeptide chain is indicated by a thin black line. Actions of glucosidases are indicated in red. Black squares, N-acetyl -D- glucosamine; green circles, mannose; blue circles, glucose; yellow circles, galactose; diamonds, N-acetyl-neuraminic acid

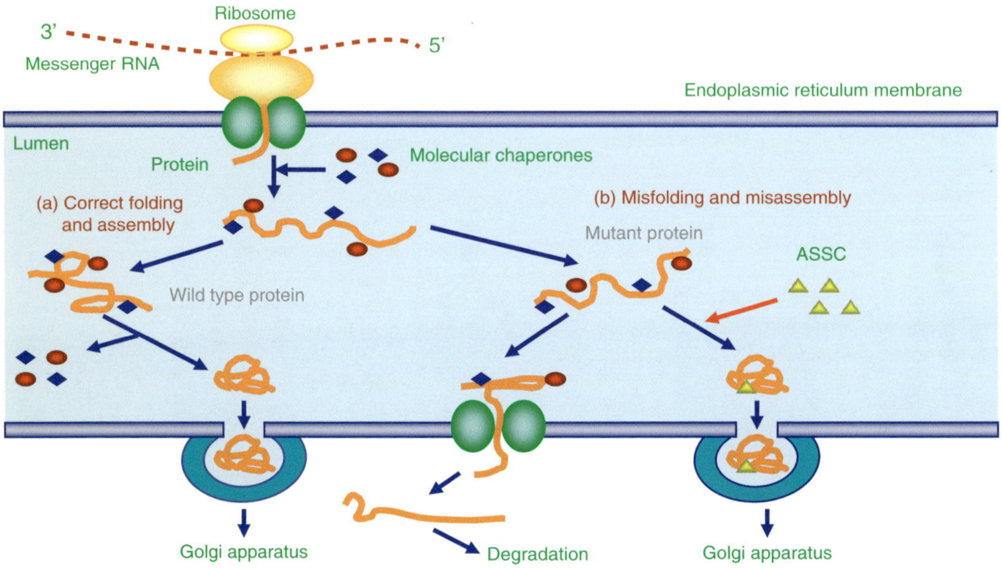

Figure 10.2 Endoplasmic reticulum (ER) quality-control system and active-site-specific chaperone (ASSC) therapy. The quality-control system for proteins in the ER is a 'proofreading' mechanism for newly synthesized proteins and is important for the fidelity of cellular function. This monitoring is enforced by several molecular chaperones and folding-assistant enzymes (red ovals and blue diamonds). (a) Appropriately folded proteins are transported out of the ER, whereas (b) misfolded and unfolded proteins are retained in the ER and are eventually degraded by ER-associated degradation [34]. Wild-type enzymes tend to fold and assemble into their native conformation and are quickly transported to the Golgi apparatus for further maturation. By contrast, mutant proteins that do not fold properly are retained and subsequently degraded. ASSCs (yellow triangles) bind to the active sites of the mutant enzymes and induce their properly folded conformation. This prevents ER-associated degradation and promotes transport of the mutant enzymes to the Golgi apparatus, thereby increasing the concentration of mutant enzymes and residual enzyme activity. Reprinted from Fan, J.-Q. (2003) *Trends Pharmacol. Sci*, **24**, 355, © (2003) with permission from Elsevier

further fuelled by anticipated favourable clinical results. ASSC therapy brings hope to the treatment of a variety of inherited metabolic disorders beyond lysosomal storage disorders, and particularly those with neuronopathic involvement. A new line of treatment, which uses traditional noninvasive oral drugs specifically acting on the target, is on the horizon for lysosomal storage disorders and other inherited inborn errors of enzymes, receptors and even channel proteins.

References

1. Winchester, B., Vellodi, A. and Young, E. (2000) The molecular basis of lysosomal storage diseases and their treatment. *Biochem. Soc.Trans.*, **28**, 150–154.
2. Meikle, P. J., Hopwood, J. J., Clague, A. E. and Carey, W. F. (1999) Prevalence of lysosomal storage disorders, *JAMA*, **281**, 249–254.
3. Rider, J. A. and Rider, D. L. (1999) Thirty years of Batten disease research: present status and future goals. *Mol.Genet.Metab.*, **66**, 231–233.
4. Santavuori, P. (1988) Neuronal ceroid-lipofuscinoses in childhood. *Brain Dev.*, **10**, 80–83.
5. Schiffmann, R. and Brady, R. O. (2002) New prospects for the treatment of lysosomal storage diseases. *Drugs*, **62**, 733–742.
6. Ellinwood, N. M., Vite, C. H. and Haskins, M. E. (2004) Gene therapy for lysosomal storage diseases: the lessons and promise of animal models. *J. Gene. Med.*, **6**, 481–506.
7. Eto, Y., Shen, J. S., Meng, X. L. and Ohashi, T. (2004) Treatment of lysosomal storage disorders: cell therapy and gene therapy. *J. Inherit. Metab. Dis.*, **27**, 411–415.
8. Hobbs, J. R., Hugh-Jones, K., Barrett, A. J., Byrom, N., Chambers, D., Henry, K., James, D. C., Lucas, C. F., Rogers, T. R., Benson, P. F., Tansley, L. R., Patrick, A. D., Mossman, J. and Young, E. P. (1981) Reversal of clinical features of Hurler's disease and biochemical improvement after treatment by bone-marrow transplantation. *Lancet*, **2**, 709–712.
9. Krivit, W., Peters, C. and Shapiro, E. G. (1999) Bone marrow transplantation as effective treatment of central nervous system disease in globoid cell leukodystrophy, metachromatic leukodystrophy, adrenoleukodystrophy, mannosidosis, fucosidosis, aspartylglucosaminuria, Hurler, Maroteaux-Lamy, and Sly syndromes, and Gaucher disease type III. *Curr. Opin. Neurol.*, **12**, 167–176.
10. Cox, T., Lachmann, R., Hollak, C., Aerts, J., van Weely, S., Hrebicek, M., Platt, F., Butters, T., Dwek, R., Moyses, C., Gow, I., Elstein, D. and Zimran, A. (2000) Novel oral treatment of Gaucher's disease with N-butyldeoxynojirimycin (OGT 918) to decrease substrate biosynthesis. *Lancet*, **355**, 1481–1485.
11. Cox, T. M., Aerts, J. M., Andria, G., Beck, M., Belmatoug, N., Bembi, B., Chertkoff, R., Vom Dahl, S., Elstein, D., Erikson, A., Giralt, M., Heitner, R., Hollak, C., Hrebicek, M., Lewis, S., Mehta, A., Pastores, G. M., Rolfs, A., Miranda, M. C. and Zimran, A. (2003) The role of the iminosugar N-butyldeoxynojirimycin (miglustat) in the management of type I (non-neuronopathic)Gaucher disease: a position statement. *J. Inherit. Metab. Dis.*, **26**, 513–526.
12. Fan, J.-Q. (2003) A contradictory treatment for lysosomal storage disorders: inhibitors enhance mutant enzyme activity. *Trends Pharmacol. Sci.*, **24**, 355–360.
13. Desnick, R. J. and Schuchman, E. H. (2002) Enzyme replacement and enhancement therapies: lessons from lysosomal disorders, *Nat. Rev. Genet.*, **3**, 954–966.
14. Fan, J.-Q., Ishii, S., Asano, N. and Suzuki, Y. (1999) Accelerated transport and maturation of lysosomal α-galactosidase A in Fabry lymphoblasts by an enzyme inhibitor. *Nature Med.*, **5**, 112–115.

15. Asano, N., Ishii, S., Kizu, H., Ikeda, K., Yasuda, K., Kato, A., Martin, O. R. and Fan, J.-Q. (2000) *In vitro* inhibition and intracellular enhancement of lysosomal α-galactosidase A activity in Fabry lymphoblasts by 1-deoxygalactonojirimycin and its derivatives. *Eur. J. Biochem.*, **267**, 4179–4186.
16. Sawkar, A. R., Cheng, W. C., Beutler, E., Wong, C. H., Balch, W. E. and Kelly, J. W. (2002) Chemical chaperones increase the cellular activity of N370S β-glucosidase: a therapeutic strategy for Gaucher disease. *Proc. Natl. Acad. Sci U S A*, **99**, 15428–15433.
17. Tropak, M. B., Reid, S. P., Guiral, M., Withers, S. G. and Mahuran, D. (2004) Pharmacological enhancement of β-hexosaminidase activity in fibroblasts from adult Tay-Sachs and Sandhoff Patients. *J. Biol. Chem.*, **279**, 13478–13487.
18. Matsuda, J., Suzuki, O., Oshima, A., Yamamoto, Y., Noguchi, A., Takimoto, K., Itoh, M., Matsuzaki, Y., Yasuda, Y., Ogawa, S., Sakata, Y., Nanba, E., Higaki, K., Ogawa, Y., Tominaga, L., Ohno, K., Iwasaki, H., Watanabe, H., Brady, R. O. and Suzuki, Y. (2003) Chemical chaperone therapy for brain pathology in G(M1)-gangliosidosis. *Proc. Natl. Acad. Sci. U S A*, **100**, 15912–15917.
19. Chang, H. H., Asano, N., Ishii, S., Ichikawa, Y. and Fan, J. Q. (2006) Hydrophilic iminosugar active-site-specific chaperones increase residual glucocerebrosidase activity in fibroblasts from Gaucher patients. *FEBS J*, **273**, 4082–4092.
20. Frustaci, A., Chimenti, C., Ricci, R., Natale, L., Russo, M. A., Pieroni, M., Eng, C. M. and Desnick, R. J. (2001) Improvement in cardiac function in the cardiac variant of Fabry's disease with galactose-infusion therapy. *N. Engl. J. Med.*, **345**, 25–32.
21. Morello, J. P., Salahpour, A., Laperriere, A., Bernier, V., Arthus, M. F., Lonergan, M., Petaja-Repo, U., Angers, S., Morin, D., Bichet, D. G. and Bouvier, M. (2000) Pharmacological chaperones rescue cell-surface expression and function of misfolded V2 vasopressin receptor mutants. *J. Clin. Invest.*, **105**, 887–895.
22. Noorwez, S. M., Kuksa, V., Imanishi, Y., Zhu, L., Filipek, S., Palczewski, K. and Kaushal, S. (2003) Pharmacological chaperone-mediated in vivo folding and stabilization of the P23H-opsin m,utant associated with autosomal dominant retinitis pigmentosa. *J. Biol. Chem.*, **278**, 14442–14450.
23. Janovick, J. A., Maya-Nunez, G. and Conn, P. M. (2002) Rescue of hypogonadotropic hypogonadism-causing and manufactured GnRH receptor mutants by a specific protein-folding template: misrouted proteins as a novel disease etiology and therapeutic target. *J. Clin. Endocrinol. Metab.*, **87**, 3255–3262.
24. Pedemonte, N., Lukacs, G. L., Du, K., Caci, E., Zegarra-Moran, O., Galietta, L. J. and Verkman, A. S. (2005) Small-molecule correctors of defective DeltaF508-CFTR cellular processing identified by high-throughput screening. *J. Clin. Invest.*, **115**, 2564–2571.
25. Ulloa-Aguirre, A., Janovick, J. A., Brothers, S. P. and Conn, P. M. (2004) Pharmacological rescue of conformationally-defective proteins: implications for the treatment of human disease. *Traffic*, **5**, 821–837.
26. Perlmutter, D. H. (2002) Chemical chaperones: a pharmacological strategy for disorders of protein folding and trafficking. *Pediatr. Res.*, **52**, 832–836.
27. Bernier, V., Lagace, M., Bichet, D. G. and Bouvier, M. (2004) Pharmacological chaperones: potential treatment for conformational diseases. *Trends Endocrinol. Metab.*, **15**, 222–228.
28. Cohen, F. E. and Kelly, J. W. (2003) Therapeutic approaches to protein-misfolding diseases. *Nature*, **426**, 905–909.
29. Beutler, E., Nguyen, N. J., Henneberger, M. W., Smolec, J. M., McPherson, R. A., West, C. and Gelbart, T. (1993) Gaucher disease: gene frequencies in the Ashkenazi Jewish population. *Am J. Hum. Genet.*, **52**, 85–88.
30. Leinekugel, P., Michel, S., Conzelmann, E. and Sandhoff, K. (1992) Quantitative correlation between the residual activity of β-hexosaminidase A and arylsulfatase A and the severity of the resulting lysosomal storage disease. *Hum. Genet.*, **88**, 513–523.

31. Schuchman, E. H. and Desnick, R. J. (1988) Mucopolysaccharidosis type I subtypes. Presence of immunologically cross-reactive material and in vitro enhancement of the residual α-L-iduronidase activities. *J. Clin. Invest.*, **81**, 98–105.
32. Schueler, U. H., Kolter, T., Kaneski, C. R., Zirzow, G. C., Sandhoff, K. and Brady, R. O. (2004) Correlation between enzyme activity and substrate storage in a cell culture model system for Gaucher disease. *J. Inherit. Metab. Dis.*, **27**, 649–658.
33. Anfinsen, C. B. and Scheraga, H. A. (1975) Experimental and theoretical aspects of protein folding. *Adv. Protein Chem.*, **29**, 205–300.
34. Ellgaard, L. and Helenius, A. (2003) Quality control in the endoplasmic reticulum. *Nat. Rev. Mol. Cell. Biol.*, **4**, 181–191.
35. Hartl, F. U. and Hayer-Hartl, M. (2002) Molecular chaperones in the cytosol: from nascent chain to folded protein. *Science*, **295**, 1852–1858.
36. Oda, Y., Hosokawa, N., Wada, I. and Nagata, K. (2003) EDEM as an acceptor of terminally misfolded glycoproteins released from calnexin. *Science*, **299**, 1394–1397.
37. Fewell, S. W., Travers, K. J., Weissman, J. S. and Brodsky, J. L. (2001) The action of molecular chaperones in the early secretory pathway. *Annu. Rev. Genet.*, **35**, 149–191.
38. Schubert, U., Anton, L. C., Gibbs, J., Norbury, C. C., Yewdell, J. W. and Bennink, J. R. (2000) Rapid degradation of a large fraction of newly synthesized proteins by proteasomes. *Nature*, **404**, 770–774.
39. Yewdell, J. W. (2001) Not such a dismal science: the economics of protein synthesis, folding, degradation and antigen processing. *Trends Cell. Biol.*, **11**, 294–297.
40. Kuznetsov, G. and Nigam, S. K. (1998) Folding of secretory and membrane proteins. *N. Engl. J. Med.*, **339**, 1688–1695.
41. Ellgaard, L., Molinari, M. and Helenius, A. (1999) Setting the standards: quality control in the secretory pathway. *Science*, **286**, 1882–1888.
42. Pasyk, E. A. and Foskett, J. K. (1995) Mutant (ΔF508) cystic fibrosis transmembrane conductance regulator Cl- channel is functional when retained in endoplasmic reticulum of mammalian cells. *J. Biol. Chem.*, **270**, 12347–12350.
43. Meacham, G. C., Patterson, C., Zhang, W., Younger, J. M. and Cyr, D. M. (2001) The Hsc70 co-chaperone CHIP targets immature CFTR for proteasomal degradation. *Nat. Cell. Biol.*, **3**, 100–105.
44. Zhu, X. X., Sheth, K. A., Li, S., Chang, H.-H. and Fan, J. Q. (2005) Rational design and synthesis of highly potent β-glucocerebrosidase inhibitors. *Angew. Chem. Int. Ed. Engl.*, **44**, 7450–7453.
45. Fan, J. Q. and Ishii, S. (2003) Cell-based screening of active-site specific chaperone for the treatment of Fabry disease. *Methods Enzymol.*, **363**, 412–420.
46. Ishii, S., Yoshioka, H., Mannen, K., Kulkarni, A. B. and Fan, J. Q. (2004) Transgenic mouse expressing human mutant α-galactosidase A in an endogenous enzyme deficient background: a biochemical animal model for studying active-site specific chaperone therapy for Fabry disease. *Biochem. Biophys. Acta*, **1690**, 250–257.
47. Brady, O. R., Gal, A. E., Bradley, R. M., Martensson, E., Warshaw, A. L. and Laster, L. (1967) Enzymatic defect in Fabry's disease: Ceramidetrihexosidase deficiency. *N. Engl. J. Med.*, **276**, 1163–1167.
48. Desnick, R. J., Ioannou, Y. A. and Eng, C. M. (2001) α-Galactosidase A deficiency: Fabry disease, in *The Metabolic and Molecular Bases of Inherited Disease* (Scriver, C. R., Beaudet, A. L., Sly, W. S. and Valle, D., eds), McGraw-Hill, New York,USA, pp. 3733–3774.
49. Remeo, G., Urso, M., Piszcane, A., Blum, E., de Falco, A. and Ruffilli, A. (1975) Residual activity of α-galactosidase A in Fabry's disease. *Biochem. Genet.*, **13**, 615–628.
50. Bishop, D. F., Grabowski, G. A. and Desnick, R. J. (1981) Fabry disease: An asymptomatic hemizygote with significant residual α-galactosidase A activity. *Am. J. Hum. Genet.*, **33**, 71A.

51. Ishii, S., Kase, R., Sakuraba, H. and Suzuki, Y. (1993) Characterization of a mutantα- galactosidase gene product for the late-onset cardiac form of Fabry disease. *Biochem. Biophys. Res. Comm.*, **197**, 1585–1589.
52. Sakuraba, H., Oshima, A., Fukuhara, Y., Shimmoto, M., Nagao, Y., Bishop, D. F., Desnick, R. J. and Suzuki, Y. (1990) Identification of point mutations in the α-galactosidase A gene in classical and atypical hemizygotes with Fabry disease. *Am. J. Hum. Genet.*, **47**, 784–789.
53. Ishii, S., Sakuraba, H. and Suzuki, Y. (1992) Point mutations in the upstream region of the α-galactosidase A gene exon 6 in an atypical variant of Fabry disease. *Hum. Genet.*, **89**, 29–32.
54. Ishii, S., Kase, R., Okumiya, T., Sakuraba, H. and Suzuki, Y. (1996) Aggregation of the inactive form of human α-galactosidase in the endoplasmic reticulum. *Biochem. Biophys. Res. Commun.*, **220,** 812–815.
55. Yam, G. H., Zuber, C. and Roth, J. (2005) A synthetic chaperone corrects the trafficking defect and disease phenotype in a protein misfolding disorder. *Faseb J.*, **19**, 12–18.
56. Yam, G. H., Bosshard, N., Zuber, C., Steinmann, B. and Roth, J. (2006) Pharmacological chaperone corrects lysosomal storage in Fabry disease caused by trafficking-incompetent variants. *Am. J. Physiol. Cell. Physiol.*, **290**, C1076–1082.
57. Beutler, E. and Grabowski, G. A. (2001) Gaucher disease, in *The Metabolic and Molecular Bases of Inherited Disease,* 8th edn (Scriver, C. R., Beaudet, A. L., Sly, W. S. and Valle, D., eds), McGraw-Hill, New York, USA, pp. 3635–3668.
58. Brady, R. O., Kanfer, J. N. and Shapiro, D. (1965) Metabolism of glucocerebrosides. II. Evidence of an enzymatic deficiency in Gaucher's disease. *Biochem. Biophys. Res. Commun.*, **18**, 221–225.
59. Grabowski, G. A. (1993) *Gaucher disease: enzymology, genetics, and treatment,* Plenum Press, New York, USA.
60. Ron, I. and Horowitz, M. (2005) ER retention and degradation as the molecular basis underlying Gaucher disease heterogeneity. *Hum. Mol. Genet.*, **14**, 2387–98.
61. Schmitz, M., Alfalah, M., Aerts, J. M., Naim, H. Y. and Zimmer, K. P. (2005) Impaired trafficking of mutants of lysosomal glucocerebrosidase in Gaucher's disease. *Int. J. Biochem. Cell Biol.*, **37**, 2310–20.
62. Ogawa, S., Kobayashi, Y., Kabayama, K., Jimbo, M. and Inokuchi, J. (1998) Chemical modification of β-glucocerebrosidase inhibitor N-octyl-β-valienamine: synthesis and biological evaluation of N-alkanoyl and N-alkyl derivatives. *Bioorg. Med. Chem.*, **6**, 1955–1962.
63. Lin, H., Sugimoto, Y., Ohsaki, Y., Ninomiya, H., Oka, A., Taniguchi, M., Ida, H., Eto, Y., Ogawa, S., Matsuzaki, Y., Sawa, M., Inoue, T., Higaki, K., Nanba, E., Ohno, K. and Suzuki, Y. (2004) N-octyl-β-valienamine up-regulates activity of F213I mutant β-glucosidase in cultured cells: a potential chemical chaperone therapy for Gaucher disease. *Biochim. Biophys. Acta*, **1689**, 219–228.
64. Sawkar, A. R., Adamski-Werner, S. L., Cheng, W. C., Wong, C. H., Beutler, E., Zimmer, K. P. and Kelly, J. W. (2005) Gaucher disease-associated glucocerebrosidases show mutation-dependent chemical chaperoning profiles. *Chem. Biol.*, **12**, 1235–1244.
65. Compain, P., Martin, O. R., Boucheron, C., Godin, G., Yu, L., Ikeda, K. and Asano, N. (2006) Design and synthesis of highly potent and selective pharmacological chaperones for the treatment of Gaucher's disease. *ChemBioChem.*, 7, 1356–1359.
66. Yu, L., Ikeda, K., Kato, A., Adachi, I., Godin, G., Compain, P., Martin, O. and Asano, N. (2006) α-1-C-Octyl-1-deoxynojirimycin as a pharmacological chaperone for Gaucher disease. *Bioorg. Med. Chem.*, **14**, 7736–7744.

67. Steet, R. A., Chung, S., Wustman, B., Powe, A., Do, H. and Kornfeld, S. A. (2006) The iminosugar isofagomine increases the activity of N370S mutant acid β-glucosidase in Gaucher fibroblasts by several mechanisms. *Proc. Natl. Acad. Sci. U S A*, **103**, 13813–13818.
68. Suzuki, Y., Oshima, A. and Nanba, E. (2001) β-Galactosidase deficiency (β-galactosidosis): G_{M1}-gangliosidosis and Morquio B disease, in *The Metabolic and Molecular Bases of Inherited Disease* (Scriver, C. R., Beaudet, A. L., Sly, W. S. and Valle, D., eds) , McGraw-Hill, New York, USA, pp. 3775–3810.
69. Ogawa, S., Matsunaga, Y. K. and Suzuki, Y. (2002) Chemical modification of the β-glucocerebrosidase inhibitor N-octyl-beta-valienamine: synthesis and biological evaluation of 4-epimeric and 4-O-(β-D-galactopyranosyl) derivatives. *Bioorg. Med. Chem.*, **10**, 1967–1972.
70. Gravel, R. A., Kaback, M. M., Proia, R. L., Sandhoff, K., Suzuki, K. and Suzuki, K. (2001) *The GM2 Gangliosidoses*, 8th edn, McGraw-Hill, New York, USA.
71. Asano, N., Oseki, K., Tomioka, E., Kizu, H. and Matsui, K. (1994) N-Containing sugars from Morus alba and their glycosidase inhibitory activities. *Carbohydr. Res.*, **259**, 243–255.
72. Asano, N., Nishiba, M., Kizu, H., Matsui, K., Watson, A. A. and Nash, R. J. (1997) Homonojirimycin isomers and glycosides from Aglaonema treubii. *J. Nat. Prod.*, **60**, 98–101.
73. Pili, R., Chang, J., Partis, R. A., Mueller, R. A., Chrest, F. J. and Passaniti, A. (1995) The α-glucosidase I inhibitor castanospermine alters endothelial cell glycosylation, prevents angiogenesis, and inhibits tumor growth. *Cancer Res.*, **55**, 2920–2926.
74. Fan, J.-Q. and Ishii, S. Method for enhancing mutant enzyme activities in lysosomal storage disorders. U.S. Patent, S/N 6,589,964, Issued on 07/08/03.
75. Ichikawa, Y., Igarashi, Y., Ichikawa, M. and Suhara, Y. (1998) 1-N-Iminosugars: Potent and selective inhibitors of β-glycosidases. *J Am Chem Soc.*, **120**, 3007–3018.
76. Dorling, P. R., Huxtable, C. R. and Colegate, S. M. (1980) Inhibition of lysosomalα- mannosidase by swainsonine, an indolizidine alkaloid isolated from Swainsona canescens. *Biochem J.*, **191**, 649–651.
77. Aoyagi, T., Yamamoto, T., Kojiri, K., Morishima, H., Nagai, M., Hamada, M., Takeuchi, T. and Umezawa, H. (1989) Mannostatins A and B: new inhibitors of α-D-mannosidase, produced by Streptoverticillium verticillus var. quintum ME3-AG3: taxonomy, production, isolation, physico-chemical properties and biological activities. *J. Antibiot. (Tokyo)*, **42**, 883–889.
78. Asano, N., Yasuda, K., Kizu, H., Kato, A., Fan, J. Q., Nash, R. J., Fleet, G. W. and Molyneux, R. J. (2001) Novel α-L-fucosidase inhibitors from the bark of Angylocalyx pynaertii (Leguminosae). *Eur. J. Biochem.*, **268**, 35–41.
79. Aoyagi, T., Suda, H., Uotani, K., Kojima, F., Aoyama, T., Horiguchi, K., Hamada, M. and Takeuchi, T. (1992) Nagstatin, a new inhibitor of N-acetyl-β-D-glucosaminidase, produced by Streptomyces amakusaensis MG846-fF3. Taxonomy, production, isolation, physico-chemical properties and biological activities. *J. Antibiot. (Tokyo)*, **45**, 1404–1408.
80. Tatsuta, K., Miura, S., Ohta, S. and Gunji, H. (1995) Syntheses and glycosidase inhibiting activities of nagstatin analogs. *J. Antibiot. (Tokyo)*, **48**, 286–288.
81. Cenci di Bello, I., Dorling, P., Fellows, L. and Winchester, B. (1984) Specific inhibition of human beta-D-glucuronidase and α-L-iduronidase by a trihydroxy pipecolic acid of plant origin. *FEBS Lett.*, **176**, 61–64.
82. Umezawa, H., Aoyagi, T., Komiyama, T., Morishima, H. and Hamada, M. (1974) Purification and characterization of a sialidase inhibitor, siastatin, produced by Streptomyces. *J. Antibiot (Tokyo)*, **27**, 963–969.

11

Iminosugar inhibitors for substrate reduction therapy for the lysosomal glycosphingolipidoses

Terry D. Butters

11.1 Glycolipid lysosomal storage diseases

The glycosphingolipidoses are a group of inherited diseases that are caused primarily by mutations in catabolic glycosidases resulting in a reduction in the hydrolysis of glycosphingolipid (GSL) oligosaccharide. If any product of the catalytic cascade fails to be sequentially hydrolysed, lysosomal storage occurs [1]. There appears to be no feedback mechanism that signals the endoplasmic reticulum and Golgi to reduce biosynthesis to match the reduced rate of hydrolysis in the lysosome and as the influx of GSL from the cell surface remains constant, lysosomal accumulation increases with time. As substrate concentration increases to inhibitory saturating levels the catalytic efficiency is reduced further, eventually generating a densely packed intralysosomal aggregate of protein and lipid that restricts access of enzyme to substrate and accelerates GSL storage.

When the catabolism of glycosphingolipids is impaired following mutation of the glycosidases, several severe pathological conditions are produced in humans. Although individually disease incidence is rare, collectively they pose a significant and challenging group of disorders to treat. The incidence of lysosomal storage disease has been estimated to occur at 1 in 18 000 live births world-wide [2] and it is the most frequent cause of paediatric neurodegenerative disease.

The most frequent lysosomal storage disease (LSD) is Gaucher disease with an estimated prevalence of 1:200 000 in the general population that predicts that over 30 000 people are affected globally. Deficiencies in the activity of ceramide β-glucosidase (glucocerebrosidase) result in the accumulation of lysosomal glucosylceramide. More than 76 gene mutations have been identified that lead to deficient enzyme activity and different clinical phenotypes (type I, type II/III) can be predicted when null, severe or mild alleles

Iminosugars: From synthesis to therapeutic applications Edited by Philippe Compain and Olivier R. Martin
 ISBN 978-0-470-03391-3

are combined [3]. Amino acid residues involved in the binding of substrate and activator have been identified and the catalytic nucleophile reported to be E340 [4]. This amino acid lies within a region of the protein where most of the human mutations are found, including the most frequent N370S substitution but an X-ray structural analysis of glucocerebrosidase reveals that N370 is located in an α-helical region too far from the active site to be involved directly in catalysis [5]. Since mutations in this region are expected to cause minor changes in the enzyme structure it is not surprising to find that this results in an impairment of catalytic activity and the most benign form (type 1) of the disease. One explanation for this is that the changes in this helical region appear to affect the interaction of enzyme with activator protein (Saposin C) and/or phospholipid-rich lysosomal membranes leading to reduced catalytic activity [6, 7].

Substrate docking experiments show that the active site of the enzyme coordinates glucose, or glucose mimics like deoxynojirimycin, castanospermine or conduritol ß-epoxide, and is surrounded by an annulus of hydrophobic residues. This conformation would predict a membrane or ceramide chain interaction that could be mimicked by iminosugar analogues containing hydrocarbon chains. In support of this, increasing the hydrophobicity of N-alkylated iminosugar analogues increases their inhibitory potency using *in vitro* assays [8, 9]. This activity has been exploited in cellular systems by increasing the amount of folded protein and enhancing the catalytic activity of mutant enzyme.

Those affected with Gaucher disease have hepatosplenomegaly, haematological disturbance and bone destruction caused by the accumulation of glucosylceramide in macrophages where the phagocytic activity of these cells induces an additional lysosomal burden of glucosylceramide. Lysosomal or extra-lysosomal GSL either acts as a primary stimulus for a disruption in cell signalling events inducing cell death or the storage lipids (or their metabolites) may be directly cytotoxic. A pro-inflammatory cascade, also demonstrated in neural tissue mediated by microglial cells, could be a mechanism for pathogenesis in Gaucher disease [10, 11] and all the glycosphingolipidoses [12].

11.2 Options for glycosphingolipidosis therapy

Despite the incidence for individual glycosphingolipidoses being very low, the Orphan Drug Act passed in 1983 has stimulated the development of novel treatments for these diseases. The first success was in providing enzyme replacement therapy (ERT) for type I Gaucher disease where glucocerebrosidase (imiglucerase, Ceredase®, Cerezyme®) has been a registered drug since 1992. Over 4300 Gaucher disease patients take a regular infusion of enzyme, recombinantly expressed in CHO cells, where the N-linked oligosaccharide has been modified to expose mannose residues that can be recognized by the macrophage mannose receptor to aid cell capture and increase internalization [13].

Two options for ERT in Fabry disease using recombinant α-galactosidase are available: Replagal and Fabrazyme, both of which were licensed for use in Europe in 2003. Only Fabrazyme has been approved in the USA [13]. Disease-specific treatments for the other glycosphingolipidoses (Tay-Sachs, Sandhoff and GM1 gangliosidosis) are lacking but enzyme replacement using iduronidase (Elaprase™, Shire Human Genetic Therapies) for mucopolysaccharidoses II, Hunter syndrome and glycogen storage disease (Pompe disease) treatment with acid glucosidase (Myozyme®, Genzyme), have both been recently approved by the FDA.

The major limitation of enzyme replacement for treating the glycosphingolipidoses is the poor access to neural tissue by the infused enzyme. Potential therapies for the diseases where neurological involvement is the main cause of disease phenotype, namely type II/III Gaucher disease, as well as the GM1 and GM2 gangliosidoses include bone marrow transplantation and gene therapy. The latter offers long-term stable expression of enzyme and although proof of principle has been achieved *in vitro* and *in vivo*, a number of concerns remain regarding safety [14]. The simplest of methods may be more practical, and a recent study in mucopolysaccharidosis VII showed that the blood–brain

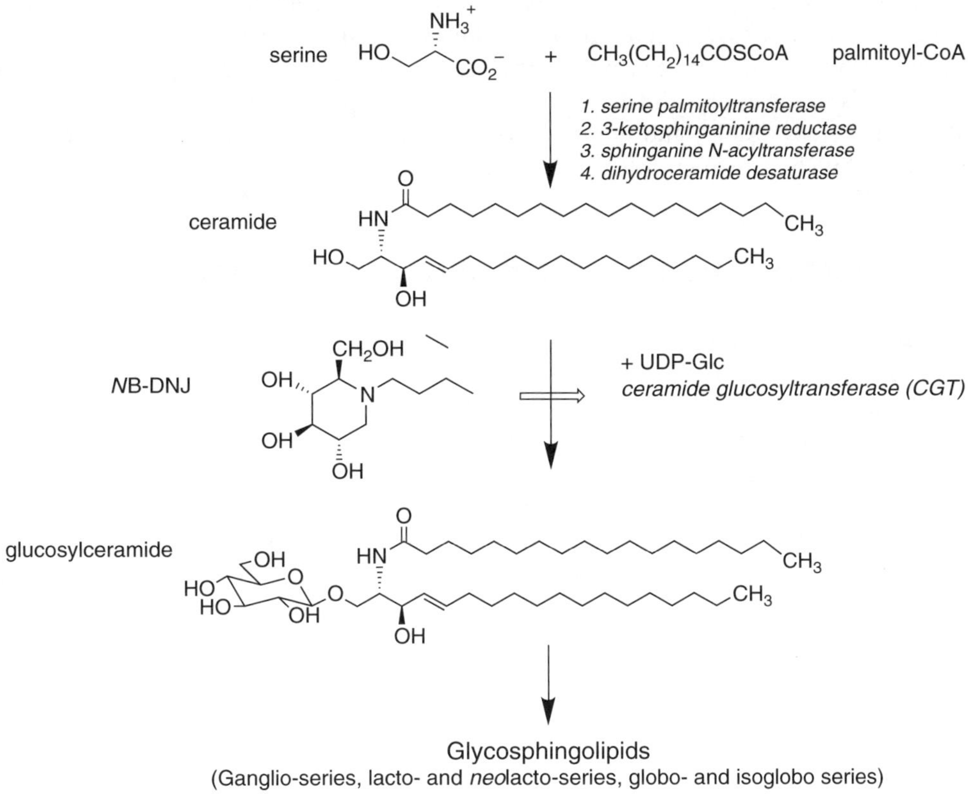

Figure 11.1 Key steps in the biosynthesis of glycosphingolipids. Serine palmitoyltransferase condenses serine and palmitoyl-CoA and via a number of enzymes, ceramide is produced. Ceramide glucosyltransferase initiates ceramide glycosylation to produce the glycosphingolipids series shown, a reaction that is inhibited by iminosugars such as *N*B-DNJ. Ceramide is also a substrate for a glycosyltransferase to make galactosylceramide, and a synthase to produce sphingomyelin. Neither of these reactions is inhibited by *N*B-DNJ. As a consequence of treating cells or tissues treated with *N*B-DNJ, no free ceramide accumulates, and the amount of sphingomyelin or galactosylceramide is increased to compensate for the increase in ceramide substrate

barrier (BBB) could be overcome by using a high-dose enzyme strategy [15]. Although these data have yet to demonstrate a reversal of the neurological deficit, storage of glycosaminoglycan (GAG) was reduced in glial and neuronal cells in a mouse model for disease. This strategy has yet to be applied in a human study but the prospect of crossing the BBB may be a possibility for all diseases where ERT is approved, despite the obvious consequences for cost. A recent study of a type III Gaucher Norrbottnian phenotype also demonstrated that over a 10 year treatment period, neuronopathic disease was slower to progress [16].

The idea that lysosomal accumulation of glycolipid could be controlled by inhibiting the biosynthesis of substrate, was promoted several years ago [17], but biologically evaluated molecules were lacking to take this strategy from the bench to the clinic. Following the discovery that N-alkylated iminosugars with the correct chirality were respectable inhibitors *in vitro*, in cultured cells and *in vivo*, of ceramide glucosyltransferase (CGT), a pivotal enzyme in the biosynthetic pathway for most glycolipids (Figure 11.1), iminosugars were rapidly exploited for clinical use.

This review will summarize some of the important experimental data that supported clinical trials of the approach to reduce lysosomal enzyme substrate synthesis, substrate reduction therapy (SRT), and approval for one compound, N-butyl-deoxynojirimycin (*N*B-DNJ) for the treatment of type I Gaucher disease.

11.3 Iminosugars for substrate reduction therapy

1-Deoxynojirimycin and N-alkylated derivatives are potent α-glucosidase inhibitors and several studies indicated the role of these enzymes in early stages of glycoprotein N-linked oligosaccharide processing in the endoplasmic reticulum (ER) [18]. Following entry and transcription of the viral genome most, if not all, enveloped viruses use the host cell pathway for glycosylation to generate virally encoded glycoproteins. The use of glucosidase inhibition to modulate the structure of the oligosaccharide chain was found to reduce viral infectivity and this approach became a pharmacological target, particularly when initiatives for HIV therapy were required. Using naturally occurring alkaloid derivatives found in plants as lead antiviral compounds, chemical libraries of iminosugars were created [19, 20] to discover therapeutically beneficial inhibitors and *N*B-DNJ was developed by Searle/Monsanto as a leading candidate for HIV treatment, culminating in clinical trials in humans as a monotherapy [21] and in combination with AZT [22].

The conventional chemical synthesis of iminosugars requires several steps due the presence of four chiral centres and of four hydroxyl groups of nearly equivalent reactivity that need protection and deprotection to preserve the desired stereochemistry. Searle/Monsanto responded to the need for commercial scale synthesis for a clinical evaluation in HIV-infected individuals by developing a rapid chemoenzymatic one-pot route for the manufacture of *N*B-DNJ (Figure 11.2. Starting from affordable materials, glucose and *n*-butylamine, the conversion to the sorbofuranose was accomplished in a single step catalysed by a *Gluconobacter* polyol dehydrogenase and without the need for product purification or protection group chemistry. A final

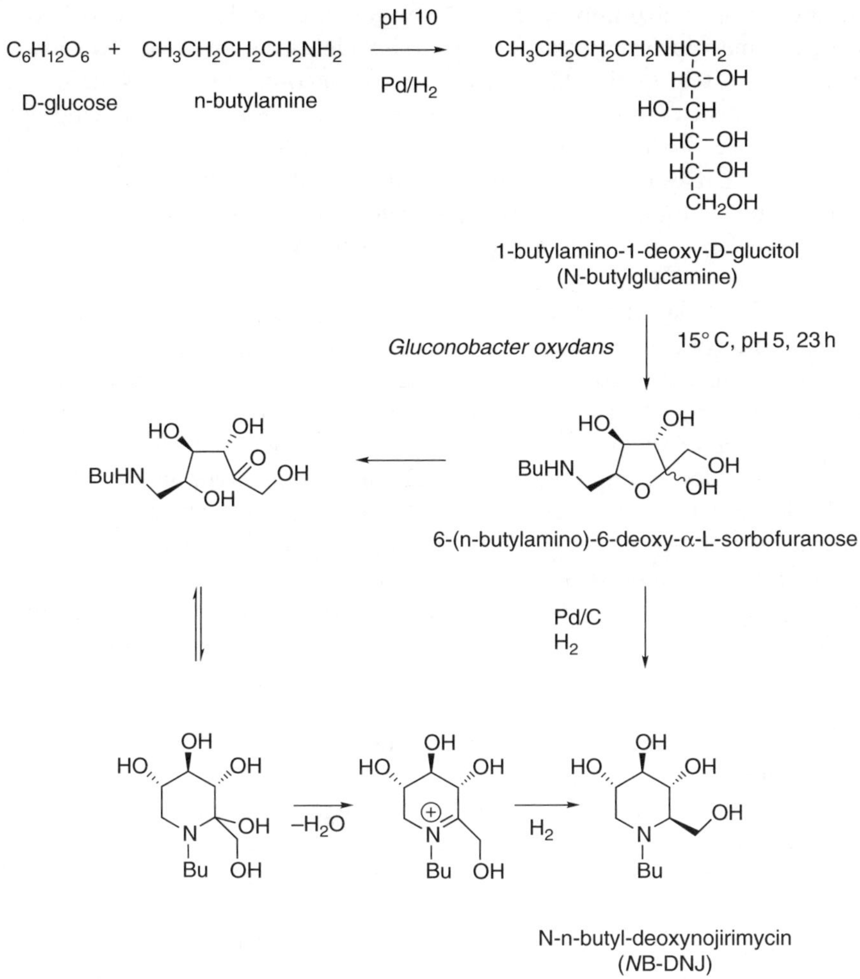

Figure 11.2 Three-step chemo-enzymatic synthesis of *N*B-DNJ. The catalytic hydrogenation of glucose with butylamine is reacted with whole Gluconobacter bacteria, or purified dehydrogenase. A further hydrogenation step affords *N*B-DNJ

catalytic hydrogenation step completed nitrogen cyclization and regenerated glucose stereochemistry [23].

Although clinical trials with *N*B-DNJ and a pro-drug derivative, perbutyrylated *N*B-DNJ [24], designed to overcome gastrointestinal side effects, were unsuccessful, and the precise mechanism by which α-glucosidase inhibitors were antiviral had not been established, these trials allowed exposure of iminosugars to humans and provided valuable biological safety information for a subsequent evaluation of SRT in treating the glycosphingolipidoses.

Until *N*B-DNJ was approved for use, N-hydroxyethyl-DNJ (Miglitol) for noninsulin dependent (NID) diabetes was the sole commercial iminosugar and there were few data indicating that iminosugars could be tolerated without overt toxicity or had a defined pharmacological profile *in vivo*. The identification of the quality control mechanism for protein folding that involves monoglucosylated glycans interacting with ER-chaperones such as calnexin [25], has revealed the mechanism for viral glycoprotein misfolding and reduction in infectivity observed following glucosidase inhibitor treatment [26, 27]. Despite the lack of a significant effect as a therapy for AIDS, the use of iminosugars to misfold viral glycoprotein as a therapeutic option has now been applied to hepatitis B (see Chapter 9) and has gained pharmaceutical industry interest [28].

The effects of GSL depletion have been experimentally determined in mouse knockout studies and a null phenotype for CGT is embryonically lethal [29]. Deletion of genes coding glycosyltransferases downstream from glucosylceramide also leads to neurological disturbance [29]. Important lessons have been learnt from these studies and serve to highlight the difference between species when using mouse models for understanding disease pathology, for example the lack of apparent phenotype in Hex A and α-Gal knockout mice. Furthermore in the GM3 synthase knockout, mice are viable and some neurological differences can be observed, whereas in humans a GM3 deficiency leads to infantile epilepsy and dramatic neurological dysfunction [30]. In recognition of the roles played by many GSL, developmentally and physiologically, the aim of SRT is to modulate biosynthesis to balance the impaired lysosomal hydrolysis of the GSL. A significant reduction in GSL composition, although tolerated in single cells, is neither warranted nor desirable *in vivo*.

The generation of genetically derived mouse models for disease have been an important acquisition for experimentalists in evaluating disease progression, and for developing therapeutic strategies ([31, 32]. Although these models have been engineered to be null for a specific enzyme activity, transgenic mice show many of the typical pathological traits observed in human disorders of lysosomal storage diseases [29]. Where lysosomal storage is a feature of the knockout model and no apparent disease phenotype can be observed, such as Fabry and Tay-Sachs, these models still offer good experimental paradigms for addressing proof of principle studies [33].

In the 10 year period since the discovery of iminosugars as therapeutic agents to treat the glycosphingolipidoses, two novel approaches have been developed. The first is a generic strategy that uses inhibition of the biosynthetic cycle to reduce GSL substrate influx to the catabolically compromised lysosome. SRT has translated from basic research to a medicinal product. The second, more experimental approach uses iminosugars to target the protein folding and trafficking pathways of glycosidases in order to assist correction of lysosomal enzyme activity. This approach has been called chaperone mediated therapy, CMT [34]. These therapeutic methods can be considered as an intervention in normal cellular processes to produce a partial effect on enzyme activity. SRT reduces the synthesis of GSL with concentrations of iminosugar that are designed to avoid complete depletion from cell surfaces of these biologically critical components. An enhancement in enzyme activity by CMT may be sufficient to reduce the threshold concentration of GSL substrate in the lysosome to nonpathological levels using noninhibitory concentrations of iminosugar.

11.4 Iminosugars as inhibitors of glycolipid biosynthesis

Iminosugars with the correct chirality and N-substituted with groups containing a minimum of three carbon atoms inhibit GSL biosynthesis [35, 36]. Further studies have shown that the molecular basis of this activity was inhibition of CGT and that *N*B-DNJ was a reversible, competitive inhibitor for ceramide and noncompetitive for the sugar donor, UDP-glucose, in the reaction scheme (Figure 11.1). The inhibitory constant (K_i) of *N*B-DNJ using ceramide as an acceptor for CGT activity is 7.4 μM. A similar value (10.6 μM) was found for *N*B-DGJ. The type of inhibition as determined by double reciprocal plots was reversible and competitive for ceramide and noncompetitive for UDP-glucose [37]. A comparison of the IC_{50} values reveals a similar trend indicating the lack of any contribution of the 4-hydroxyl group in enzyme inhibition (Figure 11.3). When the 5-hydroxymethyl epimers of DNJ and DGJ were assessed, in the L-*ido* and L-*altro* series respectively, both were weaker in inhibitory potency, but the specificity for CGT is far greater (unpublished data).

A partial explanation of the structure/function studies was obtained using molecular modelling studies of the crystal structure of ceramide and the NMR solution structure

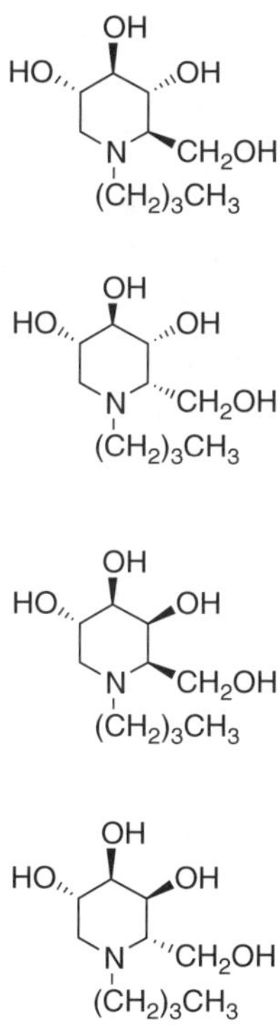

N-butyl-deoxynojirimycin (miglustat, Zavesca®) IC_{50} CGT = 20 μM

N-butyl-L-idonojirimycin IC_{50} CGT = 27 μM

N-butyl-galactonojirimycin IC_{50} CGT = 40 μM

N-butyl-L-altronojirimycin IC_{50} CGT = 70 μM

Figure 11.3 Biologically active *N*-butylated iminosugars. Iminosugars that mimic glucose and galactose, and their 6-OH epimers, are shown together with ceramide glucosyltransferase inhibition constants

of *N*B-DNJ [37]. Although not sufficient to explain all the data, it has acted as a firm base for synthesizing analogues' more potent CGT inhibitory activity or a more selective inhibition profile for CGT.

DNJ derivatives have recently been described that were synthesized to engender greater ceramide molecular mimicry based on the molecular modelling described above. Since the prediction that alignment of the N-acyl chain of ceramide and the N-alkyl chain of the iminosugar contributes significantly to potency and the partial redundancy of the 4-hydroxyl group, addition of a further alkyl chain could improve mimicry, hence inhibitory potency. This was not proved by experiment [38], the likely explanation being that the conformational space of the alkyl and acyl chains is restricted in solution phase by the DNJ cyclic structure and no close alignment with ceramide was allowed [38]. In cellular assays, C-glycosides were less potent than their N-alkylated counterparts but showed a modest increase compared with *in vitro* values, similar to the effects seen with more hydrophobic molecules (see below). It is therefore apparent that in a cellular membrane, insertion of the hydrocarbon chains is not optimal for ceramide presentation.

An increase in the hydrophobicity of the N-substituent elevates potency [39] in *in vitro* assays and increases membrane adsorption and tissue persistence, providing an additional inhibitory benefit in cells and *in vivo* [39]. Using recombinant CGT, kinetic measurements of the interaction between N-alkylated iminosugars and the native ceramide substrate reveals subtle differences between the modes of inhibition. Compounds with longer alkyl chains, for example N-nonyl-DNJ, have a noncompetitive mode of inhibition for ceramide suggesting that there may be more than one binding site on the recombinant enzyme and this may be additional to the active site that coordinates glucose binding [40].

No crystal structure of the CGT enzyme has been obtained but an *in silico* model has provided further information regarding the interaction of N-alkylated iminosugars with the enzyme [40]. Short chain N-alkylated inhibitors bind to a site with a more hydrophilic environment than a membrane associated hydrophobic site that accepts longer chain (>5C) inhibitors [40].

At least three chiral centres in the six-membered piperidines are required for inhibition of CGT and those with D-*gluco*, D-*galacto*, L-*ido* and L-*altro* configuration are active. This has allowed further compounds to be evaluated for SRT, the most notable being the *galacto* analogue, *N*B-DGJ [36, 37] (see Figure 11.3). Development of this compound has followed a route similar to *N*B-DNJ with efficacy shown in tissue culture models of Gaucher disease [36], *in vivo* tolerability [41] and efficacy in animal models for Sandhoff disease [42]. The *galacto* analogue is equally as effective as *N*B-DNJ at reducing substrate burden in the brain and increasing life expectancy [42, 43]. The lack of potential side effects noted with experiments performed with *N*B-DNJ, such as gastrointestinal distress and weight loss in mice, and inhibition of ER-glucosidases, reveals that *N*B-DGJ is both more selective and tolerated better than *N*B-DNJ. These factors make a powerful inducement for a clinical evaluation of this compound, particularly in paediatric cases where early intervention is required before significant neuropathology has occurred.

11.4.1 Cell and tissue penetration of iminosugars for SRT

The entry of iminosugars into cells appears to be by passive, nonfacilitated diffusion, or by flip-flop across the membrane [39, 44]. The active site of the Golgi membrane CGT faces the cytosol and cellular experiments have indicated that iminosugars are able to inhibit glycosyl transfer rapidly, in less than 1 minute [44]. The rate of entry is independent of N-alkyl chain length but increasing lipophilicity results in more protein and membrane binding [39]. It is therefore likely that the improved cellular potency of longer alkyl chain compounds reflects the increased deposition of compound in membranes close to the site of ceramide glucosylation. Increasing the alkyl chain length from 4 to 18 carbons leads to a 10-fold improvement in inhibition of CGT activity *in vitro*, yet a more than 50-fold increase was observed in the efficacy for reducing GSL biosynthesis in cells [44].

In vivo studies have also revealed a correlation between increasing N-alkyl-iminosugar hydrophobicity and tissue access and sequestration. Using radiolabelled compounds [39] administered by gavage to mice, slowed penetration to the liver and brain from the gut was observed as the N-alkyl chain was increased in length. The proportion of *N*N-DNJ (containing a 9-carbon chain compound) was 15-fold greater in liver and brain than the 4-carbon chain compound, *N*B-DNJ, 90 minutes post gavage. Since hydrophobic compounds in general cross the blood–brain barrier more effectively, this might have been expected, but the relative proportions of iminosugars gaining access from plasma are restricted. Approximately 25 per cent of the serum concentration was found in brain tissue from a mouse model for Sandhoff disease fed a diet of either *N*B-DNJ or *N*B-DGJ [42]. The level of *N*B-DNJ found in the brain of Sandhoff mice was 25 per cent higher than the wild-type littermate mice indicating that the blood–brain barrier is compromised in this disease [12, 41].

These factors, if extrapolated to humans, offer considerable potential benefit for treating the neuronopathic disorders where enzyme replacement therapy is of marginal efficacy due to lack of blood–brain barrier access. However, compound selection to provide a better therapeutic outcome is critical because of the many other properties that N-alkylated iminosugars display [45].

11.4.2 Reversibility of glycosphingolipid inhibition

The long-term effects of reducing GSL levels in cells and tissues is difficult to predict because of our lack of knowledge of the biological roles played by these molecules. N-alkylated iminosugar inhibitors have relatively high affinity for CGT, reach their site of action rapidly and can accumulate in tissues either transiently or for long periods, depending on hydrophobicity. All of these properties are reversible. Administration of *N*B-DNJ to cells to inhibit CGT completely, followed by removal of the inhibitor leads to a full restoration of GSL levels after 24 hours [44]. The effects of more hydrophobic compounds, by contrast, are less reversible and the deposition of these compounds in cellular lipid rich membranes could act as a reservoir enabling slow release of inhibitor.

Consequently, the inhibition of GSL by C_9-DNJ (*N*N-DNJ) was only partially reversed 24 hours following removal whereas C_{18}-DNJ, exhibited almost complete inhibition over the same period. These data point to important considerations when selecting compounds for therapy. A small molecule that is rapidly cleared by the body is much easier to regulate by dose to obtain the desired therapeutic effect than one where therapeutic efficacy is tissue- and time-dependent.

Further support for the lack of long-term effects following administration of N- alkylated iminosugars to rodents is the absence of any up-regulation of CGT that might increase GSL synthesis after compound removal [40] and the return of fertility to male mice after *N*B-DNJ withdrawal [46].

11.4.3 Cytotoxicity of iminosugars for SRT

Iminosugars appear to be resistant to metabolism and are excreted intact, mostly by the kidney. Cytotoxicity might be expected to occur given that increased hydrophobicity leads to greater tissue penetration. However, cellular studies have revealed that the compounds with the greatest therapeutic value at present, *N*B-DNJ and *N*B-DGJ, have extremely high CC_{50} values (concentration at which 50 per cent of cells become nonviable), which are in the mM range [39, 47]. As predicted, the more hydrophobic compounds are more cytotoxic and have lower CC_{50} values, again an important factor in therapeutic compound design. The toxicity associated with long chain N-alkylated iminosugars in cells is unrelated to inhibition of GSL biosynthesis, the generation of ceramide or any detergent-like effects. The major cause of cellular toxicity with these amphiphilic compounds appears to be cell lysis following increased membrane insertion and pore formation [48].

11.4.4 Side effects of iminosugars for SRT

Side effects are due to compounds having activities against other cellular enzymes and are distinct from cytotoxicity as discussed above. The inhibitory effect of iminosugars on processing α-glucosidases leads to misfolded glycoproteins that are unable to exert their biological functions. This also creates an opportunity that can be taken advantage of to treat other human diseases. Inhibition of glycoprotein processing leading to hyperglucosylation and prevention of calnexin-mediated interactions has been exploited for reducing viral infectivity [26, 47, 49, 50]. By contrast, if mutant proteins that are largely misfolded and eliminated from the ER by ER-associated degradation (ERAD) pathways are prevented from calnexin-mediated recognition and presentation necessary for ERAD, increased amounts of protein may be transported through the secretory pathway to their final destination. In this manner it is assumed that mutant cystic fibrosis transductance regulator (CFTR) has an increased ion channel activity in cells treated with *N*B-DNJ [51].

To completely block cleavage of glucosyl residues during biosynthesis of glycoproteins, concentrations of 1000- to 10 000-fold in excess of that necessary to inhibit

CGT is required. For global changes to the HIV gp120 protein conformation that are necessary to prevent gp41-assisted membrane fusion, concentrations in the mM range were required in cellular assays and, not surprisingly, this could not be achieved in humans, even at the highest possible dose (3 g/day). The observation that *N*B-DNJ is a far weaker α-glucosidase inhibitor in cells,compared with *in vitro* assays, because of it's restricted access to the ER, has been confirmed *in vivo* [40]. Using highly sensitive methods to determine the effects of *N*B-DNJ on glycosylation of proteins, we have been able to detect the degradation products of misfolded glycoproteins, free oligosaccharides (FOS), following their removal from the ER. Glucosidase inhibition by *N*B-DNJ generates glucosylated FOS [52] that can be analysed following extraction from cells, tissues and biological fluids, labelling with a fluorescence group and separation by HPLC. When plasma from a Niemann–Pick type C patient, treated with 100 mg/day *N*B- DNJ (Zavesca®) [53] was subjected to this analysis, glucosylated FOS were identified (Figure 11.4, Butters and Alonzi, unpublished).

Even at this relatively low dose, glucosidase inhibition in the ER was clearly evident, despite the failure of previous methods to detect changes in glycosylation at much higher doses *in vivo* ([54]. Further experiments are required to determine the precise amount of cellular and tissue specific glucosidase inhibition *in vivo*, but these data

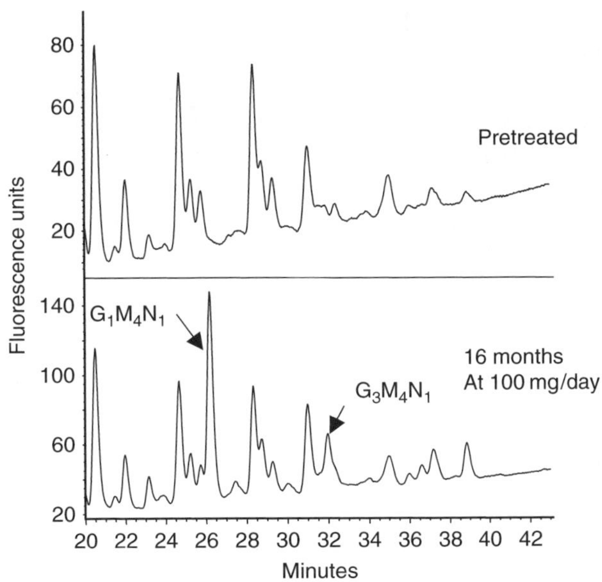

Figure 11.4 NP-HPLC of oligosaccharides detected in NP-C patient plasma. Plasma was analysed before (top panel) or following (bottom panel) treatment with *N*B-DNJ for 16 months at 100 mg/day. Free, glucose-containing oligosaccharides, $Glc_1Man_4GlcNAc_1$ ($G_1M_4N_1$) and $Glc_3Man_4GlcNAc_1$ ($G_3M_4N_1$) were detected in the treated sample only, indicating that ER-glucosidase activity was inhibited and misfolded protein removed via the ERAD pathway. These oligosaccharides are exquisite biomarkers for glucosidase inhibition and drug access and efficacy

support the potential for generating subtle effects on protein folding in the ER with low concentrations of iminosugars that may mediate an enhancement of glycosidase activity.

Other glucosidases, such as the gastrointestinal tract enzyme sucrase/isomaltase and the liver and muscle glycogenolytic enzymes [55] have also been shown to be sensitive to inhibition by *N*B-DNJ. In humans, the gastric distress induced by inhibition of disaccharidases appears to be transient and easily controlled by over-the-counter drugs [53]. Inhibition of the acid α-1, 4- and α-1, 6-glucosidase activity in mice treated with high doses of *N*B-DNJ does not lead to significant lysosomal storage of glycogen that would induce a disease phenotype characteristic of Pompe disease [55].

Galactosylceramide synthesis is not inhibited by *N*B-DNJ or *N*B-DGJ and these drugs therefore have no effects on myelin stability in neural tissue [40]. The reported peripheral neurological side effects observed in some patients treated with *N*B-DNJ, which was mild and axonal in origin, does not appear to be related to the potential of iminosugars to inhibit ceramide galactosyltransferase.

11.4.5 Preclinical studies of iminosugars for SRT

In vitro studies

The therapeutic potential of N-alkylated iminosugars with CGT inhibitory activity has been investigated using an *in vitro* model of Gaucher disease [35, 36]. To generate an authentic lipid storage macrophage cell phenotype, macrophages were treated with an irreversible inhibitor of β-glucocerebrosidase, conduritol ß-epoxide to reduce the activity of this enzyme to less than1 per cent of control. The coadministration of either *N*B-DNJ or *N*B-DGJ with CBE was able to prevent the lysosomal storage in these cells as detected by electron microscopy. Therefore, at the cellular level, the concept of SRT was experimentally valid. At 5 μM both compounds were equally effective at preventing GSL storage and correlated well with the observed K_i values for these compounds, as measured against isolated enzyme.

In vivo studies

*N*B-DNJ was evaluated in healthy mice to determine whether or not GSL depletion could be tolerated. By a number of criteria a significant tissue depletion of higher gangliosides, including GM2, (50–70 per cent in liver) was found when mice are fed on a diet of *N*B-DNJ. The compound was well tolerated, generated no osmotic diarrhoea, or outward signs of abnormal behaviour, but led to a 15 per cent weight loss after a dose of 1800 mg/kg/day at day 120 of the study [54]. Despite relatively high serum concentrations of compound (56.8 μM at 2400 mg/Kg/day) few effects due to the inhibition of α-glucosidase were evident in spleen cell membrane glycoproteins, whereas there was a nonselective decrease (60–70 per cent) of surface gangliosides.

When the drug was removed from the diet, the effects on lipid depletion were reversed and normal expression of liver GSL were found after a 2 week period. An increased conversion of ceramide to sphingomyelin followed inhibition of glucosylceramide synthesis in cells by *N*B-DNJ, as predicted by the ceramide biosynthetic pathway.

In the liver, a similar quantitative increase in sphingomyelin was observed confirming the specificity of *N*B-DNJ inhibition of GSL biosynthesis observed using *in vitro* assays and in cells.

In mouse models for lysosomal storage disease

Tay–Sachs disease results from a deficiency in β-hexosaminidase A (hex A) activity leading to the accumulation of ganglioside GM2 [56]. The accumulation is more pronounced in the brain because of the elevated synthesis of these ganglioside species in neural tissue. Mouse knockout models for this disorder have no hex A activity and accumulate GM2 in the lysosomes [57]. The presence of a sialidase activity in mouse tissue partially hydrolyses stored GM2 to asialoGM2 (AGM2) that is a substrate for ß-hexosaminidase B. Therefore in the mouse the storage levels of glycolipid never exceed the threshold required to provoke neuropathology or reduce the normal lifespan. Under conditions of stress, however, such as repeated breeding of females, clinical features can be induced, similar to those seen in the late onset variant of Tay–Sachs disease in humans [58].

The oral treatment of the Tay–Sachs mouse with *N*B-DNJ reduced ganglioside accumulation in the liver and, importantly, in the brain as demonstrated by GM2 analysis, histology and electron microscopy [33]. These were important proof of principle studies that showed following access of iminosugars via the blood–brain barrier, significant levels of drug were available to reduce the storage burden in the brain. In addition, these data made an evaluation of clinical efficacy a meaningful prospect for the neuronopathic disorders [59].

Further demonstration of the potency of iminosugars, in particular *N*B-DNJ, to prevent symptom onset dramatically was provided by the study of the effects of SRT in a symptomatic model of gangliosidosis, Sandhoff disease. In murine knockout models of disease, where both hex A and hex B activities are null, a severe phenotype is presented in more than one glycolipid species and other protein derived N-acetylhexosamine glycoconjugates. As a consequence Sandhoff mice show rapid, progressive neurodegeneration and die at 4–5 months of age. After treatment with *N*B-DNJ from 3–6 weeks of age mice were found to have reduced glycolipid storage in all tissues and showed an increased life expectancy of 40 per cent [43].

11.4.6 Clinical studies of SRT

The preclinical and clinical data from the HIV trial with *N*B-DNJ allowed a 12 month assessment for efficacy in type I Gaucher disease sponsored by Oxford GlycoSciences. Following publication of the results showing improvements of organ volumes and haematological parameters [60] (see also Chapter 13), *N*B-DNJ gained approval in Europe, USA and Israel for use for patients with this indication who were unable or unwilling to take ERT. Although the clinical benefit was deemed small and a number of issues regarding toxicity were noted [61], this proof of principle study has been extended to cover low dose administration [62] and a 3 year continuation study [63]. Statistically significant improvements in all the major efficacy changes were achieved indicating that consistent with the mechanism of action of Zavesca® as a modulator of GSL biosynthesis, treatment was increasingly effective with time. No serious adverse events have been reported; diarrhoea and weight loss decreased and no new case of peripheral neuropathy, noted in the first trial, has been found.

Given this relatively small group of patients, these data are remarkable in showing clinical improvement with a small dose of inhibitor. A treatment regime of 100–300 mg three times per day produced a stable plasma concentration of 6–8 μM, a concentration that was predicted partially to inhibit CGT leading to substrate reduction in cellular assays [35]. The safety and tolerability of long-term iminosugar treatment has been assessed and contra-indications for the use of *N*B-DNJ may be explained by underlying predilection for disease in the Gaucher population that complicates small clinical trials in seemingly monogenetic disorders [64–68]. The next challenge for iminosugar SRT will be treatment of the neuronopathogenic disorders and one recent study has demonstrated the efficacy of Zavesca® for these 'untreatable' disorders and provides further support for the translocation of iminosugars across the blood–brain barrier in species other than rodents [53]. A dose of 100 mg/day was sufficient to reduce the pathological burdens of GSL that are apparent in a Niemann–Pick C (NP-C) patient. The lipid disorder in NP-C is thought to be abnormal trafficking of cholesterol but a number of studies have identified the accumulation of GSL, and point to these as being the signal for downstream pathological events in neural tissue [31]. After 7 months therapy, approximately 20 per cent (0.6 μM) of the circulating plasma concentration (3 μM) was found in the CSF, similar to figures obtained when mice are orally administered *N*B-DNJ [53]. These data provide some optimism that other neuropathogenic disorders such as types II/III Gaucher disease and the gangliosidoses may respond to SRT. Recent evidence in support of this approach comes from SRT in the infantile form of Tay-Sachs disease [59]. The severe forms of the disease are usually fatal and are characterized by progressive neurological involvement. Management is limited to palliative care, but based on the preclinical evidence that SRT using *N*B-DNJ (Zavesca®) could successfully reduce the neuronal cell glycolipid burden in mouse models for Tay-Sachs and Sandhoff disease, where very little residual or compensating enzyme is detected, an evaluation in humans has promise. In this limited study with two patients, both failed to develop macrocephaly, a typical characteristic of the infantile disease and SRT slowed disease progression in one patient. Importantly, however, significant concentrations of *N*B-DNJ were measured in cerebrospinal fluid (approximately 3 μM) which may offer some further prospects for depleting neuronally stored GSL in all the gangliosphingolipidoses.

11.4.7 Current use and exposure of *N*B-DNJ for SRT

To date, the total number of patients enrolled in clinical trials for SRT using *N*B-DNJ as monotherapy is 200, and of these 99 are type I Gaucher disease, 30 are type III Gaucher disease, 41 are Niemann-Pick type C (29 adult and juvenile, 12 pediatric) and 30 GM2 gangliosidosis. In addition, during the period 20 October, 2004, to 19 October,

2005, 182 patients have been treated with the commercial drug formulation of *N*B-DNJ (Miglustat), Zavesca®, marketed by Actelion.

A post-marketing surveillance programme (Intensive Safety Surveillance Scheme (IS3) was launched in March, 2003, at the time of Zavesca® marketing in the EU. IS3 is a web-based tool that collects comprehensive baseline and ongoing information with the focus on neurological and skeletal symptoms. The treating physician submits these data on a voluntary basis. Of the 126 patients included in the IS3 programme up to April, 2006, 63 patients (50 per cent) had type I Gaucher disease and 63 patients (50 per cent) were treated for other indications: seven type 3 Gaucher disease, 26 Niemann- Pick C, 10 GM1 gangliosidosis, 19 GM2 gangliosidosis and one not coded.

Overall exposure to Zavesca® represents a cumulative post-marketing experience of 83 patient-years in type I Gaucher disease. This is a significant number of individuals with an exposure to *N*B-DNJ and with the continuing treatment of at least two patients from the original trial in 1999 [63], a significant body of data now supports the use of this drug for SRT in all the glycosphingolipidoses. Changing the label to allow type I Gaucher patients who are currently on ERT is an achievable goal and offers a choice of treatment regimes where previously none was available. A combination of clinical approaches may provide the most favourable outcome for successful therapy and these are discussed below.

11.5 SRT combination therapy

11.5.1 Bone marrow transplantation (BMT)

Complementation of this approach with enzyme delivered by bone marrow transplantation (BMT) provided a significant (25 per cent) synergistic outcome in the more severe Sandhoff disease mouse model [69]. Disease progression is determined by the amount of residual, or augmented enzyme in the case of BMT, and even a small increase in enzyme activity is profoundly effective. One case for using *N*B-DNJ following BMT in a sub-acute paediatric case of Tay-Sachs disease has been reported [70].

11.5.2 Antiinflammatory agents

For infantile disorders a strategy of enzyme supplementation in addition to SRT should be beneficial, but until storage levels can be reduced to levels where the inflammatory component either in the brain, or signalled from peripheral organs, has been lowered, this remains an experimental alternative.

A combination of antiinflammatory agents, NSAIDS, and SRT was effective in the Sandhoff disease mouse model at extending the presymptomatic phase from 65 days in the untreated mice to 115 days in the combination group and resulted in a significant increase in survival [71]. This synergistic effect, 6 per cent in the treatment group, can be improved further by optimization of SRT and NSAID dose but offers a further prospect for alternative therapy for the gangliosidoses.

11.5.3 ERT

The potential for coadministration of enzyme and *N*B-DNJ also offers additional benefit in adjusting the treatment regime to disease type and severity. Although there are no clinical data to support this use in Gaucher disease, where ERT and SRT are available, experimental data indicate that a combination may be complementary. At the current doses given to humans (3×100 mg/day) a plasma circulating concentration of 5–6 μM is sufficient to inhibit CGT partially ($IC_{50} = 20\,\mu M$) [1, 72], whereas inhibition of β-glucocerebrosidase requires 25-fold higher concentrations ($IC_{50} = 520\,\mu M$). When coadministered to normal mice, *N*B-DNJ increased both peak activity and half-life of β-glucocerebrosidase [73] and could indicate that some positive chaperone-mediated effects on the normal enzyme are possible with *N*B-DNJ.

11.5.4 CMT

Chaperone mediated therapy (CMT) relies on reversible tight binding constants for an iminosugar inhibitor and its enzyme substrate (see Chapter 10). Since many iminosugars are active site-directed with good affinities (10^{-6} to 10^{-9} M) co- or posttranslational binding to the enzyme may provide sufficient protection against misfolding or inactivation. CMT could be used to provide an improvement in folding of the enzyme in the endoplasmic reticulum (ER) or assist transfer to the lysosome [34]. Several β-glucosidase inhibitors have been designed to bind at noninhibitory levels to enhance enzyme activity in the lysosome [74] and a recent attempt to synthesize inhibitors that can discriminate between α- and β-glucosidases has been reported as successful [75].

*N*B-DNJ is proposed to be an active site-directed inhibitor of β-glucosidase and some data support this [76]. As such, an improvement in enzyme activity was observed in cells transfected with ß-glucosidase with Gaucher specific mutations. At low concentrations, 10 μM, *N*B-DNJ increased the activity of several mutant enzymes, including the common N370S [76]. The binding of *N*B-DNJ and similar glucose analogues to additional regions of the enzyme to the active site cannot be discounted, and may be advantageous to protocols for therapy when inhibitor and enzyme need to be dissociated in the lysosome [34]. Interestingly, longer chain DNJ-based compounds are found to be more effective at chaperoning Gaucher disease mutant ß-glucosidase in [74, 75]. The combination of SRT and CMT effect by iminosugars is therefore predicted to provide greater efficacy for reducing the lysosomally stored material.

11.6 Prospects for iminosugars as therapeutics

Published and current clinical trials with *N*B-DNJ to treat specific cases of disease provide further optimism for continued and long-term use in glycosphingolipidoses patients where neuropathology remains a major obstacle for conventional treatment. Refinements to this drug may be necessary to further decrease the side effects shown by Zavesca®, particularly where paediatric use is indicated, but more potent drugs may not be required to obtain corrective treatment. Access of *N*B-DNJ to the central nervous tissue is limited and partially effective so a further understanding for the basis of cell and organelle is necessary before small molecules can be designed for optimal use in the neuropathogenic disorders.

Several iminosugars are in preclinical or Phase I/II clinical trials for an evaluation of CMT for the glycosphingolipidoses leading to an increase in the options to treat specific diseases. Iminosugar exposure to patients with different disease phenotypes is expected to increase with both SRT and CMT and will provide more tolerable therapeutic alternatives for treating the glycosphingolipidoses.

Acknowledgements

The author thanks Raymond Dwek for reviewing the manuscript, the Glycobiology Institute Endowment for support and Actelion Pharmaceuticals for providing data relating to the clinical use of Zavesca®.

References

1. Butters, T. D., Dwek, R. A. and Platt, F. M. (2000) *Chem. Rev.*, **100**, 4683–4696.
2. Meikle, P. J., Hopwood, J. J., Clague, A. E. and Carey, W. F. (1999) *J. Am. Med. Assoc.*, **281**, 249–254.
3. Beutler, E. and Gelbart, T. (1997) *Blood Cells Mol. Dis.*, **23**, 2–7.
4. Miao, S. C., Mccarter, J. D., Grace, M. E., Grabowski, G. A., Aebersold, R. and Withers, S. G. (1994) *J. Biol. Chem.*, 269, 10975–10978.
5. Dvir, H., Harel, M., McCarthy, A. A., Toker, L., Silman, I., Futerman, A. H. and Sussman, J. L. (2003) *EMBO Rep.*, **4**, 704–709.
6. Liou, B., Kazimierczuk, A., Zhang, M., Scott, C. R., Hegde, R. S. and Grabowski, G. A. (2006) *J. Biol. Chem.*, **281**, 4242–4253.
7. Salvioli, R., Tatti, M., Scarpa, S., Moavero, S. M., Ciaffoni, F., Felicetti, F., Kaneski, C. R., Brady, R. O. and Vaccaro, A. M. (2005) *Biochem. J.*, **390**, 95–103.
8. Legler, G. and Liedtke, H. (1985) *Biol. Chem. Hoppe-Seyler*, **366**, 1113–1122.
9. Zhu, X., Sheth, K. A., Li, S., Chang, H. H. and Fan, J. Q. (2005) *Angew. Chem. Int. Ed. Engl.*, **44**, 7450–7453.
10. Aerts, J. M., Hollak, C., Boot, R. and Groener, A. (2003) *Phil. Trans. R. Soc. Lond. B Biol. Sci.*,**358**, 905–914.
11. Lichtenstein, M., Zimran, A. and Horowitz, M. (1997) *Blood Cells Mol. Dis.*, **23**, 395–401.
12. Jeyakumar, M., Thomas, R., Elliot-Smith, E., Smith, D. A., van der Spoel, A. C., d'Azzo, A., Perry, V. H., Butters, T. D., Dwek, R. A. and Platt, F. M. (2003) *Brain*, **126**, 974–987.
13. Brady, R. O. (2006) *Ann. Rev. Med.*, **57**, 283–296.
14. Sands, M. S. and Davidson, B. L. (2006) *Mol. Ther.*, **13**, 839–849.
15. Vogler, C., Levy, B., Grubb, J. H., Galvin, N., Tan, Y., Kakkis, E., Pavloff, N. and Sly, W. S. (2005) *Proc. Nat. Acad. Sci. USA*, **102**, 14777–14782.
16. Erikson, A., Forsberg, H., Nilsson, M., Astrom, M. and Mansson, J. E. (2006) *Acta Paediatr.*, **95**, 312–317.
17. Vunnam, R. R. and Radin, N.S. (1980) *Chem. Phys. Lipids*, **26**, 265–278.

18. Elbein, A. D. (1991) *FASEB J.*, **5**, 3055–3063.
19. Fleet, G. W., Karpas, A., Dwek, R. A., Fellows, L. E., Tyms, A. S., Petursson, S., Namgoong, S. K., Ramsden, N. G., Smith, P. W., Son, J. C. *et al.* (1988) *FEBS Lett.*, **237**, 128–132.
20. Karpas, A., Fleet, G. W., Dwek, R. A., Petursson, S., Namgoong, S. K., Ramsden, N. G., Jacob, G. S. and Rademacher, T. W. (1988) *Proc. Nat. Acad. Sci. USA*, **85**, 9229–9233.
21. Tierney, M., Pottage, J., Kessler, H., Fischl, M., Richman, D., Merigan, T., Powderly, W., Smith, S., Karim, A., Sherman, J. *et al.* (1995) *J. Acquir. Immune Defic. Syndr. Hum. Retrovirol.*, **10**, 549–553.
22. Fischl, M. A., Resnick, L., Coombs, R., Kremer, A. B., Pottage Jr., J. C., Fass, R. J., Fife, K. H., Powderly, W. J., Collier, A. C., Aspinall, R. L., Smith, S. L., Kowalski, K. G. and Wallemark, C. -B. (1994) *J. Acquir. Immune Defic. Syndr.*, **7**, 139–147.
23. Grabner, R. W., Landis, B. H., Wang, P. T., Prunier, M. L. and Scaros, M. G. (1992) European Patent EPO477160 Monsanto (US), Searle (US).
24. Jacob, G. S., Scudder, P., Butters, T.D., Jones, I. and Tiemeier, D.C. (1992) Aminosugar attenuation of HIV infection, in *Natural Products as Antiviral Agents* (Chu, C. K. and Cutler, H. G., eds), Plenum Press, New York, USA.
25. Helenius, A. and Aebi, M. (2001) *Science*, **291**, 2364–2369.
26. Fischer, P. B., Collin, M., Karlsson, G. B., James, W., Butters, T. D., Davis, S. J., Gordon, S., Dwek, R. A. and Platt, F. M. (1995) *J. Virol.*, **69**, 5791–5797.
27. Fischer, P. B., Karlsson, G. B., Butters, T. D., Dwek, R. A. and Platt, F. M. (1996) *J. Virol.*, **70**, 7143–7152.
28. Dwek, R. A., Butters, T. D., Platt, F. M. and Zitzmann, N. (2002) *Nat. Rev. Drug Discov.*, **1**, 65–75.
29. Proia, R. L. (2003) *Phil. Trans. R. Soc. Lond. B*, **358**, 879–883.
30. Simpson, M. A., Cross, H., Proukakis, C., Priestman, D. A., Neville, D. C., Reinkensmeier, G., Wang, H., Wiznitzer, M., Gurtz, K., Verganelaki, A., Pryde, A., Patton, M. A., Dwek, R. A., Butters, T. D., Platt, F. M. and Crosby, A. H. (2004) *Nat. Genet.*, **36**, 1225–1229.
31. Jeyakumar, M., Butters, T. D., Dwek, R. A. and Platt, F. M. (2002) *Neuropathol. Appl. Neurobiol.*, **28**, 343–357.
32. Platt, F. M., Jeyakumar, M., Andersson, U., Heare, T., Dwek, R. A. and Butters, T. D. (2003) *Phil. Trans. R. Soc. Lond. B Biol. Sci.*, **358**, 947–954.
33. Platt, F. M., Neises, G. R., Reinkensmeier, G., Townsend, M. J., Perry, V. H., Proia, R. L., Winchester, B., Dwek, R. A. and Butters, T. D. (1997) *Science*, **276**, 428–431.
34. Butters, T. D., Dwek, R. A. and Platt, F. M. (2005) *Glycobiology*, **15**, 43R-52R.
35. Platt, F. M., Neises, G. R., Dwek, R. A. and Butters, T. D. (1994) *J. Biol. Chem.*, **269**, 8362–8365.
36. Platt, F. M., Neises, G. R., Karlsson, G. B., Dwek, R. A. and Butters, T. D. (1994) *J. Biol. Chem.*, **269**, 27108–27114.
37. Butters, T. D., van den Broek, L. A. G. M., Fleet, G. W. J., Krulle, T. M., Wormald, M. R., Dwek, R. A. and Platt, F. M. (2000) *Tetrahedron: Asymmetry*, **11**, 113–124.
38. Boucheron, C., Desvergnes, V., Compain, P., Martin, O. R., Lavi, A., Mackeen, M., Wormald, M. R., Dwek, R. A. and Butters, T. D. (2005) *Tetrahedron: Asymmetry*, **16**, 1747–1756.
39. Mellor, H. R., Nolan, J., Pickering, L., Wormald, M. R., Platt, F. M., Dwek, R. A., Fleet, G. W. and Butters, T. D. (2002) *Biochem. J.*, **366**, 225–233.
40. Butters, T. D., Mellor, H. R., Narita, K., Dwek, R. A. and Platt, F. M. (2003) *Phil. Trans. R. Soc. Lond. B Biol. Sci.*, **358**, 927–945.
41. Andersson, U., Butters, T. D., Dwek, R. A. and Platt, F. M. (2000) *Biochem. Pharmacol.*, **59**, 821–829.

42. Andersson, U., Smith, D., Jeyakumar, M., Butters, T. D., Borja, M. C., Dwek, R. A. and Platt, F. M. (2004) *Neurobiol. Dis.*, **16**, 506–515.
43. Jeyakumar, M., Butters, T. D., CortinaBorja, M., Hunnam, V., Proia, R. L., Perry, V. H., Dwek, R. A. and Platt, F. M. (1999) *Proc. Nat. Acad. Sci. USA*, **96**, 6388–6393.
44. Mellor, H. R., Neville, D. C., Harvey, D. J., Platt, F. M., Dwek, R. A. and Butters, T. D. (2004) *Biochem. J.*, **381**, 861–866.
45. Andersson, U., Butters, T. D., Dwek, R. A. and Platt, F. M. (2000) *Biochem. Pharmacol.*, **59**, 821–829.
46. van der Spoel, A. C., Jeyakumar, M., Butters, T. D., Charlton, H. M., Moore, H. D., Dwek, R. A. and Platt, F. M. (2002) *Proc. Nat. Acad. Sci. USA*, **99**, 17173–17178.
47. Durantel, D., BranzaNichita, N., CarroueeDurantel, S., Butters, T. D., Dwek, R. A. and Zitzmann, N. (2001) *J. Virol.*, **75**, 8987–8998.
48. Mellor, H. R., Platt, F. M., Dwek, R. A. and Butters, T. D. (2003) *Biochem. J.*, **374**, 307–314.
49. Block, T. M., Lu, X. Y., Mehta, A. S., Blumberg, B. S., Tennant, B., Ebling, M., Korba, B., Lansky, D. M., Jacob, G. S. and Dwek, R. A. (1998) *Nature Med.*, **4**, 610–614.
50. Zitzmann, N., Mehta, A. S., Carrouee, S., Butters, T. D., Platt, F. M., McCauley, J., Blumberg, B. S., Dwek, R. A. and Block, T. M. (1999) *Proc. Nat. Acad. Sci. USA*, **96**, 11878–11882.
51. Norez, C., Noel, S., Wilke, M., Bijvelds, M., Jorna, H., Melin, P., DeJonge, H. and Becq, F. (2006) *FEBS Lett.*, **580**, 2081–2086.
52. Mellor, H. R., Neville, D. C., Harvey, D. J., Platt, F. M., Dwek, R. A. and Butters, T. D. (2004) *Biochem. J.*, **381**, 867–875.
53. Lachmann, R. H., Te Vruchte, D., Lloyd-Evans, E., Reinkensmeier, G., Sillence, D. J., Fernandez-Guillen, L., Dwek, R. A., Butters, T. D., Cox, T. M. and Platt, F. M. (2004) *Neurobiol. Dis.*, **16**, 654–658.
54. Platt, F. M., Reinkensmeier, G., Dwek, R. A. and Butters, T. D. (1997) *J. Biol. Chem.*, **272**, 19365–19372.
55. Andersson, U., Reinkensmeier, G., Butters, T. D., Dwek, R. A. and Platt, F. M. (2004) *Biochem. Pharmacol.*, **67**, 697–705.
56. Gravel, R. A., Clarke, J. T. R., Kaback, M. M., Mahuran, D., Sandhoff, K. and Suzuki, K. (1995) The GM2 Gangliosidoses, in *The Metabolic Bases of Inherited Disease*, 7th edn (Scriver, C. R., Beaudet, A. L., Sly, W. S. and Valle, D., eds), McGraw-Hill, New York, USA.
57. Yamanaka, S., Johnson, M. D., Grinberg, A., Westphal, H., Crawley, J. N., Taniike, M., Suzuki, K. and Proia, R. L. (1994) *Proc. Nat. Acad. Sci. USA*, **91**, 9975–9979.
58. Jeyakumar, M., Smith, D., Eliott-Smith, E., Cortina-Borja, M., Reinkensmeier, G., Butters, T. D., Lemm, T., Sandhoff, K., Perry, V. H., Dwek, R. A. and Platt, F. M. (2002) *Neurobiol. Dis.*, **10**, 201–210.
59. Bembi, B., Marchetti, F., Guerci, V. I., Ciana, G., Addobbati, R., Grasso, D., Barone, R., Cariati, R., Fernandez-Guillen, L., Butters, T. and Pittis, M. G. (2006) *Neurology*, **66**, 278–280.
60. Cox, T., Lachmann, R., Hollak, C., Aerts, J., van Weely, S., Hrebicek, M., Platt, F., Butters, T., Dwek, R., Moyses, C., Gow, I., Elstein, D. and Zimran, A. (2000) *Lancet*, **355**, 1481–1485.
61. Pastores, G. M., and Barnett, N. L. (2003) *Expert Opin. Investig. Drugs*, **12**, 273–281.
62. Heitner, R., Elstein, D., Aerts, J., van Weely, S. and Zimran, A. (2002) *Blood Cells Mol. Dis.*, **28**, 127–133.
63. Elstein, D., Hollak, C., Aerts, J. M., van Weely, S., Maas, M., Cox, T. M., Lachmann, R. H., Hrebicek, M., Platt, F. M., Butters, T. D., Dwek, R. A. and Zimran, A. (2004) *J. Inherit. Metab. Dis.*, **27**, 757–766.
64. Aharon-Peretz, J., Badarny, S., Rosenbaum, H. and Gershoni-Baruch, R. (2005) *Neurology*, **65**, 1460–1461.

65. Aharon-Peretz, J., Rosenbaum, H., and Gershoni-Baruch, R. (2004) New England J Med 351, 1972–1977
66. Clark, L. N., Nicolai, A., Afridi, S., Harris, J., Mejia-Santana, H., Strug, L., Cote, L. J., Louis, E. D., Andrews, H., Waters, C., Ford, B., Frucht, S., Fahn, S., Mayeux, R., Ottman, R. and Marder, K. (2005) *Mov. Disord.*, **20**, 100–103.
67. Halperin, A., Elstein, D. and Zimran, A. (2006) *Blood Cells Mol. Dis.*, **36**, 426–428.
68. Spitz, M., Rozenberg, R., Silveira, P. A. and Barbosa, E. R. (2006) *J. Neurol. Neurosurg. Psychiatry*, **77**, 709–710.
69. Jeyakumar, M., Norflus, F., Tifft, C. J., CortinaBorja, M., Butters, T. D., Proia, R. L., Perry, V. H., Dwek, R. A. and Platt, F. M. (2001) *Blood*, **97**, 327–329.
70. Jacobs, J. F., Willemsen, M. A., Groot-Loonen, J. J., Wevers, R. A. and Hoogerbrugge, P. M. (2005) *Bone Marrow Transplant*, **36**, 925–926.
71. Jeyakumar, M., Smith, D. A., Williams, I. M., Borja, M. C., Neville, D. C., Butters, T. D., Dwek, R. A. and Platt, F. M. (2004) *Ann. Neurol.*, **56**, 642–649.
72. Butters, T. D., Dwek, R. A. and Platt, F. M. (2003) *Curr. Top. Med. Chem.*, **3**, 561–574.
73. Priestman, D. A., Platt, F. M., Dwek, R. A. and Butters, T. D. (2000) *Glycobiology*, **19**, iv-vi.
74. Sawkar, A. R., Cheng, W. C., Beutler, E., Wong, C. H., Balch, W. E. and Kelly, J. W. (2002) *Proc. Nat. Acad. Sci. USA*, **99**, 15428–15433.
75. Compain, P., Martin, O. R., Boucheron, C., Godin, G., Yu, L., Ikeda, K. and Asano, N. (2006) *ChemBioChem.*, **7**, 1356–1359.
76. Alfonso, P., Pampin, S., Estrada, J., Rodriguez-Rey, J. C., Giraldo, P., Sancho, J. and Pocovi, M. (2005) *Blood Cells Mol. Dis.*, **35**, 268–276.

12

Iminosugar-based antitumoural agents

Yoshio Nishimura

12.1 Introduction

Iminosugars, carbohydrate analogues that most frequently carry nitrogen at the position of the endocyclic oxygen atom, form the most attractive class of glycomimetics. Several types of iminosugars have been discovered from natural sources or have been synthesized as probes of the mechanism of enzymatic glycoside hydrolysis. Now they have gained a remarkable importance as new therapeutic agents in a wide range of diseases.

Currently, cell-surface carbohydrates of glycoconjugates such as glycoproteins, glycolipids and proteoglycans are known as bioregulators which mediate cell–cell communication and recognition, cell adhesion and growth regulation, differentiation and transport. Ubiquitous cell-surface carbohydrates are increasingly thought to have a significant role in regulating a wide variety of biological processes, including the immune response, angiogenesis, oncogenesis and tumour metastasis. Enzymes of oligosaccharide metabolism take part in the control of these functions. Specific inhibitors of these enzymes are useful for unravelling how glycoconjugates regulate biological functions, and also in developing new drugs for serious diseases associated with metabolism of glycoconjugates, including cancer, tumour metastasis and inflammatory disorders.

Tumour cell surface oligosaccharides are remarkably distinct from those of normal cells. Malignant transformation is associated with a variety of structural alterations in the size and composition of the carbohydrates of glycoconjugates on the tumour cell surface [1–4]. Tumour cell surface oligosaccharides have been shown to play an important role in expression of the malignant phenotype and the metastatic spread of tumour cells. The N-linked oligosaccharides of malignant and transformed cells are generally larger, highly-branched and sialylated at their chain termini. Oligosaccharide

Iminosugars: From synthesis to therapeutic applications Edited by Philippe Compain and Olivier R. Martin
 ISBN 978-0-470-03391-3

synthesis in the endoplasmic reticulum and Golgi depends on carbohydrate processing enzymes such as glycosidases as well as glycosyltransferases [5, 6]. Recent biochemical studies have also revealed that cellular function and phenotype are highly influenced by heparan sulphate proteoglycans (HSPGs) of extracellular matrix (ECM), and the enzymatic degradation of ECM is involved in fundamental biological phenomena, including angiogenesis and cancer metastasis [7–11]. Proteolytic enzymes (heparanase and matrix metalloproteinases (MMPs)) secreted by tumour cells are capable of degrading extracellular matrix (ECM) and basement membrane components, and their activities are closely related to the metastatic potential of malignant cells [12–16].

The metastatic potential of cancer is an important determinant for morbidity and mortality. The metastatic process is thought to consist of a number of distinct steps: proliferation, local invasion, intravasation, migration and survival in the blood or lymphatic circulation, arrest in the endothelial basement membrane, extravasation and proliferation in the target tissue (Figure 12.1). Tumour cell invasion is a critical event in the process of metastasis and involves cell–cell interaction by cellular adhesion, chemotactic response and degradation of various constituents of the basement membrane.

New drugs aimed at the enzymes involved in the metabolism and/or catabolism of oligosaccharides of glycoconjugates using iminosugars as sugar mimics have opened up a new approach for anticancer therapy [17]. This chapter describes current progress in the chemistry, biochemistry and pharmacology of iminosugars, focusing mainly on identifying novel therapeutic agents against tumour metastasis.

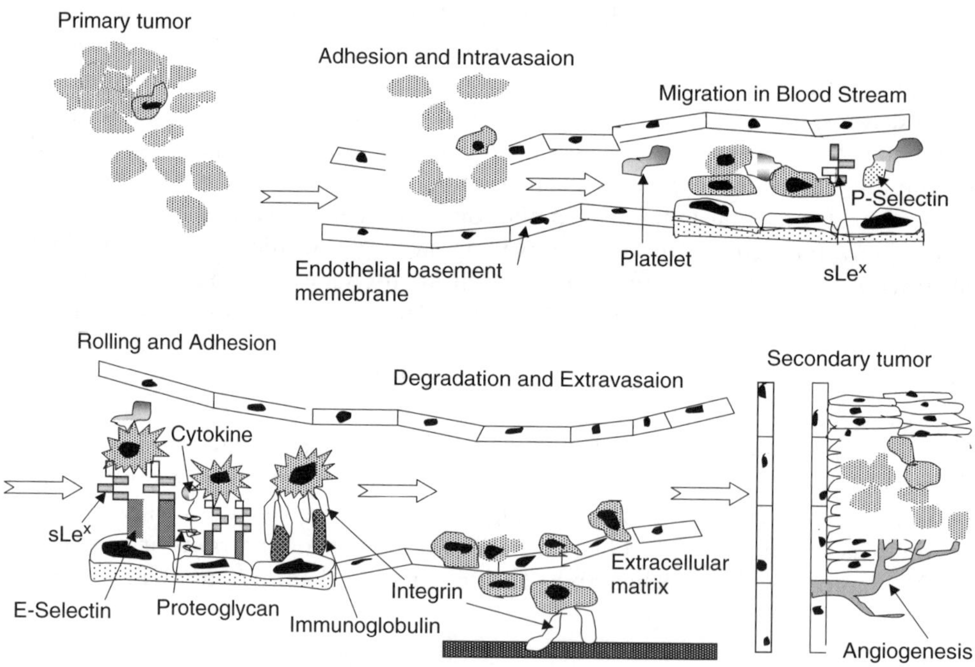

Figure 12.1 The process of tumour cell metastasis

12.2 Hexose-mimetic iminosugars

Monocyclic and bicyclic hexose-mimicking iminosugars have been isolated from microorganism cultures and plants. These compounds are well known as competitive inhibitors of glycosidases from various organisms, and several of them interfere specifically with the processing of asparagine-linked oligosaccharides [6, 18–21]. Hexose-mimicking iminosugars have a common structural feature in that they resemble the corresponding glycoses, while they specifically inhibit the corresponding glycosidase. They are useful molecular tools for probing several important biological processes, and are available as chemicals in medical applications.

Neoplastic transformation in human tumour cells often increases the degree of β1-6-N-acetylglucosamine branching at the trimannosyl core of N-linked oligosaccharides (GlcNacβ1-6Manα1-6Manβ1) (Figure 12.2) which are required for tumour cell

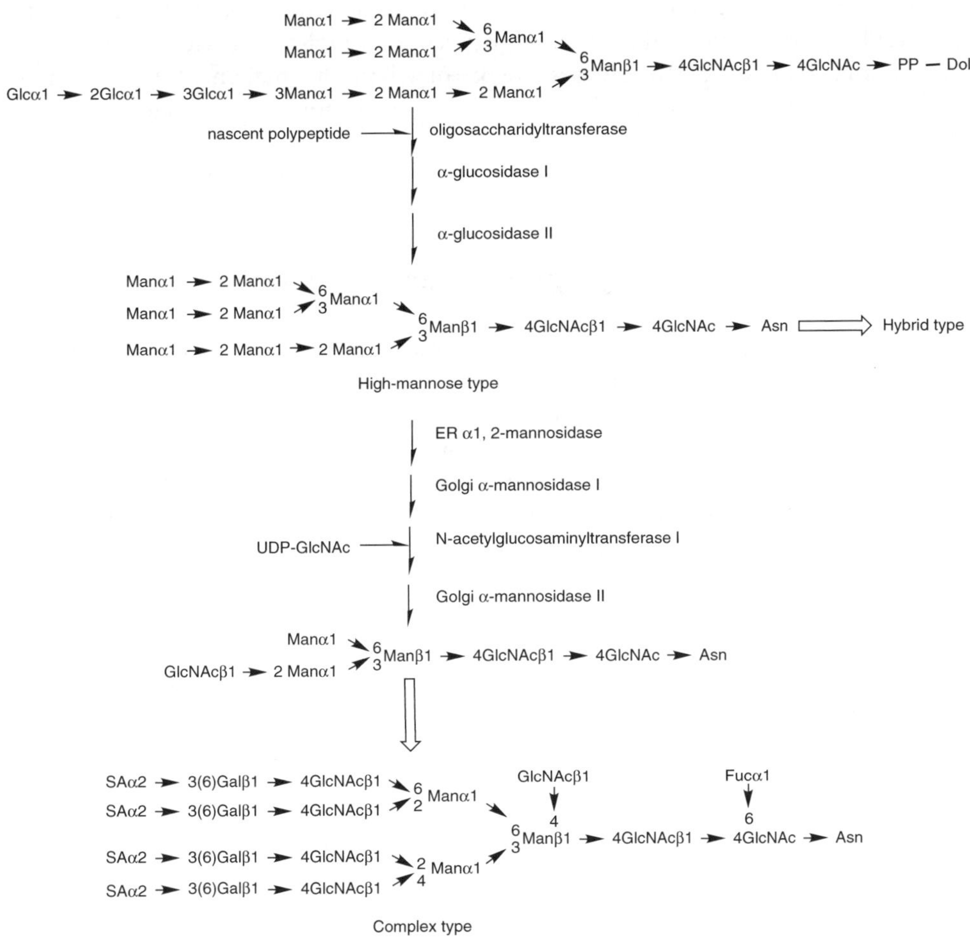

Figure 12.2 Schematic pathway of asparagine-linked oligosaccharide biosynthesis and processing

invasion and efficient tumour metastasis [22–27]. The biosynthesis of N-linked oligosaccharides begins in the lumen of the rough endoplasmic reticulum (ER) where the lipid-linked high-mannose oligosaccharide, $Glc_3Man_9GlcNAc_2$-(PP-Dol), is transferred to an asparagine residue in a nascent polypeptide chain by a protein oligosaccharidyltransferase (Figure 12.2). During transfer from the ER to the Golgi apparatus, the structure of the precursor, $Glc_3Man_9GlcNAc_2$-Asn is modified and altered by processing, such as trimming and glycosylation reactions. Glucosidase I removes the terminal α 1,2-glucose moiety, and then glucosidase II removes the two inner α 1,3-glucose residues to leave a high-mannose type oligosaccharide. ER α 1,2-mannosidase catalyses the removal of at least one α 1,2-mannose moiety. Golgi mannosidase I further removes three α 1,2-mannose residues and, after addition of GlcNAc by N-acetyglucosaminyltransferase I, α 1,3- and α 1,6-mannose residue are removed by Golgi mannosidase II to create a $GlcNAcMan_3GlcNAc_2$ oligosaccharide. Glycosyltransferases, such as N-acetylglucosaminyltransferase, fucosyltransferase, galactosyltransferase and sialyltransferase, may act sequentially at the late stages to transfer the corresponding sugar moieties to a $GlcNAcMan_3GlcNAc_2$ oligosaccharide intermediate, resulting in the formation of several types of complex oligosaccharide. As both catabolic and processing glycosidases are involved in the malignant transformation of cells, the specific glycosidase inhibitors may be promising candidates for cancer chemotherapy.

12.2.1 Nojirimycin, 1-deoxynojirimycin and mannojirimycin

Nojirimycin (NJ), 1-deoxynojirimycin (DNJ) and mannojirimycin (MJ) were isolated from microorganism cultures and/or plants [19, 21]. They all resemble monosaccharides in which the ring oxygen is replaced by nitrogen (Figure 12.3). NJ, DNJ and MJ are, due to their configurational similarity, potent and specific competitive inhibitors for D-glucosidase and D-mannosidase from various organisms. DNJ was also found to show marked inhibition against trimming glucosidase I (IC_{50} 2 μM) and II (IC_{50} 20 μM), which inhibits the formation of complex oligosaccharides from high-mannose oligosaccharides, and causes a decrease in complex oligosaccharides and an increase in high-mannose oligosaccharides [28]. NJ also shows inhibition against glucosidase I (IC_{50} 0.16 mM). Deoxymannojirimycin (DMJ), an analogue of MJ, specifically inhibits Golgi α-mannosidase IA/B, among Golgi α-mannosidases I and II, and conversion of high-mannose to complex oligosaccharide during glycoprotein processing [29]. Due to their remarkable biological properties, they have become an attractive synthetic target, reflected by various synthetic approaches, such as convergent synthesis starting from carbohydrates, asymmetric synthesis from noncarbohydrate precursors, and chiral and chemo-enzymatic synthesis (see Chapter 3 of this volume) [19, 30, 31]. NJ, DNJ and MJ inhibit experimental pulmonary metastasis of B16 variant cells in mice (Table 12.1), suggesting inhibition of the formation of complex oligosaccharides from high-mannose oligosaccharides by glucosidase I or mannosidase IA/B [32]. NJ shows 94 per cent inhibition of experimental pulmonary metastasis of

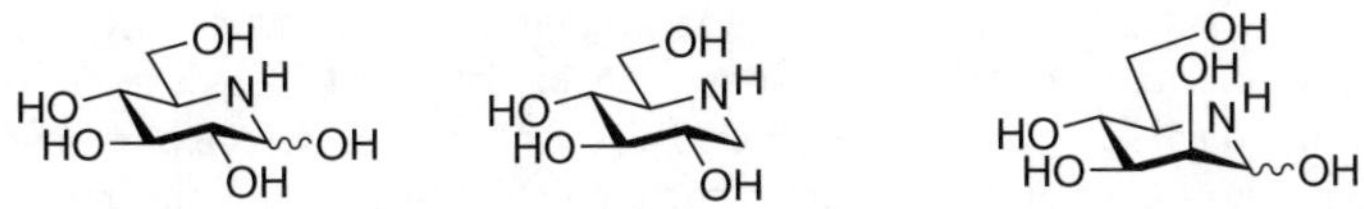

Nojirimycin (NJ) Deoxynojirimycin (DNJ) Mannojirimycin (MJ)

Figure 12.3 Structures of nojirimycin (NJ), deoxynojirimycin (DNJ) and mannojirimycin (MJ)

Table 12.1 Inhibition of experimental pulmonary metastasis of B16 variant cells by *in vitro* treatment with nojirimycin analogues. B16 variant cells were cultured with compounds for 3 days. The cells (5×10^4) were implanted i.v. into the tail vein of mice; 14 days later, the lung tumour colonies were counted

Compounds	Concentration (μg/ml)	Inhibition (%)
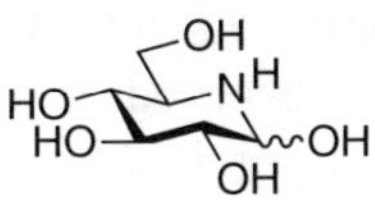 Nojirimycin (NJ)	10	98.7
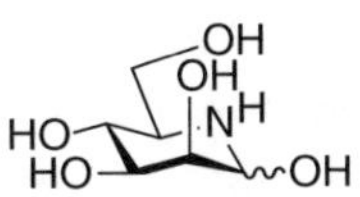 Mannojirimycin (MJ)	10	57.1
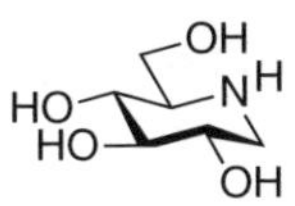 Deoxynojirimycin (DNJ)	10	80.5

B16-F10 cells in mice, by *in vitro* treatment of the cells with NJ (32 μg/ml). NJ also shows dose-dependent inhibition in the range of 65–80 per cent and 40–75 per cent by intraperitoneal administration for the complete period (day −2 to day 14) and for the early period (day −2 to day 2), respectively (Table 12.2) [33]. NJ-treated B16-F10 cells were less susceptible to concanavalin A, suggesting modification of high-mannose asparagine-linked oligosaccharides on the cell surface. NJ-treated B16-F10 cells also exhibited reduction of binding ability to endothelial cells, and NJ elicited no substantial increase in the cytotoxic activity of splenic NK cells against YAC-1 target cells [33]. These results indicate that NJ modifies specific structures of the cell surface carbohydrates, and acts on the process of adhesion, thereby inhibiting pulmonary metastasis.

Table 12.2 The effects of nojirimycin treatment on melanoma metastasis. Mice were treated with two daily injections with or without NJ at dose schedules of (1) from day −2 to 14 days after B16-F10 cell challenge (complete period), (2) from day −2 until 2 days after challenge (early period). The mice were challenged intravenously with 1×10^5 B16-F10 cells at day 0, and were examined for pulmonary metastases at day 15

Schedule	Dose (mg/kg/day)	Inhibition of metastasis (%)
1	4	80
2	4	75

12.2.2 Swainsonine and castanospermine

Swaisonine (SW) and castanospermine (CS) have often been found in plants but rarely in microorganism cultures [6, 19, 21, 34–37]. SW structurally resembles D-mannose in the furanose form. The protonated form of SW is also believed to be structurally similar to the mannosyl oxocarbenium ion, a proposed intermediate in mannosyl transfer (Figure 12.4). On the other hand, the CS structure resembles D-glucose in the pyranose form (Figure 12.5). SW and CS are potent and specific competitive inhibitors of D-mannosidase and D-glucosidase from various organisms, respectively. SW prevents the formation of complex glycoproteins by inhibition of α-1,3- and α-1,6-mannosidase activity of Golgi mannosidase II ($K_i = 40$ nM), leading to formation of hybrid-type oligosaccharides [34–37]. On the other hand, CS inhibits glycoprotein-processing glucosidase I, causing accumulation of the high mannose-type oligosaccharides, unprocessed oligosaccharide chains terminating in mannose [38–42]. Due to a wide range of interesting biological effects and to the 1-azabicyclic polyhydroxylated alkaloid structure, SW and CS have attracted considerable synthetic interest, resulting in a large number of different synthetic approaches [19, 30, 31, 43–47].

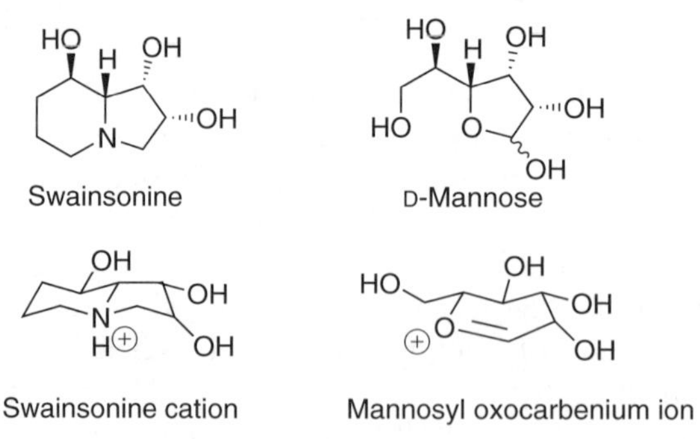

Figure 12.4 Structural resemblance of swainsonine (SW) and swainsonine cation to D-mannose and mannosyl oxocarbenium ion

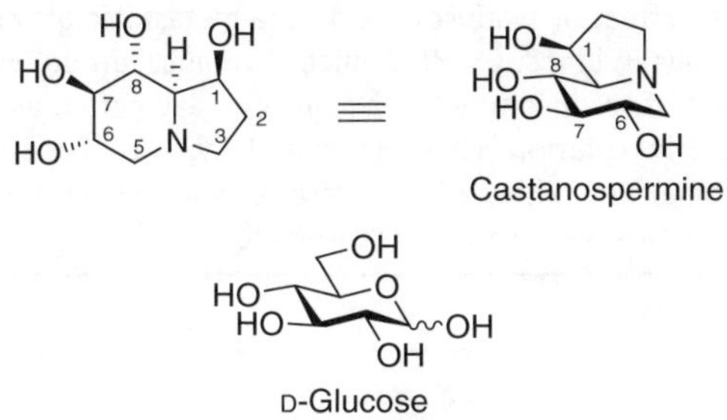

Figure 12.5 Structural resemblance of castanospermine (CS) to D-glucose

Of particular interest is the fact that SW inhibits tumour metastasis as well as tumour growth. By inhibiting Golgi α -mannosidase II, SW blocks the expression of complex-type asparagine-linked oligosaccharides, which are required for effective tumour cell invasion involved in a critical step of the metastatic cascade [34–37]. SW potently inhibits experimental metastasis of B16-F10 melanoma and MDAY-D2 lymphoid tumour cells in mice without any cytotoxic side effects (Table 12.3) [48–50], and spontaneous metastasis of B16 melanoma cells in mice (Table 12.4) [51]. SW also enhances the anti-metastatic and antiproliferative effect of interferon (Table 12.5) [49]. SW was also shown to be an antimetastatic immunomodulator in mice. The antimetastatic activity of SW is mediated primarily by the ability of the drug to augment natural killer cell reactivity [52, 53]. Concanavalin A-induced proliferation and proliferative response in mixed lymphocyte culture, which is suppressed by an immunosuppressive factor obtained from the serum of sarcoma 180 tumour-bearing mice, is restored

Table 12.3 Effect of processing inhibition on pulmonary colonization of mice by B16-F10 cells. Cultures of B16-F10 cells were treated with the indicated concentrations of castanospermine (CS) and swainsonine (SW) in growth medium for 18 hours and then aliquots of 8×10^4 (CS) or 5×10^4 (SW) cells were tested for their lung colonization capacity

Compounds	Concentration (μg/ml)	Inhibition (%)
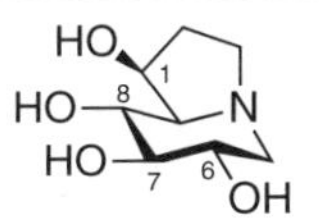 Castanospermine	1.0	63.0*
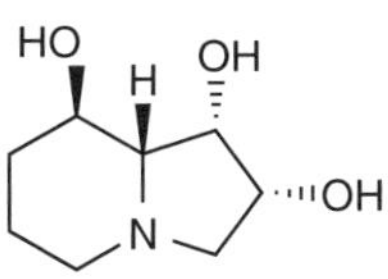 Swainsonine	1.0	64.4*

*: $p < 0.001$

Table 12.4 Effect of swainsonine on the metastatic growth of B16 melanoma cells in lungs. C57BL/6 mice (five mice/group) were given s.c. injection of swainsonine three days, one day before and one day after tumour inoculation. Mice inoculated i.v. with 1×10^5 cells of B16 melanoma, were killed 14 days after tumour inoculation and the pulmonary metastases in the lungs were counted

Dose (mg/kg/day)	Inhibition
100	73.2*
300	86.7*
1000	94.6*

*: $p < 0.05$

Table 12.5 Inhibition of lung colonization of the B16-F10 by swainsonine (SW) with poly(I). poly(C). B16-F10 cells were cultured for 48 hours with or without swainsonine (0.3 μg/ml) before injection of 10^5 cells into the lateral tail veins of mice on day 0. Mice were given swainsonine (2.5 μg/ml) 2 days before tumour cells were injected and maintained on swainsonine for 17 days. Mice were given an i.p. injection of Poly(I)·poly(C) (100 μg) the day before tumour cells were injected and again on day 3. Lung nodules were counted on day 24 and each group consisted of five mice

Treatment		Inhibition (%)
Cells	Mice	
Nil	Nil	0
Nil	Poly(I)·poly(C)	36.2
SW	Nil	72.4**
SW	Poly(I)·poly(C)	58.6*
SW	SW	91.4***
SW	SW+ Poly(I)·poly(C)	93.1***

*: $p < 0.05$ **: $p < 0.02$ ***: $p < 0.005$

to normal levels by SW [54]. SW-induced loss of complex-type oligosaccharides in malignant cells not only inhibits metastasis but also reduces tumour cell proliferation *in vitro* and *in vivo* [55, 56]. Because of its anticancer activity, oral availability and low toxicity in mice, SW was selected for Phase I clinical testing as the first iminosugar-type anticancer drug [57–61]. Based on the results of a Phase I study, utilizing oral and intravenous administration of SW in patients with advanced solid tumour and haematological malignancies, a Phase II study was conducted to determine efficacy and safety of chronic oral SW treatment and to assess the pharmacokinetics and pharmacodynamics of the drug in patients with advanced or metastatic renal cell carcinoma [62]. The therapeutic effects were limited to stabilization of the disease. Adverse events such as fatigue, anorexia, nausea and diarrhoea were common but generally mild. No evidence of antitumour activity of SW was noted in this study.

CS, as well as SW, were found to reduce pulmonary colonization of metastatic B16-F10 murine melanoma cells in C57BL/6 mice ($\geq$80 per cent reduction) without any cytotoxic side effects (Table 12.3) [48]. Frequent administration of CS during the early phase of tumour growth in mice is also effective in retarding the growth of murine tumour cells and human prostate tumour cells [63]. CS, an α-glucosidase I inhibitor, alters the structures of oligosaccharides on endothelial cells, which are required for angiogenesis, and inhibits tumour growth. In addition, CS prevents the normal glycosylation processing of the v-*fms*-transforming glycoprotein and inhibits v-*fms*-transformed tumour cell growth [64]. CS also inhibits platelet aggregation of metastatic H-ras-transformed 10T1/2 fibroblasts [65].

12.3 Hexuronic acid-mimetic iminosugars

Hexuronic acid-mimicking iminosugars have been isolated from microorganism cultures and plants, and synthesized [66–73]. These compounds form a class of competitive inhibitors of glycuronidases from various organisms, and some of them interfere with the metabolism of glycosaminoglycans of endothelial basement membranes and ECM. Hexuronic acid-mimicking iminosugars have a common structural feature in that they resemble the corresponding glycuronic acids while they specifically inhibit the corresponding glycuronidase. They have been recognized as useful molecular tools for probing several important biological processes and as potential therapeutic agents.

Heparan sulphate proteoglycans (HSPGs), distributed ubiquitously in nature, are common constituents of cell surfaces and ECM, including basement membranes. HSPGs are members of the glycosaminoglycan (GAG) family, which are unbranched sulphate-substituted polysaccharides that consist of a disaccharidic repeat unit of hexuronic acid (either D-glucuronic or L-iduronic acid) linked to glucosamine. HSPGs regulate several aspects of cancer biology including tumour progression, angiogenesis and metastasis. HSPGs are known to interact with various molecules like growth factors (e.g. bFGF, VEGF, PDGF, HGF and TGF-β), cytokines (e.g. interleukin-8), enzymes (e.g. heparanase and MMPPs) and ECM structural proteins (e.g. type IV collagen) [10, 11, 74]. Changes in both expression and structure of HSPGs on the cell surface and in ECM are important for malignant transformation of tumour cells. Adhesion to, and degradation of, ECM are required for invasion and metastasis of cancer cells [7–11]. During invasion and metastasis, tumour cells secrete enzymes that degrade both the protein (e.g. MMPs) and GAG chain (e.g. heparanase) components of the basement-membrane barrier. Heparanase is an endo-β-glucuronidase that specifically cleaves the β-1,4-linkage between glucuronic acid and N-Acetyl-glucosamine in HSPGs (Figure 12.6) [75–80]. After heparanase cleaves HSPGs, HSPG-binding growth factors, such as bFGF, PDGF, HGF, VEGF and TGF-β are released and activated, and are utilized for tumour cell growth [81, 82]. Heparanase activity was shown to correlate directly to the invasiveness of tumour cells [77, 78, 83]. On the other hand, the O-sulphonation of HSPGs creates complex heparan sulphoforms with high structural diversity. The combination of the GAG composition and the sulphonation pattern provide for diversity in cell functions [84, 85]. Heparan sulphotransferases sulphate specific hydroxyl groups in the final stages of HSPG biosynthesis. The final steps of heparan sulphate (HS)

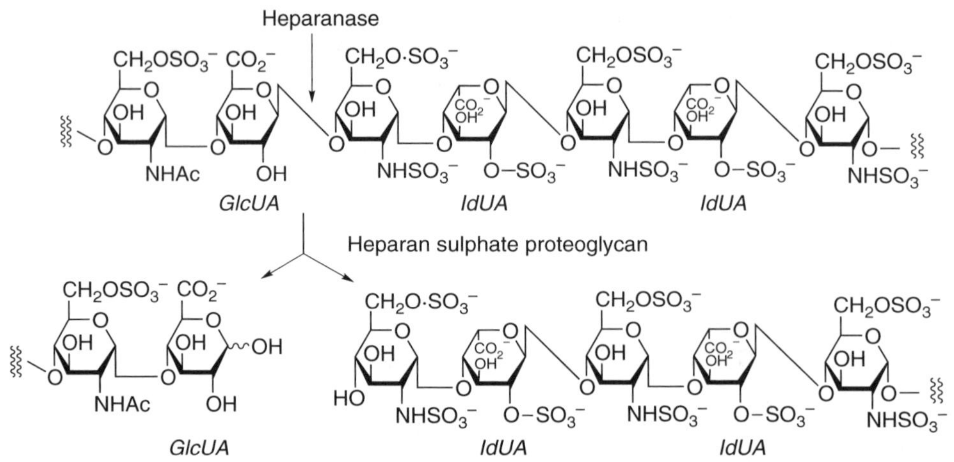

Figure 12.6 Heparanase cleaves heparan sulphate (HS)

differentiation determine the specific function of the product involved in cellular recognition, adhesion and migration. Altering HS biosynthesis could be of therapeutic benefit for treating disorders related to aberrant growth, such as tumour growth and metastasis [86, 87].

12.3.1 *Gem*-diamine 1-*N*-iminosugars

In 1974, an unusual iminosugar, siastatin B (STB) was isolated as an inhibitor of neuraminidase as well as β-D-glucuronidase and *N*-acetyl-β-D-glucosaminidase from *Streptomyces* culture [67]. STB resembles structurally D-glucuronic acid and *N*-acetyl-D-glucosamine as well as *N*-acetyneuraminic acid (Figure 12.7). STB is the first example of a 1-*N*-iminosugar, a class of monosaccharide mimic in which the anomeric carbon is replaced by nitrogen, distinct from well-known

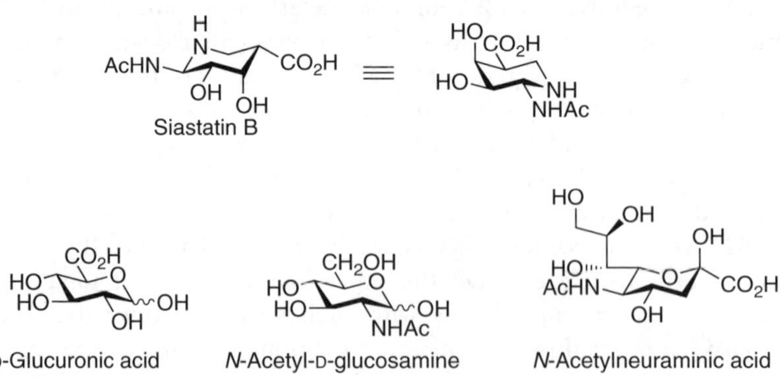

Figure 12.7 Structural resemblance of siastatin B to D-glucuronic acid, *N*-acetyl-D-glucosamine and *N*-acetyneuraminic acid

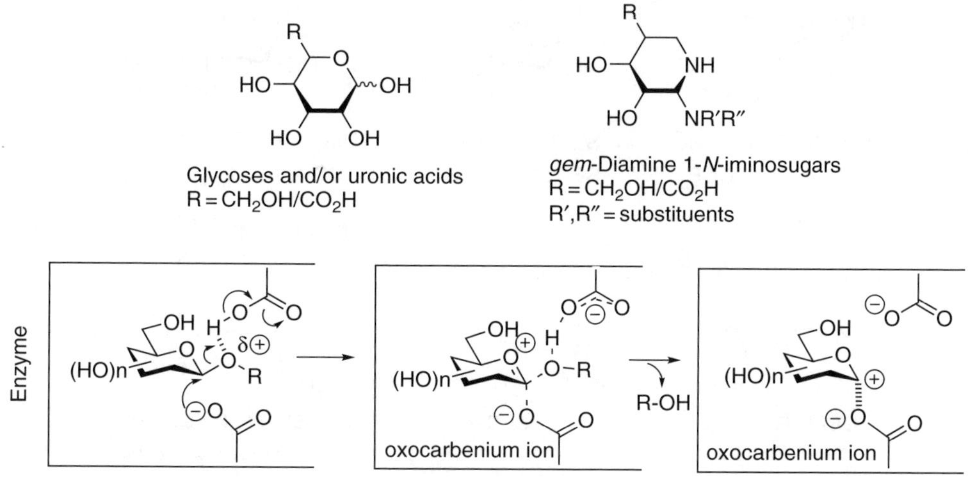

Figure 12.8 General structure of *gem*-diamine 1-*N*-iminosugars and oxocarbenium ion, the putative transition state of enzymatic glycosidic hydrolysis

nojirimycin-like glycohydrolase inhibitors with a nitrogen atom in place of the ring oxygen. Synthetic *gem*-diamine 1- *N*-iminosugars have been proposed as a new class of glycosidase inhibitors by Nishimura *et al.* [69, 73, 88]. *Gem*-diamine 1-*N*-iminosugars may mimic charge at the anomeric position in the transition state of the enzymatic glycosidic hydrolysis, and show strong and specific inhibition against glycosidases (Figure 12.8). Many kinds of uronic acid-type *gem*-diamine 1-*N*-iminosugars were synthesized by flexible synthetic strategies from natural STB or monosaccharide lactones (Figure 12.9) [73, 89–95]. D-Glucuronic acid-related *gem*-diamine 1-*N*-iminosugars show highly potent inhibition against *β*-D-glucuronidase (Table 12.6). These D-hexuronic (D-glucuronic, D-galacturonic and D-mannuronic) acid-type iminosugars inhibit recombinant heparanase from human melanoma A375M cells transfected with pBK-CMV expression vectors containing the heparanase cDNA (Table 12.6). Of these, D-galacturonic acid-type iminosugar (SF-4) shows the most potent inhibition. A typical D-galacturonic acid-type iminosugar (SF-1) also inhibits sialyltransferase in the mouse mammary carcinoma mutant cell line which shows a high level of sialic acid transfer to lactosylceramide to express ganglioside GM_3. Tumour metastatic potential is also known to correlate with the sialic acid content and/or the sialyltransferase activity of tumour cell membranes [96–99], and to be inhibited by the sialyltransferase inhibitors [100, 101]. Therefore, D-glucuronic acid-related *gem*-diamine 1-*N*-iminosugars may be capable of inhibiting tumour metastasis as both heparanase inhibitors and sialyltransferase inhibitors. On the other hand, while L-uronic acid-type *gem*-diamine 1-*N*-iminosugars show no inhibition against heparanase, the typical L-iduronic acid-related iminosugar (SDD-8) inhibits heparan sulphate2-O-sulphotransferase (HS 2-OST) with $\geq$80 per cent at 25 μM [102]. HS 2-O-ST catalyses sulphate transfer from the sulphate donor, adenosine 3'-phosphate-5'-phosphosulphate (PAPS) to the C-2 hydroxyl group of L-iduronic acid in HS

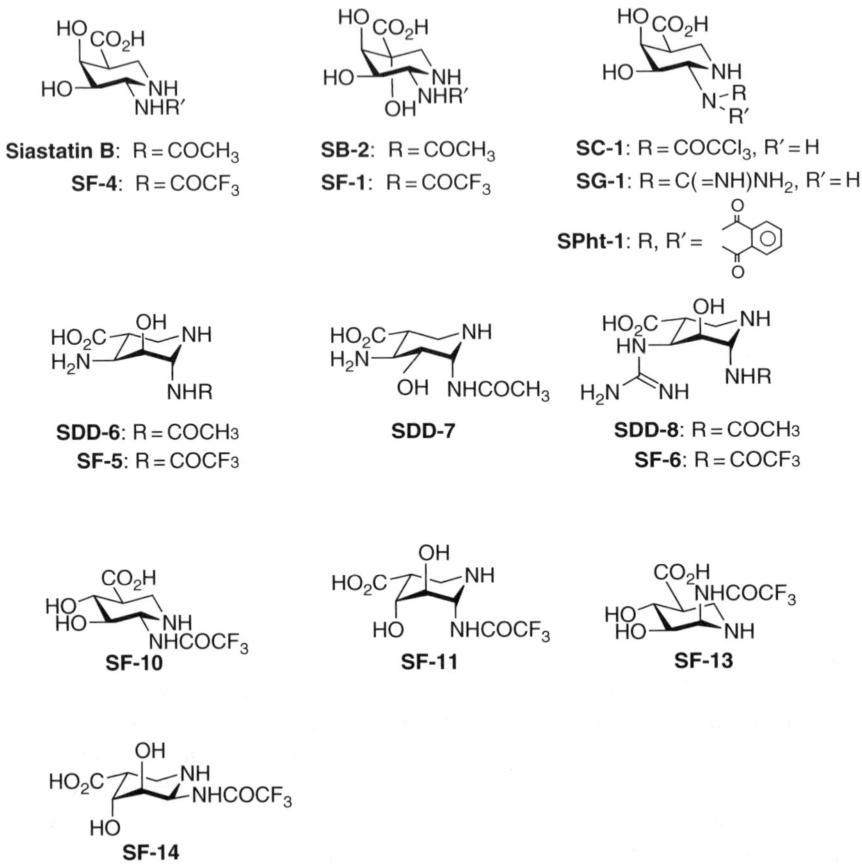

Figure 12.9 Structures of uronic acid-type *gem*-diamine 1-*N*-iminosugars

(Figure 12.10). The products of HS 2-O-ST and HS 6-O-ST are known to participate in FGF binding and activation, respectively, and are correlated with angiogenesis [103–107]. SDD-8 may mimic L-iduronic acid in HS biosynthesis, consequently altering HS structures and leading to therapeutic benefits for treating tumour growth and metastasis.

A typical D-galacturonic acid-type iminosugar (SF-4) inhibits the invasion of B16BL6 and Lewis lung carcinoma (3LL) cells through the reconstituted basement membrane Matrigel (Table 12.7) [108]. Pulmonary colonization, after i.v. transplantation of B16BL6 cells treated with D-galacturonic acid-type iminosugars (SF-1, SF-4, SC-1 and SG-1) *in vitro*, is strongly suppressed in a dose-dependent manner (Table 12.8) [90, 92]. Of these, SF-4 shows the most potent suppression with 90.8 per cent at 50 μg/ml. These iminosugars have no effects on tumour cell growth. SF-4 also potently inhibits spontaneous lung metastasis of 3LL cells in a dose-dependent manner by multiple i.v. administrations, after the surgical excision of primary tumours in mice, without any cytotoxic side effects (Table 12.9) [108].

Table 12.6 Inhibitory activities of uronic acid-type *gem*-diamine 1-*N*-iminosugars against glucuronidase and heparanase

Compounds		β-D-Glucuronidase[a] IC_{50} (M)	Heparanase[b] IC_{50} (μM)
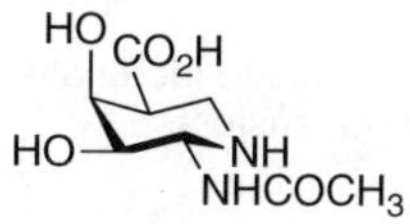	Siastatin B	7.1×10^{-5}	NT
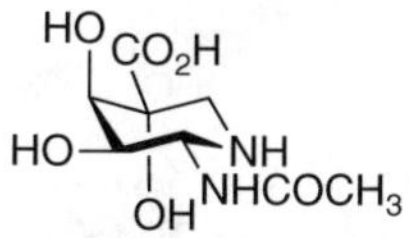	SB-2	1.2×10^{-4}	NT
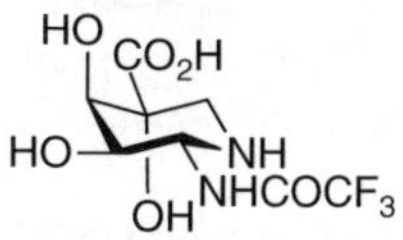	SF-1	6.2×10^{-8}	NT
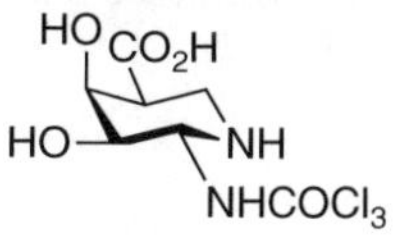	SC-1	9.2×10^{-8}	NT
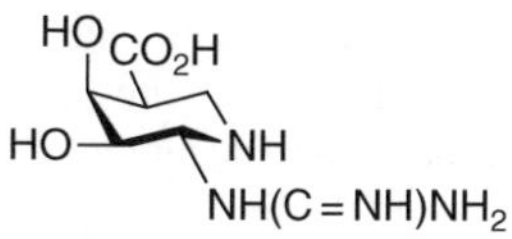	SG-1	1.3×10^{-7}	NT
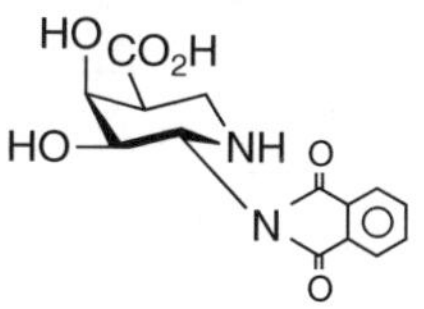	SPht-1	6.8×10^{-8}	NT
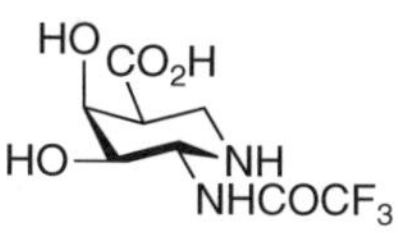	SF-4	6.5×10^{-8} [c]	1.0
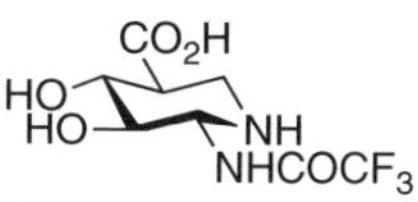	SF-10	6.5×10^{-8}	10.5
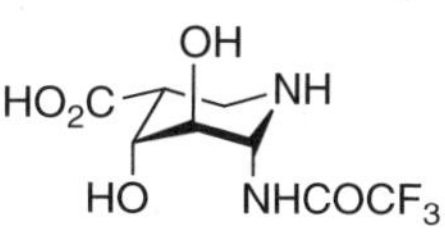	SF-11	1.3×10^{-4}	NI
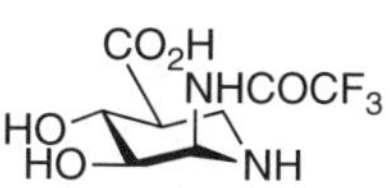	SF-13	6.5×10^{-8}	29.0

a): Bovine liver (pH 5.0); b): Human heparanase (melanoma A375M cell)(pH 4.2); c): $K_i = 6.4 \times 10^{-8}$M, NT: not tested, NI: no inhibition.

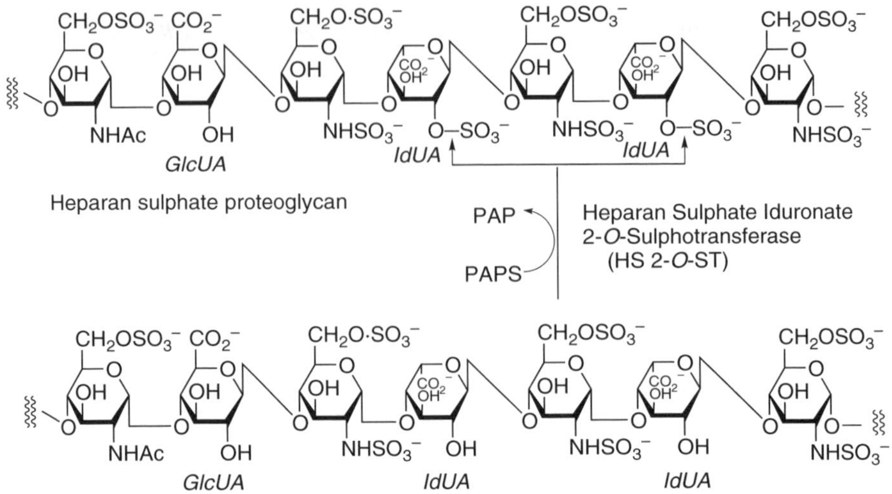

Figure 12.10 Heparan sulphate iduronate 2-*O*-sulphotransferase (HS 2-*O*-ST)-catalysed reaction with PAPS as the cosubstrate

Table 12.7 Inhibition of invasive activity of tumour cells by SF-4. Tumour cells were cultured with or without SF-4 for 72 hours (B16BL6) or 15 hours (3LL). Numbers of cells which invaded the reconstituted basement membrane Mutrigel in 3 hours (B16BL6) or 4 hours (3LL) were counted. A laminin coated under the filter surface was used as a cell attractant

Tumour cell line	Concentration (μg/ml)	Inhibition (%)
B16BL6	200	64.1
3LL	200	91.5

The typical L-iduronic acid-related iminosugars (SDD-8 and SF-6) inhibit the invasion of B16BL6 cells through the reconstituted basement membrane Matrigel (Table 12.10) [93, 94]. The experimental pulmonary metastasis of B16BL6 cells is potently inhibited in a dose-dependent manner by treatment with L-iduronic acid-related iminosugars (SDD-6, SDD-8, SF-5 and SF-6) *in vitro* (Table 12.11). Of these, SDD-8 shows the most potent inhibition with ≥97 per cent at 50 μg/ml.

The *gem*-diamine 1-*N*-iminosugars related to D-glucuronic and L-iduronic acid markedly inhibit experimentally induced lung metastasis of B16BL6 and /or 3LL cells, and also suppress spontaneous lung metastasis of 3LL cells following i.v. administration.

It is likely that the *gem*-diamine 1-*N*-iminosugars related to D-glucuronic and L-iduronic acid mimic the respective uronic acids in the metabolism of ECM and/or basement membrane involved in tumour metastasis. These iminosugars seem to modify the cell surface glycoconjugates of tumour cells simultaneously, thereby altering cell properties involved in cellular recognition and adhesion.

Table 12.8 Inhibition of pulmonary metastasis of B16BL6 cells by *in vitro* treatment with SF-1, SF-4, SC-1 and SG-1 in mice. B16BL6 cells were cultured with or without compounds for 3 days. The cells (1×10^5) were implanted i.v. into the tail vein of mice; 14 days later, the lung tumour nodules were counted

Compounds	Dose (μg/ml)	Inhibition (%)
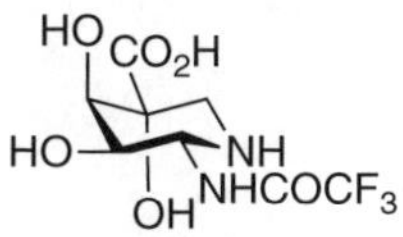 SF-1	50	80.5**
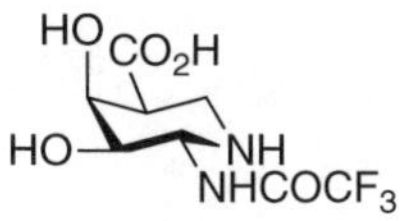 SF-4	50	90.8**
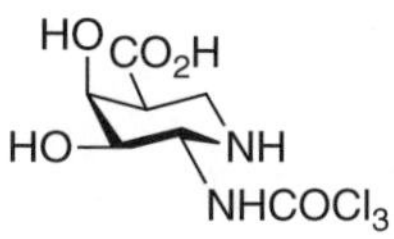 SC-1	50	67.3*
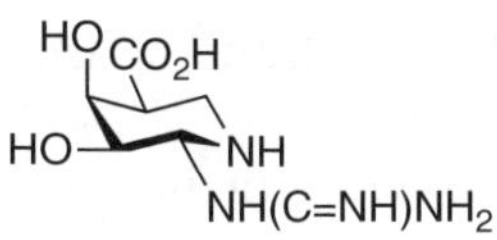 SG-1	50	87.1**

*: $p < 0.01$, **: $p < 0.001$

Table 12.9 Inhibitory effect of SF-4 on spontaneous lung metastasis of 3LL cells in mice. Five mice per group inoculated with 3LL cells (1×10^6) by s.c. intrafootpad injection were administered i.v. with SF-4 for 5 days starting on the day of the surgical excision of primary tumours on day 9. Mice were killed 10 days after tumour excision, and the lung tumour colonies were counted

Administered dose (mg/kg/day)	Inhibition (%)
50	23.5
100	57.1*

*: $p < 0.01$

Table 12.10 Inhibition of invasive activity of B16BL6 cell by SDD-8 and SF-6. The cells were cultured with SDD-8 and SF-6 for 72 hours and 24 hours, respectively. Numbers of invaded cells on the lower surface of the Mutrigel/laminin-coated filters were counted

Compounds	Concentration (μg/ml)	Inhibition (%)
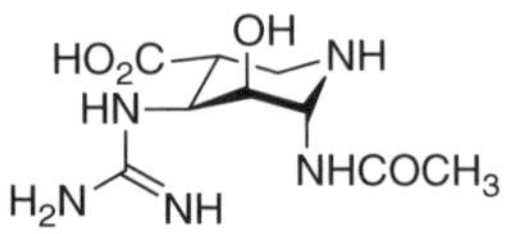 SDD-8	300	58.9*
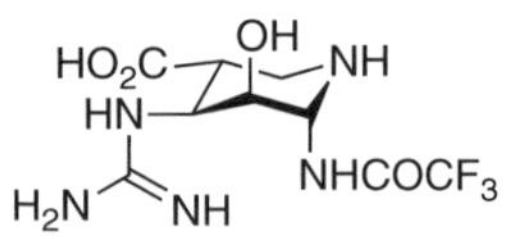 SF-6	100	61.1**

*: $p < 0.01$, **: $p < 0.001$

12.3.2 D-Glucaro-δ-lactam (lactam of 5-amino-5-deoxy-D-glucaric acid)

Sodium D-glucaro-δ-lactam (ND2001) (Figure 12.11) derived from nojirimycin is known as a potent competitive β-D-glucuronidase inhibitor *in vitro* (IC_{50} 0.18 μM, bovine liver) and *in vivo* (inhibitions against β-D-glucuronidase in mouse liver by oral administration, and in rat serum by intraperitoneal administration) [109]. Later, ND2001 was shown to inhibit the invasion and metastasis of tumour cells in a screening program of metastasis inhibitors [110]. ND2001 shows inhibition of the invasion of B16 variant and 3LL cells through the reconstituted basement membrane Matrigel by *in vitro* treatment, but no inhibition of mouse BMT-11 fibrosarcoma and rat SST-2 breast carcinoma (Table 12.12). ND2001 inhibits experimental pulmonary metastasis of the B16 variant (Table 12.13). ND2001 also shows remarkable inhibition of spontaneous metastasis of B16 variant, 3LL and rat KDH-8 liver carcinoma cells with 99.5 per cent at 30 mg/kg/day for 10 days, 98.0 per cent at 100 mg/kg/day for 4 days, and 82.5 per cent at 100 mg/kg/day for 30 days, respectively (Table 12.14). ND2001 shows neither cytocidal nor antitumour activity. ND2001 also inhibits, by 69.2 per cent, experimental pulmonary metastasis of rat hepatoma cKDH-8/11 by *in vitro* treatment with a concentration of 50 μg/ml (Table 12.15) [111]. ND2001 shows suppression of the invasive activity of cKDH-8/11 cells through the reconstituted basement membrane Matrigel (Table 12.16). While ND2001 shows no suppression of the random motility of cKDH-8/11 cells, it inhibits by 77 per cent the haptotaxis of the cells toward laminin, at a concentration of 50 μg/ml. ND2001 seems to inhibit pulmonary metastasis at the stage of basement membrane invasion, by directly changing some property of the tumour cells, such as haptotaxis to laminin.

Table 12.11 Inhibition of pulmonary metastasis of B16BL6 cell by *in vitro* treatment with SDD-6, SDD-8, SF-5 and SF-6 in mice. B16BL6 cells were cultured with SDD-6 and SDD-8 for 3 days and with SF-5 and SF-6 for 1 day. The cells were injected i.v. into the tail vein of mice; 14 days later, the mice were autopsied and the lung tumour colonies were counted

Compounds	Concentration (μg/ml)	Inhibition (%)
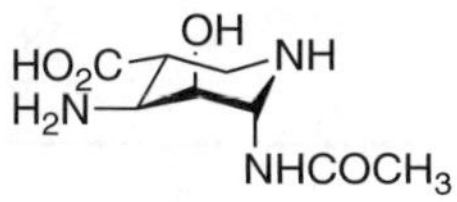 SDD-6	50	44.3*
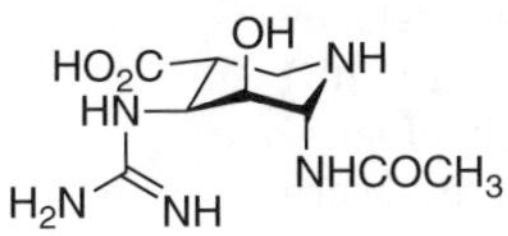 SDD-8	50	97**
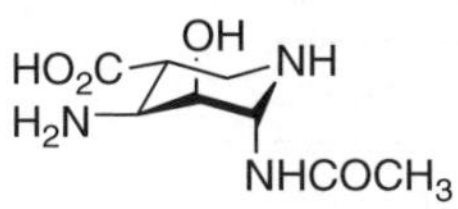 SF-5	50	75.5**
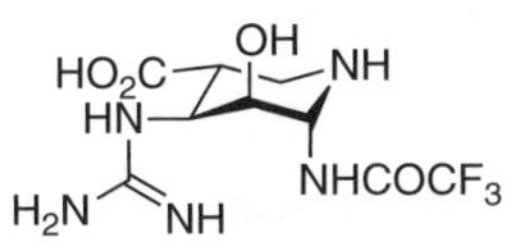 SF-6	50	81.0**

*: $p < 0.05$, **: $p < 0.001$

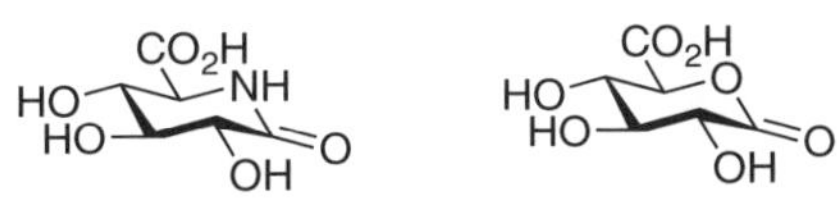

ND2001 D-glucaro-δ-lactone

Figure 12.11 Structures of ND2001 and D-glucaro-δ-lactone

12.4 Concluding remarks

The metastatic potential of cancer is an important determinant for morbidity and mortality. Inhibitors of tumour metastasis, as well as of tumour angiogenesis, are rapidly emerging as important new drug candidates for cancer therapy. Tumour invasion

Table 12.12 Inhibition of invasive activity of tumor cells by ND2001. The tumour cells were cultured with ND2001 for 3 days. Numbers of invaded cells on the lower surface of the reconstituted basement membrane Matrigel for 4 hours were counted

Tumour cell line	Concentration (μg/ml)	Inhibition (%)
B16 variant	100	83.0
3LL	100	89.9
BMT-11	100	25.8
STT-2	100	14.9

Table 12.13 Inhibition of experimental pulmonary metastasis of B16 variant cells by ND2001 in mice. B16 variant cells (1×10^5) were implanted i.v. into the tail vein of mice. ND2001 was administered i.v. to the mice twice a day from the day before the implantation (day 1) to day 4 for 5 days. Pulmonary tumour colonies were counted on day 14

Dose (mg/kg/day)	Inhibition (%)
10	67.1*
30	73.0*
100	91.4*

*: $p < 0.001$

Table 12.14 Inhibitory effect of ND2001 on spontaneous lung metastasis of tumour cells

Tumour cell	Dose (mg/kg/day)	Inhibition (%)
B16 variant[a)]	30	99.5*
3LL[b)]	50	96***
KDH-8[c)]	100	82.5**

a): Mice inoculated with B16 variant cells (5×10^5) by s.c. intrafootpad injection were administered i.v. with ND2001 at dose schedules of from day 7 to 27 after tumour cell challenge.
b): Mice inoculated with 3LL cells (1×10^6) by s.c. intrafootpad injection were administered i.v. with ND2001 for 5 days starting on the day of the surgical excision of primary tumours on day 14. Metastasized colonies were counted after fixation on day 23.
c): KDH-8 cells (5×10^6) were implanted s.c. into the right back of rats and ND2001 was administered i.v. on day 1 through 30. The rats were killed on day 31, and the metastasized colonies were counted after fixation.
*: $p < 0.05$, **: $p < 0.01$, ***: $p < 0.001$

through the ECM and basement membrane is a crucial step in the multistage process which leads to metastasis. During invasion, tumour cells detach from one another and from the ECM, and migrate through neighbouring tissue. This requires remodelling of cell-surface adhesion receptors and ligands, and the secretion of proteolytic enzymes and

Table 12.15 Inhibition of lung metastasis of rat hepatoma cKDH-8/11 cells by ND2001 in syngeneic rats. cKDH-8/11 cells (1×10^5) were treated with ND2001 *in vitro* for 24 hours and implanted i.v. into WKAH rats Rats were killed 16 days after the tumour implantation and lung metastasis colonies were counted

Concentration (μg/ml)	Inhibition (%)
10	19.4
50	69.2*

*: $p < 0.02$

Table 12.16 Inhibition of invasive activity of rat hepatoma cKDH-8/11 cells by ND2001. cKDH-8/11 cells were cultured with ND2001 for 24 hours. Numbers of cells which invaded the reconstituted basement membrane Matrigel for 3 hours were counted

Concentration (μg/ml)	Inhibition (%)
10	64.6
50	69.3

glycosidases to degrade ECM components. The biosynthetic and/or catabolic pathway of glycans involved in tumour invasion should be promising therapeutic targets for tumour metastasis.

Iminosugars were found to interact with enzymes involved in the metabolic pathway of glycans responsible for tumour cell invasion and migration. SW, CS and NJ, competitive inhibitors of *N*-linked glycan processing in the Golgi, reduced invasion and inhibited tumour metastasis. Of these, SW with low toxicity was first chosen for Phase I human cancer trials, and reached Phase II testing. However, the disadvantage of these types of inhibitors is that they affect all *N*-linked glycan biosynthesis, potentially leading to side effects. D-Uronic acid-type *gem*-diamine 1-*N*-iminosugars inhibit tumour heparanase (endo-β-glucuronidase) activity, tumour cell invasion and tumour metastasis *in vivo*, with no toxicity, an effect that probably results from their resemblance to D-glucuronic acid as a substrate for tumour heparanase. L-Uronic acid-type *gem*-diamine 1-*N*-iminosugars inhibit HS 2-O-ST activity, tumour cell invasion and tumour metastasis *in vivo*, with no toxicity, an effect that probably results from their resemblance to L-iduronic acid as a substrate for HS 2-OST. ND2001 inhibits the haptotaxis of tumour cells toward the laminin of basement membrane components, tumour cell invasion and tumour metastasis *in vivo*. ND2001 is distinct from the above iminosugars which interact with adhesion molecules or inhibit ECM degradation enzymes. That iminosugars are potent inhibitors of tumour metastasis holds promise for new drug candidates for cancer chemotherapy.

References

1. (a) Warren, L., Buck, C. A., and Tuszynski, G. P (1978) Glycopeptide changes and malignant transformation: a possible role for carbohydrate in malignant behavior. *Biochim. Biophys. Acta.*, **516**, 97-127. (b) Hakomori, S-I (1989) Aberrant glycosylation in tumors and tumor-associated carbohydrate antigens. *Adv. Cancer Res.*, **52**, 257–331.
2. Dennis, J. W. (1991) Changes in glycosylation with malignant transformation and tumor progression, in *Cell Surface Carbohydrates and Cell Development*, (ed. Fukuda, M.), CRC Press, Boca Raton, USA, pp. 161–194.
3. Varki, A. (1993) Biological roles of oligosaccharides: all of the theories are correct. *Glycobiology*, **3**, 97–130.
4. Muramatsu, T. (1993) Carbohydrate signals in metastasis and prognosis of human carcinomas. *Glycobiology*, **3**, 294–296.
5. Kornfeld, R. and Kornfeld, S. (1985) Assembly of asparagine-linked oligosaccharides. *Annu. Rev. Biochem.*, **54**, 631–664.
6. Elbein, A. D. and Molyneux, R. J. (1999) Inhibitors of glycoprotein processing, in *Iminosugars as Glycosidase Inhibitors* (Stütz, A., ed.), Wiley–VCH, Weinheim, Germany, pp. 216–251.
7. Kjellen, L. and Lindahl, U. (1991) Proteoglycans: structures and interactions. *Annu. Rev. Biochem.*, **60**, 443–475.
8. Vlodavsky, I., Bar-Shavit, R., Korner, G. and Fucks, Z. (1993) Extracellular matrix-bound growth factors, enzymes and plasma proteins, in *Basement Membranes: Cellular and Molecular Aspects* (Rohrbach, D. H. and Timpl, R., eds), Academic Press, Orland, USA, pp. 327–343.
9. Iozzo, R. V. (1998) Matrix proteoglycans: from molecular design to cellular function. *Annu. Rev. Biochem.*, **67**, 609–652.
10. Turnbull, J., Powell, A. and Guimond, S. (2001) Heparan sulfate: decoding a dynamic multifunctional cell regulator. *Trends Cell Biol.*, **11**, 75–82.
11. Esko, J. D. and Selleck, S. B. (2002) Order out of chaos: assembly of ligand binding sites in heparan sulfate. *Annu. Rev. Biochem.*, **71**, 435–471.
12. (a) Liotta, L. A., Thorgeirsson, U. P. and Garbisa, S. (1982) Role of collagenase in tumor cell invasion. *Cancer Metastasis Rev.*, **1**, 277–288. (b) Sloane, B. F. and Honn, K. V. (1984) Cysteine proteinase and metastasis. *Cancer Metastasis Rev.*, **3**, 249–263.
13. Woolley, D. E. (1984) Collagenolytic mechanisms in tumor cell invasion. *Cancer Metastasis Rev.*, **3**, 361–372.
14. Bernacki, R., Niedbala, M. J. and Korytnyk, W. (1985) Glycosidases in cancer and invasion. *Cancer Metastasis Rev.*, **4**, 81–101.
15. Nakajima, M. Irimura, T. and Nicolson, G. L. (1988) Heparanase and tumor metastasis. *J. Cell Biochem.*, **36**, 157–167.
16. Nakajima, M. Morikawa, K, Fabra, A., Bucana, C. D. and Fidler, I. J. (1990) Influence of organ environment on extracellular matrix degradative activity and metastasis of human colon carcinoma cells. *J. Natl. Cancer Inst.*, **82,** 1890–1898.
17. Wrodnigg, T. M., Steiner, A. J. and Ueberbacher, B. J. (2007) Natural and synthetic iminosugars as carbohydrate-processing enzyme inhibitors for cancer therapy. *Curr. Med. Chem. Anti-cancer Agents*. In press.
18. Elbein, A. D. (1991) Glycosidase inhibitors: inhibitors of N-linked oligosaccharide processing. *FASEB J.*, **5**, 3055–3063.
19. Nishimura, Y. (1992) Glycosidase and glycosyltransferase inhibitors, in *Studies in Natural Products Chemistry*, Volume 10, (Atta-ur-Rahman,ed.), Elsevier, Amsterdam, Holland, pp. 495–583.

20. Winchester, B. and Fleet, G. W. J. (1992) Amino-sugar glycosidase inhibitors: versatile tools for glycobiologists. *Glycobiology*, **2**, 199–210.
21. Asano, N. (2003) Natural occurring iminosugars and related compounds: structure, distribution, and biological activity. *Curr. Top. Med. Chem.*, **3**, 471–484.
22. Yagishita, K., Ohkura, T., Tachibana, Y., Takasuki, S. and Kobata, A. (1984) Comparative study of the oligosaccharides released from baby hamster kidney cells and their polyoma transformant by hydrazinolysis. *J. Biol. Chem.*, **259**, 10834–10840.
23. Collard, J. G., Van Beek, W. P., Jansen, J. W. G. and Schijiven, J. F. (1985) Transfection by human oncogenes: concomitant induction of tumorigenicity and tumor-associated membrane alterations. *Int. J. Cancer*, **35**, 207–214.
24. Pierce, M. and Arngo, J. (1986) Rous sarcoma virus-transformed baby hamster kidney cells express higher levels of asparagine-linked tri- and tetraantennary glycopeptides containing [GLcNac-β(1,6)Man-α-(1,6)Man] and poly-Nacetyllactosamine sequences than baby hamster kidney cells. *J. Biol. Chem.*, **261**, 10772–10777.
25. Debray, H., Qin, Z., Delannoy, P., Montreuil, J., Dus, D., Radikowski, C., Christensen, B. and Kieler, J. (1986) Altered glycosylation of membrane glycoproteins in human uroepithelial cell lines. *Int. J. Cancer*, **37**, 607–611.
26. Kerbel, R. S., Waghorne, C. Man, M. S., Elliott, B. and Breitman, M. L. (1987) Alteration of tumorigenic and metastatic properties of neoplastic cells is associated with the process of calcium phosphate-mediated DNA transfection. *Proc. Natl. Acad. Sci. USA*, **84**, 1263–1267.
27. Dennis, J. W. and Laferté, S. (1988) Asn-linked oligosaccharides and metastatic phenotype, in *Altered Glycosylation in Tumor Cells*, (eds Reading, C. L., Hakomori, S. and Marcus, D. M.), Alan R. Liss, New York, USA, pp. 257–267.
28. Hettkamp, H., Bause, E. and Legler, G. (1982) Inhibition by nojirimycin and 1-deoxynojirimycin of microsomal glucosidases from calf liver acting on the glycoprotein oligosaccharides Glc1-3Man9GlcNAc2. *Biosci. Rep.*, **2**, 899–906.
29. Bischoff, J. and Kornfeld, R. (1984). The effect of 1-deoxymannojirimycin on rat liver α-mannosidases. *Biochem. Biophys. Res. Commun.*, **125**, 324–331.
30. Fechter, M. H., Stütz, A. E. and Tauss, A. (1999) Chemical and chemo-enzymatic approaches to unnatural ketoses and glycosidase inhibitors with basic nitrogen in the sugar ring. *Curr. Org. Chem.*, **3**, 269–285.
31. Cipolla, L., La Ferla, B. and Nicotra, F. (2003) General methods for iminosugar synthesis. *Curr. Top. Med. Chem.*, **3**, 485–511.
32. Tsuruoka, T., Fukuyasu, H., Ishii, M., Usui, T., Shibahara, S. and Inouye, S. (1996) Inhibition of mouse tumor metastasis with nojirimycin-related compounds. *J. Antibiot.*, **49**, 155–161.
33. Tsukamoto, K., Uno, A., Kubota, Y., Shimada, S., Hori, Y. and Imokawa, G. (1992) Role of asparagine-linked carbohydrates in pulmonary metastasis of B16-F10 murine melanoma cells: implication through glycosylation inhibition by nojirimycin. *Melamona Res.*, **2**, 33–39.
34. (a) Molyneux, R. J. and James, L. F. (1991) Swaisonine, slaframine and castanospermine, in *Mycotoxins Phytoalexins*, CRC Press, Boca Raton, USA, pp. 637–656. (b) Elbein, A. D., Solf, R., Dorling, P. R. and Vosbeck, K. (1981) Swainsonine: an inhibitor of glycoprotein processing. *Proc. Natl. Acad. Sci.*,**78**, 7393–7397.
35. Tulsiani, D. R. P., Harris, T. M. and Touster, O. (1982) Swainsonine inhibits the biosynthesis of complex glycoproteins by inhibition of Golgi mannosidase II. *J. Biol. Chem.*, **257**, 7936–7939.
36. Tulsiani, D. R. P. and Touster, O. (1983) Swainsonine causes the production of hybrid glycoproteins by human skin fibroblasts and rat liver Golgi preparation. *J. Biol. Chem.*, **258**, 7578–7585.

37. Winchester, B., Al Daher, S., Carpenter, N. C., Cenci di Bello, I., Choi, S. S., Fairbanks, A. J. and Fleet, G. W. J. (1993) The structural basis of the inhibition of human α-mannosidases by azafuranose analogs of mannose. *Biochem. J.*, **290**, 743–749.
38. Sasak, V. W., Ordovas, J. M., Elbein, A. D. and Berninger, R. W. (1985) Castanospermine inhibits glucosidase I and glycoprotein secretion in human hepatoma cells. *Biochem. J.*, **232**, 759–766.
39. Palamarczyk, G. and Elbein, A. D. (1985) The effect of castanospermine on the oligosaccharide structures of glycoproteins from lymphoma cell lines. *Biochem. J.*, **227**, 795–804.
40. Merkle, R. K., Elbein, A. D. and Heifetz, A. J. (1985) The effect of swainsonine and castanospermine on the sulfation of the oligosaccharide chains of N-linked glycoproteins. *J. Biol. Chem.*, **260**, 1083–1089.
41. Gross, V., Train-Thi, T-A., Schwarz, R. T., Elbein, A. D., Decker, K. and Heinrich, P. C. (1986) Different effects of the glucosidase inhibitors 1-deoxynojirimycin, N-methyl-1-deoxynojirimycin and castanospermine on the glycosylation of rat α_1-proteinase inhibitor and α_1-acid glycoprotein. *Biochem. J.*, **236**, 853–860.
42. Foddy, L. and Hughes, C. (1988) Assembly of asparagine-linked oligosaccharides in baby hamster kidney cells treated with castanospremine, an inhibitor of processing glucosidases. *Eur. J. Biochem.*, **175**, 291–299.
43. Nemr, A. E. (2000) Synthetic methods for the stereoisomers of swainsonine and its analogues. *Tetrahedron*, **56**, 8579–8629.
44. Zhao, H., Hans, S., Cheng, X. and Mootoo, D. R. (2001) Allylated monosaccharides as precursors in triple reduction amination strategies: synthesis of castanospermine and swainsonine. *J. Org. Chem.*, **66**, 1761–1767.
45. Pyne, S. G. (2005) Recent developments on the synthesis of (-)-swainsonine and analogues. *Curr. Org. Synth.*, **2**, 39–57.
46. Martin, R., Murruzzu, C., Pericas, M. A. and Riera, A. (2005) General approach to glycosidase inhibitors. Enantioselective synthesis of deoxymannojirimycin and swainsonine. *J. Org. Chem.*, **70**, 2325–2328.
47. Michael, J. P. (2005) Indolizidine and quinolizidine alkaloids. *Nat. Prod. Rep.*, **21**, 625–649.
48. Humphries, M. J., Matsumoto, K., White, S. L. and Olden, K. (1986) Inhibition of experimental metastasis by castanospermine in mice: blockage of two distinct stages of tumor colonization by oligosaccharide processing inhibitors. *Cancer Res.*, **46**, 5215–5222.
49. Dennis, J. W. (1986) Effect of swainsonine and polyinosinic:polycytidylic acid on murine tumor cell growth and metastasis. *Cancer Res.*, **46**, 5131–5136.
50. Humphries, M. J., Matsumoto, K., White, S. L., Molyneux, R. J. and Olden, K. (1990) An assessment of the effects of swainsonine on survival of micc injected with B16-F10 melanoma cells. *Cin. Exp. Metastasis*, **8**, 89–102.
51. Kino, T., Inamura, N., Nakahara, K., Kiyota, S., Goto, T., Terano, H., Kohsaka, M., Aoki, H. and Imanaka, H. (1985) Studies of immunomodulator, swainsonine II. Effect of swainsonine on mouse immunodeficient system and experimental murine tumor. *J. Antibiot.*, **38**, 936–940.
52. Humphries, M. J., Matsumoto, K., White, S. L., Molyneux, R. J. and Olden, K. (1988) Augmentation of murine natural killer cell activity by swaisonine, a new antimetastatic immunomodulator. *Cancer Res.*, **48**, 1410–1415.
53. Olden, K., Breton, P., Grzegorzewski, K. Yasuda, Y. Gause, B. L., Oredipe, O. A., Newton, S. A. and White, S. L. (1991) The potential importance of swainsonine in therapy for cancers and immunology. *Pharm. Ther.*, **50**, 285–290.
54. Hino, M., Nakayama, O., Tsurumi, Y., Adachi, K., Shibata, T., Terano, H., Kohsaka, M., Aoki, H. and Imanaka, H. (1985) Studies of an immunomodulator, swainsonine I. Enhancement of immune response by swainsonine *in vitro*. *J. Antibiot.*, **38**, 926–935.

55. Humphries, M. J., Matsumoto, K., White, S. L. and Olden, K. (1986) Oligosaccharide modification by swainsonine treatment inhibits pulmonary colonization by B16-F10 murine melanoma cells. *Proc. Natl. Acad. Sci., USA*, **83**, 1752–1756.
56. Dennis, J. W., Koch, K., Yousefi, S. and VanderElst, I. (1990) Growth inhibition of human melanoma tumor xenografts in athymic nude mice by swainsonine. *Cancer Res.*, **50**, 1867–1872.
57. Olden, K., Breton, P. Grzegorzewski, K., Yasuda, Y., Gause, B. L., Oredipe, O. A., Newton, S. A. and White, S. L. (1991) The potential importance of swainsonine in therapy for cancers and immunology. *Pharm. Ther.*, **50**, 285–290.
58. Goss, P. E., Baptiste, J., Fernandes, B., Baker, M. and Dennis, J. W. (1994) A phase study of swaisonine in patients with advanced malignancies. *Cancer Res.*, **54**, 1450–1457.
59. Baptista, J. A., Goss, P., Nghiem, M., Krepinsky, J. J., Baler, M. and Dennis, J. W. (1994) Measuring swainsonine in serum of cancer patients: Phase I clinical trial. *Clin. Chem.*, **40**, 426–430.
60. Goss, P. E., Baker, M. A., Carver, J. P. and Dennis, J. W. (1995) Inhibition of carbohydrate processing: a new class of anticancer agents. *Clin. Cancer Res.*, **1**, 935–944.
61. Goss, P. E., Reid, C. L., Bailey, D. and Dennis, J. W. (1997) Phase IB clinical trial of the oligosaccharide processing inhibitor swainsonine in patients with advanced malignancies. *Clin. Cancer Res.*, **3**, 1077–1086.
62. Shaheen, P. E., Stadler, W., Elson, P. Knox, J. Winquist, E. and Bukowski, R. M. (2005) Phase II study of the efficacy and safety of oral GD0039 in patients with locally advanced or matastatic renal cell carcinoma. *Invest. New Drug*, **23**, 577–581.
63. Pili, R., Chang, J., Partis, R. A., Mueller, R. A., Chrest, F. J. and Passaniti, A. (1995) The α-glucosidase I inhibitor castanospermine alert endothelial cell glycosylation, prevents angiogenesis, and inhibits tumor growth. *Cancer Res.*, **55**, 2920–2926.
64. Ostrander, G. K., Scribner, N. K. and Rohrschneider L. R. (1988) Inhibition of v-fms-induced tumor growth in nude mice by castanospermine. *Cancer Res.*, **48**, 1091–1094.
65. Spearman, M. A., Ballon, B. C., Gerrard, J. M., Greenberg, A. H. and Wright, J. A. (1991) The inhibition of platelet aggregation of metastatic H-ras-transformed 10T1/2 fibroblasts with castanospermine, an N-linked glycoprotein processing inhibitor. *Cancer Lett.*, **60**, 185–191.
66. Tsuruoka, T., Niwa, T., Shomura, T., Matsuno, T., Itoh, N., Inouye, S. and Niida, T. (1973) Synthesis of D-glucaro-δ-lactam, an oxidation product of nojirimycin. *Sci. Rep. Meiji Seika Kaisha*, **13**, 80–84.
67. Umezawa, H., Aoyagi, T., Komiyama, H., Morishima, M., Hamada, M. and Takeuchi, T. (1974) Purification and characterization of a sialidase inhibitor, siastatin produced by Streptomyces. *J. Antibiot.*, **27**, 963–969.
68. Manning, K. S., Lynn, D. G., Shabanowitz, J., Fellows, L. E., Singh, M. and Schrire, B. D. J. (1985) A glucuronidase inhibitor from the seeds of Baphia racemosa: application of fast atom bombardment coupled with collision activated dissociation in natural product structure assignment. *Chem. Commun.*, 127–129.
69. Nishimura, Y. (1995) Stereoselective synthesis and transformation of siastatin B, a novel glycosidase inhibitor, directed toward new drugs for viral infection and tumor metastasis, in *Studies in Natural Products Chemistry*, Volume 16 (Atta-ur-Rahman, ed.) , Elsevier, Amsterdam, Holland, pp. 75–121.
70. Takatsu, T., Takahashi, M., Kawase, Y., Enokita, R., Okazaki, T., Matsukawa, H., Ogawa, K., Sakaida, Y., Kagasaki, T., Kinoshita, T., Nakajima, M. and Tanzawa, K. (1996) A-72363 A-1, A-2, and C, novel heparanase inhibitors from Streptomyces nobilis SANK 60192, I. Taxonomy of producing organism, fermentation, isolation and structural elucidation. *J. Antibiot.*, **49**, 54–60.

71. Kawase, Y., Takahashi, M., Takatsu, T., Arai, M., Nakajima, M. and Tanzawa, K. (1996) A-72363 A-1, A-2, and C, novel heparanase inhibitors from Streptomyces nobilis SANK 60192, II. Biological activities. *J. Antibiot.*, **49**, 61–64.
72. Kim, Y. J., Ichikawa, M. and Ichikawa, Y. J. (2000) Highly selective synthesis of 1-N-iminosugars of the D-glucose and –glucuronic acid types. *J. Org. Chem.*, **65**, 2599–2602.
73. Nishimura, Y. (2006) gem-Diamine 1-N-iminosugars, a new family of glycosidase inhibitors: synthesis and biological activity. *Heterocycles*, **67**, 461–488.
74. Yu, W-H. and Woessner, J. F. Jr. (2000) Heparan sulfate proteoglycans as extracellular docking molecules for matrilysin (matrix metalloproteinase 7). *J. Biol. Chem.*, **275**, 4183–4191.
75. Nakajima, M. Irimura, T., Di Ferrante, N. and Nicolson, G. L. (1984) Metastatic melanoma cell heparanase. Characterization of heparan sulfate degradation fragments produced by B16 melanoma endoglucuronidase. *J. Biol. Chem.*, **259**, 2283–2290.
76. Toyoshima, M. and Nakajima, M. (1999) Human heparanase. Purification, characterization, cloning, and expression. *J. Biol. Chem.*, **274**, 24153–24160.
77. Vlodavsky, I., Friedmann, Y., Elkin, M., Aingorn, H., Atzmon, R., Ishai-Michaeli, R., Bitan, M., Pappo, O., Peretz, T., Michal, I. and Speckers, I. (1999) Mammalian heparanase: gene cloning, expression, and fuction in tumor progression and matastasis. *Nat. Med.*, **5**, 793–802.
78. Hulett, M. D., Freeman, B. J., Baker, R. T., Harris, M. J. and Parish, C. R. (1999) Cloning of mammalian heparanase, an important enzyme in tumor invasion and metastasis. *Nat. Med.*, **5**, 803–809.
79. Fairbanks, M. B., Mildner, A. M., Leone, J. W., Cavey, G. S., Mathews, W. R., Drong, R. F., Slightom, J. L., Bienkowski, M. J., Smith, C. W., Bannow, C. A. and Heinrikson, R. L. (1999) Processing of the human heparanase precursor and evidence that the active enzyme is a heterodimer. *J. Biol. Chem.*, **274**, 29587–29590.
80. Kussie, P. H., Hulmers, J. D., Ludwig, D. L., Patel, S., Navarro, E. C., Seddon, A. P., Giorgio, N. A. and Bohlen, P. (1999) Cloning and functional expression of a human heparanase gene. *Biochem. Biophys. Res. Commun.*, **261**, 183–187.
81. Vlodavsky, I., Goldshmidt, O., Zcharia, E., Metzger, S., Chajek-Shaul, T., Atzmon, R., Guatta-Rangini, Z. and Friedmann, Y. (2001) Molecular properties and involvement of heparanase in cancer progression and normal development. *Biochimie*, **83**, 831–839.
82. Goldshmidt, O., Zcharia, E., Abramovitch, R., Metzger, S., Aingorn, H., Friedmann, Y., Schirrmacher, V., Mitrani, E. and Vlodavsky, I. (2002) Cell surface expression and secretion of heparanase markedly promote tumor angiogenesis and metastasis. *Proc. Natl. Acad.Sci. USA*, **99**, 10031–10036.
83. Vlodavsky, I. and Friedmann, Y. (2001) Molecular properties and involvement of heparanase in cancer metastasis and angiogenesis. *J. Clin. Invest.*, **108**, 341–347.
84. Capila, I. and Linhardt, R. J. (2002) Heparin-protein interactions. *Angew. Chem. Int. Ed.*, **41**, 391–412.
85. Chapman, E., Michael, D. B., Sarah, R. H. and Wong, C-H. (2004) Sulfotransferase: structure, mechanism, biological activity, inhibition, and synthetic utility. *Angew. Chem. Int. Ed.*, **43**, 3526–3548.
86. Fuster, M. and Esko, J. D. (2005) The sweet and sour of cancer: glycans as novel therapeutic targets. *Nat. Rev. Cancer*, **5**, 526–542.
87. Fjeldstad, K. and Kolset, S. O. (2005) Decreasing the metastatic potential in cancers – targeting the heparan sulfate proteoglycans. *Curr. Drug Targets*, **6**, 665–682.
88. Nishimura, Y. (2003) gem-Diamine 1-N-iminosugars and related iminosugars, candidate of therapeutic agents for tumor metastasis. *Curr. Top. Med. Chem.*, **3**, 575–591.

89. Kudo, T., Nishimura, Y. Kondo, S. and Takeuchi, T. (1992) Totally synthetic analogues of siastatin B. I. Optical active 2-acetamidopiperidine derivatives. *J. Antibiot.*, **45**, 954–962.
90. Nishimura, Y. Kudo, T. Kondo, S. and Takeuchi, T. (1992) Totally synthetic analogues of siastatin B. II. Optically active piperidine derivative having trifluoroacetamide and hydroxyacetamide groups at C-2'. *J. Antibiot.*, **45**, 963–970.
91. Satoh, T., Nishimura, Y. Kondo, S. and Takeuchi, T. (1996) A practical synthesis from siastatin B of (3S,4S,5R,6R)-4,5-dihydroxy-6-(trifluoroacetamido)piperidine-3-carboxylic acid having antimetastatic activity in mice. *Carbohydr. Res.*, **286**, 173–178.
92. Satoh, T., Nishimura, Y., Kondo, S. and Takeuchi, T. (1996) Synthesis and antimetastatic activity of 6-trichlorocetamido and 6-guanidino analogues of siastatin B. *J. Antibiot.*, **49**, 321–325.
93. Nishimura, Y., Satoh, T., Adachi, H, Kondo, S., Takeuchi, T, Azetaka, M., Fukuyasu, H. and Iizuka, Y. (1996) The first L-iduronic acid-type 1-N-iminosugars having inhibitory activity of experimental metastasis. *J. Am. Chem. Soc.*, **118**, 3051–3052.
94. Nishimura, Y., Satoh, T. Adachi, H., Kondo, S., Takeuchi, T. Azetaka, M., Fukuyasu, H. and Iizuka, Y. (1997) Synthesis and antimetastatic activity of L-iduronic acid-type 1-N-iminosugars. *J. Med. Chem.*, **40**, 2626–2633.
95. Nishimura, Y. Shitara, E., Adachi, H., Toyoshima, M., Nakajima, M., Okami, Y. and Takeuchi, T. (2000) Flexible synthesis and biological activity of uronic acid-type gem-diamine 1-N-iminosugars: a new family of glycosidase inhibitors. *J. Org. Chem.*, **65**, 2–11.
96. Yogeeswaran, G. and Salk, P. (1981) Metastatic potential is positively correlated with cell surface sialylation of cultured murine cell lines. *Science (Wash. DC)*, **212**, 1514–1516.
97. Fogel, M., Altevogt, P. and Schirrmacher, V. (1983) Metastatic potential severely altered by changes in tumor cell adhesiveness and cell-surface sialylation. *J. Exp. Med.*, **157**, 371–376.
98. Peracaula, R. Tabares, G. Lopez-Ferrer, A. Brossmer, R., de Bolos, C. and de Llorens, R. (2005) Role of sialyltransferases involved in the biosynthesis of Lewis antigens in human pancreatic tumor cells. *Glycoconjugate J.*, **22**, 135–144.
99. Cheung I. Y., Vickers, A. and Cheung, N. K. (2006) Sialyltransferase STX (ST8SiaII): a novel molecular marker of metastatic neuroblastoma. *Int. J. Cancer*, **119**, 152–156.
100. Kijima-Suda, I., Miyamoto, Y., Toyoshima, S., Itoh, M. and Osawa, T. (1986) Inhibition of experimental pulmonary metastasis of mouse colon adenocarcinoma 26 sublines by a sialic acid:nucleoside conjugate having sialyltransferase inhibiting activity. *Cancer Res.*, **46**, 858–862.
101. Kijima-Suda, I. Miyazawa, T., Itoh, M., Toyoshima, S. and Osawa, T. (1988) Possible mechanism of inhibition of experimental pulmonary metastasis of mouse colon adenocarcinoma 26 sublines by a sialic acid:nucleoside conjugate. *Cancer Res.*, **48**, 3728–3732.
102. Brown, J. M., Nishimura, Y. and Esko, J. D. (2006) Synthesis and biological evaluation of gem-diamine 1-N-iminosugars related to L-iduronic acid as inhibitors of heparan sulfate 2-O-sulfotransferase. *Bioorg. Med. Chem. Lett.*, **16**, 532–536.
103. Rapraeger, A. C. (1995) In the clutches of proteoglycans: how does heparin sulfate regulate FGF binding? *Chem. Biol.*, **2**, 645–649.
104. Faham, S., Hileman, R.E., Fromm, J. R., Linhardt, R. J. and Rees, D. C. (1996) Heparin structure and interactions with basic fibroblast growth factor. *Science*, **271**, 1116–1120.
105. Pye, D. A., Vives, R. R., Turnbull, J. E., Hyde, P. and Gallagher, J. T. (1998) Heparan sulfate oligosaccharides require 6-O-sulfation for promotion of basic fibroblast growth factor mitogenic activity. *J. Biol. Chem.*, **273**, 22936–22942.
106. Zhang, L. J., Beeler, D. L., Lawrence, R., Lech, M., Liu, J., Davis, J. C., Shriver, Z., Sasisekharan, R. and Rosenberg, R. D. (2001) 6-O-Sulfotransferase-1 represents a critical

enzyme in the anticoagulant heparan sulfate biosynthetic pathway. *J. Biol. Chem.*, 276, 42311–42321.
107. Wu, Z. L., Zhang, L., Yabe, T., Kuberan, B., Beeler, D. L., Love, A. and Rosenberg, R. D. (2003) The involvement of heparan sulfate (HS) in FGF1/HS/FGFR1 signaling complex. *J. Biol. Chem.*, **278**, 17121–17129.
108. Nishimura, Y., Satoh, T., Kondo, S., Takeuchi, T., Azetaka, M., Fukuyasu, H., Iizuka, Y. and Shibahara, S. (1994) Effect on spontaneous metastasis of mouse Lewis lung carcinoma by a trifluoroacetamide analogue of siastatin B. *J. Antibiot.*, **47**,. 840–842.
109. Niwa, T., Tsuruoka, T., Inouye, S., Naito, Y., Koeda, T. and Niida, T. (1972) A new potent β-glucuronidase inhibitor, D-glucaro-δ-lactam derived from nojirimycin. *J. Biochem.*, **72**, 207–211.
110. Tsuruoka, T., Fukuyasu, H., Azetaka, M., Iizuka, Y., Inouye, S., Hosokawa, M. and Kobayashi, H. (1995) Inhibition of pulmonary metastasis and tumor cell invasion in experimental tumors by sodium D-glucaro-δ-lactam (ND2001). *Jpn. J. Cancer Res.*, **86**, 41–47.
111. Kuramitsu, Y., Hamada, J-I., Tsuruoka, T., Morikawa, K., Kobayashi, H. and Hosokawa, M. (1998) A new anti-metastatic drug, ND-2001, inhibits lung metastases in rat hepatoma cells by suppressing haptotaxis of tumor cells toward laminin. *Ant-Cancer Drugs*, **9**, 88–92.

13
Medicinal use of iminosugars

Timothy M. Cox, Frances M. Platt and **Johannes M.F.G. Aerts**

13.1 Introduction

13.1.1 Definition

In understanding the medicinal uses of iminosugars, it is important for these biological compounds to be defined in terms that are acceptable to chemists and at the same time sufficiently inclusive to have meaning in fields of clinical practice and pharmaceutical research. Many iminosugars have previously been known simply as alkaloids – in other words, as molecules that occur naturally in plants and have nitrogen-containing heterocyclic structures; alkaloids have pronounced physiological actions. Other members of the family have been independently identified as microbial products, e.g. nojirimycin, originally obtained from the mould, *Streptomyces nojiriensis* (Figure 13.1). In this example, the iminosugar resembles a simple monosaccharide in which the oxygen within the pyranose ring is replaced by a nitrogen atom. Thus iminosugars consist of monocyclic or bicyclic derivatives of pyrrolidine, piperidine, pyrrolizidine, indolizidine (fusion of piperidine and pyrrolidine backbones) or nortropane ring structures [1]. A common characteristic of the monosaccharide-like iminosugars and their heterocyclic, alkaloid-type congeners, is that the biologically active members of both molecular families are almost invariably polyhydroxylated.

13.1.2 Overall mode of action

As is evident from the contributors in other sections of this volume, as a family, the iminosugars exert their biological actions principally as sugar mimics, thus inhibiting the activity of glycosidases and of glycosyltransferases. These actions form the basis of their main medicinal (and allied toxic) effects. However, the iminosugars are invested with other important biological activities which can readily be applied to medicine: they

Iminosugars: From synthesis to therapeutic applications Edited by Philippe Compain and Olivier R. Martin
 ISBN 978-0-470-03391-3

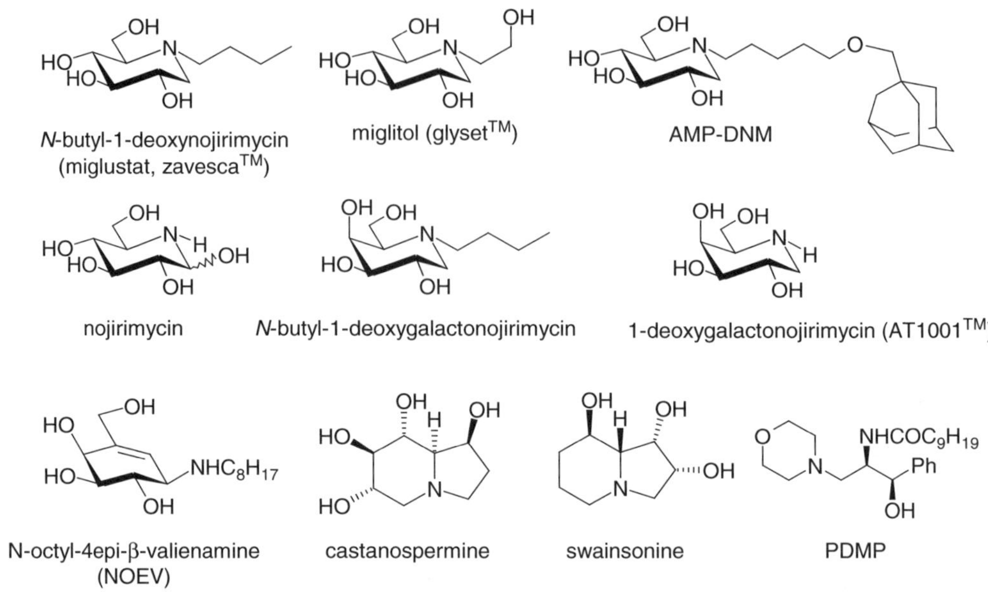

Figure 13.1 Molecular structures of the principal iminosugars discussed in the text

can serve as agents that modify the glycolipid environment and function of membrane proteins (e.g. the insulin receptor) and by binding to domains in nascent proteins, as molecular chaperones. The complex pathways of glycoconjugate biosynthesis and degradation pose challenges for medicinal applications: the key to successful development lies in establishing the specificity of iminosugars for given molecular targets and their ability selectively to inhibit particular processes at defined loci within the living cell.

13.1.3 History

'The desire to take medicine is perhaps the greatest feature which distinguishes man from animals.' Sir William Osler's statement holds true for the iminosugars, which have been consumed as phytomedicines since antiquity and are readily available today from suppliers both in the East and in the West. A preferred source, the white mulberry root-bark, *sang bai pi*, is recommended by traditional Chinese practitioners for the relief of respiratory congestion, urinary symptoms and as a hypotensive agent. Use of the mulberry has a parallel history in Western medicine: extracts are found in cosmetic applications for lightening the skin and they have also been used as laxative vermifuges. Indeed, the celebrated Calvinist physician, botanist and chemist, Herman Boeerhaave, while Professor of Medicine in the University of Leiden in the 17th and early part of the 18th centuries, had commercial interests in medicinal Haarlem oil for the purposes of weight loss, whitening of the skin and the treatment of diabetes [2]. Haarlem oil is still available as a licensed medicine for numerous, rather less precise indications. So it is that mulberry, and principally *Morus alba,* a rich source of iminosugars [3–5], has had a pharmaceutical history in the West for at least three centuries.

Thus natural products of mulberry, rich in deoxynojirimycin, its alkyl, and other more stable derivatives of nojirimycin (the first glucose analogue with a ring nitrogen), preceded the licensed use of 2- hydroxyethyldeoxynojirimycin (miglitol; Glyset) for type 2 diabetes mellitus (by Bayer, in 1996) and *N*-butyldeoxynojirimycin (miglustat; Zavesca) for Gaucher's disease (licensed by Actelion, in 2003), by many years [1, 6]. Many studies were predicated on pharmacological research using iminosugars obtained from natural sources to investigate potential agents for the treatment of diabetes mellitus in experimental systems [7].

There have been several announcements in the press and the worldwide web that Thailand and China will investigate herbal preparation for patients with human immunodeficiency virus of which mulberry root is a major component [8]. The agent is thought to enhance immunity against HIV infection including cases of drug-resistant HIV/AIDS. Use of the herbal drug is apparently based on the favourable outcome of studies by the Medical Science Department of Thailand's Public Health Ministry which conducted clinical trials in 60 patients, 40 of whom were taking the drug and 20 who did not [9].

Mulberry extracts remain in wide use but this application of a herbal remedy containing iminosugars for HIV infection is unlikely to meet the rigorous requirements for licensing a modern pharmaceutical drug.

13.2 Recent pharmaceutical developments

Contemporary development of the iminosugars within medicine has followed the more conventional path of linear thought involving biochemical study, experimental examination of the molecular cell biology of glycoconjugates in cultured cells and whole organisms – crowned finally by the translation of this research at the level of clinical trials in target diseases and the ultimate medicinal use of iminosugars, two of which are now available as modern prescription drugs. Many other iminosugars are in pharmaceutical development.

13.2.1 Discoveries in veterinary medicine

In the early modern phase of discovery, advances in the pharmaceutical use of iminosugars occurred on an eclectic background of discovery in veterinary medicine and animal nutrition [10]. Understanding of the biological actions of iminosugars has in part emerged from the discovery of toxic molecules responsible for economically important outbreaks of poisoning in domestic and farm animals.

The classic example of the toxicity of iminosugars is provided by the indolizidine toxin, swainsonine that occurs in the Darling pea, *Swainsona canescens* in Australasia and in the loco weeds, *Astragalus sp.* and related *Oxytropis sp.* in the American rangelands. These toxins give rise to locoism in cattle, horses, sheep and other grazing animals. Other ruminants eating red clover hay infected by *Rhizoctonia leguminicola* suffer 'slobber syndrome' due to the presence of a related indolizidine toxin, slaframine [10]. The toxins are powerful inhibitors of α-mannosidases and early on were shown to induce a phenocopy of the hereditary lysosomal disease, α-mannosidosis [11]. Swainsonine induces the formation of pathological hybrid glycans and the accumulation of mannose-rich oligosaccharides in lysosomes; these abnormalities are prominent in neurones, the mononuclear phagocyte system, kidney tubules and other organs, such as the thyroid gland. In pregnant sheep and cattle, swainsonine is an abortifacient; it interferes with the synthesis and secretion of progestagens that are essential for the physiological maintenance of pregnancy.

Intensive biochemical studies showed that the naturally occurring heterocyclic iminosugar, swainsonine, is not a selective inhibitor – it inhibits mannosidase II located in the Golgi as well as lysosomal acid α-mannosidase [12]. These and related studies with another naturally occurring indolizidine iminosugar, castanospermine, (from the Australian black bean and identified in the near-extinct Moreton bay chestnut, *Castanospermum australe*), that is an inhibitor of α- and β-glucosidases [13], heightened expectations that this iminosugar or more potent synthetic analogues of greater selectivity, could be useful in the investigation of cellular glycoprotein processing [14, 15]. It was also postulated that the study of these molecules might be extended to the future development of antiviral drugs with inhibitory actions on the processing of essential viral N-linked glycoproteins [16]. Investigations showing that castanospermine inhibited processing of several host and microbial glycoproteins essential for viral proliferation and growth (e.g. the haemaglutinin of influenza virus and glycoproteins essential for the production of Dengue virus) further stimulated research into the use of iminosugars as agents with activity against infection with human immunodeficiency deficiency virus [17].

13.2.2 Potential medicinal applications of iminosugars

Iminosugars potentially have the following medicinal applications:

- As inhibitors of glycosidases: with potential actions on glycoconjugate processing in the endoplasmic reticulum, glycoprocessing and glycoprotein biosynthesis in the Golgi apparatus and other membraneous intracellular compartments. Inhibition of the activity of digestive enzymes in the small intestinal brush border, and enzymes involved in the breakdown of intracellular glycogen, have a potential application in the regulation of carbohydrate metabolism. Here the physiological targets would be the intestine, liver and cells of the immune system.
- As inhibitors of glycosyltransferases: key biosynthetic reactions involved in the formation of complex glycoconjugates including glycosphingolipids. This role has been assumed for application in the glycosphingolipidoses, which result from the failure to degrade glycosphingolipids in the lysosomal compartment of many tissues, including the brain.

- As pharmacological chaperones: maintenance of glycosidases and other proteins in their active conformation during biosynthesis, by binding to the active site - a role that is being explored for lysosomal glycosidases and membrane proteins, such as the mutated cystic fibrosis transmembrane regulator (CFTR protein).
- As modifiers of lipid rafts: binding to glycan-rich domains in lipid rafts to influence receptor function (e.g. insulin receptor) and/or membrane carrier proteins. They may also modify the composition of lipid rafts by restricting glycosphingolipid biosynthesis.

13.3 Clinical application of iminosugars as medicines

13.3.1 Diabetes mellitus

Crude extracts of mulberry roots and leaves contain 1-deoxynojirimycin, *N*-methyl-1-deoxynojirimycin, 1,4-dideoxy-1,4-imino-D-arabinitol, 1,4-dideoxy-1,4-imino-D-ribitol and their congeners, as well as the nortropane iminosugars, the calystegins [4, 5].Since the 17th century, mulberry preparations have been known to have powerful metabolic effects. Aqueous and ethanolic extracts of mulberry root, including Haarlem oil, were used in the treatment of diabetes mellitus and were known to induce significant weight loss. In this context one can only marvel at the acute powers of observation of the early physicians and herbalists who determined this action, since a related and more effective nojirimycin 2-hydroxyethyldeoxynojirimycin, now licensed as miglitol, is used to inhibit α-glucosidases in the digestive tract [1], and thus serves as an oral drug for the management of type 2 diabetes mellitus - either taken alone or in synergistic combination with other antidiabetic agents (Figure 13.1).

Miglitol has been licensed as Glyset since 1996 for diabetic patients in the USA and several European countries, as well as in South America. This agent was developed by the Bayer Company and its postulated mode of action is as an inhibitor of α-1, 4 glucosidases in the intestinal brush border. It will thus reduce postprandial hyperglycaemia by decreasing the release of glucose and oligosaccharides release by digestion of luminal carbohydrate and is active as a monotherapy in type 2 diabetes mellitus [18] The naturally occurring 1,4-dideoxy-1,4-imino-D-arabinitol is a potent inhibitor of glycogen phosphorylase *in vitro* and *in vivo*; it thus reduces hepatic gluconeogenesis from glycogen [1, 19]. This effect, or a related inhibition of the α-1, 6-glucosidase activity of glycogen debranching enzyme, might also contribute to the antidiabetic properties of miglitol, which is absorbed systemically [20].

Given that type 2 diabetes is a metabolic condition approaching near epidemic proportions in many societies worldwide, the availability of antihyperglycaemic drugs with modes of action that are distinct from the biguanides (e.g. metformin) and sulphonylureas (e.g. glicazide) and the thiazolidinediones (e.g. rosiglitazone), is increasingly important. It seems likely that α-glucosidase inhibitors acarbose and miglitol will have a key role in improving the health of diabetic patients. Several studies indicate that miglitol synergizes with sulphonyurea and biguanides drugs and that type 2 diabetic patients taking the combination have improved long-term glycaemic control [21, 22]. Clearly, however, agents which inhibit intestinal glucosidases will have predictable unwanted effects, principally those related to the mode of action which can be predicted: flatulence, bloating and diarrhoea, caused by inhibition of intestinal disaccharidases with attendant delivery of increased nutrient substrates to bacteria present in the lower intestinal lumen leading to enhanced fermentation with a production of carbon dioxide, hydrogen and methane as well as osmotically active products such as lactic acid. In patients taking insulin and other medications to treat diabetes, the α-glucosidases may unexpectedly lower blood sugar concentrations. However, it is noteworthy that when given alone at therapeutic doses, miglitol does not normally appear to have an action on glycogenolysis, since it does not lower blood glucose concentrations in fasting human subjects.

Miglitol is given by mouth alone or with a sulphonylurea in initial doses of 25 mg three times daily with meals, increased if necessary to a maximum 100 mg three times daily. At a dose of 25 mg in adults, miglitol is completely absorbed but at the higher dose only 50–70 per cent of the dose is absorbed; the iminosugar is not metabolized and appears unchanged in the urine with a plasma elimination half-life estimated to be about 2 hours. Trials indicate that when taken by patients with non-insulin-dependent diabetes mellitus in conjunction with an appropriate diet and exercise programme with blood pressure management, miglitol contributes to improved glycaemic control and thereby prevents stroke, heart and kidney disease as complications.

In those patients taking insulin as part of their management of type 2 diabetes, miglitol can exacerbate the frequency of hypoglycaemia, presumably by reducing the influx of portal glucose as a result of inhibiting breakdown of starch and other dietary carbohydrates in the intestinal lumen. The drug passes into breast milk and, therefore, should not be continued in lactating women who are breastfeeding their infants.

Miglitol is marketed in the USA as Glyset and in trials reduced the mean glycosylated haemoglobin concentrations (Hb A_1C) as a biomarker of medium- to long-term diabetic control. The drug reduces plasma glucose concentrations after meals as well as postprandial concentrations of serum insulin [23]. Its principal mode of action appears to be against membrane-bound α-glucosidases including sucrase, isomaltase and glucoamylase rather than on lactase, thus milk intolerance is a rare feature of patients receiving miglitol therapy. Gastrointestinal symptoms, related to α-glucosidase inhibition in patients with diabetes receiving miglitol, are reported to improve with time.

13.3.2 Human viral infections

Hitherto, most antiviral drugs have been used to target gene products that are unique to a given viral pathogen. Although this stratagem generates highly specific and well-tolerated antiviral drugs, it is vulnerable to the emergence of escape mutants which circumvent the mode of action and lead to infection with drug-resistant strains of virus.

An alternative approach is to target a cellular process with preferential effects on the life-cycle of the virus rather than the host cell. This would minimize the potential for the development of escape mutants. In the case of infections with HIV, hepatitis B virus (HBV) and hepatitis C virus (HCV), both therapeutic avenues are being explored [24].

One cellular pathway that can be targeted in the host is the N-glycan processing pathway [25, 26]. The proper folding and assembly of many nascent glycoproteins depends on interactions with calnexin in the endoplasmic reticulum (ER) [27]. The precursor N-glycan that is added cotranslationally to the nascent polypeptide has three terminal glucose residues, which are removed in a stepwise fashion by α-glucosidases in the ER. The monoglucosylated form interacts with the chaperones, calnexin and calreticulin. Interaction with calnexin is required for the correct folding of some, but not all, glycoproteins [27]. Inhibition of the α-glucosidases can therefore be used specifically to target proteins that depend on this interaction. Viruses subvert the machinery of host cells to glycosylate their envelope glycoproteins and, as a consequence, the folding of certain viral glycoproteins is calnexin-dependent. Partial inhibition of ER α-glucosidases may interfere with the correct folding of these proteins, and thus has therapeutic potential in certain viral infections with little effect on the viability of the host. Iminosugars that inhibit α-glucosidases I and II have been evaluated in several viral systems (reviewed in [28]).

Here we discuss two viral pathogens (HIV and HBV) and the effects of iminosugars that inhibit ER α-glucosidase *in vitro* and in whole animals.

Human immunodeficiency virus (HIV) and AIDS

HIV-1 is the causative agent of acquired immunodeficiency syndrome (AIDS). The virus encodes two envelope glycoproteins (gp120 and gp41), derived from a precursor protein (gp160) by endoproteolytic cleavage in the *cis*-Golgi [29]. Although proteolytically cleaved, gp120 remains noncovalently attached to the transmembrane protein gp41, which anchors the complex. During infection, gp120 binds to its cellular receptor (CD4), and undergoes a conformational change that exposes gp41 and leads to fusion with the host cell plasma membrane and entry of the virus into the cell [30].

Glycoprotein gp120 of HIV-1 is one of the most heavily N-glycosylated viral proteins known and contains a rich complement of oligomannose and complex-type glycans [31]. It has been found that treatment of HIV-1 infected cells with *N*-butyldeoxynojirimycin (*N*B-DNJ), an ER α-glucosidase inhibitor, blocked viral infectivity and syncytium formation *in vitro* [32]. Under these circumstances, the conformational change of gp120 that normally follows binding to CD4 and leads to exposure of gp41, does not occur - and so fusion of virus with the host cell is abrogated. It was later shown that misfolding of the V1/V2 loop of gp120 was responsible for the antiviral effect of *N*B-DNJ since, in order to achieve its correct conformation, there is a critical requirement for this region of the glycoprotein to interact with calnexin [33].

N-butyldeoxynojirimycin (*N*B-DNJ) was the subject of a patent application by the company Nippon Shinyaku. As with *N*-hydroxymethyl-1 deoxynojirimycin, the initial application for *N*B-DNJ was based upon its potential use as an inhibitor of small intestinal α-glucosidases (and, possibly, hepatic glycogen debranching enzyme) in diabetes. However, studies showing that *N*B-DNJ penetrated the cell and could inhibit α-glucosidases I and II in the endoplasmic reticulum and thus disturb the folding of mature glycoproteins provided the mechanistic basis for clinical trials of this iminosugar in patients with HIV/AIDS, with or without the coadministration of the nucleotide reverse transcriptase inhibitor, zidovudine.

*N*B-DNJ was evaluated in Phase II clinical trials as an antiHIV agent: although modest effects were noted on viraemia, it was not possible to achieve sufficiently high serum concentrations of the drug [34, 35] to bring about its action in the ER. *In vitro* studies indicate that *N*B-DNJ has an inhibitory constant for α-glucosidase I in the submicromolar range but in whole cell cultures concentrations of the drug in the millimolar range are required for complete inhibition and suggest poor penetration to the site of action of α-glucosidase I. No further clinical studies in HIV infected patients were therefore conducted.

In the trial, *N*B-DNJ was generally well tolerated, and the principal unwanted effect was diarrhoea, attributable to inhibition of intestinal disaccharidase activity. The drug was used at up to 3 grams daily for 6 months in patients with HIV/AIDS: although disappointing in the failure to demonstrate antiviral properties *in vivo*, these trials yielded important information on the unwanted effects – and tolerability – of *N*B-DNJ in human subjects. After 6 months, the trials were abandoned and *N*-butyldeoxynojirimycin has not been further developed for the treatment of human retroviral infections.

Hepatitis B virus(HBV)

HBV causes liver disease [36] and hepatocellular carcinoma [37]. The HBV genome encodes three envelope proteins: large (L), middle (M) and small (S), which are derived from a single open reading frame [38]. These comprise the principal components of the viral envelope, but are also secreted as DNA-free subviral particles that are noninfectious and greatly outnumber infectious virions. Unlike the HIV envelope glycoproteins, HBV envelope proteins contain only two N-glycosylation sites. Despite this difference both HBV and HIV are sensitive to inhibitors of ER α-glucosidases [28]. The M protein of HBV is dependent on interaction with calnexin for correct folding, which in turn is required for correct and viral envelopment.

Until recently there were no replication systems for studying HBV *in vitro*. Therefore woodchucks chronically infected with woodchuck hepatitis virus (WHV), a hepadnavirus and close relative of HBV, have been commonly used to test drugs for antiviral activity. Woodchucks chronically infected with WHV were treated with *N*-nonyl-DNJ (*N*N-DNJ), a nine-carbon alkyl derivative of DNJ, which has greater potency than *N*B-DNJ against the virus *in vitro*. Secretion of enveloped virus was reduced in the woodchucks in a dose-dependent manner [39]. At the lower concentrations of *N*N-DNJ that inhibited WHV secretion, the glycosylation of most serum glycoproteins appeared to be unaffected, thus indicating that replication of this HBV-like virus in the living host is very sensitive to minimal inhibition of α-glucososidase. In another viral system (bovine viral diarrhoea virus, BVDV, a virus closely related to HCV), long alkyl chain iminosugars were found *in vitro* to have an additional mode of action (inhibition of the HCV ion channel, p7) [40, 41].

Potential applications of iminosugars as antiviral agents and their diverse mechanisms of action have recently been reviewed [42]; it is clear that the field offers rich opportunities for preclinical research and should encourage continued pharmaceutical investment in the development of innovative medicines.

13.3.3 Use of iminosugars in patients with advanced malignancy

In mice with experimental exogenous tumours, administration of heterocyclic iminosugars such as swainsonine inhibits formation of metastases (Figure 13.1). After it was shown that swainsonine inhibited spontaneous metastasis and altered glycoprotein expression implicated in cell-to-cell contact inhibition, which prevents metastatic growth following seeding of tumours [43], Phase 1 clinical studies in patients with advanced malignancy were initiated.

Patients with advanced tumours were treated with escalating doses (50–550 micrograms per kg per day) of swainsonine in an attempt to arrest the progression of their cancer by inhibiting metastasis and altering the course of their condition [44]. Administration of the drug was associated with oligosaccharidauria, as predicted by the mode of action. In several patients, tumour growth was retarded, including one head and neck tumour which within 1 week of administration decreased in size by approximately 50 per cent; two patients with advanced lymphangitis carcinomatosa also showed objective signs of tumour regression. However, swainsonine treatment was associated with significant hepatotoxicity which was particularly severe in patients with hepatic abnormalities revealed by serum tests before introduction of the drug. One patient died in the trial as a result of acute respiratory distress syndrome, for which the drug may have been responsible. So far no further trials of swainsonine have been proposed in patients with malignant conditions.

13.3.4 Iminosugars as inhibitors of glycosphingolipid biosynthesis

Biochemical studies

As explained in Chapter 11 of this volume, Norman Radin in Michigan had suggested that inhibition of glycosphingolipid biosynthesis would have potential therapeutic application in Gaucher's disease and related lysosomal disorders in which the rate of degradation of glycolipid substrates was reduced by inherited enzyme deficiency [45]. Dr Radin and colleagues investigated morpholino- and pyrrolidino compounds that were considered to be mimetics of ceramide and thus potent inhibitors of the limiting step for glycosphingolipid biosynthesis, UDP-glucosylceramide transferase [46, 47]. It was argued that in the conditions where the breakdown of glycosphingolipids is impaired but not absent, reduction of new incoming glycosphingolipid substrates by inhibition of the first biosynthetic step would, in a stratagem resembling the use of statins to inhibit the first committed biosynthetic step for cholesterol formation, enable the storage material to be cleared.

Most surviving children and adults with glycosphingolipidoses, and Gaucher's disease in particular, harbour mutations in the cognate degradative enzyme (in Gaucher's disease, glucosylceramidase) that impair rather than completely disable catalytic activity. Thus, when synthesis of the substrate is decreased, the residual activity of the glucosylceramidase will slowly degrade the stored glycolipid within the lysosomal compartment – and thus rebalance the rate of glycosphingolipid biosynthesis with its rate of degradation. A central tenet of the understanding of the lysosomal diseases is that the storage macromolecules induce the cellular pathology; hence clearance of the substrate within the lysosomal compartment would be predicted to ameliorate the manifestations of the disease.

At the University of Oxford, one of the authors (FMP) and colleagues, including Drs Dwek and Butters, demonstrated that *N*-butyldeoxynojirimycin and its congener *N*-butyldeoxygalactonojirimycin (*N*B-DGJ) inhibited the biosynthesis of glycosphingolipids by an action on UDP-glucosylceramide transferase and thus might replace the compounds developed in the University of Michigan as substrate-reducing drugs for use in glycosphingolipid storage disorders [48, 49]. Iminosugars with the appropriate chiral configuration and substituents at the ring nitrogen with groups comprising at least three carbon atoms were found to be active inhibitors. Dr Butters and colleagues have provided data from biochemical and structural modeling studies indicating, unexpectedly, that the inhibitory action of these alkyl iminosugars may in part be explained by behaviour as molecular ceramide mimetics rather than as competitive inhibitors at the sugar-binding domain on the transferase molecule [15, 50]. Formal confirmation of this postulate will require further investigation and ultimately will require a source of sufficient pure ceramide-specific UDP-glucosyl transferase for full biochemical and structural characterization.

Early investigations showed that culture of the murine macrophage cell line WHEI-3B in the presence of an irreversible inhibitor of glucocerebrosidase, conduritol-β-epoxide, would induce lysosomal abnormalities accompanied by increased glucosylceramide concentrations in whole cell extracts [48]. Addition of *N*B-DNJ and *N*B-DGJ to cultures so treated abrogated this storage. Cells cultured in the presence of 5 μM *N*B-DNJ had concentrations of glucosylceramide comparable to cells cultured in the absence of conduritol epoxide. Electron microscopy provided additional evidence that the storage material in the lysosomes of conduritol-β-epoxide-treated cells was decreased by treatment with the iminosugars.

These experiments carried out *in vitro* raised the question as to whether *N*B-DNJ was a drug that could be used as a substrate inhibitor to treat glycosphingolipidoses, including patients with Gaucher's disease, as specifically proposed by Radin [51]. Given that *N*-butyldeoxynojirimycin is a small molecule with the potential to enter the brain, it was also considered that the drug could be used to treat those glycosphingolipidoses associated with significant brain injury and accumulation of glycosphingolipids in the nervous system.

Iminosugars in experimental glycosphingolipidoses

When the action of the iminosugar *N*-butyldeoxynojirimycin on UDP-glucosylceramide transferase was identified, no satisfactory model of Gaucher disease in an animal was available for experimental purposes. Accordingly, investigations were conducted in mice genetically modified by targeted disruption of genes encoding key degradative enzymes in the breakdown of glycosphingolipids other than those which accumulate in Gaucher disease. These molecules, the GM2, GA2 gangliosides, are degraded in the same lysosomal pathway and normally give rise to glucosylceramide. The Tay–Sachs mouse, lacking hexosaminidase A, was generated by disruption of the hexosaminidase α-subunit in embryonic stem cells (the first knockout model of a glycosphingolipid storage disease to be generated).

Latterly, Sandhoff mice with clinical manifestations resulting from loss of hexosaminidase A and B activity induced by genetic disruption of the murine β-subunit gene common to both the lysosomal hexosaminidases, were used.

Unlike humans, mice lacking the α-subunit of hexosaminidase A store modest quantities of GM2 and GA2 ganglioside in the brain but have a normal lifespan and only develop mild manifestations of neurodegenerative disease beyond 1 year of age [52]. Although the indolent phenotype of this animal model precluded evaluation of the clinical efficacy of substrate reduction with iminosugars, the mice did allow assessment of the long-term biochemical and cellular effects of the treatment in a storage disorder.

Preclinical studies were undertaken first in Tay–Sachs mice and later in the clinically-affected neurodegenerative model of the glycosphingolipidosis, the Sandhoff mouse. *N*-butyldeoxynojirimycin was administered at high doses (0.6–2.4 g/kg/day) in mouse chow and the animals were investigated for reduction of the burden of glycosphingolipids in the tissues combined with optical and electron microscopy to investigate the subcellular effects on ganglioside storage within neurones. Administration of *N*B-DNJ slowed the rate of storage of glycosphingolipids in the brain and visceral organs and electron microscopy confirmed significant improvement of neuronal glycosphingolipid accumulation when compared with control animals, including affected litter mates that did not receive the drug. *N*-butyldeoxynojirimycin is excreted actively by the kidney in mice but the concentrations in the diet gave rise to serum concentrations of 18–57 μM, leading to approximately a 70 per cent reduction in tissue glycosphingolipid concentrations in several peripheral organs [53].

When the murine model of Sandhoff disease, a GM2 gangliosidosis resulting from disruption of the β-hexosaminidase A and B activities, was available further studies were undertaken to explore the potential therapeutic effects of iminosugars on an acute model which closely recapitulates human infantile and juvenile GM2 gangliosidoses. Untreated mice with Sandhoff disease develop tremor, paralysis and wasting of their hind limbs, accompanied by rigidity and ataxia; after 16 weeks of age they become moribund and so require euthanasia [52]. Oral administration of *N*B-DNJ in the chow significantly delayed the onset of these stereotypic disease manifestations and attenuated the rate at which motor function decreased. Sandhoff mice receiving *N*-butyldeoxynojirimycin had a life extended by approximately 40 per cent compared with their untreated litter mates. Pathological, biochemical and ultrastructural examination of the tissues showed that storage of GM2, GA2 and cognate glycolipids in the peripheral organs and brain was significantly less in animals receiving the iminosugar treatment [54]. Similar studies were carried out with the galactose analogue, *N*-butyldeoxygalactonojirimycin, *N*B-DGJ. Administration of *N*B-DGJ did not induce weight loss and was better tolerated in the mice with Sandhoff disease; nonetheless, a comparable or marginally improved salutary therapeutic effect was observed [55].

13.3.5 Clinical studies of *N*B-DNJ (miglustat) in Gaucher disease and other glysosphingolipid disorders affecting the lysosome

Preclinical studies in Tay–Sachs and murine Sandhoff disease combined with the extensive studies of high-dose *N*-butyldeoxynojirimycin given orally to human subjects with late-stage HIV infection and AIDS immediately suggested a clinical trial of *N*B-DNJ in patients with type 1 Gaucher's disease.

As set out in Chapter 11 of this volume, the first international study in 28 patients with mild-to-moderate type 1 (non-neuronopathic) Gaucher's disease who were unwilling or unable to take enzyme replacement therapy were enrolled in this pivotal study [56]. These patients had a confirmed diagnosis of Gaucher disease and mild-to-moderate hypersplenism or, if they had already undertaken splenectomy for the disease, greatly increased liver volume and thrombocytopenia. In the trial they were treated with 100 mg of *N*-butyldeoxynojirimycin given orally three times daily with the option to increase the dose subject to tolerance.

In Gaucher's disease the mode of action of substrate reduction therapy is indirect. In this condition, the disorder is manifest principally in the cells of the mononuclear phagocyte system resident in the spleen, liver and bone marrow, and the lungs derive their excess glycospingolipids from the digestion of glycolipids present on the surface membrane of leucocytes, erythrocytes and platelets. It was postulated that in the presence of *N*-butyldeoxynojirimycin, delivery of formed blood cells with reduced expression of glucosylceramide and other related complex glycosphingolipids would correct the imbalance between the rate of glucosylceramide degradation and the rate of release of exogenous glycosphingolipids within macrophages. Correction of this imbalance should ultimately ensure regression of glucosylceramide storage in the fixed tissue macrophages that are the pathological focus of this disorder. The therapeutic effect is thus predicated on the pathological but scientifically ill-understood relationship between lysosomal storage of glycosphingolipids and the manifestations of Gaucher's disease.

The initial clinical trial [56] and subsequent 3 year follow-up studies [57] clearly demonstrated a sustained therapeutic effect of *N*-butyldeoxynojirimycin treatment at the low doses used in patients with mild-to-moderate Gaucher's disease. Plasma concentrations of 1–2 μg (approximately 5 μM) *N*-butyldeoxynojirimycin were documented in patients receiving the drug. Under these conditions, at the 1 year time point, *N*B-DNJ treatment was associated with a decrease of 12 per cent in liver and 19 per cent in spleen volumes; there was, moreover, a significant reduction in the activity of the surrogate biomarker of Gaucher disease activity, plasma chitotriosidase, which decreased by more than 16 per cent over the year. In patients with anaemia at the start of treatment, *N*B-DNJ significantly improved blood haemoglobin concentrations with a modest but clear increase in absolute platelet count after 1 year. In the extension phase of an open-labelled prolongation of the trial, 13 patients completed treatment.

In summary, 100–300 mg of *N*-butyldeoxynojirimycin administered orally three times a day brought about a sustained therapeutic effect on all clinical, radiological and laboratory markers of disease activity, as demonstrated over a 3 year period. Evidence for a dose-related effect of *N*B-DNJ was shown in a subsequent trial where patients were given 50 mg of the drug three times daily. Under these circumstances, modest reductions in liver and spleen volumes (5.9 and 4.5 per cent respectively after 6 months of treatment) were demonstrated [58].

These promising initial studies have been recapitulated in other centres: in a noncomparative open-label study over 24 months, *N*-butyldeoxynojirimycin was administered at a dose of 100 mg three times daily. Clinical safety evaluation, assessment of neurological status and other monitoring confirmed a significant percentage reduction in spleen volume, an improvement in absolute platelet count, reduction of baseline chitotriosidase activity by more than a quarter at 24 months of administration of the drug. The treatment was well-tolerated but adverse events as predicted from all previous studies, with gastrointestinal (abdominal) pain, diarrhoea, flatulence and transient tremor were observed [59]. Thus independent studies have confirmed the salutary effects of *N*-butyldeoxynojirimycin in patients with mild Gaucher disease. As a result, the first oral treatment designed specifically to address the underlying biochemical defect of this inherited lysosomal disorder has been licensed for medicinal use.

The best standards of care for Gaucher's disease involve the administration of an expensive, parenteral, protein-replacement therapy in the form of recombinant mannose-terminated human glucocerebrosidase currently given, initially at least, at 60 iu/kg/bodyweight by intravenous infusion twice monthly. Enzyme treatment improves visceromegaly and haematological abnormalities of this lysosomal disease affecting the mononuclear phagocyte system; enzyme therapy thus corrects the underlying inherited defect in this disorder. In contrast, substrate reduction therapy with the iminosugar *N*-butyldeoxynojirimycin has a slower onset of action, as correctly predicted in the design of the first experimental trial of its use in Gaucher disease [56]. The treatment addresses key aspects of the disease and in a manner that is related to the proposed mode of drug action. In the pivotal clinical trial, the concentration of GM1 ganglioside measured in lymphocytes decreased by more than 38 per cent in five patients after treatment with the iminosugar for 1 year [56]. As predicted, baseline plasma glucosylceramide also decreased in eight subjects (from 14.2 μmol/l to 11.4 μmol/l) after 12 months of therapy.

Adverse effects

As mentioned above, several predicted unwanted effects in miglustat occurred in patients with Gaucher's disease. These included abdominal bloating, flatulence and diarrhoea and are probably related to the ability of the agent to inhibit disaccharidases and/or their biosynthesis in the intestinal brush border membrane. Fortunately the diarrhoea, though common at the outset, appears to be transient in most patients and with minor dietary alterations and the judicious use of antimotility agents, appears to be well-controlled. Transient tremor develops in about 25 per cent of the patients, examination of which indicates a physiological or sympathomimetic tremor which resolved spontaneously – it remains unexplained. Several patients develop marked weight loss, partially aggravated by self-imposed dietary changes predicated on the advice they were given to minimize the diarrhoea. In the first trial, a few patients developed a painful axonal neuropathy that improved slowly on withdrawal of the drug [56]. These patients were among those who lost the most weight during initiation of the therapeutic trial and in others peripheral neuropathy developed in the context of significant comorbidity, particularly deficiency of vitamin B12. Peripheral neuropathy does not appear to be a frequent or recurrent problem in patients receiving *N*-butyldeoxynojirimycin for Gaucher disease or other disorders in whom weight loss is not severe [57].

Although early reports of disturbed cognitive function emerged from trials conducted in Israel, and of peripheral neuropathy in the first clinical trials of the drug, this is not a proven direct toxic effect. Impaired mental function has not subsequently been noted in follow-up monitoring, further clinical trials or pharmacovigilance reports [56–59].

The first Gaucher patients to receive *N*B-DNJ received the agent that was chemically rederived from stock of a prodrug, glycovir (SC 49483) which had been stored anaerobically under argon gas since its use in the preceding human trials in HIV AIDS. The prodrug, *N*B-DNJ-tetrabutanoate, was designed to reduce local inhibition of glycosidases in the intestine, but it would be rapidly converted to the active *N*B-DNJ after absorption. In experimental animals morphological changes were reported in various tissues with this perbutylated derivative and were attributed to nonspecific inhibition of intracellular glycosidases (see [1] and references therein).

Recently, further 2 year preclinical studies required by the licensing authority, the Euopean Medicines Agency (EMEA) in mice receiving large doses of *N*B-DNJ, as required to obtain therapeutic blood concentrations of the agent in this species, have been conducted. Morphological changes in the large intestine, with cystic colitis, increased cell proliferation and focal hyperplasia and adenomatous lesions with frank carcinomas were observed in the mice – but not rats that were also studied or, in retrospect, in any other of the species, including monkeys, that were the subject of earlier preclinical toxicology data. Although the relationship of these findings to human exposure to *N*B-DNJ as a therapeutic drug is unclear,in March 2007 the Actelion company issued an appropriate warning notice and modified Summary of Product Characteristics (SPC) to treating physicians, who are advised to investigate any unexpected chronic gastrointestinal symptoms in patients receiving *N*B-DNJ (miglustat) that persist despite simple dietary measures. To the authors' knowledge, no colonic complications of this kind have been identified in human subjects.The doses of *N*B-DNJ (miglustat) used in the 2 year toxicology studies (0.21,0.42,0.5 and 0.84 g/kg body weight daily) are 16–65-fold greater than those used in humans based on body weight; they are, however, about one quarter to one half of the doses generally used in the preclinical therapeutic studies of Tay–Sachs, Sandhoff disease in experimental animals [52–55].

Miglustat is not recommended for children or adolescents with Gaucher disease and contraception is advised for all persons receiving the drug who are in the reproductive age group, based on the studies showing aberrant morphology and mobility of spermatozoa in some strains of experimental mice [60]. However, there are at least two anecdotal reports of pregnancies with the birth of normal offspring, occurring in women whose husbands were receiving the drug for Gaucher disease. This in itself, however, cannot be considered sufficient justification for avoiding contraception even though the drug has no known teratogenic properties and even in susceptible mouse strains in which aberrant sperm structures are induced, *N*B-DNJ does not appear to affect the genetic integrity of male germ cells [61].

Very recently the effects of miglustat on sperm number, sperm morphology, motility and function (acrosome reaction) have been evaluated in healthy men. For this study, miglustat was used at the same dosage as used in patients with Gaucher disease (100 mg three times daily). No abnormalities in any parameter were identified in this 8 week study [62] suggesting that the effects on male fertility in mice cannot be applied to healthy humans. Further studies have identified a striking strain dependency to the effect of miglustat on sperm maturation in inbred mouse strains [63]. In the same study, male rabbits were found to be unaffected by miglustat treatment, confirming striking intra- and interspecies variation in the action of miglustat on male fertility.

Recently, a possible explanation for the effect of *N*B-DNJ on spermatogenesis has been proposed [64, 65].Aerts and colleagues have shown that *N*B-DNJ and other hydrophobic nojirimycin derivatives are potent inhibitors of a nonlysosomal membrane-bound glucocerebrosidase that is active at neutral pH and not deficient in Gaucher disease. The cognate gene, encoding the enzyme assigned the term, GBA-2, has now been identified by molecular cloning and its sequence characterized [64]. Almost simultaneously, male mice deficient in this enzyme as a result of homozygosity for an inactivated GBA-2 locus generated by disruption in embryonic stem cells,have been reported to be infertile with the production of morphologically abnormal sperm with round heads, abnormal acrosomes and impaired motility; mild glycolipid excess was detected in the tissues but the mice of both sexes had a normal lifespan and were without overt disease [65]. The molecular basis for strain and species variation is currently under investigation but it seems probable that the enzyme is implicated in the trafficking of glycolipids from the endoplasmic reticulum to other membraneous domains, including the cell membrane and acrosome – at least of murine spermatozoa. Although *N*B-DNJ is known to reduce the rate of formation of glycosphingolipids as predicted from its proposed action as a substrate-inhibitor for Gaucher and related diseases at the level of UDP glucosyl ceramide transferase, it is formally possible that its powerful effect on the nonlysosomal glucocerebrosidase, GBA-2, contributes to its therapeutic efficacy. In any event, given the experimental evidence so far available, miglustat does not appear materially to affect reproduction in male patients with type 1 Gaucher disease; indeed, neither the drug nor its congeners seems likely to give rise to an innovative class of oral contraceptives for men, as was once thought [60].

In Gaucher's disease, several patients have had an exposure for more than 8 years to *N*-butyldeoxynojirimycin with sustained therapeutic effects and no evidence of toxicity. In two patients an impressive reduction in the abnormal fat signal was demonstrated by MRI spectroscopy using the Dixon's quantitative chemical shift imaging (QCSI) method to determine the fat signal that is pathologically disturbed in patients with active Gaucher disease presumably as a result of replacement by the normal haematopoietic bone marrow by lipid-laden cells [66]. Sustained improvement in the pathological fat signal to the normal range, which has been observed in these patients after prolonged treatment with *N*-butyldeoxynojirimycin, thus attesting to the value of the drug in key foci of the disease that prove most resistant to the first-line drug choice based on enzyme replacement therapy.

Anderson–Fabry disease

This X-linked lysosomal disease is a glycosphingolipidosis in which globotriasoylceramides accumulate as a result of deficient activity of α-galactosidase A. The disease, which is associated with widespread deposits, including in the endothelium, heart, glomerulus and medium-sized arteries, occurs in heterozygous females but is usually more severe in men; it causes renal failure, cardiomyopathy, neuralgic pains and premature stroke.

Early clinical trials of miglustat in Fabry disease were terminated as a result of untoward neurological toxicity. Of 16 men with Fabry disease who received the drug, 13 developed tremor and two complained of parasthesiae; three patients withdrew from the study [67]. However, since miglustat is principally excreted in the kidney and men with established disease are likely to have pre-existing renal failure, it seems probable that high plasma concentrations of miglustat were achieved in the conduct of the trial and this would have led to enhanced toxicity. Although the residual activity of the enzyme α-galactosidase A which is deficient in Fabry disease, is often low, a combination of miglustat with enzyme therapy might be justified in patients with a severe and a progressive lysosomal disease affecting key organs such as the kidney, heart and brain. It is notable also that iminosugars (see below) are being investigated in clinical trials as pharmacological chaperones of endogenous α-galactosidase A.

Licensing of N-butyldeoxynojirimycin

N-butyldeoxynojirimycin is registered as miglustat (Zavesca™) and licensed originally from the Searle Company who acquired it from Japan. The drug has been evaluated by the US Food and Drug Administration (FDA) and the European Agency for the Evaluation of Medicinal Products (EMEA) under the orphan drug regulatory procedures. Efficacy data from 80 patients with type 1 Gaucher's disease and safety data were examined leading to marketing approval in November 2002. The drug was launched by Actelion in the UK in March 2003 and awarded orphan status by the FDA for the treatment of Gaucher's disease in 2000 with approval in July 2003. The Israeli company TEVA filed for approval of Zavesca in January 2003.

Licensing approval for Zavesca (miglustat) is conditional upon the introduction of comprehensive surveillance programmes (pharmacovigilance): physicians prescribing the agent are encouraged to submit clinical safety data whether or not the patient so treated has Gaucher disease or another disorder for which, at present, the treatment is not licensed. Original safety data on 220 patients were available at the time of review in October 2004 and the evaluation of risk/harm effects in miglustat have remained in place.

As indicated above, the recent finding of colonic cancer and precancerous changes in mice – but not in other experimental animals used in preclinical toxicology testing – means that although there is a low probability of carcinogenicity in human patients receiving the agent, use of miglustat will require continued safety monitoring. In March 2007, the Actelion company issued a revised SPC for Zavesca (miglustat) relating to the mouse carcinogenicity study. The surveillance programme indicates that diarrhoea, tremor or unremitting progression of neurological disease in patients receiving the drug for unlicensed indications in sphingolipidoses, have been convincing and understandable reasons for drug withdrawal.

The therapeutic position of miglustat

In 2003, an advisory council comprising international experts based in the European Working Group for the Study of Gaucher Disease and patient organizations met to submit a position statement on the indications for the use of miglustat in Gaucher disease [68]. The drug is considered to be indicated for the oral treatment of mild-to-moderate Gaucher disease which is defined as patients with a haemoglobin concentration greater than 9.0 d/l and a platelet count greater than 50×10^9/l in patients without rapidly evolving bony complications of the condition.

Miglustat is at present suggested as a second-line agent for patients who are unsuitable for enzyme replacement, for example those who have poor venous access or a phobia for injections, and those who are otherwise unwilling or unable to receive enzyme therapy. In the US, the slower and less marked therapeutic responses of miglustat, when compared with enzyme therapy with imiglucerase, have been met with caution by physicians treating Gaucher disease [69].

Expanding the use of miglustat in Gaucher disease

The nature of 'unwilling or unsuitable' The suggestion that a patient with active Gaucher's disease may be considered unsuitable for enzyme replacement therapy poses genuine questions of its own. Some patients may have religious or other objections to receiving infusions of enzyme by vein but it is very unusual in practice for patients to suffer sensitivity or other adverse reactions to enzyme replacement therapy. It must be made clear, however, that in Gaucher disease, the safety profile of Cerezyme , based on more than 10 years experience in thousands of patients worldwide has been excellent, and immune reactions of a severity sufficient to require cessation of enzyme therapy are very rare (<1 per cent). Over a prolonged period, however, a slightly larger proportion of patients with Gaucher's disease receiving enzyme therapy develop a genuine distaste for – or even a horror of – needles as a result of persistent difficulties with the infusions.These difficulties usually arise as a result of either poor technique, physical disability or an inability satisfactorily to conduct infusions of enzyme at home; this is especially the case where access to appropriate centres for giving the infusions, or simply providing practical advice, is restricted. Needle phobia may be absolute or relative but some patients also develop genuine difficulties with access to exiguous peripheral veins for the administration of the drug.

A few patients with Gaucher disease, particularly adolescents and young adults in secondary education, or those who travel extensively in relation to their employment or educational requirements, find the discipline of regular infusions difficult to maintain. Once their disease has been controlled during the debulking phase of enzyme replacement therapy, the question of oral treatment in the maintenance phase may become a real one. It is thus not surprising that the Genzyme corporation, the manufacturer of imiglucerase , is conducting its own clinical trials of an orally active ceramide mimic (GENZ 112638) in Gaucher disease that is a powerful inhibitor of UDP-glucosylceramide transferase and developed 'in-house'.This interesting water-soluble compound is not an iminosugar but another aromatic pyrrolidine derivative, linked to an eight carbon fatty acid side-chain [47, 51].

Future clinical development of miglustat in Gaucher disease Miglustat appears to be quite well tolerated and represents an acceptable option for patients with Gaucher's disease who are unable or unwilling to receive enzyme replacement therapy. Further trials are underway to evaluate the potential role of miglustat in the maintenance therapy of patients with Gaucher's disease that has been stabilized by prior enzyme replacement therapy. In this open-labelled noncomparative study, patients treated with enzyme replacement therapy for several years are enrolled for a 24 month trial to examine the long-term efficacy and safety of treatment with miglustat. The trial is designed to investigate whether noninferiority to maintenance enzyme replacement therapy on the basis of clinical criteria, physiological and haematological markers as well as the determination of biomarker activity can be demonstated. It seems likely that such a study will further enhance our clinical knowledge of miglustat use; it may also lead to an enlarged therapeutic position for the agent in the long-term treatment of patients who have already secured the principal benefits of enzyme therapy in the initial phase of disease regression.

13.3.6 Miglustat in neurodegenerative disorders

Glycosphingolipids frequently accumulate in neurones and those that have a parent molecule based on glucosylceramide represent plausible targets for examining the effects of iminosugar therapy in experimental animals and humans [53–56, 70, 71].

Miglustat in neuronopathic Gaucher disease

Extensive clinical studies have shown that neurological manifestations occur in some patients who were formerly characterized as having type 1 disease (nonneuronopathic) but who later experienced neurological deficits including disordered eye movements, an extrapyramidal syndrome resembling Parkinson's disease (but responds poorly to specific agents), and neurosensorial deafness. Clearly these symptoms may develop in patients receiving enzyme replacement therapy and may serve as a stimulus for the evaluation of miglustat as a potential therapy.

Patients with established type 3 neuronopathic Gaucher's disease (i.e. non-acute neuronopathic Gaucher's disease) have been enrolled in a multicentre clinical trial as a primary endpoint to determine the effects of miglustat on the abnormal saccadic eye movements that characterize this condition. Sampling of cerebrospinal fluid has shown that miglustat does indeed penetrate into the nervous system and that concentrations of the drug reach 20-40 per cent of those achieved therapeutically in the plasma. Preliminary reports of this trial conducted at Great Ormond Street Hospital, London,

UK, and the National Institutes for Health, Bethesda, Maryland, USA, indicate that the agent did not achieve its primary therapeutic endpoint on the abnormal saccadic eye movements, The redesign of the trials in patients with type 3 Gaucher disease may reveal hitherto unknown beneficial effects of miglustat therapy but, at the time of writing, there is considerable disappointment about the poor effect of the drug in what might have appeared to be the ideal therapeutic target.

It should be recognized, however, that in this so-called 'neuronopathic' variant of Gaucher disease, gaze paralysis is usually apparent early in life and by the time patients are enrolled in clinical trials, they are likely to have sustained irretrievable neuronal injury or cell loss in mid-brain nuclei responsible for occulomotor function. This emphasizes the formidable difficulties in designing suitable studies to investigate the ability of any agent to affect clinically relevant aspects of neurodegenerative disorders, including those that affect glycosphingolipid metabolism in the lysosome. Thus it is necessary to consider a broad range of endpoints in the design of clinical trials and to take into account the slow but relentless progression of many of these conditions in the untreated state.

Further trials for patients with inherited GM2 gangliosidoses (late-onset Tay–Sachs disease and juvenile Sandhoff disease) are at late stages of completion and represent clear logical extensions on the preclinical data obtained in mice with experimental GM2 gangliosidosis due to disruption of the α- and β-subunits of hexosaminidase.

Secondary glycosphingolipid storage disorders

Several disorders of lysosomal function are associated with the accumulation of complex glycosphingolipids, especially complex gangliosides in neurons, which appear to be consequential rather than primary effects of the underlying cause of the disease and in which the enzymes of glycosphingolipid breakdown appear to be intact.

Niemann–Pick disease type C (NPC) This lysosomal disorder, associated with oculomotor palsy, dementia and cerebellar ataxia, is the paradigm of secondary glycosphingolipid disorders. The disorder accompanied by the neuronal accumulation of complex gangliosides including GM1, GM2 and GM3 and neutral glycosphingolipids, GlcCer and LacCer. These appear to accumulate as a result of a trafficking defect of cholesterol and gangliosides within neuronal and other cells. A complete block in trafficking at the level of the late endosome arises in NPC cells as a result of mutations in the cognate membrane protein, NPC1, localized in the late endosome/lysosome and, rarely, the product of its soluble lysosomal partner, NPC2. It is also notable that patients with Sanfillipo disease (mucopolysaccharidosis type 3) also accumulate glycosphingolipids in neuronal tissue [72] in a cruel, but biochemically unrelated lysosomal disorder of glycosaminoglycan breakdown that is associated with prominent neurodegeneration – often in the absence of the overt somatic features that characterize the other mucopoysaccharidoses.

For many years, it was unclear to what extent the ganglioside accumulation – as opposed to the cholesterol-trafficking defect – contributes to neuronal injury in NPC. Experimental evidence in animal models now clearly implicates lysosomal ganglioside storage in the neuropathology of NPC1. Natural mutations of the murine npc1 gene are associated with a well-characterized acute neurodegenerative disorder, which serves as an authentic model of human NPC. When homozygous NPC mice are interbred with mice lacking the capacity to synthesize GM2 and complex gangliosides due to targeted disruption of the β- 1,4 GalNAc transferase gene in embryonic cells, the neuronal pathology is markedly attenuated and outcome is improved, thus indicating a pathogenic role for glycosphingolipid storage in the brain. Further evidence for the role of glycosphingolipids in the neuronal pathology of NPC has been provided by the salutary effects of treating NPC mice (and NPC cats) with *N*–butyldeoxynojirimycin as a substrate-reducing agent. The iminosugar significantly delayed the onset and progression of the neurological manifestations in these natural disease models; this salutary effect was accompanied by decreased neuronal storage of the pathological gangliosides [71, 72].

With these preclinical studies available, a 2 year trial Phase 1, randomized for the first 12 months of miglustat, has been carried out in patients with Niemann–Pick disease type C, who were treated with up to 200 mg three times daily with the drug [73]. The interim outcome of this study has been particularly encouraging with improved cognitive function and other neuropsychological evaluations. Unlike the clinical trial of miglustat in type 3 Gaucher disease, a broad range of clinical and neurological parameters were measured, and in patients receiving the active drug, of those outcomes that changed, all were improved. In one adult patient, miglustat, which was shown to enter the cerebrospinal fluid at concentrations about 20 per cent of those in plasma, reduced pathological storage of lipids and corrected the lipid trafficking defect that occurs in peripheral blood lymphocytes [74]. Although this is merely to be regarded as a surrogate biomarker and cannot be representative of the outcome of substrate reduction therapy in the brain, it does serve as a potentially informative biomarker demonstrating the potential for the drug to correct the disturbed cell function definitively. The findings will offer encouragement to the use of the drug in other patients with Niemann–Pick disease type C.

Clinical development of N-butyldeoxygalactonojirimycin (NB-DGJ)

The galactose analogue, *N*-butyldeoxygalactonojirimycin (*N*B-DGJ) of *N*-butyldeoxynojirimycin (*N*B-DNJ) appears to have identical therapeutic efficacy on the biosynthesis of glucosylceramide but lacks the capacity to inhibit α-glucosidases [49, 55, 75]. This molecule is thus largely free of undesirable gastrointestinal effects related to inhibition of mucosal disaccharidases but is an inhibitor of the β-galactosidase, lactase-phloridzin hydrolase.

In experimental animals, the galactose analogue *N*B-DGJ does not cause atrophy of lymphoid organs, including the spleen, and is unassociated with weight loss. In this context the more selective inhibitor *N*B-DGJ is clearly extremely attractive to further clinical investigation: it has been given in gram quantities in acute exposure without ill-effect to normal human volunteers and, given its favourable chemistry and apparent lack of unwanted toxicity, were it to be chosen for therapeutic development, this analogue appears ideally-suited for clinical use in the glycosphingolipidoses; its clinical use may necessitate the imposition of a diet with a reduced lactose content. It should, however, be noted that neither *N*B-DGJ nor *N*B-DNJ affect the activity of the UDP-galactosylceramide transferase, which catalyses the first step in the biosynthesis of galactosylceramide (defective catabolism of which causes Krabbe disease) and the sulphatides (which accumulate in metachromatic leucodystrophy).

Other iminosugar inhibitors of glucosidases have been identified in biochemical and screening programmes and adamantine-pentyl-deoxynojirimycin, developed in the Netherlands by one of the authors (JMFGA) and colleagues, inhibits glycolipid biosynthesis at concentrations at least 1000 or lower in the plasma of experimental animals than those achieved with therapeutic doses of *N*B-DNJ [76].

It can thus be seen that with an important medicinal licensing programme underway, active animal and clinical research is greatly stimulated [77]. Synthetic iminosugars of greater selectivity and specificity are in preclinical development and offer hope that the concept of substrate reduction therapy can be established on a firmer basis for rare lysosomal disorders [78], for which no other treatment is in immediate prospect.

13.3.7 Iminosugars as pharmacological chaperones in lysosomal diseases

Mutations responsible for many human diseases are associated with the instability of the cognate protein leading to misfolding and aggregation, which disrupts intracellular trafficking to the site of action. Such instability activates the unfolded protein response and is accompanied, to varying degrees, by arrested transport within the endoplasmic reticulum and impaired maturation; thus disturbed folding promotes premature degradation and further contributes to the loss of protein function [79]. These considerations have stimulated research into the capacity of small molecules to bind to the newly-translated products of mutant genes that occur in inherited disorders and improve their stability by binding to an otherwise unstable protein conformer [80]. Such molecules, including the classical example of the heat-shock proteins, are described as 'molecular chaperones'.

Chaperones usually possess hydrophobic domains and several small molecular chaperones serve as weak inhibitors of enzymes as a result of low-affinity binding to the active site. Small hydrophobic molecules serving as pharmacological chaperones may also have the ability to penetrate otherwise inaccessible foci of disease, even crossing the blood–brain barrier. Since iminosugars can be considered as molecular mimics of biologically active sugars, they may serve as pharmacological chaperones for a wide range of proteins whose folding in the endoplasmic reticulum is destabilized by mutation. By binding to the target polypeptide in its nascent state, iminosugars restore function and prevent misfolding during the process of glycoprotein trimming and maturation.

Use of pharmacological chaperones in lysosomal diseases

In the special case of the lysosomal glycosidases, which are deficient in several important lysosomal storage diseases, there is a particular advantage in harnessing the molecular chaperone principle. Lysosomal glycosidases are synthesized on the endoplasmic reticulum in an environment that enjoys an approximately neutral hydrogen ion concentration: however, the final destination of the mature glycosidases is in the acidic environment of the lysosome. For most of these enzymes the organellar pH, at about 4.5, represents a condition under which they act maximally on their natural substrates and in which their biochemical stability is greatest. Thus, iminosugars with a capacity to bind at the active site of nascent acid glycosidases in the endoplasmic reticulum would be expected to maintain the polypeptide chain in a tight and stable conformation during cotranslational processing in the endoplasmic reticulum and during maturation of the glycoprotein as it traverses the Golgi. On delivery to the acidic environment of the lysosome, the avid binding of the iminosugar as a substrate mimetic to the active site of the glycosidase will be reduced, thus freeing the nascent enzyme variant from the weak inhibitor at its final cellular destination and physiological site of action. The iminosugar molecule will have served as a true chaperone, protecting its larger partner from degradation due to misfolding on the hazardous pathway of maturation and glycoprotein trimming through intracellular compartments to the delivery site in the lysosome.

Iminosugars can, on the one hand, serve as useful inhibitors of enzyme systems but, as a result of their weak inhibitory properties, they can also paradoxically be employed to improve the efficiency of protein folding – thereby operationally enhancing the activity of mutant enzymes by promoting their delivery to the site of action. Careful titration of these two competing activities will be required to identify a dose sufficient to bring about protein stabilization but not so high that it competitively inhibits activity and aggravates lysosomal storage by continuing to occupy the active site of the enzyme in the lysosome.

Fabry disease

An early clinical example of the application of this principle in lysosomal diseases is provided by studies conducted by Frustaci and colleagues from the University of Rome, Italy. These studies were preceded by experiments using lymphoblasts obtained from classical and nonclassical Fabry patients by Suzuki and colleagues who showed that galactose, and later 1-galactonojirimycin, stabilized the missense mutants of α-galactosidase A *in vitro* [81, 82]. The clinical investigators infused the sugar galactose in a patient with a cardiac variant of Fabry disease due to deficiency of acid α-galactosidase A. Like 1-galactonojirimycin, D-galactose serves as an inhibitor of this lysosomal enzyme. After *in vitro* studies, its capacity to enhance the enzyme activity was demonstrated in lymphocytes and myocardial biopsy specimens: infusions of 1 gram of galactose per kg bodyweight over 4 hours on a daily basis significantly enhanced α-galactosidase A activity, in the case of the heart tissue from less than 12 per cent to 35 per cent of human control cardiac tissue. After the galactose infusions were continued for 3 months, microscopic examination showed reduced vacuolation of myocardial cells on biopsy with reduced globotriosylceramide storage. These findings were confirmed by greatly improved ultrastructure, with significantly decreased lysosomal proliferation and distention after treatment [83].

In Fabry disease, the company, Amicus, has pursued this principle using the iminosugar, 1-deoxygalactonojirimycin (AT1001™, Figure 13.1). In Fabry fibroblasts cultured from patients with classic Fabry disease (D244 and E59K missense mutations) – a patient with a cardiac variant of Fabry disease (M296V mutation), and a patient with mild Fabry disease associated with N215S mutation in the α-galactosidase – treatment with the iminosugar at 20 μM *in vitro* significantly enhanced enzymatic activity. The degree of enzyme enhancement related to these mutations was variable but was about three- to eightfold. Over 65 missense proteins associated with Fabry disease have now been studied *in vitro,* three-quarters of which manifest at least a 20 per cent increase in α-galactosidase A activity after incubation with 1-deoxynojirimycin at concentrations in the micromolar range.

A Phase 1 study of AT1001 (1-deoxygalactonojirimycin) has been conducted by Amicus. The agent was tolerated well and the iminosugar appeared to be bioavailable. The drug has now been administered to more than 60 healthy volunteers and oral bioavailability was estimated to be greater than 80 per cent. In a dose-escalation study, administration of 50–150 mg of 1-deoxygalactonojirimycin showed a dose-response enhancement of α-galactosidase A activity in peripheral blood lymphocytes. At the time of writing, Phase 2 studies in suitable male and female patients with Fabry disease are underway using 25–250 mg of the drug and rigorous endpoints for safety and efficacy including biopsies of skin, heart and kidney and accompanied by functional monitoring. Given the intractable nature of Fabry disease and the innovative approach taken, the results of these trials are eagerly awaited.

Gaucher disease

Studies by Sawkar and colleagues have investigated the influence of alkylated deoxynojirimycins on acid β-glucosidase activity in fibroblasts obtained from patients homozygous for the N370S glucocerebrosidase allele [84], which is the most frequent to cause Gaucher disease. These studies showed that after incubating the cells for 5 days in the presence of the iminosugar at μM concentrations, and compared with assays of untreated fibroblast extracts assigned a relative activity of 1, octyl and nonyl deoxynojirimycins enhanced the enzyme activity assayed at acid pH giving values of 1.4–1.7 of normalized activity, respectively.

The optimal enhancing effect was usually detected below 30 μM iminosugar in the culture medium. Control studies conducted in fibroblasts similarly treated but obtained from healthy individuals showed no enhancement of acid β-glucosidase activity, thus further confirming the existence of an enhancing effect resulting from selective interactions of the iminosugars with the enzyme in mutant fibroblast extracts. Using cells transfected *in vitro* with several constructs that express mutant glucocerebrosidase variants, including N370S at high abundance, 10 μM *N*B-DNJ was shown to enhance activity [85].

However, it was unclear from these experiments whether the postulated chaperone effect was dependent on the conditions of enzymatic overexpression rather than on a stabilizing effect on unstable enzyme variants *in situ.* It remains possible that the reported stabilizing effect of the imnosugars is only observed in diluted cell extracts in assays conducted *in vitro* and does not relate to the activity of the N370S protein variant *in situ.* Finally, it remains uncertain as to whether the human N370S mutant glucocerebrosidase truly represents an unstable and/or misfolded protein variant.

Research conducted with the support of the Amicus Company has shown that fibroblasts obtained from Gaucher patients harbouring the N370S mutant glucocerebrosidase and incubated with a related deoxynojirimycin-based iminosugar (designated AT2101) also showed partial restoration of enzyme activity when these cell extracts were assayed *in vitro*. Clinical trials of this agent in patients with Gaucher disease are in a late stage of development.

Other lysosomal disorders

The capacity of iminosugars to bind weakly to key lysosomal acid hydrolases is clearly a rich area for pharmaceutical exploration. Use of this approach in patients suffering from glycogen storage disease type 2 (Pompe disease), due to a deficiency of acid maltase or acid α-glucosidase, is also in an active stage of development based on the by now familiar molecular mimicry of iminosugars and their capacity to enhance the stability of mutant lysosomal enzymes during intracellular synthesis and maturation.

Disordered protein folding occurs in numerous diseases, including other lysosomal disorders for which treatments are urgently needed [91]. These include GM1 gangliosidosis, Krabbe disease, metachromatic leucodystrophy, α-mannosidosis and several mucopolysaccharidosis syndromes.

Quasi-iminosugars as chaperones

Latterly Suzuki and colleagues have combined a synthetic chemical approach with enzymatic screening to identify molecules with improved properties as pharmacological chaperones for enhancing the expression and function of other lysosomal hydrolases. The quasi-iminosugar, N-octyl-β-valienamine was found to bind to mutant acid β-glucosidase at the neutral pH of the ER and Golgi, allowing the stabilized chaperone-nascent protein complex to retain its active conformation until it reaches the lysosome [92]. This agent thus has potential application for the treatment of Gaucher disease. Its synthetic analogue, N-octyl-4-epi-β–valienamine (NOEV), shows powerful enhancing activity towards the acid β-galactosidase which is deficient in GM1 gangliosidosis (Figure 13.1) [90]. Early preclinical studies in a murine model of GM1 gangliosidosis, which expresses the enhancing but disease-associated human R201C mutant β-galactosidase created by 'knock-in' genetic technology on a background engineered to lack expression of the mouse enzyme, show evidence of regression of neuronal pathology [93]. Further studies of the effects on neurological outcomes are awaited with interest.

13.3.8 Other medicinal applications of iminosugars as chaperones

Although, for reasons explained above, the very nature of intracellular delivery of lysosomal glycosidases lends itself to therapeutic application of pharmacological chaperone therapy, many other genetic diseases are associated to misfolded proteins and may also be accessible to this therapeutic stratagem.

One particular disorder that merits attention in this connection is *cystic fibrosis* due to mutations occurring in the cystic fibrosis membrane trans-regulator (CFTR) gene. The most frequent mutation, responsible for at least two-thirds of mutant alleles, is the deletion of a phenylalanine in the ATP binding cassette at position 508 of this member of the family of ABC transporters. ΔF508 and other mutant CFTR proteins are associated with misfolding of the large nascent CFTR polypeptide with a resultant deficiency of the transporter at its site of action on the cell surface membrane of epithelial cells. There is premature destruction by the ubiquitin pathway, with aggregation of the mutant CFTR protein leading to defective processing in the translocation machinery [89]. Several molecules, including myoinositol and betaine, can restore the folding defect of mutant CFTR molecules but are yet to show therapeutic effects in living animals [90]. Recently it has been reported that several iminosugars, including *N*-butyldeoxynojirimycin, bind to the nascent ΔF508 CFTR protein during cotranslational processing; this apparently enhances its delivery to the apical epithelial membrane and in the ileum of ΔF508 transgenic mice, apparently restores CFTR channel function in a mechanism postulated to represent pharmacological chaperone activity [91].

Given the frequency and severity of cystic fibrosis, as well as the need for definitive treatments to improve the outcome for affected individuals who occur with the birth frequency of approximately 1 in 2000 infants of North European descent, the potential for iminosugars such as *N*-butyldeoxynojirimycin to be developed as pharmacological chaperones for clinical use is very large. The existence of animal models, as well as cultured cells and tissues from patients with cystic fibrosis, could expedite the conduct of clinical research in this field and it is only to be hoped that the iminosugars, which offer so much therapeutic promise in otherwise intractable diseases, may be developed for this disabling condition.

13.3.9 Novel effects on the insulin receptor and in diabetes mellitus

Impaired responsiveness to insulin is reflected by increased glucose production by the liver and reduced glucose uptake by skeletal muscle and adipose tissue. Chronic insulin resistance causes diabetes mellitus type 2. The strong correlation between insulin resistance and intracellular lipid concentrations in muscle cells suggests that excessive exposure to lipids or their metabolites, so-called lipotoxicity, has a crucial effect on insulin receptor function.

Glycosphingolipids are important constituents of so-called lipid rafts, semiordered microdomains of the cell membrane in which the insulin receptor also resides. A regulatory role for glycosphingolipids in insulin sensitivity was first proposed by Tagami and coworkers [92]. They reported that addition of GM3, the simplest ganglioside, to cultured adipocytes, suppresses phosphorylation of the insulin receptor and its downstream substrate IRS-1, resulting in reduced glucose uptake. Other observations further substantiate the role of the ganglioside GM3 in responsiveness to insulin. Mutant mice lacking GM3 show an enhanced phosphorylation of the skeletal muscle insulin receptor after ligand binding and are protected from high-fat diet induced insulin resistance [93]. Conversely, GM3 concentrations are elevated in the muscle of certain strains of mice and rats that serve as models of human obesity and insulin resistance.

Inokuchi and coworkers employed the ceramide-analogue D-*threo*-1-phenyl-2-decanoylamino-3-morpholino-propanol (PDMP), an inhibitor of glucosylceramide synthase, to reduce glycosphingolipids in cultured adipocytes (Figure 13.1). They noted that PDMP counteracted the inhibitory effects of TNF-alpha on insulin receptor and IRS-1 phosphorylation [92]. More recently it was reported by Inokuchi and colleagues that increased abundance of GM3 diminished IR accumulation in detergent-resistant membrane microdomains and also decreased insulin-dependent internalization of the insulin receptor [94]. Again glycosphingolipid depletion by incubation of cells with PDMP prevented these abnormalities.

These important findings made with PDMP pose some difficulties for interpretation since this compound not only inhibits conversion of ceramide to glucosylceramide but also its transacylation to 1-O-acylceramide and consequently increases cellular ceramide concentrations [95]. To establish whether ceramide itself, or rather its glycosphingolipid metabolites, are instrumental in the development of insulin resistance, Aerts and collaborators examined in obese mice and rats the effect of reduction of glycosphingolipids [96]. The iminosugar N-(5'-adamantane-1'-yl-methoxy)-pentyl-1-deoxynojirimycin (AMP-DNM) was used for these studies (Figure 13.1) [76].

This latter compound is a 100-fold more powerful inhibitor of glycosphingolipid biosynthesis compared with *N*-butyldeoxynojirimycin; it is, moreover, freely bioavailable and apparently well tolerated [96, 97]. It was observed that pharmacological lowering of glycosphingolipids by AMP-DNM, without significant reduction of ceramide, dramatically reverses insulin resistance in all animal models. Treatment of obese animals with AMP-DNM markedly lowered circulating glucose concentrations, improved oral glucose tolerance, reduced glycated haemoglobulin and improved insulin sensitivity in muscle and liver. Of particular note, in contrast to untreated control animals, Zucker Diabetic Fat (ZDF) rats treated with 25 mg AMP-DNM/kg/day retained the ability to produce significant amounts of insulin. Thus, in this rodent model of diabetes mellitus, treatment with the iminosugar apparently exerts a protective effect on the endocrine pancreas.

Since AMP-DNM, like miglitol, also inhibits intestinal glycosidases, Aerts and Overkleeft have recently developed a mimic of AMP-DNM that more specifically inhibits glucosylceramide synthase. This molecule does not interfere with intestinal glycosidases. Notably, administration of the new compound to animals serving as models of insulin resistance also improved their metabolic parameters, although the effects were less striking than those observed with AMP-DNM. These findings suggest that the iminosugar AMP-DNM exerts its beneficial effects synergistically by inhibiting intestinal carbohydrate assimilation and glycosphingolipid reduction in tissues. Given that compounds like AMP-DNM seem to be well tolerated in living mammals, further investigations into their therapeutic potential for the treatment of human diabetes mellitus type 2 appear to be warranted.

13.4 Conclusions

Iminosugars have many properties that entice the chemist and physician for development as human medicines [98]. Unlike the conventional pantheon of candidate drugs, their wide distribution but conserved evolution as toxins in nature immediately differentiates the iminosugars into molecular classes with defined and powerful actions. Hitherto, the challenge has been to identify their pharmacological activities and define their molecular specificities within living organisms rather than the test tube; this is because subtle evolutionary changes have ensured that the site(s) of physiological action of a given biological molecule is preferentially targeted. For example, numerous iminosugars of the nojirimycin class occur in mulberry leaf extracts as alkyl and pentose derivatives, together with calystegins [5, 6]. These potent derivatives with diverse solubility properties presumably penetrate many targets in different predators, as has been ordained by evolution. Recent studies indicate that only the favoured larva of the silkworm (*Bombyx mori*), has been selected to survive this biochemical onslaught delivered in the diet [99]. Intracellular targeting by diffusion through cell membrane barriers is critical for therapeutic victory in the whole organism, as vividly illustrated by the fortunes of the alkylated iminosugar, *N*-butyl deoxynojirimycin – a resolute failure as an antiviral agent in patients with HIV/AIDS but successful as the first oral drug with specificity for Gaucher disease and with therapeutic efficacy in other glycosphingolipid disorders.

By the same token, despite their selectivity as molecular mimics and fascination for biochemists as inhibitors, it should be remembered that many of the naturally-occurring iminosugars will only become 'dirty drugs'. The presumed function of these compounds as poisons and the ambitions of survival on the part of their living manufacturers, amplified by natural selection, ensure that invincible weapons are available to strike multiple intracellular targets. Thus, swainsonine and castanospermine, with powerful effects on several α-mannosidases, or α- and β-glucosidases, respectively, may render them superficially attractive as human medicines in cancer or viral infections but, in practice, their toxicity has so far proved to be intractable.

With these drawbacks to the development of iminosugars as credible drugs come challenges to the ingenuity of the medicinal chemist. There is a need for evermore close collaboration with the molecular cell biologist, the biochemical geneticist and the clinician–scientist. Hitherto, human inventiveness in drug discovery has relied upon fastidious clinical observation by the great but unknown healers of the ancient past. The splendid mulberry tree has been the key to an important branch of empirical medicine far beyond its capacity to provide fodder for the international cash crop of antiquity; thus medical know-how, as well as the secret luxury of silk, was distributed to the west from Asia along the old trade routes.

In the new era of rational drug design, shared perspectives are crucial; the full clinical development of miglustat as a licensed agent for Gaucher disease depended on real interdisciplinary dialogue. When, in the future, progress with a new molecule becomes apparently intractable, one should recall the near-miraculous transformation of an arcane compound known to the few as *N*-butyldeoxynojirimycin; this molecule has been promoted only after surviving its long journey from struggling antidiabetic agent and ignominious failure at the late stage of clinical trials as a promising antiviral drug. The small molecule has now realised a more humble destiny as Zavesca™ (miglustat) – nonetheless, the pride of an international company and the first of several potential oral agents to be licensed for a lysosomal disease. As a small molecule with the capacity to enter the brain, *N*B-DNJ illustrates another attractive feature of many iminosugars; it continues to be the subject of clinical experimentation as an agent with the potential to ameliorate some neurodegenerative diseases which would otherwise be beyond therapeutic reach.

Development of Zavesca™ from the obscure *N*B-DNJ has stimulated other initiatives in the surprisingly exorbitant field of orphan drug development. The appearance of iminosugars as pharmacological chaperones in clinical trials, as well as the emergence of Genzyme's own competing but chemically unrelated morpholino-type inhibitor of glucosylceramide synthase, GENZ 112638, provide striking examples of such commercial 'must have' attitudes. These initiatives occur only where there has been, at the very least, a whiff of success with the lead agent – albeit in the apparently constrained and diminutive but nonetheless lucrative, orphan disease market.

Finally, we can take encouragement from the drawbacks and advantages of the iminosugars as potential medicinal agents; here again, *N*B-DNJ provides a vivid example. Although successful, miglustat (Zavesca) is itself a 'dirty' drug and shares with its licensed congener, miglitol (Glyset), an inhibitory action on intestinal disaccharidases. While it might be advantageous for miglitol to have an action on targets of glucose metabolism other than intestinal glucosidases, for example hepatic glycogen phosphorylase or debranching enzyme, we would certainly prefer miglustat not to inhibit these enzymes, even though there is evidence that it can inhibit the glucose-yielding debranching enzyme [100]

Here lies the seductive power of these molecules for the medicinal chemist, whose innocent pleasure in the creation of structural analogues with improved inhibitory specificities and biological actions, is boundless. The ring nitrogen atom common to the iminosugars endears itself to those who marvel at the rich diversity of templates supplied by nature and which render their work facile, as well as rewarding. Modern combinatorial chemistry also has an immense power to generate candidate drugs of preordained specificity [78, 101]. Already the appearance of the selective inhibitor, *N*-butyldeoxygalactonojirimycin and the highly potent adamantane pentyl deoxynojrimycin, provides convincing evidence of the opportunities that now exist for innovation and clinical discovery in the field of iminosugar research.

To temper this optimism, a cautious approach to our high expectations is required, because the path to rational drug design is littered with promising medicinal science that is lost in translation to the clinic. Our future challenge lies no longer in the identification of medicinal iminosugars with powerful and defined actions in living cells, but in our ability to determine how to use these molecules wisely in living patients.

Acknowledgements

We thank Aarnoud van der Spoel and Nicole Zitzmann for helpful discussions and Joan Grantham for preparing much of the manuscript.

References

1. Asano, N. (2003) *Glycobiology*, **13**, 93R-104R.
2. Aerts, J.M.F.G. (2006) The remarkable story of Hermann Boerhaave, the mulberry and iminosugar drugs. *Proc. European Working Group for the Study of Gaucher Disease.* http://www.ewggd.co.uk/EWGGD%20Abstracts.pdf.
3. Yagi, M., Kuono, T., Aoyagi, Y. and Murai, H. (1976) *Nippon Nogei Kagaku Kaishi*, **50**, 571–572.
4. Asano, N., Tomioka, E., Kizu, K.and Matsui, K. (1994) *Carb.Res.*, **253**, 235–245.
5. Asano, N., Oseiko, K., Tomioka, E., Kizu, H.and Matsui, K. (1994) *Carb. Res.*, **259**, 243–255.
6. Scott, L. J. and Spenser, C. M. (2000) *Drugs*, **59**, 521–549.
7. Nojima, H., Kimura, I., Chen, F. J., Sugihara, Y., Haruno, M., Kato, A. and Asano, N. (1998) *J.Nat.Prod.*, **61**, 397–400.
8. Macleod, C. (2000) Chinese doctors say herbal cures can combat AIDS. *The Independent*, London, 9 December.
9. Ministry of Science and Technology, People's Republic of China (2000) Chinese traditional herbs treating AIDS. *China Science and Technology Newsletter*, No. 232, 20 August.
10. Broquist, H.P. (1985) *Ann. Rev. Nutrition*, **5**, 311–409.
11. Dorling, P. R., Huxtable, C. R. and Colgate, S. M. (1980) *Biochem. J.*, **191**, 649–651.
12. Elbein, A. D., Solf, R., Dorling, P. R. and Vosbeck, K. (1981) *Proc. Nat. Acad. Sci. USA*, **78**, 7393–7397.
13. Saul, R., Chambers, J. P., Molyneux, R. J. and Elbein, A. D. (1983) *Arch. Biochem. Biophys.*, **232**, 593–597.
14. Asano, N. (2003) *Curr.Top.Med.Chem.*, **3**, 471–484.
15. Winchester, B. and Fleet, G. W. (1992) *Glycobiology*, **2**, 199–210.
16. Peyrieras, N., Bause, E., Legler, G., Vasilov, R., Claesson, L., Peterson, P. and Ploegh, H. (1983) *EMBO J.*, **2**, 823–832.
17. Karpas, A., Fleet, G. W., Dwek, R. A., Petrusson, S., Namgoong, S. K., Ramsden, N. G., Jacob, G. S. and Rademacher, T. W. (1988) *Proc. Nat. Acad.Sci. USA*, **85**, 9229–9233.
18. Standl, E., Schernthaner, G., Rybka, J., Hanefeld, M., Raptis, S. A. and Naditch, L. (2001) *Diabetes Res.Clin.Pract.*, **51**, 205–213.
19. Andersen, B., Rassov, A., Westergaard, N. and Lundgren, K. (1999) *Biochem. J.*, **342**, 545–550.
20. Joubert, P.H., Venter, H. L. and Fourkaridis, G. N. (1990) *Brit.J.Pharmacol.*, **30**, 391–396.
21. Chiasson, J. L. and Naditch, L. (2001) *Diabetes Care*, **24**, 989–994.
22. Van Gaal, L., Maislos, M., Scherthaner, G., Rybka, J. and Segal, P. (2001) *Diabetes Obes. Metab.*, **3**, 326–331.
23. Drent, M. L., Tollefsen, A. T., van Heusden, F. H., Hoenderdos, E. B., Jonker, J. J. and van der Veen, E.A. (2002) *Diabetes Nutr. Metab.*, **15**, 152–159.
24. Winston, A., Mallon, P. W. and Boffito, M. (2006) *Expert Opin. Drug Metab. Toxicol.*, **2**, 447–458.
25. Kornfeld, R. and Kornfeld, S. (1985) *Ann. Rev. Biochem.*, **54**, 631–664.
26. Helenius, A., Trombetta, E. S., Hebert, D. N. and Simons, J. F. (1997) *Tren. Cell Biol.*, **7**, 193–200.
27. Parodi, A. J. (2000) *Biochem J.*, **348** Pt 1,1–13.
28. Dwek, R. A., Butters, T. D., Platt, F. M. and Zitzmann, N. (2002) *Nature Drug Disc.*, **1**, 65–75.
29. Freed, E. O.and Martin, M. A. (2001) HIVs and their replication, in *Field's Virology* (B. N. Fields, P. M. Howley, D. E. Griffin, R. A. Lamb, M. A. Martin, B. Roizman, S. E. Straus and D. M. Knipe, eds), Lippincott, Williams and Willkins, Vol. 2, pp. 1971–2041.
30. Pohlmann, S. and Reeves, J. D. (2006) *Curr. Pharm. Des.*, **12**, 1963–1973.

31. Mizuochi, T., Matthews, T. J., Kato, M., Hamako, J., Titani, K., Solomon, J. and Feizi, T. (1990) *J. Biol. Chem.*, **265**, 8519–8524.
32. Fischer, P. B., Collin, M., Karlsson, G. B., James, W., Butters, T. D., Davis, S. J., Gordon, S., Dwek, R. A. and Platt, F. M. (1995) *J. Virol.*, **69**, 5791–5797.
33. Fischer, P. B., Karlsson, G. B., Butters, T. D., Dwek, R. A. and Platt, F. M. (1996) *J. Virol.*, **70**, 7143–7152.
34. Butters, T. D., Dwek, R. A. and Platt, F. M. (2005) Substrate deprivation: a new therapeutic approach for the glycosphingolipid lysosomal storage diseases. *Glycobiology*, **15**, 43R-52R.
35. Butters, T., Dwek, R. and Platt, F. (2000) *Chem. Rev.*, **100**, 4683–4696.
36. Chu, C. M., and Liaw, Y. F. (2006) *Semin. Liver Dis.*, **26**, 142–152.
37. Chan, H. L. and Sung, J. J. (2006) *Semin. Liver Dis.*, **26**, 153–161.
38. Bruss, V. (2004) *Virus. Res.*, **106**, 199–209.
39. Block, T. M., Lu, X. Y., Mehta, A. S., Blumberg, B. S., Tennant, B., Ebling, M., Korba, B., Lansky, D. M., Jacob, G. S. and Dwek, R. A. (1998) *Nature Med.*, **4**, 610–614.
40. Zitzmann, N., Mehta, A. S., Carrouee, S., Butters, T. D., Platt, F. M., McCauley, J., Blumberg, B. S., Dwek, R. A. and Block, T. M. (1999) *Proc. Nat. Acad. Sci. USA*, **96**, 11878–11882.
41. Pavlovic, D., Fischer, W., Hussey, M., Durantel, D., Durantel, S., Branza-Nichita, N., Woodhouse, S., Dwek, R. A. and Zitzmann, N. (2005) *Adv. Exp. Med. Biol.*, **564**, 3–4.
42. Zitzmann, N., Block, T., Methta, A., Rudd, P., Burton, D., Wilson, I., Platt, F., Butters, T. and Dwek, R. A. (2005) *Adv. Exp. Med. Biol.*, **564**, 1–2.
43. Newton, S. A., White, S. L., Humphries, M. J. and Olden, K. (1989) *JNCI*, **81**, 1024–1028.
44. Goss, P. E., Baptiste, J., Fernandes, B., Baker, M. and Dennis, J. W. (1994) *Cancer Res.*, **54**, 1450–1457.
45. Vunnam, R. R. and Radin, N. R. (1980) *Chem.Phys. Lipids*, **26**, 265–278.
46. Inokuchi, J. and Radin, N. R. (1987) *J. Lipid.Res.*, **28**, 565–571.
47. Radin, N. R., Shaymam, J. A. and Inokuchi, J. (1993) *Adv. Lipid. Res.*, **26**, 183–213.
48. Platt, F. M., Neises, G. R., Dwek, R. A. and Butters, T. D. (1994) *J. Biol. Chem.*, **269**, 8362–8365.
49. Platt, F. M., Neises, G. R., Karlsson, G. B., Dwek, R. A. and Butters, T. (1994) *J.Biol. Chem.*, **269**, 27108–27114.
50. (a) Butters, T., van den Broek, L. A. G., Fleet, G. W. J., Krulle, T. M., Wormald, M. R., Dwek, R. A. and Platt, F. M. (2000) *Tetrahedron: Asymmetry*, **11**, 113–124. (b) Boucheron, C., Desvergnes, V., Compain, P., Martin, O. R., Lavi, A., Mackeen, M., Wormald, M. R., Dwek, R. A. and Butters, T. D. (2005) *Tetrahedron: Asymmetry*, **16**, 1747–1756.
51. Radin, N. R.(1996) *Glycoconj.J.*, **13**, 153–157.
52. Sango, K., Yamanaka, S., Hoffmann, A., Okuda, Y., Grinberg, A., Westphal, H., McDonald, M. P., Crawley, J. N. and Sandhoff, K. (1995) *Nature Genet.*, **11**, 170–176.
53. Platt, F. M., Neises, N. R., Reinkensmeier, G., Townsend, M. J., Perry, V. H., Proia, R. L., Winchester, B., Dwek, R. A. and Butters, T.(1997) *Science*, **276**, 428–431.
54. Jeyakumar, M., Butters, T. D., Cortina-Borja, M., Proia, R. L., Perry, V. H., Dwek, R. A. and Platt, F. M. (1999) *Proc. Nat. Acad. Sci. USA*, **96**, 6388–6393.
55. Andersson, U., Butters, T. D., Dwek, R. A. and Platt, F. M. (2000) *Biochem. Pharmacol.*, **59**, 821–829.
56. Cox, T., Lachmann, R., Hollak, C., Aerts, J., van Weely, S., Hrebicek, M., Platt, F., Butters, T., Dwek, R., Moyses, C., Gow, I., Elstein, D. and Zimran, A. (2000) *Lancet*, **355**, 1481–1485.
57. Elstein, D., Hollak, C., Aerts., J., van Weely, S., Maas, M., Cox, T. M., Lachmann, R. H., Hrebicek, M., Platt, F. M., Butters, T. D., Dwek., R. A. and Zimran, A. (2004) *J. Inherit. Metab.Dis.*, **27**, 757–766.

58. Heitner, R., Elstien, D., Aerts, J., van Weely, S. and Zimran, A. (2002) *Blood Cells Mol. Dis.*, **28**, 127–133.
59. Pastores, G. M., Barnett, N. L. and Kolodny, E. H. (2005) *Clin. Ther.*, **27**, 1215–1227.
60. van der Spoel, A. C., Jeyakumar, M., Butters, T. D., Charlton, H. M., Moore, H. D., Dwek, R. A. and Platt, F. M. (2002) *Proc. Nat. Acad. Sci. USA*, **99**, 17173–17178.
61. Sugananuma, R., Walden, C., Butters, T. D., Dwek, R. A., Platt, F. M., Yanagimachi, R. and van der Spoel, A. C. (2005) *Biol.Reprod.*, **72**, 805–813.
62. Amory, J. K., Muller, C. H., Page, S. T., Leifke, E., Pagel, E. R., Bhandari, A., Subramanyam, B., Bone, W., Radlmaier, A. and Bremner, W. J. (2007) *Hum. Reprod.*, **22**, 702–707.
63. Bone, W., Walden, C. M., Fritsch, M., Voigtmann, U., Leifke, E., Gottwald, U., Boomkamp, S., Platt, F. M. and van der Spoel, A. C. (2007) *Reprod Biol Endocrinol.*, **5**, 1.
64. Yildiz, Y., Matern, H., Thompson, B., Allegood, J. C., Warren, R. L., Ramirez, D. M., Hammer, R. E., Hamra, F. K., Matern, S. and Russell, D. W. (2006) *J Clin Invest.*, **116**, 2985–2994.
65. Boot, R. G., Verhoek, M., Donker-Koopman, W., Strijland, A., van Marle, J., Overkleeft, H. S., Wennekes, T. and Aerts, J. M. (2007) *J Biol Chem.*, **282**, 1305–1312.
66. Hollak, C. E. M., Maas, M., Akkerman, E., den Heeten, A. and Aerts, J. M. F. G. (2002)*Blood Cells Mol. Dis.*, **27**, 1005–1012.
67. Data on file, Actelion Pharmaceuticals Ltd., CH 4123, Allschwil, Switzerland.
68. Cox, T. M., Aerts, J. M. F. G., Andria, G., Beck, M., Belmatoug, N., Bembi, B., Chertkoff, R., Vom Dahl, S., Elstein, D., Erikson, A., Giralt, M., Heitner, R., Hollak, C., Hrebicek, M., Lewis, S., Mehta, A., Pastores, G. M., Rolfs, A., Sa Miranda, M. C. and Zimran, A. (2003) J. Inherit. Metab. Dis., **26**, 513–526.
69. Weinreb, N. J., Barranger, J. A., Charrow, J., Grabowski, G. A., Mankin, H. J. and Mistry, P. (2005) *Am. J. Hematol.*, **80**, 223–229.
70. Kasperzyk, J. L., D'Azzo, A., Platt, F. M., Alroy, J. and Seyfried, T. N.(2005) *J. Lipid Res.*, **46**, 744–751.
71. Zervas, M., Somers, K. L., Thrall, M. A. and Walkley, S. U. (2001) *Curr. Biol.*, **11**, 1283–1287.
72. Walkley, S. (2003) *Phil. Trans. R. Soc. Lond. B. Biol. Sci.*, **358**, 893–904.
73. Patterson, M. C. and Platt, F. M. (2004) *Biochem. Biophys. Acta*, **1685**, 8362–8365.
74. Lachmann, R. H., te Vruchte, D., Lloyd-Evans, E., Reinkensmeier, G., Sillence, D. J., Fernandez-Guillen, L., Dwek, R. A., Butters, T. D., Cox, T. M. and Platt, F. M. (2004) *Neurobiol. Dis.*, **16**, 654–658.
75. Andersson, U., Butters, T. D., Dwek, R. A. and Platt, F. M. (2000) *Biochem. Pharmacol.*, **59**, 821–829.
76. Overkleeft, H. S., Renkema, G. H., Neele, J., Vianello, P., Hung, I. O., Strijland, A., van der Burg, A. M., Koomen, G. J., Pandit, U. K. and Aerts, J. M. F. G. (1998) *J. Biol. Chem.*, **273**, 26522–26527.
77. Aerts, J. M. F. G., Hollak, C. E. M., Boot, R. G., Groener, J. E. M. and Maas, M. (2006) *J. Inherit. Metab. Dis.*, **29**, 448–455.
78. Compain, P., Martin, O., Boucheron, C., Godin, G., Yu, L., Ikeda, K. and Asano, N. (2006) *Chembiochem.*, 7, 1356–1359.
79. Gregersen, N. (2006) *J.Inherit. Metab. Dis.*, **29**, 456–470.
80. Hendrick, J. P. and Hartl, F. U. (1993) *Ann. Rev. Biochem.*, **62**, 349–384.
81. Okiumiya, T., Ishii, S., Takenata, T., Kase, R., Kamei, S., Sakuraba, H. and Suzuki, Y. (1995) *Biochem. Biophys. Res. Commun.*, **214**, 1219–1224.
82. Fan, J. Q., Ishi, S., Asano, N. and Suzuki, Y. (1999) *Nature Med.*, **5**, 112–115.
83. Frustaci, A., Chimenti, C., Ricci, R., Natale, L., Russo, M. A., Pieroni, M., Eng, C. M. and Desnick, R. J. (2001) *N. Engl. J. Med.*, **345**, 25–32.

84. Sawkar, A. R., Cheng, W. C., Beutler, E., Wong, C. H., Balch, W. E. and Kelly, J. W. (2002) *Proc. Nat. Acad. Sci. USA*, **99**, 15428–15433.
85. Alfonso, P., Pampin, S., Estrada, J., Rodriguez-Rey, J. C., Sancho, J. and Pocovi, M. (2005) *Blood, Cells Mol. Dis.*, **35**, 268–276.
86. Suzuki, Y. (2006) *J. Inherit. Metab.Dis.*, **29**, 471–476.
87. Lin, H., Sugimoto, Y., Ohsaki, Y., Ninomiya, H., Oka, A., Taniguchi, M., Ida, H., Eto, Y., Ogawa, S., Matsuzaki, Y., Sawa, M., Inoue, T., Higaki, K., Nanba, E., Ohno, K. and Suzuki, Y. (2004) *Biochem. Biophys. Acta*, **1689**, 219–228.
88. Masuda, J., Suzuki, O., Oshima, A., Yamamoto, Y., Noguchi, A., Takimoto, K., Itoh., M., Matsuzaki, Y., Yasuda, Y., Ogawa, S., Sakata, Y., Nanba, E., Higaki, K., Ogawa, Y., Tominaga, L., Ohno, K., Iwasaki, H., Watanabe, H., Brady, R. O. and Suzuki, Y. (2003) *Proc. Nat. Acad. Sci. USA*, **100**, 15912–15917.
89. Cheng, S. H., Gregory, R. J., Marshall, J., Paul, S., Souza, D. W., White, G. A., O'Riordan, C. R. and Smith, A. E. (1990) *Cell*, **63**, 827–834.
90. Zhang, X. M., Wang, X. T., Yue, H., Leung, S. W., Thibodeau, P. H., Thomas, P. J. and Guggino, S. E. (2003) *J. Biol. Chem.*, **278**, 51232–51242.
91. Norez, C., Noel, S., Wilke, M., Bijvelds, M., Jovna, H., Melin, P., De Jonge, H. and Becq, F. (2006) *FEBS Lett.*, **580**, 2081–2086.
92. Tagami, S., Inokuchi, J. J., Kabayama, K., Yoshimura, H., Kitamura, F., Uemura, S., Ogawa, C., Ishii, A., Saito, M., Ohtsuka, Y., Sakue, S. and Igarashi, Y. (2002). *J. Biol. Chem.*, **277**, 3085–3092.
93. Yamashita, T., Hashirimoto, A., Haluzik, M., Mizukami, H., Beck, S., Norton, A., Kono, M., Tsuji, S., Daniotti, J. L., Werth, N., Sandhoff, R., Sandhoff, K. and Proia, R. L. (2003) *Proc. Nat. Acad. Sci. USA*, **100**, 3445–3449.
94. Kabayama, K., Sato, T., Kitamura, F., Uemura, S., Kang, B. W., Igarashi, Y. and Inokuchi, J. (2005) *Glycobiol.*, **15**, 21–29.
95. Shayman, J. A., Abe, A. and Hiraoka, M. (2004) *Glycoconj. J.*, **20**, 25–32.
96. Aerts, J. M., Ottenhoff, R., Powlson, A. S., Grefhorst, A., van Eijk, J., Dubbelhuis, P. F., Boot, R. G., van Dijk, T. H., Kuipers, F., Aten, J., Groener, J. E., Strijland, A., Groen, A. K., Serlie, M. J., Sauerwein, H. P. and Wennekes, T. (2007) *Diabetes*, Feb 7 (Epub ahead of print).
97. Aerts, J. M., Hollak, C., Boot, R. and Groener, A. (2003) *Phil. Trans. R.Soc. B. Biol. Sci.*, **358 (1433)**, 905–914.
98. Fleet, G. W., Fellows, L. E. and Winchester, B. (1990) *Ciba Found Symp.*, **154**, 112–122.
99. Konno, K., Ono, H., Nakamura, M., Tateishi, K., Hirayama, C., Tamura, Y., Hattori, M., Koyama, A. and Kono, K. (2006) *Proc. Nat. Acad. Sci. USA*, **103**, 1337–1341.
100. Andersson, U., Reinkensmeier, G., Butters, T. D., Dwek, R. A. and Platt, F. M. (2004) *Biochem Pharmacol.*, **67**, 697–705.
101. Cipolla, L., La Ferla, B. and Gregori, M. (2006) *Comb. Chem. High Throughput Screen*, **9**, 571–582.

14

Tables of iminosugars, their biological activities and their potential as therapeutic agents

Philippe Compain, Valérie Desvergnes, Virginie Liautard, Christelle Pillard and **Sylvestre Toumieux**

14.1 Introduction

From the very beginning of our project to edit a book devoted to iminosugars, our aim was to include a table which would correlate the structure of all reported iminosugars with their biological activities. Such an 'at-a-glance table' including detailed data of more than 400 iminosugar-based glycosidase inhibitors was produced by A. E. Stütz *et al.* in the late 1990s [1]. Based on this outstanding and herculean work, we at first 'naively' decided to extend its biological scope to enzymatic targets and to focus on studies reported in the literature after 1998. Our first attempts towards this goal indicated that thousands of novel iminosugars of biological interest had been published in less than 10 years! This exponential increase superbly highlights the remarkable dynamism of this field of research. We therefore decided to reduce the scope of our initial project to only iminosugars of therapeutical interest reported in the literature since the 1960s. The following tables have been designed to stimulate further research in the iminosugar area by relating structures to significant properties. Compounds have been selected for their potential as therapeutic agents, and as long as they were tested against relevant biological targets for activating or inhibitory activity. The data extracted from the literature range from preliminary *in vitro* enzymatic assays to human trials. Despite all our good intentions and the efforts made, the probability that the reader will detect oversights and mistakes inherent to this type of work is unfortunately not zero. We apologize in advance for any inconvenience caused.

The tables, which include more than 600 compounds and 200 references, have been organized by type of disease (diabetes, infectious diseases, cancer, lysosomal

Iminosugars: From synthesis to therapeutic applications Edited by Philippe Compain and Olivier R. Martin
 ISBN 978-0-470-03391-3

diseases, etc.). In each category of disease, compounds are classified by structure (mono- or bicyclic compounds, ring size, absolute configuration, etc.). The main disadvantage of this method of classification is that 'best-selling' structures, such as *N*-butyl 1-deoxynojirimycin, appeared several times in the tables since they have been evaluated against various diseases. The main advantage is nevertheless to highlight the structure–activity relationship of a family of compounds against given therapeutical targets which is the purpose of our project. The primary value of the following tables is to provide a rapid entry to key structures, biological data and guiding references. We do hope that this tool will be helpful and will stimulate further research in the field of iminosugars of therapeutic interest.

14.2 Type 2 diabetes

14.2.1 Five-membered rings

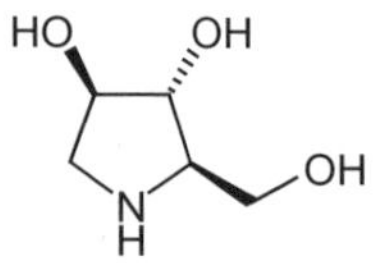	**Type 2 diabetes** *Rat intestinal maltase* (EC 3.2.1.20) – IC_{50} 55 μM [2],[3]. *Rat intestinal sucrase* (EC 3.2.1.48) – IC_{50} 16 μM [2],[3]. *Rat intestinal isomaltase* (EC 3.2.1.10) – IC_{50} 5.8 μM [2],[3]. *Mouse intestinal maltase* (EC 3.2.1.20) – IC_{50} 35 μM [4]. *Mouse intestinal sucrase* (EC 3.2.1.48) – IC_{50} 23 μM [4]. *Mouse intestinal isomaltase* (EC 3.2.1.10) – IC_{50} 4 μM [4]. *Rat intestine lactase* (EC 3.2.1.23) – IC_{50} 260 μM [5],[3]. *Trehalase (Rat intestine)* (EC 3.2.1.28) – IC_{50} 25 μM [3],[5]. *Trehalase (Mouse intestinal)* (EC 3.2.1.28) – IC_{50} 22 μM [4]. *Trehalase (Porcine kidney)* (EC 3.2.1.28) – IC_{50} 2.5 μM [2]; IC_{50} 4.8 μM [5]. *Glycogen phosphorylase* (EC 2.4.1.1) – IC_{50} 1 μM [6]. *Glycogen phosphorylase phosphorylated* (GP*a*) *rabbit muscle* (EC 2.4.1.1) K_i 0.392 ≥ M [7]. *Glycogen phosphorylase phosphorylated* (GP*a) rabbit liver* (EC 2.4.1.1) K_i 0.374 ≥ M[7]. *Glycogen phosphorylase phosphorylated* (GP*a) human liver* (EC 2.4.1.1) K_i 0.530 ≥M [7].
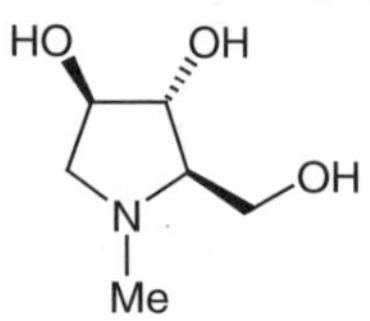	**Type 2 diabetes** *Rat intestinal maltase* (EC 3.2.1.20) – IC_{50} 500 μM [3]. *Rat intestinal sucrase* (EC 3.2.1.48) – IC_{50} 70 μM [3]. *Rat intestinal isomaltase* (EC 3.2.1.10) – IC_{50} 100 μM [3].
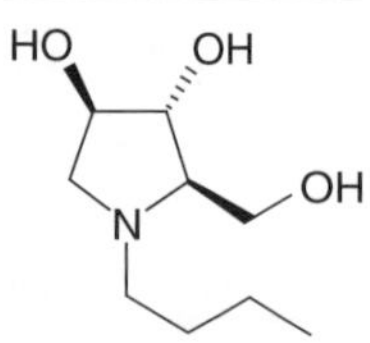	**Type 2 diabetes** *Rat intestinal sucrase* (EC 3.2.1.48) – IC_{50} 270 μM [3]. *Rat intestinal isomaltase* (EC 3.2.1.10) – IC_{50} 340 μM [3].

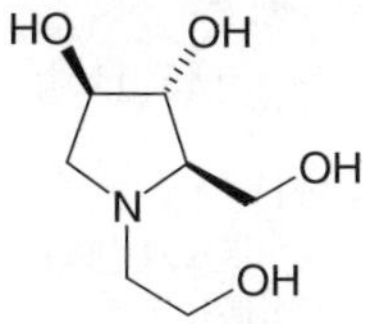	**Type 2 diabetes** *Rat intestinal sucrase* (EC 3.2.1.48) – IC_{50} 250 μM [2]. *Rat intestinal isomaltase* (EC 3.2.1.10) – IC_{50} 84 μM [2]. *Glycogen phosphorylase (Rat liver)* (EC 2.4.1.1) – IC_{50} 10 μM [8].
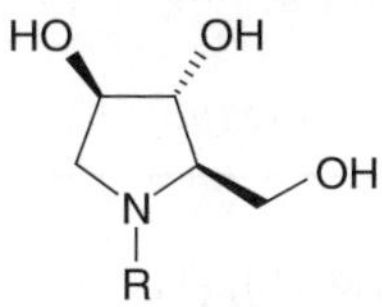	**Type 2 diabetes** *Glycogen phosphorylase (Rat liver)* (EC 2.4.1.1) R = cyclopropylmethyl IC_{50} 169 μM [8]. R = *n*-butyl IC_{50} 60 μM [8]. R = 2,2,2-trifluoroethyl IC_{50} 145 μM [8].
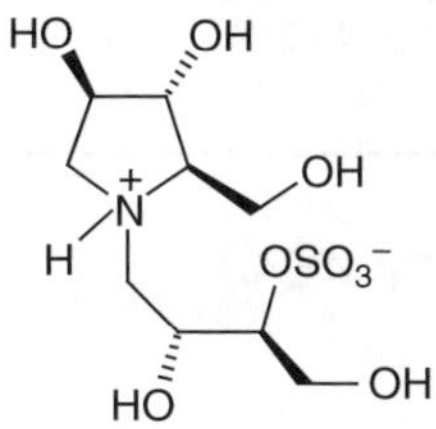	**Type 2 diabetes** *Rat intestinal maltase* (EC 3.2.1.20) – IC_{50} 306 μM [4]. *Rat intestinal sucrase* (EC 3.2.1.48) – IC_{50} 44 μM [4]. *Rat intestinal isomaltase* (EC 3.2.1.10) – IC_{50} 136 μM [4]. *Trehalase rat intestinal* (EC 3.2.1.28) – IC_{50} > 315 μM [4].
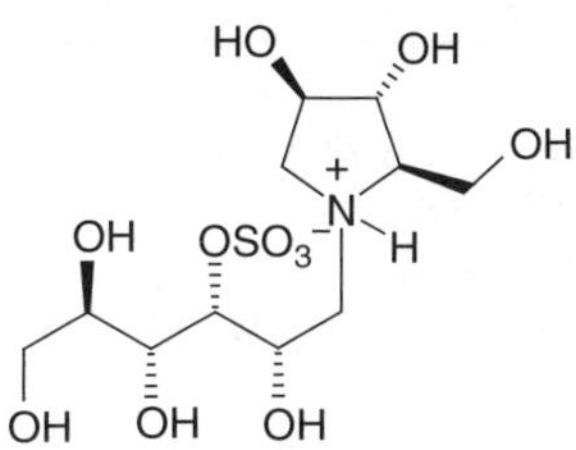	**Type 2 diabetes** *Recombinant human maltase glucoamylase* (EC 3.2.2.20 and 3.2.1.3.) – K_i 26 μM [9].
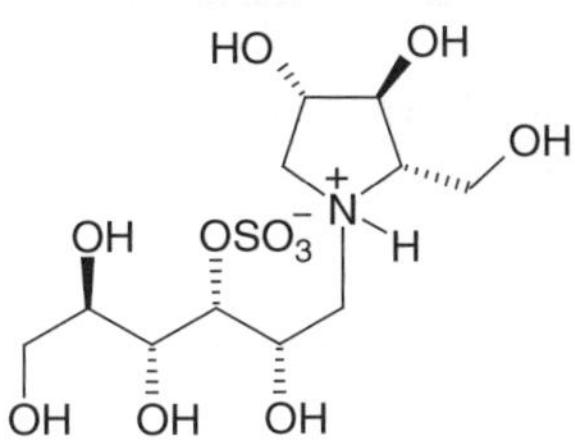	**Type 2 diabetes** *Recombinant human maltase glucoamylase* (EC 3.2.2.20 and 3.2.1.3.) – K_i 5 μM [9].
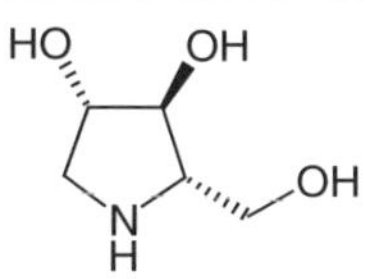	**Type 2 diabetes** *Rat intestinal maltase* (EC 3.2.1.20) – IC_{50} 1.3 μM [10]. *Rat intestinal sucrase* (EC 3.2.1.48) – IC_{50} 1.7 μM [10]. *Rat intestinal isomaltase* (EC 3.2.1.10) – IC_{50} 0.08 μM [10].
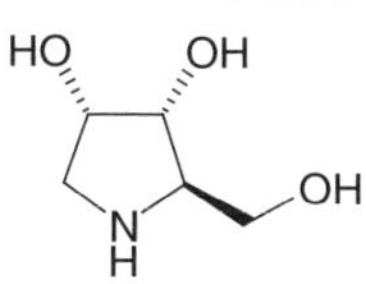	**Type 2 diabetes** *Rat intestinal sucrase* (EC 3.2.1.48) – IC_{50} 95 μM [2]. *Rat intestinal isomaltase* (EC 3.2.1.10) – IC_{50} 17 μM [2],[5]. *Rat intestinal lactase* (EC 3.2.1.23) – IC_{50} 17 μM [5].

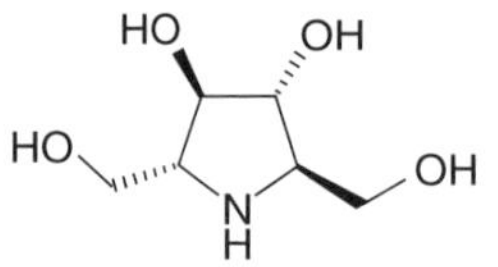	**Type 2 diabetes** *Rat intestinal isomaltase* (EC 3.2.1.10) – IC_{50} 91 μM [3],[5],[13]; IC_{50} 75 μM [11]. *Rat intestinal sucrase* (EC 3.2.1.48) – IC_{50} 55 μM [11]. *Porcine sucrase* (EC 3.2.1.48) – IC_{50} 44 μM [3]; IC_{50} 55 μM [11]. *Rat intestinal lactase* (EC 3.2.1.23) – IC_{50} 3.6 μM [3],[5],[13]. *Trehalase (Rat intestine)* (EC 3.2.1.28) – IC_{50} 360 μM [5],[13],[3]. *Trehalase (Porcine kidney)* (EC 3.2.1.28) – IC_{50} 200 μM [5],[12].
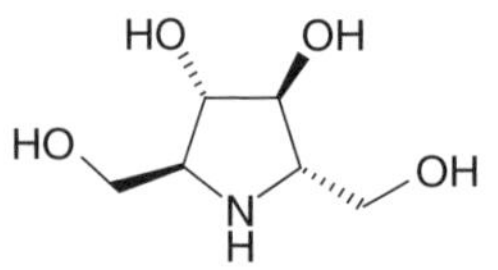	**Type 2 diabetes** *Rat intestinal maltase* (EC 3.2.1.20) – IC_{50} 1.4 μM [10],[11]. *Rat intestinal isomaltase* (EC 3.2.1.10) – IC_{50} 0.05 μM [11]. *Rat intestinal sucrase* (EC 3.2.1.48) – IC_{50} 0.1 μM [11]. *Porcine sucrase* (EC 3.2.1.48) – IC_{50} 1.5 μM [11].
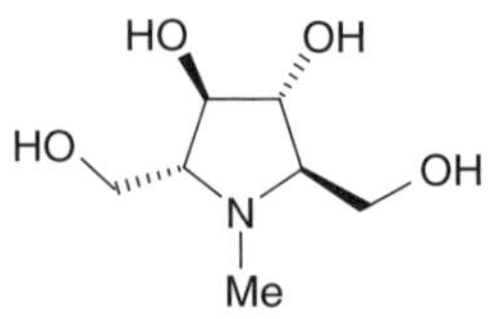	**Type 2 diabetes** *Rat intestinal sucrase* (EC 3.2.1.48) – IC_{50} 620 μM [3].
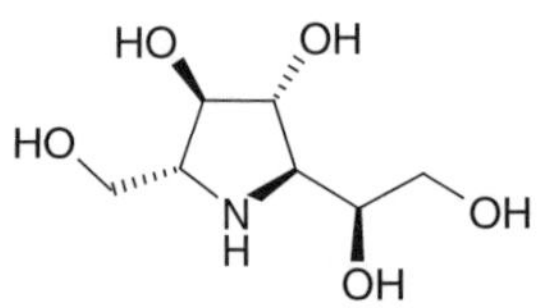	**Type 2 diabetes** *Rat intestinal maltase* (EC 3.2.1.20) – IC_{50} 400 μM [13],[14],[15]. *Rat intestinal sucrase* (EC 3.2.1.48) – IC_{50} 300 μM [13]. *Rat intestinal lactase* (EC 3.2.1.23) – IC_{50} 4.0 μM [13],[15]. *Trehalase (Porcine kidney)* (EC 3.2.1.28) – IC_{50} 5.0 μM [12], [13],[15]. *Trehalase (Rat intestine)* (EC 3.2.1.28) – IC_{50} 2.0 μM [13].
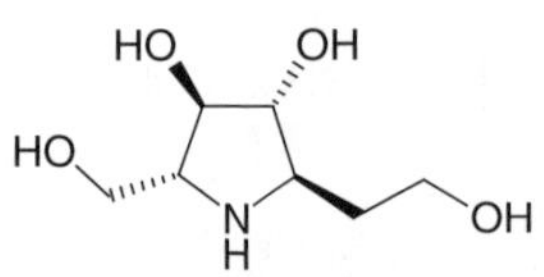	**Type 2 diabetes** *Rat intestinal maltase* (EC 3.2.1.20) – IC_{50} 2.5 μM [13]. *Rat intestinal sucrase* (EC 3.2.1.48) – IC_{50} 11 μM [13]. *Rat intestinal lactase* (EC 3.2.1.23) – IC_{50} 320 μM [13]. *Trehalase (Rat intestine)* (EC 3.2.1.28) – IC_{50} 380 μM [13].
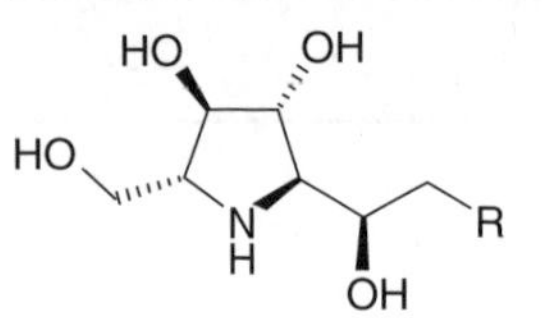	**Type 2 diabetes** *Rat intestinal lactase* (EC 3.2.1.23) R = apiosyl IC_{50} 1.6 μM [15]. R = *β-D*-xylosyl IC_{50} 0.18 μM [15].
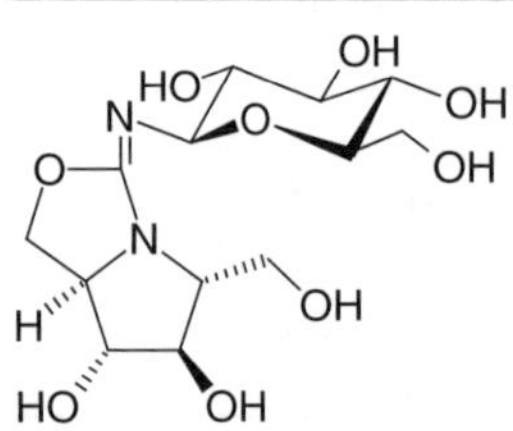	**Type 2 diabetes** *Trehalase (Porcine kidney)* (EC 3.2.1.28) – K_i 20 μM [16].

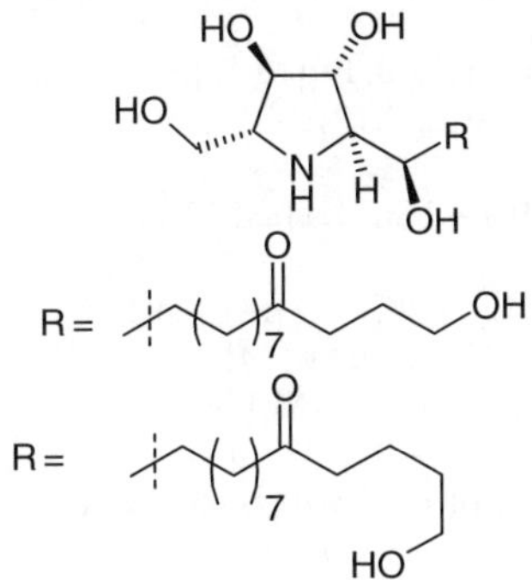

Type 2 diabetes
α – Glucosidase inhibitors [17].

14.2.2 Six-membered rings

1-Deoxynojirimycin and analogues

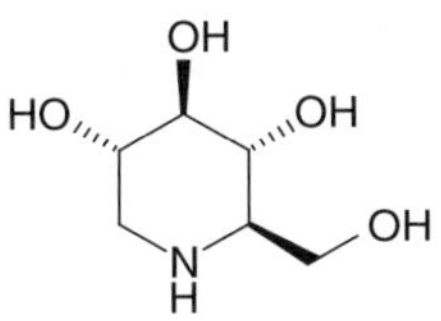

Type 2 diabetes
Rat intestinal maltase (EC 3.2.1.20)
IC_{50} 0.36 μM [3],[5],[14]; IC_{50} 0.65 μM [18].
Rat intestinal sucrase (EC 3.2.1.48) – IC_{50} 0.21 μM [3].
Rat intestinal isomaltase (EC 3.2.1.10) – IC_{50} 0.3 μM [3],[5].
α-Glucosidase type I (Calf liver) (EC 3.2.1.20) – K_i 1 μM [19].
Rat intestine lactase (EC 3.2.1.23) – IC_{50} 26 μM [5]; IC_{50} 34 μM [3].
Trehalase (Rat intestine) (EC 3.2.1.28) – IC_{50} 42 μM [3],[5].
Trehalase (Porcine kidney) (EC 3.2.1.28)
IC_{50} 41 μM [5]; IC_{50} 4.3 μM [21].
Glycogen phosphorylase (EC 2.4.1.1) – K_i 55000 μM [22].
Glycogen phosphorylase (Pig liver) (EC 2.4.1.1) – $IC_{50} > 500$ μM [23].

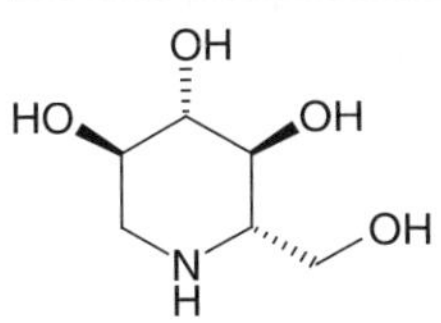

Type 2 diabetes
Rat intestinal maltase (EC 3.2.1.20) – IC_{50} 28 μM [18].

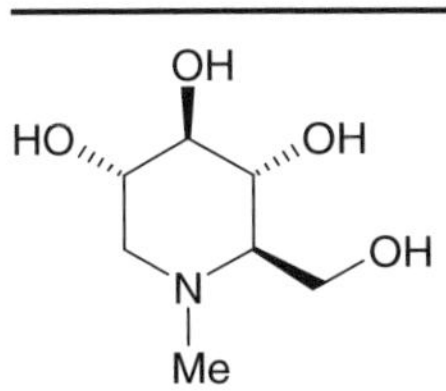

Type 2 diabetes
Rat intestinal maltase (EC 3.2.1.20) – IC_{50} 0.12 μM [3].
Rat intestinal sucrase (EC 3.2.1.48) – IC_{50} 0.04 μM [3].
Rat intestinal isomaltase (EC 3.2.1.10) – IC_{50} 4.4 μM [3].
α-Glucosidase type I (Calf liver) (EC 3.2.1.20) – K_i 0.07 μM [19].
Rat intestinal lactase (EC 3.2.1.23) – IC_{50} 4.4 μM [3].
Trehalase (Rat intestine) (EC 3.2.1.28) – IC_{50} 28 μM [3],[5].

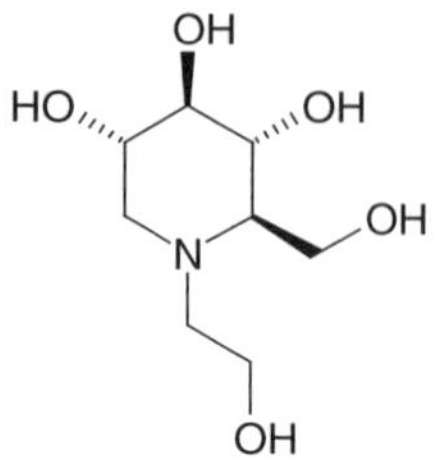	**Type 2 diabetes** *Rat intestinal maltase* (EC 3.2.1.20) – IC_{50} 0.49 μM [2]. *Rat intestinal sucrase* (EC 3.2.1.48) – IC_{50} 0.19 μM [2]; K_i 0.086 μM [24]; *Rat intestinal isomaltase* (EC 3.2.1.10) – IC_{50} 1.1 μM [2]; K_i 0.36 μM [24]. *Rat intestinal glucoamylase* (EC 3.2.1.3) – K_i 0.21 μM [24]. *Rat intestinal lactase* (EC 3.2.1.23) – K_i 4.85 μM [24]. *Trehalase (Rat intestine)* (EC 3.2.1.28) – K_i 49 μM [24]. *Trehalase (Porcine kidney)* (EC 3.2.1.28) – IC_{50} 700 μM [2]. ED_{50} 0.24 mg/kg body weight (inhibitor dose which reduces the postprandrial increase in blood glucose by 50%.) [25]. Reduce postprandrial elevations of blood glucose and plasma insulin [25]. Commercially available (GLYSET™) for the treatment of type 2 diabetes.
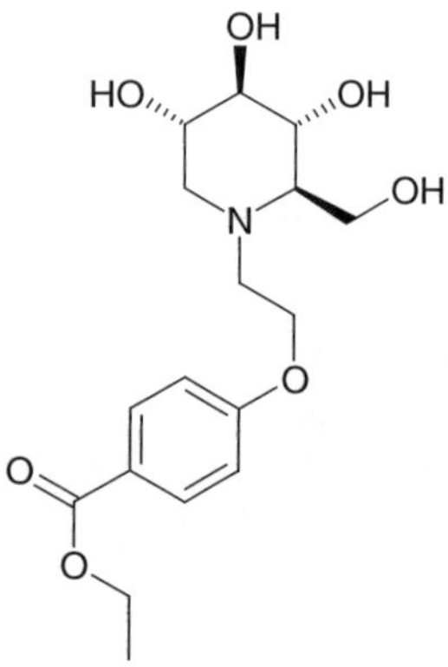	**Type 2 diabetes** ED_{50} 0.16 mg/kg body weight [25]. Reduce postprandrial elevations of blood glucose and plasma insulin [25].
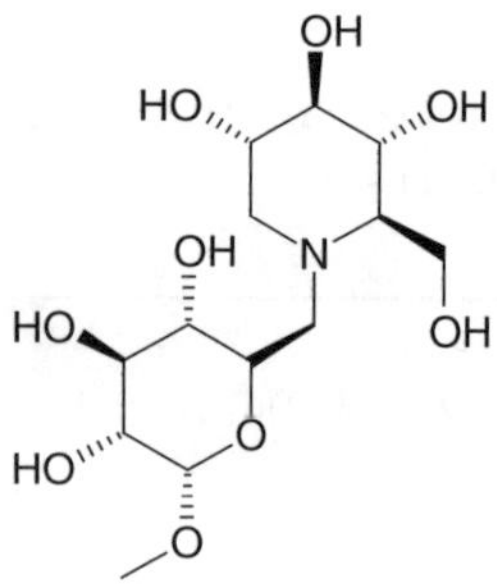	**Type 2 diabetes** *Rat intestinal sucrase* (EC 3.2.1.48) – IC_{50} 6 μM (no preincubation), IC_{50} 0.2 μM (2 h preincubation) [26]. *Rat intestinal maltase* (EC 3.2.1.20) – IC_{50} 500 μM (no preincubation), IC_{50} 1 μM (2 h preincubation) [26]. *Rat intestinal glucoamylase* (EC 3.2.1.3) – IC_{50} 50 μM (no preincubation), IC_{50} 5 μM (2 h preincubation) [26]. *Rat intestinal isomaltase* (EC 3.2.1.10) – IC_{50} 1000 μM (no preincubation), IC_{50} 8 μM (2 h preincubation) [26]. *Rat intestinal lactase* (EC 3.2.1.23) – IC_{50} 800 μM (no preincubation), IC_{50} 800 μM (2 h preincubation) [26]. Reduce postprandrial elevations of blood glucose and plasma insulin [25].
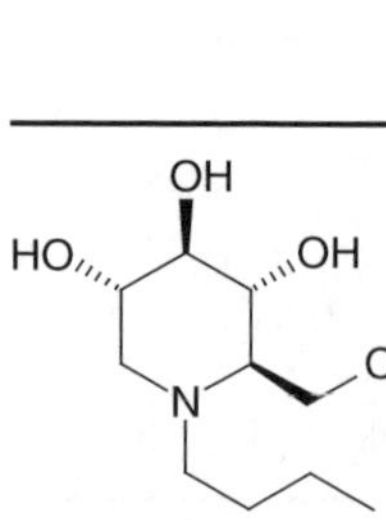	**Type 2 diabetes** *Rat intestinal maltase* (EC 3.2.1.20) – IC_{50} 2.1 μM [3]. *Rat intestinal sucrase* (EC 3.2.1.48) – IC_{50} 0.58 μM [3]. *Rat intestinal isomaltase* (EC 3.2.1.10) – IC_{50} 2.7 μM [3]. *α-Glucosidase type I (Calf liver)* (EC 3.2.1.20) – K_i 0.09 μM [19]. *Trehalase (Rat intestine)* (EC 3.2.1.28) – IC_{50} 13 μM [3].

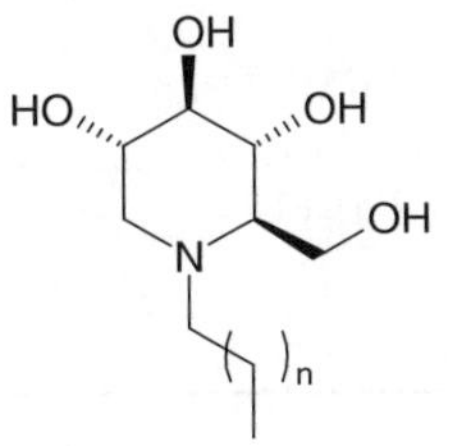

Type 2 diabetes
Rat intestinal maltase (EC 3.2.1.20)
$n=1$ IC_{50} 7.0 μM [27]; $n=4$ IC_{50} 3.7 μM [27]; $n=6$ IC_{50} 1.5 μM [27]; $n=7$ IC_{50} 1.3 μM [27].
Rat intestinal sucrase (EC 3.2.1.48)
$n=1$ IC_{50} 2.0 μM [27]; $n=4$ IC_{50} 0.45 μM [27]; $n=6$ IC_{50} 0.66 μM [27]; $n=7$ IC_{50} 0.66 μM [27].
Rat intestinal isomaltase (EC 3.2.1.10)
$n=1$ IC_{50} 1.4 μM [27]; $n=4$ IC_{50} 0.45 μM [27]; $n=6$ IC_{50} 0.28 μM [27]; $n=7$ IC_{50} 0.23 μM [27].

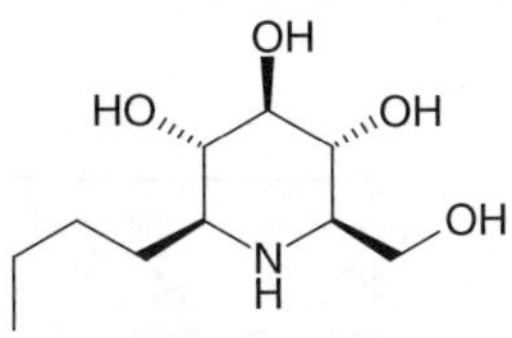

Type 2 diabetes
Rat intestinal maltase (EC 3.2.1.20) – IC_{50} 210 μM [27].
Rat intestinal sucrase (EC 3.2.1.48) – IC_{50} 155 μM [27].
Rat intestinal isomaltase (EC 3.2.1.10) – IC_{50} 115 μM [27].

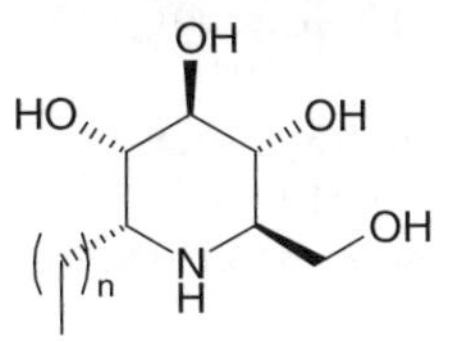

Type 2 diabetes
Rat intestinal maltase (EC 3.2.1.20)
$n=0$ IC_{50} 18 μM [14]; $n=2$ IC_{50} 43 μM [27]; $n=3$ IC_{50} 12 μM [27]; $n=5$ IC_{50} 3.5 μM [27]; $n=7$ IC_{50} 1.9 μM [27]; $n=8$ IC_{50} 4.0 μM [27].
Rat intestinal sucrase (EC 3.2.1.48)
$n=2$ IC_{50} 39 μM [27]; $n=3$ IC_{50} 3.6 μM [27]; $n=5$ IC_{50} 1.2 μM [27]; $n=7$ IC_{50} 1.5 μM [27]; $n=8$ IC_{50} 2.5 μM [27].
Rat intestinal isomaltase (EC 3.2.1.10)
$n=2$ IC_{50} 0.53 μM [27]; $n=3$ IC_{50} 0.15 μM [27]; $n=5$ IC_{50} 0.31 μM [27]; $n=7$ IC_{50} 0.09 μM [27]; $n=8$ IC_{50} 0.0035 μM [27].
Trehalase (Porcine kidney) (EC 3.2.1.28) – $n=3$ IC_{50} 780 μM [27].

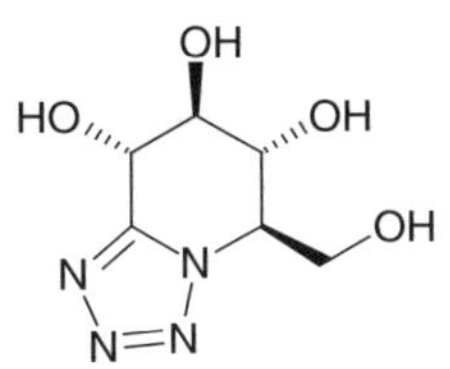

Type 2 diabetes
Glycogen phosphorylase (Rabbit muscle) (non-phosphorylated form GP*b*) (EC 2.4.1.1) – K_i 53 μM [7].

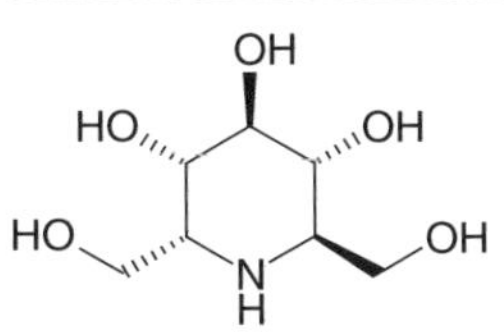

Type 2 diabetes
Rat intestinal maltase (EC 3.2.1.20) – IC_{50} 0.34 μM [28].
Rat intestinal sucrase (EC 3.2.1.48) – IC_{50} 0.17 μM [28].
Trehalase (Porcine kidney) (EC 3.2.1.28) – IC_{50} 34 μM [28].

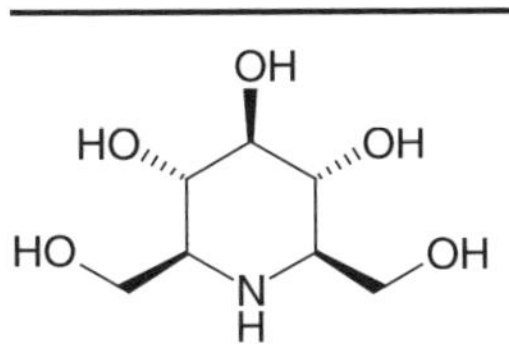

Type 2 diabetes
Rat intestinal maltase (EC 3.2.1.20) – IC_{50} 15 μM [28].
Rat intestinal sucrase (EC 3.2.1.48) – IC_{50} 7.2 μM [28].

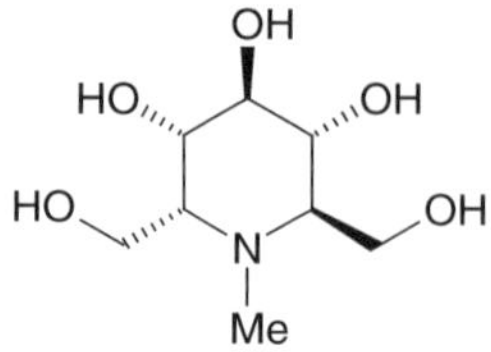	**Type 2 diabetes** *Rat intestinal maltase* (EC 3.2.1.20) – IC_{50} 0.17 μM [28]. *Rat intestinal sucrase* (EC 3.2.1.48) – IC_{50} 0.02 μM [28]. *Rat intestinal isomaltase* (EC 3.2.1.10) – IC_{50} 2.8 μM [28].
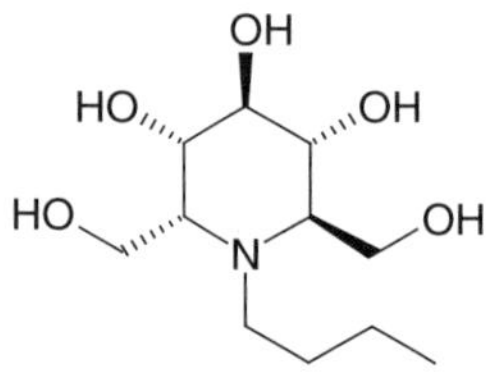	**Type 2 diabetes** *Rat intestinal maltase* (EC 3.2.1.20) – IC_{50} 4.8 μM [28]. *Rat intestinal sucrase* (EC 3.2.1.48) – IC_{50} 3.0 μM [28]. *Rat intestinal isomaltase* (EC 3.2.1.10) – IC_{50} 100 μM [28].
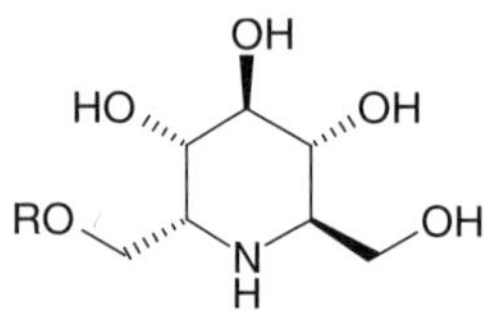 R = β-D-glucopyranosyl	**Type 2 diabetes** *Rat intestinal sucrase* (EC 3.2.1.48) – K_i 2 μM, IC_{50} 3.3 μM [29]. *Rat intestinal maltase* (EC 3.2.1.20) – IC_{50} 2.3 μM [29]. *Rat intestinal isomaltase* (EC 3.2.1.10) – IC_{50} 0.27 μM [29]. *Trehalase (Pig kidney)* (EC 3.2.1.28) – IC_{50} 0.14 μM; IC_{50} 0.011 μM without preincubation [21]; IC_{50} 0.013 μM [29]. *Trehalase (Intestinal)* (EC 3.2.1.28) – IC_{50} 0.25 μM [21]. Reduce postprandrial elevations of blood glucose and plasma insulin [25].
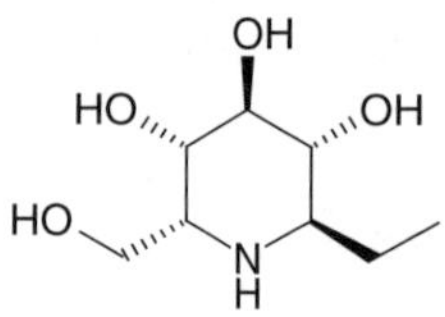	**Type 2 diabetes** *Rat intestinal sucrase* (EC 3.2.1.48) – IC_{50} 20 μM [29]. *Rat intestinal maltase* (EC 3.2.1.20) – IC_{50} 150 μM [29]. *Rat intestinal isomaltase* (EC 3.2.1.10) – IC_{50} 330 μM [29].
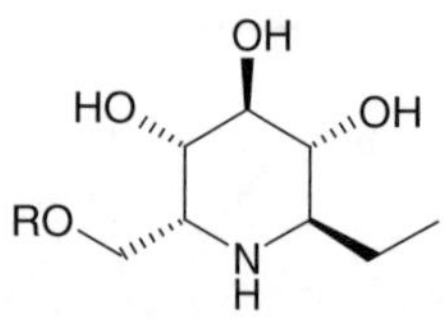 R = β-D-glucopyranosyl	**Type 2 diabetes** *Rat intestinal sucrase* (EC 3.2.1.48) – IC_{50} 1000 μM [29]. *Rat intestinal isomaltase* (EC 3.2.1.10) – IC_{50} 880 μM [29]. *Trehalase (Porcine kidney)* (EC 3.2.1.28) – IC_{50} 53 μM [29].
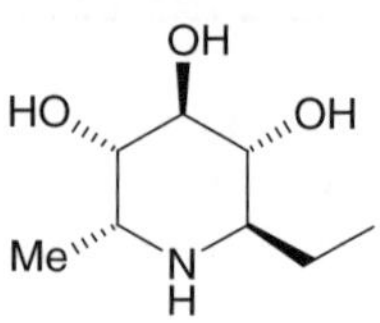	**Type 2 diabetes** *Rat intestinal sucrase* (EC 3.2.1.48) – IC_{50} 110 μM [29]. *Rat intestinal maltase* (EC 3.2.1.20) – IC_{50} 280 μM [29]. *Rat intestinal isomaltase* (EC 3.2.1.10) – IC_{50} 110 μM [29].
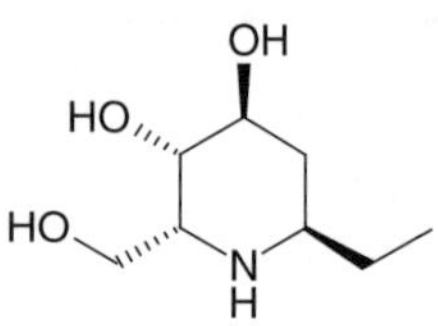	**Type 2 diabetes** *Trehalase (Porcine kidney)* (EC 3.2.1.28) – IC_{50} 1000 μM [29].

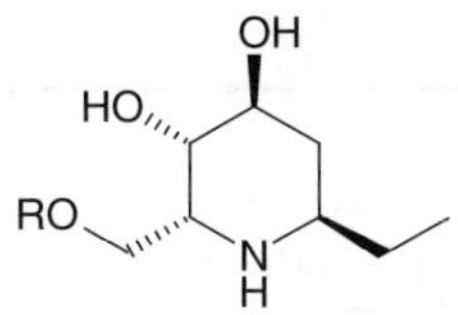

R = β-D-glucopyranosyl

Type 2 diabetes
Rat intestinal sucrase (EC 3.2.1.48) – IC_{50} 2.4 μM [29].
Rat intestinal maltase (EC 3.2.1.20) – IC_{50} 6.1 μM [29].
Rat intestinal isomaltase (EC 3.2.1.10) – IC_{50} 2.1 μM [29].

D-Galacto-configuration

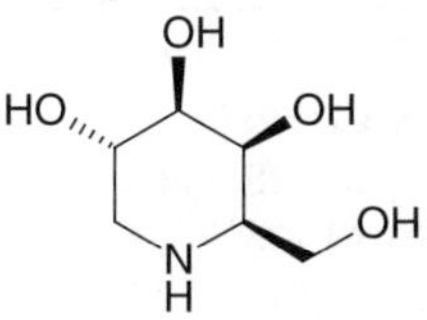

Type 2 diabetes
Rat intestinal isomaltase (EC 3.2.1.10) – IC_{50} 940 μM [5].
Rat intestinal lactase (EC 3.2.1.23) – IC_{50} 30 μM [5].

D- and L-Manno-configuration

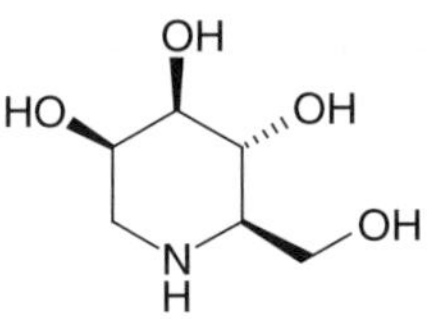

Type 2 diabetes
Rat intestinal maltase (EC 3.2.1.20) – IC_{50} 150 μM [5]; IC_{50} 110 μM [18].
Glycogen phosphorylase (Pig liver) (EC 2.4.1.1) – $IC_{50} > 500$ μM [23].

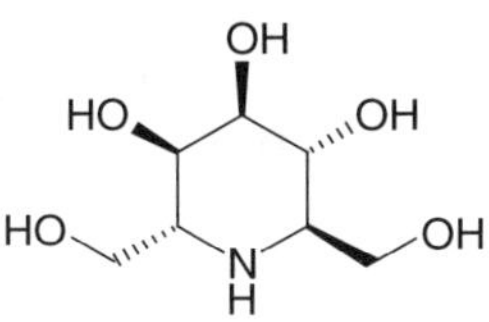

Type 2 diabetes
Rat intestinal maltase (EC 3.2.1.20) – IC_{50} 46 μM [28].
Rat intestinal sucrase (EC 3.2.1.48) – IC_{50} 27 μM [28].
Trehalase (Porcine kidney) (EC 3.2.1.28) – IC_{50} 460 μM [28].

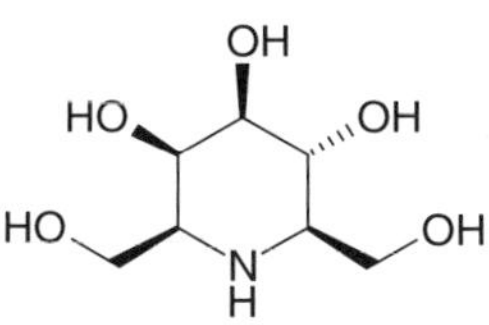

Type 2 diabetes
Rat intestinal maltase (EC 3.2.1.20) – IC_{50} 4.6 μM [28].
Rat intestinal sucrase (EC 3.2.1.48) – IC_{50} 3.0 μM [28].
Trehalase (Porcine kidney) (EC 3.2.1.28) – IC_{50} 360 μM [28].

Miscellaneous configurations

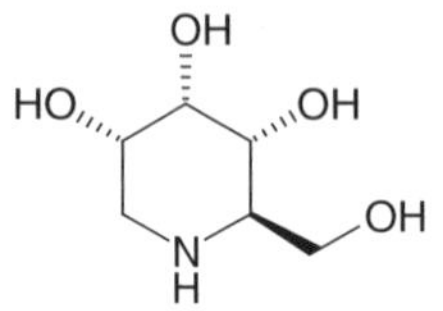

Type 2 diabetes
Rat intestinal maltase (EC 3.2.1.20) – IC_{50} 560 μM [5].
Rat intestinal isomaltase (EC 3.2.1.10) – IC_{50} 34 μM [5].
Rat intestinal lactase (EC 3.2.1.23) – IC_{50} 27 μM [5].

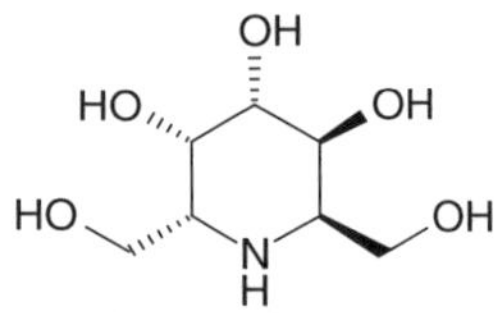

Type 2 diabetes
Rat intestinal sucrase (EC 3.2.1.48) – IC_{50} 410 μM [28].
Rat intestinal lactase (EC 3.2.1.23) – IC_{50} 290 μM [28].

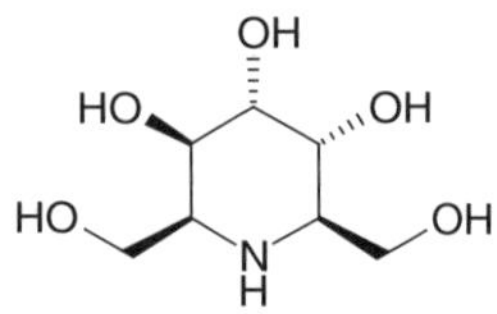

Type 2 diabetes
Rat intestinal maltase (EC 3.2.1.20) – IC_{50} 1.6 μM [28].
Rat intestinal sucrase (EC 3.2.1.48) – IC_{50} 0.8 μM [28].
Rat intestinal lactase (EC 3.2.1.23) – IC_{50} 130 μM [28].
Trehalase (Porcine kidney) (EC 3.2.1.28) – IC_{50} 140 μM [28].

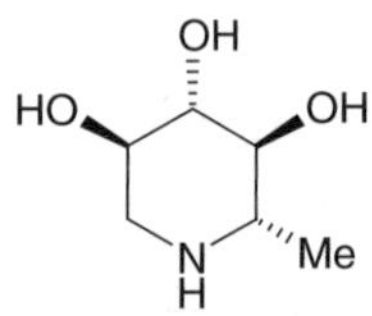

Type 2 diabetes
Rat intestinal maltase (EC 3.2.1.20) – inhibition rate at 1000 μM 45% [30].
Rat intestinal isomaltase (EC 3.2.1.10) – IC_{50} 1000 μM [30].
Rat intestinal sucrase (EC 3.2.1.48) – inhibition rate at 1000 μM 26% [30].

Fagomine and analogues

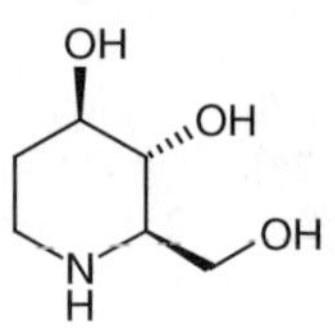

Type 2 diabetes
Rat intestinal maltase (EC 3.2.1.20) – IC_{50} 820 μM [5], [12].
Rat intestinal isomaltase (EC 3.2.1.10) – IC_{50} 460 μM [5]; IC_{50} 46 μM [31].
Rat intestinal lactase (EC 3.2.1.23) – IC_{50} 15 μM [5], [31].
Glycogen phosphorylase (Pig liver) (EC 2.4.1.1) – IC_{50} 200 μM [23].
Reduces blood glucose level 2 h after intraperitoneal administration and the hyperglycemic activity is sustained over 2–6 h after administration [25].

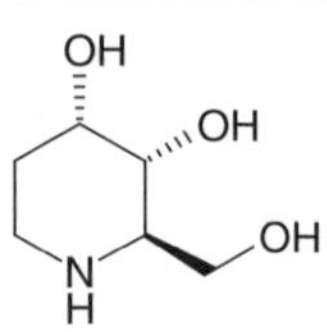

Type 2 diabetes
Rat intestinal maltase (EC 3.2.1.20) – IC_{50} 500 μM [5].
Rat intestinal isomaltase (EC 3.2.1.10) – IC_{50} 6.4 μM [5], [31].
Rat intestinal lactase (EC 3.2.1.23) – IC_{50} 4 μM [5], [31].

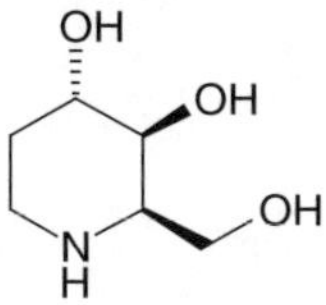

Type 2 diabetes
Glycogen phosphorylase (Pig liver) (EC 2.4.1.1) – $IC_{50} > 200\,\mu M$ [23].

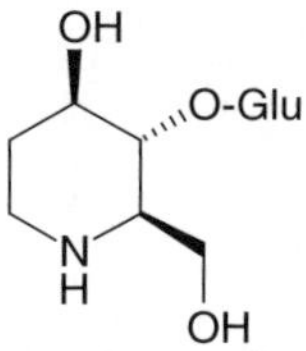

Type 2 diabetes
Glycogen phosphorylase (EC 2.4.1.1) – IC_{50} 13.5 μM [7], [22].

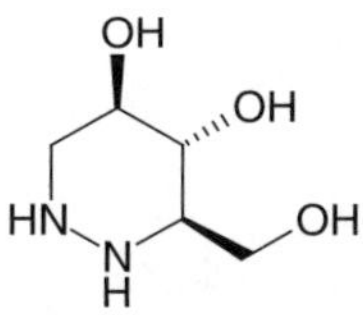

Type 2 diabetes
Glycogen phosphorylase (EC 2.4.1.1) – IC_{50} 13.5 μM [7], [22].

Isofagomine and analogues

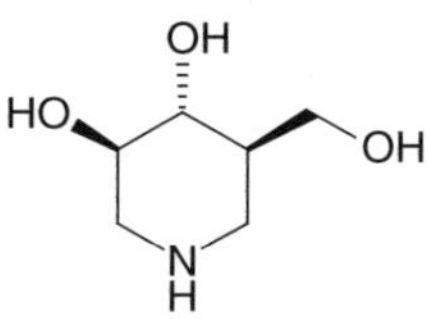

Type 2 diabetes
Rat intestinal isomaltase (EC 3.2.1.10) – IC_{50} 115 μM [10].
Glycogen phosphorylase (Pig liver) (EC 2.4.1.1) – IC_{50} 0.7 μM [23], [32].

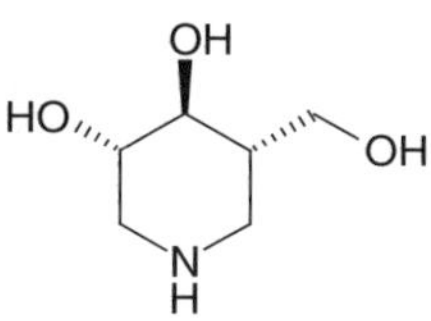

Type 2 diabetes
Glycogen phosphorylase (Pig liver) (EC 2.4.1.1) – IC_{50} 19.8 μM [23].

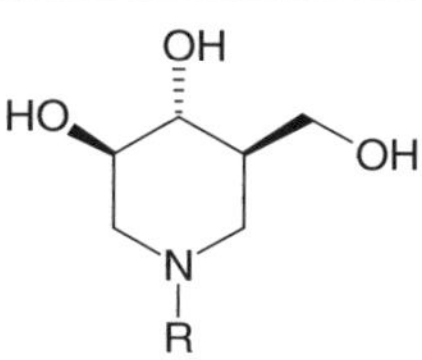

Type 2 diabetes
Glycogen phosphorylase (Pig liver) (EC 2.4.1.1)

R = methyl	IC_{50} 15 μM [23].
R = *n*-butyl	IC_{50} 20 μM [23].
R = allyl	IC_{50} 4 μM [23].
R = propyn-3-yl	$IC_{50} > 200\,\mu M$ [23].
R = 1-dodecyl	IC_{50} 8 μM [23].
R = acetyl	IC_{50} 60 μM [23].
R = 2-carboxyethyl	IC_{50} 15 μM [23].
R = benzyl	IC_{50} 46 μM [23].
R = 2-phenyl-ethyl	IC_{50} 75 μM [23].
R = 2-(4-nitro-phenyl)-ethyl	IC_{50} 16 μM [23].
R = cyclohexylprop-3-yl	IC_{50} 2 μM [23].
R = 3-phenyl-propyl	IC_{50} 1 μM [23].
R = 3-phenyl-allyl	IC_{50} 8 μM [23].

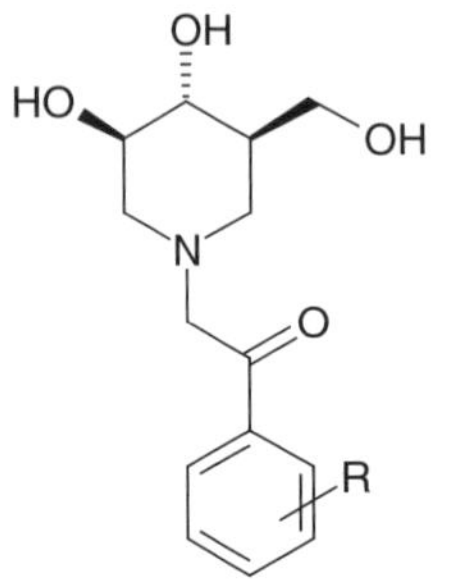

Type 2 diabetes
Glycogen phosphorylase (Pig liver) (EC 2.4.1.1)

R = H	IC_{50} 7 μM [23].
R = 4-Br	IC_{50} 40 μM [23].
R = 4-Ph	IC_{50} 70 μM [23].
R = 4-OCH_3	IC_{50} 10 μM [23].
R = 3-OCH_3	IC_{50} 8 μM [23].
R = 4-$N(CH_2CH_3)_2$	IC_{50} > 100 μM [23].
R = 4-OCF_3	IC_{50} 80 μM [23].
R = 3,4-di-F	IC_{50} 8 μM [23].
R = 3-CH_3, 4-Cl	IC_{50} 27 μM [23].
R = 2,4-di-OCH_3	IC_{50} > 100 μM [23].
R = 3-F	IC_{50} 6 μM [23].

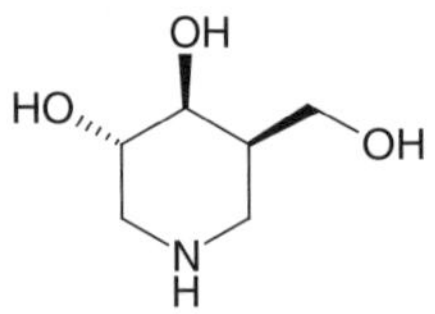

Type 2 diabetes
Glycogen phosphorylase (Pig liver) (EC 2.4.1.1) – IC_{50} > 200 μM [23].

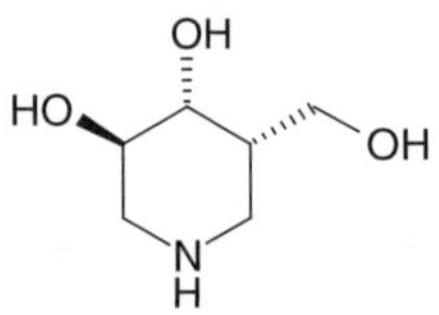

Type 2 diabetes
Glycogen phosphorylase (Pig liver) (EC 2.4.1.1) – IC_{50} > 200 μM [23].

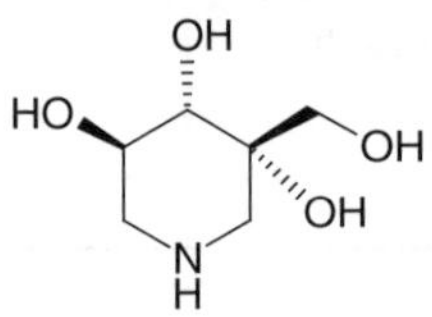

Type 2 diabetes
Glycogen phosphorylase (Pig liver) (EC 2.4.1.1) – IC_{50} > 500 μM [23].

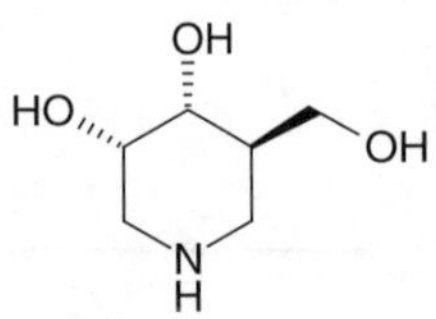

Type 2 diabetes
Glycogen phosphorylase (Pig liver) (EC 2.4.1.1) – IC_{50} > 200 μM [23].

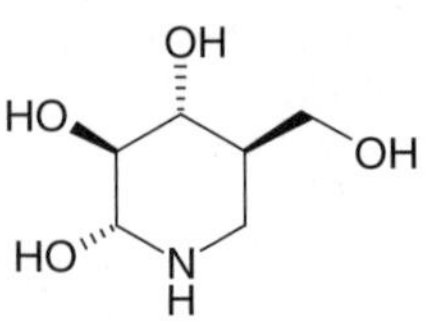

Type 2 diabetes
Glycogen phosphorylase (Rabbit liver) (EC 2.4.1.1) – IC_{50} 4 μM [7].

14.2.3 Bicyclic derivatives

Indolizidines

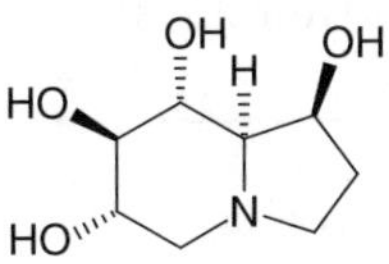

Type 2 diabetes
Trehalase (Pig liver) (EC 3.2.1.28) – IC_{50} 2.5 μM [21].

Pyrrolizidines

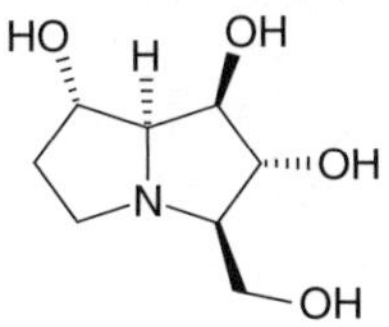

Type 2 diabetes
Rat intestinal maltase (EC 3.2.1.20) – IC_{50} 540 μM [33].
Trehalase (Porcine kidney) (EC 3.2.1.28) – IC_{50} 55 μM [33].

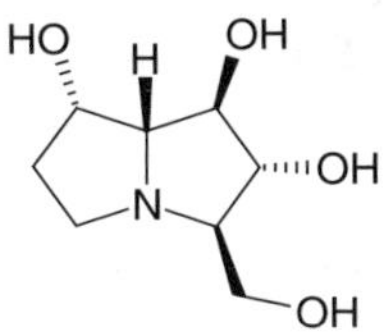

Type 2 diabetes
Rat intestinal maltase (EC 3.2.1.20) – IC_{50} 24 μM [33].
Rat intestinal isomaltase (EC 3.2.1.10) – IC_{50} 97 μM [33].
Rat intestinal sucrase (EC 3.2.1.48) – IC_{50} 4.6 μM [33].
Trehalase (Porcine kidney) (EC 3.2.1.28) – IC_{50} 160 μM [33].

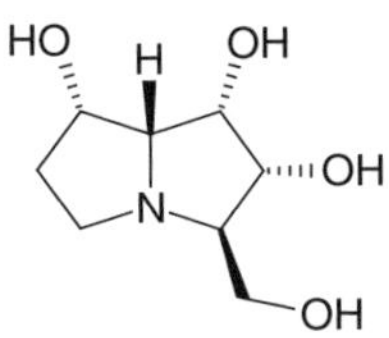

Type 2 diabetes
Rat intestinal sucrase (EC 3.2.1.48) – IC_{50} 470 μM [33].

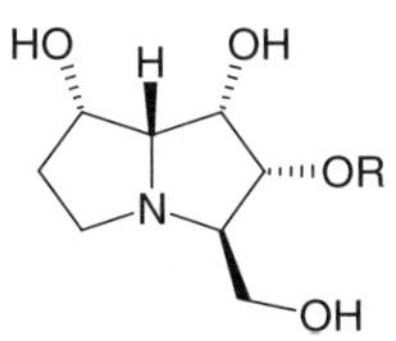

R = β-D-glucopyranosyl

Type 2 diabetes
Rat intestinal maltase (EC 3.2.1.20) – IC_{50} 41 μM [33].
Rat intestinal isomaltase (EC 3.2.1.10) – IC_{50} 2.9 μM [33].
Trehalase (Porcine kidney) (EC 3.2.1.28) – IC_{50} 450 μM [33].

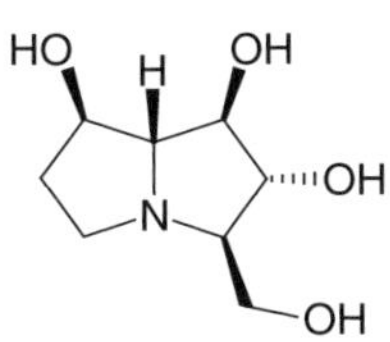

Type 2 diabetes
Rat intestinal maltase (EC 3.2.1.20) – IC_{50} 310 μM [33].
Rat intestinal isomaltase (EC 3.2.1.10) – IC_{50} 350 μM [33].
Rat intestinal sucrase (EC 3.2.1.48) – IC_{50} 100 μM [33].
Trehalase (Porcine kidney) (EC 3.2.1.28) – IC_{50} 310 μM [33].

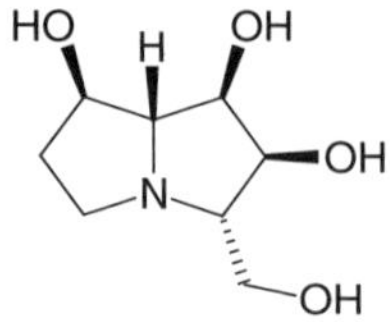	**Type 2 diabetes** *Rat intestinal maltase* (EC 3.2.1.20) – IC_{50} 130 μM [33]. *Rat intestinal isomaltase* (EC 3.2.1.10) – IC_{50} 69 μM [33]. *Rat intestinal sucrase* (EC 3.2.1.48) – IC_{50} 35 μM [33]. *Trehalase (Porcine kidney)* (EC 3.2.1.28) – IC_{50} 310 μM [33].
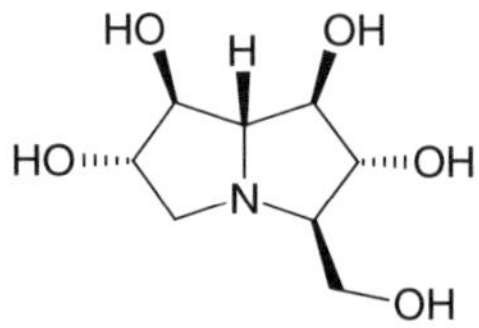	**Type 2 diabetes** *Rat intestinal maltase* (EC 3.2.1.20) – IC_{50} 0.7 μM [33]. *Rat intestinal isomaltase* (EC 3.2.1.10) – IC_{50} 3.9 μM [33]. *Rat intestinal sucrase* (EC 3.2.1.48) – IC_{50} 305 μM [33]. *Trehalase (Porcine kidney)* (EC 3.2.1.28) – IC_{50} 12 μM [33].
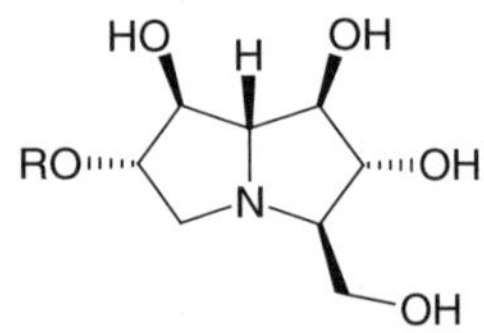 R = α-D-glucopyranosyl	**Type 2 diabetes** *Rat intestinal maltase* (EC 3.2.1.20) – IC_{50} 260 μM [33]. *Trehalase (Porcine kidney)* (EC 3.2.1.28) – IC_{50} 0.34 μM [33].
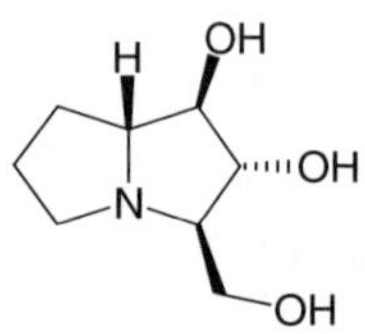	**Type 2 diabetes** *Rat intestinal lactase* (EC 3.2.1.23) – IC_{50} 73 μM [34]. *Trehalase (Porcine kidney)* (EC 3.2.1.28) – IC_{50} 260 μM [34].
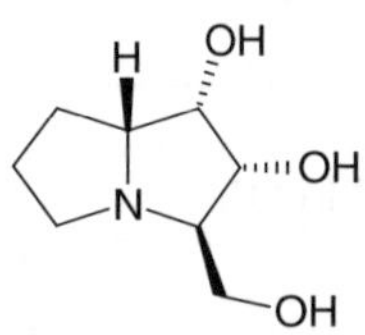	**Type 2 diabetes** *Rat intestinal lactase* (EC 3.2.1.23) – IC_{50} 4.4 μM [34].
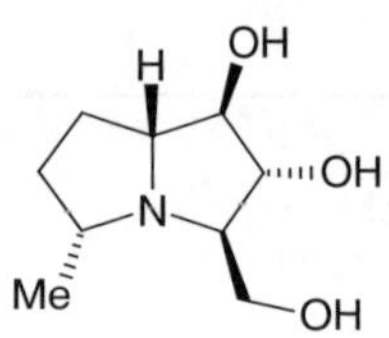	**Type 2 diabetes** *Rat intestinal lactase* (EC 3.2.1.23) – IC_{50} 160 μM [34].
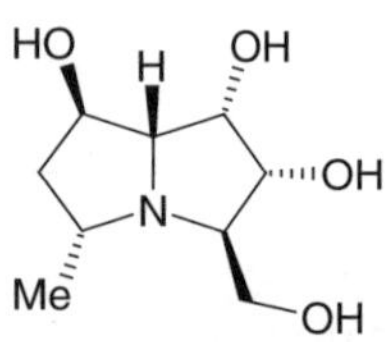	**Type 2 diabetes** *Rat intestinal lactase* (EC 3.2.1.23) – IC_{50} 18 μM [34].

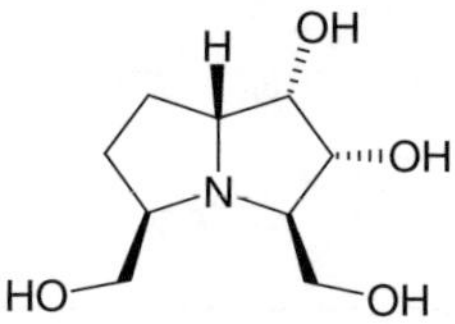

Type 2 diabetes
Rat intestinal lactase (EC 3.2.1.23) – IC_{50} 270 μM [15].

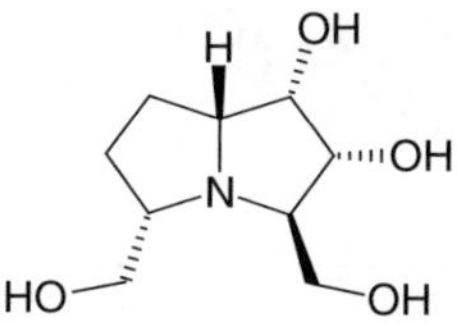

Type 2 diabetes
Rat intestinal lactase (EC 3.2.1.23) – IC_{50} 3.6 μM [15].

Nortropanes

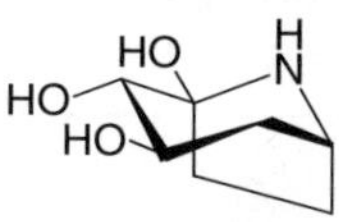

Type 2 diabetes
Rat intestinal lactase (EC 3.2.1.23) – IC_{50} 110 μM [35].
Trehalase (Rat small intestine) (EC 3.2.1.28) – IC_{50} 12 μM [35].
Trehalase (Pig kidney) (EC 3.2.1.28) – IC_{50} 13 μM [35].

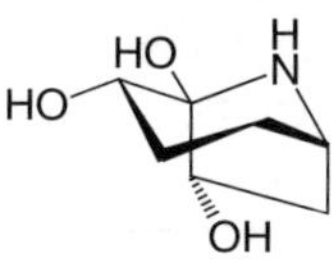

Type 2 diabetes
Trehalase (Porcine kidney) (EC 3.2.1.28) – IC_{50} 210 μM [36].

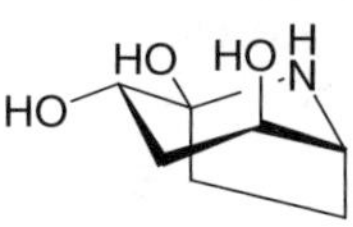

Type 2 diabetes
Moderate inhibitor of trehalase [17].

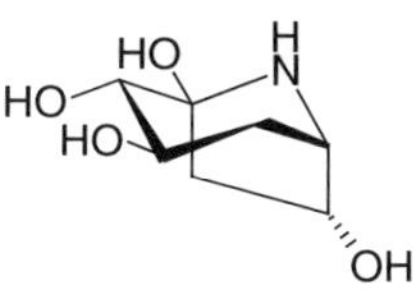

Type 2 diabetes
Rat intestinal lactase (EC 3.2.1.23) – IC_{50} 2.6 μM [35].
Trehalase (Rat small intestine) (EC 3.2.1.28) – IC_{50} 260 μM [35].

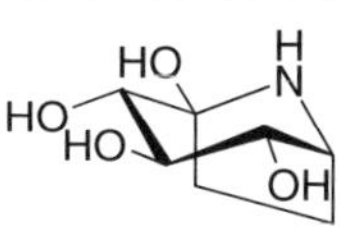

Type 2 diabetes
Rat intestinal maltase (EC 3.2.1.20) – IC_{50} 640 μM [35], [36], [38]
Rat intestinal sucrase (EC 3.2.1.48) – IC_{50} 500 μM [35], [36].
Rat intestinal lactase (EC 3.2.1.23) – IC_{50} 7.8 μM [35], [36], [38].
Trehalase (Rat intestine) (EC 3.2.1.28) – IC_{50} 9.0 μM [35], [38].
Trehalase (Porcine kidney) (EC 3.2.1.28) – IC_{50} 10 μM [35], [36], [37], [38]; K_i 5.3 μM competitive inhibition [36].

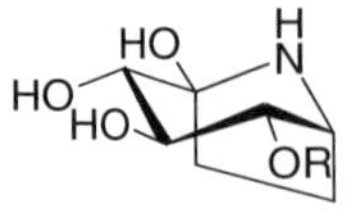 R = β-D-glucopyranosyl	**Type 2 diabetes** *Trehalase (Porcine kidney)* (EC 3.2.1.28) – IC_{50} 440 μM [37].
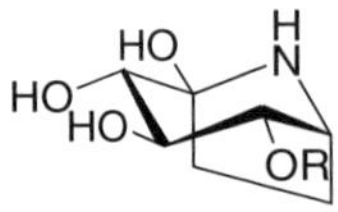 R = β-D-galactopyranosyl	**Type 2 diabetes** *Trehalase (Porcine kidney)* (EC 3.2.1.28) – IC_{50} 34 μM [37].
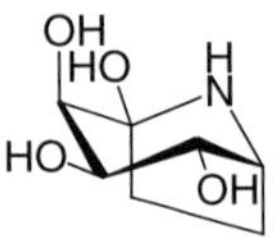	**Type 2 diabetes** *Trehalase (Rat intestine)* (EC 3.2.1.28) – IC_{50} 92 μM [35], [38]. *Trehalase (Porcine kidney)* (EC 3.2.1.28) – IC_{50} 200 μM [35], [38].
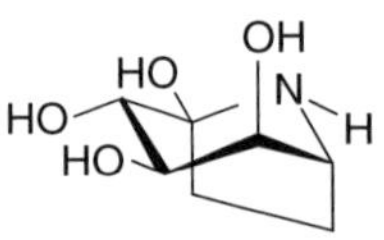	**Type 2 diabetes** *Rat intestinal lactase* (EC 3.2.1.23) – IC_{50} 110 μM [35]. *Trehalase (Rat intestine)* (EC 3.2.1.28) – IC_{50} 9.8 μM [35]. *Trehalase (Porcine kidney)* (EC 3.2.1.28) – IC_{50} 4.8 μM [35].
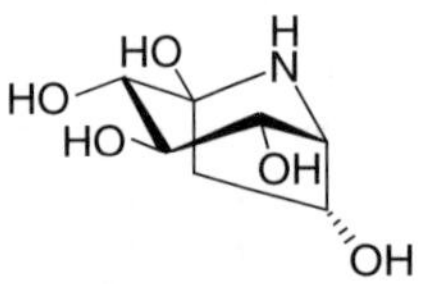	**Type 2 diabetes** *Rat intestine maltase* (EC 3.2.1.20) – IC_{50} 190 μM [35], [38]. *Rat intestine sucrase* (EC 3.2.1.48) – IC_{50} 160 μM [35]. *Rat intestine lactase* (EC 3.2.1.23) – IC_{50} 0.38 μM [35], [38]. *Trehalase (Rat intestine)* (EC 3.2.1.28) – IC_{50} 740 μM [35], [38]. *Trehalase (Porcine kidney)* (EC 3.2.1.28) – IC_{50} 270 μM [35], [38].
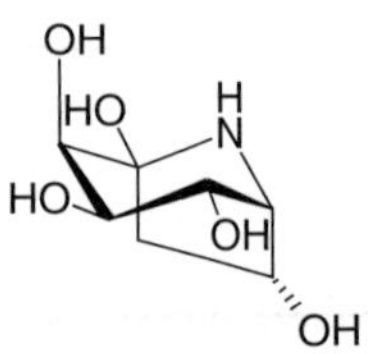	**Type 2 diabetes** *Trehalase (Rat intestine)* (EC 3.2.1.28) – IC_{50} 600 μM [38]. *Trehalase (Porcine kidney)* (EC 3.2.1.28) – IC_{50} 460 μM [38].
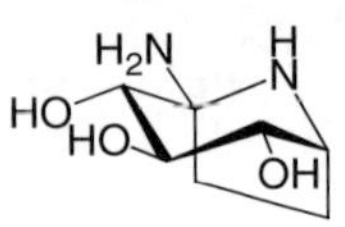	**Type 2 diabetes** *Rat intestinal sucrase* (EC 3.2.1.48) – IC_{50} 1000 μM [36]. *Rat intestinal lactase* (EC 3.2.1.23) – IC_{50} 360 μM [36]. *Trehalase (Rat intestine)* (EC 3.2.1.28) – IC_{50} 210 μM [36]. *Trehalase (Porcine kidney)* (EC 3.2.1.28) – IC_{50} 100 μM, K_i 62 μM noncompetitive inhibitor [36].

14.2.4 Oligosaccharide mimics

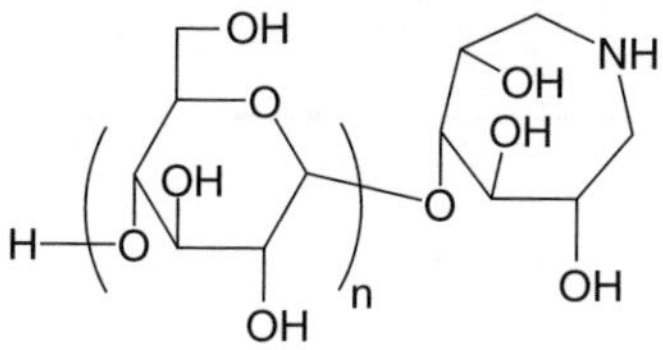

Type 2 diabetes
Human pancreatic α-amylase (EC 3.2.1.1)
$n=1$ IC_{50} 2500 μM [20].
$n=2$ IC_{50} 43 μM [20].
$n=3$ IC_{50} 30 μM [20].
$n=4$ IC_{50} 230 μM [20].
Human salivary α-amylase (EC 3.2.1.1)
$n=1$ IC_{50} 2000 μM [20].
$n=2$ IC_{50} 82 μM [20].
$n=3$ IC_{50} 49 μM [20].
$n=4$ IC_{50} 360 μM [20].
$n=2$ administration to ICR mice suppressed postprandial hyperglycemia [20].

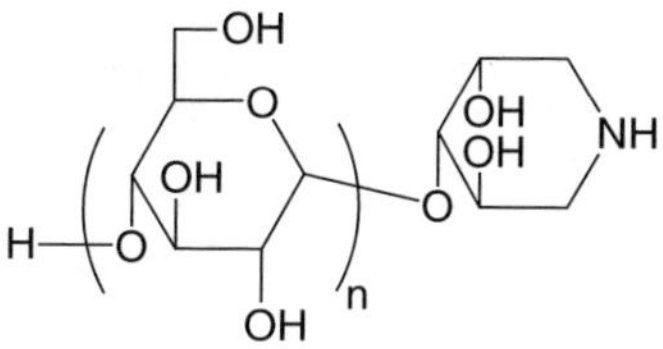

Type 2 diabetes
Human pancreatic α-amylase (EC 3.2.1.1)
$n=1$ IC_{50} 3000 μM [20].
$n=2$ IC_{50} 34 μM [20].
$n=3$ IC_{50} 25 μM [20].
$n=4$ IC_{50} 70 μM [20].
Human salivary α-amylase (EC 3.2.1.1)
$n=1$ IC_{50} 2000 μM [20].
$n=2$ IC_{50} 46 μM [20].
$n=3$ IC_{50} 42 μM [20].
$n=4$ IC_{50} 70 μM [20].
$n=2$ administration to ICR mice suppressed postprandial hyperglycemia [20].

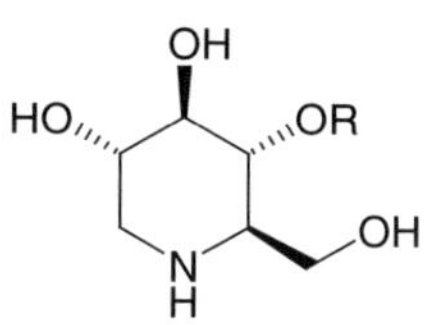

R = β-D-glucopyranosyl

Type 2 diabetes
Rabbit sucrase (EC 3.2.1.48) – IC_{50} 22 μM [39].
Rabbit maltase (EC 3.2.1.20) – IC_{50} 0.61 mM [39].
Sucrose load: 93% inhibition [39].
ED_{50} 3.9 mg/kg (dogs given 2 g/kg of starch) [39].

14.3 Antiviral activities

14.3.1 Five-membered rings

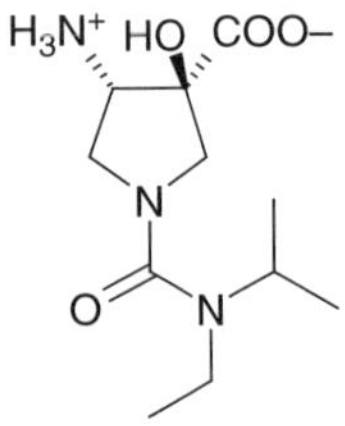	**Influenza virus** *neuraminidases* (EC 3.2.1.18) $K_i = 210\,\mu M$ for B/Memphis/3/89 [40]. $K_i = 0.36\,\mu M$ for A/Tokyo/3/67 [40]. *Inhibition of influenza virus replication* EC_{50} $2.0 \pm 1.8\,\mu M$ against A/Victoria/3/75 (H3N2) [40]. EC_{50} $305 \pm 15\,\mu M$ against B/Hong Kong/5/72 [40].
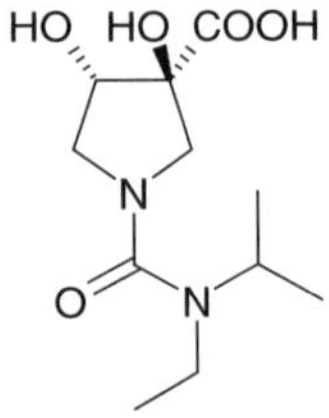	**Influenza virus** *neuraminidases* (EC 3.2.1.18) $K_i = 940\,\mu M$ for A/Tokyo/3/67 [40].
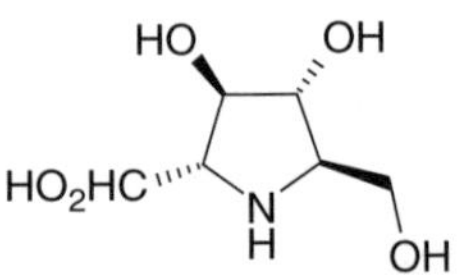	**Human immunodeficiency syndrome** *Cytotoxic activity* 35% of cell death at 0.32 mg/ml [41]. *Anti-HIV activity* 25% of CPE reduction at 0.32 mg/ml [41].
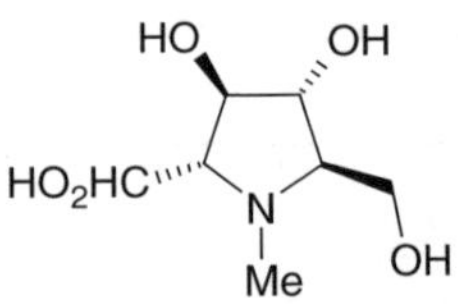	**Human immunodeficiency syndrome** *Cytotoxic activity* No cell death at 0.1 mg/ml [41]. *Anti-HIV activity* 25% of CPE reduction at 0.1 mg/ml [41].
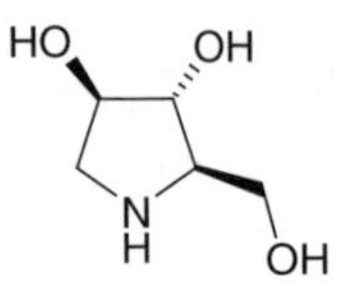	**Human immunodeficiency syndrome** *Cytotoxic activity* No cell death at 0.17 mg/ml [41]. *Anti-HIV activity* 25% of CPE reduction at 0.17 mg/ml [41]. *Jack Bean α-mannosidases (EC 3.2.1.24)* IC_{50} $100\,\mu M$ [42].
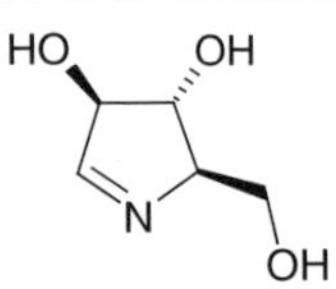	**Human immunodeficiency syndrome** Inhibitory effect on the cytopathic effect of HIV: at $10\,\mu g/ml$ 246.5 giant cells, at $100\,\mu g/ml$ 9.5 giant cells [43]. **Friend leukaemia virus** Inhibitory effect (intraveneously administered) on the induction of splenomegaly by friend leukaemia virus (mouse) 29% of inhibition for 20 mg/kg/administration [43].

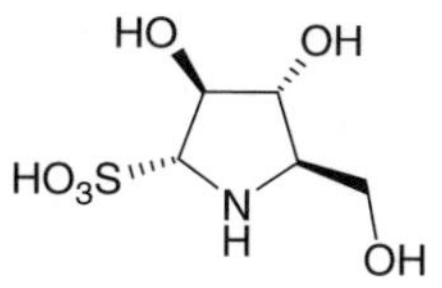	**Human immunodeficiency syndrome** Inhibitory effect on the cytopathic effect of HIV: at 10 μg/ml 89 ± 10 giant cells, at 100 μg/ml 26.5 ± 3.5 giant cells [43]. **Friend leukaemia virus** Inhibitory effect (oral administered) on the induction of splenomegaly by friend leukaemia virus (mouse) 20% of inhibition for 10 mg/kg/administration [43].
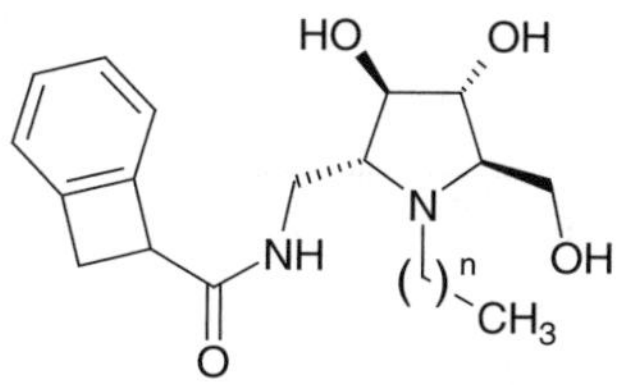	**Japanese encephalitis virus** $n = 8$, IC_{50} 11.3 ± 1.9 μM (MOI = 0.1) [44]. $n = 11$, IC_{50} 7.6 ± 0.1 μM (MOI = 0.1) [44]. **Dengue virus serotype 2** $n = 8$, IC_{50} 11.8 ± 0.2 μM (MOI = 0.1) [44]. $n = 11$, IC_{50} 6.0 ± 0.5 μM (MOI = 0.1) [44].
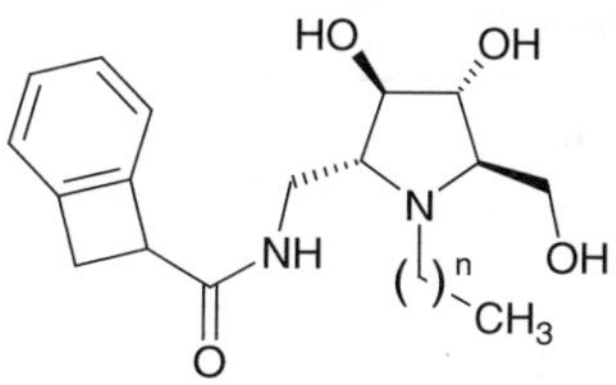	**Japanese encephalitis virus** $n = 9$, IC_{50} 9.6 ± 0.8 μM (MOI = 0.1) [44]. **Dengue virus serotype 2** $n = 9$, IC_{50} 4.7 ± 1.5 μM (MOI = 0.1) [44]. **Severe acute respiratory syndrome human coronavirus** $n = 9$, IC_{50} 3.3-10 μM [44].
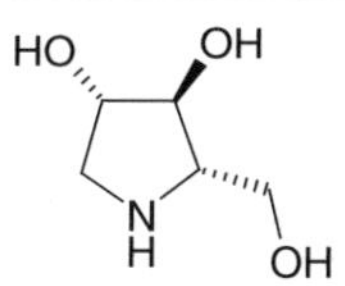	**Human immunodeficiency syndrome** *Cytotoxic activity* No cell death at 0.1 and 0.5 mg/ml [41]. *Anti-HIV activity* 50% of CPE reduction at 0.1 mg/ml and 100% at 0.5 mg/ml [41]. *T-cell growth* 1.0 10^6 at virus-uninfected-day 7 at 0.5 mg/ml [45].
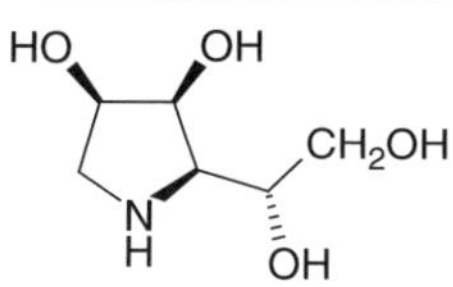	**Human immunodeficiency syndrome** *Cytotoxic activity* No cell death at 0.3 mg/ml [41]. *Anti-HIV activity* 10% of CPE reduction at 0.3 mg/ml [41]. *Jack Bean α-mannosidases (EC 3.2.1.24)* IC_{50} 0.5 μM [42].
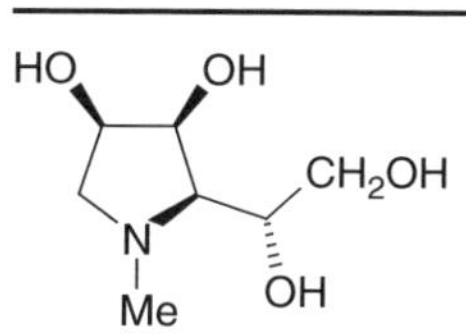	**Human immunodeficiency syndrome** *Cytotoxic activity* No cell death at 0.3 mg/ml [41]. *Anti-HIV activity* 10% of CPE reduction at 0.3 mg/ml [41].

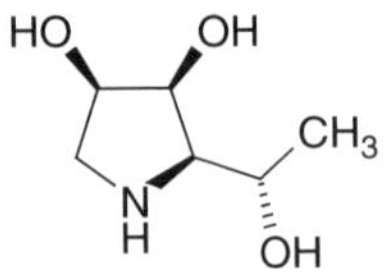	**Human immunodeficiency syndrome** *Jack Bean α-mannosidases (EC 3.2.1.24)* IC_{50} 0.6 μM [42].
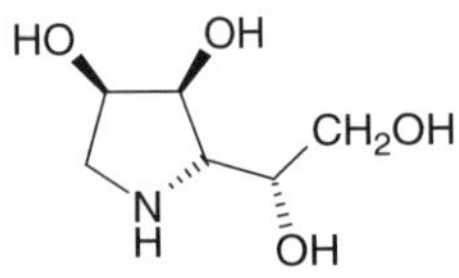	**Human immunodeficiency syndrome** *Cytotoxic activity* No cell death at 0.16 mg/ml [41]. *Anti-HIV activity* 25% of CPE reduction at 0.16 mg/ml [41].
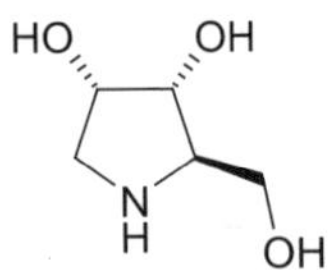	**Human immunodeficiency syndrome** *Cytotoxic activity* No cell death at 0.13 mg/ml [41]. *Anti-HIV activity* 50% of CPE reduction at 0.13 mg/ml [41].
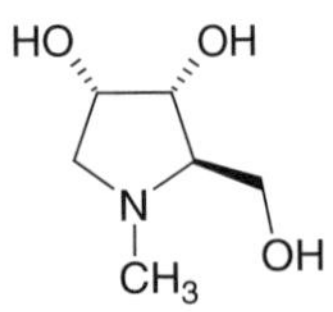	**Human immunodeficiency syndrome** *Cytotoxic activity* No cell death at 0.1 mg/ml [41]. *Anti-HIV activity* 25% of CPE reduction at 0.1 mg/ml [41].
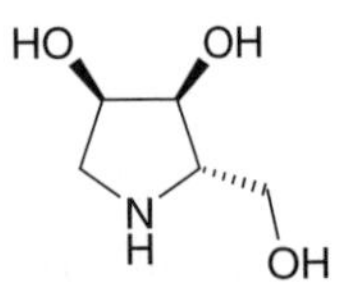	**Human immunodeficiency syndrome** *Cytotoxic activity* No cell death at 0.13 mg/ml [41]. *Anti-HIV activity* 25% of CPE reduction at 0.13 mg/ml [41].
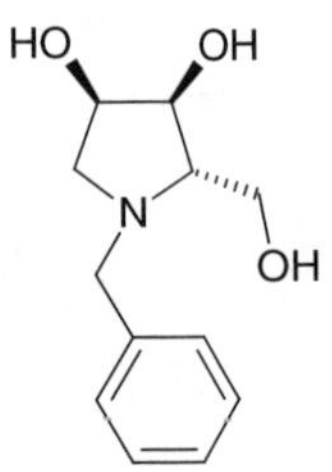	**Human immunodeficiency syndrome** *Cytotoxic activity* No cell death at 0.1 mg/ml [41]. *Anti-HIV activity* 25% of CPE reduction at 0.1 mg/ml [41].
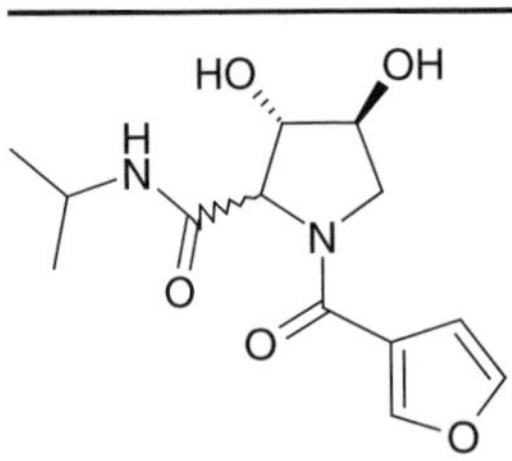	**Bovine viral diarrhoea virus** IC_{50} 30 μM (MOI = 0.5) [46].

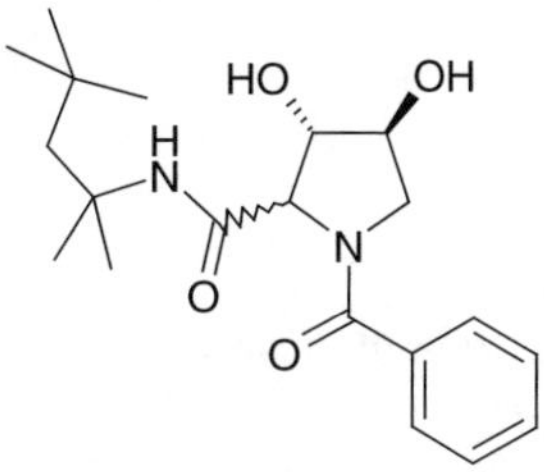

Bovine viral diarrhoea virus
IC_{50} 25 μM (MOI = 0.5) [46].

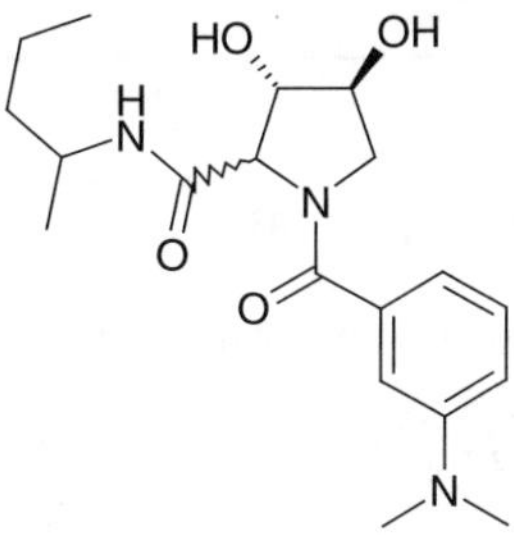

Bovine viral diarrhoea virus
IC_{50} 30 μM (MOI = 0.5) [46].

14.3.2 Six-membered rings

1-Deoxynojirimycin and analogues

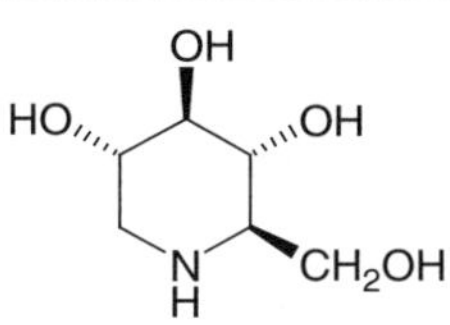

Human immunodeficiency syndrome
Cytotoxic activity
No cell death at 0.5 mg/ml [41].
Anti-HIV activity
50% of CPE reduction at 0.5 mg/ml [41].
Estimated T-cell growth
1.4 10^6 at virus-uninfected-day 7 at 0.5 mg/ml [45].
Major alterations of the properties of Env [47].
- Immunoreactivity changes of regions involved in interactions of Env with CXCR4 (V1, V2, V3 and V4).
- Altered capacity to bind CXCR4 ligands.
- Inability to mediate membrane fusion at post-CD4 binding step.

Measles virus
Viral titer from treated HEp-2 and EBV transformed human lymphoid cell line was decreased compared to untreated cells [48].
Decrease of the syncytium formation in measles virus HEp-2 cells (3.7 ± 1.6 Syncytia/cm *vs* 15.5 ± 3.1 for untreated cells) [48].

Bovine viral diarrhoea virus
MDBK cells
IC_{50} 22.5 ± 2.5 μM at low MOI (0.01), $CC_{50} > 5$ μM [49].
IC_{50} 150-300 μM, $CC_{50} > 5000$ μM [50].

(Continued)

Hepatitis B virus
HepG2 2.2.15 cells
IC_{50} 100-500 μM, $CC_{50} > 5000$ μM [50].

Dengue virus type 1
FGA/89-Infected Neuro 2a cells
Dose-dependent reduction of the production of infective particles. 20% reduction at 500 μM [52].
Prevent first step of DEN virus envelope gycoprotein processing [52].

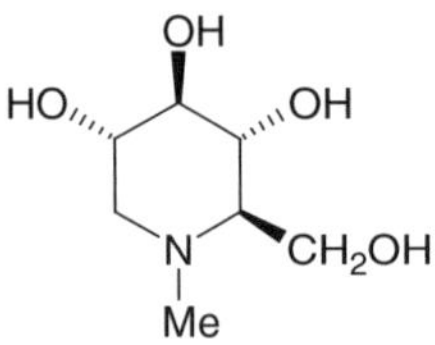

Human immunodeficiency syndrome
Cytotoxic activity
No cell death at 0.1 mg/ml and 40% at 0.5 mg/ml [41].
Anti-HIV activity
100% of CPE reduction at 0.1 mg/ml and 40% at 0.5 mg/ml [41].
Estimated T-cell growth
5 10^5 at virus-uninfected-day 7 at 0.5 mg/ml [45].

Herpes simplex type II
88% of virus control at 1.12 mM. (Viral strain HG52, HEF cells Strain MRC-5) [53].

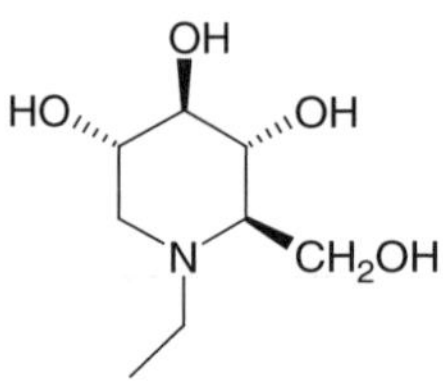

Human immunodeficiency syndrome
Cytotoxic activity
No cell death at 0.1 mg/ml and 25% at 0.5 mg/ml [41].
Anti-HIV activity
100% of CPE reduction at 0.1 mg/ml and 75% at 0.5 mg/ml [41].
Estimated T-cell growth
1.3 10^6 at virus-uninfected-day 7 at 0.1 mg/ml [45].

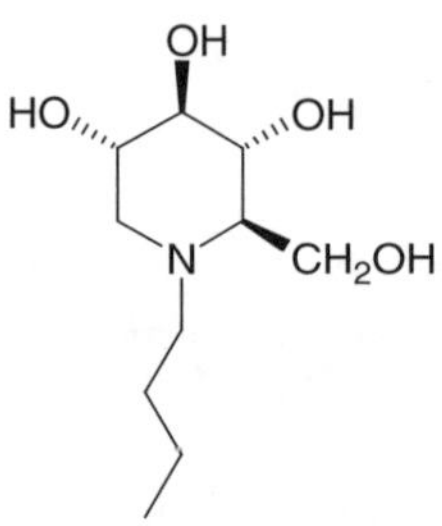

Human immunodeficiency syndrome
Anti-HIV activity
100% of CPE reduction at 0.1 mg/ml and 0.5 mg/ml [41].
Cytotoxic activity
No cell death at 0.1 mg/ml and 0.5 mg/ml [41].
Estimated T-cell growth
1.4 10^6 at virus-uninfected-day 7 at 0.1 mg/ml [45].
Virions generated in H9 cells
7.4-fold reduction in CD4-induced gp120 shedding at 2 mM [56].
4.0-fold decrease in the exposure of virion-associated gp41 at 2 mM [56].
No effect on the kinetics of the interaction between gp120 and CD4 or on the binding of virus particles to H9 cells [54].
For synergistic inhibition with Zidovudine *in vitro* see [55].
In vivo
For phase 1 dose-escalating tolerance trial see [57].

Herpes simplex type II
85% of virus control at 0.22 mM, 51% at 1.12 mM. (Viral strain HG52, HEF cells Strain MRC-5) [53].

	(Continued) **Cytomegalovirus** 87% of virus control at 0.22 mM, 32% at 0.90 mM. (Viral strain Towne, HEF cells Strain MRC-5) [53]. **Hepatitis B virus** Dose-dependent reduction of the amount of virion-associated HBV DNA released into the medium by HepG2 2.2.15 cells. Decrease of 90% at 500 μg/ml and of 99% at 1000 μg/ml [58]. *HepG2 2.2.15 cells* IC_{50} 100-500 μM, CC_{50} > 5000 μM [50]. Relative amount of HBV viral DNA secreted: 37.0 ± 13 at 200 μg/ml [51]. **Bovine diarrhoea virus** *MDBK cells* IC_{50} 27.5 ± 2.5 μM (MOI=0.01), CC_{50} > 10000 μM [49]. IC_{50} 115 μM at MOI 0.005 [59]. IC_{50} 226 μM, CC_{50} > 9000 μM. [60]. CC_{50} > 38 mM [61] IC_{50} 125-200 μ M, CC_{50} > 5000 μM [50]. Reduction of virus yields by 70-fold (n = 27) at 4.5 mM with only a 2-fold reduction in genomic RNA synthesis [61].
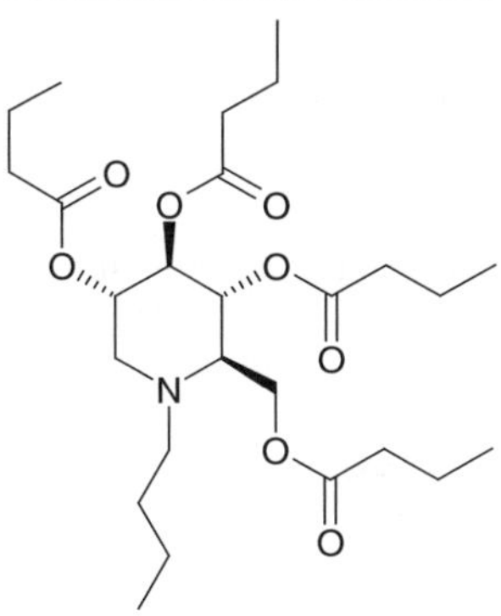	**Human immunodeficiency syndrome** Administration as prodrug led to the following relative bioavailabilities of N-Butyl DNJ (rat: 99%, dog: 15%, monkey: 42%, man: 37%) [63].
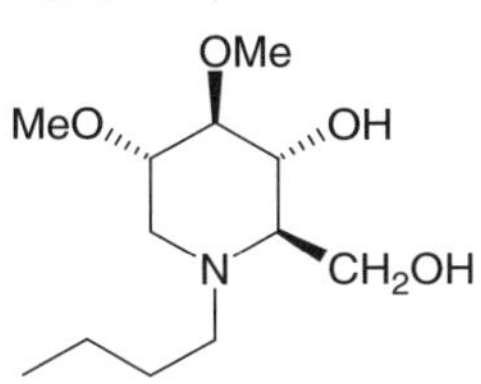	**Hepatitis B virus** 90% of HBV secretion inhibition at 200 μg/ml [51].

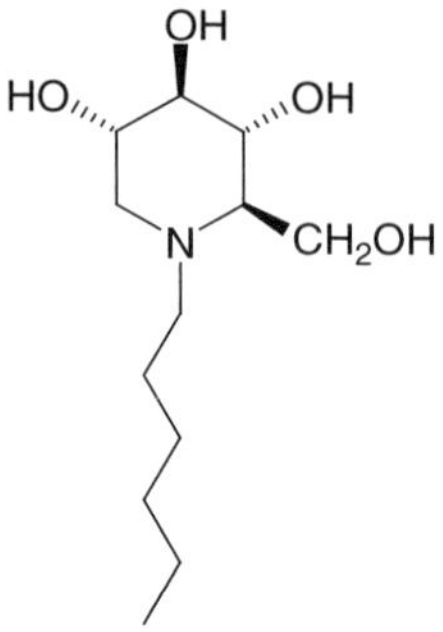	**Herpes simplex type II** 31.4% of virus control at 0.56 mM. (Viral strain HG52, HEF cells Strain MRC-5) [53].
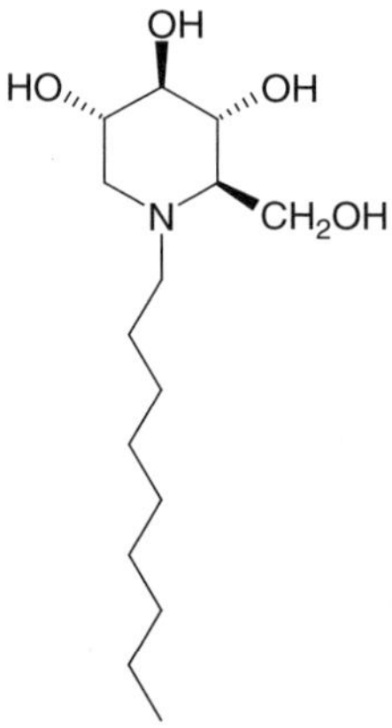	**Bovine viral diarrhoea virus** *MDBK cells plaque essays* IC_{50} 2.5 ± 0.5 μM (MOI=0.01), CC_{50} 237.5 ± 12.5 μM [49]. IC_{50} 2.5 μM (MOI=0.005) [59]. IC_{50} 6 μM, CC_{50} 175 μM [50]. **Hepatitis B virus** *Hep G2 2.2.15 cells* 8.9-fold inhibited enveloped virion secretion [64]. No inhibition of the endogenous HBV polymerase activity at 250 μmol/l [64]. IC_{50} 1-10 μM, CC_{50} 175 μM [50]. 90% HBV secretion inhibition at 0.5-1.0 μg/ml [51]. **Japanese encephalitis virus** *BHK-21 cells* Reduction of virus yields in a dose-dependent manner at 100 μM (irrespective of the MOI used) [65]. At a MOI of 5, great reduction of JEV-induced CPE [65]. In vivo, *mouse challenge model* 200 mg/kg/day orally delivered dose increased the survival rate to 47% [65]. **Dengue virus serotype 2** *BHK-21 cells* Great suppression of spreading ability of DEN-2 [65]. Dose-dependent decrease of intracellular levels of DEN-2 glycoproteins E and NS1. Reduction of DEN-2 RNA replication (16-fold less with 100 μM) [65]. **Hepatitis C virus** *H strain* Complete blockage of P7 channel activity at 105 μM [62]. **Kunjin virus** Significant decrease of the infectivity of secreted particles [66].

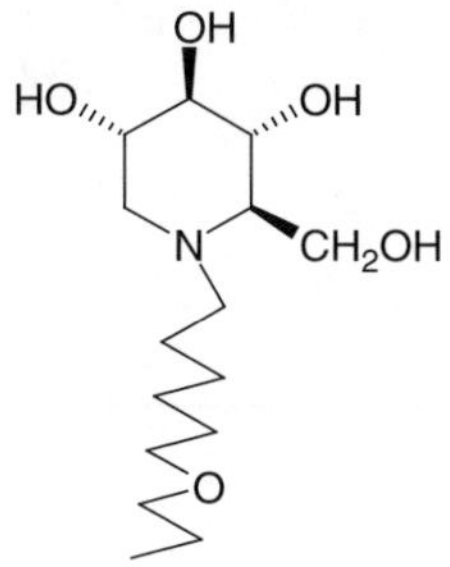	**Bovine viral diarrhoea virus** *MDBK cell plaque assays* IC_{50} 17.5 ± 2.5 μM at MOI of 0.01, CC_{50} > 5 μM [49].
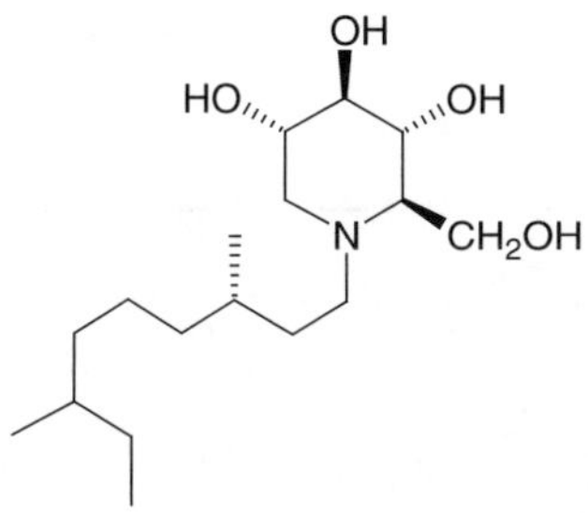	**Bovine viral diarrhoea virus** *MDBK cell plaque assays* IC_{50} 12 μM, CC_{50} 500 μM [50]. **Hepatitis B virus** *Hep G2 2.2.15 cells* IC_{50} 30-100 μM, CC_{50} 250 μM [50].
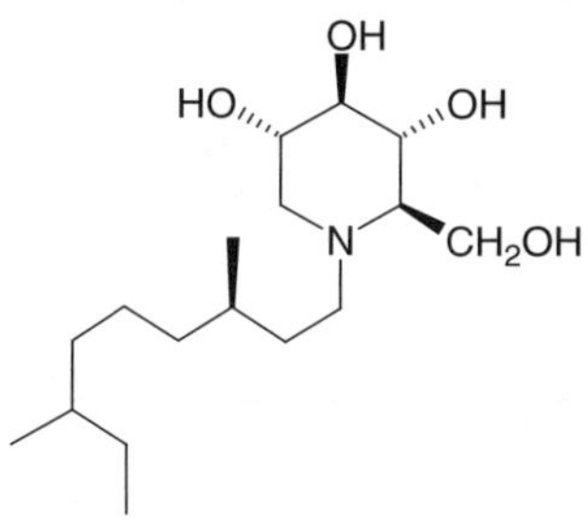	**Bovine viral diarrhoea virus** *MDBK cell plaque assays* IC_{50} 8 μM, CC_{50} 500 μM [50]. **Hepatitis B virus** *Hep G2 2.2.15 cells* IC_{50} 100 μM, CC_{50} 100 μM [50].
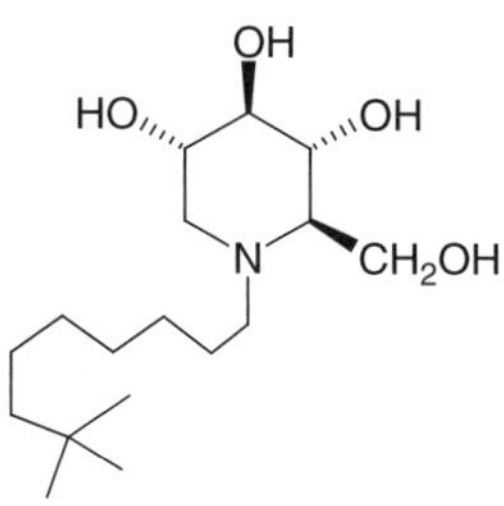	**Bovine viral diarrhoea virus** *MDBK cell plaque assays* IC_{50} 30 μM, CC_{50} 500 μM [50]. **Hepatitis B virus** *Hep G2 2.2.15 cells* IC_{50} 30 μM, CC_{50} 600 μM [50].
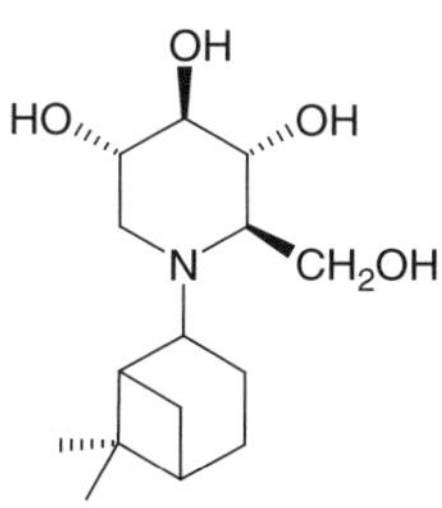	**Bovine viral diarrhoea virus** *MDBK cell plaque assays* IC_{50} 150 μM, CC_{50} >1000 μM [50]. **Hepatitis B virus** *Hep G2 2.2.15 cells* IC_{50} >100 μM, CC_{50} 800 μM [50].

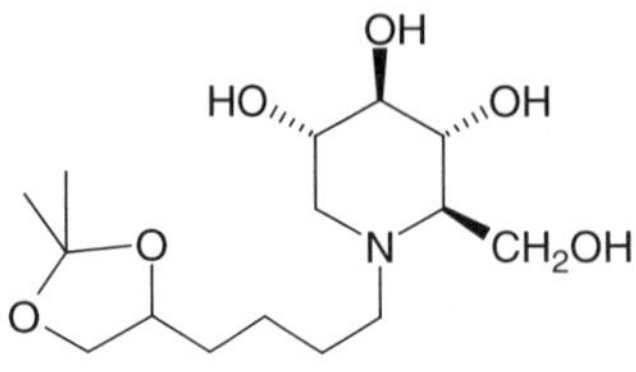	**Bovine viral diarrhoea virus** *MDBK cell plaque assays* IC_{50} 30 μM, CC_{50} >1000 μM [50]. **Hepatitis B virus** *Hep G2 2.2.15 cells* IC_{50} >100 μM, CC_{50} > 1000 μM [50].
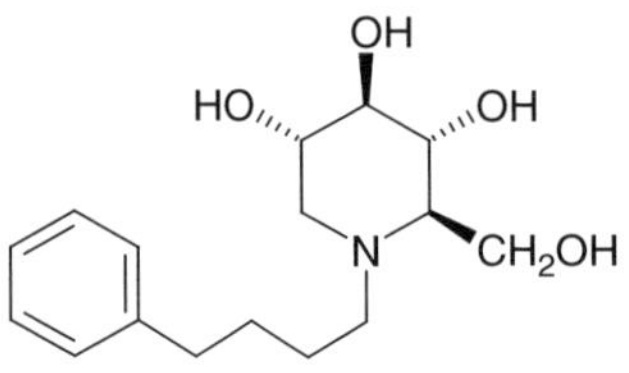	**Bovine viral diarrhoea virus** *MDBK cell plaque assays* IC_{50} 40 μM, CC_{50} 1000 μM [50]. **Hepatitis B virus** *Hep G2 2.2.15 cells* CC_{50} > 1000 μM [50].
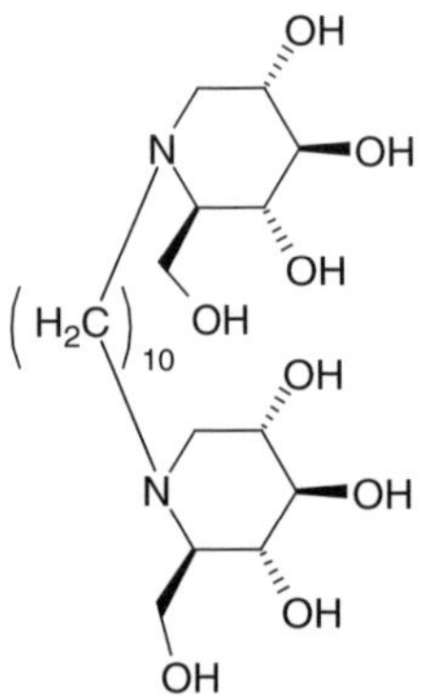	**Bovine viral diarrhoea virus** *MDBK cell plaque assays* IC_{50} 25 μM, CC_{50} 1000 μM [50]. **Hepatitis B virus** *Hep G2 2.2.15 cells* IC_{50} > 100 μM, CC_{50} > 1000 μM [50].
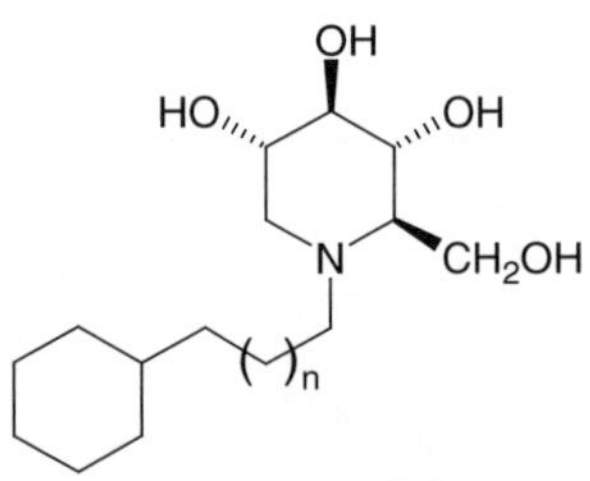	**Bovine viral diarrhoea virus** *MDBK cell plaque assays* $n=1$ IC_{50} 3 μM, CC_{50} 300 μM [50]. $n=2$ IC_{50} 3 μM, CC_{50}400 μM [50]. **Hepatitis B virus** *Hep G2 2.2.15 cells* $n=1$ CC_{50} 300 μM [50]. $n=2$ IC_{50} 30 μM, CC_{50} 500 μM [50].
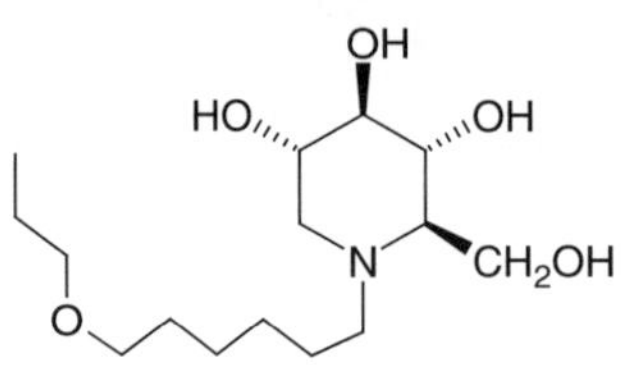	**Bovine viral diarrhoea virus** *MDBK cell plaque assays* IC_{50} 30 μM, CC_{50} > 2000 μM [50]. **Hepatitis B virus** *Hep G2 2.2.15 cells* IC_{50} 100 μM, CC_{50} > 2000 μM [50].

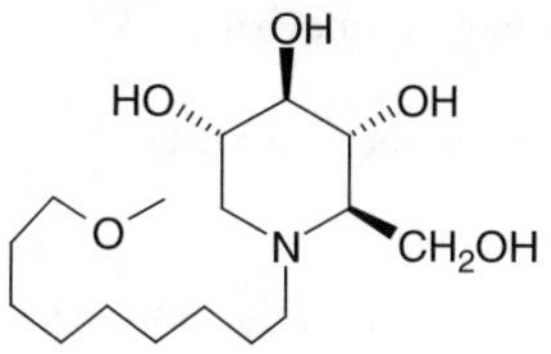

Bovine viral diarrhoea virus
MDBK cell plaque assays
IC_{50} 3 μM, $CC_{50} > 2000$ μM [50].

Hepatitis B virus
Hep G2 2.2.15 cells
IC_{50} 3 μM, $CC_{50} > 2000$ μM [50].

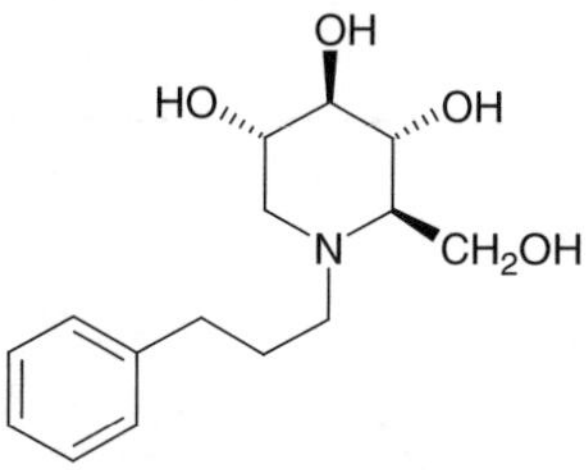

Hepatitis B virus
Hep G2 2.2.15 cells
Relative amount of HBV viral DNA secreted: 85.0 ± 5 at 200 μg/ml [51].

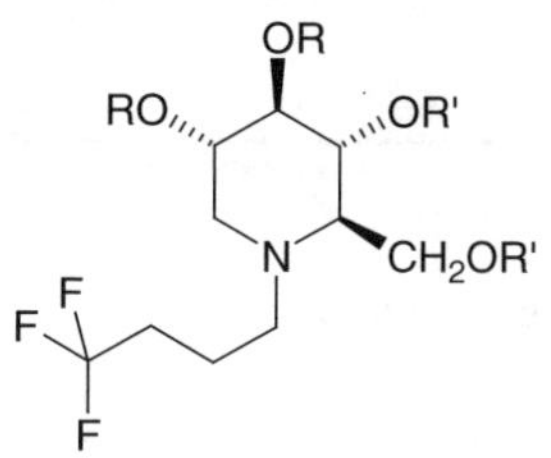

Hepatitis B virus
Hep G2 2.2.15 cells
R = Me, R' = Ac 90% HBV inhibition >200 μg/ml for [51].
R, R' = H Relative amount of HBV viral DNA secreted: 93 ± 1 at 200 μg/ml [51].

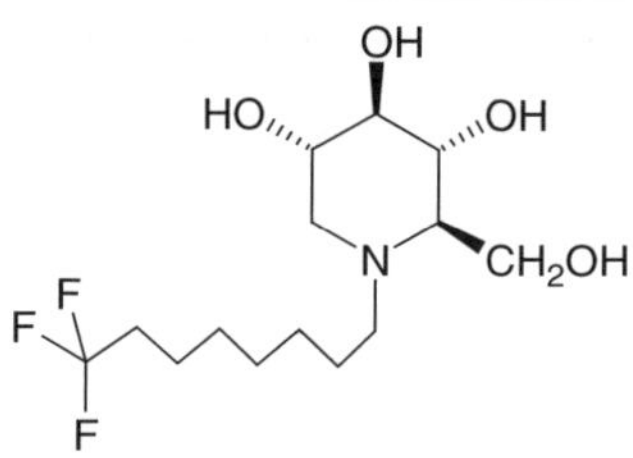

Hepatitis B virus
Hep G2 2.2.15 cells
Relative amount of HBV viral DNA secreted: 58 ± 20 at 200 μg/ml [51].
90% HBV inhibition at >200 μg/ml [51].

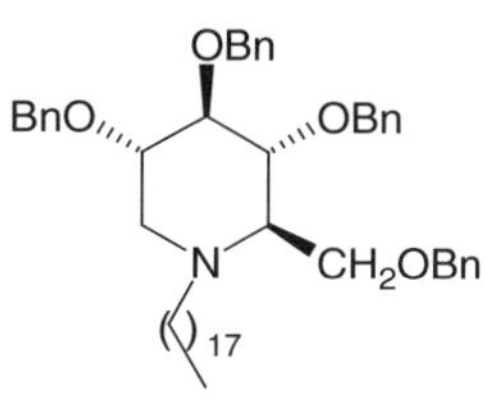

Severe acute respiratory syndrome human coronavirus
Anti-SARS activity
Protective effect at 10 μM. Noninhibitory at the anti-SARS concentrations [68].
Cytotoxic activity
20% inhibition of cell growth (Vero E6) at 2 times the anti-SARS concentrations [68].

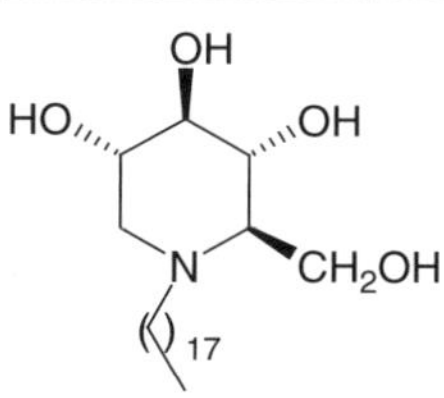

Human immunodeficiency syndrome
Potent and specific affinity to gp120 equal to GalCer itself [69].

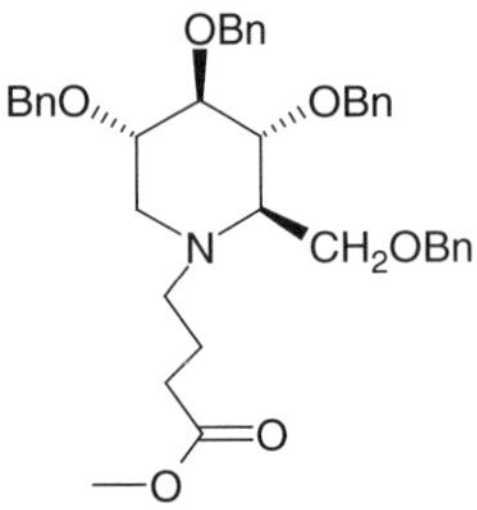

Severe acute respiratory syndrome human coronavirus
Anti-SARS activity
Protective effect at 10 μM. Noninhibitory at the anti-SARS concentrations [68].
Cytotoxic activity
20% inhibition of cell growth (Vero E6) at 2 times the anti-SARS concentrations [68].

D-Galacto-configuration

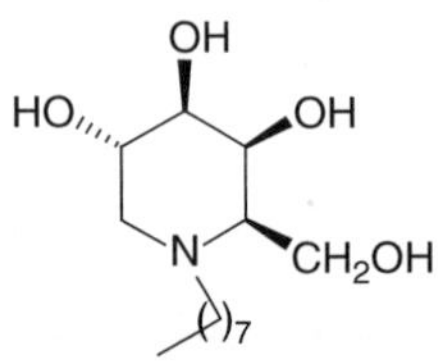

Hepatitis B virus
Hep G2 2.2.15 cells
Inhibition of virus secretion at 20 μM but no inhibition at 2 μM [70].

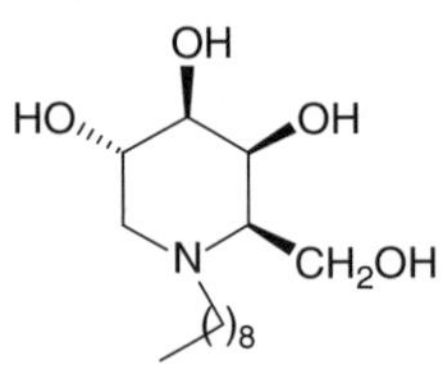

Bovine viral diarrhoea virus
IC_{50} 2.5 ± 0.5 μM at MOI (0.01), CC_{50} 237.5 ± 12.5 μM [49].

Hepatitis B virus
Hep G2 2.2.15 cells
15-fold decrease in the amount of secreted viral DNA. IC_{50} 1-2 μM of virus secretion [64].
No inhibition of the endogenous HBV polymerase activity at 250 μM [64].
Reduction of the amout of HBV nucleocapsid in tissue culture [71].
12-fold reduction in HBV secretion at 20 μM [70].

Hepatitis C virus
H strain
Complete blockage of P7 channel activity at 140 μM [62].

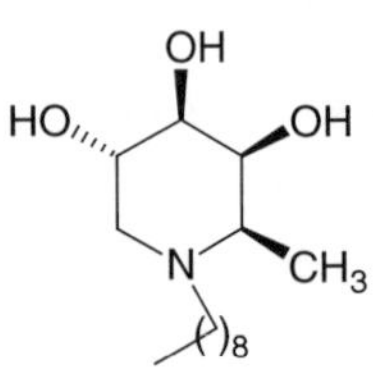

Bovine viral diarrhoea virus
MDBK cells plaque essays
IC_{50} 2 ± 0.5 μM at MOI (0.01) [49].
CC_{50} 187.5 ± 12.5 μM [49].

Hepatitis B virus
Hep G2 2.2.15 cells
Inhibition of virus secretion at 20 μM [70].

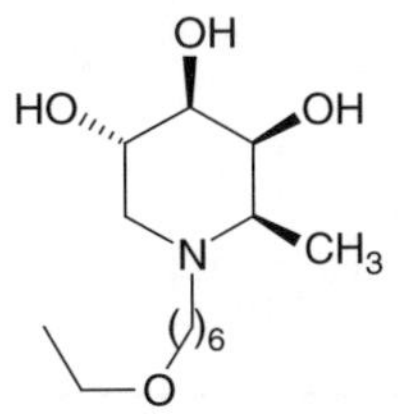	**Hepatitis C virus** *H strain* Complete blockage of P7 channel activity at 180 μM [62]. *In vivo* Entered phase 1 clinical trials in July 2002 [62]. **Bovine Viral Diarrhoea Virus** ***MDBK cells plaque essays*** IC_{50} 2.5 ± 0.5 μM at low MOI (0.01), CC_{50} > 4000 μM [67], [49].
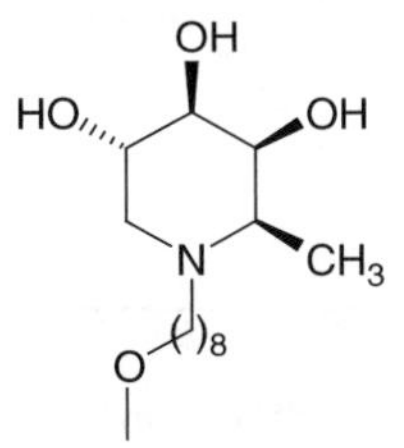	**Hepatitis B virus** *Hep G2 2.2.15 cells* IC_{50} ~ 1 μM (50% inhibitory concentration of the secretion of enveloped HBV), CC_{50} > 2000 μM [72]. **Hepatitis C virus** *9-13 cells* IC_{50} 1.5 ± 4.0 μM of HCV RNA amount [72]. Dose dependent reduction of HCV NS5A protein was observed [72].
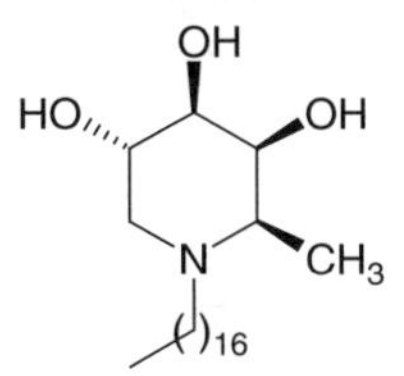	**Human immunodeficiency syndrome** Significantly higher affinity than GalCer for gp 120 [74].

Miscelleanous configurations

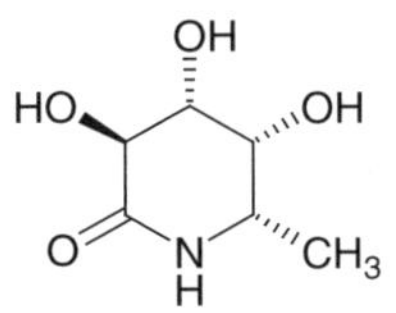	**Human immunodeficiency syndrome** *Cytotoxic activity* 25% of cell death at 0.1 mg/ml [41]. *Anti-HIV activity* 50% of CPE reduction at 0.1 mg/ml [41].
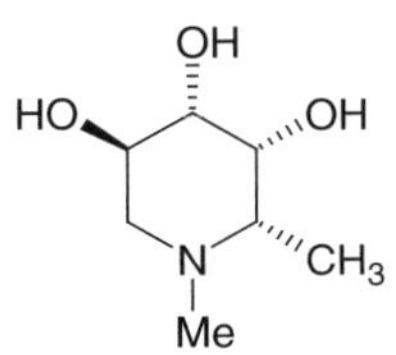	**Human immunodeficiency syndrome** *Cytotoxic activity* No cell death at 0.1 mg/ml [41]. *Anti-HIV activity* 25% of CPE reduction at 0.1 mg/ml [41].

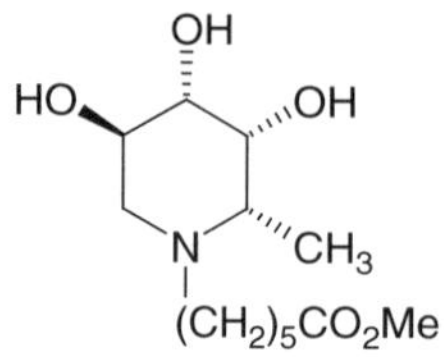	**Human immunodeficiency syndrome** *Cytotoxic activity* No cell death at 0.1 and 0.25 mg/ml [41]. *Anti-HIV activity* 75% of CPE reduction at 0.1 and 90% at 0.25 mg/ml [41].
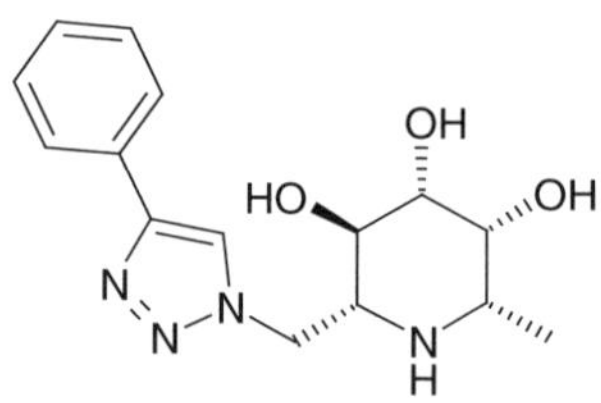	**Severe acute respiratory syndrome human coronavirus** *Anti-SARS activity* Protective effect at 10 μM. *Cytotoxic activity* Noninhibitory to Vero E6 at concentrations greater than 4 times the anti-SARS concentrations [68].
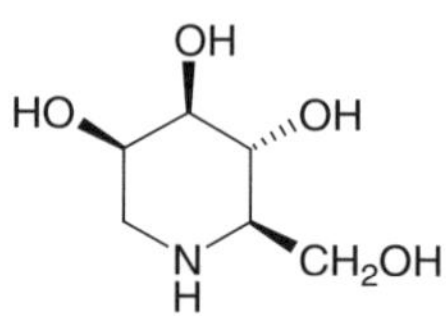	**Measles virus** Viral titer from treated Hep-2 and EBV transformed human lymphoid cell line was decreased compared with untreated cells [48]. No significant influence on the amount of syncytia in measles virus Hep-2 cells [48]. **Human immunodeficiency syndrome** *Jack Bean α-mannosidases* (EC 3.2.1.24) IC_{50} 150 μM [42], [75].
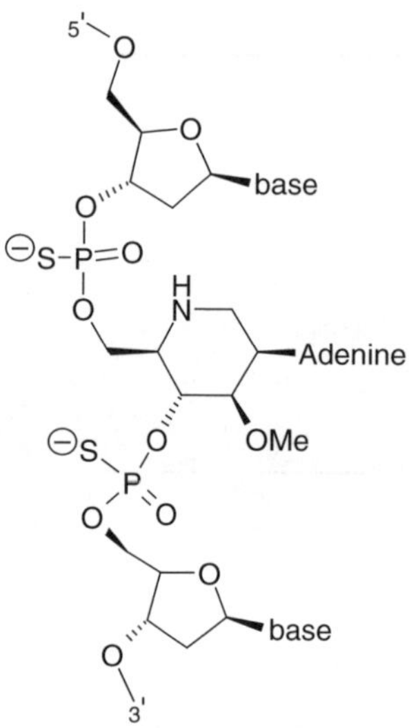	**Human immunodeficiency syndrome** Potent anti-HIV-1 activity: TI (therapeutic index) was at least higher than 400 [76].

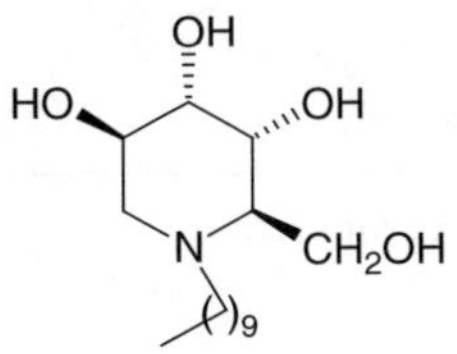

Hepatitis B virus
Hep G2 2.2.15 cells
12-fold reduction in HBV secretion at 20 μM [70].

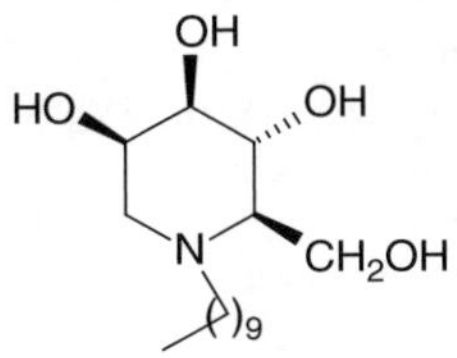

Hepatitis B virus
Hep G2 2.2.15 cells
Antiviral activity similar to that of *N*-Nonyl DFJ or DGJ [70].

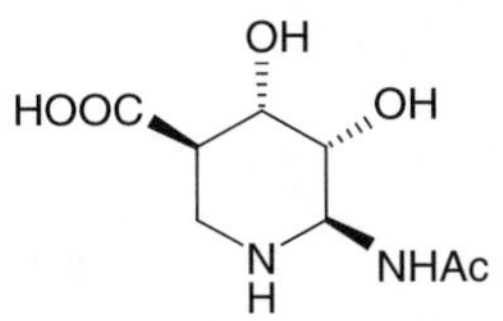

Influenza virus
N-Acetyl neuraminidase (EC 3.2.1.18)
IC_{50} 74 μM for A/FM/1/47 (H1N1) [77].
88.9% inhibition against influenza virus A/FM/1/47 (H1N1) infection in MDCK cells at 40 μM [77].
$IC_{50} > 10$ μM A/Kayano/57 (H2N2) [77].
IC_{50} 42 μM B/Lee/40 [77].

Sendai virus
N-Acetyl neuraminidase (EC 3.2.1.18)
No inhibition at 100 μM of (HVJ)/Fushimi [77].

Newcastle disease virus
N-Acetyl neuraminidase (EC 3.2.1.18)
No inhibition at 100 μM of (ND)/Miyadera [77].

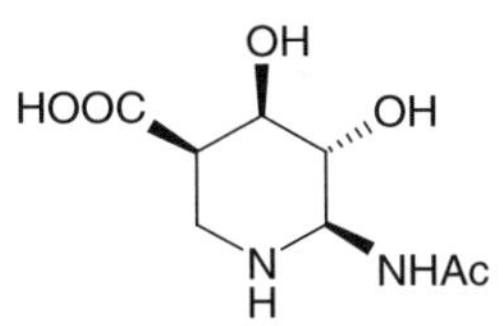

Influenza virus
N-Acetyl neuraminidase (EC 3.2.1.18)
No inhibition at 100 μM for A/FM/1/47 (H1N1) [77].
2% inhibition at 100 μM for A/Kayano/57 (H2N2) [77].
19.8% inhibition at 100 μM for B/Lee/40 [77].

Sendai virus
N-Acetyl neuraminidase (EC 3.2.1.18)
No inhibition at 100 μM of (HVJ)/Fushimi [77].

Newcastle disease virus
N-Acetyl neuraminidase (EC 3.2.1.18)
22.4% inhibition at 100 μM for (ND)/Miyadera [77].

14.3.3 Seven-membered rings

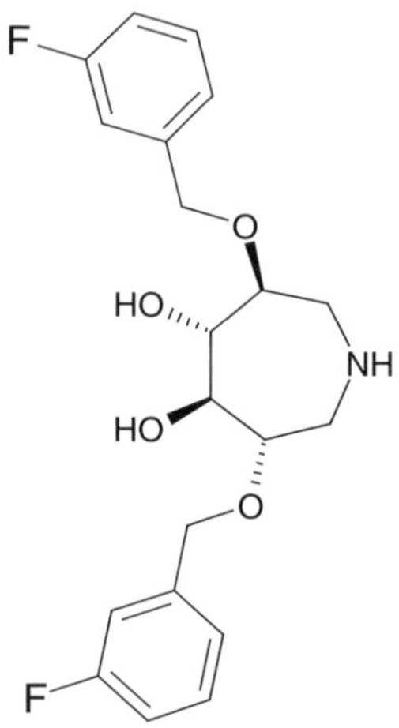

Severe acute respiratory syndrome human coronavirus
Anti-SARS activity
Protective effect at 10 μM [68].
Cytotoxic activity
Noninhibitory at the anti-SARS concentrations.
Growth retardation, to ~ 80% of the control level (Vero E6 cells treated 2 times at the anti-SARS concentration) [68].

14.3.4 Bicyclic derivatives

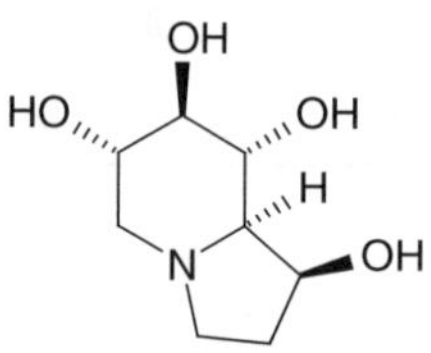

Human immunodeficiency syndrome
Cytotoxic activity
No cell death at 0.1 mg/ml and 40% at 0.4 mg/ml [41].
Anti-HIV activity
75% of CPE reduction at 0.1 mg/ml and 60% at 0.4 mg/ml [41].
Estimated T-cell growth
8 10^5 at virus uninfected-day 7 at 0.35 mg/ml [45].
H9/HTLV-IIIB cell
Inhibition of HIV-induced syncytium formation [78].
Inhibition of virus replication
6-fold reduction of infectious virus produced at 200 μg/ml [78].
Reduction in HIV infectivity
IC_{50} 7 μg/ml [79], [80] , [86].
For synergistic inhibition study with AZT *in vitro* see [81].

Moloney leukaemia virus
IC_{50} 1–2 μg/ml [79], [82].

Rausher leukemia virus
IC_{50} 1–2 μg/ml [79], [82], [86].

Herpes Simplex Type II
Viral strain HG52, HEF cells strain MRC-5
56% of virus control at 0.25 mM [53].

Cytomegalovirus
Viral strain Towne, HEF cells strain MRC-5
81% of virus control at 0.25 mM [53].

(Continued)

Feline immunodeficiency syndrome
Crandell feline kidney cells
Significant decrease of FIV infectivity compared with untreated virus [83].

Dengue virus
Neuro 2a cells (strain FGA/89) DEN-1
Dose dependent reduction in the amount of infectious virus, down to 5% of the control value at 500 μM [52].
Prevent first steps of DEN virus envelope gycoprotein processing [52].
Huh-7 human hepatoma cell line
Inhibition of the production of infectious DEN-2 in Huh-7 human hepatoma cell line [84].
IC_{50} 85.7 μM [84].
In BHK-21 cells
IC_{50} 1 μM [84].
20 to 40% reduction of marker gene expression or propagation of DEN replicon in BHK-21 cells when treatment for 48h with 15 to 500 μM range.
~20-fold reduction of viral RNA, 3000-fold reduction of infectious virus [84].
In vivo
AJ Mice treated with 250 mg/kg day had survival rate of 85% [84].
For West Nile Virus (WNV) and Yellow Fever Virus (YFV) serotypes see [84].

Bovine viral diarrhoea virus
MDBK cells
Dose-dependent inhibition of BVDV production
IC_{50} 47 μM, CC_{50} 6500 μM [60].

Measles virus
Hep-2 cells
Significant reduction of the production of viral particles in MV-infected cells.
In treated cells, association of ER chaperone calnexin with the Fusion protein was diminished, much more than with Haemagglutinin protein [85].

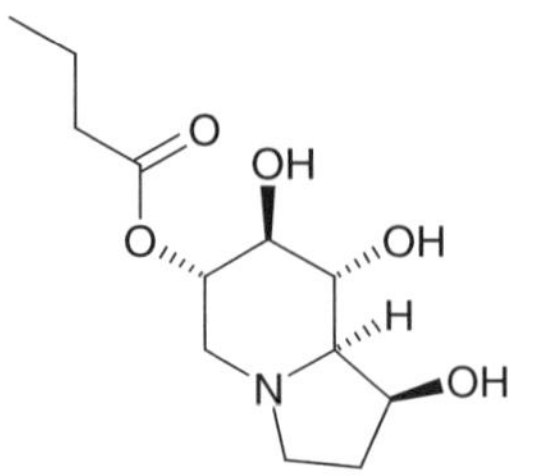	**Human immunodeficiency syndrome** ***$B_{16}F_{10}$ and JM1-cells*** Low inhibitory activity against sucrase (IC_{50} 12 µg/ml) [86]. *In vivo: Mice* Rapidly converted to catanospermine in plasma; low potential gastrointestinal toxicity [86]. **Herpes simplex type I** *Strain SC16* IC_{50} 37 ± 5.5 µM and 15 ± 4.8 µM with a pretreatment [87]. *Strain 17 i* IC_{50} 30 ± 8.5 µM and >100 µM with a pretreatment [87]. *In vivo* Virus load in the brains of mice was reduced ~100-fold [87]. **Herpes simplex virus type 2** IC_{50} 105-110 µM when treatment was postinfective. IC_{50} < 4 µM when the exposition was pre- as well as post-infection [88].
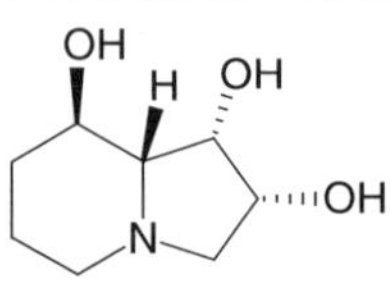	**Measles virus** Increase of viral titer from treated Hep-2 cultures compared with untreated cells [48]. No significant influence on the amount of syncytia in measles virus infected Hep-2 cells [48]. **Human immunodeficiency syndrome** *Jack Bean α-mannosidases (EC 3.2.1.24)* IC_{50} 1.75 µM [42].
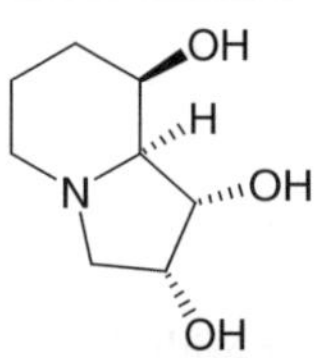	**Human immunodeficiency syndrome** *Rat liver α-mannosidases (EC 3.2.1.24)* IC_{50} ~10 µM [42].
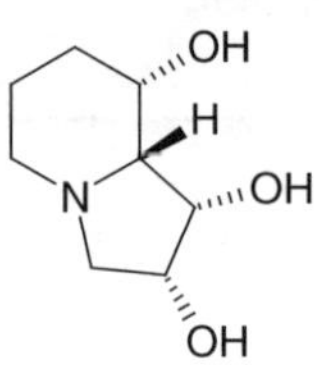	**Human immunodeficiency syndrome** *Rat liver α-mannosidases (EC 3.2.1.24)* IC_{50} ~50 µM [42].
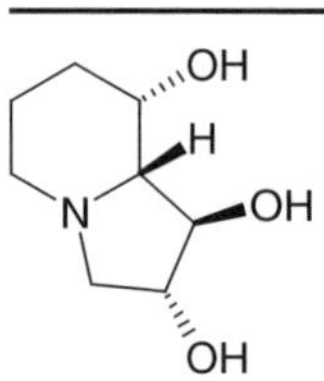	**Human immunodeficiency syndrome** *Rat liver α-mannosidases (EC 3.2.1.24)* IC_{50} ~50 µM [42].

14.4 Antibacterial activities

14.4.1 Five-membered rings

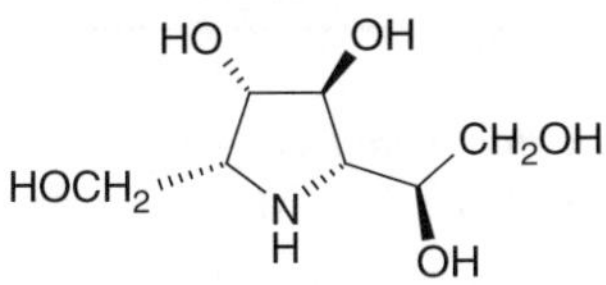	**Tuberculosis** *Mycobacterium tuberculosis* 63% Inhibition of mycobacterial galactan biosynthesis [89]. *E. Coli K12 UDP-Galactopyranose mutase (EC 5.4.99.9)* 64% Inhibition of UDP-Gal*p* to UDP-Gal*f* at 200 μg/ml [89]. 67% Inhibition of UDP-Gal*f* to UDP-Gal*p* at 200 μg/ml [89].
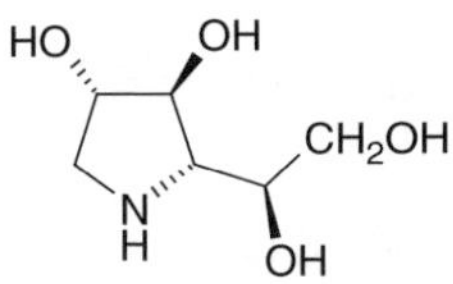	**Tuberculosis** *Mycobacterium tuberculosis* 56% inhibition of mycobacterial galactan biosynthesis [89]. *E. Coli K12 UDP-Galactopyranose mutase (EC 5.4.99.9)* 36% inhibition of UDP-Gal*p* to UDP-Gal*f* at 200 μg/ml [89]. 81% inhibition of UDP-Gal*f* to UDP-Gal*p* at 200 μg/ml [89].
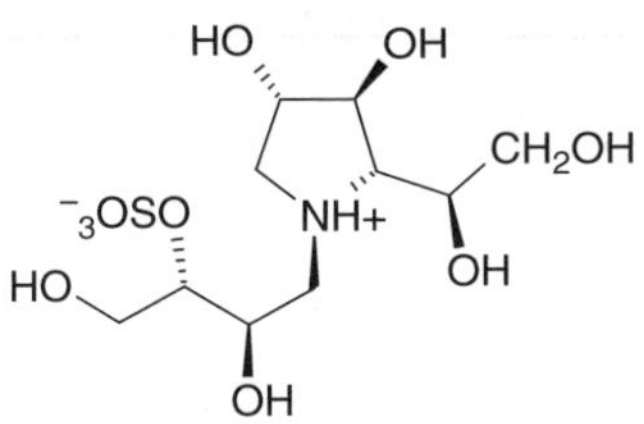	**Tuberculosis** *Mycobacterium tuberculosis* *UDP-Galactopyranose mutase (EC 5.4.99.9) from Klebsiella pneumonia* 40.7% inhibition at 10 mM [90].
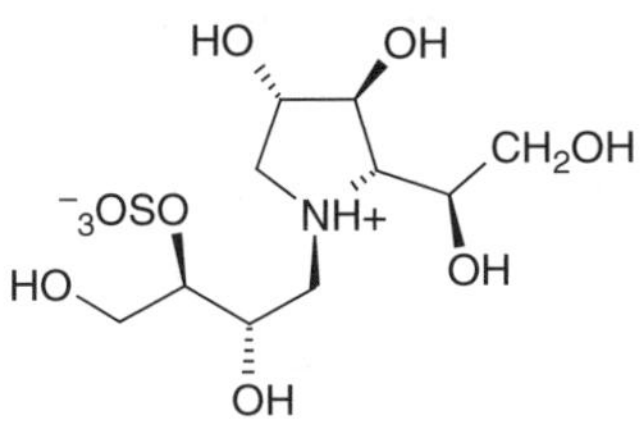	**Tuberculosis** *Mycobacterium tuberculosis* *UDP-Galactopyranose mutase (EC 5.4.99.9) from Klebsiella pneumonia* 16.7% inhibition at 10 mM [90].
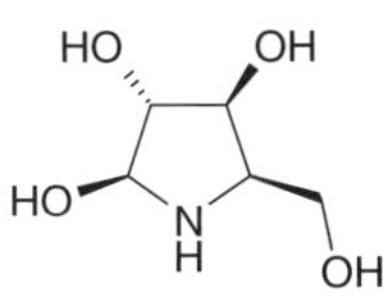	**Tuberculosis** *Mycobacterium tuberculosis* 16% inhibition of mycobacterial galactan biosynthesis at 200 μg/ml [89].
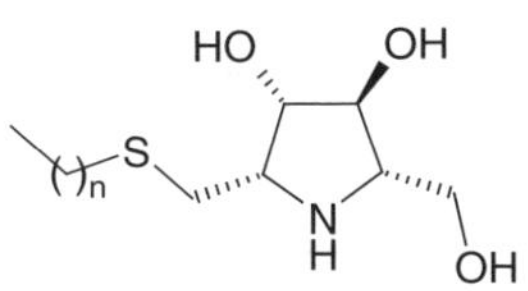	**Tuberculosis** *Mycobacterium tuberculosis Mycobacterium avium Complex* *M Tuberculosis strain H37Ra, MAC strain NJ 168, NJ 211, NJ 1854, NJ 3009 and NJ 3404* $n = 9$ MIC 32-128 μg/ml; MBC >128 μg/ml [91]. $n = 11$ MIC 16-32 μg/ml; MBC 64 – >128 μg/ml [91]. $n = 15$ MIC 16-64 μg/ml; MBC 64 – >128 μg/ml [91].

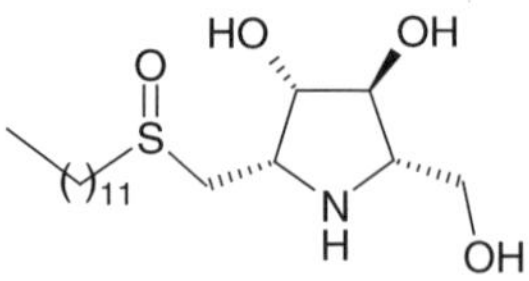	**Tuberculosis** *Mycobacterium tuberculosis Mycobacterium avium Complex M Tuberculosis strain H37Ra, MAC strain NJ 168, NJ 211, NJ 1854, NJ 3009 and NJ 3404* MIC 32-64 μg/ml, MBC >128 μg/ml [91].
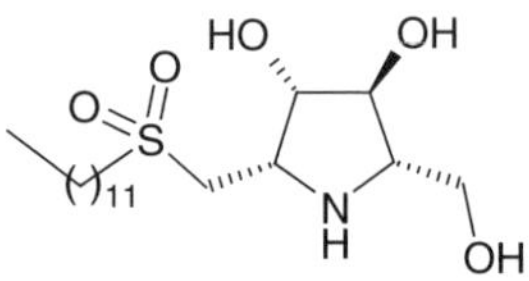	**Tuberculosis** *Mycobacterium tuberculosis Mycobacterium avium Complex M Tuberculosis strain H37Ra, MAC strain NJ 168, NJ 211, NJ 1854, NJ 3009 and NJ 3404* MIC 32-128 μg/ml, MBC >128 μg/ml [91].
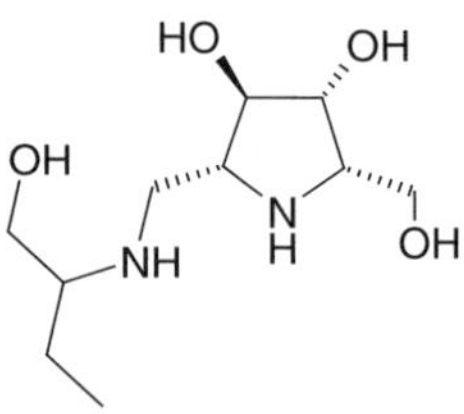	**Tuberculosis** *M. Tuberculosis strain H37Ra and 3 to 5 clinical isolates of M. Avium* MIC >128 μg/ml [92].
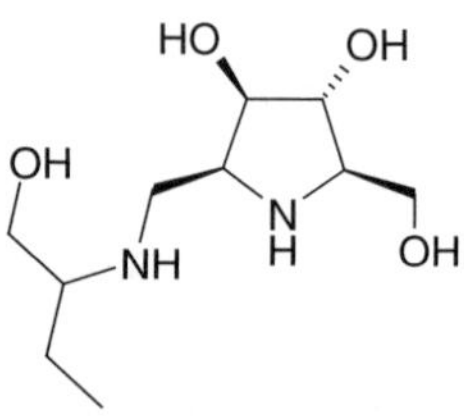	**Tuberculosis** *M. Tuberculosis strain H37Ra and 3 to 5 clinical isolates of M. Avium* MIC >128 μg/ml [92].
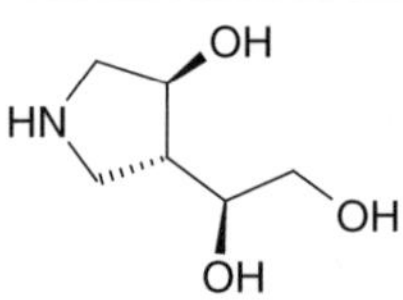	**Tuberculosis** *M. smegmatis* 40% inhibition of mycobacterial galactan biosynthesis at 8 mM [93].
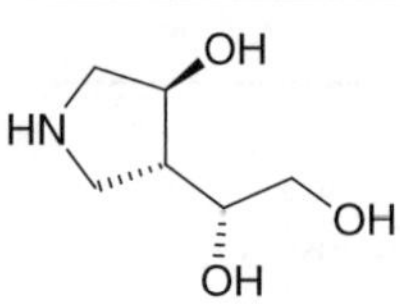	**Tuberculosis** *M. smegmatis* IC_{50} 4.8 mM (mycobacterial galactan biosynthesis) [93].
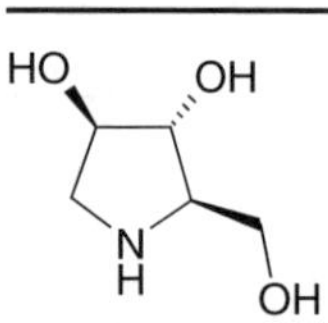	**Tuberculosis, leprosis** *Mycobacterial Arabinosyltransferase. Donor: [1-^{14}C]-β-D-arabinofuranosyl-1-decaprenyl phosphate* 22% inhibition at 4 mM [94].

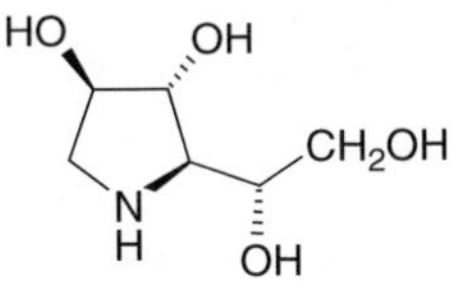

Tuberculosis, leprosis
Mycobacterial Arabinosyltransferase. Donor :
[1-^{14}C]-β-D-arabinofuranosyl-1-decaprenyl phosphate
46% inhibition at 4mM [94].

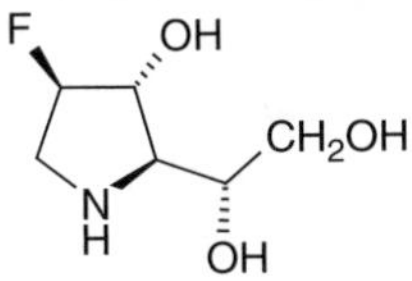

Tuberculosis, leprosis
Mycobacterial Arabinosyltransferase. Donor :
[1-^{14}C]-β-D-arabinofuranosyl-1-decaprenyl phosphate
34% inhibition at 4 mM [94].

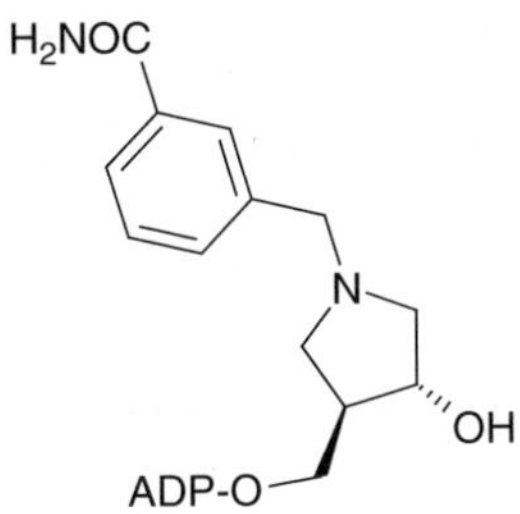

Cholera toxin
Inhibition of NAD^+ hydrolysis by the catalytic A chain of cholera toxin K_i 17.4 μM [95].
Inhibition of ATP-ribosylation of ArgOMe by the catalytic A chain of cholera toxin K_i 10.9 μM [95].

Pertussis toxin
Inhibition of NAD^+ hydrolysis by the catalytic A chain of Pertussis toxin K_i 24.4 μM [95].

Diphtheria toxin
Inhibition of NAD^+ hydrolysis by the catalytic A chain of diphtheria toxin K_i 48.2 μM [95].
Inhibition of ATP-ribosylation of EF-2 by the catalytic A chain of diphtheria toxin K_i 30.5 μM [95].

14.4.2 Six-membered rings

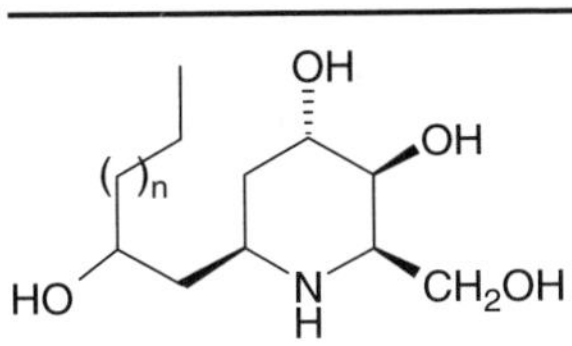

Staphylococcus epidermidis
$n = 8$ MIC ≤ 6.3 μg/ml [96].
$n = 7$ MIC ≤ 6.3 μg/ml [96].
$n = 9$ MIC ≤ 6.3 μg/ml [96].

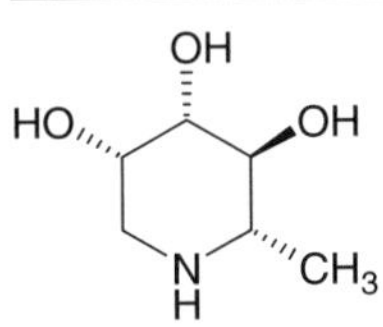

Tuberculosis
Naringinase from Penicillium decumbens (EC 3.2.1.40)
No inhibition at 770 μM [97] and at 0.1 mM [98].
dTDP-L-rhamnose biosynthesis
No inhibition at 0.1 mM [98].

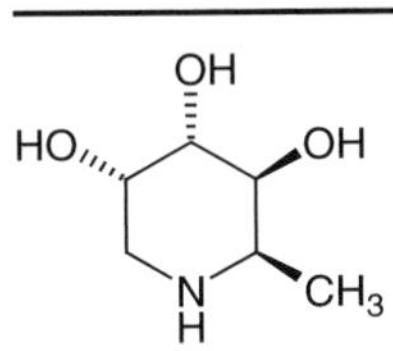

Tuberculosis
Naringinase from Penicillium decumbens (EC 3.2.1.40)
IC_{50} 5 μM, K_i 1.0 μM. [97], [98].
dTDP-L-rhamnose biosynthesis
21% at 0.1 mM [98].

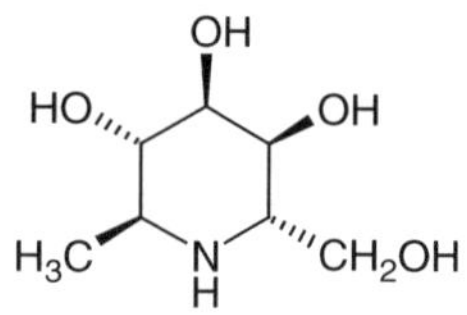	**Tuberculosis** *Naringinase from Penicillium decumbens (EC 3.2.1.40)* IC_{50} 15 μM, K_i 5.3 μM. [97], [98].
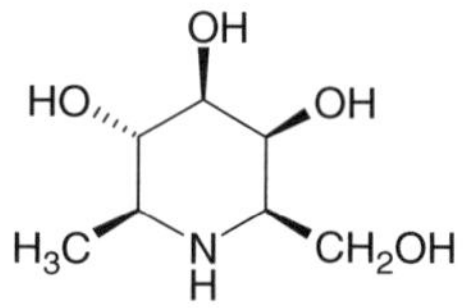	**Tuberculosis** *Naringinase from Penicillium decumbens (EC 3.2.1.40)* IC_{50} 730 μM. [97], [98].
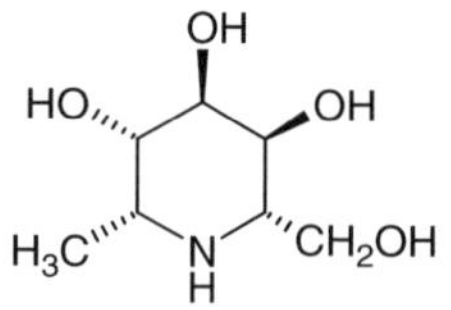	**Tuberculosis** *Naringinase from Penicillium decumbens (EC 3.2.1.40)* IC_{50} 850 μM. [97], [98]. *dTDP-L-rhamnose biosynthesis* 22% at 0.5 mM [98].
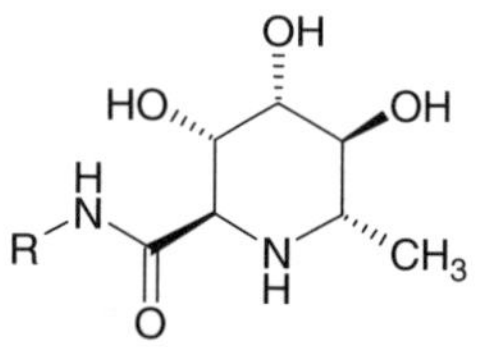 R = a: H b: CH_3 c: $CH_2CH_2CH_2CH_3$ d: CH_2Ph	**Tuberculosis** *Naringinase from Penicillium decumbens (EC 3.2.1.40)* a: IC_{50} 30 μM [98]. b: IC_{50} 21 μM [98]. c: IC_{50} 104 μM [98]. d: IC_{50} 100 μM [98]. *dTDP-L-rhamnose biosynthesis* a: 69% at 10 μM [98]. b: 39% at 100 μM [98]. c: No inhibition at 0.1 mM [98]. d: No inhibition at 0.1 mM [98].
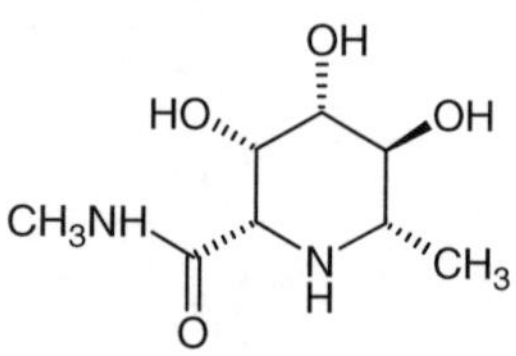	**Tuberculosis** *Naringinase from Penicillium decumbens (EC 3.2.1.40)* IC_{50} 350 μM [98]. *dTDP-L-rhamnose biosynthesis* 18% at 0.1 mM [98].
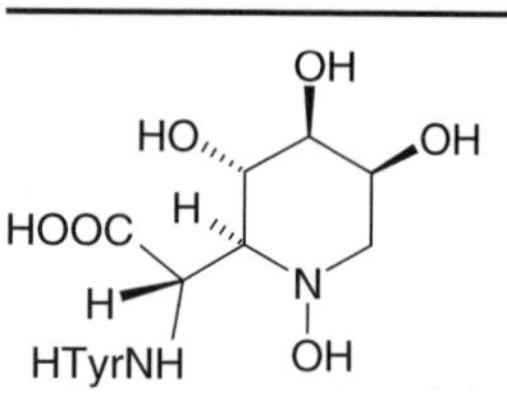	**Bacterial tyrosyl tRNA synthase** IC_{50} 1.2 nM with good selectivity over the mammalian enzyme (11% inhibition at 3 μM) [99].

14.4.3 Seven-membered rings

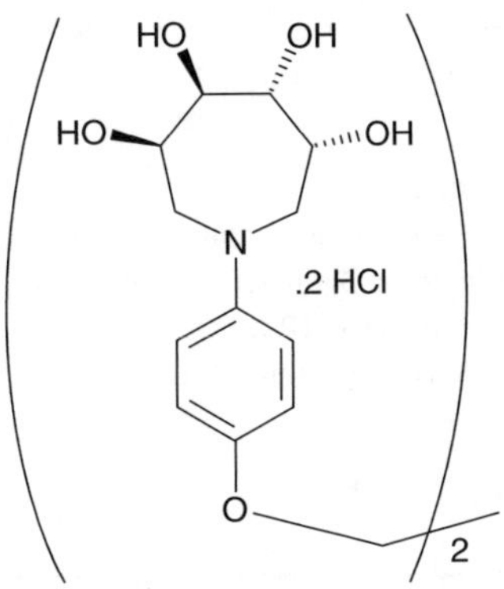

Antibiotic activity
DNA-binding affinity
Change in helix melting temperature: +0.7°C at pH 5.0 after 12 h of incubation [100].

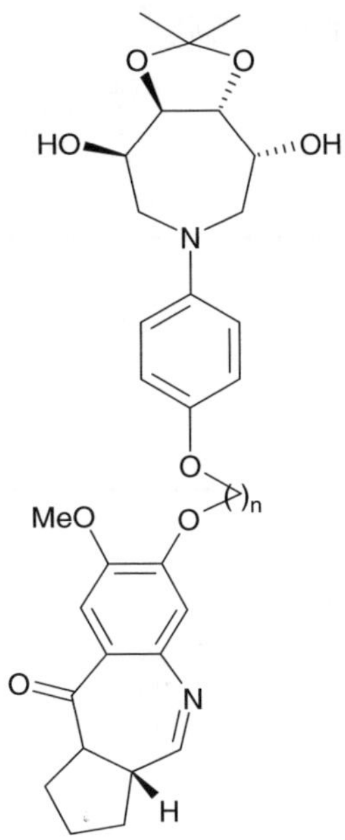

Antibiotic activity
DNA-binding affinity, thermal denaturation data for
$n = 3$, ΔTm 0.0°C after incubation at 37°C for 18 h, conjugates with calf thymus DNA [101].
$n = 4$, ΔTm 1.0°C after incubation at 37°C for 18 h, conjugates with calf thymus DNA [101].
$n = 5$, ΔTm 1.0°C after incubation at 37°C for 18 h, conjugates with calf thymus DNA [101].
$n = 8$, ΔTm 1.1°C after incubation at 37°C for 18 h, conjugates with calf thymus DNA [101].

14.4.4 Bicyclic derivatives

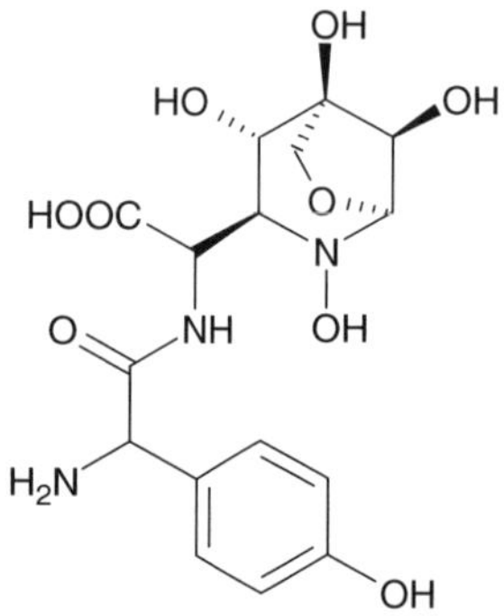	**Bacterial tyrosyl tRNA synthetase** IC_{50} < 1nM against staphylococcus tyrosyl tRNA synthetase [102]. **Activity against Gram positive bacteria** No whole cell antibacterial activity (MIC > 64 μg/ml) against *Escherichia coli, Haemophilius influenzae, Moraxella catarrhalis, Enterococcus faecalis, Staphylococcus aureus* [102]. MIC = 32 μg/ml against some *Streptococcus* strains [102].
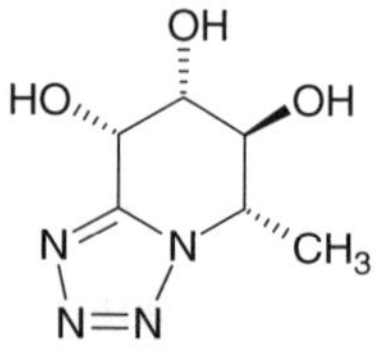	**Tuberculosis** *Naringinase from Penicillium decumbens (EC 3.2.1.40)* 25% inhibition at 770 μM [97].

14.4.5 Polysaccharides

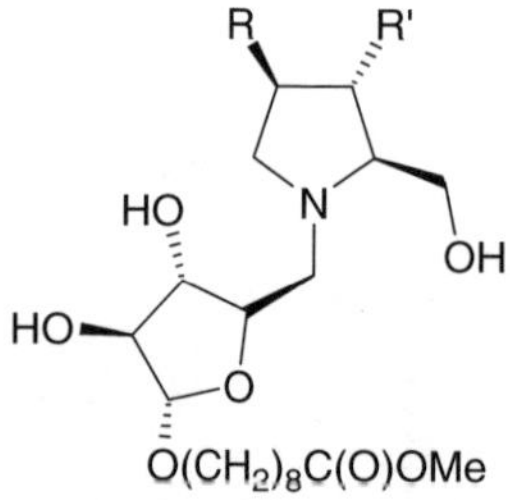	**Tuberculosis, leprosis** *Mycobacterial Arabinosyltransferase. Donor :* *[1-^{14}C]-β-D-arabinofuranosyl-1-decaprenyl phosphate* Compound a: R = H, R' = H 64% inhibition at 4 mM [94]. Compound b: R = OH, R' = OH 80% inhibition at 4 mM [94]. Compound c: R = F, R' = OH 58% inhibition at 4 mM [94].
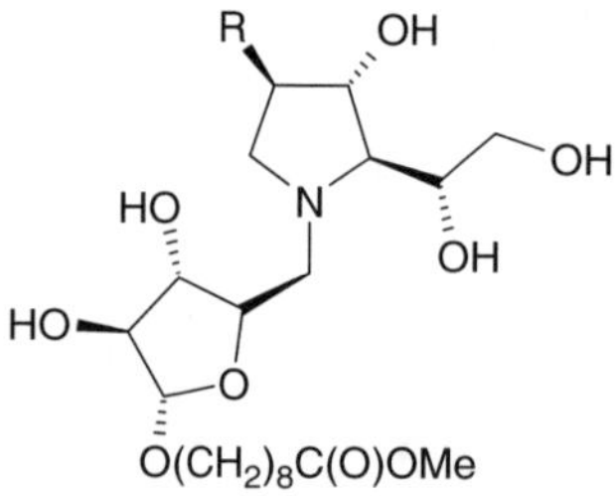	**Tuberculosis, leprosis** *Mycobacterial Arabinosyltransferase. Donor :* *[1-^{14}C]-β-D-arabinofuranosyl-1-decaprenyl phosphate* R = OH 47% inhibition at 4 mM [94]. R = F 59% inhibition at 4 mM [94].

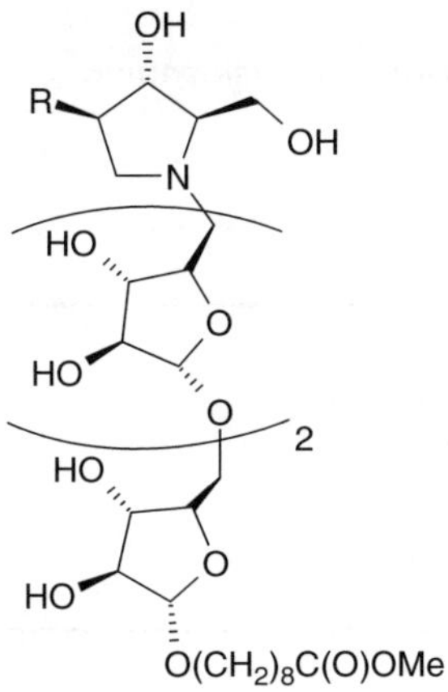

Tuberculosis, leprosis
Mycobacterial Arabinosyltransferase. Donor : [1-^{14}C]-β-D-arabinofuranosyl-1-decaprenyl phosphate
R = OH 34% inhibition at 4 mM [94].
R = F 5% inhibition at 4 mM [94].

14.5 Antiprotozoal activities

14.5.1 Five-membered rings

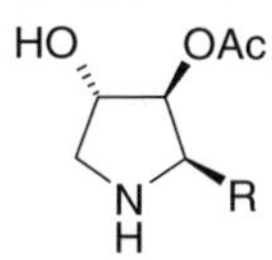

a: R = -CH_2-4-OCH_3-(C_6H_5)
b: R = -CH_2-4-CH_3-(C_6H_5)
c: R = -CH_2-(C_6H_5)
d: R = -CH_2-3-OCH_3-(C_6H_5)
e: R = -CH_2-2-OCH_3-(C_6H_5)
f: R = -CHMe-4-OMe-(C_6H_5)
g: R = -CH(C_6H_5)-4-OCH_3-(C_6H_5)

Compounds b, c, d, e, f, g were evaluated in their racemic form [103].
Trichomonas vaginalis
Compound a: 100% inhibition at 2 μg/ml, compound b: 100% inhibition at 8 μg/ml, compound c: 100% inhibition at 8 μg/ml, compound d: 100% inhibition at 32 μg/ml.
Trichomonas foetus
Compound a: 100% inhibition at 4 μg/ml, compound b: 100% inhibition at 16 μg/ml, compound c: 100% inhibition at 32 μg/ml, compound d: <25% inhibition at 8 μg/ml.
Entamoeba histolytica
Compound a: 100% inhibition at 2 μg/ml, compound b: 100% inhibition at 4 μg/ml, compound c: 100% inhibition at 16 μg/ml, compound d: 100% inhibition at 32 μg/ml.

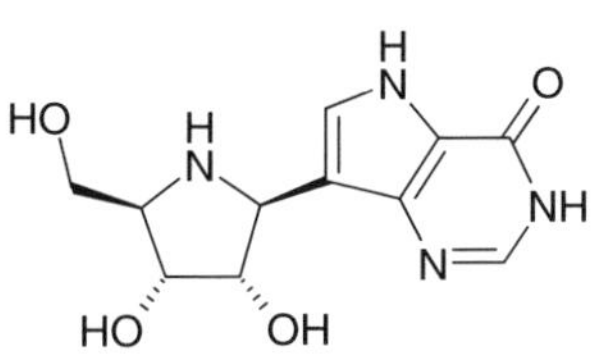

Malaria
Purine nucleoside phosphorylase from plasmodium falciparum
IC_{50} 35 ± 7 nM for *in vitro* culture [104].
K_i 29 ± 8 nM, K_i^* 0.6 ± 0.1 nM [105]; K_i 0.86 ± 0.08 nM [109].

Antiprotozoal
IU-nucleoside hydrolase from Crithidia fasciculata (EC 3.2.2.1)
K_i 0.042 μM [106].
IU-nucleoside hydrolases from Leishmania Major and Crithidia fasciculata (EC 3.2.2.1)
K_i 5.4 μM [107].
IAG-nucleoside hydrolase from Trypanosoma brucei brucei (EC 3.2.2.1)
K_i 3.6 μM [107]; K_i 0.024 μM [106].

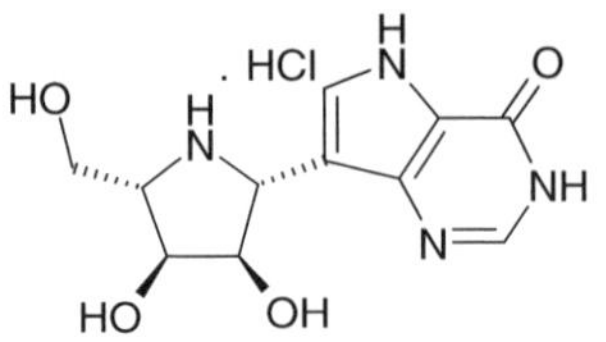

Malaria
Purine nucleoside phosphorylase from plasmodium falciparum
K_i 32 ± 9 nM [108].

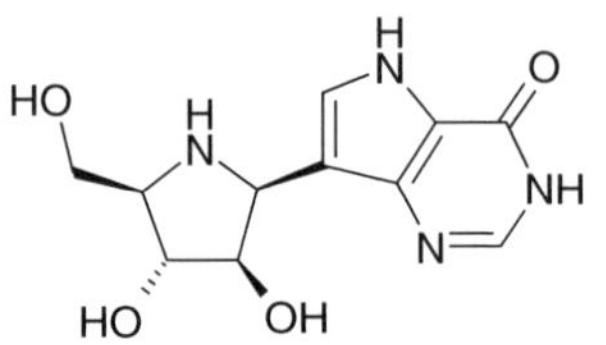

Malaria
Purine nucleoside phosphorylase from plasmodium falciparum
K_i 9.1 ± 2.3 nM [109]; K_i^* 1.2 ± 0.3 nM [105].

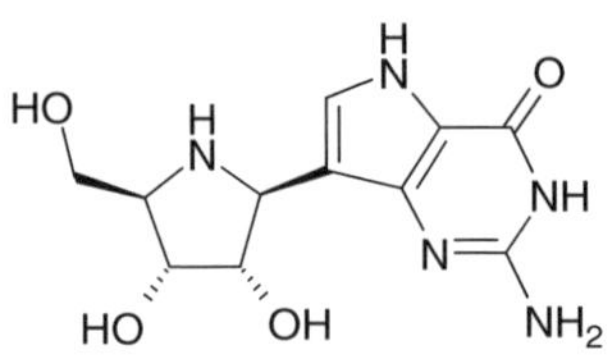

Malaria
Purine nucleoside phosphorylase from plasmodium falciparum
IC_{50} 50 ± 11 nM for *in vitro* culture [104].
K_i 0.9 ± 0.2 nM [109]; K_i^*0.9 ± 0.2 nM [105].

Antiprotozoal
IU-nucleoside hydrolase from Crithidia fasciculata (EC 3.2.2.1)
K_i 0.084 μM [106].
IU-nucleoside hydrolase from Leishmania Major and Crithidia fasciculata (EC 3.2.2.1)
K_i 22 μM [107].
IAG-nucleoside hydrolase from Trypanosoma brucei brucei (EC 3.2.2.1)
K_i 5 μM [107]; K_i 0.11 μM [106].

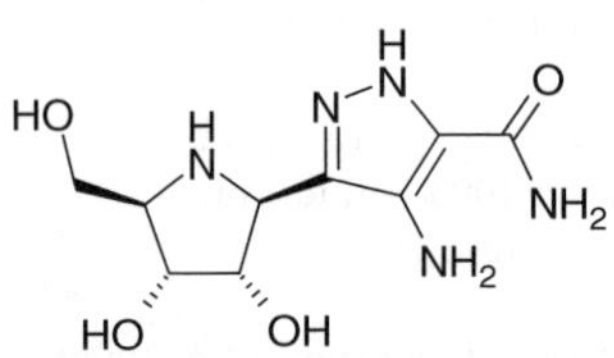

Malaria
Purine nucleoside phosphorylase from plasmodium falciparum
K_i^* 2.7 ± 0.4 nM [105].

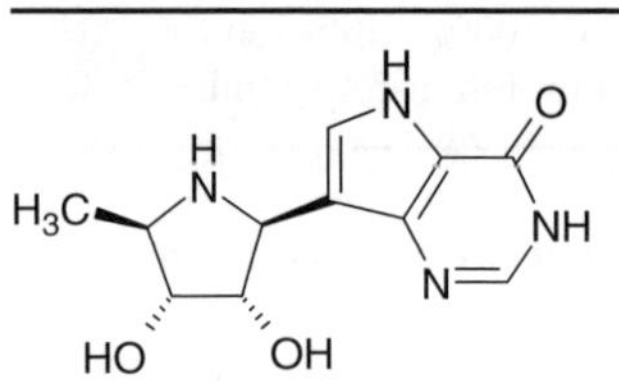

Malaria
Purine nucleoside phosphorylase from plasmodium falciparum
IC_{50} 4000 ± 1100 nM for *in vitro* culture [104].
K_i 130 ± 35 nM [105]; K_i^* 7.4 ± 1.4 nM [105]; K_i^*53 ± 6 nM [109].

Antiprotozoal
IU-nucleoside hydrolase from Crithidia fasciculata (EC 3.2.2.1)
$K_i > 50$ μM [106].
IAG-nucleoside hydrolase from Trypanosoma brucei brucei (EC 3.2.2.1)
$K_i > 50$ μM [106].

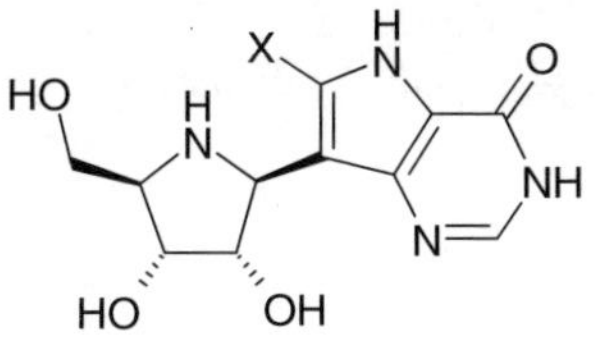

Malaria
Purine nucleoside phosphorylase from plasmodium falciparum
Compound a: X = F K_i^* 19.6 ± 0.6 nM [105];
K_i 9.1 ± 0.5 nM [109].
Compound b: X = CH_3 K_i and $K_i^* > 10\,\mu M$ [105], [109].

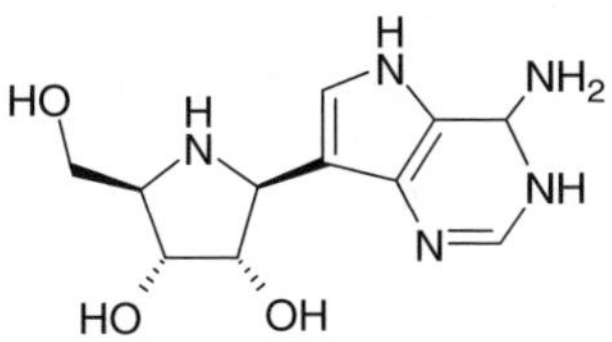

Malaria
Purine nucleoside phosphorylase from plasmodium falciparum
K_i and $K_i^* > 10\,\mu M$ [105].
$IC_{50} > 10\,\mu M$ for *in vitro* culture [104].

Antiprotozoal
IU-nucleoside hydrolase from Crithidia fasciculata (EC 3.2.2.1)
K_i 0.007 μM [106].
IU-nucleoside hydrolases from Leishmania Major and Crithidia fasciculata (EC 3.2.2.1)
K_i 21 μM [107].
IAG-nucleoside hydrolase from Trypanosoma brucei brucei (EC 3.2.2.1)
K_i 13 μM [107]; K_i^* 0.0009 μM [106].

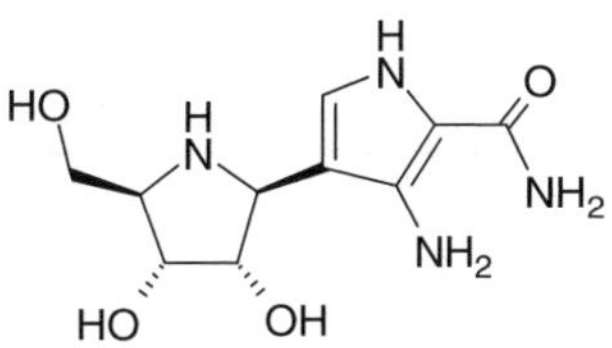

Antiprotozoal
IU-nucleoside hydrolase from Crithidia fasciculata (EC 3.2.2.1)
K_i^* 0.003 μM [106].
IAG-nucleoside hydrolase from Trypanosoma brucei brucei (EC 3.2.2.1)
K_i 0.023 μM [106].

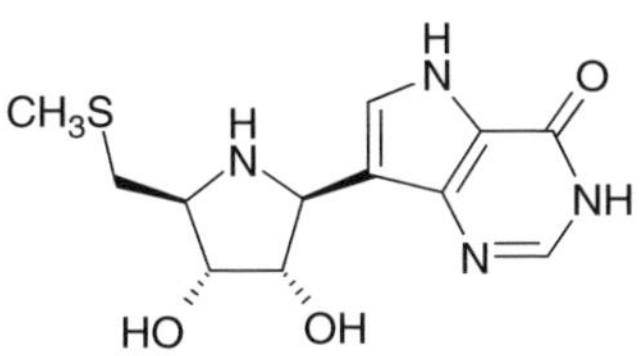

Malaria
Purine nucleoside phosphorylase from plasmodium falciparum
K_i 22 ± 3 nM [109]; K_i^* 2.7 ± 0.4 nM [109].

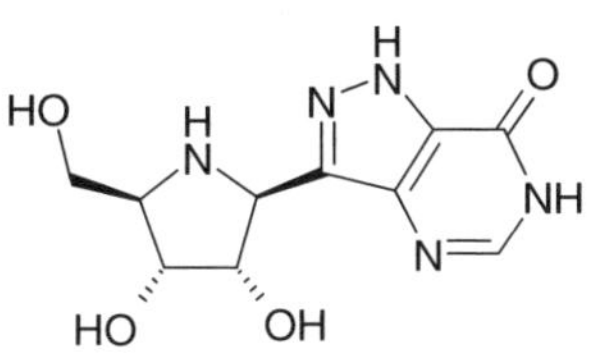

Malaria
Purine nucleoside phosphorylase from plasmodium falciparum
K_i 110 ± 10 nM [109]; K_i^*13 ± 2 nM [109].

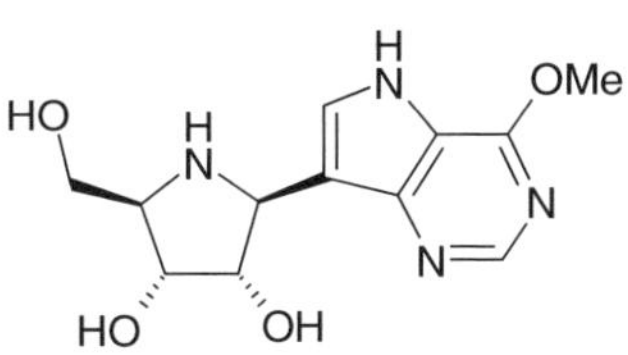

Malaria
Purine nucleoside phosphorylase from plasmodium falciparum
K_i 450 ± 30 nM [109]; K_i^*26 ± 7 nM [109].

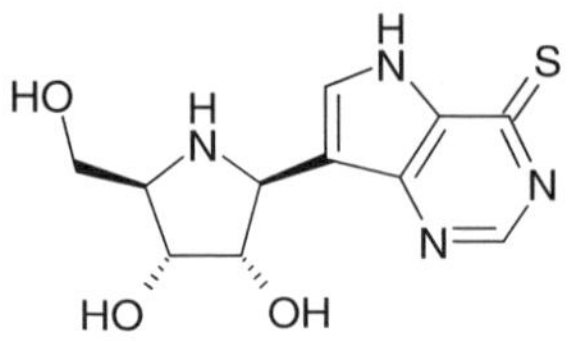

Malaria
Purine nucleoside phosphorylase from plasmodium falciparum
K_i 58 ± 6 nM [109].

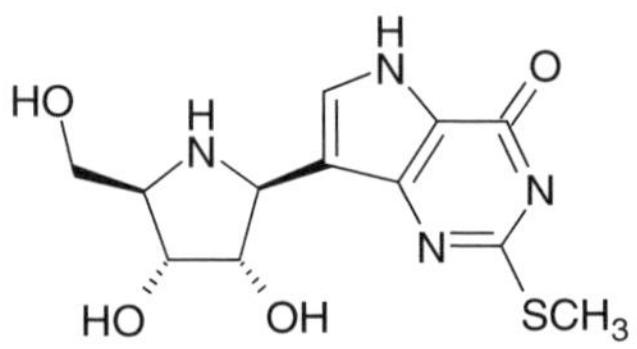

Antiprotozoal
IU-nucleoside hydrolase from Crithidia fasciculata (EC 3.2.2.1)
K_i 0.23 μM [106].
IAG-nucleoside hydrolase from Trypanosoma brucei brucei (EC 3.2.2.1)
K_i > 50 μM [106].

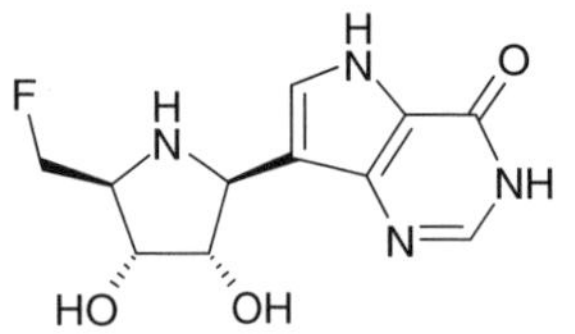

Malaria
Purine nucleoside phosphorylase from plasmodium falciparum
K_i 240 ± 20 nM [109]; K_i^* 60 ± 11 nM [109].

Antiprotozoal
IU-nucleoside hydrolase from Crithidia fasciculata (EC 3.2.2.1)
K_i 0.97 μM [106].
IAG-nucleoside hydrolase from Trypanosoma brucei brucei (EC 3.2.2.1)
K_i 0.19 μM [106].

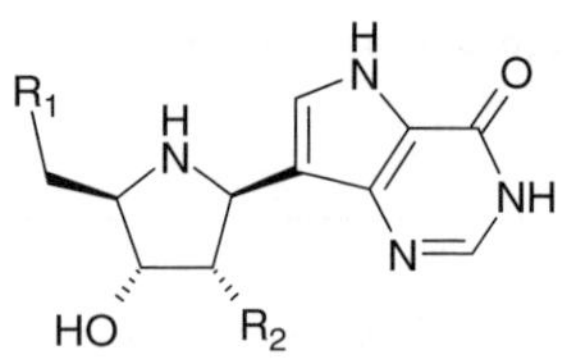

a: R_1 = SPh, R_2 = OH
b: R_1 = OH, R_2 = OMe

Malaria
Purine nucleoside phosphorylase from plasmodium falciparum
a: K_i 150 ± 8 nM [109].
b: K_i 840 ± 70 nM [109].

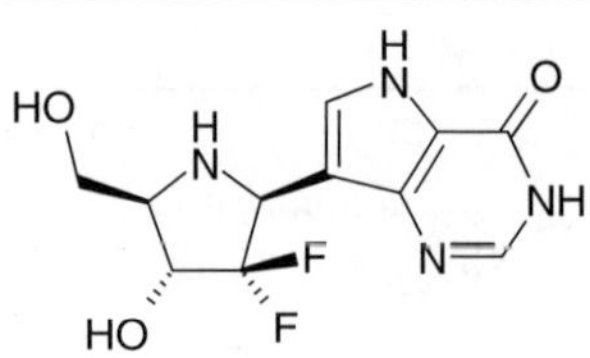

Malaria
Purine nucleoside phosphorylase from plasmodium falciparum
K_i > 15 000 nM [109].

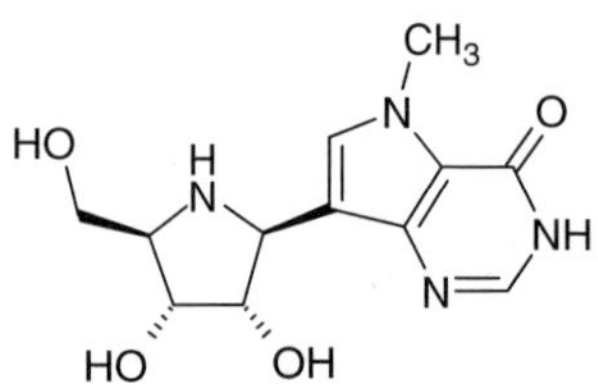

Malaria
Purine nucleoside phosphorylase from plasmodium falciparum
K_i 900 ± 160 nM [109].

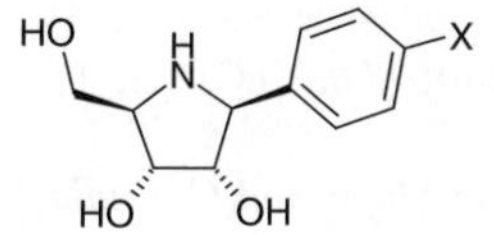

a: X = H f: X = COOH
b: X = Br g: X = NO_2
c: X = Cl h: X = NH_2
d: X = F i: X = CH_2NH_2
e: X = OH j: X = CN

Antiprotozoal
IU-nucleoside hydrolase from Crithidia fasciculata (EC 3.2.2.1)
Compound a: K_i 300 ± 27 nM, compound c: K_i 30 ± 1.4 nM, compound d: K_i 57 ± 5 nM, compound e: K_i 75 ± 4 nM, compound f: K_i 96 ± 7 nM, compound g: K_i 1.1 ± 0.1 μM [110].

IAG-nucleoside hydrolases from Trypanosoma brucei brucei, Trypanosoma congolense, Trypanosoma vivax (EC 3.2.2.1)
Compound a: K_i 180 ± 15 μM, compound c: K_i 190 ± 8 μM, compound d: K_i 205 ± 14 μM, compound e: K_i 35 ± 2 μM, compound f: K_i > 480 μM, compound g: K_i > 360 μM [110].

IU-nucleoside hydrolases from Leishmania Major and Crithidia fasciculata (EC 3.2.2.1)
Compound b: K_i 0.028 μM [107], [110]; compound h: K_i 0.03 μM [107], [110]; compound i: K_i 5.0 μM [107]; compound j: K_i 0.92 μM. [107]

IAG-nucleoside hydrolase from Trypanosoma brucei brucei (EC 3.2.2.1)
Compound b: K_i 115 μM [107], [110]; compound h: K_i 12 μM [107], [110]; compound i: K_i > 2 μM [107]; compound j: K_i 3.4 μM. [107].

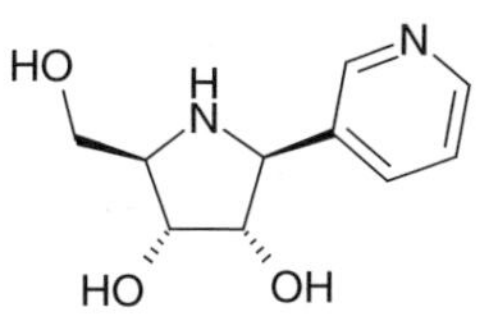

Antiprotozoal
IU-nucleoside hydrolase from Crithidia fasciculata (EC 3.2.2.1)
K_i 7.9 ± 0.6 μM [106].
IAG-nucleoside hydrolases from Trypanosoma brucei brucei, Trypanosoma congolense, Trypanosoma vivax (EC 3.2.2.1)
K_i > 240 μM [106].

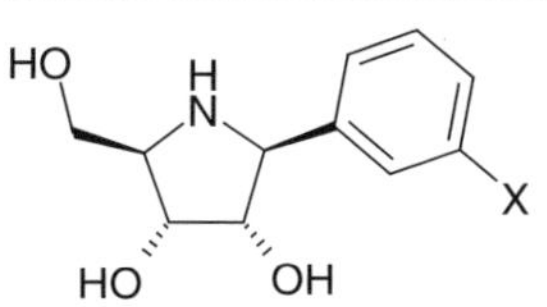

a: X = NO_2
b: X = NH_2

Antiprotozoal
IU-nucleoside hydrolase from Crithidia fasciculata (EC 3.2.2.1)
Coupound a: K_i 7.5 ± 0.5 μ M [110], compound b: K_i 51 ± 4 nM [110].
IAG-nucleoside hydrolases from Trypanosoma brucei brucei, Trypanosoma congolense, Trypanosoma vivax (EC 3.2.2.1)
Coupound a: K_i > 360 μM [110], compound b: K_i 38 ± 4 μM [110].

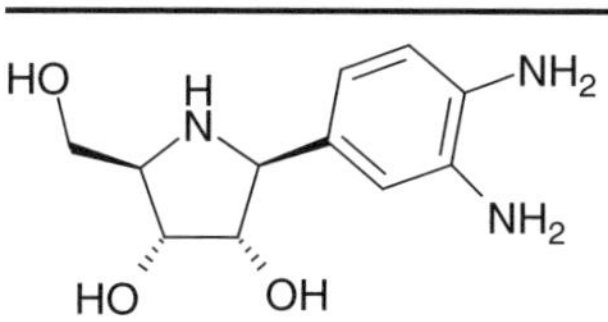

Antiprotozoal
IU-nucleoside hydrolases from Leishmania Major and Crithidia fasciculata (EC 3.2.2.1)
K_i 0.21 μM [107].
IAG-nucleoside hydrolase from Trypanosoma brucei brucei (EC 3.2.2.1)
K_i > 2 μM [107].

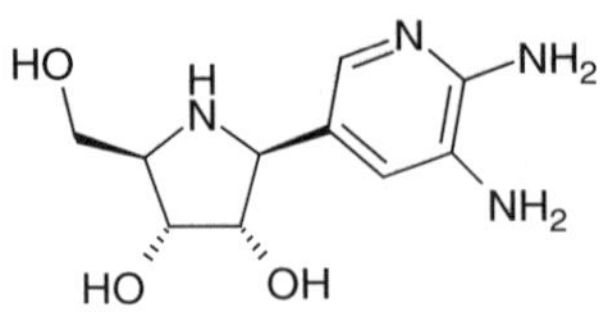	**Antiprotozoal** *IU-nucleoside hydrolase from Crithidia fasciculata* (EC 3.2.2.1) K_i 0.77 μM [106]. *IU-nucleoside hydrolases from Leishmania Major and Crithidia fasciculata* (EC 3.2.2.1) K_i 0.94 μM [107]. *IAG-nucleoside hydrolase from Trypanosoma brucei brucei* (EC 3.2.2.1) $K_i > 30$ μM [107], [106].
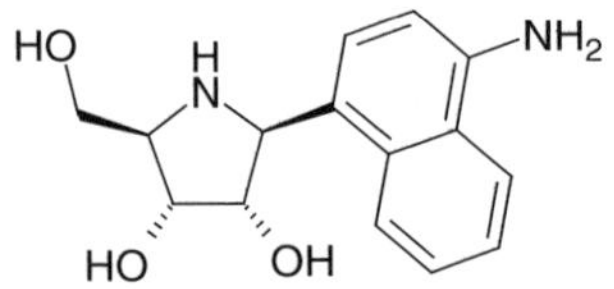	**Antiprotozoal** *IU-nucleoside hydrolases from Leishmania Major and Crithidia fasciculata* (EC 3.2.2.1) K_i 0.43 μM [107], [106]. *IAG-nucleoside hydrolase from Trypanosoma brucei brucei* (EC 3.2.2.1) $K_i > 50$ μM [107], [106].
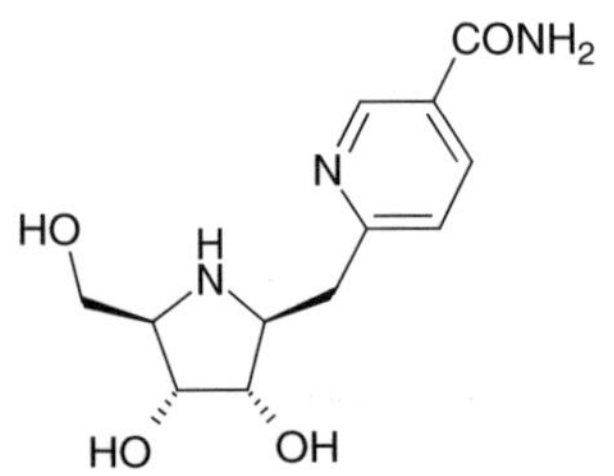	**Antiprotozoal** *IU-nucleoside hydrolases from Leishmania Major and Crithidia fasciculata* (EC 3.2.2.1) K_i 5.2 μM [107], [106]. *IAG-nucleoside hydrolase from Trypanosoma brucei brucei* (EC 3.2.2.1) K_i 2.3 μM [107], [106].
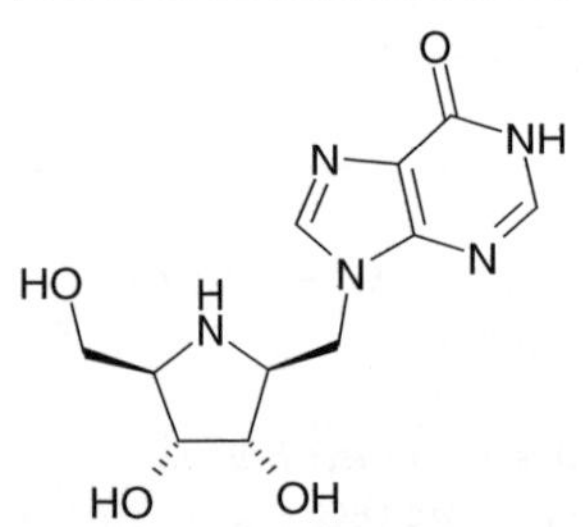	**Antiprotozoal** *IU-nucleoside hydrolase from Crithidia fasciculata* (EC 3.2.2.1) K_i 5.4 μM [106]. *IAG-nucleoside hydrolase from Trypanosoma brucei brucei* (EC 3.2.2.1) K_i 3.6 μM [106].
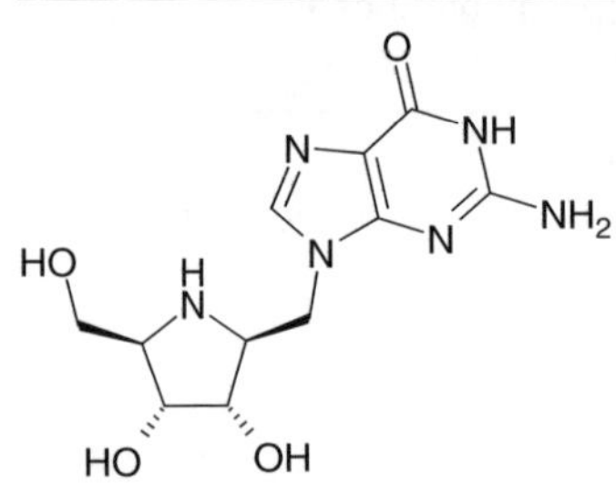	**Antiprotozoal** *IU-nucleoside hydrolase from Crithidia fasciculata* (EC 3.2.2.1) K_i 22 μM [106]. *IAG-nucleoside hydrolase from Trypanosoma brucei brucei* (EC 3.2.2.1) K_i 5 μM [106].

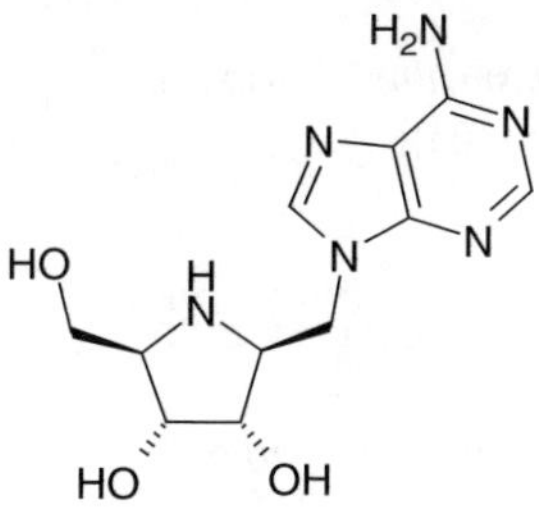

Antiprotozoal
IU-nucleoside hydrolase from Crithidia fasciculata (EC 3.2.2.1)
K_i 21 μM [106].
IAG-nucleoside hydrolase from Trypanosoma brucei brucei (EC 3.2.2.1)
K_i 13 μM [106].

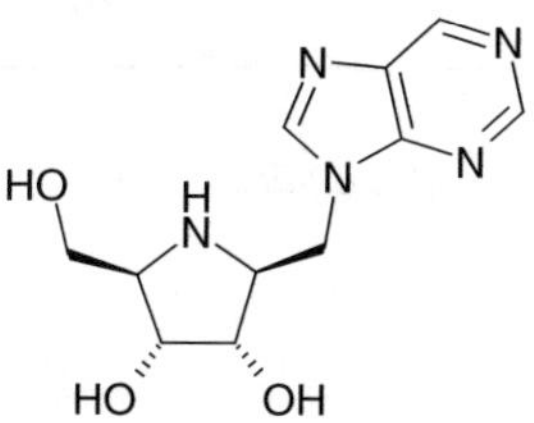

Antiprotozoal
IU-nucleoside hydrolase from Crithidia fasciculata (EC 3.2.2.1)
$K_i > 50$ μM [106].
IAG-nucleoside hydrolase from Trypanosoma brucei brucei (EC 3.2.2.1)
$K_i > 50$ μM [106].

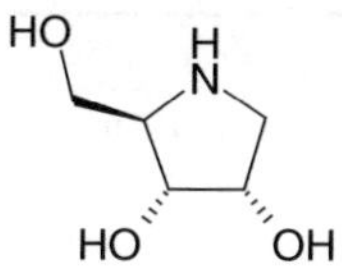

Antiprotozoal
IU-nucleoside hydrolase from Crithidia fasciculata (EC 3.2.2.1)
K_i 4.5 ± 0.4 μM [110].
IAG-nucleoside hydrolases from Trypanosoma brucei brucei, Trypanosoma congolense, Trypanosoma vivax (EC 3.2.2.1)
K_i 44 ± 4 μM [110].

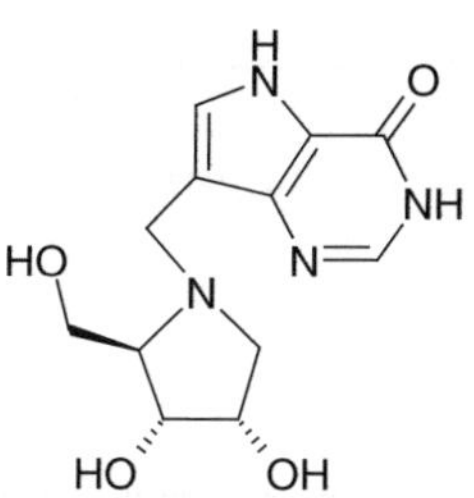

Malaria
Purine nucleoside phosphorylase from plasmodium falciparum
K_i 530 ± 70 nM [109].

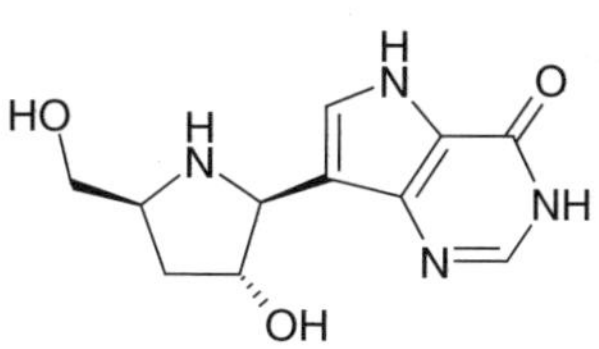

Malaria
Purine nucleoside phosphorylase from plasmodium falciparum
K_i^* 1.7 ± 0.4 nM [105]; K_i 42 ± 2 nM [109].

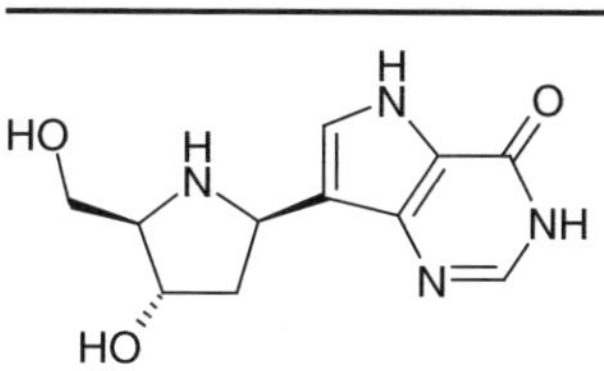

Malaria
Purine nucleoside phosphorylase from plasmodium falciparum
K_i^* 2.2 ± 0.1 nM [105]; K_i 2.2 ± 0.1 nM [109].
IC_{50} 500 ± 50 nM for *in vitro* culture [104].

Antiprotozoal
IU-nucleoside hydrolase from Crithidia fasciculata (EC 3.2.2.1)
$K_i > 50$ μM [106].
IAG-nucleoside hydrolase from Trypanosoma brucei brucei (EC 3.2.2.1)
K_i 0.66 μM [106].

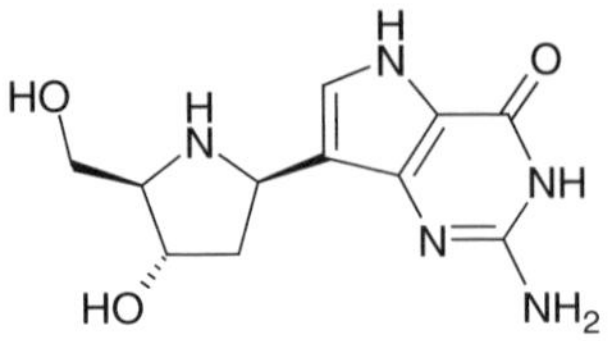

Malaria
Purine nucleoside phosphorylase from plasmodium falciparum
IC_{50} 1600 ± 400 nM for *in vitro* culture [104];
K_i 8.5 ± 0.9 nM [105], [109].

Antiprotozoal
IU-nucleoside hydrolase from Crithidia fasciculata (EC 3.2.2.1)
K_i > 50 μM [106].
IAG-nucleoside hydrolase from Trypanosoma brucei brucei (EC 3.2.2.1)
K_i 0.69 μM [106].

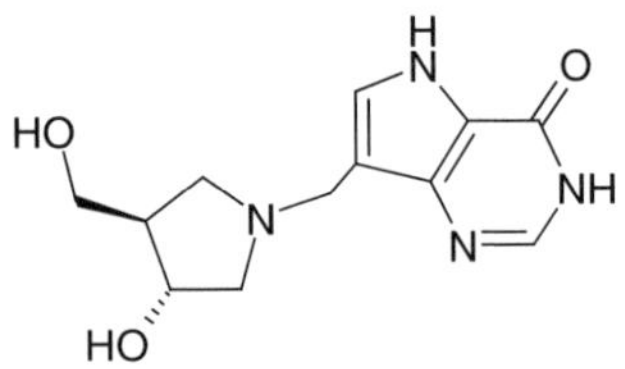

Malaria
Purine nucleoside phosphorylase from plasmodium falciparum
K_i 0.50 ± 0.04 nM [109], [108].

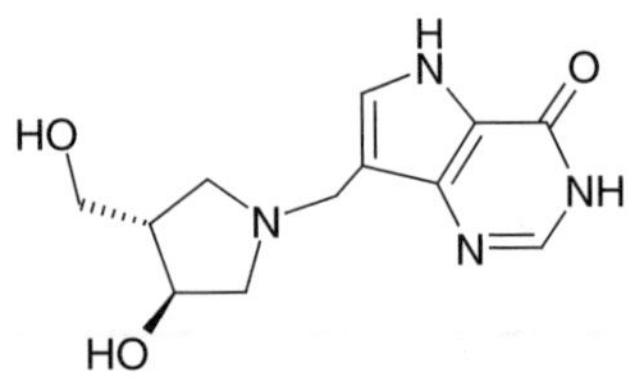

Malaria
Purine Nucleoside Phosphorylase from Plasmodium falciparum
K_i 1700 ± 300 nM [108].

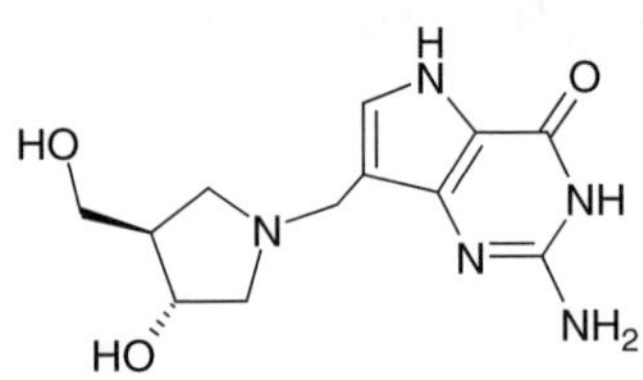

Malaria
Purine nucleoside phosphorylase from plasmodium falciparum
K_i 0.89 ± 0.06 nM [109].

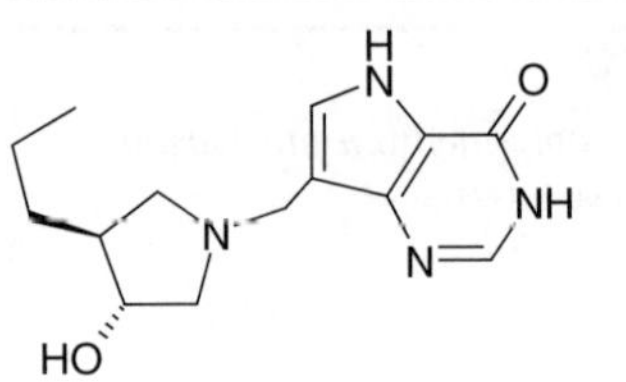

Malaria
Purine nucleoside phosphorylase from plasmodium falciparum
K_i 11.13 ± 1.7 nM [109].
K_i^* 1.6 + 0.2 nM [109].

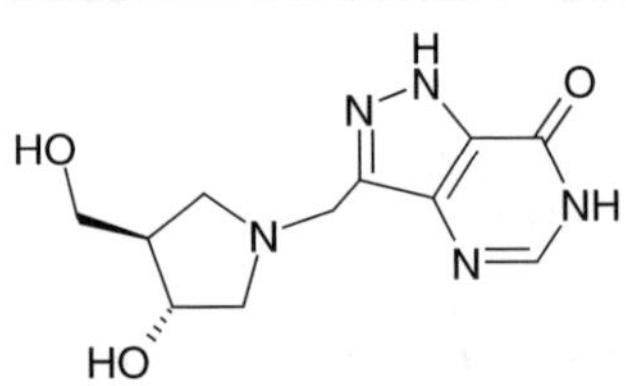

Malaria
Purine nucleoside phosphorylase from plasmodium falciparum
K_i 5.5 ± 0.2 nM [109].

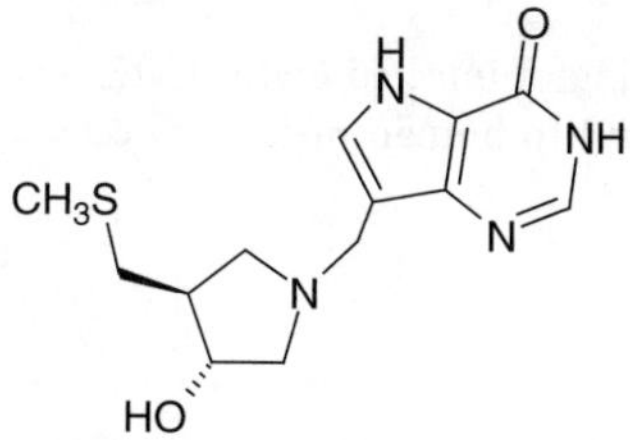

Malaria
Purine nucleoside phosphorylase from plasmodium falciparum
K_i 11 ± 4 nM [109].
K_i^* 0.9 ± 0.1 nM [109].

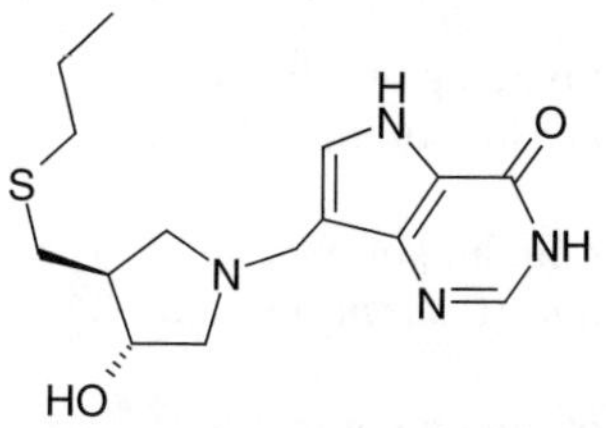

Malaria
Purine nucleoside phosphorylase from plasmodium falciparum
K_i 160 ± 40 nM [109].

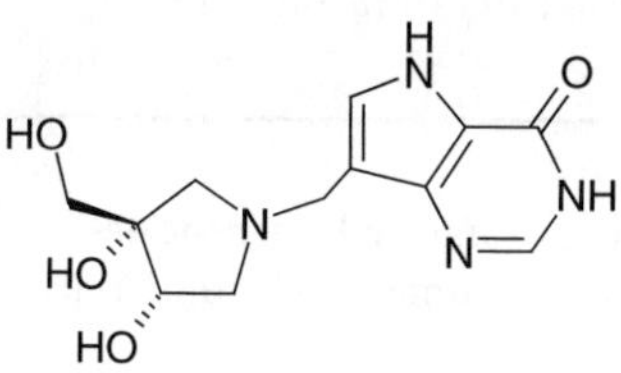

Malaria
Purine nucleoside phosphorylase from plasmodium falciparum
K_i 4300 ± 800 nM [109].

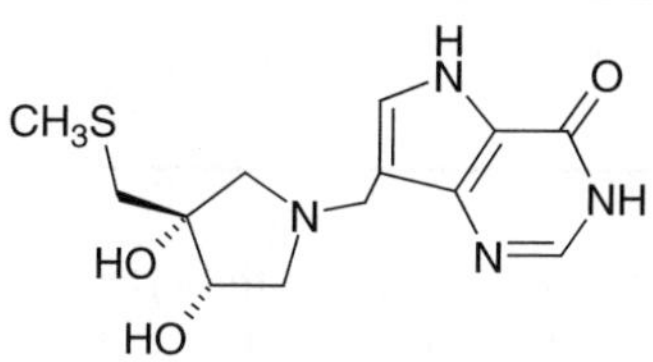

Malaria
Purine nucleoside phosphorylase from plasmodium falciparum
K_i 45000 ± 3000 nM [109].

14.5.2 Bicyclic derivatives

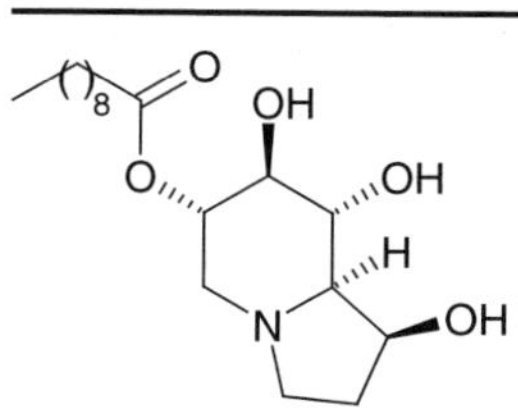

Malaria
Plasmodium falciparum (ItG strain)
59% of control of human melanoma cells growth at 25 μM [111].

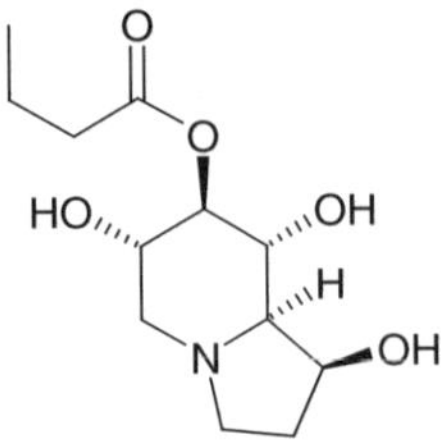	**Malaria** At 25 μM, 26% inhibition of binding of infected erythrocytes (*Plasmodium falciparum ItG strain*) to human melanoma cells *in vitro* [111].
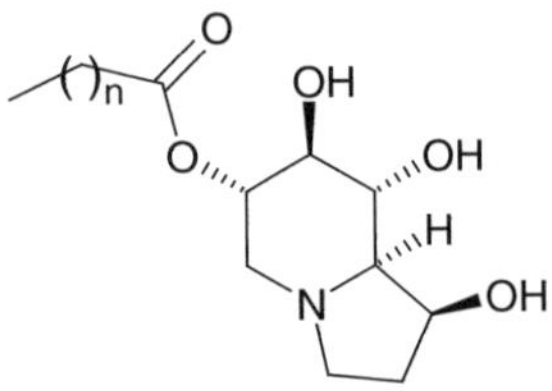	**Malaria** *n* = 4 : At 25 μM, 44% inhibition of binding of infected erythrocytes (*Plasmodium falciparum ItG strain*) to human melanoma cells *in vitro*. *n* = 3 : At 25 μM, 27% inhibition of binding of infected erythrocytes (*Plasmodium falciparum ItG strain*) to human melanoma cells *in vitro*. *n* = 14 : At 25 μM, 50% inhibition of binding of infected erythrocytes (*Plasmodium falciparum ItG strain*) to human melanoma cells *in vitro* [111].
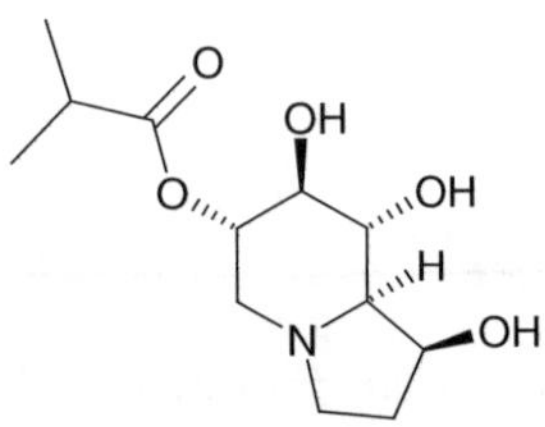	**Malaria** At 25 μM, 78% inhibition of binding of infected erythrocytes (*Plasmodium falciparum ItG strain*) to human melanoma cells *in vitro* [111].
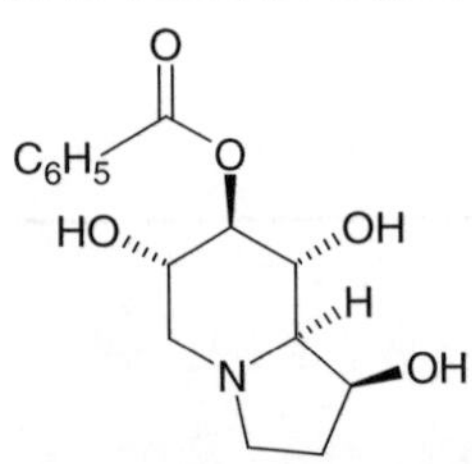	**Malaria** At 25 μM, 92% inhibition of binding of infected erythrocytes (*Plasmodium falciparum ItG strain*) to human melanoma cells *in vitro* [111].
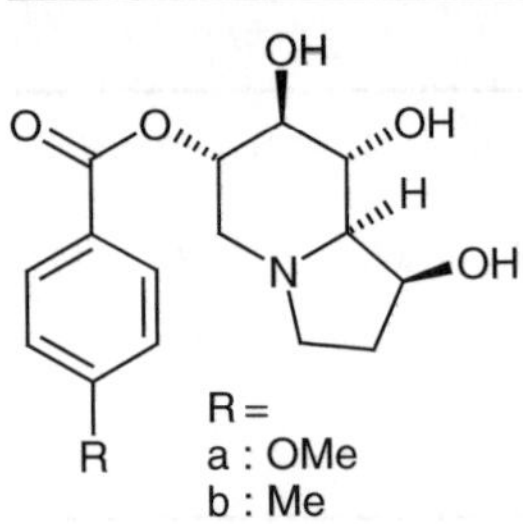	**Malaria** Compound a: At 25 μM, 43% inhibition of binding of infected erythrocytes (*Plasmodium falciparum ItG strain*) to human melanoma cells *in vitro* [111]. Compound b: At 25 μM, 25% inhibition of binding of infected erythrocytes (*Plasmodium falciparum ItG strain*) to human melanoma cells *in vitro* [111].

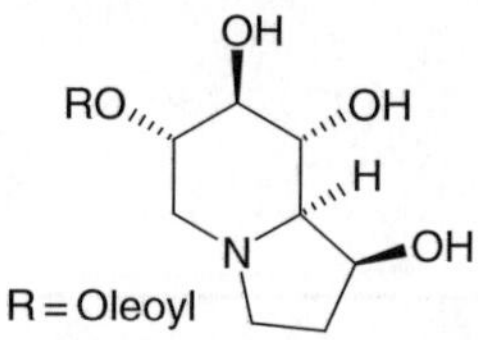

Malaria
At 25 μM, 7% inhibition of binding of infected erythrocytes (*Plasmodium falciparum ItG strain*) to human melanoma cells *in vitro* [111].

14.6 Antiparasitarial activities

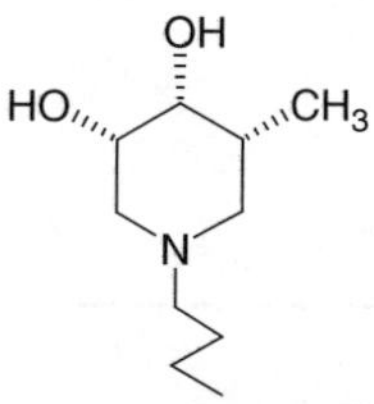

Schistosoma Mansoni
S. Mansoni fucosyltransferase (EC 2.4.1.65)
IC_{50} 15.1 ± 4.8 mM [73].
K_i 11.6 ± 2.8 mM [73].

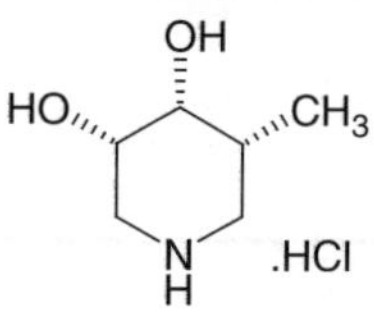

Schistosoma Mansoni
S. Mansoni fucosyltransferase (EC 2.4.1.65)
IC_{50} 10.4 ± 2.8 mM [73].
K_i 7.8 ± 1.1 mM [73].

14.7 Cancer

14.7.1 Five-membered rings

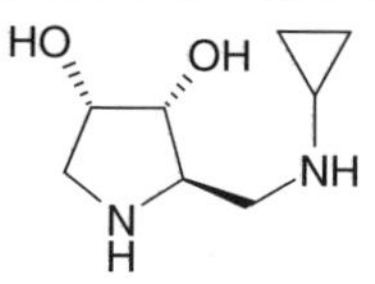

Potential anticancer agent
Jack beans α-mannosidase (EC 3.2.1.24)
IC_{50} 650 μM [122].

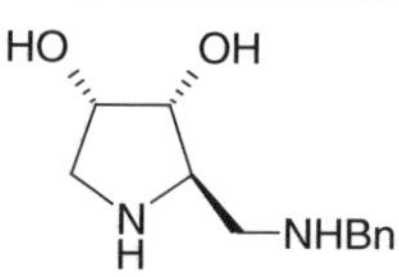

Potential anticancer agent
Jack beans α-mannosidase (EC 3.2.1.24)
IC_{50} 60 μM; K_i 7.4 μM [121].
No activity on LN18 and LNZ308 glioblastoma cells [120].

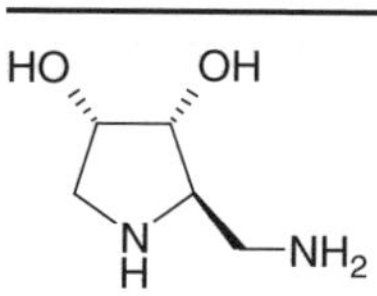

Potential anticancer agent
Jack beans α-mannosidase (EC 3.2.1.24)
IC_{50} 170 μM; K_i 53 μM [121].

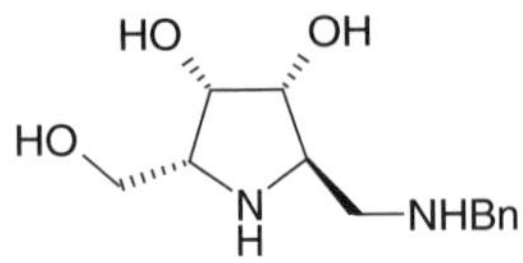	**Potential anticancer agent** *Jack beans α-mannosidase* (EC 3.2.1.24) IC_{50} 6.2 μM; K_i 1.2 μM [121].
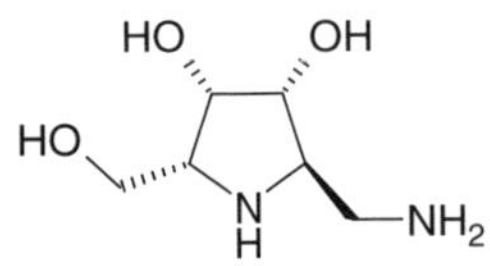	**Potential anticancer agent** *Jack beans α-mannosidase* (EC 3.2.1.24) 97% inhibition at 1 mM [121].
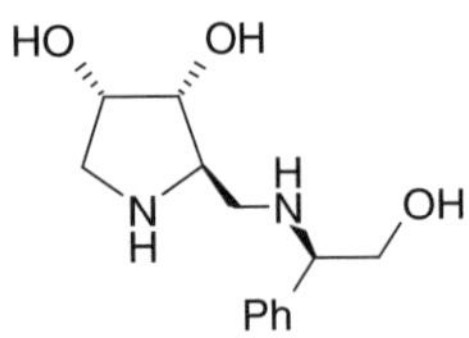	**Potential anticancer agent** *Jack beans α-mannosidase* (EC 3.2.1.24) IC_{50} 700 nM; K_i 135 nM [120].
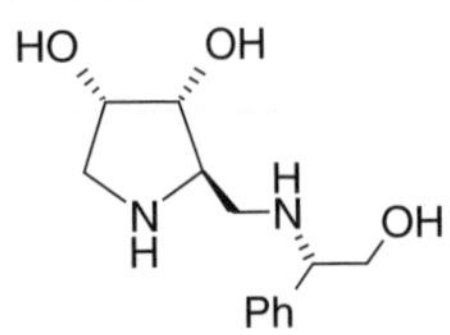	**Potential anticancer agent** *Jack beans α-mannosidase* (EC 3.2.1.24) IC_{50} 100 μM [120].
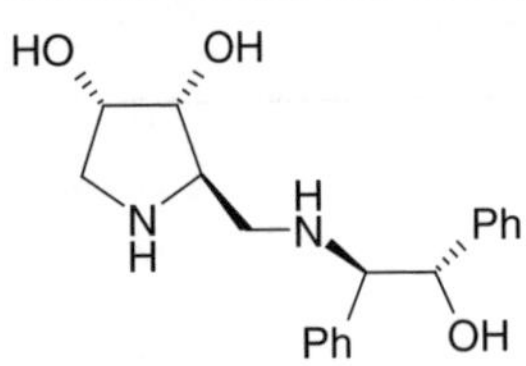	**Potential anticancer agent** *Jack beans α-mannosidase* (EC 3.2.1.24) IC_{50} 110 μM [120].
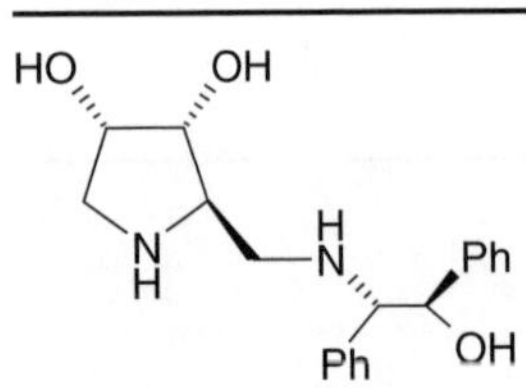	**Potential anticancer agent** *Jack beans α-mannosidase* (EC 3.2.1.24) IC_{50} 128 μM [120].
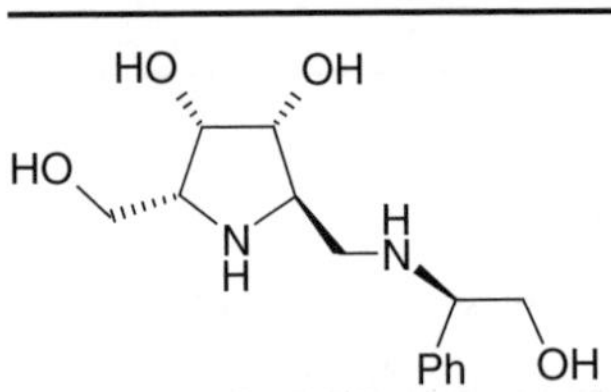	**Potential anticancer agent** *Jack beans α-mannosidase* (EC 3.2.1.24) IC_{50} 4.2 μM; K_i 1.35 μM [120].

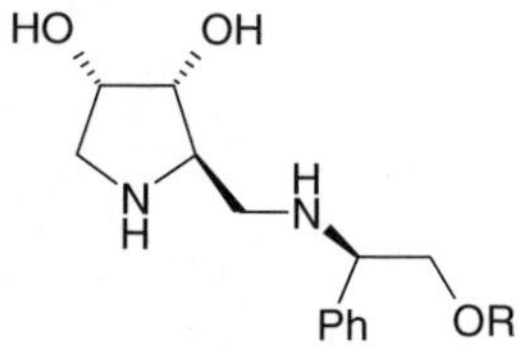	**Potential anticancer agent** *Jack beans α-mannosidase* (EC 3.2.1.24) R = Me 77% inhibition at 1 mM [120]. R = Bn IC_{50} 58 μM; K_i 16 μM [120].
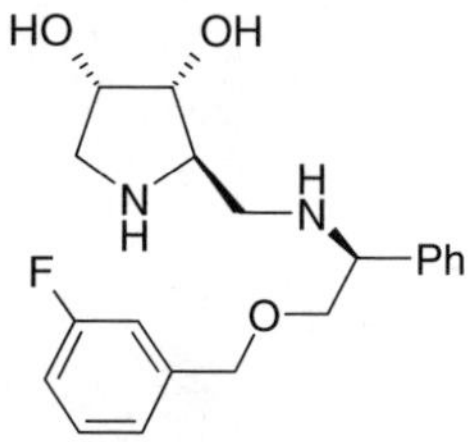	**Potential anticancer agent** *Jack beans α-mannosidase* (EC 3.2.1.24) IC_{50} 30 μM; K_i 7 μM [120].
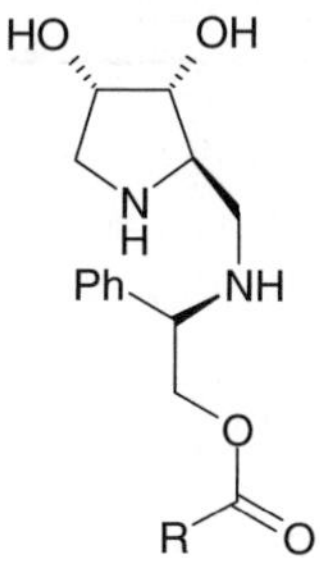	**Potential anticancer agent** *Jack beans α-mannosidase* (EC 3.2.1.24) R = 2-fluorophenyl IC_{50} 63 μM; K_i 24 μM [120]. R = 4-fluorophenyl IC_{50} 60 μM; K_i 19 μM [120]. R = 3-bromophenyl 79% inhibition at 1 mM [120]. R = 4-bromophenyl IC_{50} 55 μM [120]. **Inhibition of DNA in human glioblastoma:** R = 4-Bromophenyl 93% inhibition at 300 μM in LN18 [120]. Rate of protein synthesis decrease: 82% at 300 μM [120].
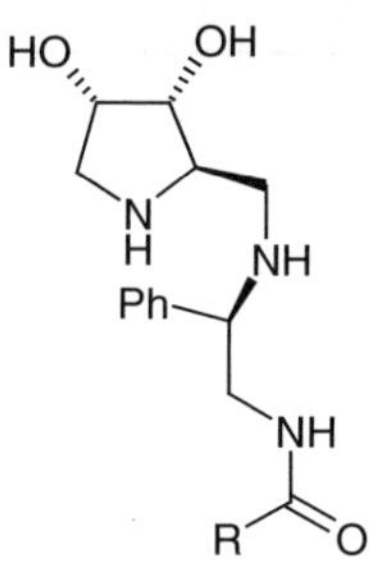	**Potential anticancer agent** *Jack beans α-mannosidase* (EC 3.2.1.24) R = 3-Bromophenyl IC_{50} 89 μM [120]. Human glioblastoma cells (LN18 and LNZ308) R = 3-Bromophenyl IC_{50} 225 μM [120].
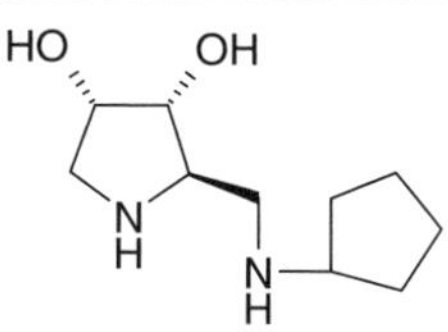	**Potential anticancer agent** *Jack beans α-mannosidase* (EC 3.2.1.24) IC_{50} 560 μM [122].
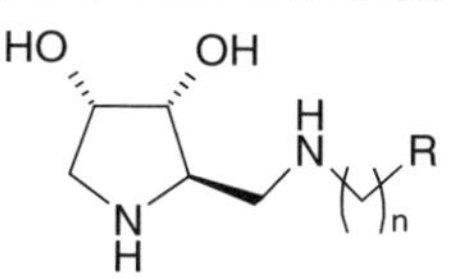	**Potential anticancer agent** *Jack beans α-mannosidase* (EC 3.2.1.24) $n = 5$ R = OH 44% inhibition at 1 mM [122]. $n = 1$ R = Ph IC_{50} 60 μM; K_i 7.4 μM [122]. $n = 2$ R = Ph IC_{50} 300 μM [122]. $n = 3$ R = Ph IC_{50} 200 μM [122]. $n = 4$ R = Ph IC_{50} 380 μM; K_i 120 μM [122].

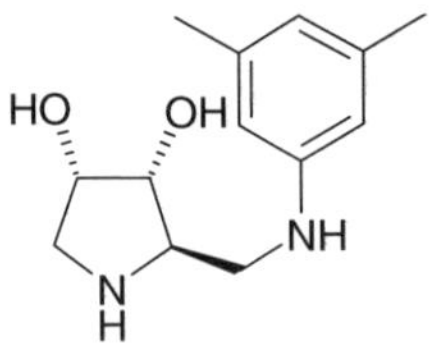

Potential anticancer agent
Jack beans α-mannosidase (EC 3.2.1.24)
IC_{50} 560 μM; K_i 190 μM [122].

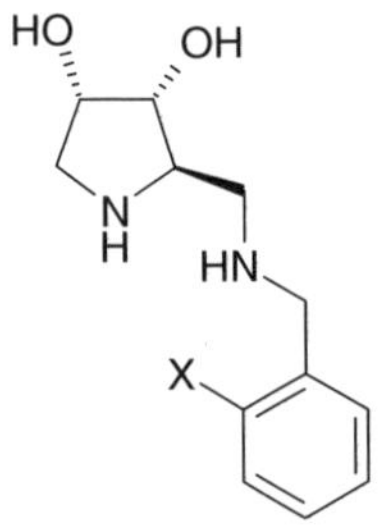

Potential anticancer agent
Jack beans α-mannosidase (EC 3.2.1.24)

X = Me	IC_{50} 45 μM; K_i 11 μM [122].
X = OMe	IC_{50} 105 μM; K_i 20 μM [122].
X = F	IC_{50} 75 μM; K_i 20 μM [122].
X = Cl	IC_{50} 29 μM; K_i 13 μM [122].
X = Br	IC_{50} 36 μM; K_i 11 μM [122].

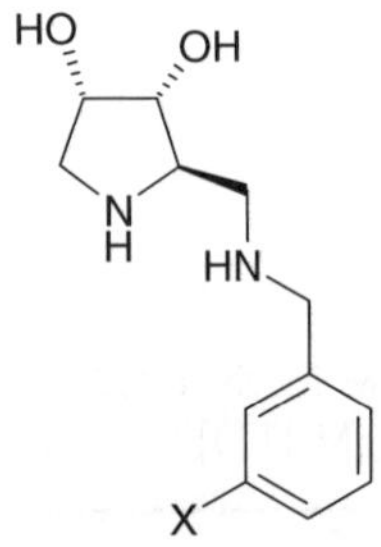

Potential anticancer agent
Jack beans α-mannosidase (EC 3.2.1.24)

X = Me	IC_{50} 60 μM; K_i 10 μM [122].
X = OH	IC_{50} 34 μM; K_i 8.4 μM [122].
X = OMe	IC_{50} 35 μM; K_i 9.2 μM [122].
X = F	IC_{50} 140 μM; K_i 36 μM [122].
X = Cl	IC_{50} 38 μM; K_i 12 μM [122].
X = Br	IC_{50} 35 μM; K_i 10 μM [122].

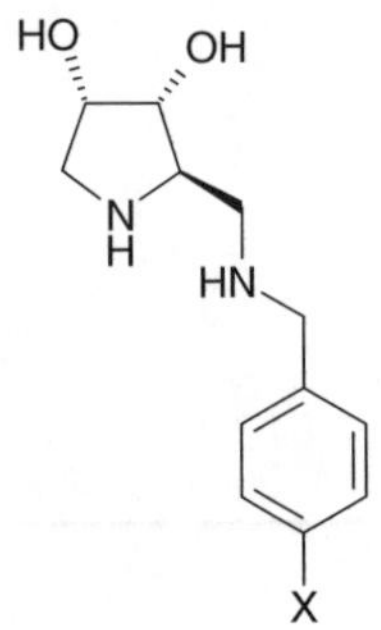

Potential anticancer agent
Jack beans α-mannosidase (EC 3.2.21.24)

X = Me	IC_{50} 20 μM; K_i 3 μM [122].
X = OH	IC_{50} 23 μM; K_i 5 μM [122].
X = OMe	IC_{50} 10 μM; K_i 7 μM [122].
X = F	IC_{50} 36 μM; K_i 9.5 μM [122].
X = Cl	IC_{50} 35 μM; K_i 10 μM [122].
X = Br	IC_{50} 30 μM; K_i 8.5 μM [122].
X = C_6H_5	IC_{50} 5 μM; K_i 2.5 μM [122].

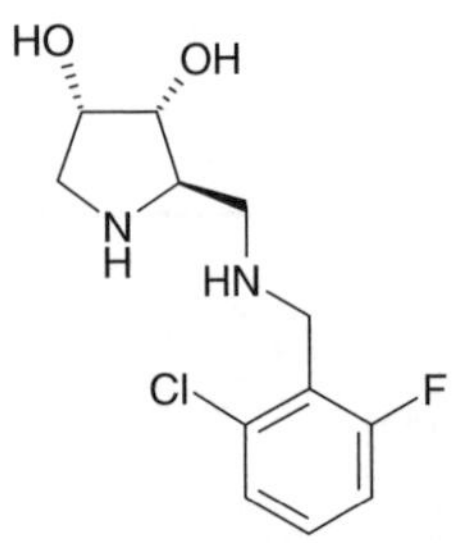

Potential anticancer agent
Jack beans α-mannosidase (EC 3.2.1.24)
IC_{50} 255 μM [122].

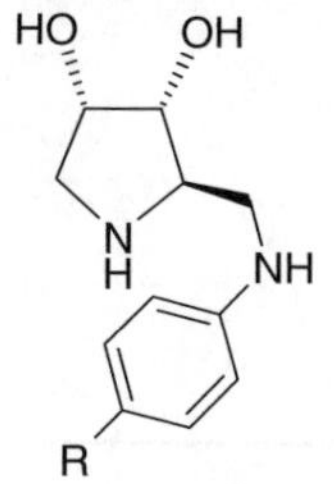	**Potential anticancer agent** *Jack beans α-mannosidase* (EC 3.2.1.24) R = Ph IC_{50} 250 μM; K_i 56 μM [122]. R = OPh IC_{50} 130 μM; K_i 66 μM [122].
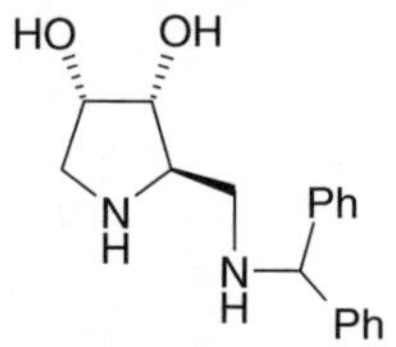	**Potential anticancer agent** *Jack beans α-mannosidase* (EC 3.2.1.24) IC_{50} 375 μM [122].
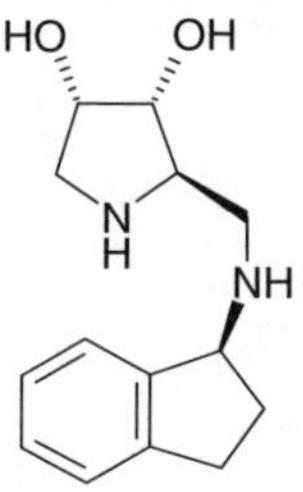	**Potential anticancer agent** *Jack beans α-mannosidase* (EC 3.2.1.24) IC_{50} 17 μM; K_i 2.3 μM [122].
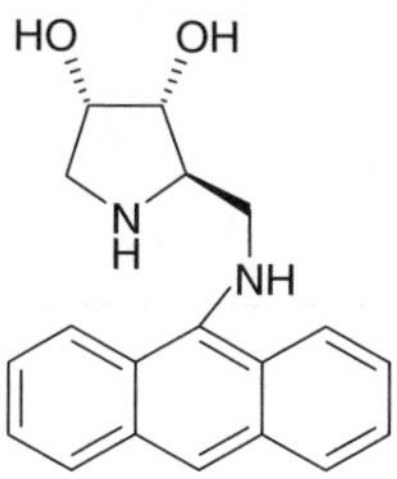	**Potential anticancer agent** *Jack beans α-mannosidase* (EC 3.2.1.24) IC_{50} 220 μM [122].
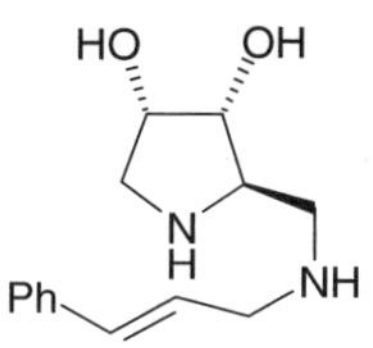	**Potential anticancer agent** *Jack beans α-mannosidase* (EC 3.2.1.24) IC_{50} 16 μM; K_i 9.5 μM [122].
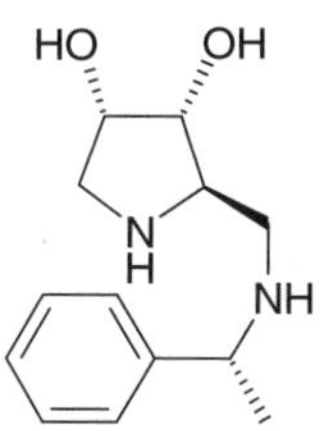	**Potential anticancer agent** *Jack beans α-mannosidase* (EC 3.2.1.24) IC_{50} 440 μM; K_i 170 μM [122].

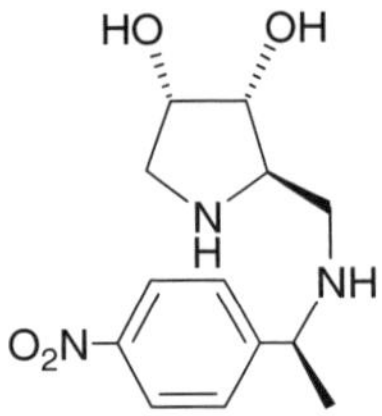	**Potential anticancer agent** *Jack beans α-mannosidase* (EC 3.2.1.24) IC_{50} 70 μM; K_i 26 μM [122].
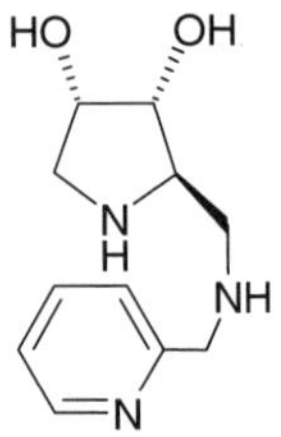	**Potential anticancer agent** *Jack beans α-mannosidase* (EC 3.2.1.24) IC_{50} 174 μM; K_i 34 μM [122].
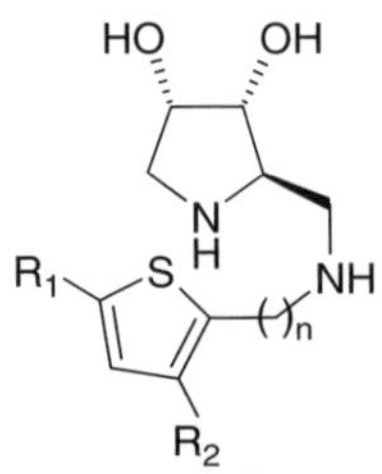	**Potential anticancer agent** *Jack beans α-mannosidase* (EC 3.2.1.24) $n = 1$ $R_1 = H$; $R_2 = H$ IC_{50} 85 μM; K_i 26 μM [122]. $n = 1$ $R_1 = Me$; $R_2 = H$ IC_{50} 10 μM; K_i 3.5 μM [122]. $n = 1$ $R_1 = H$; $R_2 = Me$ IC_{50} 44 μM; K_i 18 μM [122]. $n = 1$ $R_1 = Et$; $R_2 = H$ IC_{50} 26 μM; K_i 6.4 μM [122]. $n = 2$ $R_1 = H$; $R_2 = H$ IC_{50} 260 μM [122].
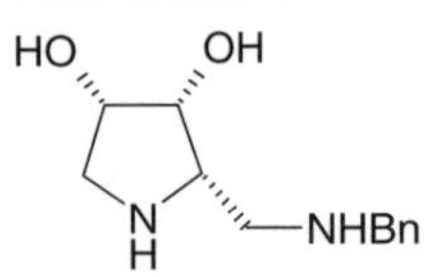	**Potential anticancer agent** *Jack beans α-mannosidase* (EC 3.2.1.24) K_i 40 μM [121].
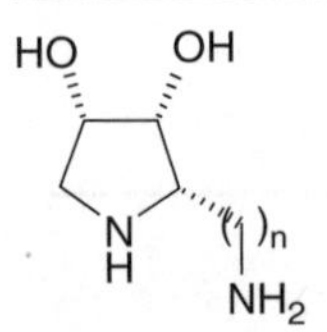	**Potential anticancer agent** *Jack beans α-mannosidase* (EC 3.2.1.24) $n = 1$ IC_{50} 30 μM; K_i 92 μM [122]. $n = 2$ IC_{50} 1 mM [123].
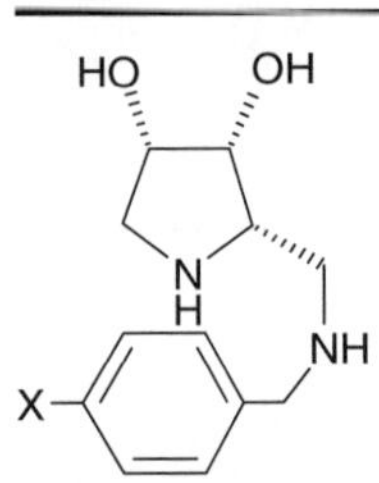	**Potential anticancer agent** *Jack beans α-mannosidase* (EC 3.2.1.24) X = H IC_{50} 361 μM; K_i 105 μM [122]; K_i 8.7 μM [123]. X = OMe IC_{50} 110 μM; K_i 56 μM [122]. X = Ph IC_{50} 800 μM [122].

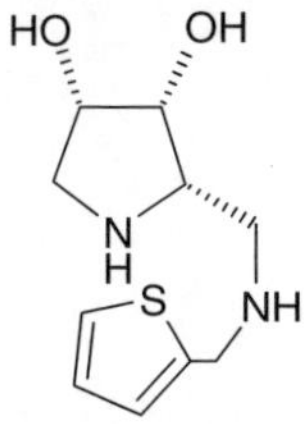

Potential anticancer agent
Jack beans α-mannosidase (EC 3.2.1.24)
IC_{50} 105 μM; K_i 40 μM [122].

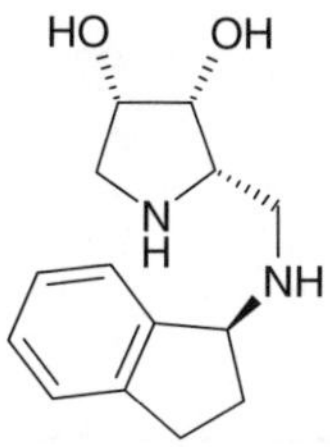

Potential anticancer agent
Jack beans α-mannosidase (EC 3.2.1.24)
IC_{50} 397 μM; K_i 12 μM [122].

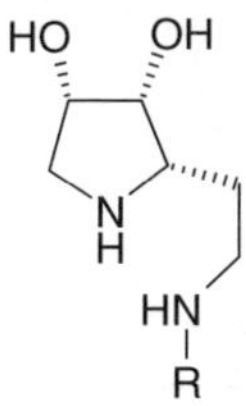

Potential anticancer agent
Jack beans α-mannosidase (EC 3.2.1.24)
R = Bn IC_{50} 620 μM; K_i 360 μM [123].
R = *i*-Pr 25% inhibition at 1mM [123].
R = $CH_2CH(OH)Me$ 54% inhibition at 1mM, IC_{50} 1 mM [123].
R = $CH_2CH(OH)Et$ 55% inhibition at 1mM, IC_{50} 830 μM [123].

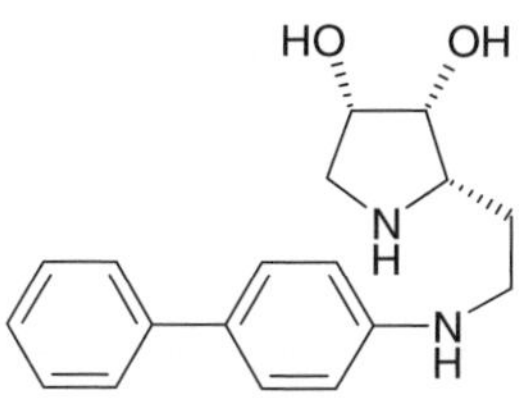

Potential anticancer agent
Jack beans α-mannosidase (EC 3.2.1.24)
70% inhibition at 1mM [123].
IC_{50} 360 μM; K_i 102 μM [123].

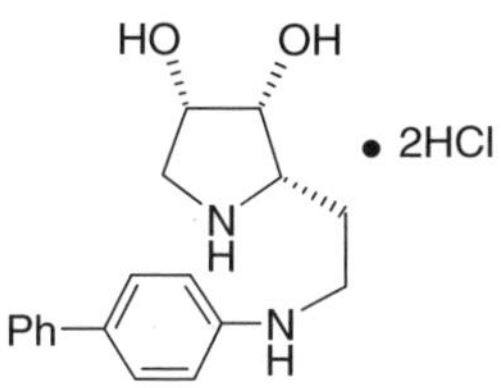

Potential anticancer agent
Jack beans α-mannosidase (EC 3.2.1.24)
59% inhibition at 1mM [123].

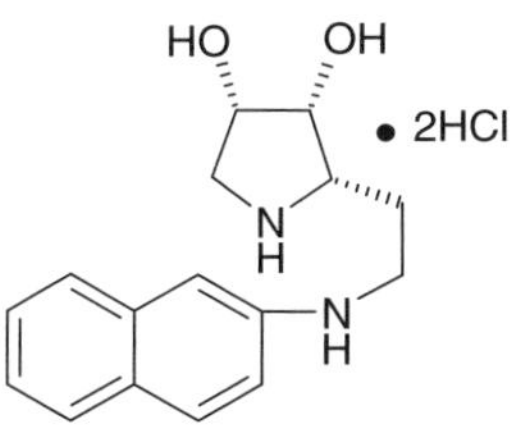

Potential anticancer agent
Jack beans α-mannosidase (EC 3.2.21.24)
53% inhibition at 1mM [123].

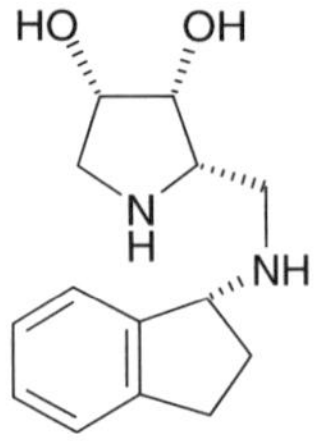

Potential anticancer agent
Jack beans α-mannosidase (EC 3.2.1.24)
IC_{50} 397 μM; K_i 12 μM [123].

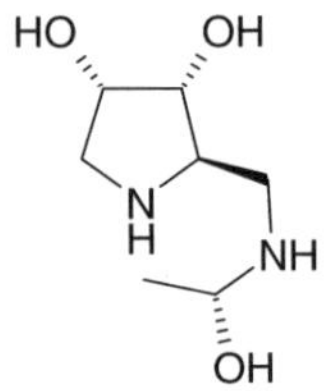

Potential anticancer agent
Jack beans α-mannosidase (EC 3.2.1.24)
IC_{50} 300 μM; K_i 59 μM [118].

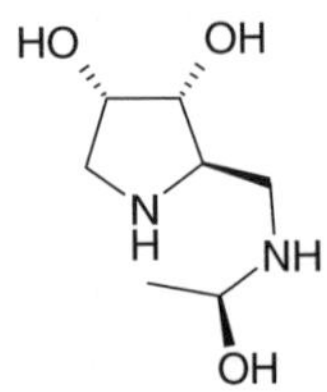

Potential anticancer agent
Jack beans α-mannosidase (EC 3.2.1.24)
IC_{50} 490 μM; K_i 120 μM [118].

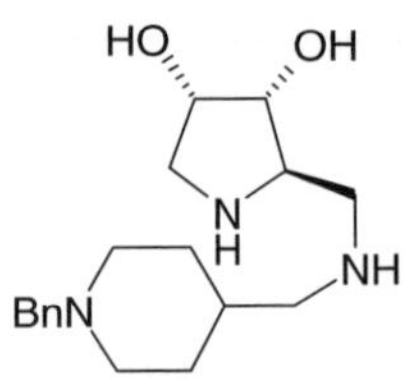

Potential anticancer agent
Jack beans α-mannosidase (EC 3.2.1.24)
IC_{50} 700 μM [118].

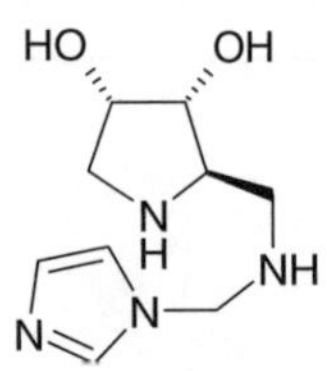

Potential anticancer agent
Jack beans α-mannosidase (EC 3.2.1.24)
IC_{50} 700 μM [118].

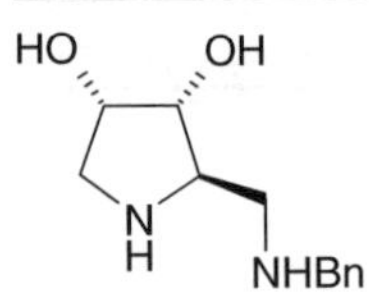

Potential anticancer agent
Jack beans α-mannosidase (EC 3.2.1.24)
IC_{50} 60 μM; K_i 7.4 μM [118].

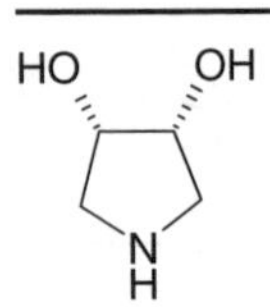

Potential anticancer agent
Jack beans α-mannosidase (EC 3.2.1.24)
IC_{50} 400 μM [118].

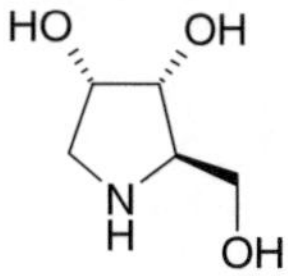	**Potential anticancer agent** *Jack beans α-mannosidase* (EC 3.2.1.24) IC_{50} 800 μM [118].
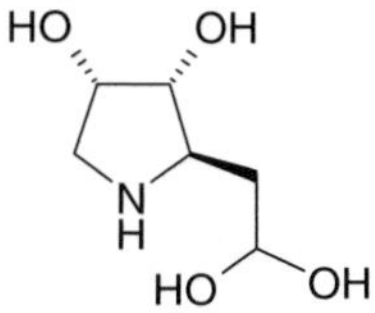	**Potential anticancer agent** *Jack beans α-mannosidase* (EC 3.2.1.24) IC_{50} 180 μM; K_i 80 μM [118].
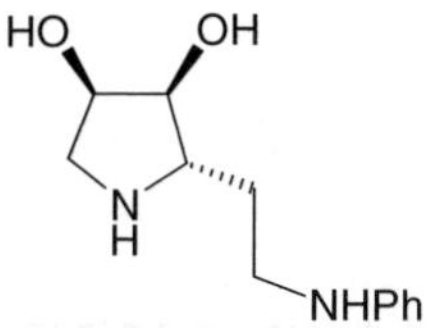	**Potential anticancer agent** *Jack beans α-mannosidase* (EC 3.2.1.24) 58% inhibition at 1 mM; IC_{50} 613 μM [123].
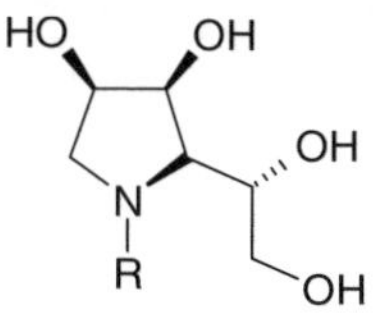	**Potential anticancer agent** *Human liver α-mannosidase* (EC 3.2.1.24) R = H K_i 13 μM [112]. *Golgi α-mannosidase II* (EC 3.2.1.114) R = H IC_{50} 100 μM [113]. R = Me 42% at 1 mM [113].
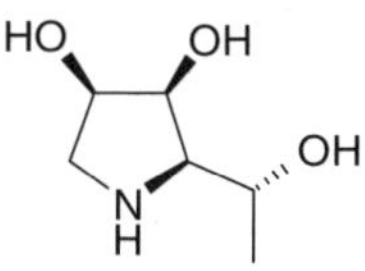	**Potential anticancer agent** *Lysosomal α-mannosidase* (EC 3.2.1.24) K_i 1.3 μM [113]. *Golgi α-mannosidase II* (EC 3.2.1.114) IC_{50} 60 μM [113].
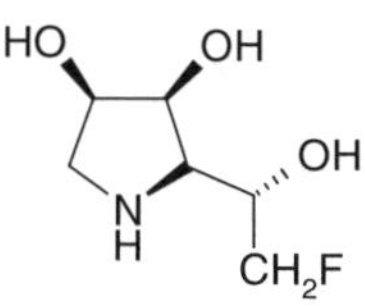	**Potential anticancer agent** *Lysosomal α-mannosidase* (EC 3.2.1.24) K_i 1.5 μM [113]. *Golgi α-mannosidase II* (EC 3.2.1.114) IC_{50} 30 μM [113].
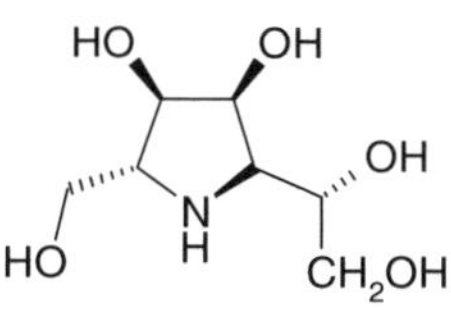	**Potential anticancer agent** *Lysosomal α-mannosidase* (EC 3.2.1.24) 71% at 1 mM [113]. *Golgi α-mannosidase II* (EC 3.2.1.114) IC_{50} 35 μM [113].
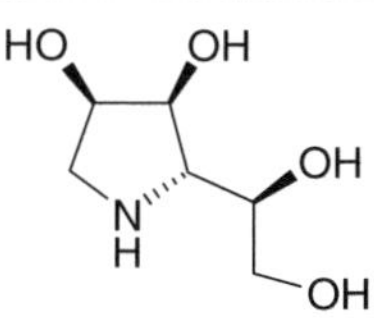	**Potential anticancer agent** *Human liver α-mannosidase* (EC 3.2.1.24) K_i 120 μM [112].

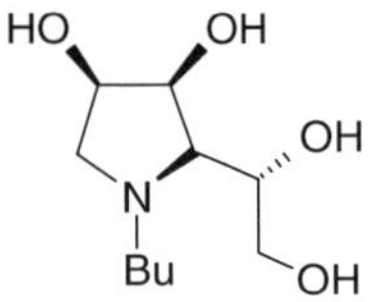	**Potential anticancer agent** *Golgi α-mannosidase II* (EC 3.2.1.114) 61% at 1mM [113].
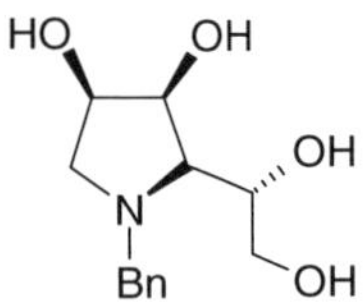	**Potential anticancer agent** *Golgi α-mannosidase II* (EC 3.2.1.114) 72% inhibition at 1mM [113].
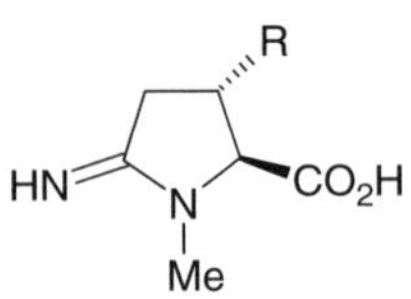	**Potential anticancer agent** *N-acetyl-β-D-glucosaminidase* (EC 3.2.1.30) R = H K_i 2.0 μM [157]. R = OH K_i 1.7 μM [157].

14.7.2 Six-membered rings

Gluco configuration

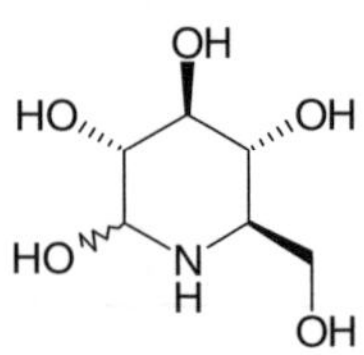	**Potential anticancer agent** *Jack beans α-mannosidase* (EC 3.2.1.24) 19% inhibition at 1 mM [132a]. *Bovine liver β-glucuronidase* (EC 3.2.1.31) 37.2% inhibition at 1 mM [132a]. **Pulmonary metastases** Inhibition of pulmonary metastasis of B16 variant cells: 98.7% at 10 μg/ml [132a]. Inhibition of pulmonary colonization of B16-F10 cells: 94% at 32 μg/ml [168].
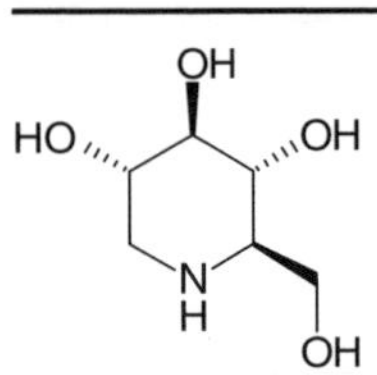	**Potential anticancer agent** ***α-mannosidase*** **(EC 3.2.1.24)** 54.3% inhibition at 1 mM [132a]. **Pulmonary metastases** Inhibition of pulmonary metastasis of B16 variant cells: 80.5% at 10 μg/ml [132a]. **Head and neck carcinoma** Sensitivity to Cisplatin of IMC-3 cells: IC_{50} 26.36 μg/ml [142].

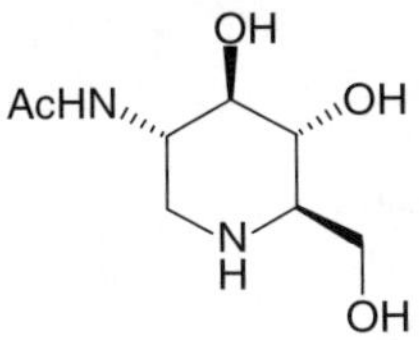

Potential anticancer agent
Human placenta N-acetyl-β-glucosaminidase (EC 3.2.1.30)
50% inhibition at 6 μM [150].
K_i 0.9 μM [150].

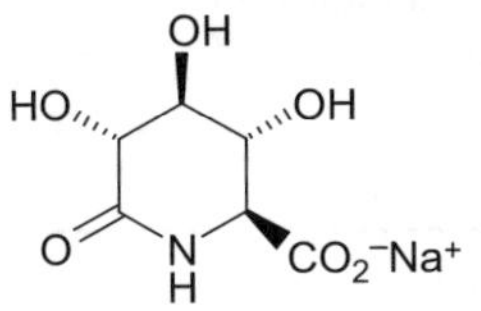

Potential anticancer agent
Bovine liver β-glucuronidase (EC 3.2.1.31)
99.6% inhibition at 1 mM [132a].

Pulmonary metastases
Inhibition of pulmonary metastasis of B16 variant cells: 91.8% at 10 μg/ml (Ca salt used) [132a].
Inhibition of pulmonary metastases of B16 melanoma variant cells: 99.5% at 30 mg/kg/day (day 7-27) [132b].
Inhibition of pulmonary metasatases of Lewis lung carcinoma (3LL): 98% at 100 mg/kg/day [132b].
Invasive activity of B16 variant cells: 17% invasion at 100 μg/ml [132b].
Invasive activity of 3LL variant cells: 10.1% invasion at 100 μg/ml [132b].
Invasive activity on 3LL cells: 89.9% inhibition rate at 100 μg/ml [132b].
Inhibition of lung metastases of rat hepatoma cKDH-811 cells: 69.2% at 50 μg/ml [171].
Inhibition of pulmonary metastases from human renal cell carcinoma
SN12Cpm6: 78.1% at 50 μg/ml [172].
Haptotaxis of rat hepatoma cKDH-811 cells toward laminin: 77% inhibition rate at 50 μg/ml [171].
Invasive activities of rat hepatoma cKDH-811 cells: 69.3% inhibition rate at 50 μg/ml [171].

Renal carcinoma
Invasive activity of human NC 65 cells: 33.1% invasion at 100 μg/ml [173].
Invasive activity of human renal cells carcinoma SN12Cpm6: 72.1% inhibition rate at 100 μg/ml [172].
Invasive activity of human ACHN cells: 61.2% invasion at 300 μg/ml [173].

Human melanoma
Invasive activity of human G 361 cells: 42.2% invasion at 300 μg/ml [173].

Liver metastases
Invasive activity of LMFS-1 cells: 1% invasion at 100 μg/ml [163].

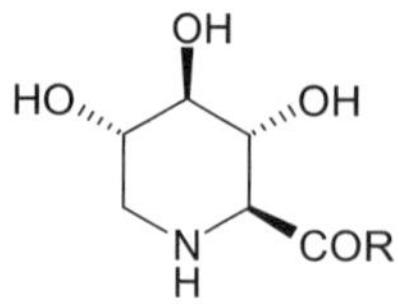

Potential anticancer agent
Bovine liver β-D-glucuronidase (EC 3.2.1.31)
R = OH 89.4% inhibition at 100 μg/ml [132a]
R = OMe 68.1% inhibition at 100 μg/ml [132a].
R = NH_2 19.5% inhibition at 100 μg/ml [132a].

Pulmonary metastases
Inhibition of pulmonary metastases of B16 variant cells:
R = OH 73.8% at 10 μg/ml [132a].
R = OMe 46.4% at 10 μg/ml [132a].
R = NH_2 75.5% at 10 μg/ml [132a].

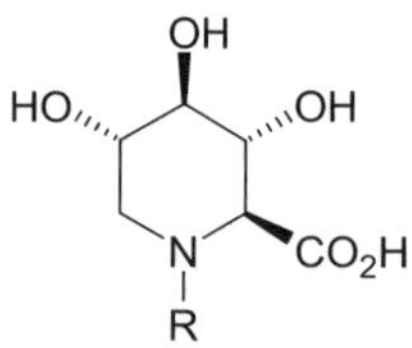

Potential anticancer agent
Bovine liver β-D-glucuronidase (EC 3.2.1.31)
R = *n*-Bu 20.9% inhibition at 100 μg/ml [132a].

Pulmonary metastases
Inhibition of pulmonary metastases of B16 variant cells:
R = *n*-Bu 48.8% at 10 μg/ml [132a].
R = Ac 52.7% at 10 μg/ml [132a].

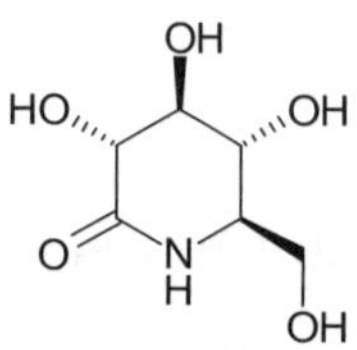

Pulmonary metastases
Inhibition of pulmonary metastases of B16 variant cells: 41.2% at 10 μ g/ml [132a].

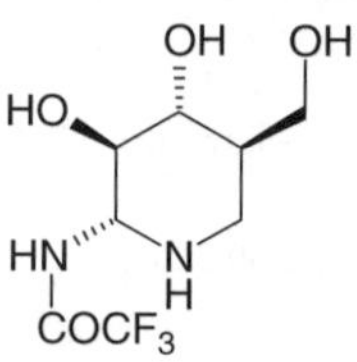

Potential anticancer agent
Jack beans α-mannosidase (EC 3.2.1.24)
IC_{50} 0.22 μM [125].

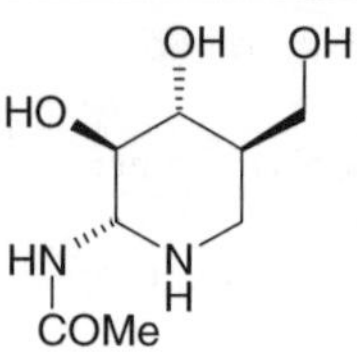

Potential anticancer agent
Jack beans α-mannosidase (EC 3.2.1.24)
IC_{50} 0.25 mM [125].
Bovine epididymis N-acetyl-β-D-glucosaminidase (EC 3.2.1.30)
IC_{50} 1.2 μM [125].

D-Galacto-configuration

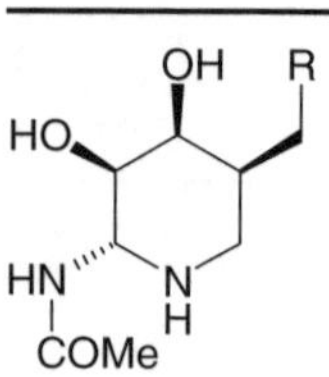

Potential anticancer agent
Chicken liver α-N-acetylgalactosaminidase (EC 3.2.1.49)
R = N_3 IC_{50} 0.6 μg/ml [128].
R = NH_2 IC_{50} 38 μg/ml [128].
R = OH IC_{50} 0.27 μg/ml [128]; IC_{50} 0.08 μg/ml [129].
R = SCH_3 IC_{50} 20 μg/ml [128].
R = $S(O)CH_3$ IC_{50} 56 μg/ml [128].

(Continued)
Bovine epididymis N-acetyl-β-D-glucosaminidase (EC 3.2.1.30)
R = N_3 IC_{50} 45 μg/ml [128].
R = OH IC_{50} 0.42 μg/ml [128], IC_{50} 0.65 μg/ml [129].
R = SCH_3 IC_{50} 0.28 μg/ml [128].

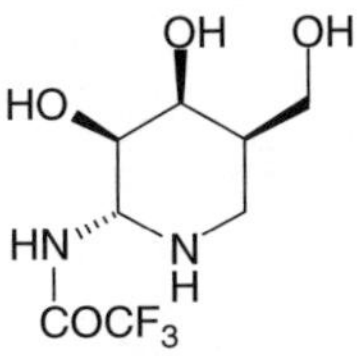	**Potential anticancer agent** *Chicken liver α-N-acetylgalactosaminidase* (EC 3.2.1.49) IC_{50} 0.65 μg/ml [129].
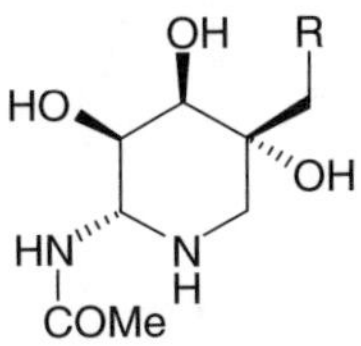	**Potential anticancer agent** *β-D-glucuronidase* (EC 3.2.1.31) R = NH_2 38% inhibition at 100 μg/ml [133]. R = NO_2 3% inhibition at 100 μg/ml [133].
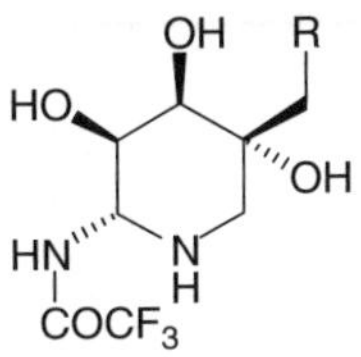	**Potential anticancer agent** *β-D-glucuronidase* (EC 3.2.1.31) R = NH_2 13% inhibition at 100 μg/ml [134]. R = NO_2 14% inhibition at 100 μg/ml [133].

D-Manno-configuration

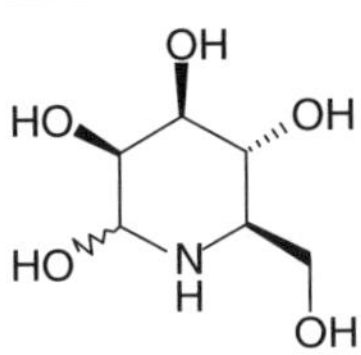	**Potential anticancer agent** *Jack beans α-mannosidase* (EC 3.2.1.24) 90.3% inhibition at 1 mM [132a]. **Pulmonary metastases** Inhibition of pulmonary metastases of B16 variant cells: 57.1% inhibition at 10 μg/ml [132a].
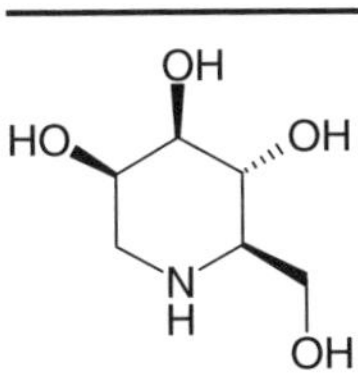	**Head and neck carcinoma** Sensitivity to cisplatin of IMC-3 cells IC_{50} 36.65 μg/ml [142]. **Potential anticancer agent** *Golgi α-mannosidase II* (EC 3.2.1.114) IC_{50} 400 μM [126].

Miscelleanous configurations

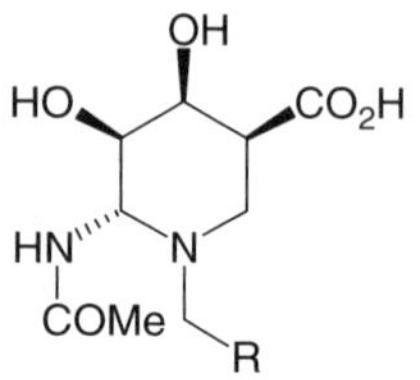

Potential anticancer agent
Streptococcus sp. N-acetylneuraminidase (EC 3.2.1.18)
R = Ph IC_{50} 3.96 μg/ml [135].
R = 2-F-C_6H_4 IC_{50} 4.93 μg/ml [135].
R = 2-CH_3-C_6H_4 IC_{50} 1.39 μg/ml [135].
R = 2-CF_3-C_6H_4 IC_{50} 16.82 μg/ml [135].
R = Et IC_{50} 0.78 μg/ml [135].
R = *i*-Pr IC_{50} 5.35 μg/ml [135].
R = *t*-Bu IC_{50} 15.14 μg/ml [135].
C. Perfringens N-acetylneuraminidase (EC 3.2.1.18)
R = Ph IC_{50} 3.5 μg/ml [135].
R = 2-F-C_6H_4 IC_{50} 1.5 μg/ml [135].
R = 2-CH_3-C_6H_4 IC_{50} 11 μg/ml [135].
R = 2-CF_3-C_6H_4 IC_{50} 68 μg/ml [135].
R = Et IC_{50} 20 μg/ml [135].
R = *i*Pr IC_{50} 25 μg/ml [135].
R = *t*Bu IC_{50} 240 μg/ml [135].

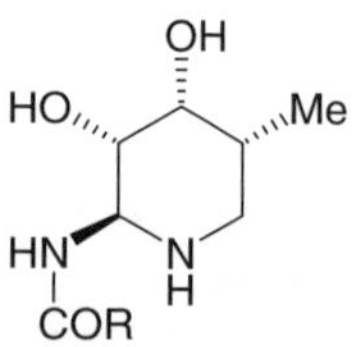

Invasive human ovarian carcinoma
Bovine kidney α-L-fucosidase (EC 3.2.1.51)
R = Me IC_{50} 0.48 μM [127].
R = CF_3 IC_{50} 11 nM [127].
R = CCl_3 IC_{50} 9 nM [127].

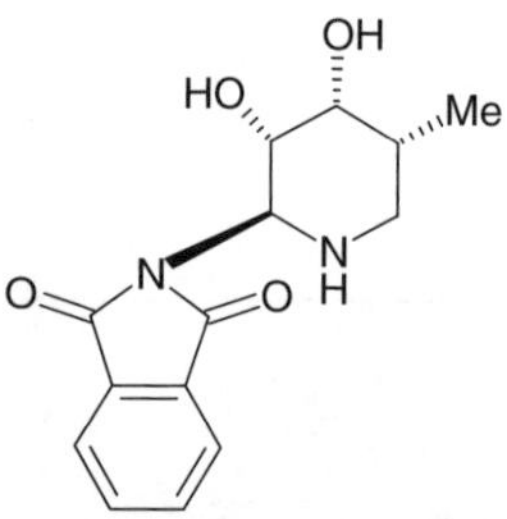

Invasive human ovarian carcinoma
Bovine kidney α-L-fucosidase (EC 3.2.1.51)
IC_{50} 13 nM [127].

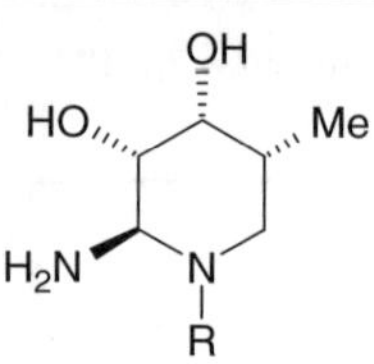

Invasive human ovarian carcinoma
Bovine kidney α-L-fucosidase (EC 3.2.1.51)
R = H IC_{50} 16 nM [127].
R = $COCH_3$ IC_{50} > 0.27 mM [127].

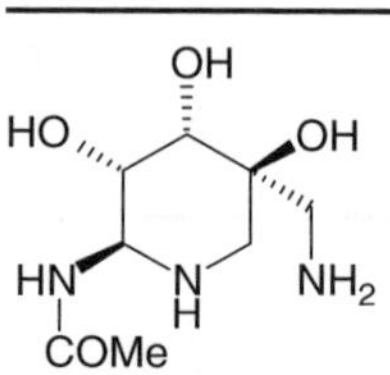

Potential anticancer agent
Bovine liver β-D-glucuronidase (EC 3.2.1.31)
24% inhibition at 100 μg/ml [133].

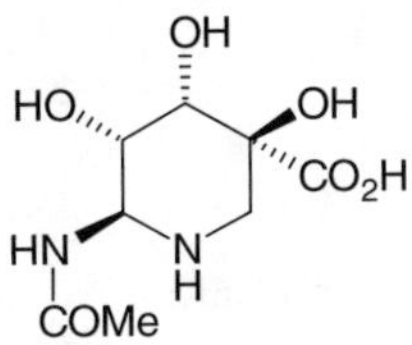

Potential anticancer agent
Bovine liver β-D-glucuronidase (EC 3.2.1.31)
2% inhibition at 100 μg/ml [133].

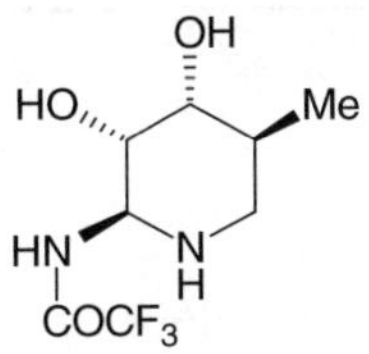

Invasive human ovarian carcinoma
Bovine kidney α-L-fucosidase (EC 3.2.1.51)
IC_{50} 1.8 μM [127].

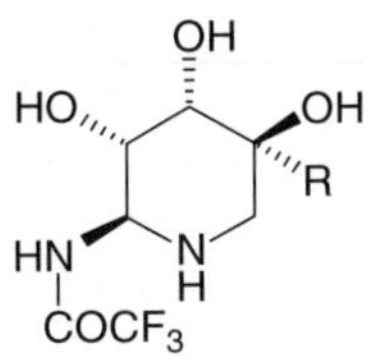

Potential anticancer agent
Bovine liver β-D-Glucuronidase (EC 3.2.1.31)
R = CH_2NH_2 43% inhibition at 100 μg/ml [134].
R = CO_2H 74% inhibition at 100 μg/ml [134].

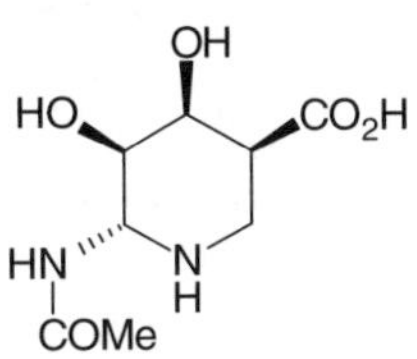

Human melanoma
Bovine liver β-D-glucuronidase (EC 3.2.1.31)
IC_{50} 71 μM [124]; IC_{50} 39 μM [138].
Human heparanase (EC 3.2.1.31)
$IC_{50} > 350$ μM [138].
C. Perfringens N-acetylneuraminidase (EC 3.2.1.18)
IC_{50} 150 μM [138].

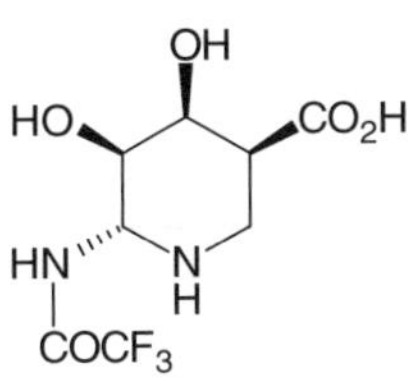

Potential anticancer agent
Human heparanase (EC 3.2.1.31)
IC_{50} 1.2 μM [124].
Bovine liver β-D-glucuronidase (EC 3.2.1.31)
IC_{50} 65 nM [124].

Pulmonary metastases
Inhibition of pulmonary metastases of B16BL6 cells (mice): 90.8% inhibition rate at 162 μM [124].
Invasive activity of B16BL6 cells: 64.1% inhibition at 1.1 mM [137].
Invasive activity of 3LL cells: 79.3% inhibition at 1.1 mM [137].
Inhibition of lung metastases of 3LL in mice: 57.1% inhibition at 100 mg/kg [124].

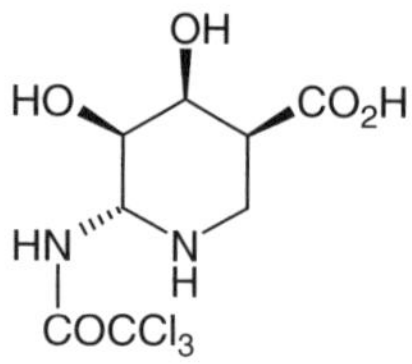	**Potential anticancer agent** *Bovine liver β-D-glucuronidase* (EC 3.2.1.31) IC_{50} 92 nM [124]. **Pulmonary metastases** Inhibition of pulmonary metastases of B16BL6 cells (mice): 67.3% inhibition rate at 140 μM [124].
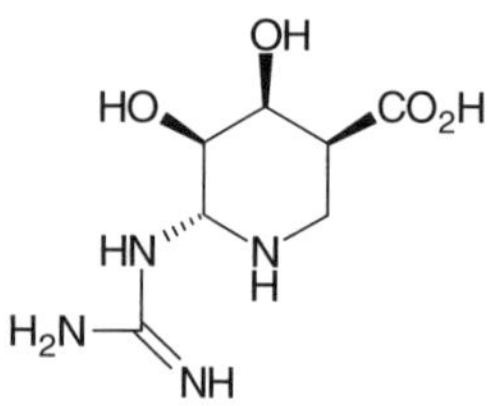	**Potential anticancer agent** *Bovine liver β-D-glucuronidase* (EC 3.2.1.31) IC_{50} 0.13 μM [124]. **Pulmonary metastases** Inhibition of pulmonary metastases of B16BL6 cells (mice): 87.1% inhibition rate at 170 μM [124].
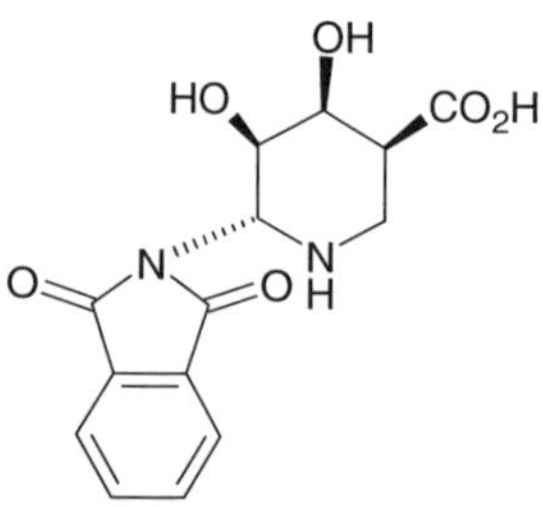	**Potential anticancer agent** *Bovine liver β-D-glucuronidase* (EC 3.2.1.31) IC_{50} 68 nM [124].
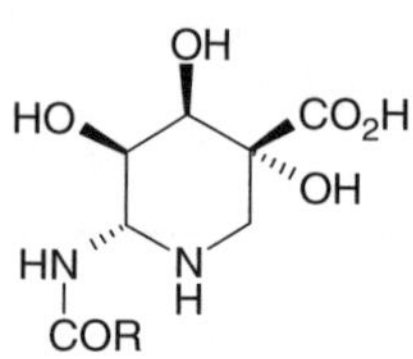	**Potential anticancer agent** *Bovine liver β-D-glucuronidase* (EC 3.2.1.31) R = Me IC_{50} 120 μM [124]; 28.5 μg/ml [133]. R = CF_3 IC_{50} 62 nM [124]. **Pulmonary metastases** R = CF_3 Inhibition of pulmonary metastases of B16BL6 cells (mice): 90.4% inhibition rate at 310 μM [124].
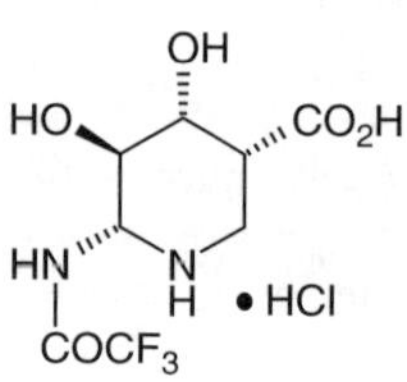	**Potential anticancer agent** *Bovine liver β-D-Glucuronidase* (EC 3.2.1.31) IC_{50} 65 nM [139]. *Human heparanase* (EC 3.2.1.31) IC_{50} 10.5 μM [124]; IC_{50} 29 μM (hydrochloride salt) [139].
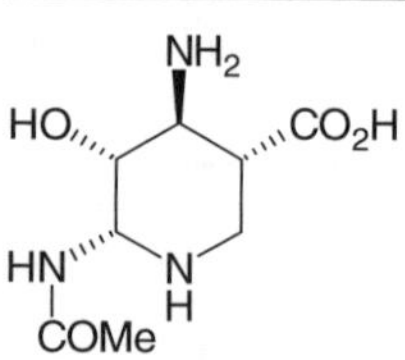	**Pulmonary metastases** Inhibition of pulmonary metastases of B16BL6 cells (mice): 30.9% inhibition rate at 172 μM [131].

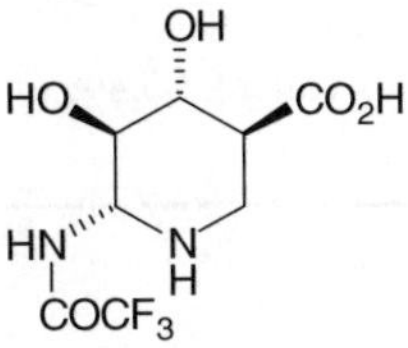	**Human melanoma** *Bovine liver β-D-glucuronidase* (EC 3.2.1.31) IC_{50} 65 nM [124]. *Human heparanase* (EC 3.2.1.31) IC_{50} 29 μM [124].
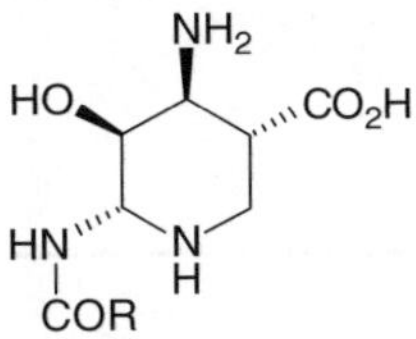	**Pulmonary metastases** Inhibition of pulmonary metastases of B16BL6 cells (mice): R = CF_3 75.5% inhibition rate at 145 μM [124]. R = Me 44.3% inhibition rate at 172 μM [124].
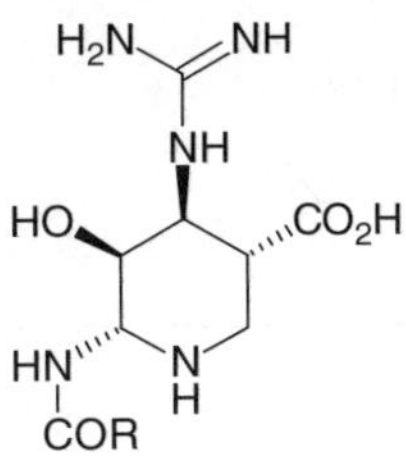	**Pulmonary metastases** Inhibition of pulmonary metastases of B16BL6 cells (mice): R = CH_3 97% inhibition rate at 150 μM [124]. R = CF_3 81% inhibition rate at 129 μM [124]. Invasive activity of B16BL6 cells: R = CH_3 58.9% inhibition at 870 μM [124]; R = CF_3 63.9% inhibition at 518 μM [124].
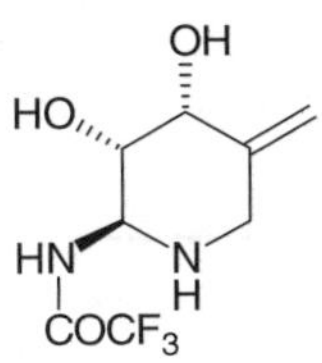	**Invasive human ovarian carcinoma** *Bovine kidney α-L-fucosidase* (EC 3.2.1.51) IC_{50} 700 nM [127].
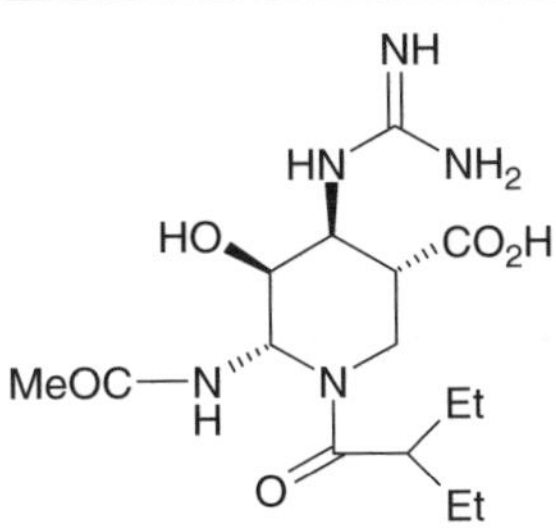	**Potential anticancer agent** *Heparan sulphate 2-O-sulphotransferase* 80% inhibition at 25 μM [130].
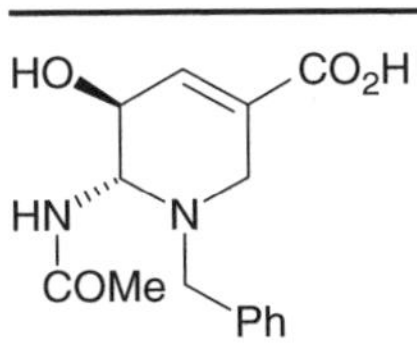	**Potential anticancer agent** *Streptococcus sp. N-acetylneuraminidase* (EC 3.2.1.18) IC_{50} 14.5 μg/ml [135]. *C. Perfringens N-acetylneuraminidase* (EC 3.2.1.18) IC_{50} 50 μg/ml [135].
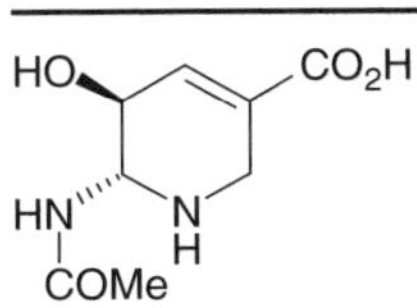	**Potential anticancer agent** *Streptococcus sp. N-acetylneuraminidase* (EC 3.2.1.18) IC_{50} 70% inhibition at 100 μg/ml [136]. *Bovine liver β-D-glucuronidase* (EC 3.2.1.31) 81% inhibition at 100 μg/ml [136].

(Continued)

Pulmonary metastases

Inhibition of pulmonary metastases of B16 cells (mice): 35.4% at 30 μg/ml [136].

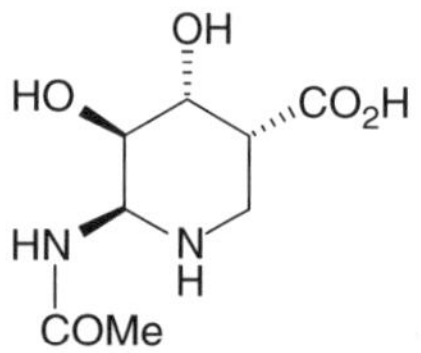

Potential anticancer agent

Human heparanase (EC 3.2.1.31)

IC_{50} > 350 μM [138].

C. Perfringens N-acetylneuraminidase (EC 3.2.1.18)

IC_{50} > 350 μM [138].

Bovine liver β-D-glucuronidase (EC 3.2.1.31)

IC_{50} 50 μM [138].

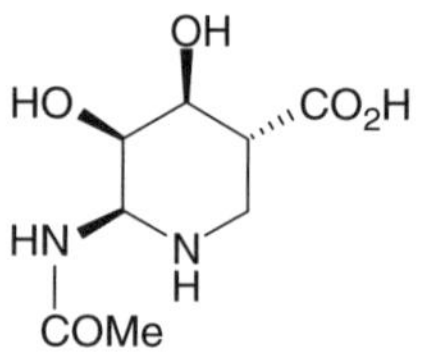

Potential anticancer agent

Human heparanase (EC 3.2.1.31)

IC_{50} > 350 μM [138].

C. Perfringens N-acetylneuraminidase (EC 3.2.1.18)

IC_{50} 250 μM [138].

Bovine liver β-D-glucuronidase (EC 3.2.1.31)

IC_{50} > 350 μM [138].

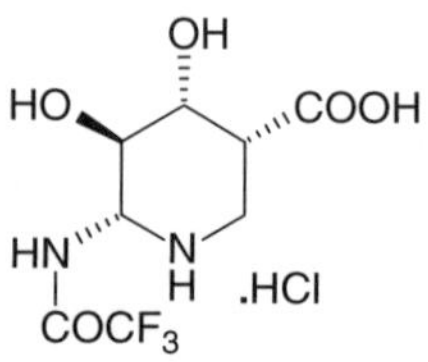

Potential anticancer agent

Bovine liver β-D-glucuronidase (EC 3.2.1.31)

IC_{50} 0.13 mM [139].

14.7.3 Bicyclic derivatives

Swainsonine and analogues

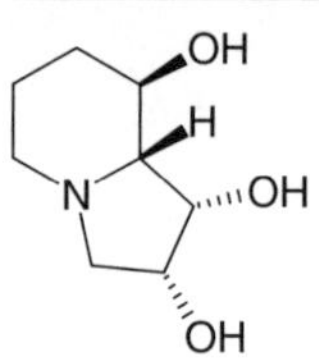

Head and neck carcinoma

Sensitivity to Cisplatin of IMC-3 cells

IC_{50} 21.12 μg/ml [142].

Antitumoural

Murine transplantable tumour growth: 100% inhibition at 30 mg/kg/day [164].

MDAY-D2 tumours cells: 30% of mice treated cells tumour free growth (mice): 30% of long-term survivor at 2.5 μg/ml in drinking water (tumour free after 100 days) [165].

Colorectal carcinoma

Inhibition of human tumour cells proliferation (HT29m): 59% at 1 μg/ml after 3 days [167].

Inhibition of human tumour cells proliferation (HT29p): 68% at 1 μg/ml after 3 days [167].

(Continued)

Renal carcinoma
Inhibition of human tumour cells proliferation (SN12L1): 82% at 1 μg/ml after 3 days [167].

Human melanoma
Inhibition of human tumour cells proliferation (A375p): 59% at 1 μg/ml after 3 days [167].

Pulmonary metastases
Inhibition of the number of B16 pulmonary metastases: 73 to 95% at 100 to 1000 mg/kg [164].
Lung colonization by B16F10 melanoma cells: 7% of lung nodules remaining at 2.5 μg/ml in drinking water (after 17 days) [165].
Inhibition of pulmonary colonization by B16F10 cells (mice): 79.4% at 0.5 μg/ml [166].
Lung colonization by B16F10 melanoma cells:
3 out of 105 B16F10 cells/mouse remaining at 5.8 μM [169].

Gastric cancer
Inhibition of tumour growth (mice with human xenograft): 73.8% with 3 μg/ml in drinking water [170].

Potential anticancer agent
Human liver α-mannosidase (EC 3.2.1.24)
K_i 70 nM at pH 4 [112].
Golgi α-mannosidase II (EC 3.2.1.114)
IC_{50} 40 nM [113].
IC_{50} 20 nM [126].
Rat liver endoplasmic reticulum mannosidase II (EC 3.2.1.113)
34% inhibition at 0.1 mM [160].
Jack beans α-mannosidase (EC 3.2.1.24)
IC_{50} 100 nM [114].

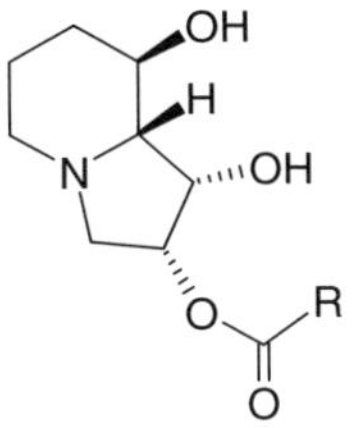

Antimetastatic
MDAY-D2 *α-mannosidase* (EC 3.2.1.24)
R = 4-Me-C_4H_6 IC_{50} 69 μM [169].
R = 4-NO_2-C_4H_6 IC_{50} 35 μM [169].
R = C_7H_{15} IC_{50} 35 μM [169].
R = Propyl IC_{50} 35 μM [169].
Lung colonization by B16F10 melanoma cells:
R = C_7H_{15} 1 out of 105 B16F10 cells/mouse remaining at 3.3 μM [169].

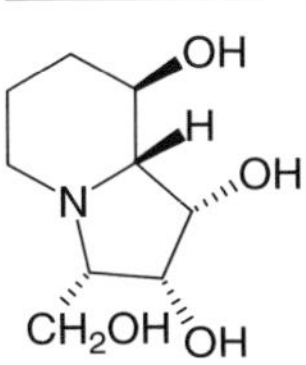

Potential anticancer agent
Jack bean α-mannosidase (EC 3.2.1.24)
IC_{50} 45 μM [115].

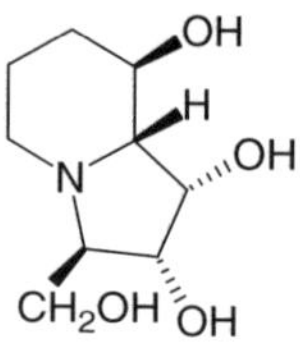	**Potential anticancer agent** *Jack bean α-mannosidase* (EC 3.2.1.24) IC_{50} 1.2 μM [115].
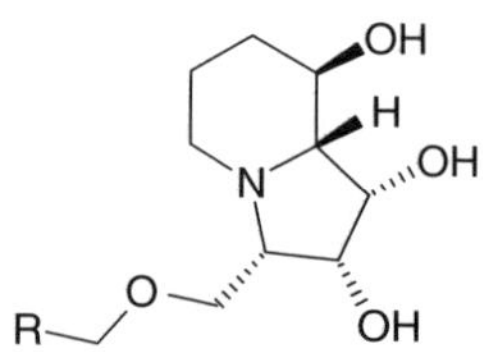	**Potential anticancer agent** *Jack beans α-mannosidase* (EC 3.2.1.24) R = Ph IC_{50} 100 μM [114]. R = 4-*t*Bu-Ph IC_{50} > 200 μM [114]. R = 4-Ph-Ph IC_{50} > 200 μM [114]. R = 2-naphthyl IC_{50} 17 μM [114].
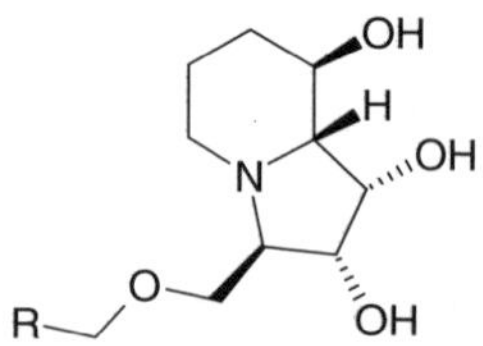	**Potential anticancer agent** *Jack beans α-mannosidase* (EC 3.2.1.24) R = Ph IC_{50} 0.39 μM [114]. R = 4-*t*Bu-Ph IC_{50} 0.05 μM [114]. R = 4-Ph-Ph IC_{50} 0.08 μM [114]. R = 2-naphthyl IC_{50} 0.11 μM [114]. R = 4-Me-Ph IC_{50} 0.07 μM [114].
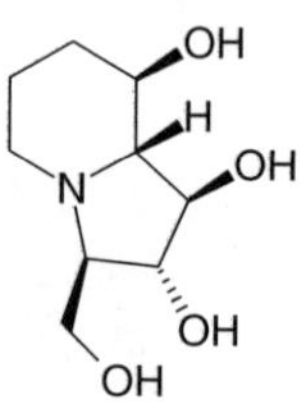	**Potential anticancer agent** *Jack beans α-mannosidase* (EC 3.2.1.24) IC_{50} 150 μM [116].
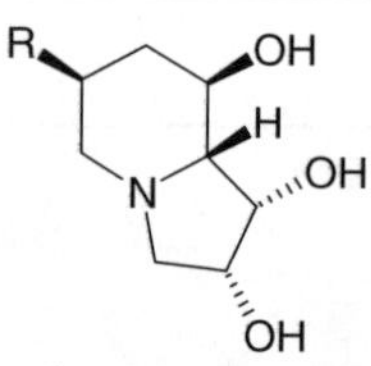	**Potential anticancer agent** *Jack beans α-mannosidase* (EC 3.2.1.24) R = Et IC_{50} 30 μM [117]. R = $(CH_2)_2OBn$ IC_{50} 230 μM [117].
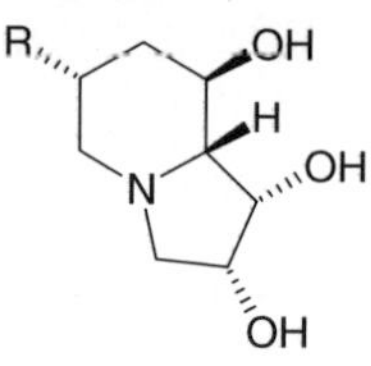	**Potential anticancer agent** *Jack beans α-mannosidase* (EC 3.2.1.24) R = Et IC_{50} 70 μM [117]. R = $(CH_2)_2OBn$ IC_{50} 275 μM [117].

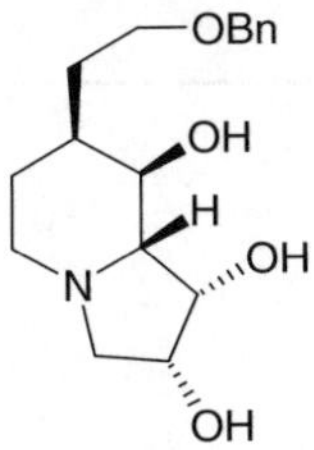

Potential anticancer agent
Jack beans α-mannosidase (EC 3.2.1.24)
IC_{50} 890 μM [117].

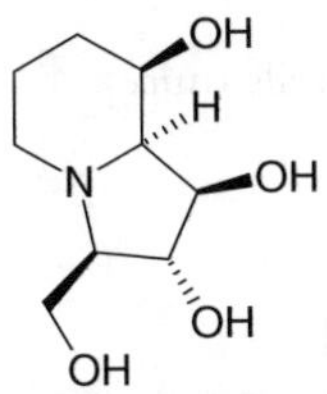

Potential anticancer agent
Jack beans α-mannosidase (EC 3.2.1.24)
IC_{50} 480 μM [116].

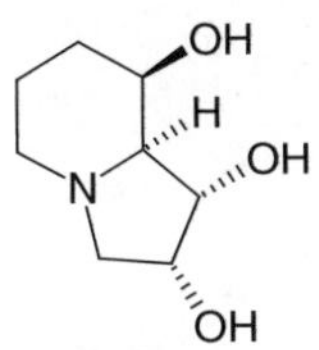

Potential anticancer agent
Human liver α-mannosidase (EC: 3.2.1.24)
K_i 75 μM [112].

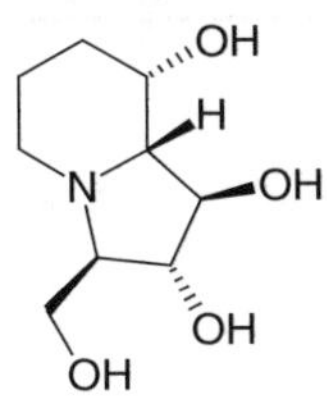

Potential anticancer agent
Jack beans α-mannosidase (EC 3.2.1.24)
IC_{50} 190 μM [116].

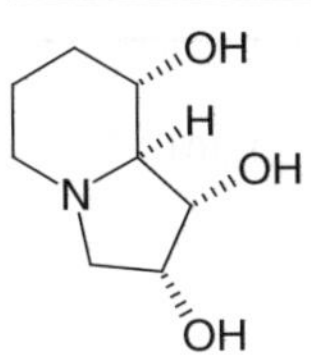

Potential anticancer agent
Human liver α-mannosidase (EC 3.2.1.24)
K_i 2 μM [112].

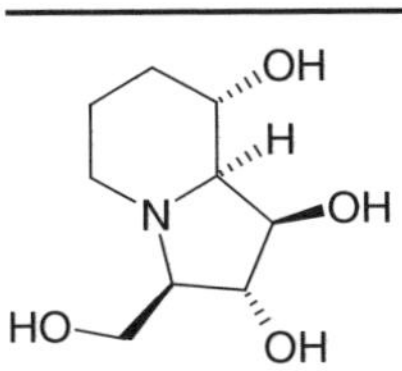

Potential anticancer agent
Jack beans α-mannosidase (EC 3.2.1.24)
IC_{50} 530 μM [116].

Castanospermine and analogues

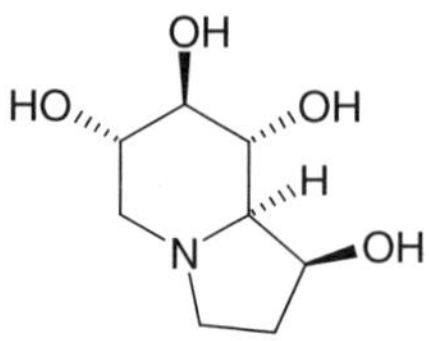

Colorectal carcinoma
Sucrase (EC 3.2.1.48)
100 % at inhibition at 25 μM (Caco-2 cells) [140].

Head and neck carcinoma
Sensitivity to cisplatin of IMC-3 cells
IC_{50} 30 μg/ml [142].

Pulmonary metastases
Inhibition of pulmonary colonization by B16F10 cells (mice): 79.4% at 0.5 μg/ml [166].

Potential anticancer agent
Human liver lysosomal α-glucosidase (EC 3.2.1.20)
100% inhibition at 1mM (pH 4); K_i 0.1 μM [143].
Human liver lysosomal β-glucosidase (EC 3.2.1.21)
96% inhibition at 1mM (pH 4); K_i 7 μM [143].
Bovine kidney β-N-acetylhexosaminidase (EC 3.2.1.52)
98% remaining activity at 250 μg/ml [148].
Bovine liver β-glucuronidase (EC 3.2.1.31)
99% remaining activity at 100 μg/ml [148].
Retard gowth of the SM-FeSV-transformed cells in soft agar [141].

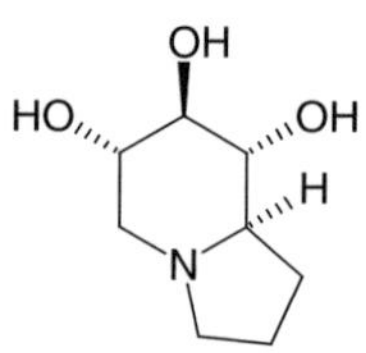

Potential anticancer agent
Human liver lysosomal α-glucosidase (EC 3.2.1.20)
7% inhibition at 1mM (pH 4) [143].
Human liver lysosomal β-glucosidase (EC 3.2.1.21)
31% inhibition at 1mM (pH 4) [143].
Human liver lysosomal α-mannosidase (EC 3.2.1.24)
21% inhibition at 1mM (pH 4) [143].

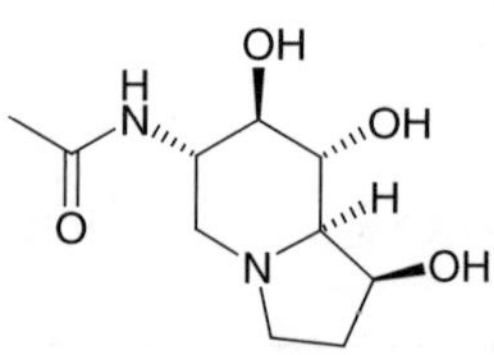

Potential anticancer agent
Human placenta N-acetyl-β-glucosaminidase (EC 3.2.1.30)
IC_{50} 0.5 μM [151].

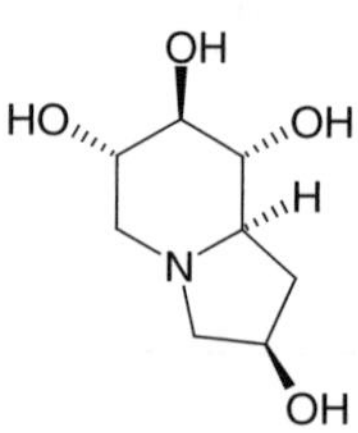

Potential anticancer agent
Jack beans α-mannosidase (EC 3.2.1.24)
IC_{50} 36.2 mM [147].

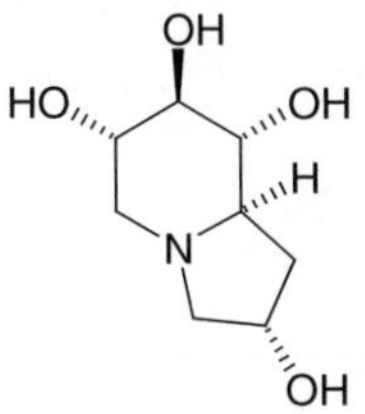

Potential anticancer agent
Jack beans α-mannosidase (EC 3.2.1.24)
IC_{50} 20 mM [147].

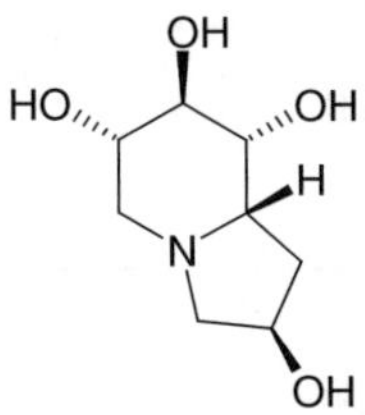

Potential anticancer agent
Jack beans α-mannosidase (EC 3.2.1.24)
IC_{50} 4.2 mM [147].

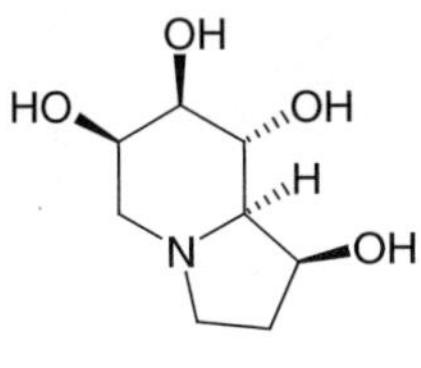

Potential anticancer agent
Human liver lysosomal α-glucosidase (EC 3.2.1.20)
28% inhibition at 1mM (pH 4) [143].
Human liver lysosomal β-glucosidase (EC 3.2.1.21)
14% inhibition at 1mM (pH 4) [143].
Human liver α-fucosidase (EC 3.2.1.51)
15% inhibition at 1mM (pH 5.5) [143].

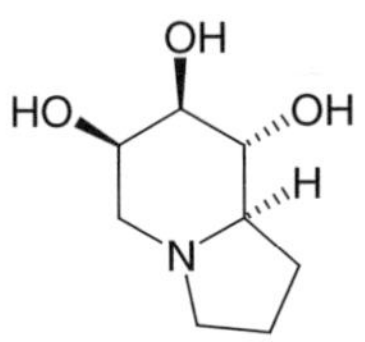

Potential anticancer agent
Human liver lysosomal α-glucosidase (EC 3.2.1.20)
7% inhibition at 1mM (pH 4) [143].
Human liver lysosomal β-glucosidase (EC 3.2.1.21)
12% inhibition at 1mM (pH 4) [143].
Human liver lysosomal α-mannosidase (EC 3.2.1.24)
75% inhibition at 1mM (pH 4); K_i 0.5 mM [143].
Human liver α-fucosidase (EC 3.2.1.51)
55% inhibition at 1mM (pH 5.5) [143].

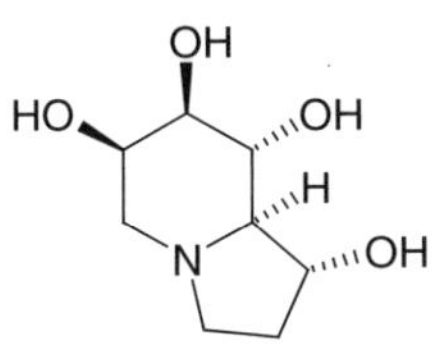

Potential anticancer agent
Human liver lysosomal α-glucosidase (EC 3.2.1.20)
3% inhibition at 1mM (pH 4) [143].
Human liver lysosomal β-glucosidase (EC 3.2.1.21)
9% inhibition at 1mM (pH 4) [143].
Human liver α-fucosidase (EC 3.2.1.51)
15% inhibition at 1mM (pH 5.5) [143].

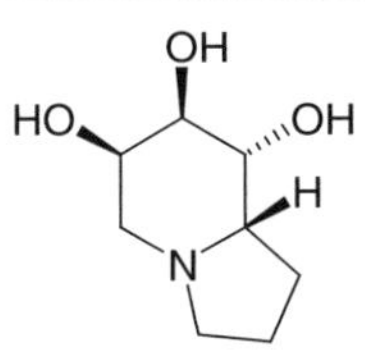

Potential anticancer agent
Human liver lysosomal α-glucosidase (EC 3.2.1.20)
20% inhibition at 1mM (pH 4) [143].
Human liver lysosomal β-glucosidase (EC 3.2.1.21)
44% inhibition at 1mM (pH 4) [143].
Human liver lysosomal α-mannosidase (EC 3.2.1.24)
22% inhibition at 1mM (pH 4) [143].
Human liver α-fucosidase (EC 3.2.1.51)
100% inhibition at 1mM (pH 5.5); K_i 1.3μM [143].

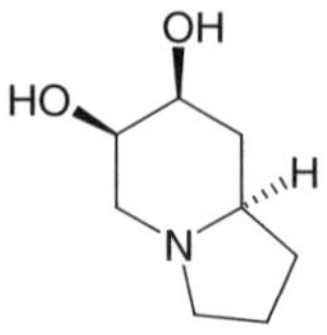

Potential anticancer agent
Human liver lysosomal β-glucosidase (EC 3.2.1.21)
29% inhibition at 1mM (pH 4) [143].
Human liver lysosomal α-mannosidase (EC 3.2.1.24)
10% inhibition at 1mM (pH 4) [143].

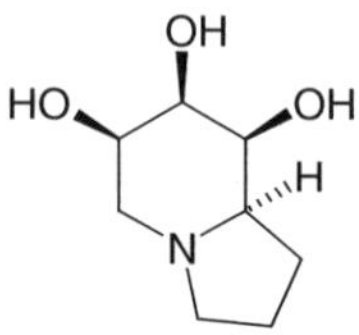

Potential anticancer agent
Human liver lysosomal α-glucosidase (EC 3.2.1.20)
6% inhibition at 1mM (pH 4) [143].
Human liver lysosomal β-glucosidase (EC 3.2.1.21)
9% inhibition at 1mM (pH 4) [143].

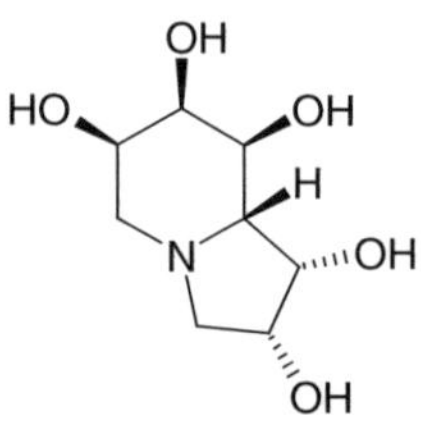

Antimetastatic
MDAY-D2 *α-mannosidase* (EC 3.2.1.24)
IC_{50} 10 μM [169].

Quinolizidine

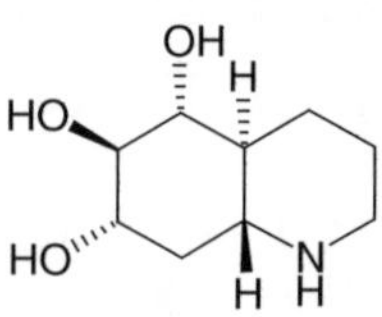

Potential anticancer agent
Bovine kidney α-fucosidase (EC 3.2.1.51)
26% inhibition at 1mM [144].

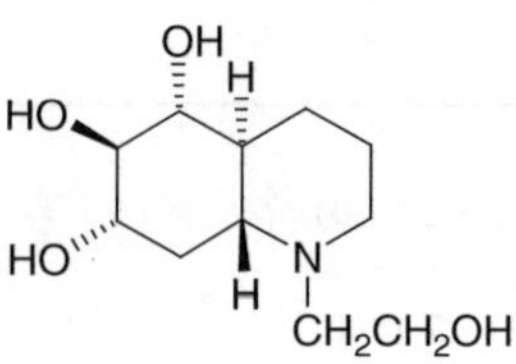

Potential anticancer agent
Bovine kidney α-fucosidase (EC 3.2.1.51)
39% inhibition at 1mM [144].

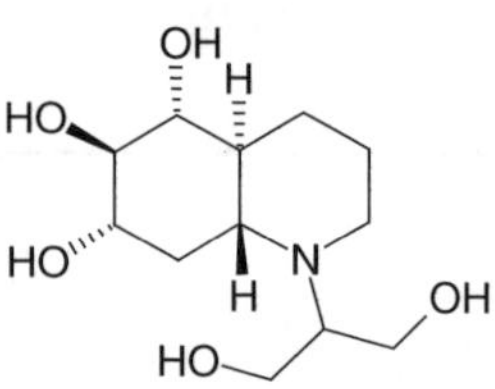

Potential anticancer agent
Bovine kidney α-fucosidase (EC 3.2.1.51)
K_i 70 μM [144].

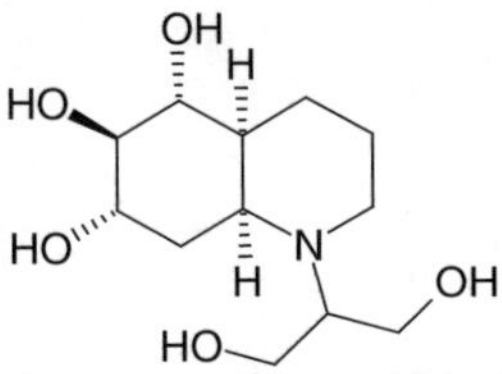

Potential anticancer agent
Bovine kidney α-fucosidase (EC 3.2.1.51)
K_i 18 μM [144].

Nagstatin and analogues

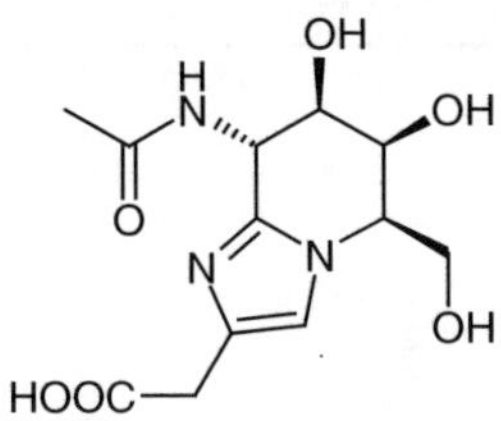

Potential anticancer agent
Bovine kidney N-acetyl-β-glucosaminidase (EC 3.2.1.30)
IC_{50} 13.4 nM [156].
Hog kidney N-acetyl-β-glucosaminidase (EC 3.2.1.30)
IC_{50} 0.0012 μg/ml [152].

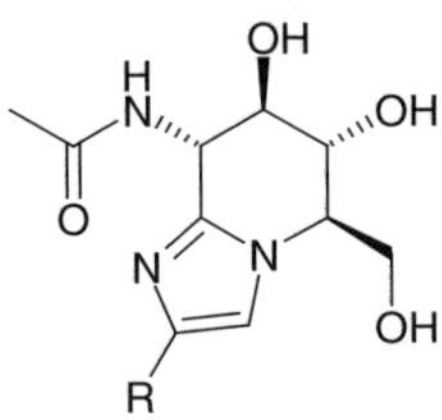

Potential anticancer agent
Snail β-mannosidase (EC 3.2.21.25)
R = CH_2CO_2Me K_i 20 μM [156].
R = CH_2COOH K_i 145 μM [156].
Bovine kidney N-acetyl-β-glucosaminidase (EC 3.2.1.30)
R = CH_2CO_2Me IC_{50} 4.5 nM; K_i 4 nM [156].
R = CH_2COOH IC_{50} 8.1 nM; K_i 5.2 nM [156].
R = H IC_{50} 7.0 nM [156]; IC_{50} 0.0017 μg/ml [153].

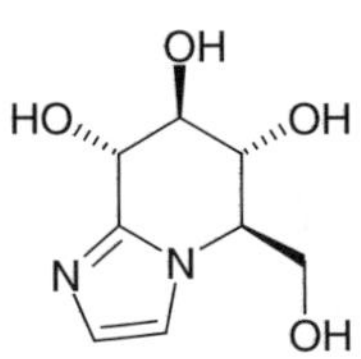

Potential anticancer agent
Bovine kidney N-acetyl-β-glucosaminidase (EC 3.2.1.30)
$IC_{50} >$ 100 μg/ml [153].
Jack beans α-D-mannosidase (EC 3.2.1.24)
IC_{50} 2.5 μg/ml [153].

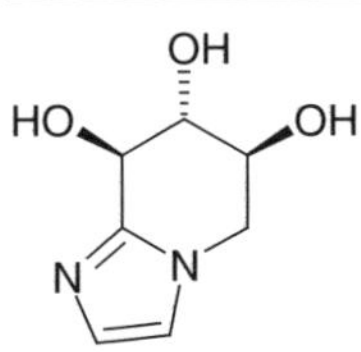

Potential anticancer agent
Jack beans α-D-mannosidase (EC 3.2.1.24)
K_i 280 μM [155].

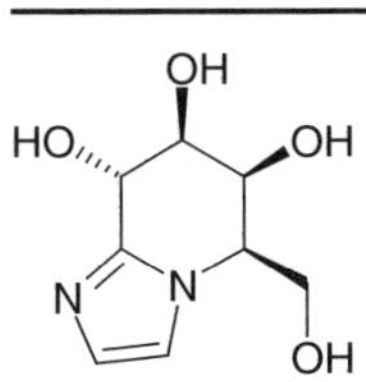

Potential anticancer agent
Bovine kidney N-acetyl-β-glucosaminidase (EC 3.2.1.30)
IC_{50} 12 μg/ml [153].

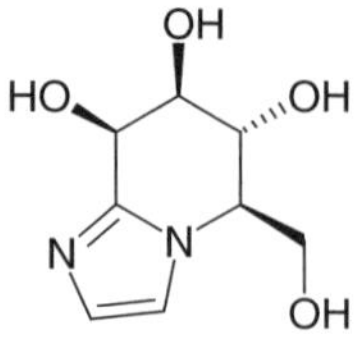

Potential anticancer agent
Jack beans α-mannosidase (EC 3.2.1.24)
IC_{50} 0.12 μg/ml [153].

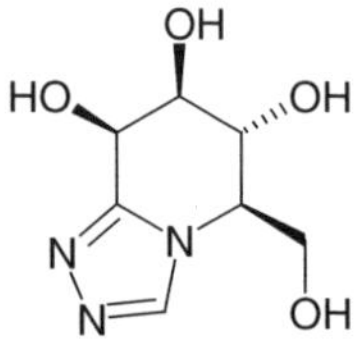

Potential anticancer agent
Jack beans α-mannosidase (EC 3.2.1.24)
IC_{50} 13 μg/ml [154].

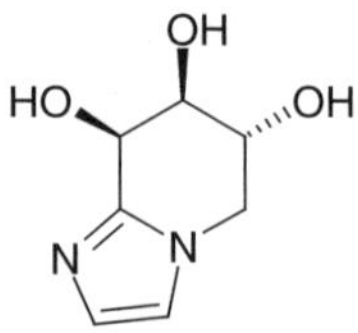

Potential anticancer agent
Jack beans α-mannosidase (EC 3.2.1.24)
K_i 360 μM [155].

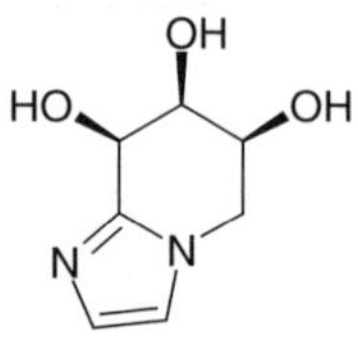

Potential anticancer agent
Jack beans α-mannosidase (EC 3.2.1.24)
K_i 285 μM [155].

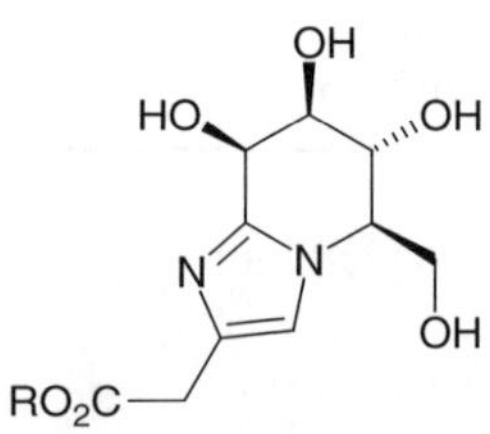

Potential anticancer agent
Snail β-mannosidase (EC 3.2.1.25)
R = Me K_i 117 nM [156].
R = H K_i 810 nM [156].

Other bicyclic derivatives

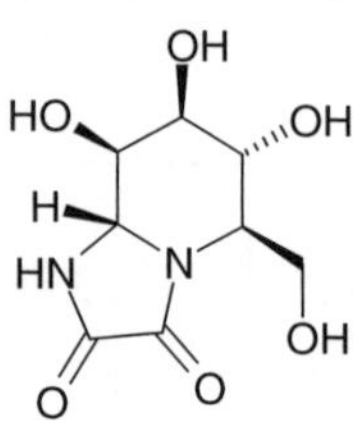

Potential anticancer agent
Jack beans α-mannosidase (EC 3.2.1.24)
IC_{50} 0.12 mM [158].
Mannosidase I (EC 3.2.1.24)
IC_{50} 20 nM [159].
Rat liver endoplasmic reticulum mannosidase I (EC 3.2.1.113)
100% inhibition at 5.0 μM [160].
Drosophilia golgi α-mannosidase II (EC 3.2.1.114)
K_i 5.2 mM [161].

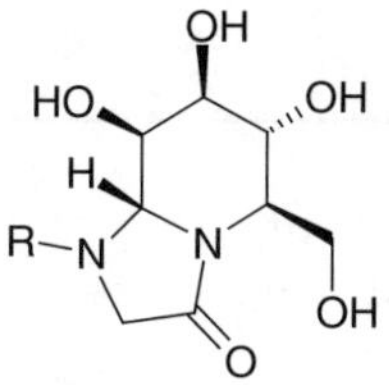

Potential anticancer agent
Human endoplasmic reticulum mannosidase I (EC 3.2.1.113)
R = H IC_{50} 0.03 mM [162].
R = Me IC_{50} 0.9 mM [162].

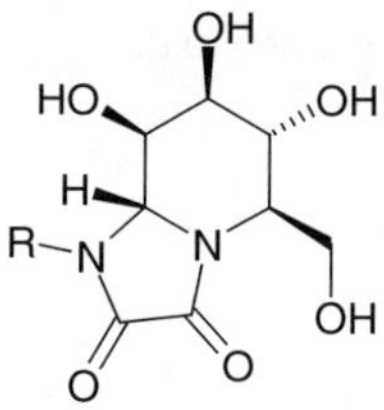

Potential anticancer agent
Human endoplasmic reticulum mannosidase I (EC 3.2.1.113)
R = Me IC_{50} 0.08 mM [162].
R = Cyclohexyl IC_{50} 2.6 mM [162].
R = $CH(CH_2OH)_2$ IC_{50} 0.2 mM [162].

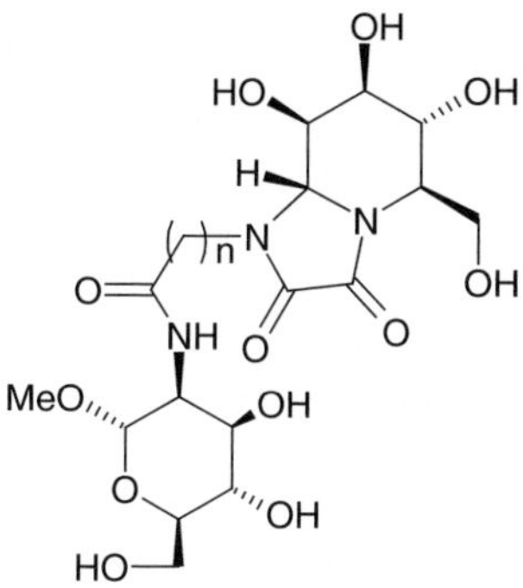

Potential anticancer agent
Human endoplasmic reticulum mannosidase I (EC 3.2.1.113)
$n = 1$ $IC_{50} > 5$ mM [162].
$n = 2$ IC_{50} 2.1 mM [162].
$n = 3$ IC_{50} 2.2 mM [162].

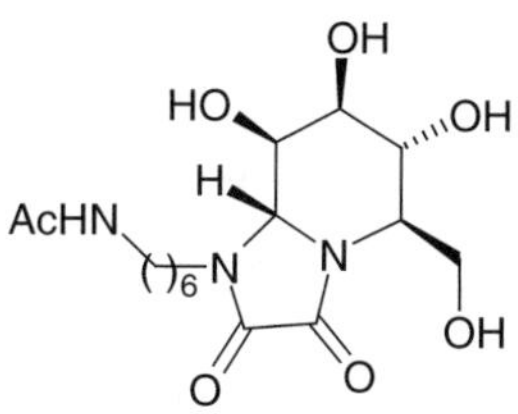

Potential anticancer agent
Human endoplasmic reticulum mannosidase I (EC 3.2.1.113)
IC_{50} 5.0 mM [162].

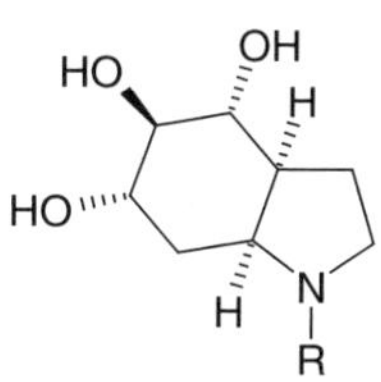

Potential anticancer agent
Jack beans α-mannosidase (EC 3.2.1.24)
R = CH_2CH_2OH 10% inhibition at 1 mM [144].
R = $CH(CH_2OH)_2$ 29% inhibition at 1 mM [144].
Bovine kidney α-fucosidase (EC 3.2.1.51)
R = CH_2CH_2OH 15% inhibition at 1 mM [144].
R = $CH(CH_2OH)_2$ 23% inhibition at 1 mM [144].
R = H 24% inhibition at 1 mM [144].

14.7.4 Seven-membered rings

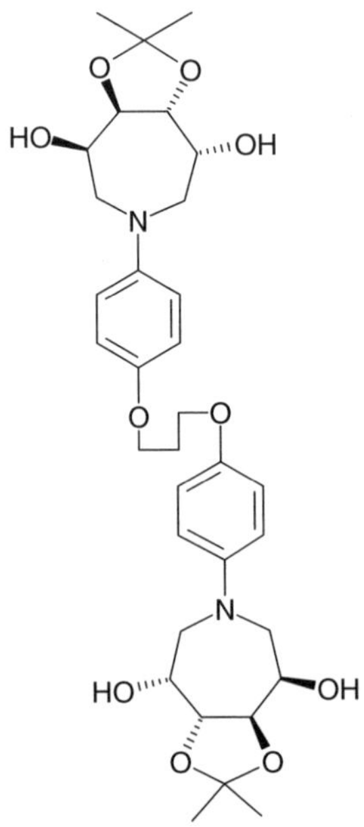

Human colon cancer
Growth inhibition of HCT116 cells: GI_{50} 32 μM [100].
Growth inhibition of HT29 cells: GI_{50} > 40 μM [100].

Human breast cancer
Growth inhibition of MCF7 cells: GI_{50} > 50 μM [100].
Growth inhibition of MCF7ADR cells: GI_{50} > 50 μM [100].

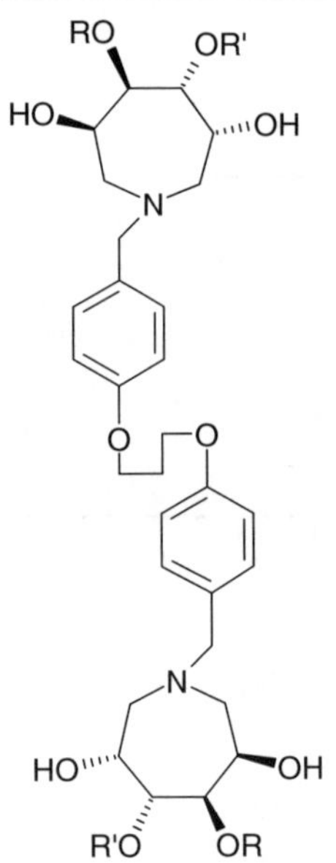

R, R' = -$C(CH_3)_2$
Human colon cancer
Growth inhibition of HCT116 cells: GI_{50} 6 μM [100].
Growth inhibition of HT29 cells: GI_{50} 23 μM [100].

Human breast cancer
Growth inhibition of MCF7 cells: GI_{50} 24 μM [100].
Growth inhibition of MCF7ADR cells: GI_{50} 10 μM [100].

R = R' = H
Human colon cancer
Growth inhibition of HCT116 cells: GI_{50} > 50 μM [100].
Growth inhibition of HT29 cells: GI_{50} 24 μM [100].

Human breast cancer
Growth inhibition of MCF7 cells: GI_{50} > 50 μM [100].
Growth inhibition of MCF7ADR cells: GI_{50} > 50 μM [100].

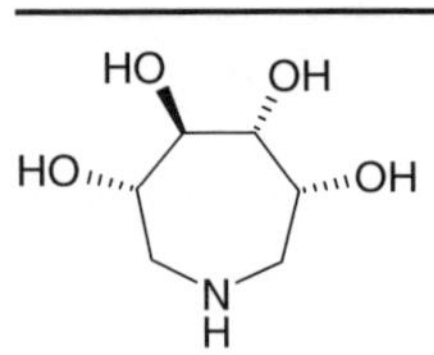

Potential anticancer agent
Jack beans α-mannosidase (EC 3.2.1.24)
31% inhibition at 120 μM; K_i 187 μM [149].
Bovine kidney α-fucosidase (EC 3.2.1.51)
95% inhibition at 120 μM; K_i 4.6 μM [149].

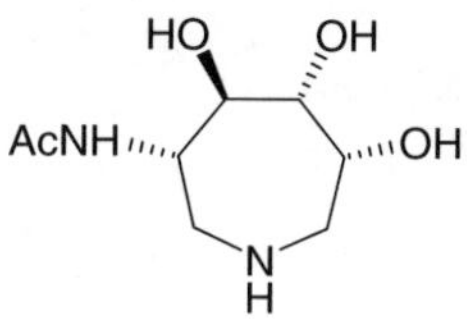	**Potential anticancer agent** *Jack beans β-N-acetylglucosaminidase* (EC 3.2.1.30) 22% inhibition at 240 μM [149]. *Jack beans α-mannosidase* (EC 3.2.1.24) 30% inhibition at 240 μM; K_i 364 μM [149]. *Bovine kidney α-fucosidase* (EC 3.2.1.51) 89% inhibition at 240 μM; K_i 10.6 μM [149].
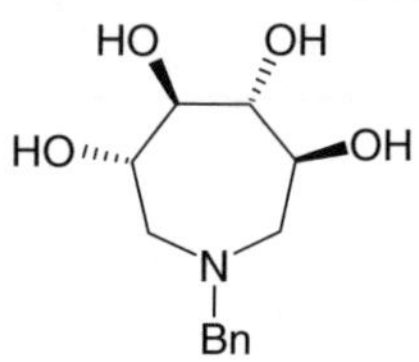	**Potential anticancer agent** *Jack beans α-mannosidase* (EC 3.2.1.24) 48% inhibition at 200 μM [149].
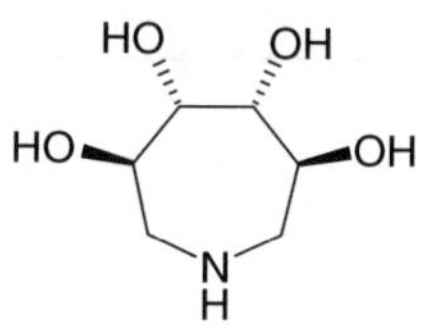	**Potential anticancer agent** *Jack beans β-N-acetylglucosaminidase* (EC 3.2.1.30) 35% inhibition at 160 μM [149]. *Jack beans α-mannosidase* (EC 3.2.1.24) 18% inhibition at 160 μM [149]. *Bovine kidney α-fucosidase* (EC 3.2.1.51) 57% inhibition at 160 μM [149].
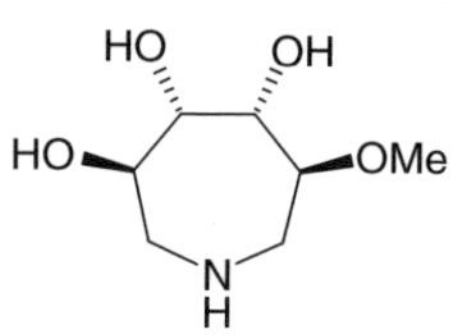	**Potential anticancer agent** *Jack beans β-N-acetylglucosaminidase* (EC 3.2.1.30) 41% inhibition at 224 μM; K_i 269 μM [149]. *Jack beans α-mannosidase* (EC 3.2.1.24) 14% inhibition at 224 μM [149]. *Bovine kidney α-fucosidase* (EC 3.2.1.51) 42% inhibition at 224 μM [149].
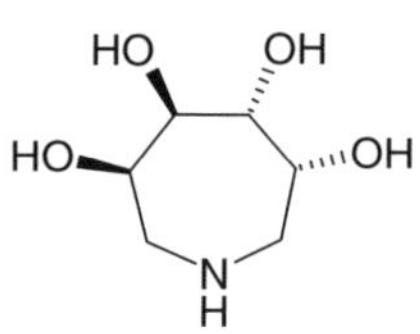	**Potential anticancer agent** *Jack beans β-N-acetylglucosaminidase* (EC 3.2.1.30) 94% inhibition at 240 μM; K_i 4.6 μM [149]; 28% inhibition at 1.67 mM [145]. *Bovine kidney α-fucosidase* (EC 3.2.1.51) 16% inhibition at 240 μM [149]; 97% inhibition at 1.67mM [145].
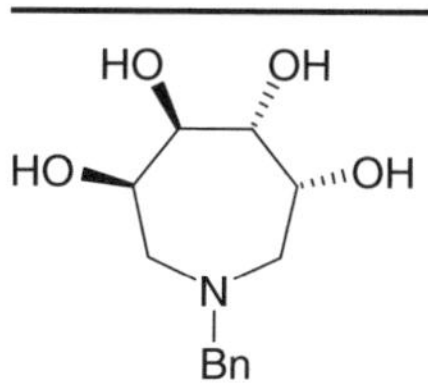	**Potential anticancer agent** *Jack beans β-N-acetylglucosaminidase* (EC 3.2.1.30) 6% inhibition at 160 μM [149]. *Jack beans α-mannosidase* (EC 3.2.1.24) 11% inhibition at 160 μM [149]. *Bovine kidney α-fucosidase* (EC 3.2.1.51) 88% inhibition at 160 μM; K_i 23.4 μM [149].

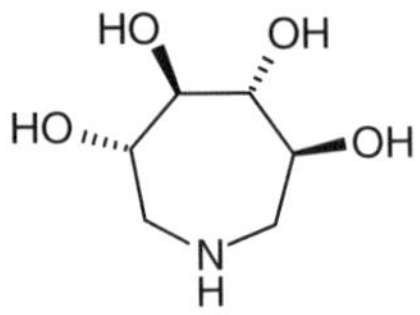	**Potential anticancer agent** *Jack beans β-N-acetylglucosaminidase* (EC 3.2.1.30) 89% inhibition at 200 μM; K_i 22.7 μM [149]. *Jack beans α-mannosidase* (EC 3.2.1.24) 81% inhibition at 200 μM; IC_{50} 25.7 μM [149]. 99% inhibition at 1.67 mM; IC_{50} 17 μM [145]. *Bovine kidney α-fucosidase* (EC 3.2.1.51) 44% inhibition at 200 μM [149]; 78% inhibition at 1.67 mM [145].
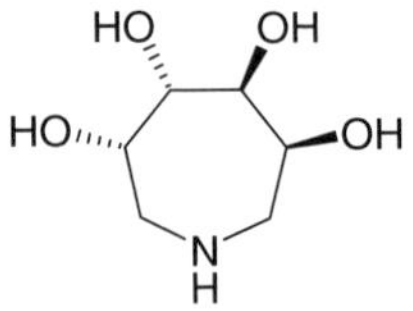	**Potential anticancer agent** *Bovine kidney α-L-fucosidase* (EC 3.2.1.51) 39% inhibition at 1.67 mM [145].
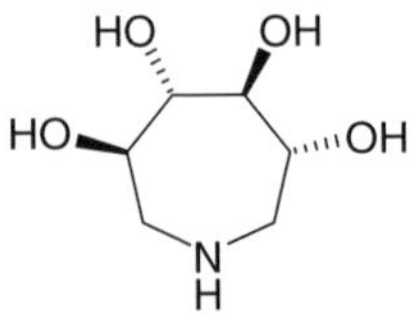	**Potential anticancer agent** *Jack beans α-mannosidase* (EC 3.2.1.24) IC_{50} 165 μM [145]. *Bovine kidney α-L-fucosidase* (EC 3.2.1.51) 15% inhibition at 1.67 mM [145].
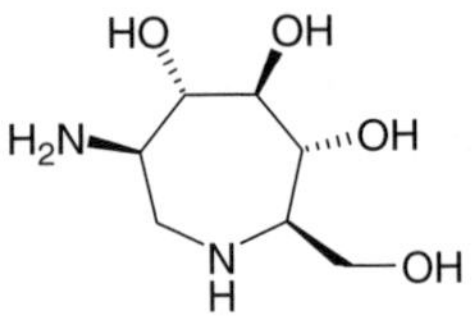	**Potential anticancer agent** *Bovine epididymis α-L-fucosidase* (EC 3.2.1.51) 76% inhibition at 1 mM [146]. *Jack beans α-mannosidase* (EC 3.2.1.24) 30% inhibition at 1 mM [146].
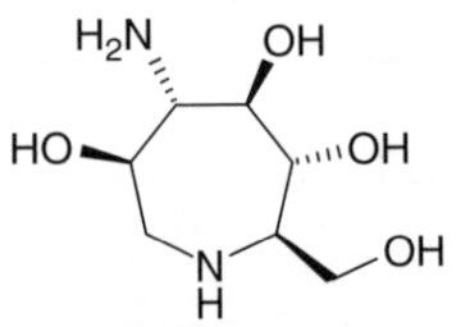	**Potential anticancer agent** *Bovine epididymis α-L-fucosidase* (EC 3.2.1.51) 42% inhibition at 1mM [146].
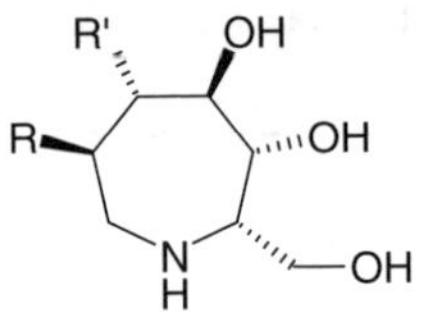	**Potential anticancer agent** *Bovine epididymis α-L-fucosidase* (EC 3.2.1.51) R = NH_2 R' = OH 48% inhibition at 1 mM [146]. R = OH R' = NH_2 57% inhibition at 1 mM [146].
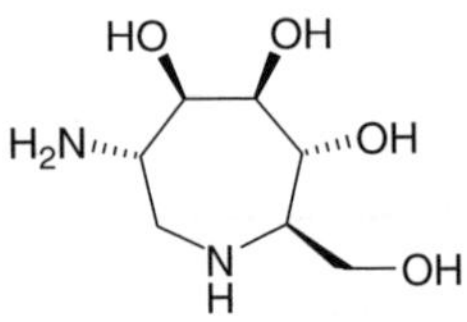	**Potential anticancer agent** *Bovine epididymis α-L-fucosidase* (EC 3.2.1.51) 82% inhibition at 1 mM; IC_{50} 100 μM [146]. *Jack beans α-mannosidase* (EC 3.2.1.24) 33% inhibition at 1 mM [146].

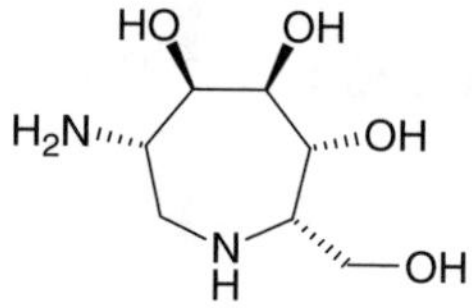

Potential anticancer agent
Bovine epididymis α-L-fucosidase (EC 3.2.1.51)
67% inhibition at 1 mM [146].

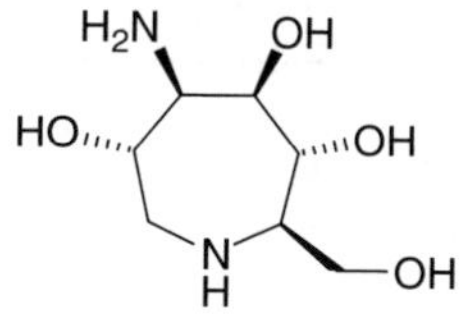

Potential anticancer agent
Bovine epididymis α-L-fucosidase (EC 3.2.1.51)
25% inhibition at 1 mM [146].

14.7.5 Nucleosides

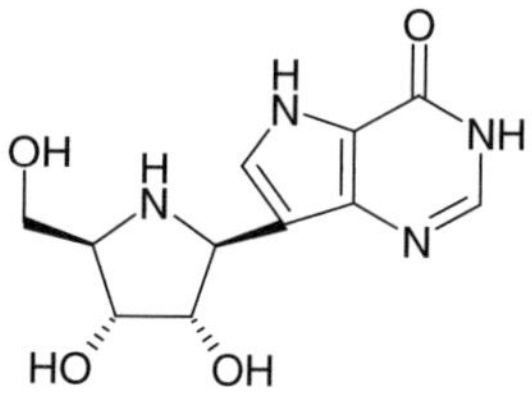

Human T-cell leukaemia
Purine nucleoside phosphorylase (EC 2.4.2.1)
IC_{50} 1.19 nM [175].
Inhibition of proliferation of CCRF-CEM cells: IC_{50} 5 nM [174].
Inhibition of proliferation of MOLT-4 cells: IC_{50} 5 nM [174].
Inhibition of proliferation of CEM-SS cells: IC_{50} 15 nM [176].

Toxic effects against cultured cell lines
Human A2780 and OVCAR-3 cells (ovarian cancer): $IC_{50} > 150\,\mu M$ [174].
Human MCF-7 and SKBR3 cells (breast cancer): $IC_{50} > 150\,\mu M$ [174].
Human HCT116, LS174t, MIP and CACO-2 (colon cancer): $IC_{50} > 150\,\mu M$ [174].
Human A549 and LX-1 cells (lung cancer): $IC_{50} > 150\,\mu M$ [174].
Human HL60 cells (leukaemia): $IC_{50} > 150\,\mu M$ [174].
Human HS27 cells (fibroblast): $IC_{50} > 150\,\mu M$ [174].

Peripheral blood lymphoma
IL-2-induced proliferation of human peripheral blood lymphoma (PBL)
IC_{50} greater than 100 μM [175].

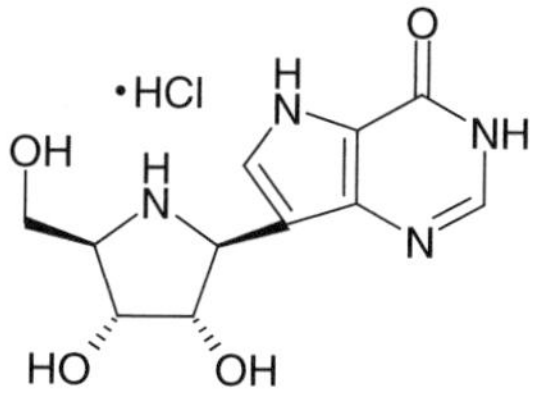

Human T-cell leukaemia
Human purine nucleoside phosphorylase (EC 2.4.2.1)
K_i 3.3 nM [108].

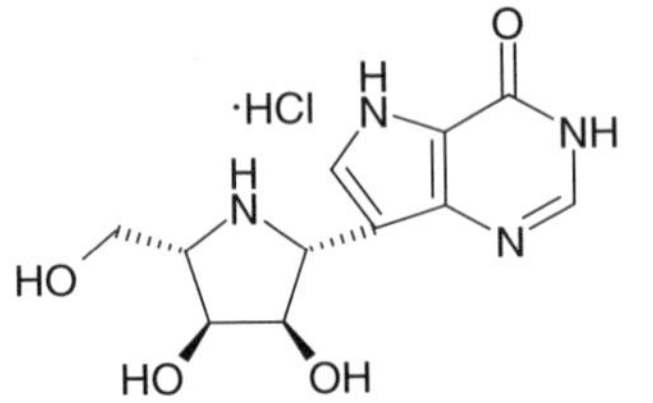

Human T-cell leukaemia
Human purine nucleoside phosphorylase (EC 2.4.2.1)
K_i 18 nM [108].

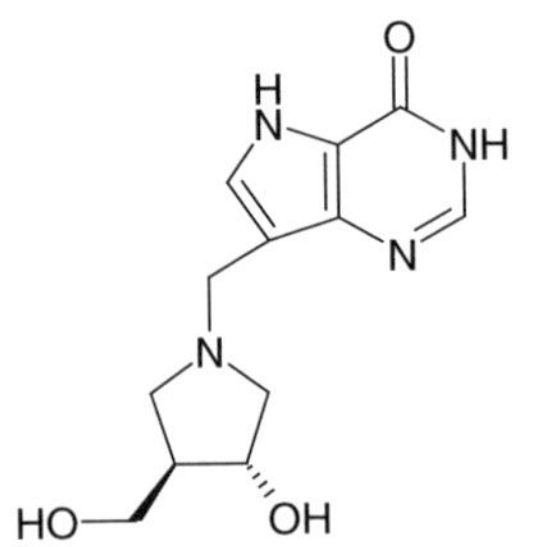

Human T-cell leukaemia
Human purine nucleoside phosphorylase (EC 2.4.2.1)
K_i 1.1 nM [108].

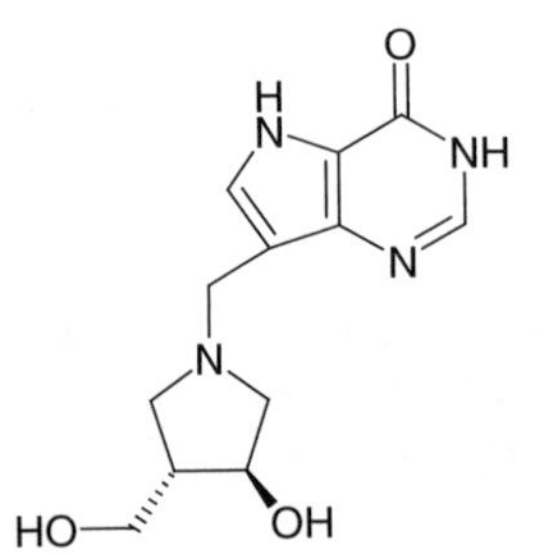

Human T-cell leukaemia
Human purine nucleoside phosphorylase (EC 2.4.2.1)
K_i 1.5 nM [108].

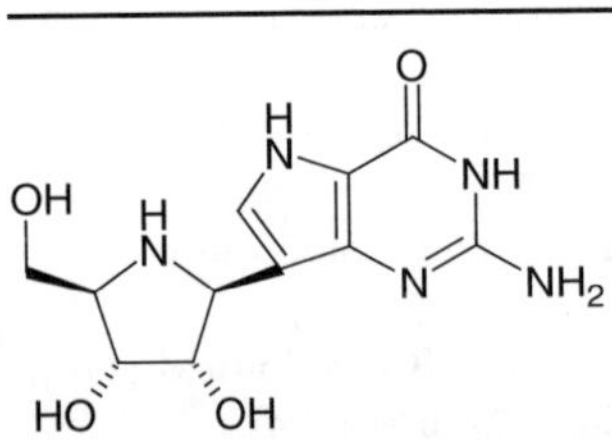

Human T-cell leukaemia
Human purine nucleoside phosphorylase (EC 2.4.2.1)
K_i 0.029 nM [177].

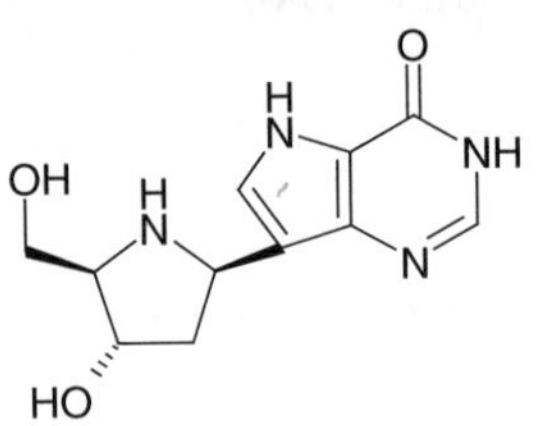

Human T-cell leukaemia
Human purine nucleoside phosphorylase (EC 2.4.2.1)
K_i 0.14 nM [109].
K_i 0.041 nM [104].

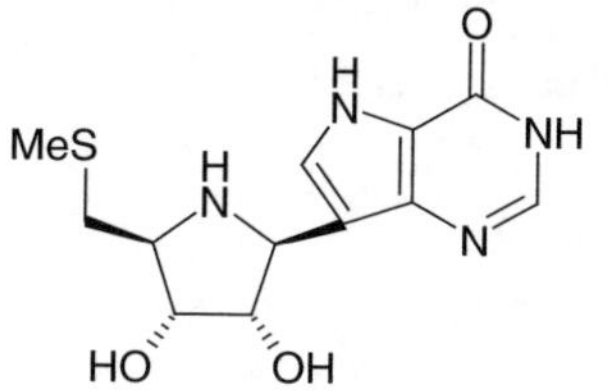

Human T-cell leukaemia
Human purine nucleoside phosphorylase (EC 2.4.2.1)
K_i 300 nM [109].

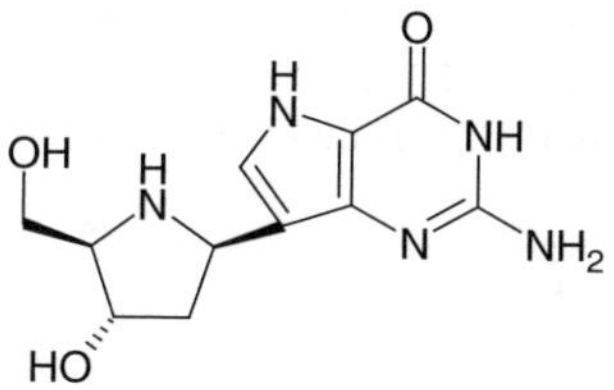

Human T-cell leukaemia
Human purine nucleoside phosphorylase (EC 2.4.2.1)
K_i 0.18 nM [109].
K_i 0.045 nM [105].
K_i* 0.045 nM [104].

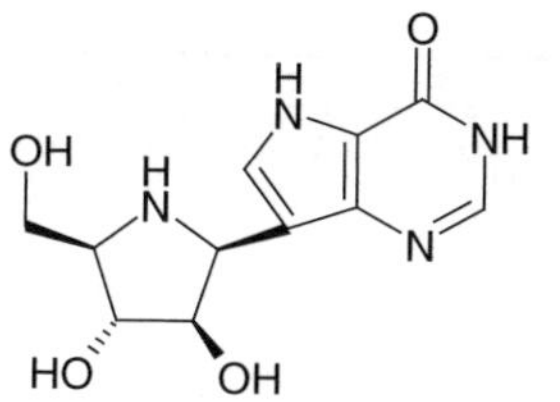

Human T-cell leukaemia
Human purine nucleoside phosphorylase (EC 2.4.2.1)
K_i 3.5 nM [109].
K_i 0.019 nM [105].

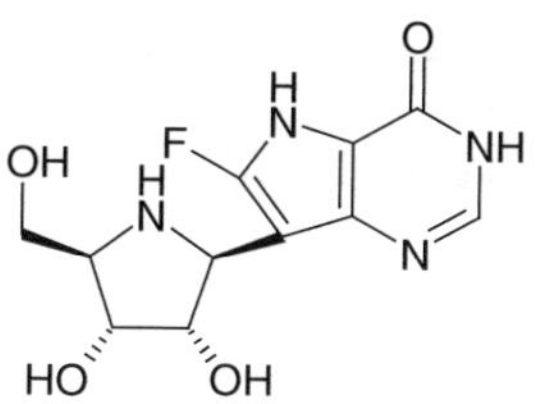

Human T-cell leukaemia
Human purine nucleoside phosphorylase (EC 2.4.2.1)
K_i* 0.39 nM [109].
K_i 0.0025 nM [109].

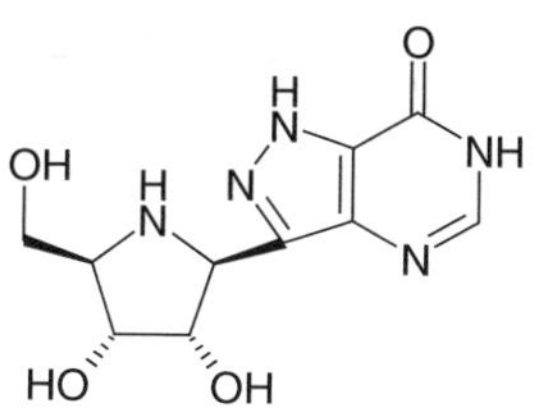

Human T-cell leukaemia
Human purine nucleoside phosphorylase (EC 2.4.2.1)
K_i* 0.18 nM [109].

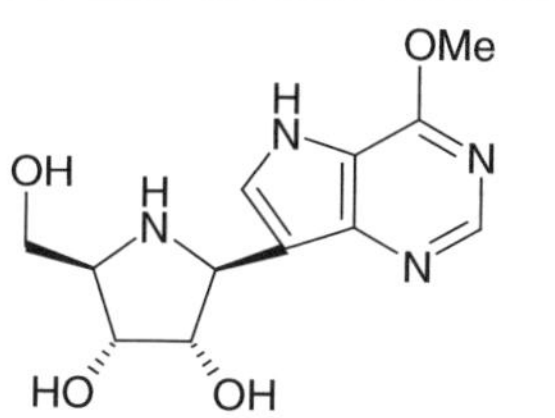

Human T-cell leukaemia
Human purine nucleoside phosphorylase (EC 2.4.2.1)
K_i 4.7 nM [109].

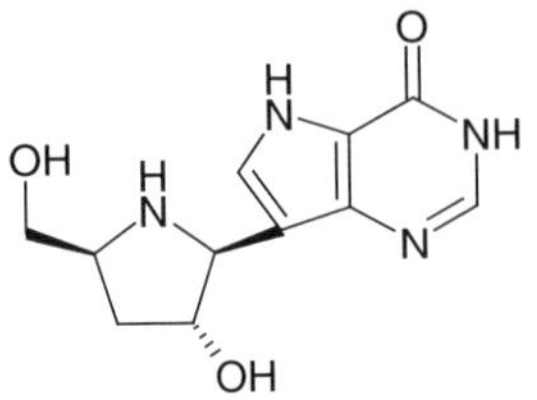	**Human T-cell leukaemia** *Human purine nucleoside phosphorylase* (EC 2.4.2.1) K_i 7.5 nM [109]. K_i* 0.0025 nM [105].
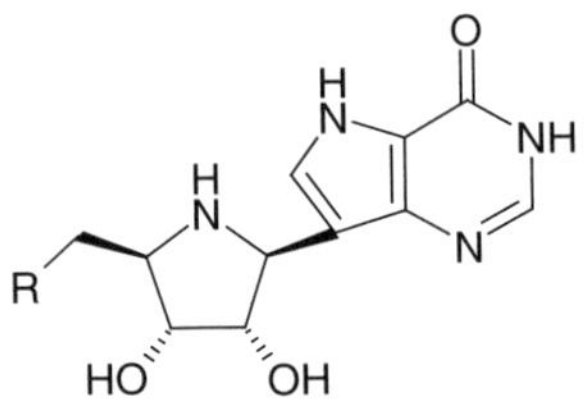	**Human T-cell leukaemia** *Human purine nucleoside phosphorylase* (EC 2.4.2.1) R = H K_i 25 nM [109]; K_i 9.4 nM [104]; K_i* 0.0094 nM [105]. **Potential anticancer agent** *5'methylthioadenosine phosphorylase* (EC 2.4.2.28) R = H K_i 720 nM [178]. R = OMe K_i 134 nM [178]. R = Ethyl K_i 44 nM [178].
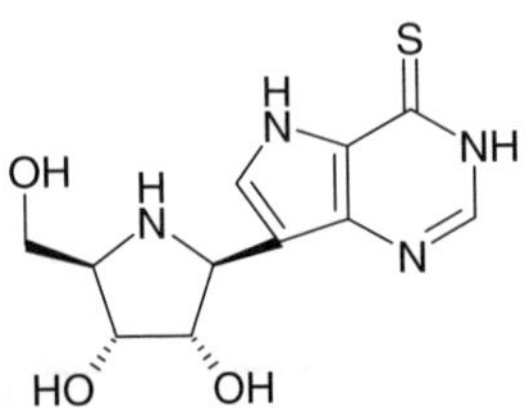	**Human T-cell leukaemia** *Human purine nucleoside phosphorylase* (EC 2.4.2.1) K_i 25 nM [109].
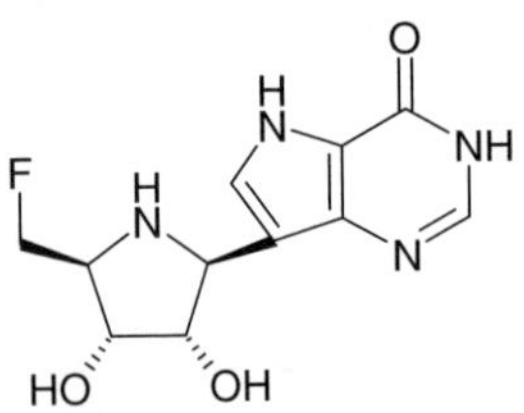	**Human T-cell leukaemia** *Human purine nucleoside phosphorylase* (EC 2.4.2.1) K_i* 6.8 nM [109]. K_i 81 nM [109].
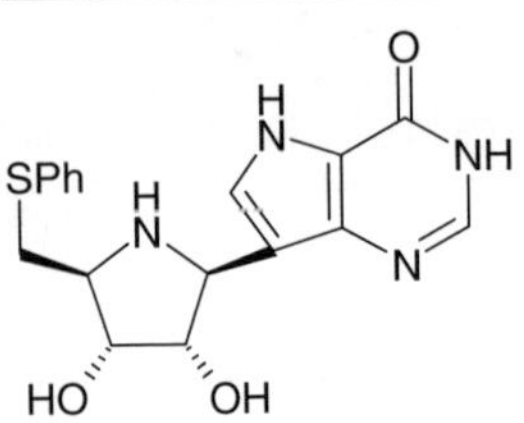	**Human T-cell leukaemia** *Human purine nucleoside phosphorylase* (EC 2.4.2.1) K_i 250 nM [109]. **Potential anticancer agent** *5'methylthioadenosine phosphorylase* (EC 2.4.2.28) K_i 3.6 nM [178].
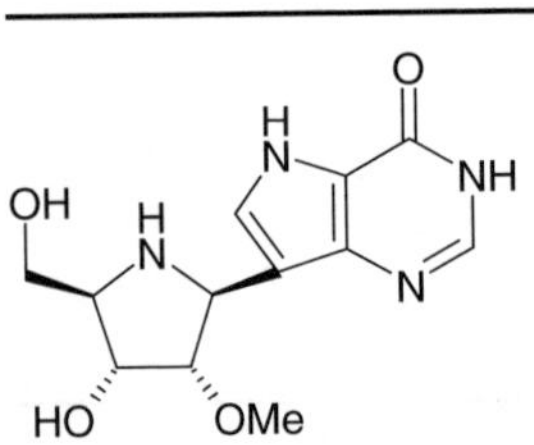	**Human T-cell leukaemia** *Human purine nucleoside phosphorylase* (EC 2.4.2.1) K_i 6 nM [109].

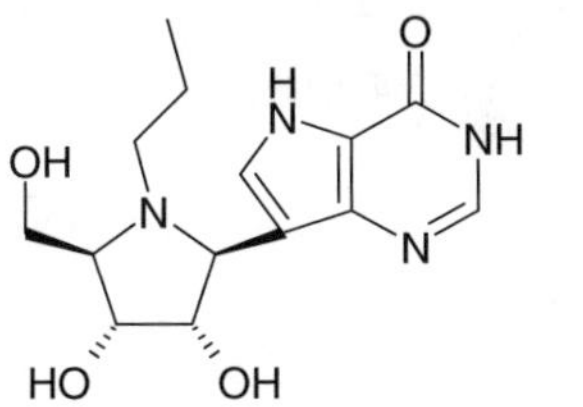

Human T-cell leukaemia
Human purine nucleoside phosphorylase (EC 2.4.2.1)
K_i 410 nM [109].

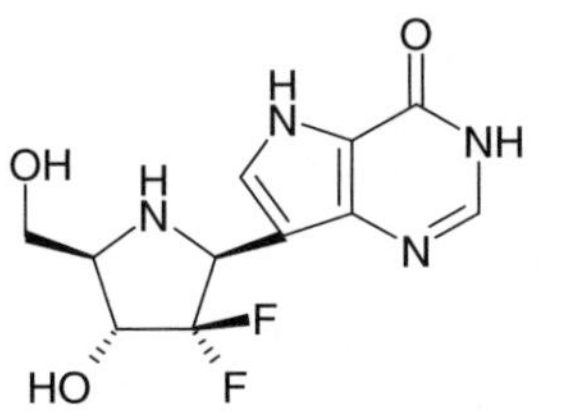

Human T-cell leukaemia
Human purine nucleoside phosphorylase (EC 2.4.2.1)
K_i 1.4 nM [109].

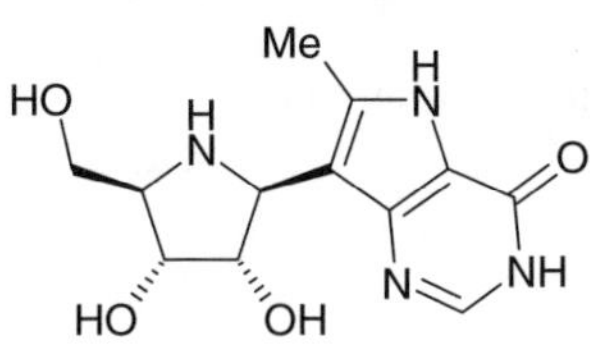

Human T-cell leukaemia
Human purine nucleoside phosphorylase (EC 2.4.2.1)
K_i 20 nM [109].
K_i* 14 nM [105].

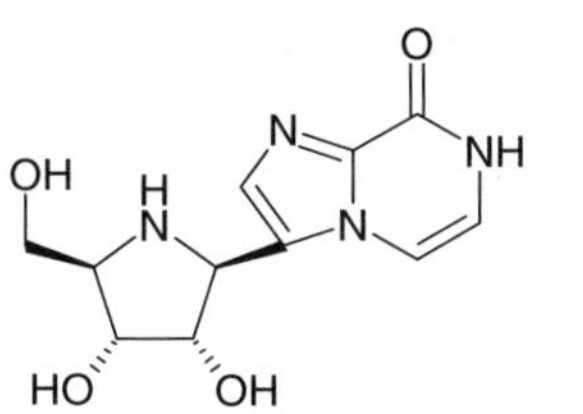

Human T-cell leukaemia
Human purine nucleoside phosphorylase (EC 2.4.2.1)
K_i 1.9 μM [109].

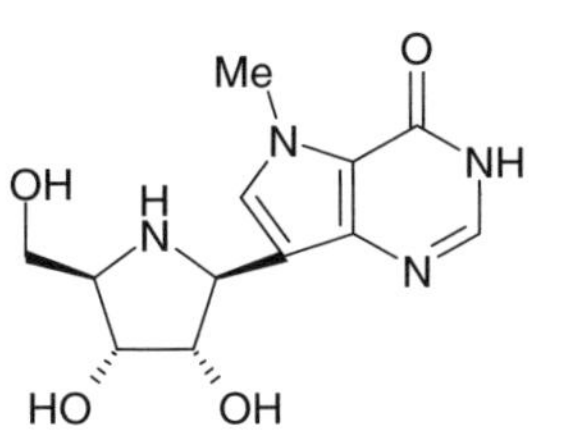

Human T-cell leukaemia
Human purine nucleoside phosphorylase (EC 2.4.2.1)
K_i 100 nM [109].

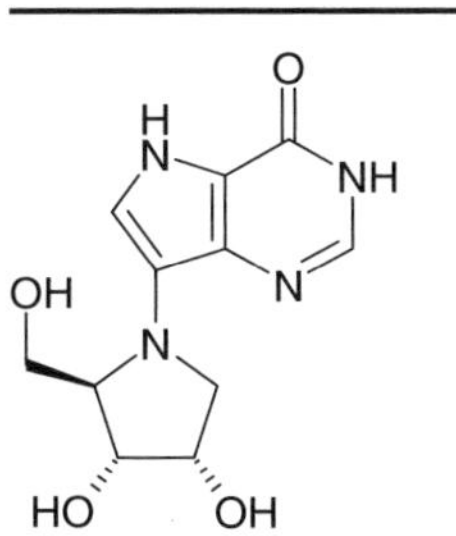

Human T-cell leukaemia
Human purine nucleoside phosphorylase (EC 2.4.2.1)
K_i > 38 μM [109].

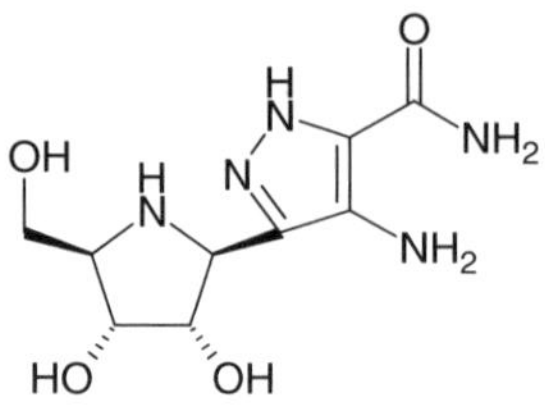

Human T-cell leukaemia
Human purine nucleoside phosphorylase (EC 2.4.2.1)
K_i* 96 pM [105].

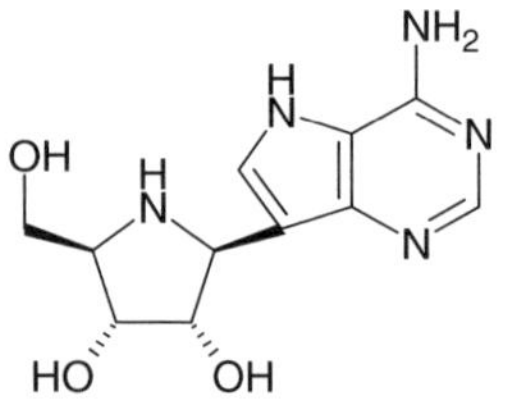

Human T-cell leukaemia
Human purine nucleoside phosphorylase (EC 2.4.2.1)
$K_i > 10\,\mu$M [104].
K_i* 33 nM [105].

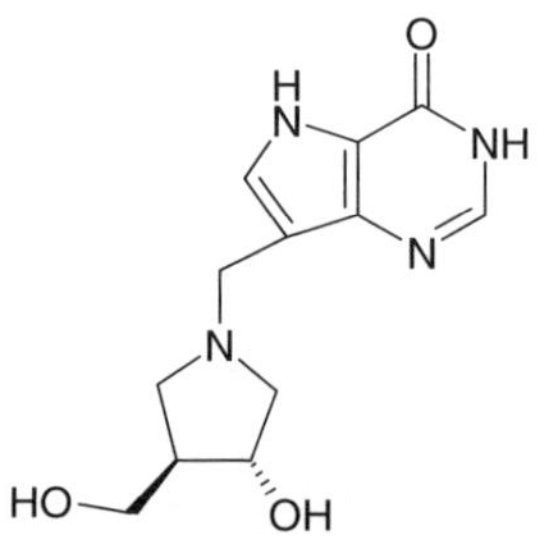

Human T-cell leukaemia
Human purine nucleoside phosphorylase (EC 2.4.2.1)
K_i 1.1 nM [109].
K_i* 16 pM [109].

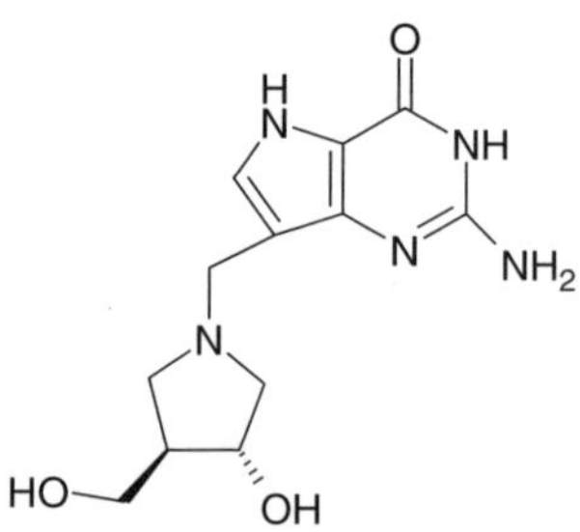

Human T-cell leukaemia
Human purine nucleoside phosphorylase (EC 2.4.2.1)
K_i 0.16 nM [109].
K_i* 0.007 nM [109].

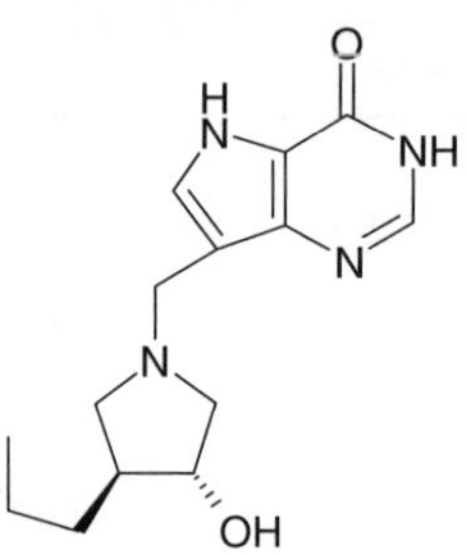

Human T-cell leukaemia
Human purine nucleoside phosphorylase (EC 2.4.2.1)
K_i 1.48 nM [109].
K_i* 0.36 nM [109].

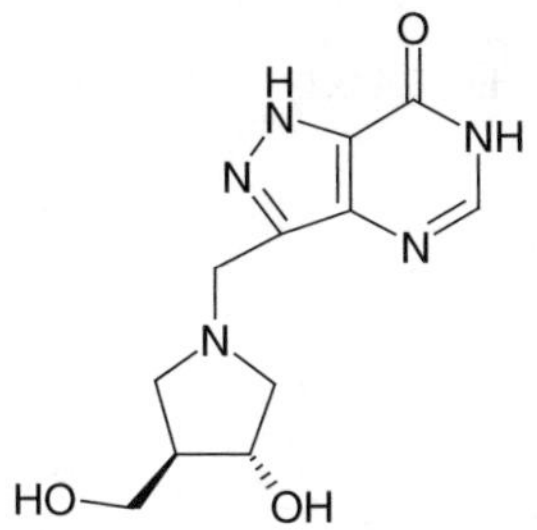

Human T-cell leukaemia
Human purine nucleoside phosphorylase (EC 2.4.2.1)
K_i 2.0 nM [109].

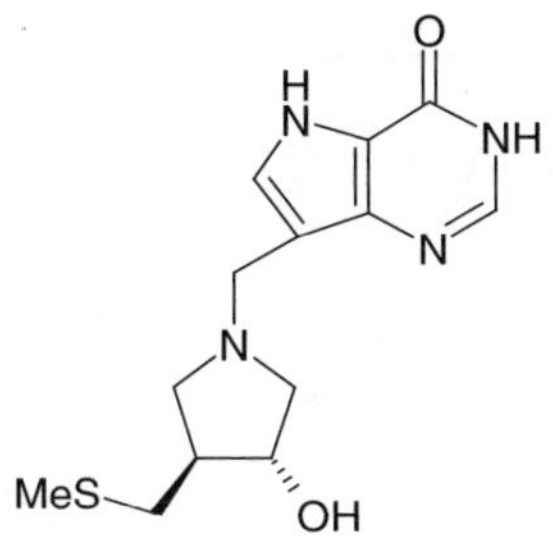

Human T-cell leukaemia
Human purine nucleoside phosphorylase (EC 2.4.2.1)
K_i 0.30 nM [109].
K_i^* 0.07 nM [109].

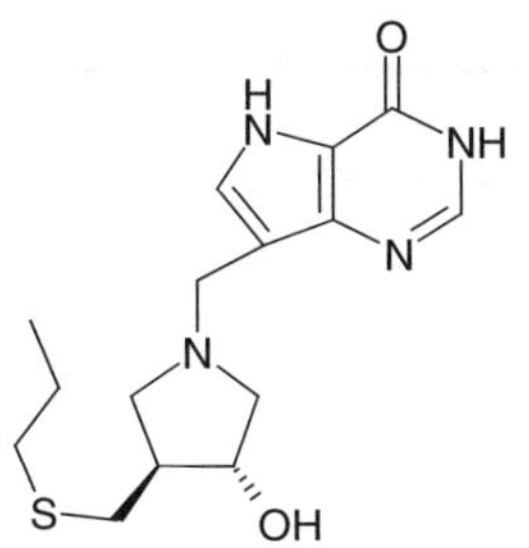

Human T-cell leukaemia
Human purine nucleoside phosphorylase (EC 2.4.2.1)
K_i 0.39 nM [109].
K_i^* 0.041 nM [109].

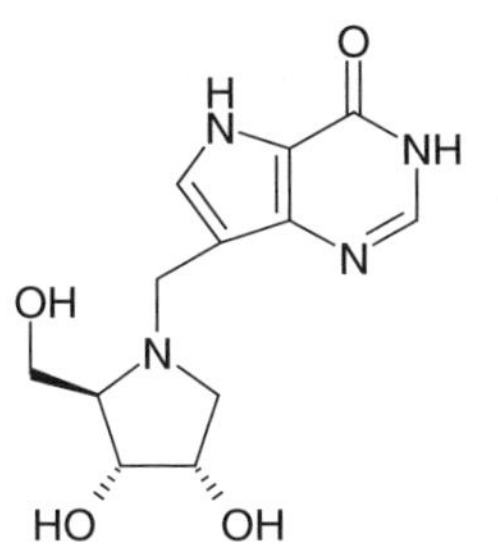

Human T-cell leukaemia
Human purine nucleoside phosphorylase (EC 2.4.2.1)
K_i 2.8 nM [109].

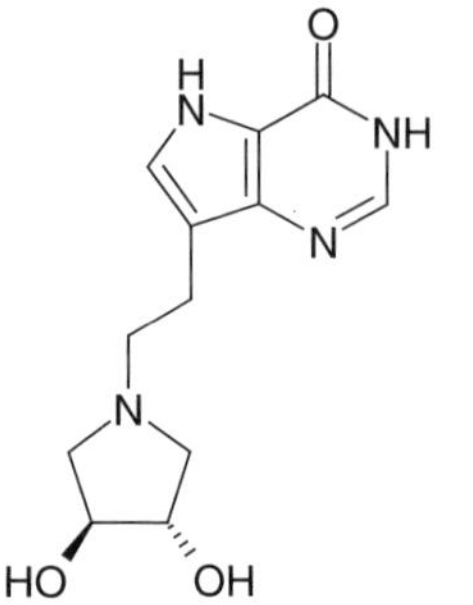

Human T-cell leukaemia
Human purine nucleoside phosphorylase (EC 2.4.2.1)
K_i 1.6 nM [109].
K_i* 0.46 nM [109].

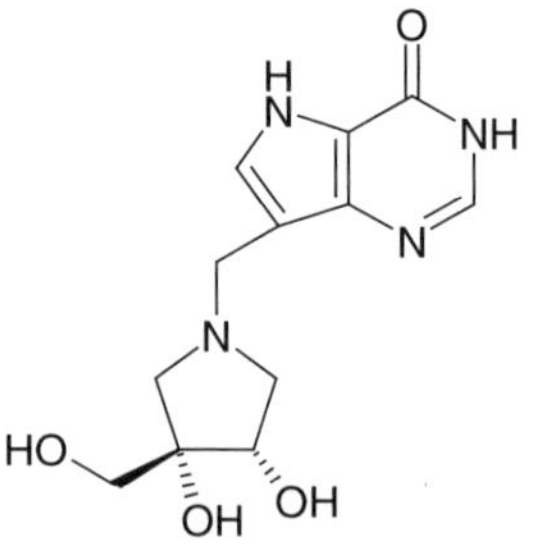

Human T-cell leukaemia
Human purine nucleoside phosphorylase (EC 2.4.2.1)
K_i 14 nM [109].

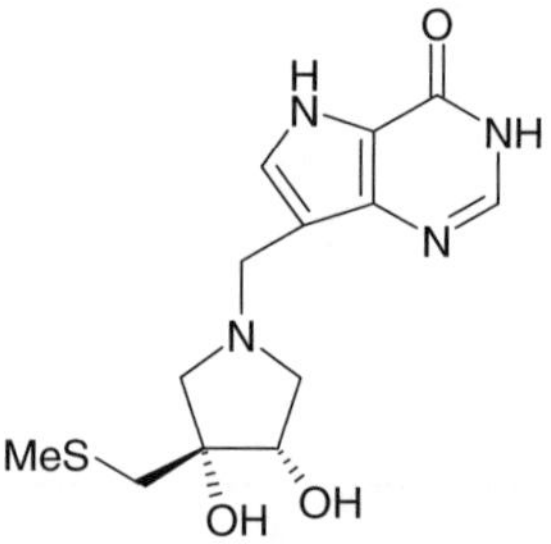

Human T-cell leukaemia
Human purine nucleoside phosphorylase (EC 2.4.2.1)
K_i 430 nM [109].

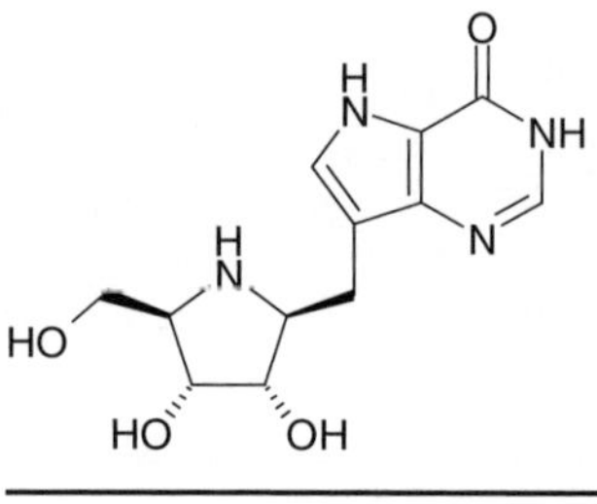

Human T-cell leukaemia
Human purine nucleoside phosphorylase (EC 2.4.2.1)
K_i 250 nM [109].

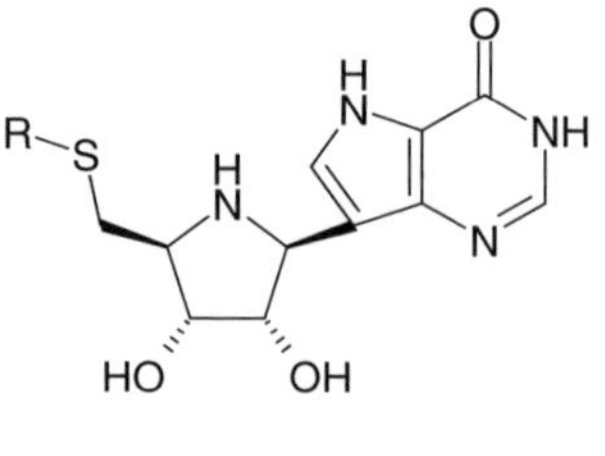

Potential anticancer agent
5'methylthioadenosine phosphorylase (EC 2.4.2.28)
R = Et K_i 9.2 nM [178].
R = *n*-propyl K_i 4 nM [178].
R = 2-hydroxyethyl K_i 56 nM [178].
R = 4-fluorophenyl K_i 6.4 nM [178].
R = 4-chlorophenyl K_i 0.576 nM [178].
R = 4-methylphenyl K_i 4.4 nM [178].

(Continued)
R = 3-chlorophenyl K_i 6.4 nM [178].
R = 3-methylphenyl K_i 1.39 nM [178].
R = benzyl K_i 26 nM [178].
R = 1-naphthyl K_i 90 nM [178].
R = 2-fluoroethyl K_i 20 nM [178].

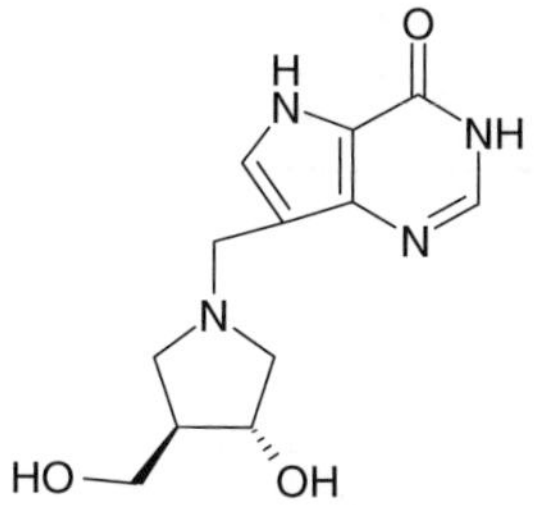	**Human T-cell leukaemia** *Human purine nucleoside phosphorylase* (EC 2.4.2.1) K_i 6.6 μM [109].
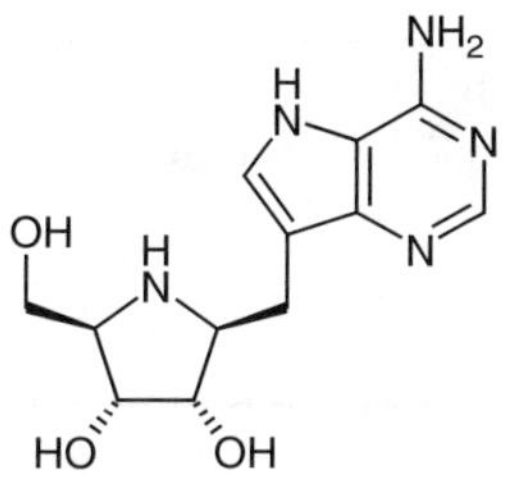	**Human T-cell leukaemia** *Human purine nucleoside phosphorylase* (EC 2.4.2.1) K_i 440 nM [109].
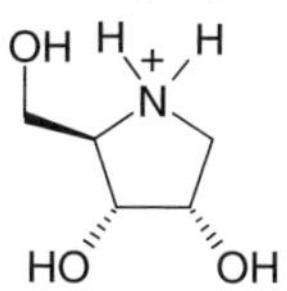	**Nucleoside hydrolase from *Crithidia fasciculata*** K_i 10 μM [119].
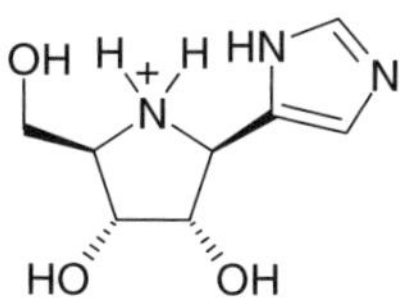	**Nucleoside hydrolase from *Crithidia fasciculata*** K_i 3 μM [119].
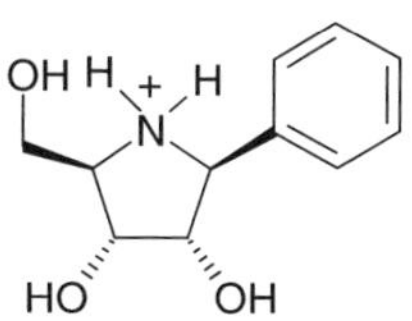	**Nucleoside hydrolase from *Crithidia fasciculata*** K_i 0.03 μM [119].

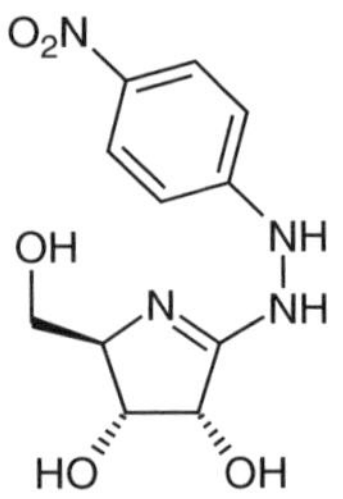	**Nucleoside hydrolase from** ***Crithidia fasciculata*** K_i 2 nM [179].

14.8 Lysosomal diseases

14.8.1 Five-membered rings

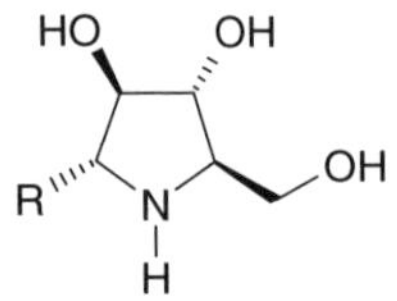	**Gaucher disease** *β-Glucocerebrosidase* (EC 3.2.1.45) – Chemical Chaperone Therapy R = H IC_{50} 160 μM [10]. R = CH_2OH IC_{50} 340 μM [10].
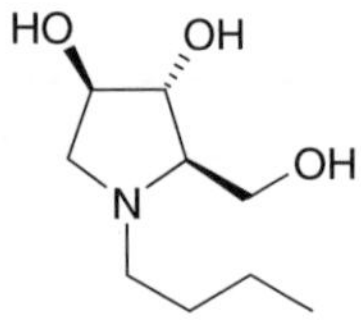	**Gaucher disease** *Glucosylceramide synthase* (EC 2.4.1.80) – Substrate Reduction Therapy 21% inhibition at 200 μM [180].
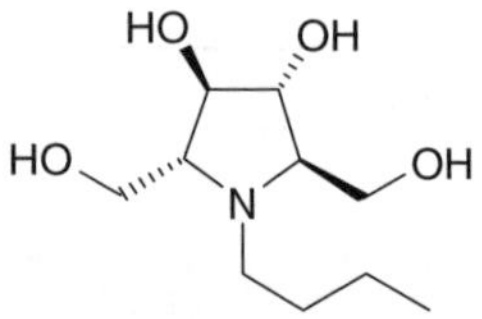	**Gaucher disease** *Glucosylceramide synthase* (EC 2.4.1.80) – Substrate Reduction Therapy 86% inhibition at 200 μM [180].
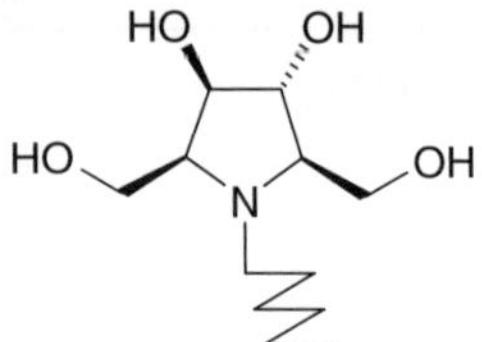	**Gaucher disease** *β-Glucocerebrosidase* (EC 3.2.1.45) – Chemical Chaperone Therapy 1.25-fold increase in activity at 0.01-100 μM of β-glucocerebrosidase (wild-type) [181]. 1.55-fold increase in activity at 15-30 μM of N370S β-glucocerebrosidase (type 1) [181]. 3.2-fold increase in activity at 150 μM of G202R β-glucocerebrosidase [181]. No significant increase in activity at 0.01-100 μM of L444P β-glucocerebrosidase (type 2) [181].

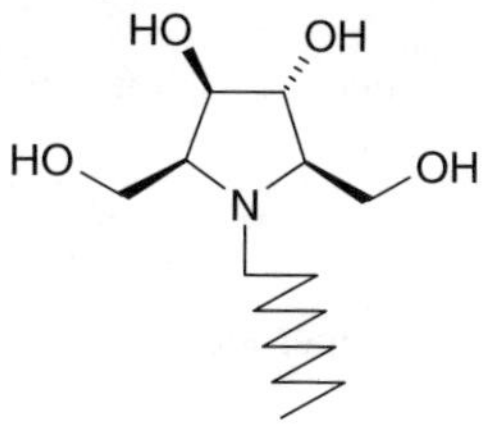	**Gaucher disease** *β-Glucocerebrosidase* (EC 3.2.1.45) – Chemical Chaperone Therapy IC_{50} 100 μM (wild-type); IC_{50} 330 μM (N370S mutation) [182]. 1.10-fold increase at 20 μM in activity of N370S β-glucocerebrosidase (type 1) [182].
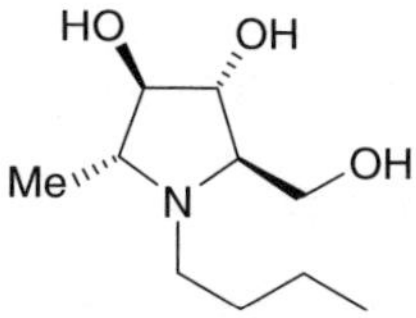	**Gaucher disease** *Glucosylceramide synthase* (EC 2.4.1.80) – Substrate Reduction Therapy 11% inhibition at 200 μM [180].
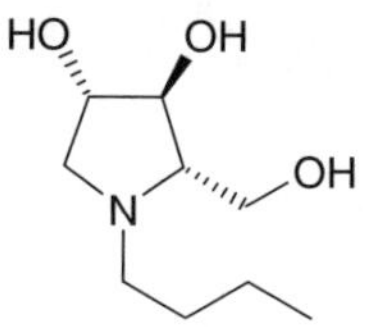	**Gaucher disease** *Glucosylceramide synthase* (EC 2.4.1.80) – Substrate Reduction Therapy 25% inhibition at 200 μM [180].
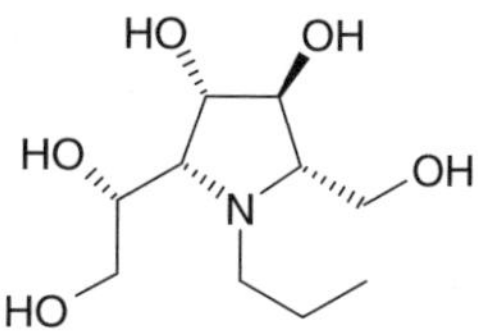	**Gaucher disease** *Glucosylceramide synthase* (EC 2.4.1.80) – Substrate Reduction Therapy 6% inhibition at 200 μM [180].
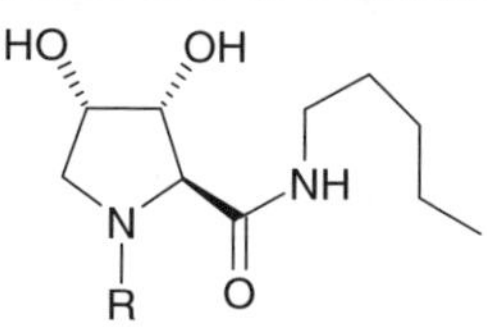	**Gaucher disease** *Glucosylceramide synthase* (EC 2.4.1.80) – Substrate Reduction Therapy R = H IC_{50} 140 μM [46]. R = *n*-Bu IC_{50} 117 μM [46].

14.8.2 Six-membered rings

1-deoxynojirimycin and analogues

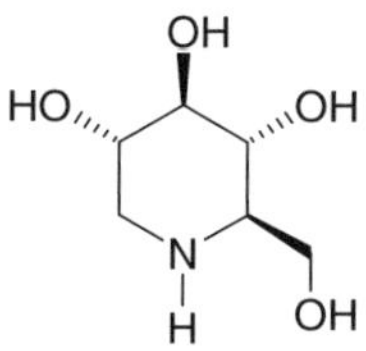	**Gaucher disease** *Glucosylceramide synthase* (EC 2.4.1.80) – Substrate Reduction Therapy no inhibition at 2 mM [180], [183]. *β-Glucocerebrosidase* (EC 3.2.1.45) – Chemical Chaperone Therapy IC_{50} 240 μM [184], [185]; IC_{50} 506 μM [186]; K_i 79 μM [184], [185].

(Continued)

Fabry disease
Lysosomal α-galactosidase A (EC 3.2.1.22) – Chemical Chaperone Therapy
IC_{50} 830 μM [187].

Tay–Sachs and Sandhoff diseases
β-Hexosaminidase (EC 3.2.1.52) – Chemical Chaperone Therapy
No increase in activity [188].

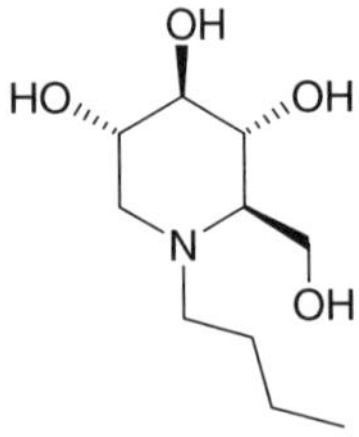

Gaucher disease
Glucosylceramide synthase (EC 2.4.1.80) – Substrate Reduction Therapy [189,190].
87% inhibition at 200 μM [180]; IC_{50} 20.4 μM [183]; IC_{50} 34.4 μM [194]; K_i 7 μM [180].
β-glucocerebrosidase (EC 3.2.1.45) – Chemical Chaperone Therapy
IC_{50} 520 μM [184]; IC_{50} 270 μM [185]; IC_{50} 912 μM [186]; K_i 116 μM [184], [185].
2.1-fold increase in activity at 10 μM of wild type β-glucocerebrosidase [191].
1.2-fold increase in activity at 0.5 μM of β-glucocerebrosidase [186].
2.3-fold increase in activity at 10 μM of N370S β-glucocerebrosidase (type 1) [191].
1.3-fold increase in activity at 10 μM of S364R β-glucocerebrosidase [191].
3.6-fold increase in activity at 10 μM of V15M β-glucocerebrosidase [191].
9.9-fold increase in activity at 10 μM of M123T β-glucocerebrosidase [191].
Small decrease in the rare P266L [191].
No increase in activity at 10 μM of L444P (type 2), L336P, S465del β-glucocerebrosidase [191].
No increase in activity at 0.01-100 μM of wild-type, N370S, L444P and G202R β-glucocerebrosidase [181].

Niemann–Pick disease
Glucosylceramide synthase (EC 2.4.1.80) – Substrate Reduction Therapy [192]
K_i 7 μM [180].

Tay–Sachs and Sandhoff diseases
Glucosylceramide synthase (EC 2.4.1.80) – Substrate Reduction Therapy [193].
K_i 7 μM [180].

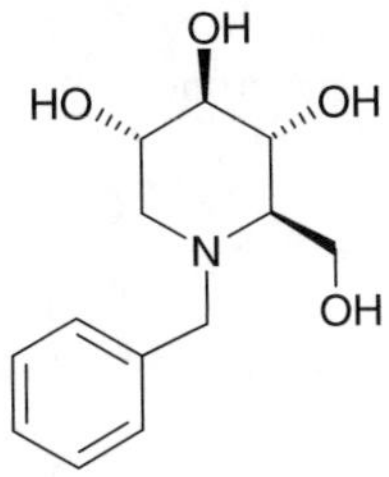

Gaucher disease
Glucosylceramide synthase (EC 2.4.1.80) – Substrate Reduction Therapy
73% inhibition at 200 μM [180].

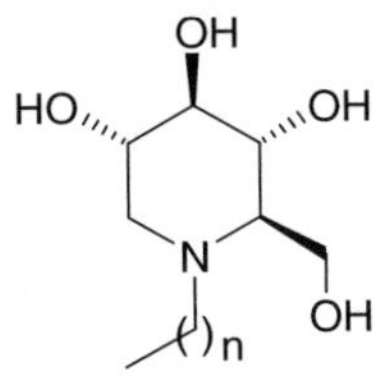

Gaucher disease
β-Glucocerebrosidase (EC 3.2.1.45) – Chemical Chaperone Therapy
$n = 0$ IC_{50} 150 μM [185].
$n = 2$ IC_{50} 700 μM [185]; IC_{50} 3546 μM [186].

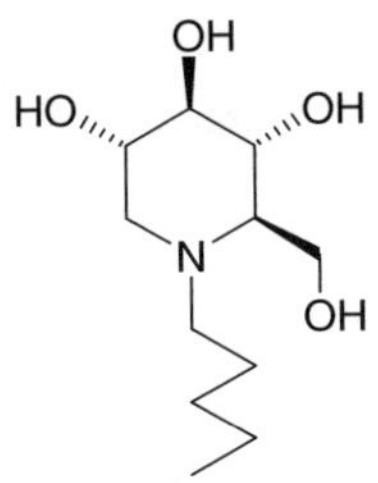

Gaucher disease
Glucosylceramide synthase (EC 2.4.1.80) – Substrate Reduction Therapy
IC_{50} 26.8 μM [194]; IC_{50} 249 μM [186].

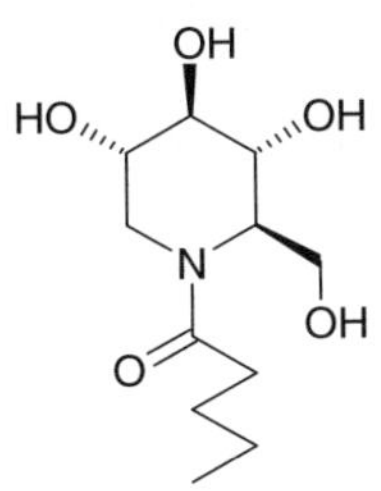

Gaucher disease
β-Glucocerebrosidase (EC 3.2.1.45) – Chemical Chaperone Therapy
IC_{50} 670 μM [186].

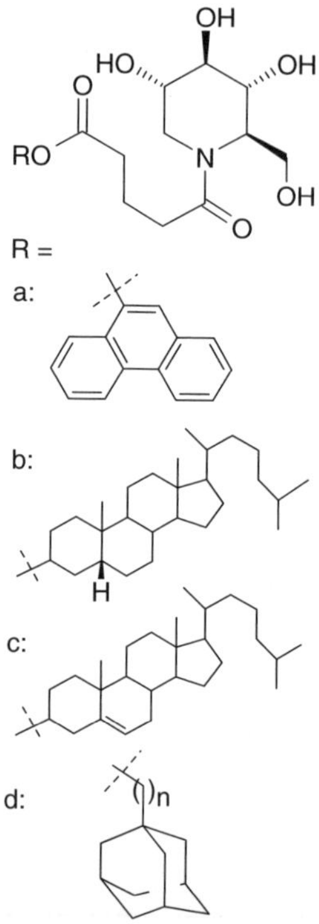 	**Gaucher disease** *β-Glucocerebrosidase* (EC 3.2.1.45) – Chemical Chaperone Therapy a: IC_{50} 11.6 μM [186]. b: IC_{50} 11.2 μM [186]. c: IC_{50} 51.6 μM [186]. d: $n=1$ IC_{50} 19.7 μM [186]; $n=0$ IC_{50} 113 μM [186].
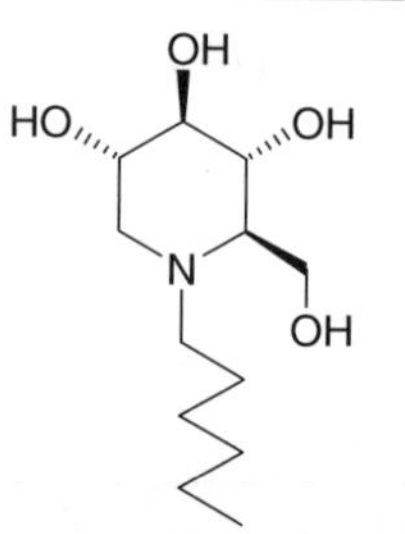 	**Gaucher disease** *Glucosylceramide synthase* (EC 2.4.1.80) – Substrate Reduction Therapy IC_{50} 23.8 μM [194]. *β-Glucocerebrosidase* (EC 3.2.1.45) – Chemical Chaperone Therapy IC_{50} 13 μM; K_i 5.5 μM [185].
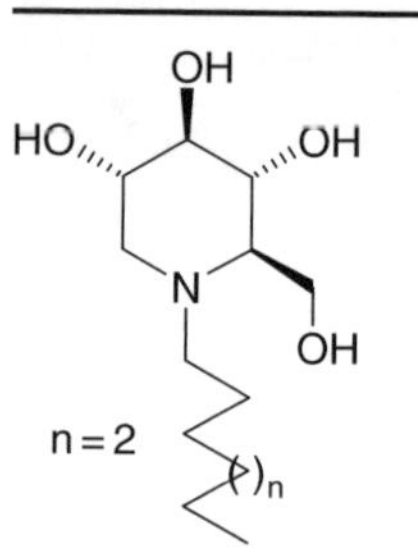 	**Gaucher disease** *β-Glucocerebrosidase* (EC 3.2.1.45) – Chemical Chaperone Therapy IC_{50} 9 μM [186].

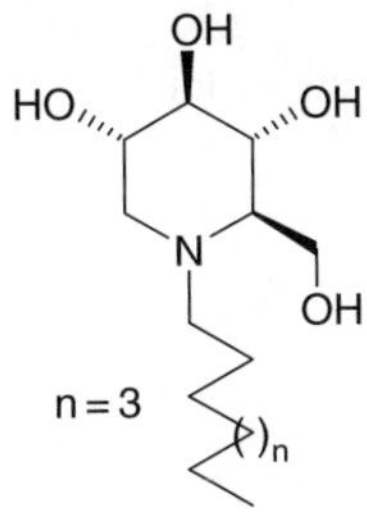

Gaucher disease

Glucosylceramide synthase (EC 2.4.1.80) – Substrate Reduction Therapy

IC_{50} 16.6 μM [194].

β-Glucocerebrosidase (EC 3.2.1.45) – Chemical Chaperone Therapy

IC_{50} 0.82 μM [185]; K_i 0.42 μM [185].

2.3-fold increase in activity at 50 μM of N370S β-glucocerebrosidase (type 1) [185].

1.2 fold increase in activity at 5 μM of L444P β-glucocerebrosidase (type 2) [185].

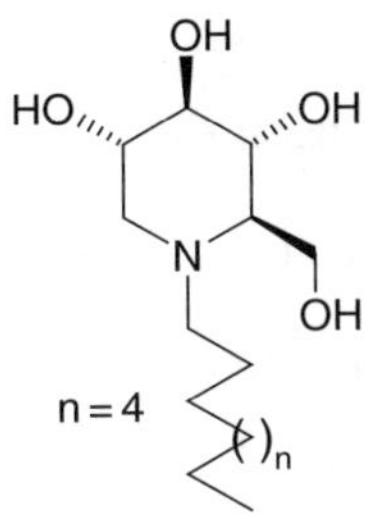

Gaucher disease

Glucosylceramide synthase (EC 2.4.1.80) – Substrate Reduction Therapy

96% inhibition at 200 μM [180]; IC_{50} 1.6 μM [183]; IC_{50} 7.4 μM [194].

β-glucocerebrosidase (EC 3.2.1.45) – Chemical Chaperone Therapy

IC_{50} 1 μM (wild-type) [182]; IC_{50} 0.66 μM [184-185]; K_i 0.3 μM [184-185]; IC_{50} 5 μM (N370S mutation) [182].

1.25-fold increase in activity at 0.5-10 μM of β-glucocerebrosidase (wild-type) [181].

1.65-fold increase in activity at 0.5-10 μM of N370S β-glucocerebrosidase (type 1) [181].

2-fold increase in activity at 10 μM of N370S β-glucocerebrosidase (type 1) [182].

2.4-fold increase in activity at 10 μM of N370S β-glucocerebrosidase (type 1) [185].

1.6-fold increase in activity at 0.5-10 μM of G202R β-glucocerebrosidase [181].

No significant increase in activity at 0.5-10 μM of L444P β-glucocerebrosidase (type 2) [181], [185].

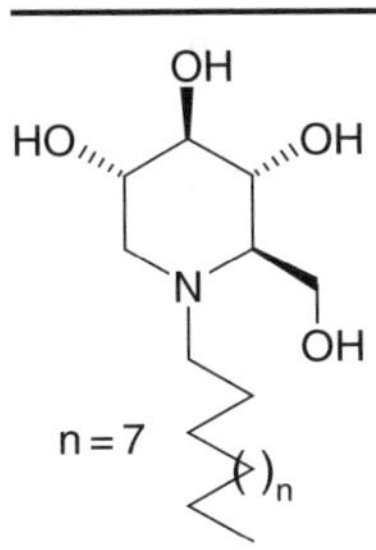

Gaucher disease

Glucosylceramide synthase (EC 2.4.1.80) – Substrate Reduction Therapy

IC_{50} 5.2 μM [194].

β-glucocerebrosidase (EC 3.2.1.45) – Chemical Chaperone Therapy

1.2-fold increase in activity at 0.01-0.1 μM of N370S β-glucocerebrosidase (type 1) [181].

1.3 fold increase in activity at 0.05-0.1 μM of G202R β-glucocerebrosidase [181].

No increase in activity at 0.01-100 μM of wild-type and L444P β-glucocerebrosidase [181].

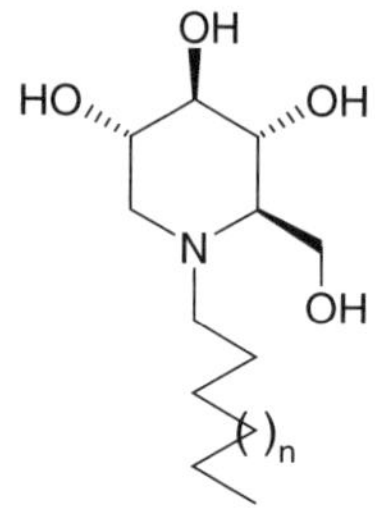

Gaucher disease
Glucosylceramide synthase (EC 2.4.1.80) – Substrate Reduction Therapy
$n = 5$ IC_{50} 3.1 μM [194].
$n = 11$ IC_{50} 3.4 μM [194].
$n = 13$ IC_{50} 4.0 μM [194].

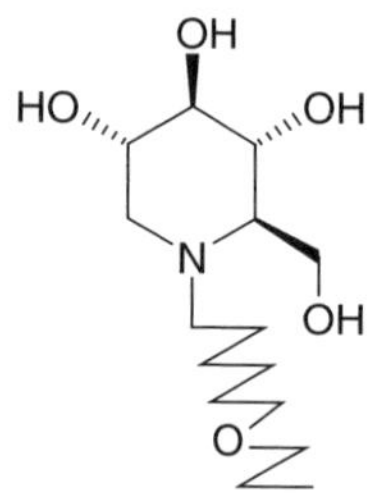

Gaucher disease
Glucosylceramide synthase (EC 2.4.1.80) – Substrate Reduction Therapy
97% inhibition at 200 μM [180]; IC_{50} 3.2 μM [183].
β-Glucocerebrosidase (EC 3.2.1.45) – Chemical Chaperone Therapy
1.25-fold increase in activity at 25 μM of β-glucocerebrosidase (wild-type) [181].
1.6-fold increase in activity at 20-40 μM of N370S β-glucocerebrosidase (type 1) [181].
1.7-fold increase in activity at 25-50 μM of G202R β-glucocerebrosidase [181].
No significant increase in activity at 0.01-1 μM of L444P β-glucocerebrosidase (type 2) [181].

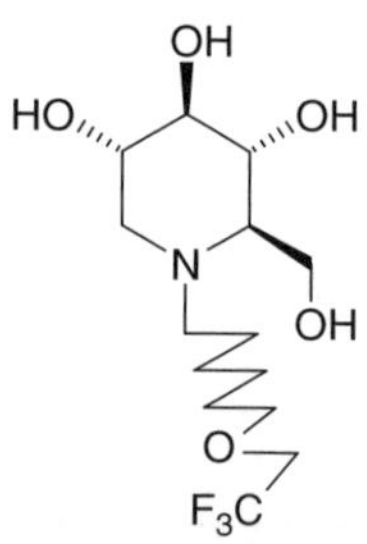

Gaucher disease
β-Glucocerebrosidase (EC 3.2.1.45) – Chemical Chaperone Therapy
1.3-fold increase in activity at 0.01-100 μM of β-glucocerebrosidase (wild-type) [181].
1.8-fold increase in activity at 10-20 μM of N370S β-glucocerebrosidase (type 1) [181].
1.7-fold increase in activity at 25-50 μM of G202R β-glucocerebrosidase [181].
No significant increase in activity at 0.01-1 μM of L444P β-glucocerebrosidase (type 2) [181].

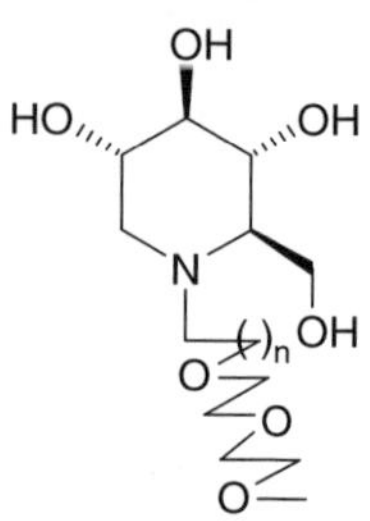

Gaucher disease
Glucosylceramide synthase (EC 2.4.1.80) – Substrate Reduction Therapy
$n = 1$ 20% inhibition at 200 μM [180].
$n = 5$ 93% inhibition at 200 μM [180].

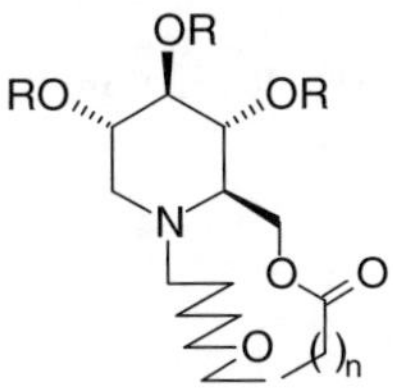

Gaucher disease
Glucosylceramide synthase (EC 2.4.1.80) – Substrate Reduction Therapy
R = Ac, *n* = 0 18% inhibition at 200 μM [180].
R = H, *n* = 2 25% inhibition at 200 μM [180].

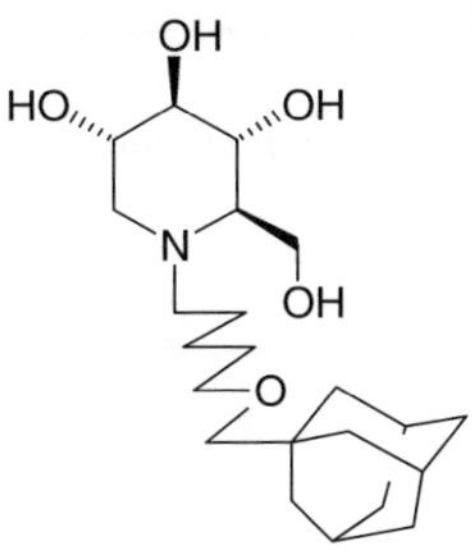

Gaucher disease
β-Glucocerebrosidase (EC 3.2.1.45) – Chemical Chaperone Therapy
IC_{50} 0.16 μM [186].
No significant increase in activity at 0.01-1 μM of β-glucocerebrosidase (wild-type) [181].
1.2-fold increase in activity at 0.01-0.5 μM of N370S β-glucocerebrosidase (type 1) [181].
1.2-fold increase in activity at 0.01-0.5 μM of G202R β-glucocerebrosidase [181].
No significant increase in activity at 0.01-1 μM of L444P β-glucocerebrosidase (type 2) [181].

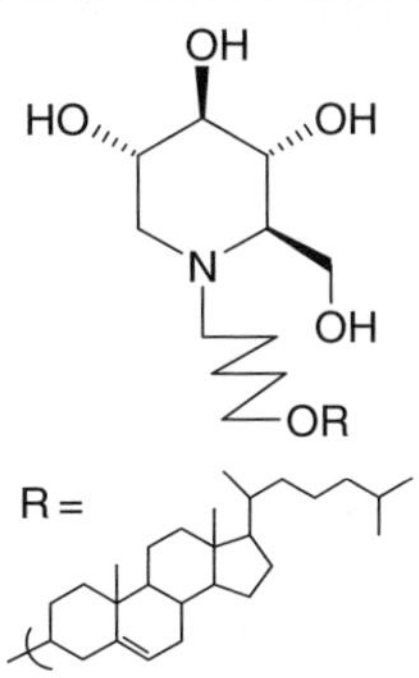

Gaucher disease
β-Glucocerebrosidase (EC 3.2.1.45) – Chemical Chaperone Therapy
IC_{50} 0.96 μM [186].

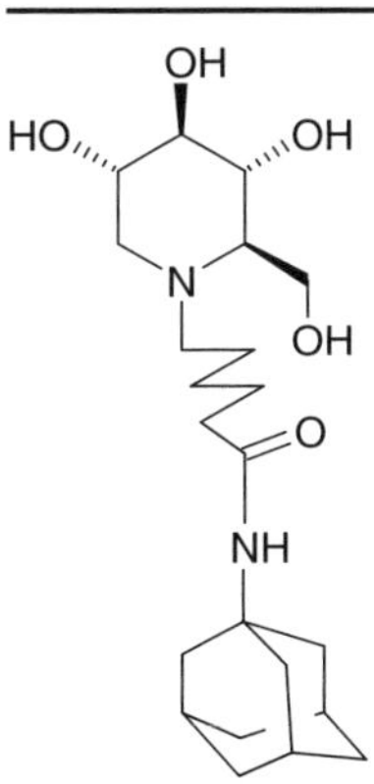

Gaucher disease
β-Glucocerebrosidase (EC 3.2.1.45) – Chemical Chaperone Therapy
1.3-fold increase in activity at 1-10 μM of β-glucocerebrosidase (wild-type) [181].
2.4-fold increase in activity at 40-60 μM of N370S β-glucocerebrosidase (type 1) [181].
3.7-fold increase in activity at 60-80 μM of G202R β-glucocerebrosidase [181].
No significant increase in activity at 0.01-5 μM of L444P β-glucocerebrosidase (type 2) [181].

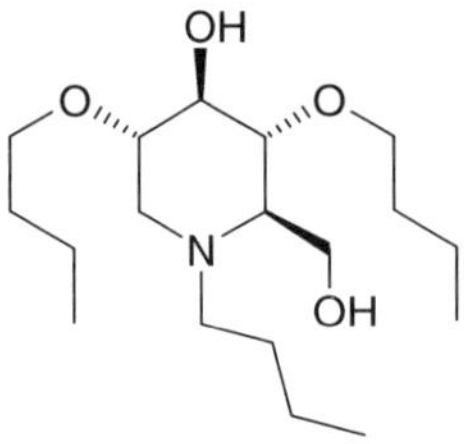	**Gaucher disease** *Glucosylceramide synthase* (EC 2.4.1.80) – Substrate Reduction Therapy 29% inhibition at 1 mM [195].
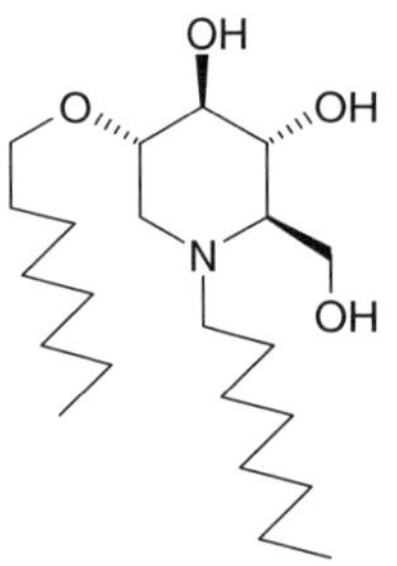	**Gaucher disease** *Glucosylceramide synthase* (EC 2.4.1.80) – Substrate Reduction Therapy IC_{50} 164 μM [195].
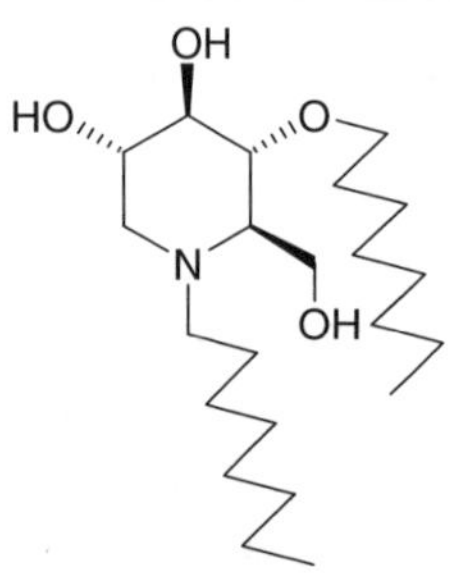	**Gaucher disease** *Glucosylceramide synthase* (EC 2.4.1.80) – Substrate Reduction Therapy IC_{50} 134 μM [195].
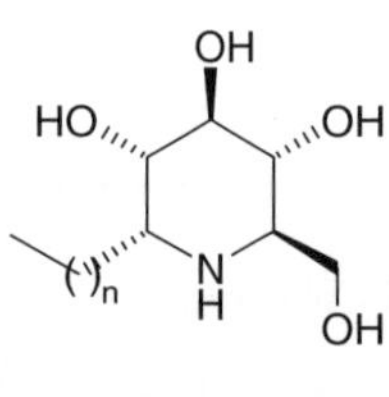	**Gaucher disease** *β-Glucocerebrosidase* (EC 3.2.1.45) – Chemical Chaperone Therapy $n=0$ IC_{50} 900 μM [185]. $n=2$ IC_{50} 400 μM [185]. $n=3$ IC_{50} 100 μM [184-185]; K_i 110 μM [184-185]. $n=5$ IC_{50} 4.2 μM [185]; K_i 2.3 μM [185].
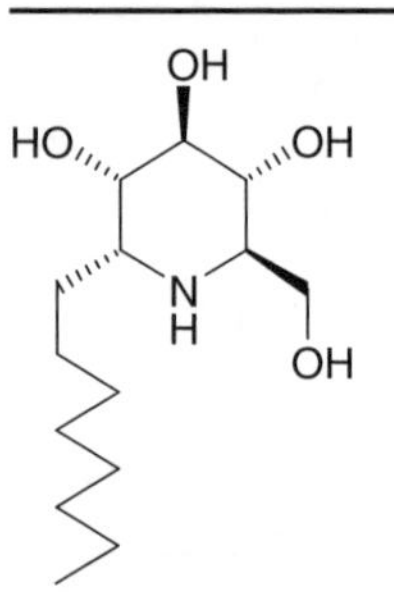	**Gaucher disease** *β-Glucocerebrosidase* (EC 3.2.1.45) – Chemical Chaperone Therapy IC_{50} 0.50 μM [185]; K_i 0.28 μM [185]. 1.7-fold increase in activity at 20 μM of N370S β-glucocerebrosidase (type 1) [185]. 1.2-fold increase in activity at 5 μM of L444P β-glucocerebrosidase (type 2) [185].

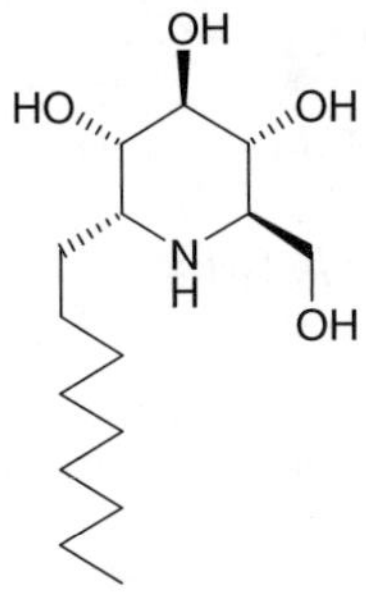

Gaucher disease
β-Glucocerebrosidase (EC 3.2.1.45) – Chemical Chaperone Therapy
IC_{50} 0.27 μM [184-185]; K_i 0.20 μM [184-185].
1.7-fold increase in activity at 2.5 μM of N370S β-glucocerebrosidase (type 1) [185].
1.1-fold increase in activity at 1 μM of L444P β-glucocerebrosidase (type 2) [185].

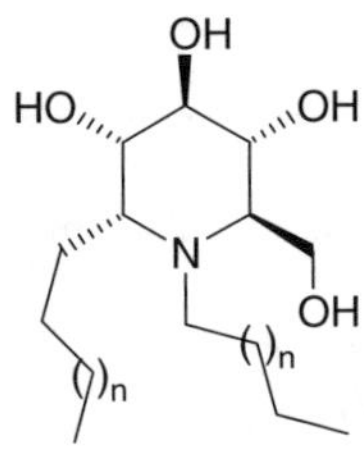

Gaucher disease
Glucosylceramide synthase (EC 2.4.1.80) – Substrate Reduction Therapy
$n = 1$ IC_{50} 609 μM [195].
$n = 5$ IC_{50} 174 μM [195].

D-Galacto-configuration

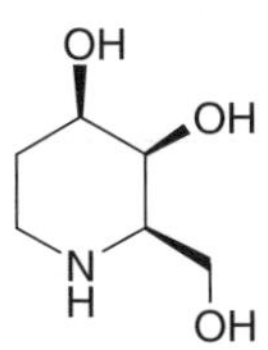

Fabry disease
α-Galactosidase A (EC 3.2.1.22) – Chemical Chaperone Therapy
IC_{50} 250 μM [187].

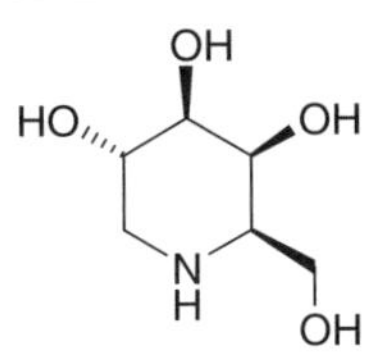

Gaucher disease
Glucosylceramide synthase (EC 2.4.1.80) – Substrate Reduction Therapy no inhibition at 2 mM [183].

Fabry disease
α-Galactosidase A (EC 3.2.1.22) – Chemical Chaperone Therapy [196]
IC_{50} 0.04 μM; K_i 0.04 μM [187].
14-fold increase in activity at 100 μM of R301Q *Lysosomal α-Galactosidase A* [187].
4-fold increase in activity at 20 μM of R301Q *Lysosomal α-galactosidase A* [196a].

G_{M1}-gangliosidis
β-Galactosidase (EC 3.2.1.23) – Chemical Chaperone Therapy
IC_{50} 25 μM [197].
1.8-fold increase in activity at 500 μM of GP8 *β-galactosidase* (normal) [197].
5.4-fold increase in activity at 500 μM of R201C *β-galactosidase* (juvenile type) [197].

(Continued)

2.2-fold increase in activity at 500 μM of I51T *β-galactosidase* (adult type) [197].

2.6-fold increase in activity at 500 μM of R201H *β-galactosidase* (adult type) [197].

6.0-fold increase in activity at 500 μM of R457Q *β-galactosidase* (adult type) [197].

Morquio B disease

β-Galactosidase (EC 3.2.1.23) – Chemical Chaperone Therapy

1.8-fold increase in activity at 500 μM of W273L *β-galactosidase* [197].

1.7-fold increase in activity at 500 μM of Y83H *β-galactosidase* [197].

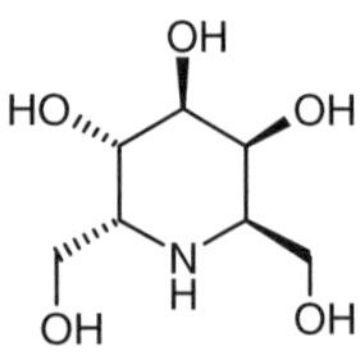

Fabry disease

α-Galactosidase A (EC 3.2.1.22) – Chemical Chaperone Therapy

IC_{50} 0.21 μM; K_i 0.17 μM [187].

5.2-fold increase in activity at 100 μM of R301Q *Lysosomal α-Galactosidase A* (12.5-fold at 1000 μM) [187].

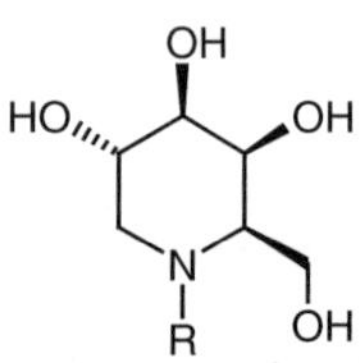

Fabry disease

α-Galactosidase A (EC 3.2.1.22) – Chemical Chaperone Therapy

R = Me IC_{50} 96 μM [187].

R = Et IC_{50} 306 μM [187].

R = *n*-Pr IC_{50} 301 μM [187].

R = $(CH_2)_2OH$ IC_{50} 520 μM [187].

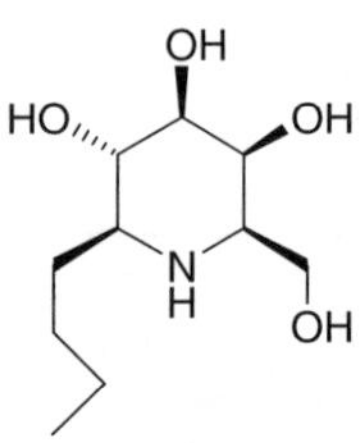

Fabry disease

α-Galactosidase A (EC 3.2.1.22) – Chemical Chaperone Therapy

IC_{50} 24 μM, K_i 16 μM [187].

2.3-fold increase in activity at 100 μM of R301Q *α-galactosidase A* (6.3-fold at 1000 μM) [187].

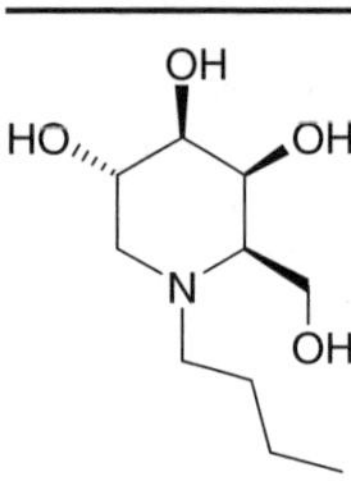

Gaucher disease

Glucosylceramide synthase (EC 2.4.1.80) – Substrate Reduction Therapy

71% inhibition at 200 μM [180]; IC_{50} 30 μM [183]; K_i 10 μM [198].

Fabry disease

α-Galactosidase A (EC 3.2.1.22) – Chemical Chaperone Therapy

IC_{50} 300 μM [187].

Sandhoff diseases

Glucosylceramide synthase (EC 2.4.1.80) – Substrate Reduction Therapy [199].

Better therapeutic efficacy than N-Butyl DNJ (evaluation in the Sandhoff disease mouse model).

(Continued)

G_{M1}-gangliosidis disease
Glucosylceramide synthase (EC 2.4.1.80) – Substrate Reduction Therapy [200].
K_i 10 μM [198].
Significantly reduces total ganglioside and GM1 content in cerebrum-brainstem and cerebellum of acid β-galactosidase knockout mice treated from postnatal day 9 to 15 (600 mg/kg) [201].
β-galactosidase (EC 3.2.1.23) – Chemical Chaperone Therapy
IC_{50} 25 μM [197].
1.4-fold increase in activity at 500 μM of GP8 *β-galactosidase* (normal) [197].
4.8-fold increase in activity at 500 μM of R201C *β-galactosidase* (juvenile type) [197].
6.1-fold increase in activity at 500 μM of I51T *β-galactosidase* (adult type) [197].
2.1-fold increase in activity at 500 μM of R201H *β-galactosidase* (adult type) [197].
5.4-fold increase in activity at 500 μM of R457Q *β-galactosidase* (adult type) [197].

Morquio B disease
β-Galactosidase (EC 3.2.1.23) – Chemical Chaperone Therapy
1.8-fold increase in activity at 500 μM of W273L *β-galactosidase* [197].
1.1-fold increase in activity at 500 μM of Y83H *β-galactosidase* [197].

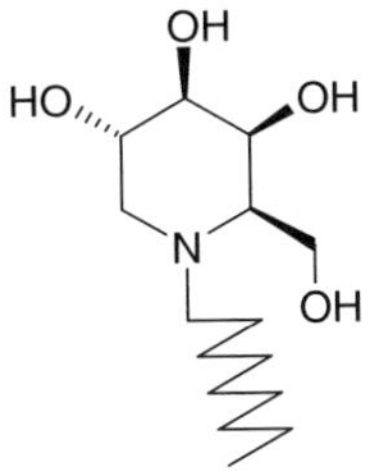

Gaucher disease
Glucosylceramide synthase (EC 2.4.1.80) – Substrate Reduction Therapy
IC_{50} 10.6 μM [183].

D-GlcNAc mimetics

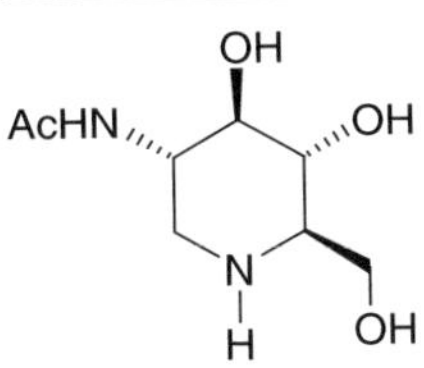

Tay–Sachs and Sandhoff diseases
β-Hexosaminidase (EC 3.2.1.52) – Chemical Chaperone Therapy
K_i 0.70 μM (Jack Bean) [188]; IC_{50} 16 μM (homogenate of human spleen *β*-hexosaminidase) [202].
2-fold increase in activity at 490 μM of α-G269S *β-hexosaminidase* [188].

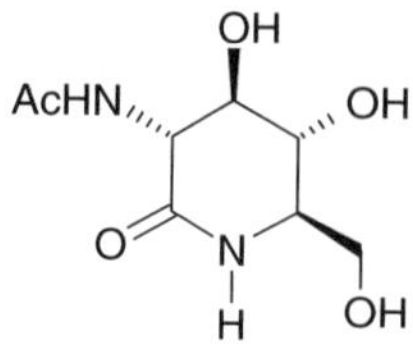	**Tay–Sachs and Sandhoff diseases** *β-Hexosaminidase* (EC 3.2.1.52) – Chemical Chaperone Therapy IC_{50} 2.5 μM (homogenate of human spleen β-hexosaminidase) [202].
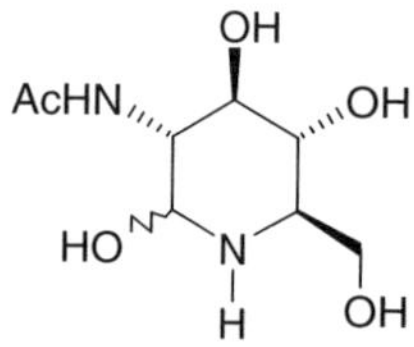	**Tay–Sachs and Sandhoff diseases** *β-Hexosaminidase* (EC 3.2.1.52) – Chemical Chaperone Therapy K_i 0.005 μM (Jack Bean) [188]. 2.5-fold increase in activity at 200 μM of α-G269S *β-hexosaminidase* [188].

D-Xylo-configuration

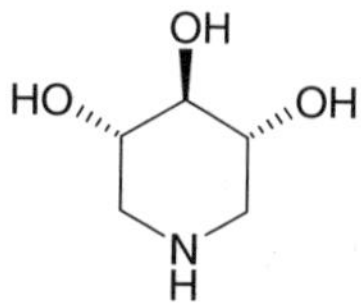	**Gaucher disease** *β-glucocerebrosidase* (EC 3.2.1.45) – Chemical Chaperone Therapy IC_{50} 2.3 μM [184]; K_i 1.9 μM [184].
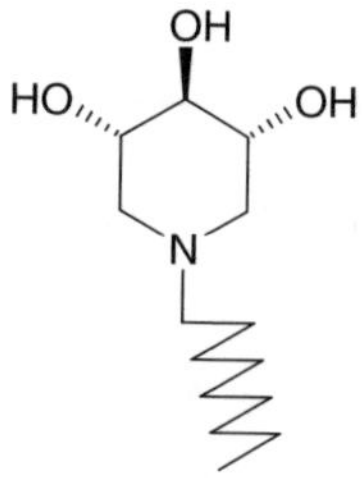	**Gaucher disease** *β-glucocerebrosidase* (EC 3.2.1.45) – Chemical Chaperone Therapy IC_{50} 1.5 μM [184]; K_i 0.5 μM [184].
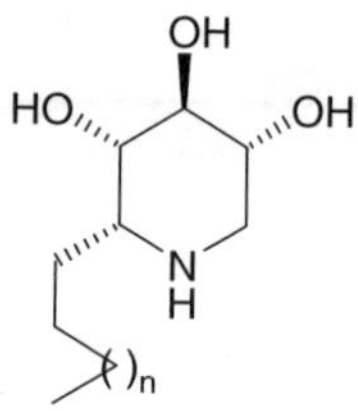	**Gaucher disease** *β-glucocerebrosidase* (EC 3.2.1.45) – Chemical Chaperone Therapy $n = 6$ IC_{50} 0.0068 μM [184]; K_i 0.0022 μM [184]. 1.8-fold increase in activity at 0.01 μM of N370S β-glucocerebrosidase (type 1) [184]. Specific inhibitor: no inhibition observed for α-glucosidases [184].
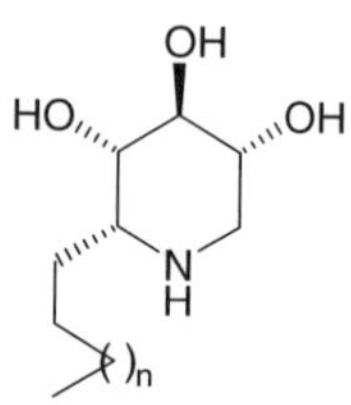	**Gaucher disease** *β-glucocerebrosidase* (EC 3.2.1.45) – Chemical Chaperone Therapy $n = 9$ IC_{50} 0.014 μM [184]; K_i 0.031 μM [184].

Miscellaneous configurations

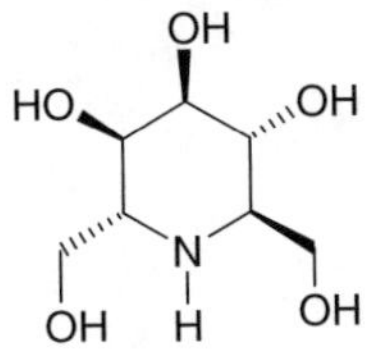

Fabry disease
α-Galactosidase A (EC 3.2.1.22) – Chemical Chaperone Therapy
IC_{50} 464 μM [187].

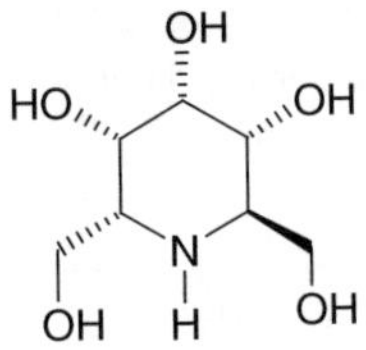

Fabry disease
α-Galactosidase A (EC 3.2.1.22) – Chemical Chaperone Therapy
IC_{50} 4.3 μM; K_i 2.6 μM [187].
2.4-fold increase in activity at 100 μM of R301Q *lysosomal α-galactosidase A* (3.9 at 1000 μM) [187].

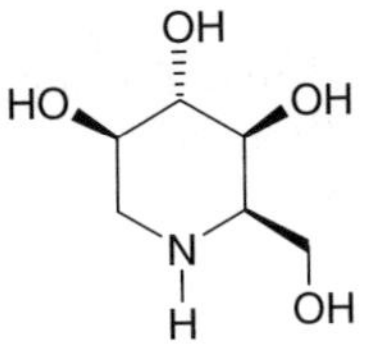

Gaucher disease
β-Glucocerebrosidase (EC 3.2.1.45) – Chemical Chaperone Therapy
IC_{50} 66 μM [18].

Isofagomine and analogues

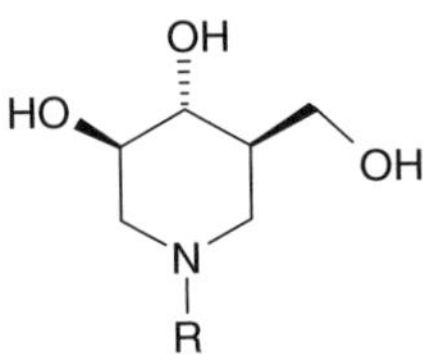

Gaucher disease
β-Glucocerebrosidase (EC 3.2.1.45) – Chemical Chaperone Therapy
R = H IC_{50} 0.056 μM; K_i 0.025 μM [203]; K_i 0.016 μM [10].
R = *n*-Bu 44 μM [203].
R = *n*-Nonyl IC_{50} > 100 μM [203].

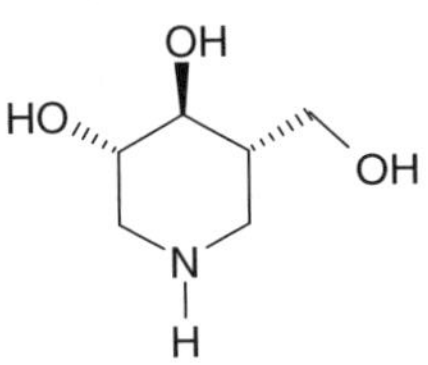

Gaucher disease
β-Glucocerebrosidase (EC 3.2.1.45) – Chemical Chaperone Therapy
K_i 5.7 μM [10].

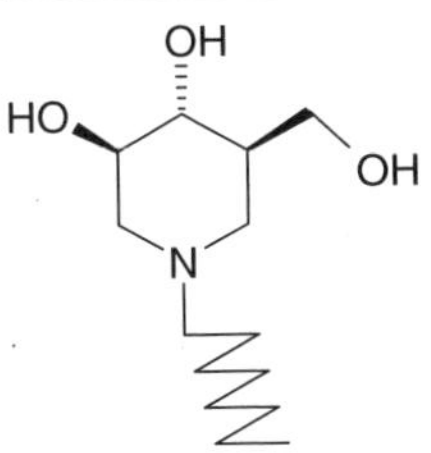

Gaucher disease
β-Glucocerebrosidase (EC 3.2.1.45) – Chemical Chaperone Therapy
1.45-fold increase in activity at 15 μM of N370S β-glucocerebrosidase (type 1) [181].
3.2-fold increase in activity at 25 μM of G202R β-glucocerebrosidase [181].

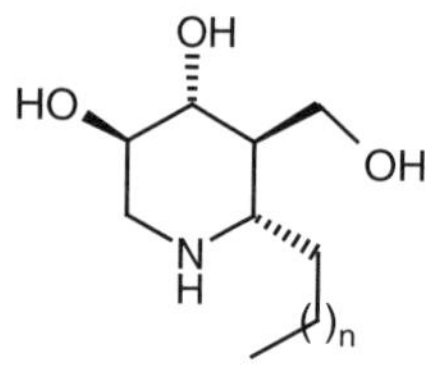

Gaucher disease
β-Glucocerebrosidase (EC 3.2.1.45) – Chemical Chaperone Therapy
$n=2$ IC_{50} 0.16 μM; K_i 0.12 μM [203].
$n=4$ IC_{50} 0.0042 μM [203].
$n=5$ IC_{50} 0.0018 μM [203].
$n=6$ IC_{50} 0.0008 μM [203].
$n=7$ IC_{50} 0.0006 μM [203].

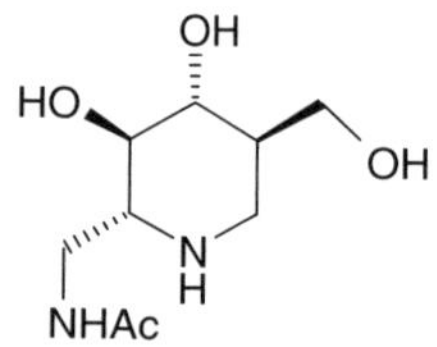

Tay–Sachs and Sandhoff diseases
β-Hexosaminidase (EC 3.2.1.52) – Chemical Chaperone Therapy
IC_{50} 5 μM; K_i 2.4 μM (homogenate of human spleen *β*-hexosaminidase) [202].

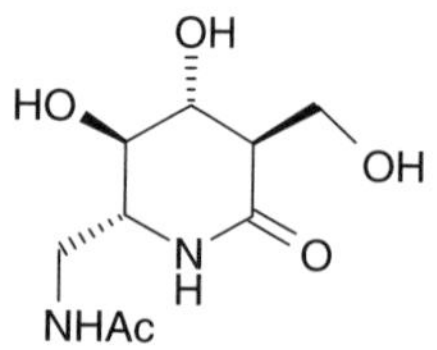

Tay–Sachs and Sandhoff diseases
β-Hexosaminidase (EC 3.2.1.52) – Chemical Chaperone Therapy
20% inhibition at 1200 μM (homogenate of human spleen *β*-hexosaminidase) [202].

14.8.3 Bicyclic derivatives

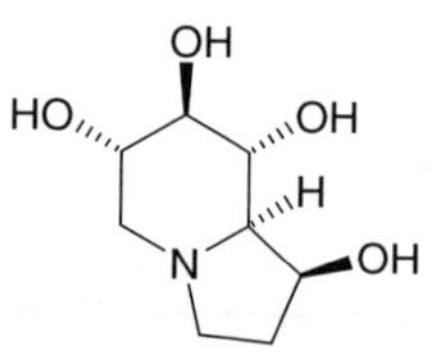

Gaucher disease
β-Glucocerebrosidase (EC 3.2.1.45) – Chemical Chaperone Therapy
No significant increase in activity at 0.01-100 μM of wild-type, N370S, G202R and L444P β-glucocerebrosidases [181].

Tay–Sachs and Sandhoff diseases
β-Hexosaminidase (EC 3.2.1.52) – Chemical Chaperone Therapy
No increase in activity [188].

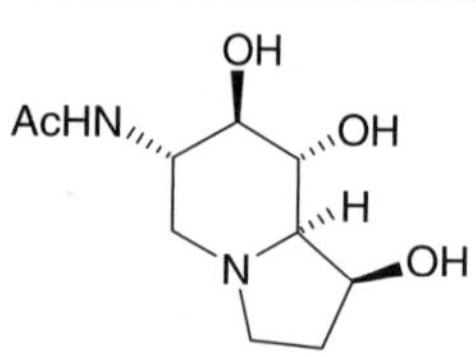

Tay–Sachs and Sandhoff diseases
β-Hexosaminidase (EC 3.2.1.52) – Chemical Chaperone Therapy
IC_{50} 0.50 μM [188].
3.5-fold increase in activity at 200 μM of *α*-G269S *β-hexosaminidase* [188].

14.9 Immunomodulating agents

14.9.1 Five-membered rings

D-Ribo-configuration

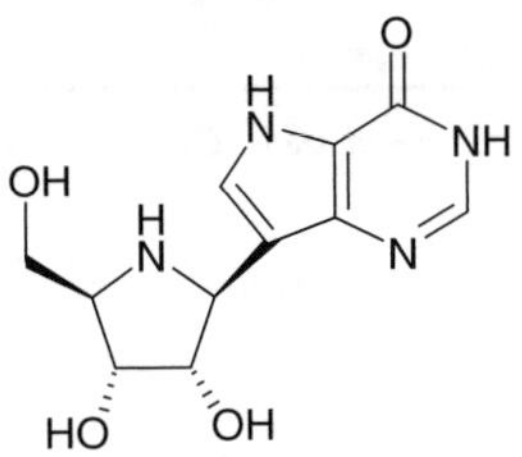

Human Purine Nucleoside Phosphorylase expressed in E. Coli (EC 2.4.2.1)

IC_{50} 1.19 ± 0.21 nM [175], K_i 3.3 ± 0.2 nM, K_i^* 0.056 ± 0.015 nM [204].

Dose-dependent inhibition of the proliferative response to IL-2 in the presence of deoxyguanosine dGuo (10 μM), IC_{50} 0.06 μM [175], [205].

Dose-dependent inhibition of PHA-stimulated lymphoproliferation of human T-cells in the presence of dGuo (10 μM), IC_{50} 0.387 μM [175].

Inhibition of activated normal peripheral blood T lymphocytes (95%), IC_{50} 5 nM [174].

Selective inhibition of CCRF-CEM and MOLT-4 cell lines proliferation

IC_{50} 5 nM [174].

Thymidine incorporation

IC_{50} 4×10^{-10} M [174].

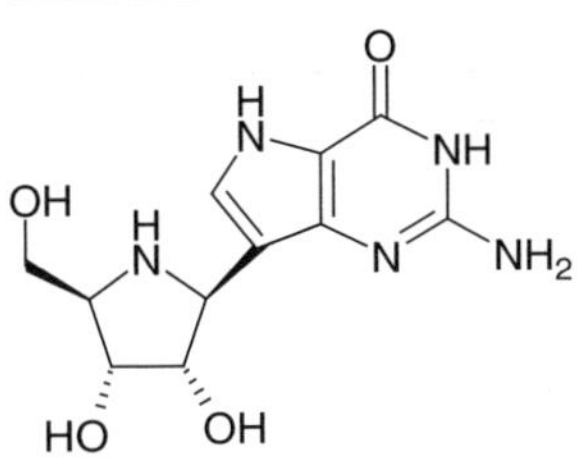

Human purine nucleoside phosphorylase expressed in E. Coli (EC 2.4.2.1)

K_i 0.54 ± 0.1 nM, K_i^* 0.042 ± 0.006 nM [204].

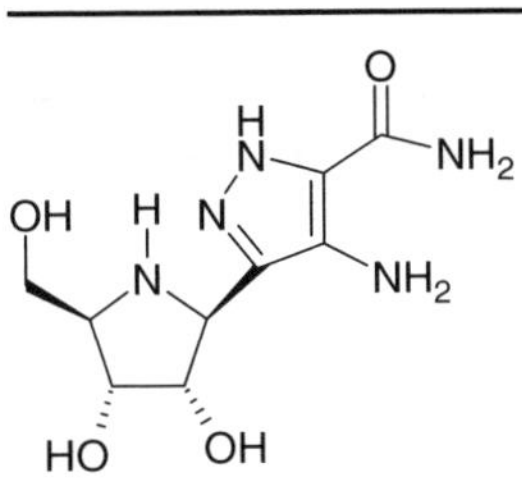

Human purine nucleoside phosphorylase expressed in E. Coli (EC 2.4.2.1)

K_i^* 96 ± 16 pM [206].

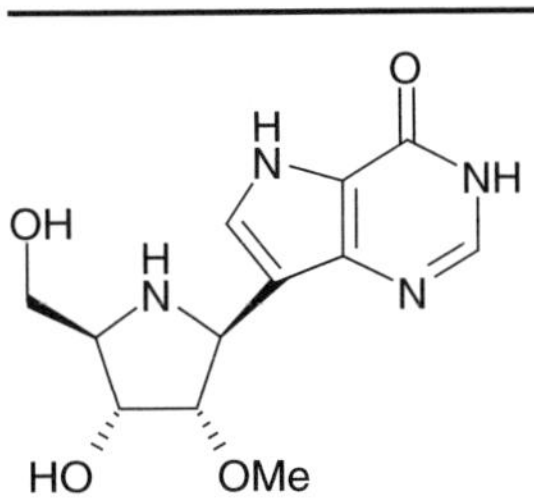

Human purine nucleoside phosphorylase expressed in E. Coli (EC 2.4.2.1)

K_i 5.9 ± 1.4 nM [204].

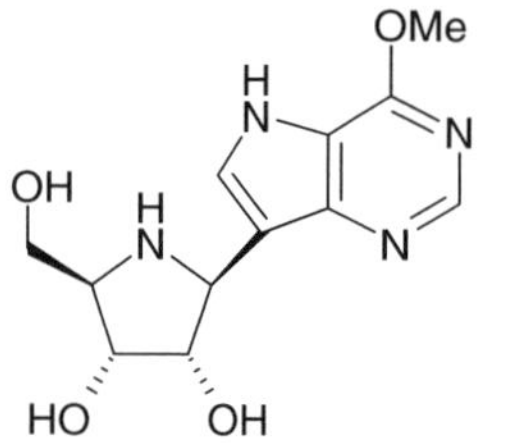	*Human purine nucleoside phosphorylase expressed in E. Coli* (EC 2.4.2.1) K_i 4.7 ± 0.5 nM [204].
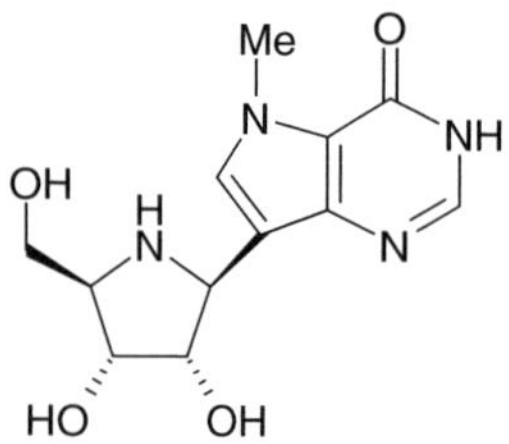	*Human purine nucleoside phosphorylase expressed in E. Coli* (EC 2.4.2.1) K_i 4.7 ± 0.3 nM [204].
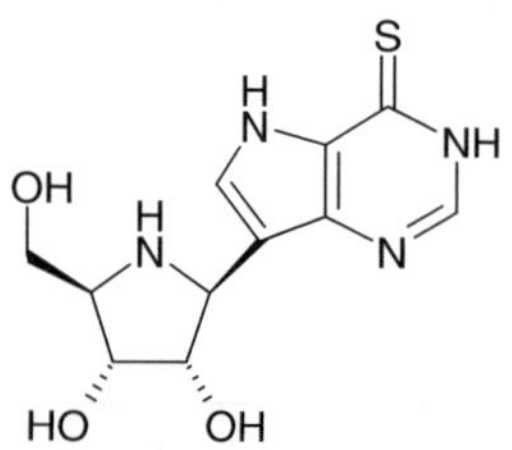	*Human purine nucleoside phosphorylase expressed in E. Coli* (EC 2.4.2.1) K_i 24.7 ± 1.1 nM [204].
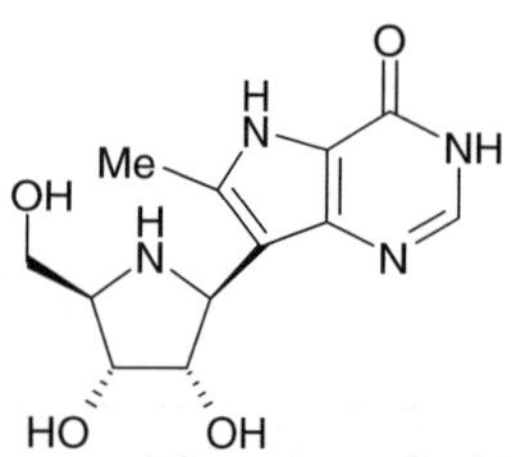	*Human purine nucleoside phosphorylase expressed in E. Coli* (EC 2.4.2.1) K_i 10.1 ± 1.6 nM [204].
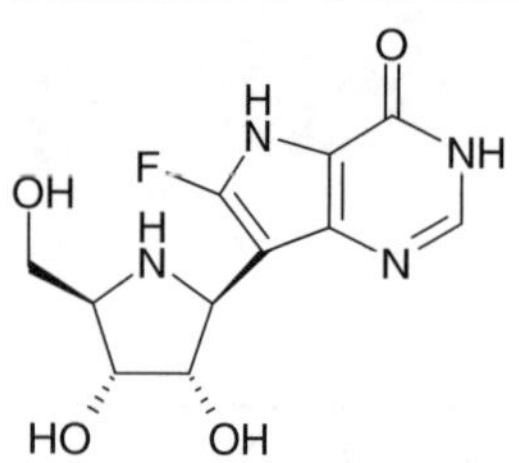	*Human purine nucleoside phosphorylase expressed in E. Coli* (EC 2.4.2.1) K_i 5.6 ± 0.7 nM, K_i* 0.39 ± 0.02 nM [204].

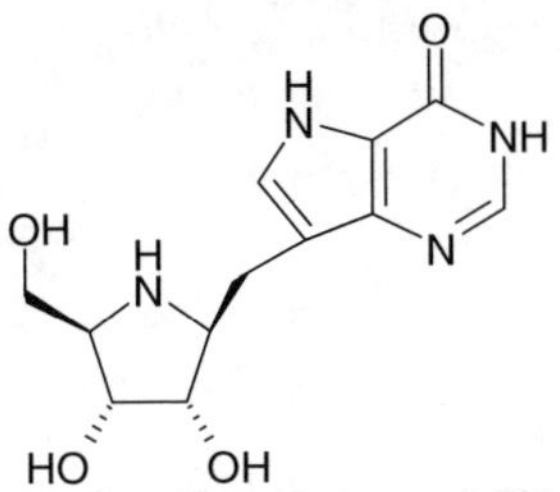

Human purine nucleoside phosphorylase expressed in E. Coli
(EC 2.4.2.1)
K_i 250 ± 7 nM [204].

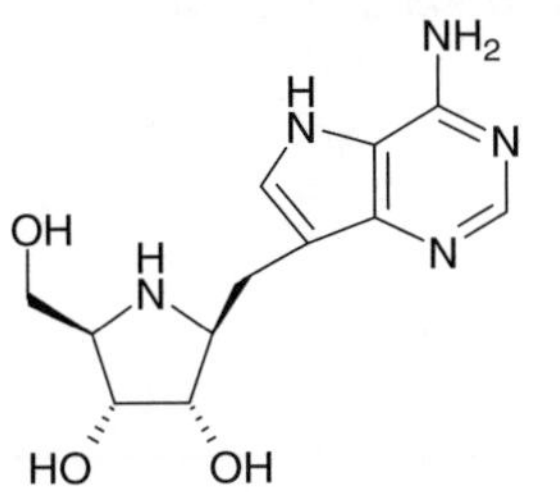

Human purine nucleoside phosphorylase expressed in E. Coli
(EC 2.4.2.1)
$K_i > 30000$ nM [204].

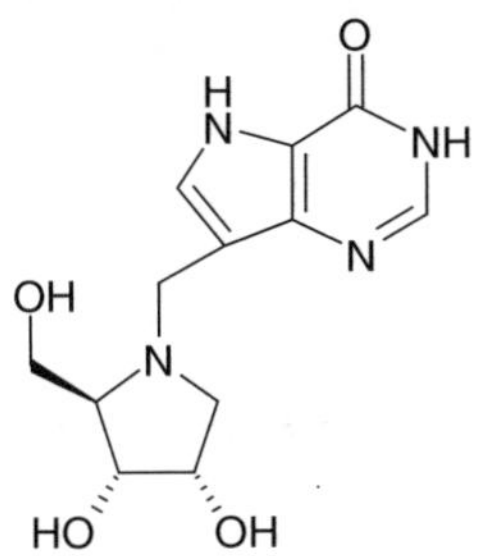

Human purine nucleoside phosphorylase expressed in E. Coli
(EC 2.4.2.1)
K_i 2.84 ± 0.06 nM [204].

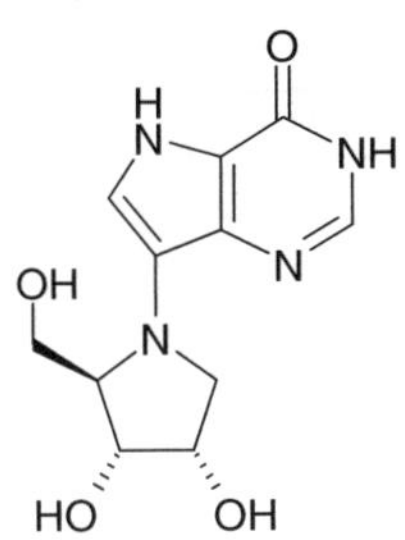

Human purine nucleoside phosphorylase expressed in E. Coli
(EC 2.4.2.1)
K_i 12700 ± 1700 nM [204].

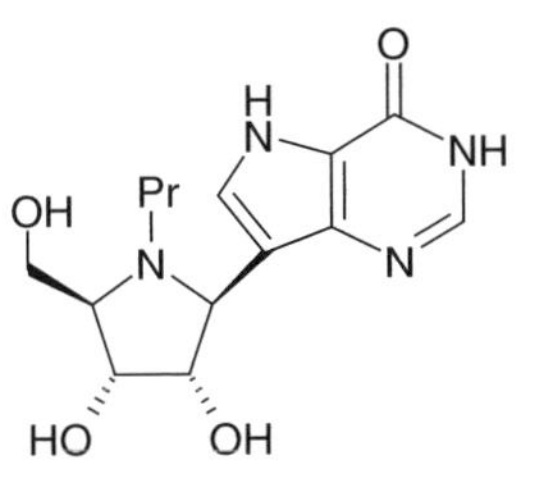

Human purine nucleoside phosphorylase expressed in E. Coli
(EC 2.4.2.1)
K_i 415 ± 86 nM [204].

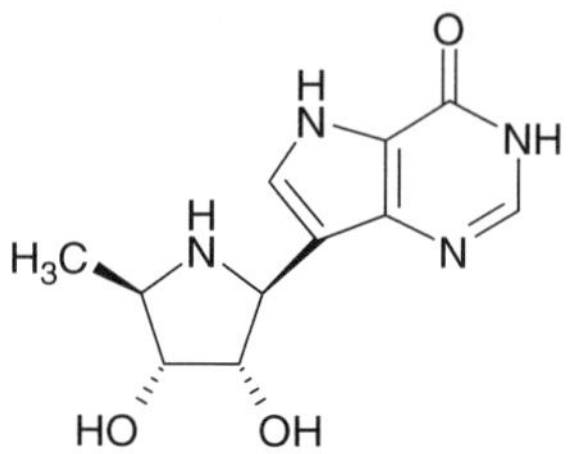	*Human purine nucleoside phosphorylase expressed in E. Coli* (EC 2.4.2.1) K_i 25 ± 7 nM [204].
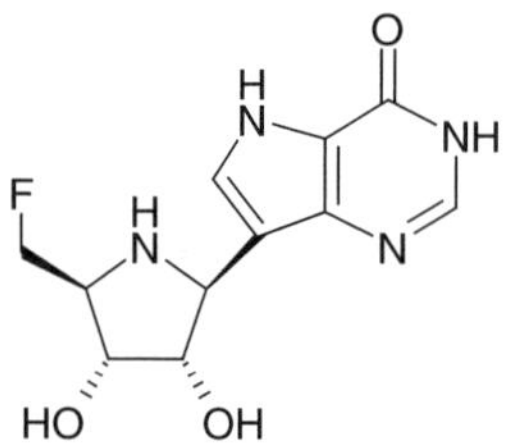	*Human purine nucleoside phosphorylase expressed in E. Coli* (EC 2.4.2.1) K_i 81 ± 5 nM, K_i* 7 ± 2 nM [204].
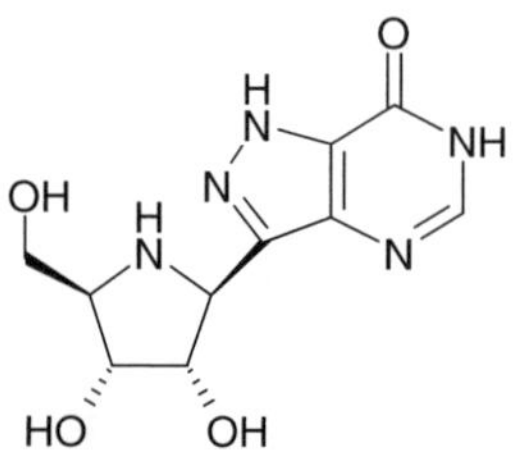	*Human purine nucleoside phosphorylase expressed in E. Coli* (EC 2.4.2.1) K_i 1.4 ± 0.2 nM, K_i* 0.18 ± 0.02 nM [204].

D-Arabino-configuration

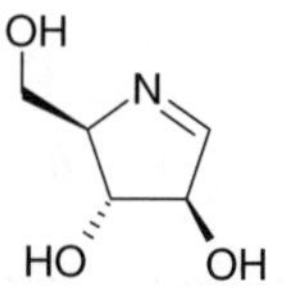 **A**	*Effect on antibody formation to SRBC in the immunodeficient mice* [207] Reversion of the depressed immune responses in mitomycin C treated mice. Control (saline): anti-SRBC pfc (×10^4 pfc spleen) 142 ± 0.55. Compound A (500 mg/kg) 150 ± 5.32. Mitomycin C (1 mg/kg) + saline 99.6 ± 7.12. Mitomycin C (1 mg/kg) + compound A (500 mg/kg) 208 ± 9.39. *Suppression of Con. A-induced mouse spleen cell proliferation by immunosuppressive factor and its restoration by iminosugar* [207] Non-treated control: [^{3}H] Thymidine uptake (cpm): 931 ± 46. Con. A (2 μg/well): 103.425 ± 5.285. Immunosuppressive factor (5 μg/well): 307 ± 83. *Compound A* (10 μg/well): 348 ± 86, (1 μg/well): 705 ± 106, (0.1 μg/well): 924 ± 182 [207].

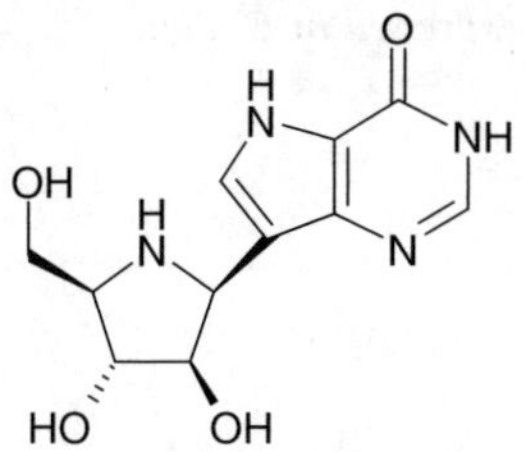

Human purine nucleoside phosphorylase expressed in E. Coli
(EC 2.4.2.1)
K_i 3.5 ± 0.3 nM [204].

Miscellaneous configurations

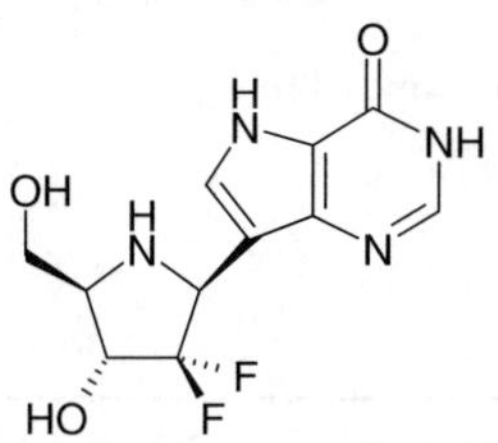

Human purine nucleoside phosphorylase expressed in E. Coli
(EC 2.4.2.1)
K_i 1.4 ± 0.2 nM [204].

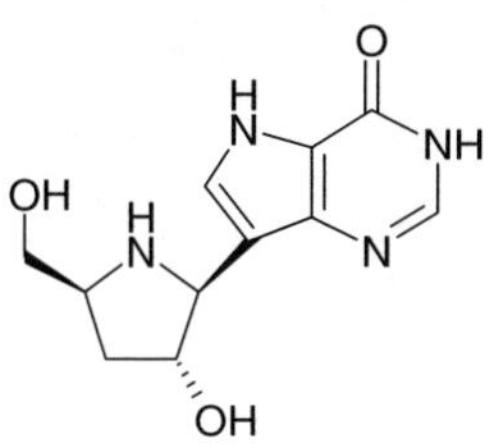

Human purine nucleoside phosphorylase expressed in E. Coli
(EC 2.4.2.1)
K_i 9.5 ± 1.6 nM [204].

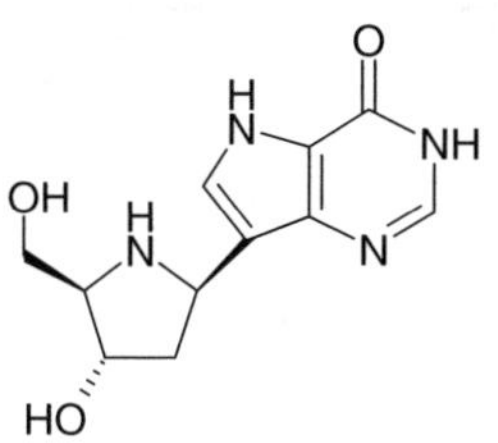

Human purine nucleoside phosphorylase expressed in E. Coli
(EC 2.4.2.1)
K_i 0.25 ± 0.06 nM [204].

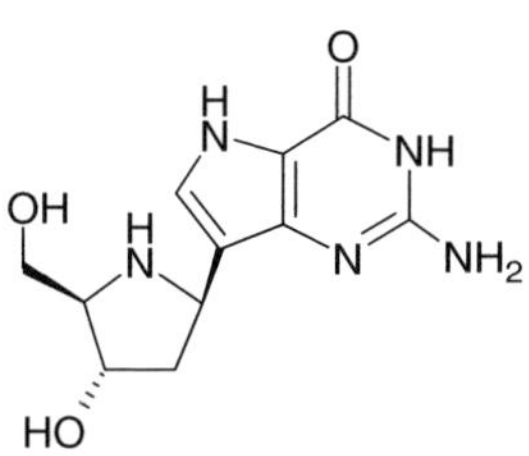

Human purine nucleoside phosphorylase expressed in E. Coli
(EC 2.4.2.1)
K_i 0.22 ± 0.02 nM [204].

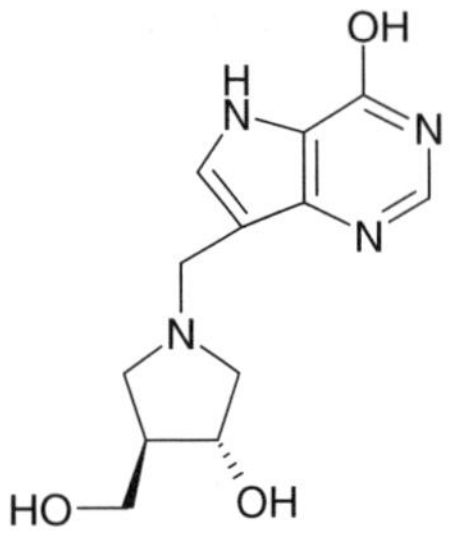	*Human purine nucleoside phosphorylase expressed in E. Coli* (EC 2.4.2.1) [208] K_i 1100 ± 120 pM, K_i* 16 ± 1.4 pM. IC_{50} 16 pM.
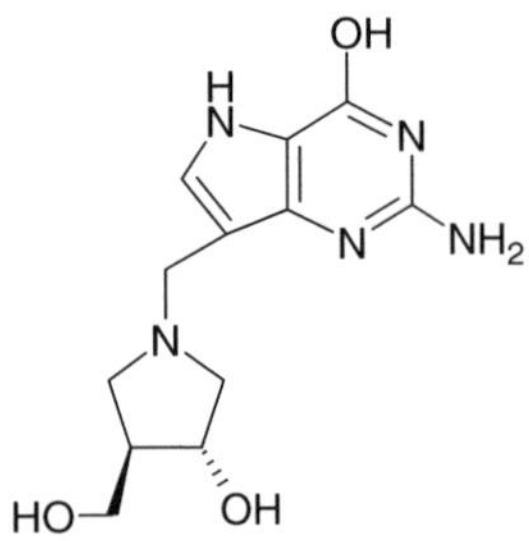	*Human purine nucleoside phosphorylase expressed in E. Coli* (EC 2.4.2.1) [208] K_i 163 ± 25 pM, K_i* 6.8 ± 1.2 pM. IC_{50} 7 pM. 8-fold increase in affinity compared with Immucillin H.
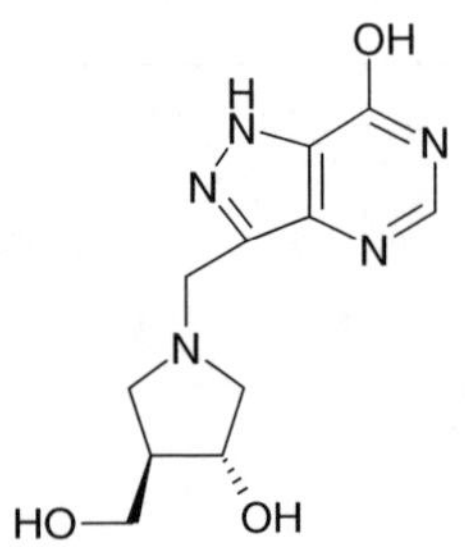	*Human purine nucleoside phosphorylase expressed in E. Coli* (EC 2.4.2.1) [208] K_i 2000 ± 50 pM, K_i* no slow onset. IC_{50} 2000 pM.

14.9.2 Six-membered rings

1-Deoxynojirimycin and analogues

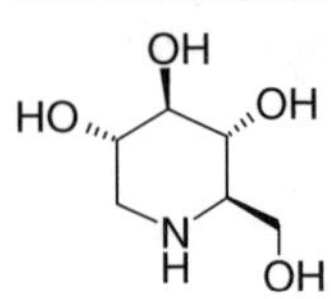	Human PBMCs proliferation for 72 h MTT assay *Mitochondrial dehydrogenase activity of surviving cells* IC_{50} 25.6 μM [209]. Less toxic than CyA to human PBMCs.

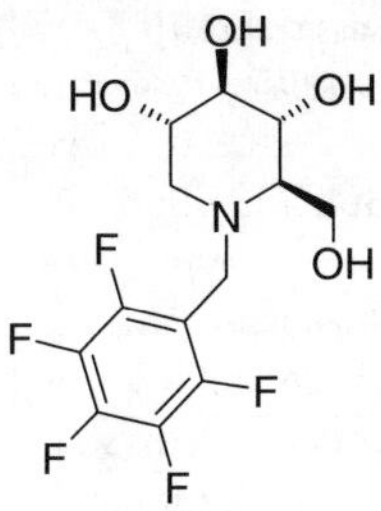	Human PBMCs proliferation for 72 h MTT assay [209] *Mitochondrial dehydrogenase activity of surviving cells* IC_{50} 20.2 μM. At 10 μM, levels of IL-4 secretion decreased 6-fold lower than the control cells. Less toxic than CyA to human PBMCs. Specific inhibition effect on the expression of CD4.

D-Galacto-configuration

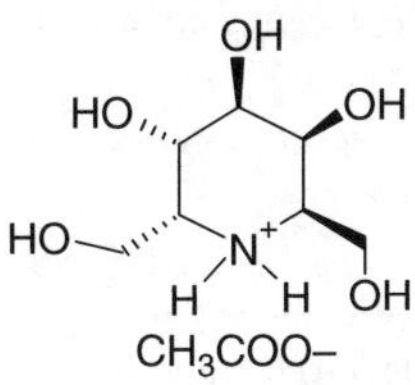	Mouse splenocytes proliferation for 72 h MTT assay [210] *Mitochondrial dehydrogenase activity of surviving cells* IC_{50} 519 μM. At 25 μM, level of IL-4 secretion reduction 69.2%. Level of INF-γ secretion reduction 75.9%.
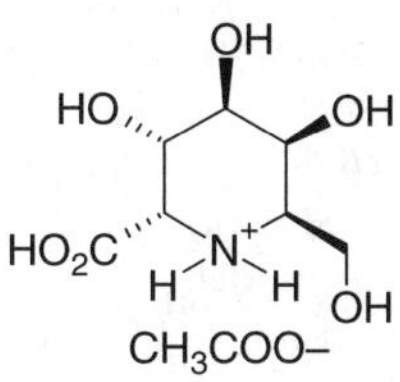	Mouse splenocytes proliferation for 72 h MTT assay [210] *Mitochondrial dehydrogenase activity of surviving cells* IC_{50} 340 μM. At 25 μM, levels of IL-4 secretion reduction 30.8%. Levels of INF-γ secretion reduction 31.2%.

L-Altro-configuration

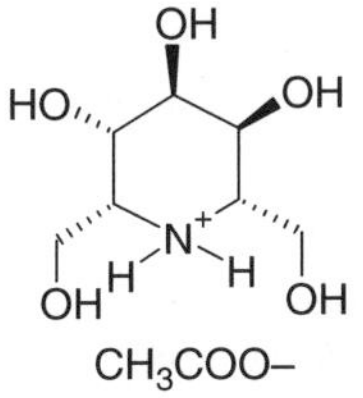	Mouse splenocytes proliferation for 72 h MTT assay [210] *Mitochondrial dehydrogenase activity of surviving cells* IC_{50} 332 μM. At 25 μM, level of IL-4 secretion reduction 92.3%. Level of INF-γ secretion reduction 86.2%.
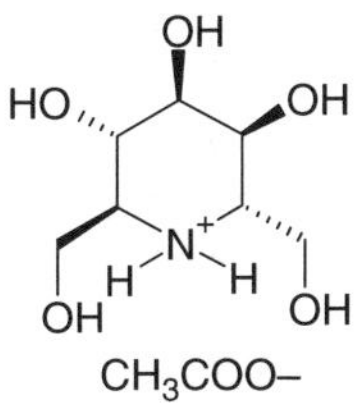	Mouse splenocytes proliferation for 72 h MTT assay [210] *Mitochondrial dehydrogenase activity of surviving cells* IC_{50} 272 μM. At 25 μM level of IL-4 secretion reduction 84.6%. Level of INF-γ secretion reduction 51.6%.

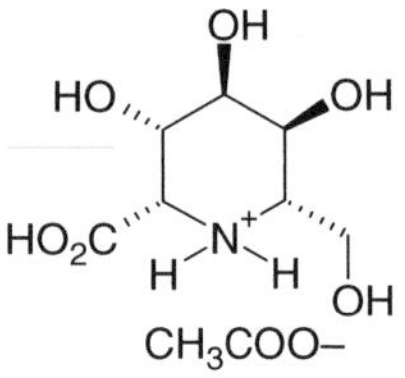	Mouse splenocytes proliferation for 72 h MTT assay [210] *Mitochondrial dehydrogenase activity of surviving cells* IC_{50} 346 μM. At 25 μM level of IL-4 secretion reduction 92.3%. Level of INF-γ secretion reduction 24.2%.

14.9.3 Bicyclic derivatives

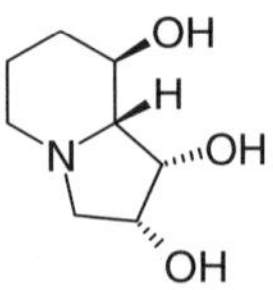	*Enhancement of the proliferative response of the T-cell clone to antigen and to the mitogen concanavalin A* [211] *Substantial decrease of proliferative response to exogenous IL-2 factor* [211].
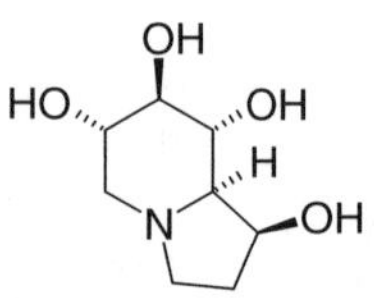	*Enhancement of the proliferative response to the T-cell clone to mitogen concanavalin A* [211] *Substantial decrease of proliferative response to exogenous IL-2 factor* [211] *Inhibition of the proliferative response to the T-cell to antigen* [211] For immunosuppression in allotransplantation see [212]. Immunosuppressive agent in a rat heart allograft model. In grafted rats, significant reduction of the expression of LFA-1α, ICAM-1, class I and II MHC, CD4, CD8, CD45. [212]. Prolongation of rat heart allograft survival in a dose-dependent manner [212].

14.10 Antipsoriatic agents

14.10.1 Five-membered rings

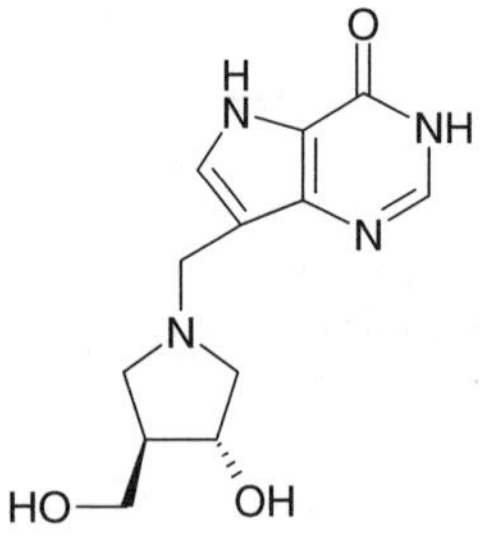	Entered phase I clinical trials as an antipsoriasis drug [108].

14.10.2 Six-membered rings

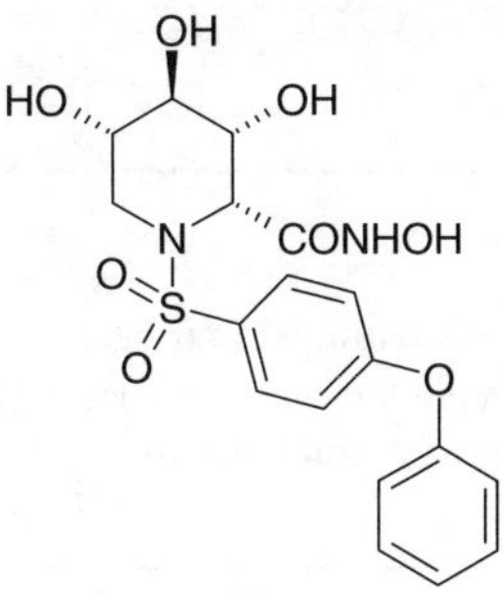

Matrix Metalloproteinase
rMMP-1 (EC 3.4.24.7) K_i 8 nM; rMMP-3 (EC 3.4.24.17) K_i 0.51 nM; rMMP-9 (EC 3.4.24.35) K_i 0.06 nM [213].
TNF-α converting enzyme, ADAM17 (TACE)
K_i 2.3 nM [213].
Heparin-binding EGF-like growth factor (shedding of HB-EGF)
IC_{50} 0.35 μM [213].

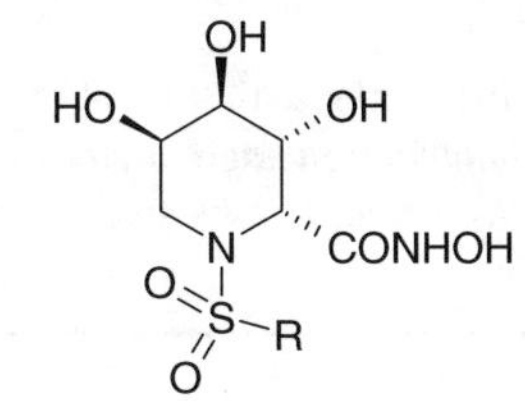

a: R = -4-O(C_6H_5)-(C_6H_5)
b: R = -4-OCH_3-(C_6H_5)
c: R = biphenyl
d: R = -4-OBn-(C_6H_5)
e: R = -4-$OCH_2C{\equiv}CCH_3$-(C_6H_5)

Matrix Metalloproteinase
a: rMMP-1 (EC 3.4.24.7) K_i 5.3 nM; rMMP-3 (EC 3.4.24.17) K_i 0.35 nM; rMMP-9 (EC 3.4.24.35) K_i 0.097 nM [213].
Effects on mice TPA-induced Epidermal Hyperplasia (psoriatic model)
Topical application of this compound dose-dependently suppressed TPA-induced inflammation at a dose of 1-100 μg/site. [213].
b: rMMP-1 K_i 26 nM; rMMP-3 K_i 2.0 nM; rMMP-9 K_i 2.0 nM.
c: rMMP-1 K_i 162 nM; rMMP-3 K_i 50 nM; rMMP-9 K_i 47 nM.
d: rMMP-1 K_i > 850 nM; rMMP-3 K_i 2.1 nM; rMMP-9 K_i 7.4 nM.
e: rMMP-1 K_i 128 nM; rMMP-3 K_i 3.3 nM; rMMP-9 K_i 14 nM.
TNF-α converting enzyme, ADAM17(TACE) [213]
a: K_i 6.2 nM, b: K_i 15 nM, c: K_i 21 nM, d: K_i 1.7 nM, e: K_i 0.53 nM
Heparin-binding EGF-like growth factor (shedding of HB-EGF) IC_{50}:
a: 0.34 μM, b: 0.45 μM, c: 21 μM, d: 0.084 μM, e: 0.028 μM [213]

14.11 Antifungal agents

14.11.1 Five-membered rings

D-Arabinose-configuration

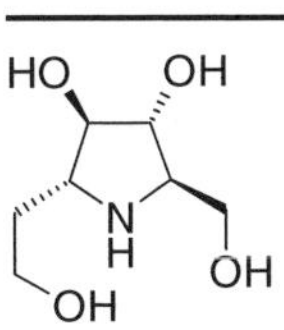

Chitin synthase (EC 2.4.1.16)
IC_{50} 65 μM, K_i 38 μM (behave as an uncompetitive inhibitor) [214].

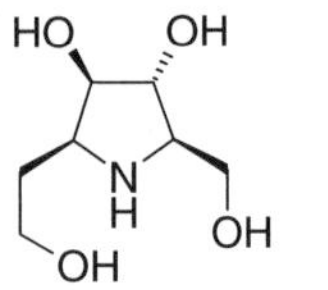

Chitin synthase (EC 2.4.1.16)
IC_{50} 2.6 mM [214].

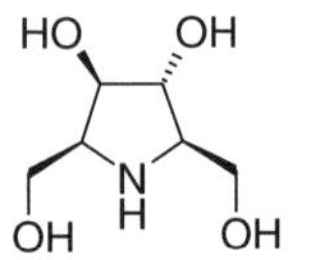

Chitin synthase (EC 2.4.1.16)
IC_{50} 5.7 ± 0.8 mM [215].
No antifungal activity was observed against pathogenic fungal strains *(Cryptococcus neoformans, Trichophyton metagrophytes, Microsporum canis, Aspergillus fumigatus, Candida albicans, Saccharomyces cerevisiae)* [215].

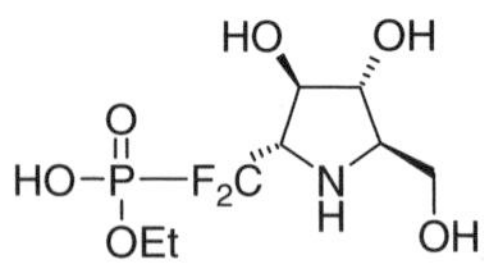

Chitin synthase (EC 2.4.1.16)
IC_{50} 38 ± 4 mM [215].
No antifungal activity was observed against pathogenic fungal strains *(Cryptococcus neoformans, Trichophyton metagrophytes, Microsporum canis, Aspergillus fumigatus, Candida albicans, Saccharomyces cerevisiae)* [215].

Miscellaneous configurations

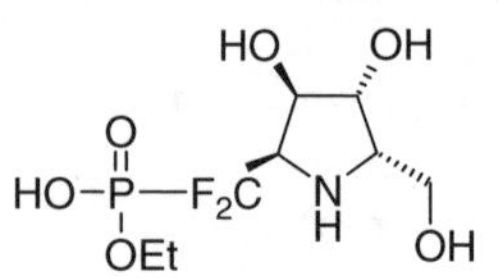

Chitin synthase (EC 2.4.1.16)
IC_{50} 4.0 ± 0.5 mM [215].
No antifungal activity was observed against pathogenic fungal strains *(Cryptococcus neoformans, Trichophyton metagrophytes, Microsporum canis, Aspergillus fumigatus, Candida albicans, Saccharomyces cerevisiae)* [215].

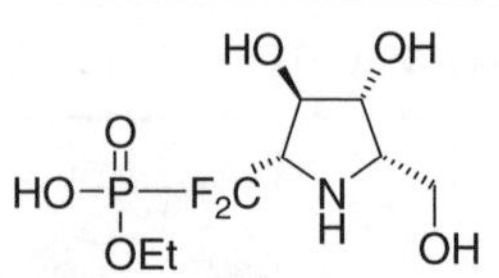

Chitin synthase (EC 2.4.1.16)
IC_{50} 1.6 ± 0.2 mM [215]
No antifungal activity was observed against pathogenic fungal strains *(Cryptococcus neoformans, Trichophyton metagrophytes, Microsporum canis, Aspergillus fumigatus, Candida albicans, Saccharomyces cerevisiae)* [215].

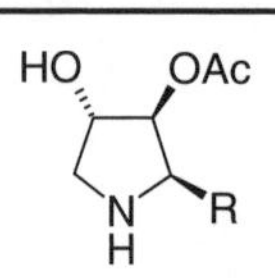

a: R = $-CH_2$-4-OCH_3-(C_6H_4)
b: R = $-CH_2$-4-CH_3-(C_6H_4)
c: R = $-CH_2$-(C_6H_5)
d: R = $-CH_2$-3-OCH_3-(C_6H_4)
e: R = $-CH_2$-2-OCH_3-(C_6H_4)
f: R = -CHMe-4-OCH_3-(C_6H_4)
g: R = -CH(C_6H_5)-4-OCH_3-(C_6H_4)

Minimum inhibitory concentration in μg/ml [103]:
compounds b, c, d, e, f, g were evaluated as racemic form
Saccharomyces cerevisiae.
Compound a: 0.25, compound b: 1, compound c: 2, compound d: 8, compound e: >32, compound f: >32, compound g: >32.
Trichophyton mentagrophytes
Compound a: 2, compound b: 4, compound c: 16, compound d: >32, compound e: >32, compound f: >32, compound g: >32.
Candida albicans
compound a: 4, compound b: 8, compound c: 32, compound d: >32, compound e: >32, compound f: >32, compound g: >32.

(Continued)

Epidermophyton floccosum

Compound a: 16, compound b: 16, compound c: 32, compound d: >32, compound e: >32, compound f: >32, compound g: >32.

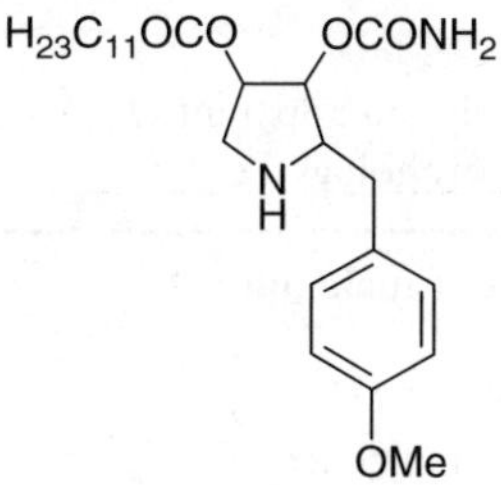

Minimum inhibitory concentration in μg/ml [216]

Candida albicans 50 μg/ml.

Saccharomyces cerevisiae 25 μg/ml.

Cryptococcus neoformans 3.1 μg/ml.

Aspergillus fumigatus 50 μg/ml.

Mucor rouxii 50 μg/ml.

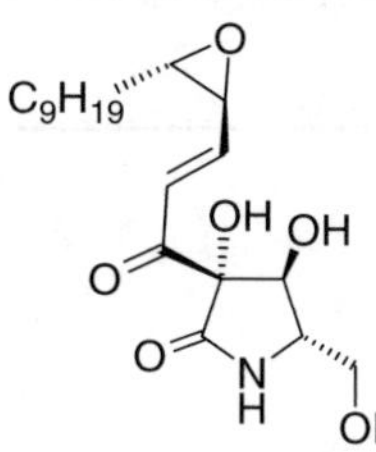

Minimum inhibitory concentration μM [217]

Candida albicans 20 μM.

Candida parapsilosis 4 μM.

Candida tropicalis 20 μM.

Candida krusei >100 μM.

Cryptococcus neoformans 0.062 μM.

Aspergillus fumigatus 100 μM.

Bacillus subtilis 4 μM.

14.12 Cystic fibrosis

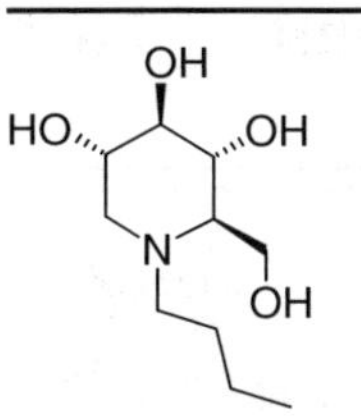

Rescue of functional delF508-CFTR channels in cystic fibrosis epithelial cells [218]

Prevention of delF508-CFTRcalnexin interaction by 55% to 75% and restoration of cAMP-activated chloride current in epithelial CF cells. Rescue of a mature and functional delF508-CFTR in the intestinal crypts of ileal mucosa from delF508 mice [218].

14.13 Infertility inducing agents

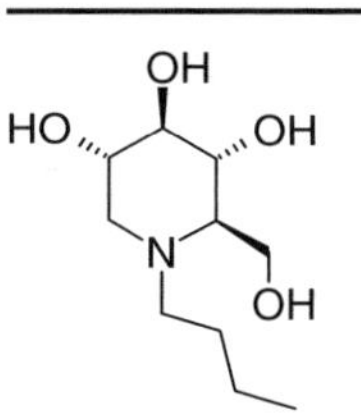

Nonhormonal reversible infertility agent of male mice

1 mg/kg/day: no affection of the male fertility. 5 mg/kg/day: significant reduction of their fertility. 15 mg/kg/day: complete infertility. Mice male treated for 6 months with 15 mg/kg/day of *N*B-DNJ regained their fertility after cessation of drug administration. [219].

The results indicated that *N*B-DNJ induced infertility was not mediated by a depression of reproductive hormones [219].

*N*B-DNJ treatment at 1.2 mg/kg/day did not cause female to become infertile [219].

	(Continued) Long-term low-dose (15 mg/kg/day) *N*B-DNJ treatment does not affect body or gonads weights, reproductive endocrinology, serum chemistry or animal behaviour. No evidence for cumulative toxicity caused by prolonged low-dose *N*B-DNJ treatment [220]. **Human reproduction** After 6 weeks therapy at 100 mg twice daily, no apparent effect on sperm concentration, mobility or morphology [230].
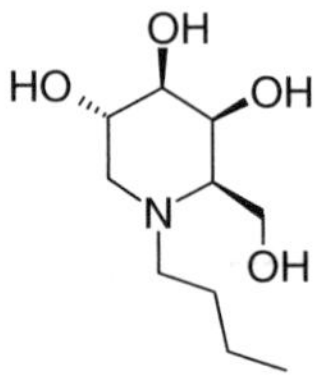	**Nonhormonal reversible infertility agent of male mice** 1200 mg/kg/day: complete infertility [219]. 15 mg/kg/day: no obvious effect on fertility [221]. 150 mg/kg/day: dramatic reduction of the incidence of spermatozoa with normal nuclei and normal acrosomes [221].

14.14 Chondroprotective agents

14.14.1 Five-membered rings

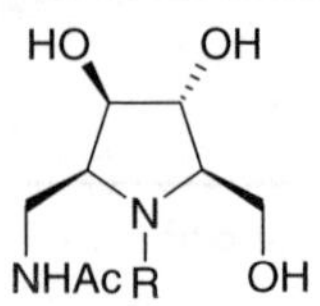	**Osteoarthritis** *β-N-Acetylhexosaminidase* (EC 3.2.1.52) *from human placenta* R = H, K_i 240 nM [222]. R = CH_3, K_i 65 nM [222].
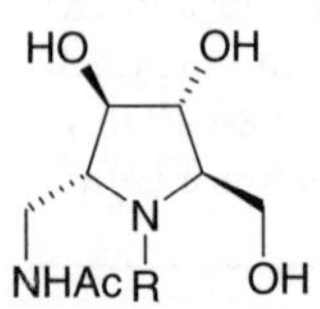	**Osteoarthritis** *β-N-Acetylhexosaminidase* (EC 3.2.1.52) *from human placenta* R = H, K_i 40 nM [222]. R = CH_3, K_i 24 nM; specific inhibition towards human hexosaminidase *in vitro*: glycosaminoglycans accumulation in cultured human articular chondrocytes and cartilage [222].
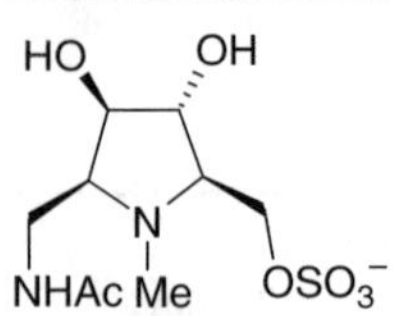	**Osteoarthritis** *β-N-Acetylhexosaminidase* (EC 3.2.1.52) *from human placenta* No inhibition at 100 μM [222].
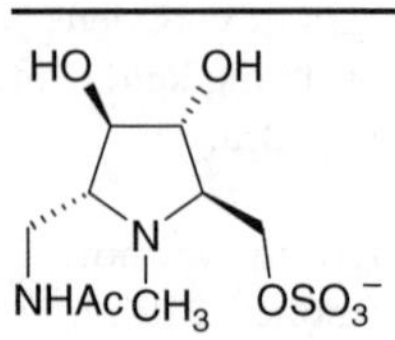	**Osteoarthritis** *β-N-Acetylhexosaminidase* (EC 3.2.1.52) *from human placenta* K_i 5000 nM. Selective inhibition of HexA, IC_{50} 100 nM [222].

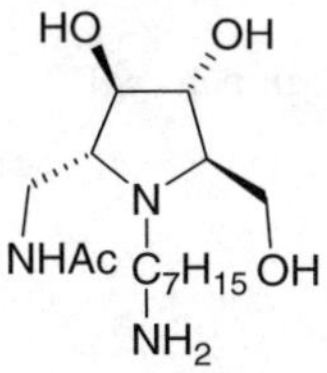	**Osteoarthritis** *β-N-Acetylhexosaminidase* (EC 3.2.1.52) *from human placenta* K_i 2.6 nM [44].
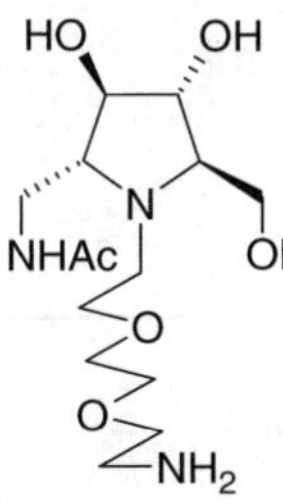	**Osteoarthritis** *β-N-Acetylhexosaminidase* (EC 3.2.1.52) *from human placenta* K_i 60 nM [44].
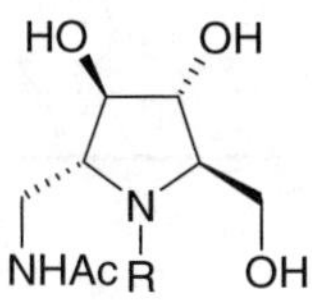	**Osteoarthritis** *β-N-Acetylhexosaminidase* (EC 3.2.1.52) *from human placenta* [44] $R = C_6H_{13}$, K_i 180 nM. $R = C_7H_{15}$, K_i 250 nM. $R = C_8H_{17}$, K_i 160 nM.
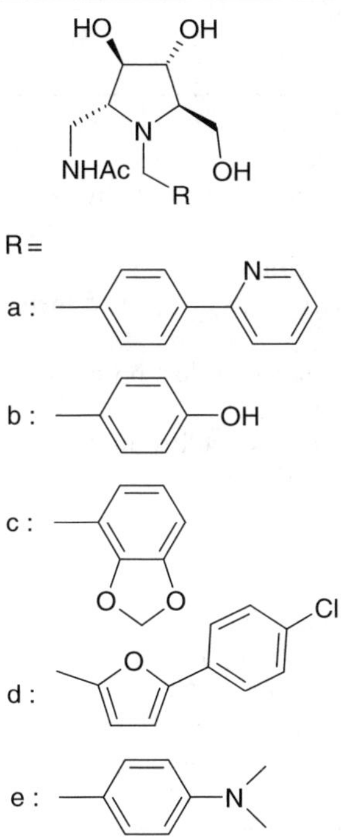	**Osteoarthritis** *β-N-Acetylhexosaminidase* (EC 3.2.1.52) *from human placenta* [223] a: K_i 9.5 μM b: K_i 4.1 μM c: K_i 38 μM d: K_i 10 μM e: K_i 18 μM

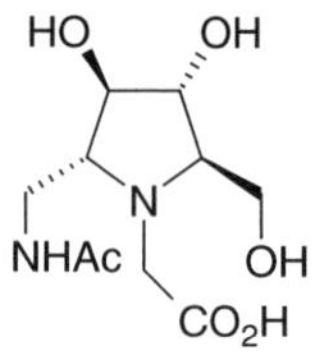

Osteoarthritis
β-N-Acetylhexosaminidase (EC 3.2.1.52) *from human placenta*
92% inhibition at 25 μM [223].

14.14.2 Six-membered rings

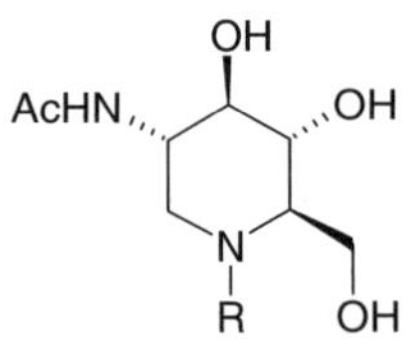

Osteoarthritis
β-N-Acetylhexosaminidase (EC 3.2.1.52) *from human placenta* [222]
R = H, K_i 1.2 μM.
R = CH_3, K_i 0.86 μM.

14.15 Inflammatory diseases

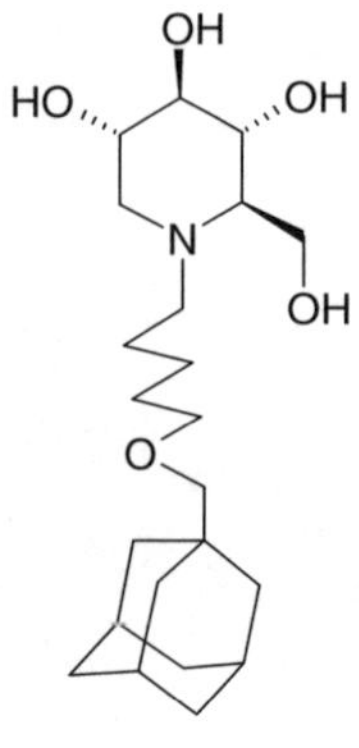

Hapten-induced colitis [224]
Test in two hapten-induced colitis models.
Strong anti-inflammatory and immune suppressive activity on both TNBS- and Oxa-induced colitis.
Effect on body weight after hapten administration
Animals treated with 0.2 mg (low dose, 10 mg/kg) regained their original body weight by day 6 or 7.
Animals treated with 1 mg (high dose, 50 mg/kg) lost no more than 10% of their original body weight.
Protective effect even more evident in the Oxa-colitis model.
Effect on macroscopic and microscopic disease scores
50% reduction in average macroscopic score (3.8 ± 1.2 *vs* 7.8 ± 0.8 in untreated mice).
Effect on tissue MPO activity
At high dose, 3-fold reduction of MPO activity.
Effect of oral AMP-DNM on antihapten antibody production
AntiOxa antibody (IgG_{2a}) lower in treated mice (OD 326.5 ± 72.7 *vs* OD 857.9 ± 191.0 in untreated mice).
AntiTNSB antibody (IgG_{2a}) reduced from OD 469.4 ± 114.5 to OD 157.2 ± 30.4.

14.16 Appetite suppressing agents

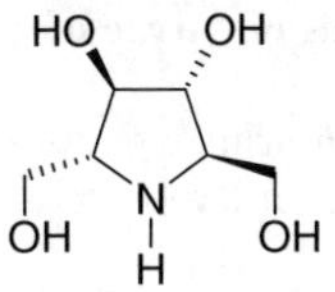

Antifeedant activity [229]

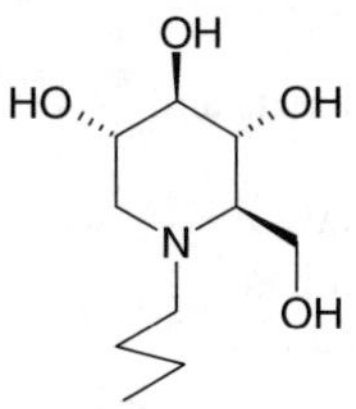

Obesity [225]

Anorectic effect

2 nmol directly injected into the lateral cerebral ventricle of lean control mouse brain. Injected mice ate 20% less.

Effect of iminosugar on mouse growth

Well-defined effect on total body weight only seen in doses of 600 mg/kg/day and above. Reduction in fat pad weight apparent at doses as low as 15mg/kg/day.

Effect on control mouse adipose tissue

40% reduction in mouse skin thickness.

Effect on obese mice

7 week-old group of obese mice treated weighed 35g after six months (40% less than the control).

Dietary intake in obese mice

Significant 30% reduction in food intake.

14.17 Nematicidal activity

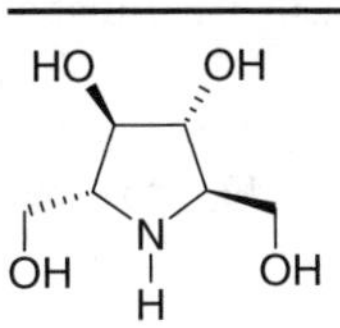

Activity against several plant parasitic nematodes [226]

Inhibits cyst hatch of *Globodera pallida* by 41% at 25 μg/ml, immobilise 53% of *G. rostochiensis* juveniles after 72 h exposure at 1000 μg/ml.

As a foliar spray, systemic activity and decrease of root galling of tomato by *Meloidogyne spp*. A 30 μg/ml soil drench inhibits virus acquisition, transmission and root galling by *Xiphinema diversicaudatum* on *Petunia*.

14.18 Insecticidal activity

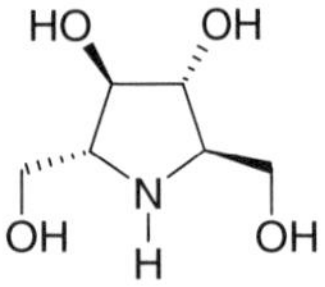	**Detrimental effects in some tropical legume seeds on larvae of the bruchid *Callosobruchus maculatus*** [227] Lethal to the larvae of the bruchid beetle *Callosobruchus maculatus* when incorporated into artificial diets at levels > 0.03%. From 0.003 to 0.03%, reduction of the larval survival in a dose-dependent manner. **Inhibition of digestive glucosidases of aquatic insects and crustacean grazers** [228] *Crustacean zooplankton α-glucosidase* – IC_{50} 19 nM. *Crustacean zooplankton β-glucosidase* – IC_{50} 49 nM. Digestive enzymes of macrozoobenthos (chironomids, trichoptera, ephemeroptera) were less sensitive. The insect digestive β-glucosidase was more effectively inhibited than the α-glucosidase.

References

1. C. W. Ekhart, M. H. Fechter, P. Hadwiger, E. Mlaker, A. E. Stütz, A. Tauss and T. M. Wrodnigg (1999) In *Iminosugars as Glycosidase Inhibitors: Nojirimycin and Beyond* (Stütz, A. E., ed.), Wiley-VCH, New York, USA, pp. 253–390.
2. K. Yasuda, H. Kizu, T. Yamashita, Y. Kameda, A. Kato, R. J. Nash, G. W. J. Fleet, R. J. Molyneux and N. Asano (2002) *J. Nat. Prod.*, **65**, 198–202.
3. N. Asano, H. Kizu, K. Oseki, E. Tomioka, K. Matsui, M. Okamoto and M. Baba (1995) *J. Med. Chem.*, **38**, 2349–2356.
4. O. Muraoka, S. Ying, K. Yoshikai, Y. Matsuura, E. Yamada, T. Minematsu, G. Tanabe, H. Matsuda and M. Yoshikawa (2001) *Chem. Pharm. Bull.*, **49**, 1503–1505.
5. N. Asano, K. Oseki, H. Kizu and K. Matsui (1994) *J. Med. Chem.*, **37**, 3701–3706.
6. B. Andersen, A. Rassov, N. Westergaard and K. Lundgren (1999) *Biochem. J.*, **342**, 545–550.
7. L. Somsak, V. Nagy, Z. Hadady, T. Docsa and P. Gergely (2003) *Curr. Pharm. Des.*, **9**, 1177–1189.
8. L. Karsten, J. Palle, K. Marit, N.-L. Leif and N. Lars (1997) Patent WO 9709040, *Chem. Abstr.*, **126**, 277–385.
9. H. Liu, L. Sim, D. R. Rose and B. M. Pinto (2006) *J. Org. Chem.*, **71**, 3007–3013.
10. N. Asano, K. Ikeda, L. Yu, A. Kato, K. Takebayashi, I. Adachi, I. Kato, H. Ouchi, H. Takahata and G. W. J. Fleet (2005) *Tetrahedron: Asymmetry*, **16**, 223–229.
11. C.-Y. Yu, N. Asano, K. Ikeda, M.-X. Wang, T. D. Butters, M. R. Wormald, R. A. Dwek, A. L. Winters, R. J. Nash and G. W. J. Fleet (2004) *Chem. Commun.*, 1936–1937.
12. N. Asano, M. Nishida, M. Miyauchi, K. Ikeda, M. Yamamoto, H. Kizu, Y. Kameda, A. A. Watson, R. J. Nash and G. W. J. Fleet (2000) *Phytochemistry*, **53**, 379–382.
13. N. Asano, A. Kato, M. Miyauchi, H. Kizu, Y. Kameda, A. A. Watson, R. J. Nash and G. W. J. Fleet (1998) *J. Nat. Prod.*, **61**, 625–628.
14. T. Yamashita, K. Yasuda, H. Kizu, Y. Kameda, A. A. Watson, R. J. Nash, G. W. J. Fleet and N. Asano (2002) *J. Nat. Prod.*, **65**, 1875–1881.

15. A. Kato, I. Adachi, M. Miyauchi, K. Ikeda, T. Komae, H. Kizu, Y. Kameda, A. A. Watson, R. J. Nash, M. R. Wormald, G. W. J. Fleet and N. Asano (1999) *Carbohydr. Res.*, **316**, 95–103.
16. M. I. Garcia-Moreno, D. Rodriguez-Lucena, C. Ortiz Mellet and J. M. Garcia Fernandez (2004) *J. Org. Chem.*, **69**, 3578–3581.
17. A. A. Watson, G. W. J. Fleet, N. Asano, R. J. Molyneux and R. J. Nash (2001) *Phytochemistry*, **56**, 265–295.
18. A. Kato, N. Kato, E. Kano, I. Adachi, K. Ikeda, L. Yu, T. Okamoto, Y. Banda, H. Ouchi, H. Takahata and N. Asano (2005) *J. Med. Chem.*, **48**, 2036–2044.
19. T. Kajimoto, K. Liu, R. L. Pederson, Z. Zhong, Y. Ichikawa, J. A. Porco and C.-H. Wong (1991) *J. Am. Chem. Soc.*, **113**, 6187–6196.
20. R. Uchida, A. Nasu, S. Tokutake, K. Kasai, K. Tobe and N. Yamaji (1999) *Chem. Pharm. Bull.*, **47**, 187–193.
21. S. V. Kyosseva, Z. N. Kyossev and A. D. Elbein (1995) *Arch. Biochem. Biophys.*, **316**, 821–826.
22. V. H. Lillelund, H. H. Jensen, X. Liang and M. Bols (2002) *Chem. Rev.*, **102**, 515–553.
23. P. Jakobsen, J. M. Lundbeck, M. Kristiansen, J. Breinholt, H. Demuth, J. Pawlas, M. P. Torres Candela, B. Andersen, N. Westergaard, K. Lundgren and N. Asano (2001) *Bioorg. Med. Chem.*, **9**, 733–744.
24. L. J. Scott and C. M. Spencer (2000) *Drugs*, **59**, 521–549.
25. N. Asano, R. J. Nash, R. J. Molyneux and G. W. J. Fleet (2000) *Tetrahedron: Asymmetry*, **11**, 1645–1680.
26. K. M. Robinson, M. E. Begovic, B. L. Rhinehart, E. W. Heineke, J. B. Ducep, P. R. Kastner, F. N. Marshall and C. Danzin (1991) *Diabetes*, **40**, 825–830.
27. G. Godin, P. Compain, O. R. Martin, K. Ikeda, L. Yu and N. Asano (2004) *Bioorg. Med. Chem. Lett.*, **14**, 5991–5995.
28. N. Asano, M. Nishida, A. Kato, H. Kizu, K. Matsui, Y. Shimada, T. Itoh, M. Baba, A. A. Watson, R. J. Nash, P. M. de Q. Lilley, D. J. Watkin and G. W. J. Fleet (1998) *J. Med. Chem.*, **41**, 2565–2571.
29. K. Ikeda, M. Takahashi, M. Nishida, M. Miyauchi, H. Kizu, Y. Kameda, M. Arisawa, A. A. Watson, R. J. Nash, G. W. J. Fleet and N. Asano (2000) *Carbohydr. Res.*, **323**, 73–80.
30. A. Bordier, P. Compain, O. R. Martin, K. Ikeda and N. Asano (2003) *Tetrahedron: Asymmetry*, **14**, 47–51.
31. A. Kato, N. Asano, H. Kizu and K. Matsui (1997) *J. Nat. Prod.*, **60**, 312–314.
32. N. Asano (2003) *Curr. Top. Med. Chem.*, **3**, 471–484.
33. A. Kato, E. Kano, I. Adachi, R. J. Molyneux, A. A. Watson, R. J. Nash, G. W. J. Fleet, M. R. Wormald, H. Kizu, K. Ikeda and N. Asano (2003) *Tetrahedron: Asymmetry*, **14**, 325–331.
34. N. Asano, H. Kuroi, K. Ikeda, H. Kizu, Y. Kameda, A. Kato, I. Adachi, A. A. Watson, R. J. Nash and G. W. J. Fleet (2000) *Tetrahedron: Asymmetry*, **11**, 1–8.
35. N. Asano, A. Kato, H. Kizu, K. Matsui, A. A. Watson and R. J. Nash (1996) *Carbohydr. Res.*, **293**, 195–204.
36. N. Asano, A. Kato, Y. Yokoyama, M. Miyauchi, M. Yamamoto, H. Kizu and K. Matsui (1996) *Carbohydr. Res.*, **284**, 169–178.
37. N. Asano, A. Kato, H. Kizu, K. Matsui, R. C. Griffiths, M. G. Jones, A. A. Watson and R. J. Nash (1997) *Carbohydr. Res.*, **304**, 173–178.
38. A. Kato, N. Asano, H. Kizu, K. Matsui, S. Suzuki and M. Arisawa (1997) *Phytochemistry*, **45**, 425–429.
39. Y. Yoshikuni, Y. Ezure, T. Seto, K. Mori, M. Watanabe and H. Enomoto (1989) *Chem. Pharm. Bull.*, **37**, 106–109.
40. W. M. Kati, D. Montgomery, C. Maring, V. S. Stoll, V. Giranda, X. Chen, W. G. Laver, W. Kohlbrenner and D. W. Norbeck (2001) *Antimicrob. Agents Chemother.*, **45**, 2563–2570.

41. G. W. J. Fleet, A. Karpas, R. A. Dwek, L. E. Fellows, A. S. Tyms, S. Petursson, S. K. Namgoong, N. G. Ramsden, P. W. Smith, J. Chan Son, F. Wilson, D. R. Witty, G. S. Jacob and T. W. Rademacher (1988) *FEBS Letters*, **237**, 128–132.
42. D. A Winkler and G. Holan (1989) *J. Med. Chem.*, **32**, 2084–2089.
43. T. Kiyoshi, M. Yohsuke, H. Toshio, K. Tsutomu, O. Masanori, Y. Yoshiko, N. Katsuya and K. Hiroshi (1991) Antiretroviral pyrroline and pyrrolidine sulfonic acid derivatives. *Eur. Pat. Appl.*, EP 0407 701 A2.
44. P.-H. Liang, W.-C. Cheng, Y.-L. Lee, H.-P. Yu, Y.-T. Wu, Y.-L. Lin and C.-H. Wong (2006) *ChemBioChem*, 7, 165–173.
45. A. Karpas, G. W. J. Fleet, R. A. Dwek, S. Petursson, S. K. Namgoong, N. G. Ramsden, G. S. Jacob and T. W. Rademacher (1988) *Proc. Nat. Acad. Sci. USA*, **85**, 9229–9233.
46. T. M. Chapman, I. G. Davies, B. Gu, T. M. Block, D. I. C. Scopes, P. A. Hay, S. M. Courtney, L. A. McNeill, C. J. Schofield and B. G. Davies (2005) *J. Am. Chem. Soc.*, **127**, 506–507.
47. M.-J. Papandréou, R. Barbouche, R. Guieu, M. P. Kieny and E. Fenouillet (2002) *Mol. Pharmacol.*, **61**, 186–193.
48. G. Bolt, I. R. Pedersen and M. Blixenkrone-Moller (1999) *Virus Res.*, **61**, 43–51.
49. D. Durantel, N. Branza-Nichita, S. Carrouée-Durantel, T. D. Butters, R. A. Dwek and N. Zitzmann (2001) *J. Virol.*, **75**, 8987–8998.
50. A. Metha, S. Ouzounov, R. Jordan, E. Simsek, X. Lu, R. M. Moriarty, G. Jacob, R. A. Dwek and T. M. Block (2002) *Antimicrob. Agents Chemother.*, **13**, 315–320.
51. R. A. Mueller, M. L. Bryant and R. A. Partis (2000) Use of substituted-1,5-dideoxy-1,5-imino-*D*-glucitol compounds for treating hepatitis virus infections. *PCT Int. Appl.*, WO 2000047198 A2.
52. M.-P. Courageot, M.-P. Frenkiel, C. Duarte Dos Santos, V. Deubel and P. Desprès (2000) *J. Virol.*, **74**, 564–572.
53. G. S. Jacob, T. W. Rademacher, A. S. Tyms and R. A. Dwek (1990) Use of alkyl derivatives of deoxynojirimycin in the treatment of herpes *Eur. Pat. Appl.*, EP 0378 984 A2.
54. P. B. Fischer, M. Collin, G. B. Karlsson, W. James, T. D. Butters, S. J. Davis, S. Gordon, R. A. Dwek and F. M. Platt (1995) *J. Virol*, **69**, 5791–5797.
55. L. Ratner (1993) *AIDS Res. Hum. Retroviruses*, **9**, 291–297.
56. P. B. Fischer, G. B. Karlsson, T. D. Butters, R. A. Dwek and F. M. Platt (1996) *J. Virol.*, **70**, 7153–7160.
57. M. Tierney, J. Pottage, H. Kessler, M. Fischl, D. Richman, T. Merigan, W. Powderly, S. Smith, A. Karim, J. Sherman and M. Hirsch (1995) *J. Acqui. Immune Defic. Syndr. Hum. Retrovirol.*, **10**, 549–553.
58. T. M. Block, X. Lu, F. M. Platt, G. R. Foster, W. H. Gerlich, B. S. Blumberg and R. A. Dwek (1994) *Proc. Nat. Acad. Sci. USA*, **91**, 2235–2239.
59. N. Zitzmann, A. S. Metha, S. Carrouée, T. D. Butters, F. M. Platt, J. McCauley, B. S. Blumberg, R. A. Dwek and T. M. Block (1999) *Proc. Nat. Acad. Sci. USA*, **96**, 11878–11882.
60. S. Ouzounov, A. Metha, R. A. Dwek, T. M. Block and R. Jordan (2002) *Antiviral Res.*, **55**, 425–435.
61. R. Jordan, O. V. Nikolaeva, L. Wang, B. Conyers, A. Mehta, R. A. Dwek and T. M. Block (2002) *Virology*, **295**, 10–19.
62. D. Pavlovic, D. C. A. Neville, O. Argaud, B. Blumberg, R. A. Dwek, W. B. Fischer and N. Zitzmann (2003) *Proc. Nat. Acad. Sci. USA*, **100**, 6104–6108.
63. C. S. Cook, P. J. Karabatsos, G. L. Schoenhard and A. Karim (1995) *Pharmaceutical Research.*, **12**, 1158–1163.
64. A. Metha, S. Carrouée, B. Conyers, R. Jordan, T. Butters, R. A. Dewk and T. M. Block (2001) *Hepathology*, **33**, 1488–1495.

65. S.-F. Wu, C.-J. Lee, C.-L. Liao, R. A. Dwek, N. Zitzmann and Y.-L. Lin (2002) *J. Virol.*, **76**, 3596–3604.
66. J. M. Mackenzie and E.G. Westaway (2001) *J. Virol.*, **75**, 10787–10799.
67. D. Durantel, S. Carrouée-Durantel, N. Branza-Nichita, R. A. Dwek and N. Zitzmann (2004) *Antimicrob. Agents Chemother.*, **48**, 497–504.
68. C.-Y. Wu, J.-T. Jan, S.-H. Ma, C.-J. Kuo, H.-F. Juan, Y.-S. E. Cheng, H.-H. Hsu, H.-C. Huang, D. Wu, A. Brik, F.-S. Liang, R.-S. Liu, J.-M. Fang, S.-T. Chen, P.-H. Liang and C. -H. Wong (2004) *Proc. Nat. Acad. Sci. USA*, **101**, 10012–10017.
69. K. T. Weber, D. Hammache, J. Fantini and B. Ganem (2000) *Bioorg. Med. Chem. Lett.*, **10**, 1011–1014.
70. A. S. Metha, B. Conyers, D. L. J. Tyrrell, K.-A. Walters, G. A. Tipples, R. A. Dwek and T. M. Block (2002) *Antimicrob. Agents Chemother.*, **46**, 4004–4008.
71. X. Lu, T. Tran, E. Simsek and T. M. Block (2003) *J. Virol.*, **77**, 11933–11940.
72. A. S. Metha, B. Gu, B. Conyers, S. Ouzounov, L. Wang, R. M. Moriarty, R. A. Dwek and T. M. Block (2004) *Antimicrob. Agents Chemother.*, **48**, 2085–2090.
73. E. T. A. Marques Jr., Y. Ichikawa, M. Strand, J. T. August, G. W. Hart and R. L. Schnaar (2001) *Glycobiology*, **11**, 249–259.
74. L. A. Augustin, J. Fantini and D. R. Mootoo (2006) *Bioorg. Med. Chem.*, **14**, 1182–1188.
75. D. R. P. Tulsiani, T. M. Harris and O.Touster (1982) *J. Biol. Chem*, **257**, 7936–7939.
76. D.-S. Lee, K.-E. Jung, C.-H. Yoon, H. Lim and Y.-S. Bae (2005) *Antimicrob. Agents Chemother.*, **49**, 4110–4120.
77. Y. Nishimura, Y. Umezawa, S. Kondo and T. Takeuchi (1993) *J. Antibiotics*, **46**, 1883–1889.
78. B. D. Walker, M. Kowalski, W. Chun Goh, K. Kozarsky, M. Krieger, C. Rosen, L. Rohrscheneider, W. A. Haseltine and J. Sodroski (1987) *Proc. Nat. Acad. Sci. USA*, **84**, 8120–8124.
79. P. S. Sunkara, D. L. Taylor, M. S. Kang, T. L. Bowlin, P. S. Liu, A. S. Tyms and A. Sjoerdsma (1989) *Lancet.*, **1**, 1206.
80. D. L. Taylor, M. S. Kang, T. M. Brennen, C. G. Bridges, P. S. Sunkara and A. S. Tyms (1994) *Antimicrob. Agents Chemother.*, **383**, 1780–1787.
81. V. A. Johnson, B. D. Walker, M. A. Barlow, T. J. Paradis, T.-C. Chou and M. S. Hirsch (1989) *Antimicrob. Agents Chemother.*, **33**, 53–57.
82. R. M. Ruprecht, S. Mullaney, J. Anderson and R. Bronson (1989) *J. AIDS*, **2**, 149–157.
83. E. B. Stephens, E. Monck, K. Reppas and E. J. Butfiloski (2001) *J. Virol.*, **65**, 1114–1123.
84. K. Whitby, T. C. Pierson, B. Geiss, K. Lane, M. Engle, Y. Zhou, R. W. Doms and M. S. Diamond (2005) *J. Virol.*, **79**, 8698–8706.
85. G. Bolt (2001) *Arch. Virol.*, *1***46**, 2055–2068.
86. M. S. Kang (1996) *Glycobiology*, **6**, 209–216.
87. C. G. Bridges, S. P. Ahmed, M. S. Kang, R. J. Nash, E. A. Porter and A. S. Tyms (1995) *Glycobiology*, **5**, 249–253.
88. S. P. Ahmed, R. J. Nash, C. G. Bridges, D. L. Taylor, M. S. Kang, E. A. Porter and A. S. Tyms (1995) *Biochem. Biophys. Res. Commun.*, **208**, 267–273.
89. R. E. Lee, M. D. Smith, R. J. Nash, R. C. Griffiths, M. McNeil, R. K. Grewal, W. Yan, G. S. Besra, P. J. Brennan and G. W. J. Fleet (1997) *Tetrahedron Lett.*, **38**, 6733–6736.
90. N. Veerapen, Y. Yuan, D. A. R. Sanders and B. M. Pinto (2004) *Carbohydr. Res.*, **339**, 2205–2217.
91. J. A. Maddry, N. Bansal, L. E. Bermudez, R. N. Comber, I. M. Orme, W. J. Suling, L. N. Wilson and R. C. Reynolds (1998) *Bioorg. Med. Chem. Lett.*, **8**, 237–242.
92. R. C. Reynolds, N. Bansal, J. Rose, J. Friedrich, W. J. Suling and J. A. Maddry (1999) *Carbohydr. Res.*, **317**, 164–179.

93. S. Cren, S. S. Gurcha, A. J. Blake, G. S. Besra and N. R. Thomas (2004) *Org. Biomol. Chem.*, **2**, 2418–2420.
94. K. Marotte, T. Ayad, Y. Génisson, G. S. Besra, M. Baltas and J. Prandi (2003) *Eur. J. Org. Chem.*, 2557–2565.
95. G.-C. Zhou, S. L. Parikh, P. C. Tyler, G. B. Evans, R. H. Furneaux, O. V. Zubkova, P. A. Benjes and V. L. Schramm (2004) *J. Am. Chem. Soc.*, **126**, 5690–5698.
96. N. L. Segraves and P. Crews (2005) *J. Nat. Prod.*, **68**, 118–121.
97. J. P. Shilvock, J. R. Wheatley, B. Davis, R. J. Nash, R. C. Griffiths, M. G. Jones, M. Müller, S. Crook, D. J. Watkin, C. Smith, G. S. Besra, P. J. Brennan and G. W. J. Fleet (1996) *Tetrahedron Lett.*, **47**, 8569–8572.
98. J. P. Shilvock, J. R. Wheatley, R. J. Nash, A. A. Watson, R. C. Griffiths, T. D. Butters, M. Müller, D. J. Watkin, D. A. Winkler and G. W. J. Fleet (1999) *J. Chem. Soc., Perkin Trans. 1*, **1**, 2735–2745.
99. J. M. Berge, R. C. B. Copley, D. S. Eggleston, D. W. Hamprecht, R. L. Jarvest, L. M. Menash, P. J. O'Hanlon and A. J. Pope (2000) *Bioorg. Med. Chem. Lett.*, **10**, 1811–1814.
100. H. A. Johnson and N. R. Thomas (2002) *Bioorg. Med. Chem. Lett.*, **12**, 237–241.
101. A. Kamal, D. Rajasekhar Reddy, P. S. Muraly Mohan Reddy and Rajendar (2006) *Bioorg. Med. Chem. Lett.*, **16**, 1160–1163.
102. A. L. Stefanska, N. J. Coates, L. M. Mensah, A. J. Pope, S. J. Ready and S. R. Warr (2000) *J. Antibiotics*, **53**, 345–350.
103. S. S. Hall, D. Loebenberg and D. P. Schumacher (1983) *J. Med. Chem.*, **26**, 469–475.
104. G. A. Kicska, P. C. Tyler, G. B. Evans, R. H. Furneaux, V. L. Schramm and K. Kim (2002) *J. Biol. Chem.*, **277**, 3226–3231.
105. G. A. Kicska, P. C. Tyler, G. B. Evans, R. H. Furneaux, K. Kim and V. L. Schramm (2002) *J. Biol. Chem.*, **277**, 3219–3225.
106. R. W. Miles, P. C. Tyler, G. B. Evans, R. H. Furneaux, D. W. Parkin and V. L. Schramm (1999) *Biochemistry*, **38**, 13147–13154.
107. R. H. Furneaux, V. L. Schramm and P. C. Tyler (1999) *Bioorg. Med. Chem.*, **7**, 2599–2606.
108. K. Clinch, G. B. Evans, G. W. J. Fleet, R. H. Furneaux, S. W. Johnson, D. H. Lenz, S. P. H. Mee, P. R. Rands, V. L. Schramm, E. A. Taylor Ringia and P. C. Tyler (2006) *Org. Biomol. Chem.*, **4**, 1131–1139.
109. A. Lewandowicz, E. A. Talor Ringia, L.-M Ting, K. Kim, P. C. Tyler, G. B. Evans, O. V. Zubkova, S. Mee, G. F. Painter D. H. Lenz, R. H. Furneaux and V. L. Schramm (2005) *J. Biol. Chem.*, **280**, 30320–30328.
110. D. W. Parkin, G. Limberg, P. C. Tyler, R. H. Furneaux, X.-Y. Chen and V. L. Schramm (1997) *Biochemistry*, **36**, 3528–3534.
111. A. J. Bitoni, P.P. McCann and A. Sjoerdsma (1991) Preparation of esters of castanospermine for the treatment of cerebral malaria. *Eur. Pat. Appl.*, EP 90–119842 19901016.
112. I. Cenci Di Bello, G. Fleet, S. K. Namgoong, K. Tadano and B. Winchester (1989) *Biochem. J.*, **259**, 855–861.
113. B. Winchester, S. Al Daher, N. C. Carpenter, I. Cenci di Bello, S. S. Choi, A. J. Fairbanks and G. W. J. Fleet (1993) *Biochem. J.*, **290**, 743–749.
114. W. H. Pearson and L. Guo (2001) *Tetrahedron Lett.*, **42**, 8267–8271.
115. E. J. Hembre and W. H. Pearson (1997) *Tetrahedron*, **53**, 11021–11032.
116. W. H. Pearson and E. J. Hembre (1996) *J. Org. Chem.*, **61**, 5546–5556.
117. W. H. Pearson and E. J. Hembre (2001) *Tetrahedron Lett.*, **42**, 8273–8276.
118. F. Popowycz, S. Gerber-Lemaire, R. Demange, E. Rodriguez-Garcia, A. T. Carmona Asenjo, I. Robina and P. Vogel (2001) *Bioorg. Med. Chem. Lett.*, **11**, 2489–2493.
119. V. L. Schramm and P. C. Tyler (2003) *Curr. Top. Med. Chem.*, **3**, 525–540.

120. H. Fiaux, F. Popowycz, S. Favre, C. Schütz, P. Vogel, S. Gerber-Lemaire and L. Juillerat-Jeanneret (2005) *J. Med. Chem.*, **48**, 4237–4246.
121. F. Popowycz, S. Gerber-Lemaire, C. Schütz and P. Vogel (2004) *Helv. Chim. Acta*, **87**, 800–810.
122. F. Popowycz, S. Gerber-Lemaire, E. Rodriguez-García, C. Schütz and P. Vogel (2003) *Helv. Chim. Acta*, **86**, 1914–1948.
123. A. Carmona, F. Popowycz, S. Gerber-Lemaire, E. Rodriguez-García, C. Schütz, P. Vogel and I. Robina (2003) *Bioorg. Med. Chem.*, **11**, 4897–4911.
124. Y. Nishimura (2003) *Curr. Top. Med. Chem.*, **3**, 575–591.
125. Y. Nishimura (2006) *Heterocycles*, **67**, 461–488.
126. J. M. H. van Den Elsen, D. A. Kuntz and D. R. Rose (2001) *EMBO J.*, **20**, 3008–3017.
127. E. Shitara, Y. Nishimura, F. Kojima and T. Takeuchi (2000) *Bioorg. Med. Chem.*, **8**, 343–352.
128. Y. Nishimura, T. Satoh, T. Kudo, S. Kondo and T. Takeuchi (1996) *Bioorg. Med. Chem.*, **4**, 91–96.
129. E. Shitara, Y. Nishimura, F. Kojima and T. Takeuchi (1999) *J. Antibiotics*, **52**, 348–350.
130. J. R. Brown, Y. Nishimura and J. D. Esko (2006) *Bioorg. Med. Chem. Lett.*, **16**, 532–536.
131. Y. Nishimura, T. Satoh, H. Adachi, S. Kondo, T. Takeuchi, M. Azetaka, H. Fukuyasu and Y. Iizuka (1997) *J. Med. Chem.*, **40**, 2626–2633.
132. (a) T. Tsuruoka, H. Fukuyasu, M. Ishii, T. Usui, S. Shibahara and S. Inouye (1996) *J. Antibiotics*, **49**, 155–161; (b) T. Tsuruoka, H. Fukuyasu, M. Azetaka, Y. Iizuka, S. Inouye, M. Hosokawa and H. Kobayashi (1995) *Jpn. J. Cancer Res.*, **86**, 41–47.
133. T. Kudo, Y. Nishimura, S. Kondo and T. Takeuchi (1992) *J. Antibiotics*, **45**, 954–962.
134. Y. Nishimura, T. Kudo, S. Kondo and T. Takeuchi (1992) *J. Antibiotics*, **45**, 963–970.
135. T. Kudo, Y. Nishimura, S. Kondo and T. Takeuchi (1992) *J. Antibiotics*, **45**, 1662–1668.
136. Y. Nishimura, T. Kudo, S. Kondo and T. Takeuchi (1994) *J. Antibiotics*, **47**, 101–107.
137. Y. Nishimura, T. Satoh, S. Kondo, T. Takeuchi, M. Azetaka, H. Fukuyasu, Y. Iizuka and S. Shibahara (1994) *J. Antibiotics*, **47**, 840–842.
138. Y. Kawase, M. Takahashi, T. Takastu, M. Arai, M. Nakajima and K. Tanzawa (1996) *J. Antibiotics*, **49**, 61–64.
139. Y. Nishimura, E. Shitara, H. Adachi, M. Toyoshima, M. Nakajima, Y. Okami and T. Takeuchi (2000)*J. Org. Chem.*, **65**, 2–11.
140. G. Trugnan, M. Rousset and A. Zweibaum (1986) *FEBS Lett.*, **195**, 28–32.
141. A. Hadwiger, H. Niemann, A. Käbisch, H. Bauer and T. Tamura (1986) *EMBO J.*, **5**, 689–694.
142. I. Noda, S. Fujieda, M. Seki, N. Tanaka, H. Sunaga, T. Ohtsubo, H. Tsuzuki, G. K. Fan and H. Saito (1999) *Int. J. Cancer*, **80**, 279–284.
143. B. G. Winchester, I. Cenci di Bello, A. C. Richardson, R. J. Nash, L. E. Fellows, N. G. Ramsden and G. Fleet (1990) *Biochem. J.*, **269**, 227–231.
144. C. Gravier-Pelletier, W. Maton, G. Bertho and Y. Le Merrer (2003) *Tetrahedron*, **59**, 8721–8730.
145. G. F. Painter, P. J. Eldridge and A. Falshaw (2004) *Bioorg. Med. Chem.*, **12**, 225–232.
146. H. Li, Y. Blériot, J-M. Mallet, E. Rodriguez-Garcia, P. Vogel, Y. Zhang and P. Sinaÿ (2005) *Tetrahedron: Asymmetry*, **16**, 313–319.
147. N. S. Karanjule, S. D. Markad, T. Sharma, S. G. Sabharwal, V. G. Puranik and D. D. Dhavale (2005) *J. Org. Chem*, **70**, 1356–1363.
148. R. Saul, J. P. Chambers, R. J. Molyneux and A. D. Elbein (1983) *Arch. Biochem. Biophys.*, **221**, 593–597.
149. F. Moris-Varas, X-H Qian and C-H Wong (1996) *J. Am. Chem. Soc.*, **118**, 7647–7652.
150. G. W. J. Fleet, P. W. Smith, R. Nash, L. E. Fellows, R. B. Parekh and T. W. Rademacher (1986) *Chem. Lett.*, **15**, 1051–1054.

151. H. G. Selnick, E. M. Radzilowski and G. S. Ponticello (1991) *Tetrahedron Lett.*, **32**, 721–724.
152. T. Aoyagi, H. Suda, K. Uotani, F. Kojima, T. Aoyama, K. Horiguchi, M. Hamada and T. Takeuchi (1992) *J. Antibiotics*, **45**, 1404–1408.
153. K. Tatsuta, S. Miura, S. Ohta and H. Gunji (1995) *J. Antibiotics*, **48**, 286–288.
154. K. Tatsuta, Y. Ikeda and S. Miura (1996) *J. Antibiotics*, **49**, 836–838.
155. F. Gessier, T. Tschamber, C. Tarnus, M. Neuburger, W. Huber and J. Streith (2001) *Eur. J. Org. Chem.*, 4111–4125.
156. M. Terinek and A. Vasella (2005) *Helv. Chim. Acta.*, **88**, 10–22.
157. T. Aoyama, F. Kojima, C. Imada, Y. Muraoka, H. Naganawa, Y. Okami, T. Takeuchi and T. Aoyagi (1995) *J. Enzym. Inhib.*, **8**, 223–232.
158. H. Kayakiri, S. Takase, T. Shibata, M. Okamoto, H. Terano, M. Hashimoto, T. Tada and S. Koda (1989) *J. Org. Chem.*, **54**, 4015–4016.
159. A. D. Elbein, J. E. Tropea, M. Mitchell and G. P. Kaushal (1990) *J. Biol. Chem.*, **265**, 15599–15605.
160. S. Weng and R. G. Spiro (1996) *Arch. Biochem. Biophys.*, **325**, 113–123.
161. N. Shah, D. A. Kuntz and D. R. Rose (2003) *Biochemistry*, **42**, 13812–13816.
162. K. W. Hering, K. Karaveg, K. W. Moremen and W. H. Pearson (2005) *J. Org. Chem.*, **70**, 9892–9904.
163. Y. Fujii, H. Iwasa, J. Hirai, Y. Takahama, K. Hasumi and M. Eriguchi (2000) *Biomed. Pharmacother*, **54**, 85–92.
164. T. Kino, N. Inamura, K. Nakahara, S. Kiyoto, T. Goto, H. Terano, M. Kohsaka, H. Aoki and H. Imanaka (1985) *J. Antibiotics*, **38**, 936–940.
165. J. W. Dennis (1986) *Cancer Res.*, **46**, 5131–5136.
166. M. J. Humphries, K. Matsumoto, S. L. White and K. Olden (1986) *Cancer Res.*, **46**, 5215–5222.
167. J. W. Dennis, K. Koch and D. Beckner (1989) *J. Nat. Cancer Inst.*, **81**, 1028–1033.
168. K. Tsukamoto, A. Uno, Y. Kubota, S. Shimada, Y. Hori and G. Imokawa (1992) *Melanoma Res.*, **2**, 33–39.
169. J. W. Dennis, S. L. White, A. M. Freer and D. Dime (1993) *Biochem. Pharmacol.*, **46**, 1459–1466.
170. B. Liu, Y. Lin and H. Yin (1998) *Zhonghua Zhong Liu Za Zhi*, **20**, 168–170.
171. Y. Kuramitsu, J Hamada, T. Tsuruoka, K. Morikawa, H. Kobayashi and M. Hosokawa (1998) *Anti-Cancer Drugs*, **9**, 88–92.
172. Y. Kuramitsu, J Hamada, T. Tsuruoka, K. Morikawa, S. Naito, H. Kobayashi and M. Hosokawa (1998) *Anti-Cancer Drugs*, **9**, 739–741.
173. S. Nakatsugawa, T. Okuda, H. Muramoto, K. Koyama, T. Ishigaki, T. Tsuruoka, M. Hosokawa and H. Kobayashi (1999) *Anti-Cancer Drugs*, **10**, 229–233.
174. G. A. Kicska, L. Long, H. Hörig, C. Fairchild, P. C. Tyler, R. H. Furneaux, V. L. Schramm and H. L. Kaufman (2001) *Proc. Nat. Acad. Sci USA*, **98**, 4593–4598.
175. S. Bantia, P. J. Miller, C. D. Parker, S. L. Ananth, L. L. Horn, J. M. Kilpatrick, P. E. Morris, T. L. Hutchison, J. A. Montgomery and J. S. Sandhu (2001) *Int. Immunopharmacol.*, **1**, 1199–1210.
176. S. Bantia, S. L. Ananth, C. D. Parker, L. L. Horn and R. Upshaw (2003) *Int. Immunopharmacol.*, **3**, 879–887.
177. G. B. Evans, R. H. Furneaux, G. J. Gainsford, V. L. Schramm and P. C. Tyler (2000) *Tetrahedron*, **56**, 3053–3062.
178. G. B. Evans, R. H. Furneaux, V. L. Schramm, V. Singh and P. C. Tyler (2004) *J. Med. Chem.*, **47**, 3275–3281.
179. M. Boutellier, B. Horenstein, A. Semenyaka, V. Schramm and B. Ganem (1994) *Biochemistry*, **33**, 3994–4000.

180. T. D. Butters, L. A. G. M. van den Broek, G. W. J. Fleet, T. M. Krulle, M. R. Wormald, R. A. Dwek and F. M. Platt (2000) *Tetrahedron: Asymmetry*, **11**, 113–124.
181. A. R. Sawkar, S. L. Adamski-Werner, W.-C. Cheng, C.-H. Wong, E. Beutler, K.-P. Zimmer and J. W. Kelly (2005) *Chem. Biol.*, **12**, 1235–1244.
182. A. R. Sawkar, W.-C. Cheng, E. Beutler, C.-H. Wong, W. E. Balch and J. W. Kelly (2002) *Proc. Nat. Acad. Sci USA*, **99**, 15428–15433.
183. T. D. Butters, R. A. Dwek and F. M. Platt (2003) *Curr. Top. Med. Chem.*, **3**, 561–574.
184. P. Compain, O. R. Martin, C. Boucheron, G. Godin, L. Yu, K. Ikeda and N. Asano (2006) *ChemBioChem*, **7**, 1356–1359.
185. L. Yu, K. Ikeda, A. Kato, I. Adachi, G. Godin, P. Compain, O. R. Martin and N. Asano (2006) *Bioorg. Med. Chem.*, **14**, 7736–7744.
186. H. S. Overkleeft, G. H. Renkema, J. Neele, P. Vianello, I. O. Hung, A. Strijland, A. M. van der Burg, G.-J. Koomen, U. K. Pandit and J. M. F. G. Aerts (1998) *J. Biol. Chem.*, **273**, 26522–26527.
187. (a) J.-Q. Fan, S. Ishii, N. Asano and Y. Suzuki (1999) *Nature Medicine*, **5**, 112-115; (b) N. Asano, S. Ishii, H. Kizu, K. Ikeda, K. Yasuda, A. Kato, O. R. Martin and J-Q. Fan (2000) *Eur. J. Biochem.*, **267**, 4179–4186.
188. M. B. Tropak, S. P. Reid, M. Guiral, S. G. Withers and D. Mahuran (2004) *J. Biol. Chem.*, **279**, 13478–13487.
189. (a) T. M. Cox, R. Lachmann, C. Hollak, J. M. F. G. Aerts, S. van Weely, M. Hrebicek, F. M. Platt, T. D. Butters, R. A. Dwek, C. Moyses, I. Gow, D. Elstein and A. Zimran (2000) *Lancet*, **355**, 1481-1485; (b) T. M. Cox, J. M. F. G. Aerts, G. Andria, M. Beck, N. Belmatoug, B. Bembi, R. Chertkoff, S. Vom Dahl, D. Elstein, A. Erikson, M. Giralt, R. Heitner, C. Hollak, M. Hrebicek, S. Lewis, A. Mehta, G. M. Pastores, A. Rolfs, M. C. Sa Miranda and A. Zimran (2003) *J. Inherit. Metab. Dis.*, **26**, 513-766; (c) R. H. Lachmann (2003) *Curr. Opin. Invest. Drugs*, **4**, 472–479.
190. F. M. Platt, G. R. Neises, R. A. Dwek and T. D. Butters (1994) *J. Biol. Chem.*, **269**, 8362–8365.
191. P. Alfonso, S. Pampin, J. Estrada, J. C. Rodriguez-Rey, P. Giraldo, J. Sancho and M. Pocovi (2005) *Blood Cells Mol. Dis.*, **35**, 268–276.
192. (a) M. Zervas, K. L. Somers, M. A. Thrall and S. U. Walkleys (2001) *Curr. Biol.*, **11**, 1283–1287; (b) R. H. Lachmann, D. Te Vruchte, E. Lloyd-Evans, G. Reinkensmeier, D. J. Sillence, L. Fernandez-Guillen, R. A. Dwek, T. D. Butters, T. M. Cox and F. M. Platt (2004) *Neurobiol. Dis.*, **16**, 654–658.
193. (a) F. M. Platt, G. R. Neises, G. Reinkensmeier, M. J. Townsend, V. H. Perry, R. L. Proia, B. Winchester, R. A. Dwek and T. D. Butters (1997) *Science*, **276**, 428-431; (b) M. Jeyakumar, T. D. Butters, M. Cortina Borja, V. Hunnam, R. L. Proia, V. H. Perry, R. A. Dwek and F. M. Platt (1999) *Proc. Nat. Acad. Sci. USA*, **96**, 6388-6393; (c) B. Bembi, F. Marchetti, V. I. Guerci, G. Ciana, R. Addobbati, D. Grasso, R. Barone, R. Cariati, L. Fernandez-Guillen, T. D. Butters and M. G. Pittis (2006) *Neurology*, **66**, 278–280.
194. H. R. Mellor, J. Nolan, L. Pickering, M. R. Wormald, F. M. Platt, R. A. Dwek, G. W. J. Fleet and T. D. Butters (2002) *Biochem. J.*, **366**, 225–233.
195. C. Boucheron, V. Desvergnes, P. Compain, O. R. Martin, A. Lavi, M. Mackeen, M. R. Wormald, R. A. Dwek and T. D. Butters (2005) *Tetrahedron: Asymmetry*, **16**, 1747–1756.
196. (a) G. H.-F. Yam, C. Zuber and J. Roth (2005) *FASEB J.*, **19**, 12-18; (b) S. Ishii, H. Yoshioka, K. Mannen, A. B. Kulkarni and J.-Q. Fan (2004) *Biochem. Biophys. Acta*, **20**, 250–257.
197. L. Tominaga, Y. Ogawa, M. Taniguchi, K. Ohno, J. Matsuda, A. Oshima, Y. Suzuki and E. Nanba (2001) *Brain Dev.*, **23**, 284–287.
198. U. Andersson, T. D. Butters, R. A. Dwek and F. M. Platt (2000) *Biochem. Pharmacol.*, **59**, 821–829.

199. U. Andersson, D. Smith, M. Jeyakumar, T. D. Butters, M. Cortina Borja, R. A. Dwek and F. M. Platt (2004) *Neurobiol. Dis.*, **16**, 506–515.
200. J. L. Kasperzyk, M. M. El-Abbadi, E. C. Hauser, A. d'Azzo, F. M. Platt and T. N. Seyfried (2004) *J. Neurochem.*, **89**, 645–653.
201. J. L. Kasperzyk, A. d'Azzo, F. M. Platt, J. Alroy and T. N. Seyfried (2005) *J. Lipid Res.*, **46**, 744–751.
202. R. J. B. H. N. van den Berg, W. Donker-Koopman, J. H. van Boom, H. M. F. G. Aerts and D. Noort (2004) *Bioorg. Med. Chem.*, **12**, 891–902.
203. X. Zhu, K. A. Sheth, S. Li, H.-H. Chang and J.-Q. Fan (2005) *Angew. Chem. Int. Ed.*, **44**, 7450–7453.
204. G. B. Evans, R. H. Furneaux, A. Lewandowicz, V. L. Schramm and P. C. Tyler (2003) *J. Med. Chem.*, **46**, 3412–3423.
205. S. Bantia, P. J. Miller, C. D. Parker, S. L. Ananth, L. L. Horn, Y. S. Babu and J. S. Sandhu (2002) *Int. Immunopharmacol.*, **2**, 913–923.
206. G. B. Evans, R. H. Furneaux, G. J. Gainsford, J. C. Hanson, G. A. Kicska, A. A. Sauve, V. L. Schramm and P. C. Tyler (2003) *J. Med. Chem.*, **46**, 155–160.
207. T. Shibata, O. Nakayama, Y. Tsurumi, M. Okuhara, H. Terano and M. Kohsaka (1988) *J. Antibiotics*, **41**, 296–301.
208. G. B. Evans, R. H. Furneaux, A. Lewandowicz, V. L. Schramm and P. C. Tyler (2003) *J. Med. Chem.*, **46**, 5271–5276.
209. X.-L. Zhang, M. Liu, P. Xie, S. Wan, J. T. Ye, X. Zhou and J. Wu (2004) *Bioorg. Med. Chem. Lett.*, **14**, 3789–3792.
210. X.-S. Ye, F. Sun, M. Liu, Q. Li, Y. Wang, G. Zhang, L.-H. Zhang and X.-L. Zhang (2005) *J. Med. Chem.*, **48**, 3688–3691.
211. K. A. Wall, J. D. Pierce and A. D. Elbein (1988) *Proc. Nat. Acad. Sci. USA*, **85**, 5644–5648.
212. P. M. Grochowicz, A. D. Hibberd, Y. C. Smart, K. M. Bowen, D. A. Clark, W. B. Cowden and D. O. Willenborg (1996) *Transplant Immun.*, **4**, 275–285.
213. H. Moriyama, T. Tsukida, Y. Inoue, K. Yokota, K. Yoshino, H. Kondo, N. Miura and S-I. Nishimura (2004) *J. Med. Chem.*, **47**, 1930–1938.
214. M. Djebaili and J-B. Behr (2005) *J. Enzym. Inhib. and Med. Chem.*, **20** (**2**), 123–127.
215. I. Gautier-Lefebvre, J-B. Behr, G. Guillerm and M. Muzard (2005) *Eur. J. Med. Chem.* **40**, 1255–1261.
216. I. Yukio, I. Makoto, O. Koji and T. Takashi (1994) Anisomycin derivatives and anticancer agents, antifungal agents and antiprotozoan agents containing them. *Eur. Pat. Appl.*, EP 0 612 728 A2.
217. R. E. Schwartz, G. L. Helms, E. A. Bolessa, K. E. Wilson, R. A. Giacobbe, J. S. Tkacz, G. F. Bills, J. M. Liesch, D. L. Zink, J. E. Curotto, B. Pramanik and J. C. Onishi (1994) *Tetrahedron*, **50**, 1675–1686.
218. C. Norez, S. Noel, M. Wilke, M. Bijvelds, H. Jorna, P. Melin, H. DeJonge and F. Becq (2006) *FEBS Letters*, **580**, 2081–2086.
219. A. C. van der Spoel, M. Jeyakumar, T. D. Butters, H. M. Charlton, H. D. Moore, R. A. Dwek and F. M. Platt (2002) *Proc. Nat. Acad. Sci. USA*, **99**, 17173–17178.
220. C. M. Walden, T. D. Butters, R. A. Dwek, F. M. Platt and A.C. van der Spoel (2006) *Human Reprod.*, **21**, 1309–1315.
221. R. Suganuma, C. M. Walden, T. D. Butters, F. M. Platt, R. A. Dwek, R. Yanagimachi and A.C. van der Spoel (2005) *Biol. Reprod.*, **72**, 805–813.
222. J. Liu, A. R. Shikhman, M. K. Lotz and C-H. Wong (2001) *Chem. Biol.*, **8**, 701–711.
223. J. Liu, M. M. D. Numa, H. Liu, S-J. Huang, P. Sears, A. R. Shikhman and C-H. Wong (2004) *J. Org. Chem.*, **69**, 6273–6283.

224. C. Shen, D. Bullens, A. Kasran, P. Maerten, L. Boon, J. M. F. G. Aerts, G. van Assche, K. Geboes, P. Rutgeerts and J. L. Ceuppens (2004) *Int. Immunopharmacol.*, **4**, 939–951.
225. D. Priestman, A. van der Spoel, T. D. Butters, R. A. Dwek and F. M. Platt (2005) Use of N-substituted imino sugars for appetite suppression. *PCT Int. Appl.*, WO 2005/092334A2.
226. A. N. E. Birch, W. M. Robertson, I. E. Geoghegan, W. J. McGavin, T. J. W. Alphey, M. S. Phillips, L. E. Fellows, A. A. Watson, M. S. J. Simmonds and E. A. Porter (1993) *Nematologica*, **39**, 521–535.
227. S. V. Evans, A. M. R. Gatehouse and L. E. Fellows (1985) *Entomol. Exp. Appl.*, **37**, 257–261.
228. F. Jüttner and H. P. Wessel (2003) *J. Phycol.*, **39**, 26–32.
229. J-B. Behr and G. Guillerm (2002) *Tetrahedron: Asymmetry*, **13**, 111–113.
230. J.K. Amory, C.H. Muller, S.T. Page, E. Leifke, E.R. Pagel, A. Bhandari, B. Subramanyam, W. Bone, A. Radlmaier and W.J. Bremner (2007) *Hum. Reprod.*, **22**, 702–707.

Index

Note: page numbers in *italics* refer to figures and tables

Iminosugars: From synthesis to therapeutic applications Edited by Philippe Compain and Olivier R. Martin
 ISBN 978-0-470-03391-3

Index compiled by Jill Halliday